BirdLife Conservation Series No. 6

HABITATS FOR BIRDS IN EUROPE
A Conservation Strategy for the Wider Environment

compiled by

Graham ___ker and Michael I. Evans

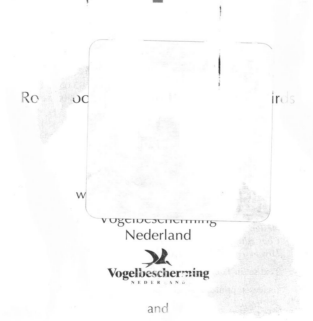

Ro___oc_____irds

w___

Vogelbescherming
Nederland

Vogelbescherming
NEDERLAND

and

Fondation Hans Wilsdorf (Montres Rolex)

Recommended citation   Tucker, G. M. and Evans, M. I. (1997) *Habitats for birds in Europe: a conservation strategy for the wider environment*. Cambridge, U.K.: BirdLife International (BirdLife Conservation Series no. 6).

© 1997 BirdLife International

Wellbrook Court, Girton Road, Cambridge CB3 0NA, UK

tel. +44-(0)1223-277318   fax +44-(0)1223-277200   email birdlife@birdlife.org.uk

BirdLife International is a UK registered charity

ISBN 0 946888 32 9

A catalogue record for this book is available from the British Library

Series editor   Duncan Brooks
Design   CBA (Cambridge) and Duncan Brooks
Copy editor   Brian Hillcoat
Layout, text preparation and graphics   Michelle Berry, Duncan Brooks, Regina Pfaff

Text set in Times (9/11 pt) and Optima

Imageset, printed and bound in Great Britain by The Burlington Press (Cambridge) Ltd.

Cover   Peregrine *Falco peregrinus* (photograph by Piers Cavendish/Ardea). This species, highly dispersed in Europe, suffered large population declines following widespread pollution of the environment by persistent agricultural pesticides. The ensuing implementation of regulations against the pesticides in most European countries has resulted in excellent population recovery over most of its European range, and is a classic success story of bird conservation in the wider environment.

CONTENTS

■

■

HABITAT CONSERVATION STRATEGIES

■

Appendices

FOREWORD

by Domingo Jiménez-Beltrán
Executive Director, European Environment Agency

SINCE 1992—when more than 150 parties or states agreed on the United Nations Convention on Biological Diversity in Rio de Janeiro—biodiversity and the conservation of biodiversity have become concepts used again and again.

Today we can see a slowly growing and spreading understanding and acceptance of the importance to all of us of biodiversity. In the words of Thomas E. Lovejoy, we begin to understand that 'As the planet becomes simpler biologically it becomes more expensive economically. The planet is also more vulnerable to disaster and the quality of life inevitably declines. It is up to science to spread the understanding that the choice is not between wild places or people. Rather it is between a rich or impoverished existence for man.'

But we are far from reaching our goals. In fact, despite our understanding, clear goals and strategies are still largely missing. Every day we are losing species and habitats; every day we still see large areas of habitats being degraded or changing and starting to malfunction.

We ourselves are responsible for the degradation and mis-use of biodiversity just as we are responsible for seeing that biodiversity in its many aspects is wisely used, well managed and, where possible, restored. This calls for understanding, for knowledge and for involvement, and that is not easily gained. It also calls for continuous follow-up and monitoring, which is costly and often seen as beyond the capabilities of the authorities.

Therefore, the work done by organizations such as BirdLife International and all its local network of organizations in separate countries is of the greatest importance in contributing to our knowledge and in provoking debate and action. The work involves a vast number of interested and concerned persons and is based on intimate local knowledge.

With this latest publication BirdLife International embarks on yet another large enterprise, showing the relationships between birds and the conditions of habitats. This is important because of the now-consolidated understanding that sustainable and wise use of habitats forms the only basis for maintaining biodiversity.

There is a very long way yet to go before the goals of the Convention on Biological Diversity are reached, but good information is emerging and good results can be seen. The information in this book contributes to this process.

4

ACKNOWLEDGEMENTS

THIS book is the result of the work and cooperation of a large number of people united by their dedication to the cause of nature conservation. We are especially grateful to all of these people, and their associated organizations, as their work was unpaid and often carried out in their spare time or squeezed into already very busy schedules.

First and foremost we are indebted to the members of the Habitat Working Groups who attended workshops and provided text and other data that formed the basis of each of the Habitat Conservation Strategies. In particular we acknowledge the contributions by the chapter compilers who assembled or wrote the individual strategies with varying degrees of help from other members of the working groups. Among the Habitat Working Group members we also wish to thank especially those who helped with the preparation, hosting and running of the workshops at which the groundwork for most of the strategies was developed (listed individually under 'Acknowledgements' at the end of each strategy).

We also particularly wish to thank the staff of the BirdLife International Partners, Partners Designate and Representatives in Europe, and other experts outside the BirdLife Network, who provided very valuable additional information and comments on the strategies during the review process (listed individually under 'Acknowledgements' at the end of each strategy). Many in the BirdLife Network helped by providing or tracing the contact details of relevant national experts.

This project was only made possible by the funding provided principally by the Royal Society for the Protection of Birds (RSPB) and by Vogelbescherming Nederland, the UK and Dutch BirdLife Partners respectively, with additional support from Fondation Hans Wilsdorf (Montres Rolex).

We wish to acknowledge the valuable guidance given by the project steering committee: Alistair Gammell (chairman), Mark Avery, Olivier Biber, Rob Fuller, Eduardo de Juana, Torsten Larsson, Frank Saris, Janine van Vessem and Tomasz Wesolowski. From the BirdLife Task Forces, we thank E. Osieck (Coordinator, Birds and Habitats Directives Task Force) for his careful and detailed review of all major chapters, and B. Briggs for reviewing the 'Overview' chapter and providing material on transport and climate-change issues.

All of the staff at the BirdLife Secretariat in Cambridge helped in some way through their contribution to the running of the office. Nevertheless, we would particularly like to thank Richard Grimmett who recognized the need for the development of a conservation strategy for the wider environment, instigated this project and provided considerable encouragement, advice and motivation for its completion.

We are indebted to 'The Producers'—Duncan Brooks, Michelle Berry and Regina Pfaff—for their hard work, skill, devotion to high standards, and attention to detail in the production of this book, happily for us accompanied by good humour even in the face of our worst failings and lapses as compilers. MB laid out most of the book and originated all graphics, while RP word-processed (repeatedly) all manuscripts and laid out the bulk of the tables. Brian Hillcoat bravely undertook the copy-editing.

Past and present BirdLife staff who we also wish to identify for their help include Colin Bibby and Zoltán Waliczky, especially for their contributions towards the initial chapters on 'Opportunities', 'Overview' and 'Main Conclusions and Recommendations', as well as Beverley Childs, Melanie Heath, Borja Heredia, Jenny Loughlin, David Chandler, Carlos Martín-Novella and Anne Collins. At BirdLife's European Community Office in Brussels, we are grateful for the prompt and accurate support provided by Nelly Paleologou, Barbara Coghlan and Mark Ferris on all questions relating to the European Union.

We particularly appreciate the valuable help we have received from Robin Standring, Hugo Rainey, Caroline Pilkington and James Lowen, who donated their time and skills as volunteers during various stages of the project.

Translations of the 'Summary' chapter were carried out by the BirdLife Partners, and we are particularly grateful to the following for their swift work on this: Isabelle Chesnot, Bernard Deceuninck and Alison Duncan (Ligue pour la Protection des Oiseaux, France), Ann Grösch, Peter Herkenrath and Claus Mayr (Naturschutzbund Deutschland), Juan Criado (Sociedad Española de Ornitología) and Elena Lebedeva (Russian Bird Conservation Union).

We are also grateful to others who contributed in many and important ways, including David Hill (Ecoscope Applied Ecologists) who provided office facilities and work-time for GMT to assist with the final production of the book after leaving BirdLife, Gwyn Williams (RSPB) who provided helpful advice on the development process for the strategies, Laurence Rose (RSPB) and Johanna Winkelman (Vogelbescherming Nederland). We also thank Olivier Biber and Luc Schifferli (Schweizerische Vogelwarte, Sempach, Switzerland) for organizing and hosting an initial planning workshop for the project (15–16 March 1993).

Lastly, we wish to thank our friends, relatives and partners who were affected in a variety of ways during the years that it has taken to prepare this book but who were always patient and understanding and provided encouragement and support throughout.

SUMMARY

BIODIVERSITY conservation in Europe rightly pursues the protection of important species and sites. More than 90% of the continent lies outside protected areas, however, and the conservation of biodiversity in this wider environment still receives far too little emphasis from government or in society as a whole.

This book is a contribution by BirdLife International towards remedying this neglect, and forms the third and final part of a 10-year research effort to promote the conservation of Europe's birds, building upon the previously published *Important Bird Areas in Europe* (Grimmett and Jones 1989) and *Birds in Europe: their conservation status* (Tucker and Heath 1994).

These and other continental-scale reviews have demonstrated that the greatest threats to birds in Europe, and to biodiversity in general, lie in the continuing erosion of the quality and extent of habitats. This loss and degradation is driven by the increasing intensity of human uses of the environment. Almost 40% of the birds of Europe now have an Unfavourable Conservation Status, because their populations are small, declining or localized.

A habitat-based approach to the conservation of the wider environment is therefore taken in this book, which aims to identify the most important measures that are needed for the conservation of Europe's birds on a wide scale. Wherever possible, the analysis of priorities is quantitative, in order to provide maximum justification for broad policies and actions that may be costly in current economic terms.

The eight major habitat-types that are important for birds in Europe are identified as being: Marine Habitats; Coastal Habitats; Inland Wetlands; Tundra, Mires and Moorland; Lowland Atlantic Heathland; Boreal and Temperate Forests; Mediterranean Forest, Shrubland and Rocky Habitats; and Agricultural and Grassland Habitats.

Thirteen workshops were held across the continent, at which relevant experts were brought together in Habitat Working Groups to prepare conservation strategies for each of the major habitat-types. Each draft strategy was then circulated for review within the Habitat Working Group, as well as throughout the wider BirdLife Network in Europe. Based on the feedback from this process, the strategies were finalized by the compilers.

The compilation of this book has therefore been a major effort by more than 190 experts on bird- and habitat-conservation issues from all over Europe.

For each habitat-type, essential background information is provided on its current distribution and extent (including trends), its origin and history, its important physical and ecological processes, and its flora and fauna. The values of the habitat to humanity and the most important human uses are also described, as well as the current socio-political factors that most influence such uses.

The bird species that will most benefit from broad habitat-conservation measures are then identified. These 'priority' species have an Unfavourable Conservation Status in Europe and/or a significant proportion of their European population depends upon the habitat-type for its survival.

The book stresses the need to conserve habitat 'quality' as well as extent. The important aspects of quality are defined for each major habitat-type by identifying the particular features or attributes of the habitat that are required by each priority bird species, and then summarizing these in order to identify those features that are most widely required among all priority species.

The most important threats to habitat quality and extent are identified, and their cumulative impacts on the European populations of the priority species are quantified.

Before addressing these threats, the available channels for broad conservation action are reviewed, in terms of the relevant international legislation, economic instruments, policy initiatives and non-statutory opportunities that currently exist within Europe.

Broad conservation objectives for the habitat-type are then identified, as well as the general biological targets that define good-quality habitat for priority birds. Building upon these, and bearing in mind strategic principles such as 'sustainable use' and 'the precautionary principle', the most important measures that need to be followed for the wide-scale conservation of the habitat-type are listed, in terms of broad policies and actions.

In conclusion, the book confirms that the main threat to birds in Europe's wider environment is the continuing intensification of land- and sea-uses by man. This is largely due to the lack of environmental objectives within the policies and regulations of use-sectors such as agriculture, forestry and tourism.

The lack of coordination between these sectors also tends to aggravate the threats to biodiversity that are posed by such activities. A single such use can have impacts on several different habitat-types, while several different uses can combine to affect one particular habitat-type.

It is also clear that the continent-wide impact of such land- and sea-uses on levels of biodiversity is inadequately monitored, and that the mechanisms of impact are poorly understood. There is an urgent need to address this situation if the sustainability of current and future policies of land- and sea-use is to be reliably assessed.

Clear environmental objectives, including targets for biodiversity conservation, should be integrated into the policies and regulations of all socio-economic sectors, at all levels. The European Union, in particular, provides a unique opportunity to influence national policy-making. A major review is needed of all EU legislation related to the Common Agricultural Policy, Structural Funds, Cohesion Policy and Common Fisheries Policy, so that these will ultimately benefit both nature conservation and regional/rural economies.

The integration of objectives can be facilitated through the development of new initiatives and concepts, such as Strategic Environmental Assessment of policies, plans and programmes, Integrated Coastal Zone Management, Catchment Management Planning, and sustainable agriculture and forestry. Pilot projects to demonstrate the feasibility of these approaches should be financed, and their results should be widely disseminated.

Biodiversity conservation in the wider environment will only succeed if it is demanded by the public at large. In addition, however, all members of society will have to act, not just conservation organizations and governments. Everyone is a stakeholder in the biodiversity conservation process, and more effort should go into promoting the widespread acceptance by society of this recent concept.

Ultimately, new forms of land- and sea-use will need to be agreed and acted on by society if conservation of the wider environment is to succeed in the long term in Europe.

ZUSAMMENFASSUNG

MAßNAHMEN zum Erhalt der biologischen Vielfalt in Europa schließen folgerichtig den Schutz bedeutsamer Arten und Habitate ein. Allerdings sind nur knapp 10% des europäischen Kontinents als Schutzgebiete ausgewiesen. Auf dem Kontinent verbleiben also 90% der Fläche zu bearbeiten, wenn man sich der Herausforderung stellen will, die biologische Vielfalt in Europa zu erhalten. Aus dieser Perspektive heraus bietet sich ein Handlungspotential an, das bisher weder von den einzelnen Regierungen noch von der Gesellschaft selbst ausreichend wahrgenommen wurde.

Einen Impuls in diese Richtung liefert das von BirdLife International herausgebrachte Buch, das der dritte und letzte Abschnitt einer auf 10 Jahre angesetzten Studie zur Förderung des Schutzes der Vögel Europas darstellt und auf die in *Important Bird Areas in Europe* (Grimmett und Jones 1989) und *Birds in Europe: their conservation status* (Tucker und Heath 1994) bereits veröffentlichten Ergebnisse aufbaut.

Schleichende Qualitätsminderung und ständige Reduzierung der Lebensräume stellen, wie diese und andere europaweit angelegten wissenschaftlichen Arbeiten gezeigt haben, die stärkste Gefährdung der Vögel Europas und der damit verbundenen biologischen Vielfalt dar. Verstärkt wird dieser Prozeß durch die zunehmend intensivere Nutzung der Umwelt durch den Menschen. Aufgrund einer kleinen, abnehmenden oder regional begrenzten Population fallen fast 40% der Vögel Europas in eine Schutzkategorie, die ihnen nur geringe Überlebenschancen in Aussicht stellt.

Das Prinzip 'Habitate als Bausteine' wird in diesem Buch als Wegbereiter in Richtung Schutz und Erhaltung der biologischen Vielfalt auf den 90% der Fläche Europas vorgestellt, die bislang nicht unter Schutz stehen und ein großes Handlungspotential bieten. Die wichtigsten und dringendsten

Maßnahmen zum Schutz der Vögel europaweit werden hervorgehoben. Die Analyse der Prioritäten erfolgt dabei auf quantitativer Basis, um die Argumentationsgrundlage maximal zu untermauern.

Zu den acht für die Vögel Europas wichtigsten Habitattypen zählen: Meereshabitate; Küstenhabitate; Feuchtgebiete des Binnenlandes; Tundra, Hoch- und Niedermoore; atlantische Tiefland-Heidelandschaften; Wälder der borealen und gemäßigten Zonen; mediterrane Wälder, Gebüsche und Felsenhabitate; landwirtschaftlich genutzte und Graslandhabitate.

An verschiedenen Orten Europas wurden dreizehn Workshops organisiert, die Experten in Arbeitsgruppen zusammenbrachten, um Schutzstrategien für jeden Habitattyp auszuarbeiten. Alle in der Gruppe Beteiligten erhielten den Entwurf der Schutzstrategien zur Überarbeitung. Der Entwurf wurde auch innerhalb Europas über das BirdLife-Netzwerk in Umlauf gebracht. Aus den Rückmeldungen heraus wurden die endgültigen Strategien zusammengestellt.

Das Buch zeichnet sich auch durch die einmalige Zusammenarbeit zwischen mehr als 190 Experten aus ganz Europa aus, die Beiträge zu Vogel-und Habitatschutzthemen geliefert haben.

Jede Habitattypbeschreibung enthält wichtige Hintergrundinformationen zu Verbreitung, Verbreitungsgebiet (inklusive Trends), Ursprung und Geschichte, zu den wichtigsten physikalischen und ökologischen Prozessen und zu der habitatspezifischen Flora und Fauna. Ebenso beschrieben wird der ethische Wert des Habitats sowie die Nutzung durch den Menschen, aber auch aktuelle soziopolitische Faktoren, die die Nutzung am stärksten beeinflußen.

Davon ausgehend werden die Vogelarten, die von breit angelegten Habitatschutzmaßnahmen am meisten profitieren, vorgestellt. Diese sog. prioritären Arten weisen einen ungünstigen Erhaltungszustand auf, und/oder ein bedeutender Teil ihrer europäischen Populationen hängt von einem spezifischen Habitattyp ab.

Das Buch unterstreicht die Notwendigkeit, sowohl die 'Qualität' als auch die Größe eines Habitats zu schützen. Um aufzuzeigen, welche lebensnotwendige Rolle die Qualität eines Habitats für die einzelne prioritäre Art spielt, werden die charakteristischen Merkmale und Eigenschaften jeder der acht Habitattypen beschrieben. Daraus werden die am häufigsten vorkommenden und zum Überleben notwendigen Qualitätsmerkmale zusammengefaßt.

Aufgezeigt werden die Faktoren, die die Qualität eines Habitats am stärksten beeinträchtigen und gefährden. Die kumulative Wirkung auf die europäischen Populationen der prioritären Arten wird quantifiziert.

Bevor auf die einzelnen Gefährdungstypen eingegangen wird, untersuchen die Autoren unter dem Aspekt internationaler Gesetzgebung, wirtschaftlicher Instrumente, Strategien und anderer vom Gesetz her nicht vorgesehenen Lösungsmöglichkeiten die gängigen Modalitäten der Umsetzung von Naturschutzmaßnahmen in Europa.

Definiert werden dann die auf das Habitat zugeschnittenen Ziele, die eine große Breitenwirkung nach sich ziehen, sowie spezifische arterhaltende Maßnahmen zum Schutz der prioritären Arten. Daraus werden die wichtigsten Maßnahmen zusammengestellt, die aufbauend auf den Grundsätzen der Nachhaltigkeit und des Vorsorgeprinzips für einen umfassenden Schutz der Habitattypen erforderlich sind und Maßstäbe bei der politischen Umsetzung setzen.

Was den Status der Vögel Europas angeht, so stellen die Autoren fest, daß die größte Bedrohung von der zunehmenden Intensivierung der Landnutzung und Ausbeutung der Meere ausgeht. Dazu tragen die Land- und Forstwirtschaft sowie der Tourismus bei, Bereiche, in denen der Schutz der Natur als Ziel gesetzlich noch nicht verankert ist.

Mangelnde Koordination zwischen diesen Bereichen verstärkt zudem das Ausmaß der Gefährdung der biologischen Vielfalt. Menschliche Aktivitäten in einem einzigen Bereich können eine Streuwirkung auf verschiedene Habitattypen auslösen, verschiedene Nutzungsarten auf einen einzigen Habitattyp gebündelt nachhaltige Veränderungen verursachen.

Es wird außerdem deutlich, daß europaweit die Auswirkungen der Land- und Meeresnutzung auf die Ab- bzw. Zunahme der biologischen Vielfalt unzureichend erfaßt werden und daß die durch menschliche Aktivitäten verursachte Wechselwirkung noch zu wenig erforscht ist. Um eine zuverlässige Aussage über die Nachhaltigkeit der zur Zeit gültigen und zukünftigen Verordnungen zu machen, ist es dringend erforderlich, auf diesen Zustand hinzuweisen.

Klar definierte Naturschutzziele, die die Erhaltung der biologischen Vielfalt beinhalten, sollten auf allen Ebenen Bestandteil aller sozio-politischen Bereiche sein und Eingang in die Gesetzgebung finden. Insbesondere bietet die Europäische Union eine einmalige Gelegenheit, Einfluß auf die nationale Politik zu nehmen. Eine Reform sämtlicher EU-Gesetze bezüglich der EU-Agrarpolitik, der Strukturfonds, der Kohäsionsfonds und der EU-Fischereipolitik ist notwendig, damit diese Bereiche letztendlich dem Naturschutz wie auch der regionalen Wirtschaft und der ländlichen Entwicklung Vorteile bringen.

Neue Initiativen und Konzepte wie z. B. eine strategische Umweltverträglichkeitsprüfung für alle

Politikfelder, Pläne und Programme, integriertes Küstengebietsmanagement, integriertes Management der Wasserressourcen sowie nachhaltige Land- und Forstwirtschaft können dazu beitragen, Naturschutzziele zu integrieren. Pilotprojekte, die die Anwendbarkeit dieser Konzepte demonstrieren, sollten finanziert und ihre Ergebnisse breit gestreut werden.

Der Schutz der biologischen Vielfalt europaweit wird nur dann gelingen, wenn der Bürger das will und das auch verlangt. Nicht nur die Naturschutzverbände und Regierungen müssen handeln—wir

alle sind zum Handeln aufgefordert. Jeder ist ein Teil des Prozesses zur Erhaltung der biologischen Vielfalt. Wir müssen uns viel mehr anstrengen, um jedem beizubringen, daß auch er oder sie letztendlich betroffen ist.

Langfristig gesehen sind neue Formen der Land- und Meeresnutzung zwingend erforderlich und müssen umgesetzt werden, wenn wir den Naturschutz als gesellschaftspolitisches Ziel aller Europäer verankern wollen!

RÉSUMÉ

LA CONSERVATION de la biodiversité en Europe vise à assurer la protection des espèces et des sites importants. Pourtant, plus de 90 % du continent ne bénéficie d'aucune zone de protection, et la conservation de la biodiversité, dans son intégralité et dans un contexte environnemental, bénéficie encore de bien peu d'attention de la part des Etats ou de la société.

Ce livre représente la contribution de BirdLife International pour remédier à cette négligence et constitue la troisième et dernière partie d'un effort de recherche de 10 ans visant à promouvoir la conservation des oiseaux en Europe. Il a été élaboré d'après les ouvrages précédemment parus *Important Bird Areas in Europe* (Grimmett et Jones 1989) et *Birds in Europe: their conservation status* (Tucker et Heath 1994).

Ces ouvrages, ainsi que d'autres bilans au niveau continental, démontrent que les menaces les plus sérieuses qui pèsent sur les oiseaux en Europe, et sur la biodiversité en générale, se révèlent être l'érosion persistante de la qualité et de la superficie des habitats. Cette érosion et cette dégradation sont liées à l'augmentation croissante des activités humaines exercées sur l'environnement. Actuellement, près de 40% des oiseaux d'Europe souffrent d'un statut de conservation défavorable en raison de la faiblesse ou de déclin de leur population.

Dans ce livre, la conservation de l'environnement se fonde sur l'approche au niveau de l'habitat. Cette dernière a pour objectif d'identifier les mesures les plus importantes, nécessaires à la conservation à

grande échelle des oiseaux en Europe. Partout où c'est possible, l'analyse des priorités est d'ordre quantitatif, afin de fournir des arguments optimaux en faveur de politiques et d'actions globales pouvant se révéler coûteuses en termes économiques actuels.

Les huit principaux types d'habitats, identifiés comme étant les plus importants pour les oiseaux en Europe, sont : les habitats marins, les habitats côtiers, les marais continentaux, la toundra, les tourbières et landes, les landes atlantiques de plaine, les forêts boréales et tempérées, la forêt méditerranéenne, les habitats rocheux et garrigues, les habitats agricoles et prairies.

A l'occasion de treize ateliers organisés en Europe, des experts compétents se sont réunis en groupes de travail, par type d'habitat, afin de préparer des stratégies de conservation pour les principaux. Chaque projet de stratégie réalisé a ainsi circulé pour relecture au sein du groupe de travail concerné, ainsi qu'à travers le réseau élargi de BirdLife International en Europe. Différentes stratégies, basées sur les résultats de cette démarche, ont été ensuite finalisées.

La compilation de ce livre est le fruit de l'effort de plus de 190 personnes originaires de toute l'Europe et expertes dans le domaine de la conservation des oiseaux et des habitats.

Pour chaque type d'habitat, des informations essentielles sont fournies sur la distribution actuelle et la superficie (y compris les tendances), l'origine et l'histoire, les processus physiques et écologiques importants, et sur la flore et la faune. Les valeurs que revêtent les habitats pour l'humanité, les exploitations les plus importantes exercées par l'homme sur l'environnement sont également décrites, ainsi que les facteurs socio-politiques qui influent principalement sur elles.

Les espèces d'oiseaux qui bénéficieront le plus des mesures globales de conservation d'habitat sont alors identifiées. Ces espèces prioritaires ont un statut de conservation défavorable en Europe et/ou une proportion significative de leurs effectifs européens dont la survie dépend de certains types d'habitats.

Ce livre souligne la nécessité de conserver la «qualité» des habitats ainsi que leur superficie. Les aspects qualitatifs importants sont définis pour chaque type d'habitat principal en identifiant leurs caractéristiques particulières ou attributs nécessaires à chaque espèce d'oiseau prioritaire. Ces caractéristiques sont ensuite résumées afin de distinguer celles qui sont les plus généralement requises pour toutes les espèces prioritaires.

Les menaces les plus importantes qui pèsent sur la qualité et la superficie des habitats sont identifiées et les impacts cumulés sur les populations européennes des espèces prioritaires sont quantifiés.

Avant d'énumérer ces menaces, les éventuelles directions relatives aux actions globales de conservation sont étudiées en termes de législation internationale, d'instruments économiques, d'initiatives politiques et d'opportunités non réglementaires qui existent actuellement en Europe.

Les objectifs généraux de conservation en faveur des types d'habitats sont ensuite identifiés ainsi que les indicateurs biologiques généraux qui définissent la bonne qualité des habitats pour les espèces d'oiseaux prioritaires. En se basant sur ceux-ci et en ayant bien à l'esprit des principes stratégiques, tels que le développement durable et le principes de précaution, les mesures les plus importantes qui ont besoin d'être suivies pour la conservation des types d'habitats à grande échelle ont été définies en termes de politiques globales et d'actions.

En conclusion, ce livre confirme que la principale menace pesant sur les oiseaux dans l'environnement européen au sens large est la poursuite de l'intensification des activités humaines sur la terre et la mer. Ceci est principalement dû à une absence d'objectifs environnementaux dans les politiques et les réglementations des secteurs concernés, tels que l'agriculture, la sylviculture et le tourisme.

Le manque de coordination entre ces secteurs tend également à aggraver les menaces que présentent de telles activités pour la biodiversité. Il faut savoir qu'une seule activité humaine sur l'environnement peut avoir des impacts sur plusieurs types d'habitats différents, tandis que plusieurs activités différentes peuvent se combiner et avoir un impact sur un seul type d'habitat particulier.

Il est également évident que l'impact, à l'échelle continentale, de telles exploitations de la terre et de la mer sur le plan de la biodiversité est insuffisamment suivi et que les mécanismes de leur impact sont mal compris. Il est nécessaire de reconnaître cette situation si l'on veut vérifier objectivement la viabilité à long terme des politiques actuelles et futures d'exploitation de l'espace terrestre et marin.

Des objectifs environnementaux clairs ainsi que des cibles pour la conservation de la biodiversité devraient être intégrés dans les politiques et réglementations de tous les secteurs socio-économiques, à tous les niveaux. L'Union Européenne, en particulier, fournit une opportunité unique d'influencer les décideurs nationaux. Un examen important de toutes les législations européennes ayant trait à la Politique Agricole Commune, aux Fonds structurels, à la politique de Cohésion et à la Politique commune de la pêche est nécessaire afin que la conservation de la nature et les économies régionales/rurales puissent en bénéficier.

L'intégration des objectifs peut être facilitée par le biais du développement de nouvelles initiatives et concepts, tels que: l'évaluation des incidences de certains projets publics et privés sur l'environnement, de politiques, et de plans et programmes; d'une gestion intégrée des zones littorales; d'une planification de la gestion intégrée des bassins versants; d'une agriculture et sylviculture durables. Des projets pilotes pour démontrer la faisabilité de ces approches devraient être financés et leurs résultats devraient être largement diffusés.

La conservation de la biodiversité dans un environnement global ne pourra réussir que si le grand public le demande. De plus, tous les membres de la société devront agir et non pas seulement les organisations de conservation et les gouvernements.

Chacun d'entre nous détient une responsabilité dans le processus de conservation de la biodiversité et davantage d'effort devrait être consacré à promouvoir l'acceptation générale de ce concept récent.

Enfin, de nouvelles formes d'exploitation de la terre et de la mer devront être approuvées et appliquées par la société si l'on souhaite que la conservation d'un environnement global réussisse à long terme en Europe.

RESUMEN

L A CONSERVACIÓN de la biodiversidad en Europa persigue la protección de lugares y especies importantes. Sin embargo, más del 90% del continente queda fuera de los Espacios Protegidos, por lo que la conservación de la biodiversidad en este ámbito más amplio, todavía es objeto de muy poca atención, tanto por parte de los gobiernos como de toda la sociedad.

Este libro es una contribución de BirdLife International para poner fin a este abandono, y constituye la tercera y última parte de 10 años de investigación para promover la conservación de las aves en Europa, apoyándose en las publicaciones anteriores *Important Bird Areas in Europe* (Grimmett y Jones 1989) y *Birds in Europe: their conservation status* (Tucker y Heath 1994).

Estas y otras revisiones a escala europea han demostrado que las mayores amenazas para las aves en Europa, y sobre la biodiversidad en general, son la disminución de la calidad y la superficie de los hábitats. El continuo aumento de la actividad humana sobre el medio ambiente es la principal causa de la pérdida y deterioro de los hábitats. Casi el 40% de las aves europeas presentan un estado de conservación desfavorable debido a que sus poblaciones o son muy pequeñas o están en declive.

La conservación del medio ambiente a través de la conservación de los hábitats es la aproximación seguida en este libro, que pretende identificar las medidas necesarias más importantes para la conservación de las aves en un sentido amplio. Siempre que ha sido posible, el análisis de prioridades ha sido cuantitativo, con la intención de justificar lo más posible acciones y políticas que puedan resultar costosas en los términos económicos actuales.

Los ocho tipos de hábitats de mayor importancia para la conservación de las aves en Europa son: Hábitats marinos; Hábitats costeros; Humedales interiores; Tundra, «mires» y páramos; Brezales atlánticos; Bosques templados y boreales; Bosque mediterráneo, hábitats de roquedos y matorral; y Hábitats esteparios y agrícolas.

Se celebraron trece seminarios en el continente en los que numerosos expertos trabajaron juntos en Grupos de Trabajo de Hábitats para la preparación una Estrategia de Conservación para cada uno de los tipos de hábitat. El borrador de cada estrategia fue distribuido a los miembros de cada Grupo de Trabajo, así como a la amplia red de BirdLife en Europa. Las contribuciones durante este proceso de revisión permitieron a los compiladores finalizar las estrategias. Por tanto, los trabajos incluidos en este libro representan el esfuerzo de más de 190 expertos en conservación de las aves (y sus hábitats) de toda Europa.

Para cada tipo de hábitat se facilitan los antecedentes sobre la distribución actual y la extensión (incluyendo la tendencia), su origen e historia, sus procesos físicos y ecológicos, y su flora y fauna. También se describen la importancia del hábitat para el hombre y los usos que desarrolla sobre éstos, así como los factores socio-políticos actuales que afectan a dichos usos.

A continuación, se identifican las especies de aves que más se beneficiarían de la aplicación de medidas generales de conservación del hábitat. Estas especies «prioritarias» tienen o bien un estado de conservación desfavorable en Europa, o bien una proporción significativa de la población europea depende de un tipo de hábitat para su supervivencia. El libro resalta la necesidad de conservar los hábitats tanto en «calidad» como en extensión. Se definen los aspectos más importantes de la calidad de cada tipo de hábitat, a través de la identificación de características o particularidades del hábitat requeridas por cada especie de ave prioritaria, para, finalmente, identificar aquellas características que son más ampliamente requeridas por todas las especies prioritarias.

Se identifican las principales amenazas para la calidad y extensión del hábitat y se cuantifican los impactos sinérgicos sobre las poblaciones europeas de las especies prioritarias.

Antes de señalar estas amenazas, se revisan las posibles actuaciones de conservación, en el sentido de la legislación internacional relevante, los instrumentos económicos, o las oportunidades políticas o de otro tipo que actualmente existen en Europa.

Se identifican objetivos amplios de conservación para cada tipo de hábitat, así como los principios biológicos generales que definen la buena calidad del hábitat para las aves prioritarias. Sobre todo ello y considerando principios generales como «desarrollo sostenible» y el «principio de precaución», se detallan las medidas más importantes para la conservación a gran escala de los tipos de hábitat, a nivel de grandes políticas y actuaciones.

En conclusión, el libro confirma que la principal

amenaza para las aves en el medio ambiente en Europa, es la continua intensificación por el hombre de los usos del suelo y el aprovechamiento de los recursos marinos. Esto se debe, fundamentalmente, a la ausencia de objetivos medioambientales en la regulación y las políticas de sectores productivos tales como la agricultura, el sector forestal y el turismo.

La falta de coordinación entre estos sectores tiende también a agravar las amenazas que suponen dichas actividades para la biodiversidad. Un único uso puede impactar sobre diferentes tipos de hábitats, mientras que la combinación de diferentes actividades puede impactar sobre un tipo concreto de hábitat. El seguimiento de los impactos producidos por los usos del suelo y el aprovechamiento de los recursos marinos sobre la biodiversidad en Europa tampoco es adecuado, y los mecanismos del impacto son poco conocidos. Hay una necesidad urgente de afrontar esta situación para evaluar seriamente la sostenibilidad de futuras políticas sobre los usos del suelo y el aprovechamiento de los recursos marinos.

Se deben integrar claros objetivos medioambientales, incluyendo los de conservación de la biodiversidad, en la regulación y las políticas de todos los sectores socioeconómicos, a todos los niveles. La Unión Europea, en particular, proporciona una oportunidad única para influir en la formulación de políticas nacionales. Es necesaria una revisión de toda la legislación comunitaria relativa a la Política Agraria Común, los Fondos Estructurales, la Política de Cohesión y la Política Pesquera Común, de manera que en último término resulten beneficiadas tanto la conservación de la naturaleza como las economías rurales y regionales.

El desarrollo de nuevas iniciativas y conceptos como la Evaluación Estratégica Ambiental de Planes, Políticas y Programas, la Gestión Integrada de las Costas, la Gestión Integral de las Cuencas y el Aprovechamiento Agrícola y forestal sostenible, facilitan la integración de esos objetivos. Se deben financiar proyectos piloto que demuestren la viabilidad de esta aproximación, y sus resultados deben ser diseminados ampliamente.

La conservación de la biodiversidad en sentido amplio sólo tendría éxito si la solicita el gran público. Además, todos los sectores de la sociedad, y no únicamente las organizaciones conservacionistas y los gobiernos, tendrán que actuar. Todos y cada uno de nosotros somos responsables en el proceso de conservación de la biodiversidad, por lo que debería realizare un mayor esfuerzo en promover una acogida cada vez más amplia de este reciente concepto por parte de la sociedad.

Por último, la sociedad debe adoptar y desarrollar nuevas formas de uso del suelo y de aprovechamiento de los recursos marinos si a largo plazo se quiere la conservación del medio ambiente en Europa.

RESUMEN

РЕЗЮМЕ

СОХРАНЕНИЕ биоразнообразия в Европе справедливо преследует цели охраны видов и мест их обитания. Более 90% суши на континенте не относится к особо охраняемым природным территориям, и сохранению биоразнообразия этих обширных экологических комплексов—среды обитания в широком смысле слова—пока уделяется крайне мало внимания как на государственном уровне, так и в обществе в целом.

Настоящая книга представляет собой вклад Международной ассоциации защиты птиц и природы BirdLife International в решение этой проблемы. Это третья, завершающая часть десятилетних исследований, направленных на совершенствование охраны птиц Европы. Она составляет единое целое с опубликованными ранее книгами 'Важные для птиц территории Европы' (Grimmett and Jones 1989) и 'Птицы Европы: их охранный статус' (Tucker and Heath 1994).

Эти и подобные обзоры, выполненные в общеевропейском масштабе, показывают, что основная угроза птицам, и биоразнообразию в целом, проистекает в Европе от снижения качества и уменьшения площадей природных местообитаний. Потери и деградация местообитаний происходят в связи с возрастающей интенсивностью использования среды человеком. Почти 40% птиц Европы оказались сейчас в числе видов с неблагоприятным статусом из-за малой или снижающейся численности их популяций.

Используемый в данной книге биотопический подход к сохранению природной среды направлен на выделение первоочередных мер, необходимых для сохранения птиц Европы в масштабах всего континента. Во всех возможных случаях при анализе использовались количественные данные, необходимые для максимальной обоснованности обширных природосберегающих стратегий и конкретных действий, выполнение которых может быть достаточно дорого в современной экономической ситуации.

Выделено восемь основных типов местообитаний, представляющих особую ценность для птиц в Европе: морские местообитания; прибрежные местообитания; внутренние водоемы; тундры и болота; атлантические вересковые низменности; таежные леса и леса умеренных широт; средиземноморские леса, кустарники и каменистые ландшафты; сельскохозяйственные и луговые местообитания.

Проведены тринадцать общеевропейских совещаний, на которых приглашенные эксперты образовали Рабочие группы по местообитаниям для разработки стратегий охраны птиц в каждом из выделенных типов угодий. Все варианты стратегий в дальнейшем распространялись для комментариев как среди участников Рабочих групп, так и более широко среди Европейской сети BirdLife International. На заключительном этапе все комментарии были обобщены составителями стратегий. Таким образом, настоящая книга явилась результатом работы более чем 190 экспертов по охране птиц и охране местообитаний из множества европейских стран.

Для каждого типа местообитаний в книге приводится исчерпывающая информация по их современному распространению и площадям (включая тенденции изменений), истории формирования, важнейшим физико-географическим и экологическим характеристикам, флоре и фауне. Рассмотрены также значимость этих местообитаний для человека, основные формы их хозяйственного использования, современные социально-политические факторы, влияющие на характер землепользования в этих угодьях.

Были определены виды птиц, существование которых зависит в первую очередь от широкомасштабных мер по сохранению местообитаний. Это 'приоритетные' виды, имеющие в Европе неблагоприятный охранный статус и (или) те виды, у которых значительная часть Европейских популяций напрямую зависит от соответствующего типа местообитаний.

В книге подчеркивается необходимость сохранения как 'качества', так и площадей местообитаний. Важнейшие качественные характеристики местообитаний для каждого из соответствующих типов угодий выделены в зависимости от экологических требований каждого из приоритетных видов птиц: чем больше видов птиц зависит от конкретной характеристики местообитания, тем выше оказывается при обобщении ее значимость.

Выделены основные факторы, угрожающие состоянию местообитаний, и количественно оценено их совокупное влияние на Европейские популяции приоритетных видов птиц. Перед рассмотрением способов снижения действия этих угрожающих факторов, проведен обзор существующих механизмов широкомасштабных природоохранных акций. Проанализировано соответствующее международное законодательство, экономические подходы, политико-стратегические инициативы и неформальные возможности деятельности, существующие в масштабах Европы.

Затем определены конкретные широкомасштабные задачи по сохранению каждого из типов местообитаний, а также общие биологические показатели, соответствующие высокому "качеству" местообитания в отношении приоритетных видов птиц. На основании этого и с учетом общестратегических принципов (таких как 'устойчивое использование' и 'упреждающий подход') перечислены наиболее важные меры, которые необходимо предпринимать для сохранения местообитаний с точки зрения общих стратегий и направлений деятельности.

В заключение необходимо еще раз подчеркнуть, что основная угроза птицам на территории Европы заключается в продолжающейся интенсификации использования водных и земельных ресурсов человеком. Это главным образом связано с тем, что в стратегиях и положениях о развитии таких секторов экономики как сельское хозяйство, лесопользование и туризм практически отсутствуют задачи сохранения окружающей природной среды. Недостаток координации деятельности между этими секторами также увеличивает риск нарушения биологического разнообразия. Один и тот же тип землепользования может наносить вред сразу в нескольких типах местообитаний, и в равной степени различные виды землепользования могут совместно поставить под угрозу один и тот же тип местообитания.

Столь же очевидно, что воздействие разных форм землепользования и использования морских ресурсов в общеевропейском масштабе контролируется недостаточно, и что механизмы этих воздействий осмыслены слабо. Необходима срочная и достоверная оценка как существующих, так и будущих стратегий использования наземных и водных ресурсов с точки зрения их устойчивости по отношению к сохранению биоразнообразия.

Четкие экологические задачи, включая задачи сохранения биоразнообразия, должны быть интегрированы в общие стратегии и регуляторные положения по всем социо-экономическим секторам и на всех уровнях. В частности, Европейский Союз представляет уникальные возможности влияния на разработку национальных стратегий. Необходимо значительно пересмотреть все законодательные механизмы Европейского Союза, имеющие отношение к Единой сельскохозяйственной политике, Структурным фондам, Политике объединения и Единой политике в области рыболовства—с тем, чтобы они приносили пользу как природной среде, так и развитию региональной или местной экономики.

Интеграция природоохранных задач может быть облегчена через развитие новых инициатив и концепций, например Стратегической экологической экспертизы стратегий, планов и программ, Интегрированного управления береговой зоной, Интегрированного планирования выловов, и через устойчивое развитие сельского и лесного хозяйства. Должны быть профинансированы пилотные проекты, демонстрирующие осуществимость этих подходов, и результаты этих проектов должны быть широко распространены.

Сохранение биологического разнообразия природной среды в широком смысле слова может иметь успех, только если это востребовано всеми слоями населения. И действия должны быть предприняты не только природоохранными организациями или правительствами, но всеми членами общества в целом. В процессе сохранения биоразнообразия каждый житель выступает как участник, и на массовое распространение этого современного мировоззрения также необходимо затрачивать больше усилий.

В конечном счете необходимо, чтобы новые формы использования земельных и водных ресурсов согласовывались и претворялись в действие на уровне общества в целом, и тогда охрана окружающей среды в Европе может иметь долговременный успех.

INTRODUCTION

THE PUBLICATION of *Important Bird Areas in Europe* by BirdLife International (then the International Council for Bird Preservation) and Wetlands International (then the International Waterfowl and Wetlands Research Bureau) represented a major step towards bird conservation on a truly European scale (Grimmett and Jones 1989). For the first time, individual sites were evaluated in a standard way and a continent-wide network of sites was identified that, if protected and managed properly, would safeguard a significant proportion of Europe's birds (Box 1).

However, the majority of Europe's threatened and declining bird species have populations that are dispersed throughout, or during critical periods of, their annual cycle, and thus they cannot be adequately conserved by a network of protected areas alone. For example, although many seabirds congregate in colonies to breed and these can be protected, they are highly dispersed when feeding, and outside the breeding season. In contrast, many important wintering sites for waterbirds receive protection, but these species receive little consideration when dispersed in their extensive breeding habitats, e.g. tun-

---

Box 1. The conservation of Important Bird Areas in Europe.

In 1989, BirdLife International (then the International Council for Bird Preservation) and Wetlands International (then the International Waterfowl and Wetlands Research Bureau) identified 2,444 Important Bird Areas (IBAs) in Europe (Grimmett and Jones 1989).

Important Bird Areas

- Are places of international significance for the conservation of wild birds.
- Offer practical means for the conservation of congregatory and localized bird species.
- Are chosen using standardized criteria.
- Should support self-sustaining populations of wild birds.
- Should be delimitable from their surroundings.
- Should form part of a wider, integrated approach to biodiversity conservation.

IBAs are **selected** for one or more of the following groups of bird species which they may support:
- Species of global conservation concern (Collar *et al*. 1994).
- Restricted-range species (Stattersfield *et al*. in press).
- Biome-restricted species (BirdLife International 1995a).
- Congregatory species (Rose and Scott 1994).
- Species of European Conservation Concern (Tucker and Heath 1994).

- Species in Annex I of the Birds Directive of the European Union (CEC 1994c).

These criteria are a development of those used in the first IBA inventory for Europe (Grimmett and Jones 1989), and are currently being applied by BirdLife International under the first periodic review of IBAs in Europe (Heath and Payne in prep.).

BirdLife International's **European IBA Programme** aims to:
- Promote the provision of legal protection for IBAs at local, national and international levels.
- Influence relevant national and international legislation and policies affecting the protection of sites.
- Safeguard IBAs against any adverse effects through site-specific conservation action.
- Restore, where possible, the ecological value of IBAs which have been destroyed or degraded by human activity.
- Promote the adequate management of all IBAs, protected or not.
- Disseminate information about the value of IBAs to decision-makers and the general public.
- Involve a wide range of interest groups and local communities in the conservation of IBAs.

BirdLife Partner organizations in 26 countries are currently operating national IBA programmes.

dra, mires and moorland in northern Europe. Some groups, such as raptors (Accipitriformes), owls (Strigiformes), woodpeckers (Picidae) and most passerines, tend to be dispersed all year round.

Furthermore, even when sites are protected and managed for conservation (as nature reserves, for example), they are not independent of the wider environment and may be influenced by external factors such as upstream alterations to river-drainage basins, changes in water quality and in erosion and sedimentation rates, air and water pollution, disturbance from adjacent human activities, and the quality of surrounding habitats. Birds share the landscape with man's economic interests and activities.

In addition to the protection and management of key species and sites, therefore, any conservation strategy for Europe's birds must also include broad measures for the conservation of habitats and landscapes, which are fully integrated into land-use policies and regulations.

To tackle this, the Dispersed Species Project was launched by BirdLife International in September 1990. Within Europe, including Turkey and Greenland, the project aimed to develop wide-scale habitat conservation measures for bird species that are in need of conservation action, thereby addressing the needs of dispersed species in particular. The project had the following principal objectives.

- To identify bird Species of European Conservation Concern (SPECs).

- To identify the priority species in each habitat (based on SPECs), and to summarize current knowledge of their habitat requirements, their relationships with land-uses, and the potential land-use changes which may affect them and their habitats.

- To identify the most important broad measures that are needed for the conservation of the habitats of these priority species.

The project was divided into two phases. Phase I collated national population data on all regularly occurring birds in Europe, in order to assess each species' conservation status and to identify SPECs (according to the criteria given in Box 2). The results of this phase were published in *Birds in Europe: their conservation status* (Tucker and Heath 1994). That publication provided clear and quantitative evidence that Europe's avifauna has changed fundamentally over the past few decades, with 195 species (38%) identified as having an Unfavourable Conservation Status in Europe, due to their European population being particularly small, declining or localized (see Box 2).

The majority of these 195 species are declining, although many are (or once were) common and widespread, thus indicating the huge scale of the problem. Furthermore, most declines were considered to be due to the generally increasing intensity of human uses of the environment over the continent. Thus, the review supported the justification and urgency for the targeting and development of wide-ranging conservation measures planned for Phase II. It was concluded that, in future, much greater emphasis must be given to incorporating environmental considerations into the management of the wider countryside.

This species-led approach to biodiversity conservation has the advantage that it is practical and allows targeting of measures to identifiable priorities, a requirement that is essential when conservation resources are as limited as at present. It also allows the quantification of threats, which in turn allows justification for actions that may be costly in economic terms, and provides the means to set objective targets for habitat conservation, and thereby to monitor the success or otherwise of actions, something for which birds are particularly well suited.

Although a bird-focused strategy may be regarded by some as too narrow in the current climate of ecosystem-orientated conservation, this publication has much wider implications than purely for bird protection. Birds are widespread and important components of biodiversity and ecosystems, and therefore patterns of bird decline and their causes can indicate broader areas of environmental concern (Box 3). Thus although this book focuses on habitat conservation for birds, it is likely that implementation of the recommended policies and actions would also contribute substantially to the conservation of other animals and plants and of ecosystems in general.

It might have been desirable to include accounts in this book of the importance of habitats for flora and non-bird fauna, their habitat requirements and the threats to these as well. This would have added new dimensions to the analysis and would have allowed a comparison between conservation priorities and requirements for other taxa and those for birds. However, the treatment of other groups of taxa which would be necessary to produce such an assessment in a reliable form would have been very time consuming, as much of the required data are dispersed or not readily accessible. For example, a simple breakdown of globally threatened taxa in Europe by habitat-type is still not readily accessible for any group other than birds. As an incomplete and superficial treatment of other taxa could be misleading, these subjects have been considered to be outside the scope of this book.

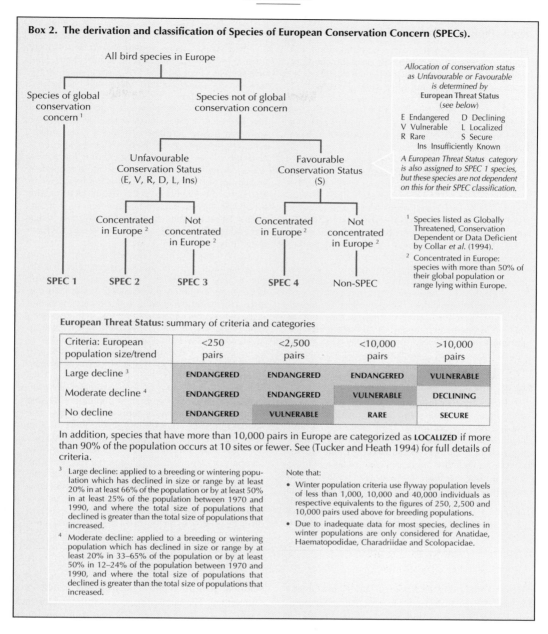

Box 2. The derivation and classification of Species of European Conservation Concern (SPECs).

All bird species in Europe

Species of global conservation concern[1]

Species not of global conservation concern

Unfavourable Conservation Status (E, V, R, D, L, Ins)

Favourable Conservation Status (S)

Concentrated in Europe[2]

Not concentrated in Europe[2]

Concentrated in Europe[2]

Not concentrated in Europe[2]

SPEC 1   SPEC 2   SPEC 3   SPEC 4   Non-SPEC

Allocation of conservation status as Unfavourable or Favourable is determined by European Threat Status (see below)

E Endangered   D Declining
V Vulnerable   L Localized
R Rare   S Secure
            Ins Insufficiently Known

A European Threat Status category is also assigned to SPEC 1 species, but these species are not dependent on this for their SPEC classification.

[1] Species listed as Globally Threatened, Conservation Dependent or Data Deficient by Collar et al. (1994).

[2] Concentrated in Europe: species with more than 50% of their global population or range lying within Europe.

European Threat Status: summary of criteria and categories

Criteria: European population size/trend	<250 pairs	<2,500 pairs	<10,000 pairs	>10,000 pairs
Large decline[3]	ENDANGERED	ENDANGERED	ENDANGERED	VULNERABLE
Moderate decline[4]	ENDANGERED	ENDANGERED	VULNERABLE	DECLINING
No decline	ENDANGERED	VULNERABLE	RARE	SECURE

In addition, species that have more than 10,000 pairs in Europe are categorized as **LOCALIZED** if more than 90% of the population occurs at 10 sites or fewer. See (Tucker and Heath 1994) for full details of criteria.

[3] Large decline: applied to a breeding or wintering population which has declined in size or range by at least 20% in at least 66% of the population or by at least 50% in at least 25% of the population between 1970 and 1990, and where the total size of populations that declined is greater than the total size of populations that increased.

[4] Moderate decline: applied to a breeding or wintering population which has declined in size or range by at least 20% in 33–65% of the population or by at least 50% in 12–24% of the population between 1970 and 1990, and where the total size of populations that declined is greater than the total size of populations that increased.

Note that:
• Winter population criteria use flyway population levels of less than 1,000, 10,000 and 40,000 individuals as respective equivalents to the figures of 250, 2,500 and 10,000 pairs used above for breeding populations.
• Due to inadequate data for most species, declines in winter populations are only considered for Anatidae, Haematopodidae, Charadriidae and Scolopacidae.

RATIONALE AND METHODOLOGY

Each habitat conservation strategy was developed by a working group, established solely for this purpose. Experts on the habitat and its birds were selected for each Habitat Working Group from representative countries, particularly from those with important areas of the habitat. These experts are listed, together with their titles and institutional affiliations, in Appendix 6 (p. 426). A workshop was then held with each Habitat Working Group at which an outline strategy was produced (according to the pattern described below).

After the workshop, further contributions were then provided by members of the Habitat Working Group and by other experts, and a complete draft produced. Each draft was then distributed for checking by the Habitat Working Group and for comments from other experts and from the BirdLife Network in Europe, including BirdLife Task Forces and other working groups where appropriate.

Each strategy covers the whole of each habitat's extent within Europe (i.e. east to the Ural mountains

Box 3. Birds as environmental indicators.

Birds are probably better researched and monitored than any other group of animals or plants, and are thus well placed to indicate the overall health of our environment (Peakall and Boyd 1987, Furness et al. 1993, Greenwood et al. 1995). Changes in the Europe-wide status of birds can warn of habitat loss and modification and can indicate the likely impact of these threats on other animals and plants. Birds have been shown to be good indicators of a variety of environmental conditions or problems, such as inappropriate pesticide use (Hardy et al. 1987), water quality (Eriksson 1987, Ormerod and Tyler 1993), the condition of the marine environment (Furness 1987a, Montevecchi 1993) and forests (Helle and Järvinen 1986, Welsh 1987, Angelstam and Mikusinski 1994), and changes in agricultural practice (Fuller et al. 1995). Furthermore, birds can be effective in gaining public support for the less tangible goals of maintaining biological diversity and functional ecosystems, through being visible 'flagships' for conservation.

Birds may not, however, be the most sensitive of environmental indicators, as many species appear relatively resilient to change. Although they tend to be high in the food-chain and can therefore indicate disruption to food-webs, responses may be slow and difficult to interpret. Bird numbers also tend to be regulated by density-dependent processes, such that their population sizes are often buffered against environmental impacts (Furness et al. 1993).

The migratory movements of birds may make it difficult to determine the actual location where impacts may be occurring. Their high mobility also facilitates the relocation of populations when conditions change, or their recolonization if habitats recover. On the other hand, evidence of past changes in range suggests that some bird species (those that are directly limited by climate) are likely to be particularly sensitive indicators of climate change. Range changes by birds may therefore be among the first observable symptoms of global warming (Huntley 1994).

Many bird species appear also to be relatively general in their habitat and food requirements. Species such as Great Bustard Otis tarda adapted in the past to agricultural habitats where these resembled the broad structure of their original habitats (in this case steppe grassland) and where practices remained non-intensive. However, some species have more restricted habitat requirements and are less adaptable. For example, Pallid Harrier Circus macrourus, Sociable Plover Chettusia gregaria and Black Lark Melanocorypha yeltoniensis appear to be dependent on natural steppe grassland and are now restricted in Europe to the fragments of this habitat that remain, along the Russia–Kazakhstan border (Tucker and Heath 1994). Furthermore, even those species that have adapted to changes in agricultural habitats have their limits.

Consequently, many species are now declining as agricultural intensification leads to habitat changes that exceed these limits. Populations of Otis tarda have gone extinct over much of western Europe during the twentieth century and continue to decline elsewhere (Tucker and Heath 1994). Similarly, many other species that initially adapted to semi-natural and low-intensity agricultural habitats, such as White Stork Ciconia ciconia, Corncrake Crex crex and Chough Pyrrhocorax pyrrhocorax, have also shown recent declines.

Because birds may be less sensitive to environmental change, declines in their populations arising from agricultural intensification or other wide-scale habitat alteration may be less severe than in other taxa. Indeed, wild plant populations have already been devastated in most agricultural landscapes in Europe (e.g. Kornas 1983, Hodgson 1987, van Dijk 1991). We must therefore assume that, although the effects of habitat change on birds also reflect the effects on other taxa, these are often likely to be very much underestimated. In other words, the loss of bird diversity is likely to be the 'tip of the iceberg' in terms of overall biodiversity loss in Europe's habitats.

---

of Russia), including the whole of Turkey and Greenland, as well as European Macaronesia[1]. Armenia, Azerbaijan, Georgia and the European part of Kazakhstan are not covered, as their participation in Phase I of the Dispersed Species Project was not possible, and the Caspian Sea is also excluded.

The process followed in the development of each habitat conservation strategy is summarized in Figure 1, and described below. The process follows a step-by-step approach to problem solving, whereby the problem (bird declines through habitat loss and degradation) is first described, then opportunities for

solutions identified and finally priority actions recommended. The same process was followed for each strategy in order to provide a consistent approach which allows direct comparisons between habitats, for example, in terms of the number of priority species and the impacts of threats.

Habitat definition and distribution

A schematic diagram indicating the division of habitats into eight broad categories (presented as separate strategies/chapters in this book) is presented in Figure 2. Each strategy starts by defining the habitat covered. Habitats were distinguished mainly on the basis of broad biogeographical factors, vegetation-types, bird communities and predominant land-uses. Existing European habitat classifications were not followed

---

[1] Macaronesia is the biogeographical region composed of the archipelagos of Madeira, the Canary Islands, the Azores and (outside Europe) the Cape Verde Islands.

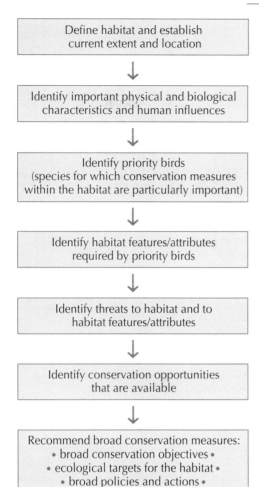

Define habitat and establish
current extent and location

↓

Identify important physical and biological
characteristics and human influences

↓

Identify priority birds
(species for which conservation measures
within the habitat are particularly important)

↓

Identify habitat features/attributes
required by priority birds

↓

Identify threats to habitat and to
habitat features/attributes

↓

Identify conservation opportunities
that are available

↓

Recommend broad conservation measures:
• broad conservation objectives •
• ecological targets for the habitat •
• broad policies and actions •

Figure 1. The process that was consistently followed in the preparation of each habitat conservation strategy.

as they were either too broad, or were restricted geographically or to certain habitats, or were primarily based on botanical criteria, e.g. CORINE (Devilliers and Devilliers-Terschuren 1996).

Most habitats in Europe have been covered by this project, but it is not entirely comprehensive. Urban habitats including parks and gardens, etc., have not been covered as they are not priority habitats at the continental scale covered here, although they are often of local importance for maintaining biodiversity. Some rare and highly important habitats, such as the inland sand-dunes or laurel forests of the Canary Islands, are not covered as these are often already subject to adequate site protection measures or other conservation initiatives.

It has also proved difficult to deal adequately

with some transitional habitats, such as the intergrade between boreal forest and tundra, and between forest and steppe. Consequently, it has been necessary to draw rather artificial dividing lines between some habitats. In other cases, for practical purposes, some overlap has been intentionally allowed. For example, to keep the habitat requirements and threats to seabirds within one strategy, sea-cliff and island breeding sites of seabirds are covered primarily under 'Marine Habitats'. However, these may also be defined as coastal habitats, and the issues and recommended policies and actions that are covered under 'Coastal Habitats' are clearly also of relevance.

The first section of the strategy also reviews the location and extent of the habitat. The aim is to establish which countries and regions have substantial areas of the habitat and, therefore, special responsibility for its conservation. This is often difficult due to inconsistencies between countries and regions in the definition of habitats and in the methods of grouping them. In some cases, maps of the breeding distribution of selected bird species (which can indicate the presence of a habitat) are used to supplement broad habitat statistics. These maps are based on data collected for *The EBCC atlas of European breeding birds* (Hagemeijer and Blair 1997).

Ecological characteristics

The important physical and biological characteristics of the habitat, and human influences, are then reviewed as the second stage of the strategy. This is necessary because such factors influence birds and their conservation. For example, oceanographic factors have a fundamental effect on the productivity of the seas, which in turn influences the number and diversity of seabirds that can be supported. Similarly, vegetation is profoundly influenced by climate, soil-type and topography, while semi-natural and artificial vegetation associations are affected by man's activities, such as wood-cutting, fire, cultivation, water management and grazing.

However, this is not a natural history text book and does not attempt to provide a comprehensive or detailed review of each habitat or ecological process.

Priority birds

The next stage of the strategy is to identify the most important bird species for habitat-conservation measures, hereafter termed 'priority' species[2]. Four classes

---

[2] This term should not be confused with the 'priority species' of the Habitats Directive of the European Union (EU), nor does it necessarily encompass the same selection of bird species which, according to BirdLife International, should be considered as such in the framework of the Birds Directive of the EU.

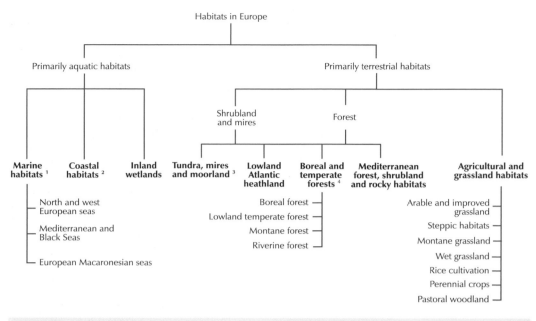

Figure 2. The division of habitats covered by this project. Habitat conservation strategies (each forming a separate chapter of this book) are indicated in bold. For identification of priority bird species and assessment of threats, some habitats are subdivided.

[1] Includes sea-cliffs and offshore islands where these are used by priority species for nesting.

[2] Includes coastal sand-dunes.

[3] Includes all wetlands within these habitats, as well as coastal habitats north of Arctic Circle.

[4] Includes forested mires.

of priority species (A–D) are identified on the basis of the species' European Threat Status as defined by Tucker and Heath (1994) and the importance of the habitat to its European population, i.e. the maximum percentage that uses the habitat during any part of the species' annual cycle (see Table 1). Thus, priority species include all birds that regularly occur in the habitat and have an Unfavourable Conservation Status in Europe (Box 2).

Birds that are highly dependent on the habitat at some point in their annual cycle (irrespective of their

conservation status) are also treated as priority species. This is intended to take account of the wider importance of habitats for biodiversity and of the need to conserve representative and characteristic birds as well as threatened species.

Species which have an Unfavourable Conservation Status and are highly dependent on the habitat should be regarded as being of particularly high priority, and the prioritization system has been designed to reflect this. *However, the prioritization of species does not take into account the threats to the*

Table 1. The assignment of priority categories (A, B, C, D) to bird species for each habitat. 'Habitat importance' is the maximum percentage of the European population using the habitat for any activity and at any stage of the annual cycle (i.e. for feeding, nesting or roosting—during the breeding season, on passage or in winter). Species which have an Unfavourable Conservation Status in Europe (SPEC categories 1–3) only because they are 'Localized' in winter (see Box 2) are treated as having a Favourable Conservation Status in Europe (SPEC 4 or Non-SPEC) within breeding habitats.

Habitat importance (% of European population using habitat)	SPEC category				
	1	2	3	4	Non-SPEC
<10%	B	C	D	—	—
10–75%	A	B	C	D	—
>75%	A	A	B	C	D

species within the habitat, and thus should not be considered as indicating by itself the urgency for conservation action. Instead, the classification primarily identifies species for which conservation measures within the habitat are particularly important for their survival.

The process of identifying priority species provides a quantitative way of assessing a habitat's relative importance compared with others. However, the absolute numbers of priority species per habitat will depend on the way that the habitats in Europe have been divided up, i.e. a large and diverse habitat will generally have more priority species. The allocation of species that has resulted here is appropriate for a European-scale conservation strategy which is focused on broad land-use policies. For example, it indicates that concentrating efforts on measures that will have broad impacts on boreal and temperate forests or arable farmland could have wider impacts than similar measures relating to riverine forests or lowland Atlantic heathlands.

However, this prioritization should *not* be applied at the regional or local scale, where priorities may be very different. Indeed, habitats such as riverine forests and lowland Atlantic heathlands have been greatly reduced in Europe and remaining areas are of considerable importance.

The aim of the prioritization is to assess the overall importance in Europe of the habitat and the threats to it, in order to identify broad priority actions. Species-by-species actions are not advocated in this publication. Thus, it is anticipated that non-priority bird species will also benefit from the broad measures identified in the strategy.

A broad analysis of the breeding distribution of priority species is also carried out at this stage, in order to identify those countries with particular responsibilities for these species. This analysis is based on the breeding bird population data held in the European Bird Database of BirdLife International and the European Bird Census Council. It should, however, be noted that because these analyses do not take into account non-breeding birds, the importance of countries that hold wintering or migrating populations may be underestimated. This is likely to be particularly true for inland wetlands and coastal habitats.

Habitat needs of priority birds

This review provides a broad picture of the habitat features required by priority bird species, based on a more detailed analysis that is given in Appendix 3 (p. 354). Its purpose is to identify those necessary features that are widely shared between species within the habitat. Particularly important requirements of individual priority species are highlighted where

appropriate.

This assessment is based on the personal experience of the Habitat Working Groups, together with inputs from other relevant experts, and on numerous accounts of species' ecology in the literature, although reference details are not normally given due to the complexity of the analysis (see Appendix 3).

Quantifying these habitat needs has proved challenging, because of the huge overall range and extent of the major habitat-types in Europe, their varied spatial characteristics, and the interactions between habitats and other environmental factors. Whilst information on habitat use is given individually for priority birds in such works as *The birds of the western Palearctic* (Cramp *et al.* 1977–1994) or the *Handbuch der Vögel Mitteleuropas* (Glutz von Blotzheim *et al.* 1966–1997) and in more specialist texts, Europe-wide summaries have never been presented for any broad habitat-types. Appendix 3 is thus particularly valuable in that it reflects the personal experience of field-workers who have researched many of the priority species in different parts of Europe.

However, again due to the potential complexity of the subject, particularly with regard to regional variations in habitats and in species' ecology, this section cannot attempt to provide a detailed account of habitat requirements.

Threats to the habitat

Threats to the habitat, and to the habitat features required by the priority species, are then reviewed, and an assessment is provided of the past scale of these threats over the last 20 years and their predicted scale over the next 20 years. Threats are considered as factors which reduce the extent or ecological quality of the habitat, thus causing in turn significant population declines of the priority species within the habitat.

By definition, a particular habitat is important for the survival of its priority species, thus net loss of habitat extent reduces the resource base upon which the priority species depend (in Europe), and usually leads to declines in European populations.

On the other hand, threats to habitat quality operate at a more subtle scale, by destroying or changing the particular features that are required within a habitat by priority species (as summarized in the previous section). Although frequently less obvious and more difficult to measure than habitat loss, reductions in habitat quality may be of equal and more immediate significance for bird populations on a continental scale. Thus, although a priority species may be able to persist in degraded low-quality habitat at a much reduced density, substantial population densities may only be sustained in habitat of the highest quality, and thus steep declines in bird populations may take place over large regions long before the habitat is even recognizably degraded to a non-informed observer.

The Habitat Working Group, with inputs from other relevant experts in the BirdLife network, predicted the impact of each threat on the European population of each priority species (within the habitat) over the next 20 years. This complex and detailed assessment is presented here in Appendix 4 (p. 401).

There are very few field studies of threats to birds at the level of the entire European population. Most studies have been local or, rarely, regional. Thus the impacts of most threats are largely inferred, and based on the majority opinion of the Habitat Working Group and the other consulted experts.

Importantly, where a threat is not listed in Appendix 4 as having an impact on individual species, this does not indicate that there is definitely no impact, but merely that no harmful effects have been reported so far. It has also not been possible in this assessment to take into full account the interactions between threats on a broad, continental scale, although these can certainly occur in a manner that magnifies the total impact beyond the sum of the individual threats. This is due to the high complexity and largely unstudied character of this field.

The direct effect on populations of taking birds for sport, food, sale or other purposes is not assessed in these strategies as it is not a component of habitat quality and is outside the scope of the project. Such an assessment would also be difficult since the actual impact resulting from such hunting and trapping is itself dependent on habitat quality, through the latter's effects on recruitment and mortality.

Conservation opportunities

Before making recommendations to alleviate the major threats to habitats and their priority bird species, a review of the main opportunities for implementing conservation actions in the habitat is made in this section. This covers international legislation (such as conventions and EU Directives), key policies and initiatives (such as the EU Common Agricultural Policy or the Helsinki Guidelines for Sustainable Forestry) and economic instruments, such as subsidies, taxes, and funds for actions (e.g. LIFE funds of the European Union).

The background to, and aims of, all of these laws, policies, funding sources and other initiatives are summarized in the introductory chapter 'Opportunities for Conserving the Wider Environment' (p. 24), since many have shared relevance and applicability across habitats. Only information that is specific or relevant to the particular habitat is included in each

strategy, together with an explanation of how conservation benefits may be obtained.

Conservation recommendations

Finally, the recommendations for conservation measures, based on the preceding analysis, are provided in the last section of the strategy.

Broad conservation objectives

Broad conservation objectives are established for the habitat and its priority species, on the basis of the priority species' habitat requirements and the broad nature of the threats to them and to the habitat in general.

Ecological targets for the habitat

The ecological needs of the priority species are then summarized in terms of targets for the habitat (an 'ideal picture' of what the habitat should be like), drawing on the previous assessment of habitat requirements. Although 'targets' should usually be quantitative and time-limited, this has not normally been possible here, because of inadequate information and the wide geographical scale of this study and the variation in the habitats covered. Strategies compiled at the national scale present a better opportunity for defining such targets.

Broad policies and actions

To meet the ecological targets for the habitat, priority policies and actions are then identified, based firstly on strategic principles for conservation in the wider environment. Since these principles are shared between all habitats, they are not elaborated upon in each strategy but are simply stated in the following chapter (under 'Strategic principles', p. 24).

Because these strategies aim to contribute towards the development of wide-scale habitat conservation measures for birds across Europe, the recommended policies and actions concentrate on broad measures that are applicable to large regions of Europe or to populations of priority species that are particularly important at the European scale. Such actions, although general, clearly have the potential to have widespread and substantial effects on bird populations if implemented. It is, however, realized that habitats, species requirements and threats can vary considerably at national and local levels, and that most opportunities to influence policy also occur at the national level. Therefore, although the treatment of these issues is also beyond the scope of this publication, BirdLife International intends through this book to set out a generic European strategy within which more detailed action plans can be developed for particular habitats, or for habitats at national, regional or local scales.

The importance of implementing each action could potentially be related to the number and severity of the threats that it addresses. However, in many cases the actions are general and potentially relate to more than one habitat, and may address many or even all of the threats to varying degrees. In addition, the priority for implementing actions should take into account the opportunities available, the likelihood of success and the potential impact of the action on the threat. These are difficult to assess or predict in the long term. Opportunities, such as reform of sections of the Common Agricultural Policy, may come and go in a relatively short time as a result of political initiatives and timetables. Although certain actions may have a low likelihood of success, their potential impact may be enormous. And, of course, national variations in conservation priorities, threats and opportunities may well be substantial. A formal prioritization of actions listed in each strategy is therefore not appropriate but, again, should be carried out as part of the planning of national or habitat-specific actions where a more detailed analysis can be undertaken.

OVERVIEW AND CONCLUSIONS

As a result of this process, each habitat conservation strategy (i.e. chapter) provides a comprehensive account of the issues and actions relating to the habitat (although in some cases cross-references are given to issues which have been treated in greater detail elsewhere). However, it is also essential to consider the relationships between habitats and to compare priorities among them in the development of an overall strategy for conserving habitats in Europe. Consistent methods have been used to identify priority species and the impacts of threats on them. It is, therefore, possible to compare threats across habitats and to produce a broad assessment of their relative importance in relation to the needs of all habitats. This is carried out in the summary chapters on 'Overview of the Strategies' (p. 41) and 'Main Conclusions and Recommendations' (p. 53). Similarly the conservation actions are reviewed to identify those which are beneficial to more than one habitat.

OPPORTUNITIES FOR CONSERVING THE WIDER ENVIRONMENT

THIS CHAPTER outlines the strategic principles and current opportunities for conserving the wider environment in Europe, in terms of international legislation, economic instruments, and broad policies and initiatives. It ends with a discussion of the opportunity presented by the publication of this book itself, particularly with regard to the development of Habitat Action Plans.

STRATEGIC PRINCIPLES

The eight main chapters in this book (the habitat conservation strategies) identify threats to habitats. In moving from these threats to identifying the broad conservation policies and actions that are needed, a number of themes or principles arise commonly among the strategies. Rather than repeat such material across chapters, these strategic principles are elaborated on below.

The principle of sustainability

The term 'sustainable development' was defined in the Brundtland Report (WCED 1987) as a development that 'meets the needs of the present without compromising the ability of future generations to meet their own needs'. If taken out of its context, this definition is sufficiently vague to appeal to many.

Although originally put forward as an alternative to 'perpetual growth', sustainability is often (mistakenly) promoted with the catchphrase 'sustainable growth'. Growth, however, cannot be sustained indefinitely in a limited system, but development (quality) probably can. In order to reduce ambiguity, a more practical definition of sustainability is needed.

A more useful definition is the amount of consumption that can be continued indefinitely without degrading capital stocks, including 'natural capital' (Costanza 1991). Although it has more practical value in driving real-life policies, this definition remains very much in the realms of mainstream economics, in that it reduces people to the level of consumers only, while nature is reduced to capital.

Within this definition, two kinds of sustainability can be distinguished (following Hicks 1946): 'weak' or 'strong'. 'Weak sustainability' strives to maintain total capital intact, without distinguishing between man-made or natural (or any other, e.g. social) capital. This assumes perfect substitutability between various types of capital. The impossibility of this claim has been widely accepted among ecologists, though less so among economists.

'Strong sustainability' on the other hand maintains the different types of capital intact and separate. This appears to be the only workable definition which actually leads to a sustainable use of natural resources.

The Ramsar Convention (see p. 27) also includes the concept of sustainability, described by the more traditional phrase of 'wise use'. Wise use is defined as sustainable utilization for the benefit of mankind in a way that is compatible with the maintenance of the natural processes of the ecosystem. Sustainable utilization is defined as the human use of a wetland so that it may yield the greatest continuous benefit to present generations while maintaining its potential for the needs and aspirations of future generations. In other words, contracting parties have agreed to a definition that makes the maintenance of biodiversity (a natural property of an ecosystem) a key test of sustainability.

Under this principle, it is also important to mention the concept of the carrying capacity of the natural environment in terms of human population. Natural resources being limited in supply, and therefore able to provide only a limited amount of services to the economic system within their bound, there must also be a limit to man's consumption of these services without disrupting the whole system. Total human consumption can be defined by two factors: the size of the population and the consumption level per capita. If total consumption increases beyond a certain threshold, the natural system will be disrupted to such an extent that the services it provides will decline. Eventually this will lead to an inevitable reduction in the total consumption level through reduced per-capita consumption, or through a reduced human population level, or both.

This implies that at every population level there is a threshold in per-capita consumption and, inversely, for every given consumption level there is also a limit to the human population. An important goal for humanity today should therefore be to identify the human carrying capacity of the Earth's natural system, and to keep our numbers and consumption well within limits.

Although it is necessary to define sustainability in terms of material consumption, this is by no means the end of the story. In the long term, we should seek to maintain the quality of all life (human or otherwise) in the widest sense, including aesthetic and cultural values. Bringing such non-material and non-monetary values into the equation, it is likely that human consumption and corresponding population levels will have to be further reduced if sustainability is to be achieved.

As a final word of caution, note however that the concept of sustainable development *does not* negate the need for protected areas, as is claimed by some in society. Biodiversity is an important indicator of sustainability. Maintaining the biodiversity and natural resources of Europe requires the setting aside of areas where human interference with species, habitats and natural ecosystem processes is regulated to some extent. Protected areas have important functions, including education and research, and have non-material and non-monetary values as well as providing the focus for national or local pride. These functions will not disappear in a sustainable society; on the contrary, they will be increasingly highly valued. The number and total extent of protected areas will need to be increased greatly if sustainability is to be achieved.

The precautionary principle

Many decisions affecting the environment are made in ignorance of sufficient knowledge to predict the impacts. The precautionary principle insists that, in such cases, the benefit of the doubt is given to the environment rather than to the case for the development. This is expressed in the Bergen Ministerial Declaration on Sustainable Development (Anon. 1990) as follows: 'where there are threats of serious or irreversible damage, lack of full scientific certainty should not be used as a reason for postponing measures to prevent environmental degradation'. In a stronger sense, no activity should be permitted unless it can be shown that it will not cause unacceptable harm to the environment.

The 'polluter pays' principle

In its simplest form, this says that the costs of measures to prevent, control, monitor and reduce environmental damage should be borne by the responsible party. In addition, all external costs of production (i.e. those costs that are imposed upon a third party, or in this case, society at large, by the parties engaged in a transaction—in this case, production), not just pollution, should be internalized and be borne by the parties responsible for the generation of these external effects.

The use of regulatory mechanisms

Regulation has the potential to set minimum standards for environmental management, and to prevent losses and damage below critical levels. Within the EU, the Birds and Habitats Directives set standards which have only partially been implemented in national legislation and on the ground. At national and sub-national levels, a range of regulations affects the maintenance of habitats by supporting protected areas, by protecting individual species, by controlling pollution, by formalizing the planning of built developments, and through other approaches.

The enormous scale of biodiversity loss in recent decades is a clear indication that regulation alone has not been sufficient to ensure sustainable use of the environment. The minimum standard that regulation can achieve is clearly too low, but few people would accept the desirability of more regulation. Clearly, society needs to find alternative approaches to the restoration of previous losses, and to the curtailment of further unnecessary damage. The most obvious measures are the provision of financial incentives.

Economic measures

In most countries, the state has a large impact on the economic attractiveness of various land-uses through a range of financial instruments. In the most obvious cases, some activities are directly subsidized. It is quite wrong for subsidy to support activities which are damaging to the environment. Such subsidies are perverse, in that they run directly counter to the intentions of environmental departments and regulations striving for contrary effect.

Alternative financial mechanisms can influence land-use, by adjusting the markets for products or by influencing the support or otherwise provided through the taxation system. Nowhere in Europe are the opportunities for such influence greater than in agriculture.

Greater use needs to be made of such economic instruments to achieve defined environmental targets. In particular, a significant extension of the 'management agreement' concept is needed (e.g. as embodied in the EU Agri-environment Regulation).

These instruments acknowledge the fact that society not only expects farmers to provide food but also, increasingly, to manage and shape the majority of Europe's habitats and landscapes. It is reasonable to strive for better returns on such subsidies, in terms of targeted and quantified environmental benefits.

'Structural' funds (for infrastructural development), whether provided by the EU, by other international or multilateral sources, or bilaterally by national governments, provide another opportunity for public funding, both to avoid environmental damage and also to provide benefit to the environment. The current perversity of such funding is frequently due to the omission of proper environmental assessment, which often fails to take into account or 'internalize' the costs incurred by non-sustainable development.

The 'polluter pays' principle (see above) is another example of the use of economic instruments to control environmental impact. Again, by internalizing environmental costs, this approach properly appraises projects or policies whose economic attractiveness might otherwise be misjudged.

INTERNATIONAL LEGISLATION

The various international and regional directives and conventions represent a strong basis for international cooperation in the conservation of the wider environment and shared natural resources such as biodiversity. These forms of international legislation are described below if they cover more than one broad habitat-type or are otherwise particularly broad-based. More habitat-specific or otherwise restricted legal instruments are described in the relevant chapters, under 'Conservation opportunities'.

International environmental legislation has two notable disadvantages (Biber-Klemm 1991), in that the potential geographical coverage, etc., is often incomplete (e.g. EU Directives only cover member states), and the contractual obligations are often not well implemented in practice, e.g. because conventions tend to be vaguely formulated. Furthermore, although the obligations of such legislation have to be fulfilled within the territory of the contracting parties, good implementation by national governments cannot be taken for granted, and yet direct inter-party control of implementation is not possible either (Biber-Klemm 1991).

A number of the conventions listed below deal with the conservation of particular habitat-types at a broad level in Europe or more globally, but it is notable that there are none aimed specifically at the two most extensive habitats in Europe, agricultural and forest habitats.

All of these instruments need to be strengthened, and more and clearer biodiversity-conservation meas-

ures and targets need to be incorporated into most. Efforts at implementation should also be stepped up by national governments and relevant international institutions.

In addition, with regard to individual sites, the greatest protection can only be achieved under proper national (as opposed to international) legislation. National governments should increase the number and area of their statutory protected areas and should make sure that all habitat-types are adequately represented in protected-area networks. Protected areas require effective implementation, management planning, staffing, funding, local involvement and public support if they are to fulfil their role in conserving biodiversity and promoting the conservation of species and habitats.

Global instruments

'Biodiversity Convention' or 'CBD': Convention on Biological Diversity

Coverage[1] Worldwide; there were 133 contracting parties by February 1996, as well as 41 signatories. Of these, 26 contracting parties (including the EU) and 12 signatories were from Europe.

Adopted May 1992, entered into force in November 1994.

Objectives To ensure the conservation of biological diversity, the sustainable use of its components and the fair and equitable sharing of the benefits arising out of the utilization of genetic resources.

Obligations The most relevant obligations, in relation to conservation of the wider environment, are as follows.

● To prepare national strategies, plans or programmes for the conservation and sustainable use of the nation's biological resources, and to integrate these into other relevant sectoral or cross-sectoral plans, programmes and policies.

● To promote the protection of ecosystems, natural habitats and the maintenance of viable populations of species in natural surroundings, and to rehabilitate and restore degraded ecosystems and promote the recovery of threatened species, *inter alia*, through the development and implementation of plans or other management strategies.

Comments This is probably the most important broad legislation for maintaining biodiversity in the European environment. It was a product of the United Nations Conference on Environment and Development (the 'Earth Summit') held in Rio de Janeiro in 1992.

---

[1] 'Coverage' indicates potential geographical coverage.

The Convention provides a framework for conserving biodiversity, mainly by setting out policies that parties should follow. Although many of the objectives and prescribed measures are not new and are imprecisely defined, the Convention has raised awareness of the needs of biodiversity conservation and placed them on the political agenda. National biodiversity conservation strategies have been developed or are under development in at least seven European countries.

The obliged national strategies, plans or programmes for the conservation and sustainable use of biodiversity are important mechanisms for integrating environmental considerations into all socio-economic sectors: they set the national priorities for biodiversity conservation, assess the resources that are required for the implementation of such priorities, and evaluate the economic, social and environmental impacts of the proposed measures. In Europe, such national strategies have already been prepared by three countries, and are under preparation in a further six countries.

The European Union (EU), as a contracting party to the Biodiversity Convention, is also obliged to prepare such a strategy. The preparation of the EU Biodiversity Strategy started in late 1996, and wide public consultation and debate of the first version is expected in 1997.

In order to coordinate implementation of the Biodiversity Convention beyond the EU and throughout Europe, a Pan-European Biological and Landscape Diversity Strategy (PEBLanDS) has been developed by the Council of Europe (see 'Broad policies and initiatives', p. 38).

'Ramsar Convention': Convention on Wetlands of International Importance especially as Waterfowl Habitat

Coverage Worldwide; there were 87 contracting parties by June 1995, including 32 European states.

Adopted February 1971, entered into force in December 1975 (the 'Regina amendments' came into force in June 1994).

Objectives To stem the loss of wetlands and to ensure their conservation and wise use.

Obligations

- To designate suitable wetlands within Contracting Parties' territories for inclusion in a 'List of Wetlands of International Importance' (Article 2,1).
- To formulate and implement Contracting Parties' planning so as to promote the conservation of the wetlands included in the List and, as far as possible, the wise use of wetlands in their territory (Article 3,1).
- To promote the conservation of wetlands and waterfowl by establishing nature reserves on wetlands whether they are included in the List or not, and provide adequately for their wardening (Article 4,1).

Comments The Convention is the most important international legal instrument for the protection of wetlands. In Europe, about 88 coastal/marine wetlands and more than 167 inland wetlands were listed as 'Wetlands of International Importance' as of December 1995 (Frazier 1996). The Ramsar classification of wetlands (Ramsar Convention Bureau 1990) is wider than the definitions of coastal and inland wetlands that have been adopted in this book (see relevant chapters, p. 93 and p. 125).

The Convention introduced the now widely used concept of '1% flyway population levels' for assessing the international importance of sites and, although this was framed around discrete sites where large numbers of wintering waterbirds occur in relatively small areas, the 1% approach is equally relevant for breeding species, albeit that sites usually cover larger geographic areas, reflecting the much more dispersed nature of many breeding waterbird populations.

'Bonn Convention' or 'CMS': Convention on the Conservation of Migratory Species of Wild Animals

Coverage Worldwide; there were 49 contracting parties by May 1996, including 20 European states and the European Community, as well as eight signatories, including one European state.

Adopted June 1979, entered into force in November 1983.

Objectives The conservation and effective management of terrestrial, marine and avian species over the whole of their migratory range.

Obligations

- To undertake research activities relating to migratory species.
- To adopt strict protection measures for migratory species that have been categorized as 'endangered' (listed in Appendix I of the Convention).
- To conclude Agreements for the conservation of migratory species which have an unfavourable conservation status or which would benefit significantly from international cooperation (listed in Appendix II of the Convention).

Comments The 'Agreement on the Conservation

of African–Eurasian Migratory Waterbirds' (AEWA) aims to create the legal basis for a concerted conservation policy among the range states of all migratory waterbird species and populations which migrate in the western Palearctic and Africa, irrespective of their current conservation status. AEWA includes 170 species within its remit, and covers an area of 60 million square kilometres, encompassing 120 range states in Africa and Europe as well as parts of Asia.

The Agreement provides a framework for conservation action, monitoring, research and management of several globally important bird-migration systems. It has been open for ratification since October 1995, had been signed by seven states by January 1997, and is due to enter into force in 1998.

'World Heritage Convention': Convention concerning the Protection of the World Cultural and Natural Heritage

Coverage Worldwide; 137 countries were contracting parties by July 1994, including 31 European states.

Adopted November 1972, came into force in December 1975.

Objectives To ensure the protection of natural and cultural areas of outstanding universal values as a duty of the international community as a whole, by granting collective assistance.

Obligations To ensure the identification, protection, conservation, presentation and transmission to future generations of the cultural and natural heritage situated on the territory of each party state (Article 4).

Comments This convention offers the opportunity to protect unique wildlife habitats and representative examples of the most important ecosystems. It offers financial and technical assistance for the protection of sites and imposes strict obligations on its contracting parties (Lyster 1985).

'Law of the Sea' or 'UNCLOS': United Nations Convention on the Law of the Sea

Coverage Worldwide; there were 81 contracting parties and 83 signatories by March 1996, of which there were 12 and 19 respectively in Europe.

Adopted December 1982, entered into force in November 1994.

Objectives A legal order for the seas and oceans which will facilitate international communication and will promote peaceful uses of the seas and oceans, the equitable and efficient utilization

of their resources, the conservation of their living resources and the study, protection and preservation of the marine environment.

Obligations
- Coastal states exercise sovereignty over their territorial seas in an area up to 12 nautical miles in width.
- Archipelagic states have sovereignty over a sea-area enclosed by straight lines drawn between the outermost points of the islands.
- Coastal states have sovereign rights in a 200-nautical-mile exclusive economic zone (EEZ) with respect to natural resources and certain economic activities.
- Establishment of the deep sea-bed as outside national jurisdiction and of its resources as 'the common heritage of mankind', and establishment of an International Sea-Bed Authority to oversee its exploitation.

Comments UNCLOS consists of 320 articles and 9 annexes dealing with almost all aspects of ocean-use. Exploitation and protection of living resources and anti-pollution measures are covered, albeit only generally. States are not allowed to make any reservations when they join UNCLOS.

'MARPOL 73/78': International Convention for the Prevention of Pollution from Ships, 1973, as modified by the Protocol of 1978 relating thereto

Coverage Worldwide; 88 states had ratified the Convention by January 1995, 33 of them being European states.

Adopted February 1973, modified in February 1978, entered into force in October 1983.

Objectives To prevent and control deliberate discharges and accidental spills from shipping into the marine environment.

Obligations
- To take measures to prevent pollution by:
  - oil;
  - noxious liquid substances in bulk;
  - harmful substances carried by sea in packaged forms, or in freight containers, portable tanks or road and rail wagons;
  - sewage;
  - garbage.

Comments This is the principal international agreement for regulating unnecessary discharges of oil from ships. The measures on sewage have not yet entered into force.

'Climate Convention': United Nations Framework Convention on Climate Change

Coverage Worldwide; 165 states or supranational organizations had ratified the Convention by early 1997, including at least 34 European states plus the EU.

Adopted December 1992, entered into force in 1994.

Objectives The Convention recognizes that man-made climate change is a problem. It sets out to achieve stabilization of greenhouse-gas concentrations in the atmosphere at a level that would prevent dangerous anthropogenic interference with the climate system. Such a level should be achieved within a time-frame sufficient to allow ecosystems to adapt naturally to climate change, to ensure that food production is not threatened, and to enable economic development to proceed in a sustainable manner.

Obligations

- To develop national inventories of greenhouse gases.
- To formulate policies to mitigate climate change.
- To cooperate in preparing for adaptation to climate change.
- To promote international cooperation in the transfer of technology.

  Developed countries also signed up to return their emissions of greenhouse gases to 1990 levels by the year 2000, and are required to adopt policies and measures to limit greenhouse-gas emissions, in order to lead the way in modifying long-term trends.

Comments Some limited progress has been made towards the objectives of the Convention since 1992. National commitments to emission-reduction targets for 2005 and the longer-term are now needed, and are the subject of intense negotiations.

Pan-European instruments

'Bern Convention': Convention on the Conservation of European Wildlife and Natural Habitats

Coverage Member states of the Council of Europe (currently 38 countries) and invited non-member states in Europe and North and West Africa; 29 European countries, as well as the EU, had ratified the Convention by December 1996.

Adopted September 1979, entered into force in June 1982.

Objectives To conserve wild flora and fauna and their natural habitats, especially those species and habitats whose conservation requires the cooperation of several states.

Obligations To take appropriate and necessary legislative and administrative measures to ensure the conservation of the habitats of the wild flora and fauna, especially those specified in Appendices I and II, and the conservation of endangered natural habitats.

Comments The need to implement the Bern Convention in the European Union was the motivation for the Habitats Directive (see below), which has extended and improved the obligations of the Convention into EU law.

The EMERALD network is a recent initiative, under this Convention, to extend the Natura 2000 network of protected areas (see 'Habitats Directive', below) beyond the EU to cover the whole of Europe.

'Espoo Convention': Convention on Environmental Impact Assessment in a Transboundary Context

Coverage Member states of the UN Economic Commission for Europe, and associated states or organizations. A total of 28 European countries were signatories as of January 1996.

Adopted February 1991, but not yet entered into force.

Objectives

- To promote the environmental assessment of proposed activities whose impacts have the potential to affect more than one state.
- To improve the methodology of such assessments, and improve mechanisms of cooperation between states, so as to minimize and mitigate the transboundary environmental impacts of economic activities.

Obligations

- As a minimum requirement, contracting parties should undertake environmental impact assessments (EIAs) at the project-level where a proposed activity is likely to cause significant adverse transboundary impacts.
- The general public of any affected areas of other contracting parties' territory should be allowed equal access to the consultation process as the general public of the state which originates the proposed activity.

Comments 'Proposed activities' are described in Appendix I. Appendix II describes the minimum information-content that is required of the documentation resulting from the EIA.

Regional instruments

'OSPAR Convention': Convention for the Protection of the Marine Environment of the North-East Atlantic

Coverage Belgium, Denmark, Finland, France, Germany, Iceland, Ireland, Luxembourg, Netherlands, Norway, Portugal, Spain, Sweden, Switzerland, United Kingdom and the Commission of the European Communities.

Adopted September 1992, but not yet entered into force.

Objectives To counteract the introduction of certain pollutants into the north-east Atlantic sea-area, and to monitor and assess the effects of certain activities.

Obligations
- To take measures to prevent and eliminate pollution from:
  - land-based sources;
  - dumping or incineration;
  - offshore sources.
- To take measures to assess the quality of the marine environment.

Comments This Convention is the result of revising and merging the Oslo and Paris Conventions (see below). The new Convention will enter into force when it has been ratified by all of the signatories to the Oslo and Paris Conventions. Meanwhile, the Oslo and Paris Conventions still apply.

'Oslo Convention': Convention for the Prevention of Marine Pollution by Dumping from Ships and Aircraft

Coverage Belgium, Denmark, Finland, France, Germany, Iceland, Republic of Ireland, Netherlands, Norway, Portugal, Spain, Sweden and UK.

Adopted February 1972, entered into force April 1974.

Objectives To regulate dumping operations involving industrial wastes, dredged material and sewage-sludge in the north-east Atlantic sea-area.

'Paris Convention': Convention for the Prevention of Marine Pollution from Land-based Sources

Coverage Belgium, Denmark, France, Germany, Iceland, Republic of Ireland, Netherlands, Norway, Portugal, Spain, Sweden, UK, and European Union.

Adopted June 1974, entered into force in May 1978.

Objectives To prevent, reduce and, as appropriate, eliminate pollution from land-based and offshore sources in the north-east Atlantic sea-area.

'Barcelona Convention': Convention for the Protection of the Mediterranean Sea against Pollution

Coverage The 20 states riparian to the Mediterranean Sea, and the European Union; all are contracting parties.

Adopted February 1976, entered into force February 1978 (reviewed and amended 1995).

Objectives
- To counteract the introduction of certain pollutants into the Mediterranean Sea area (including internal waters).
- To conserve natural species and habitats.

Obligations
- To take measures to prevent, reduce and, as appropriate, eliminate pollution:
  - by oil and other harmful substances;
  - by dumping from ships and aircraft;
  - from land-based sources;
  - resulting from exploration and exploitation of the continental shelf and seabed.
- To address problems caused by the transboundary movement of wastes and their disposal.
- To address the conservation of specially protected areas and biodiversity in the marine and coastal zone.

Comments The 'Action Plan for the Protection of the Mediterranean Environment and the Sustainable Development of the Coastal Areas of the Mediterranean' (Mediterranean Action Plan) was adopted at the Inter-governmental Meeting on the Protection of the Mediterranean in 1975, and was significantly revised in 1995 so that its objectives adhered more closely to the obligations of the Biodiversity Convention (see above).

The Regional Activity Centre for Specially Protected Areas (RAC/SPA) has compiled much information about potentially important marine protected areas as well as existing ones that need improved management.

'[New] Helsinki Convention': 1992 Helsinki Convention on the Protection of the Marine Environment of the Baltic Sea Area

Coverage Czech Republic, Denmark, Estonia, Finland, Germany, Latvia, Lithuania, Norway, Poland, Russia, Slovakia, Sweden, Ukraine, and European Union. All have signed, but not all have ratified, the Convention.

Adopted   April 1992, but not yet entered into force.

Objectives   To counteract the introduction of certain pollutants into the Baltic Sea area (including internal waters), and to conserve natural species and habitats.

Obligations

- To take measures to prevent and eliminate land-based input of the 16 noxious substances listed in Annex II.
- To implement MARPOL 73/78, and apply uniform standards for port-reception facilities.
- To maintain an ability to combat spillages of oil and other harmful substances.
- To prohibit incineration.
- To prohibit dumping (with some exceptions) at sea.
- To take measures to prevent pollution from the offshore oil and gas industry.
- To take measures to conserve natural habitats and biological diversity and to protect ecological processes.

Comments   This Convention is the result of revising the 1974 Helsinki Convention. The new Convention will not enter into force until it has been ratified by all of the signatories. This may require at least two more years. Meanwhile, the 1974 Helsinki Convention still applies.

'Helsinki Convention': 1974 Helsinki Convention on the Protection of the Marine Environment of the Baltic Sea Area

Coverage   Denmark, Estonia, Finland, Germany, Latvia, Lithuania, Poland, Russia and Sweden.

Adopted   March 1974, entered into force in May 1980.

Objectives   To counteract the introduction of certain pollutants into the Baltic Sea area (excluding internal waters).

Obligations

- To take measures to control and minimise land based input of the 16 noxious substances listed in Annex II.
- To implement MARPOL 73/78 (see above).
- To maintain an ability to combat spillages of oil and other harmful substances at sea.
- To prohibit dumping (with some exceptions) at sea.
- To take measures to prevent pollution from the offshore oil and gas industry.

'Bucharest Convention': Convention on the Protection of the Black Sea against Pollution

Coverage   All Black Sea riparian states; all states have ratified the Convention.

Adopted   1992, entered into force 1994.

Objectives

- To counteract the introduction of certain pollutants into the Baltic Sea area (including internal waters).
- To conserve natural species and habitats.

Obligations

- To take measures to prevent, reduce and, as appropriate, eliminate pollution:
  - by oil and other harmful substances;
  - by dumping from ships and aircraft;
  - from land-based sources.

Comments   The Convention forms part of the Black Sea Environmental Programme (BSEP), which was adopted in May 1993, in response to a Ministerial Declaration on the Protection of the Black Sea, with the aims of capacity building and institutional strengthening, technical assistance, and the promotion of environmentally sound investment activities.

In order to provide overall coordination to BSEP, the Black Sea Strategic Action Plan was adopted in October 1996. The development of a Protocol on Biological Diversity and Landscape Protection has been proposed, to address problems with over-exploitation of commercial species, loss of endangered species in the Black Sea and its wetlands, and lack of protection for habitats and landscapes. Some expected products and milestones by 2010 are:

- a regional network of conservation areas;
- new legal and regulatory regimes for these areas;
- recovery of high-value fisheries;
- regional licensing and quota systems for fisheries, based on national allocations.

There is a national NGO secretariat in each country, as well as an international NGO forum. By the end of 1996, the Istanbul Commission will be established and a Secretariat set up in Istanbul (Turkey) to coordinate BSEP, funded by the participating governments.

European Union instruments

There are probably no other instruments in the world which provide such high-level, legally binding measures, complete with some financial support, as the Directives of the EU. The 15 member states of the

European Union are Austria, Belgium, Denmark, Finland, France, Germany, Greece, Ireland, Italy, Luxembourg, The Netherlands, Portugal, Spain, Sweden and the United Kingdom. EU legislation generally applies to all 15 member states, unless otherwise stated.

'Birds Directive': Council Directive 79/409/EEC on the conservation of wild birds

Coverage All European Union member states.

Adopted April 1979, entered into force April 1981.

Objectives Conservation of all species of naturally occurring birds in the wild state in the European territory of the member states (excluding Greenland). It covers the protection, management and control of these species and lays down rules concerning their exploitation.

Obligations

- To preserve, maintain or re-establish a sufficient diversity and area of habitats for all the species of birds to which the Directive applies. Measures should include (a) creation of protected areas; (b) upkeep and management of habitats inside and outside the protected zones, in accordance with ecological needs; (c) re-establishment of destroyed biotopes; (d) creation of biotopes.

- To classify as Special Protection Areas (SPAs) the most suitable sites for the conservation of bird species listed in Annex I of the Directive, taking into account their protection requirements in the geographical sea- and land-area where the Directive applies.

- To take similar measures for regularly occurring migratory species not listed in Annex I, paying particular attention to the protection of wetlands, and particularly to wetlands of international importance.

- To ensure that the SPAs form a coherent whole which meets the protection requirements of the species in the geographical sea- and land-area where the Directive applies.

- To avoid significant disturbance and habitat deterioration in SPAs with respect to the Annex I species for which the areas have been designated.

- To assess the implications for the SPA of any plan or project likely to have a significant effect thereon, and to ensure that the competent national authorities agree to the plan or project only after having ascertained that it will not adversely affect the integrity of the site and, if appropriate, after having obtained the opinion of the general public.

- To take all compensatory measures necessary to ensure that the overall coherence of Natura 2000 is protected, if a plan or project must be carried out at or near an SPA for imperative reasons of overriding public interest, including those of a social or economic nature.

- To strive to avoid pollution or deterioration of habitats outside SPAs.

Comments Annex I of the Directive lists 182 species, subspecies or populations of birds in Europe (seven were added upon the accession of Austria, Finland and Sweden) that are considered in danger of extinction, vulnerable to specific changes in their habitat, rare, or requiring particular attention for reason of the specific nature of their habitat. Those Annex I species which are of priority for habitat conservation in Europe, as defined in this publication, are indicated in Appendix 1 of this book (p. 327).

Annex I has not yet been amended to take account of *Birds in Europe: their conservation status* (Tucker and Heath 1994), which constitutes the first quantitative review of the conservation status of all European birds, but many of the species considered to have an Unfavourable Conservation Status in Europe in that work are already included in Annex I.

By November 1996, the member states of the EU had designated 1,462 SPAs, which together cover 79,419 km^2 (Anon. 1996). According to the European Commission, implementation of the obligations concerning SPAs has been highly uneven between member states, being notably insufficient in France, Greece and Luxembourg, while being considered complete in Denmark and Belgium (Anon. 1996). However, most Important Bird Areas (Grimmett and Jones 1989) in the European Union that qualify as SPAs have still not received any such designation; in 1994, only about 23% of the area of such potential SPAs had actually been designated thus (CEC 1994c, Waliczky 1994), and this figure has not significantly improved since then.

'Habitats Directive': Council Directive 92/43/EEC on the conservation of natural habitats of wild fauna and flora

Coverage All European Union member states.

Adopted May 1992.

Objectives To promote the maintenance of biodiversity, taking account of economic, social, cultural and regional requirements, through measures that maintain or restore, at favourable conservation status, natural habitats and species of wild fauna and flora of Community interest in the European territory of the member states. The maintenance of such biodiversity may in certain cases require the maintenance, or indeed the encouragement, of human activities.

Obligations

- To create a coherent ecological network ('Natura 2000') of Special Areas for Conservation (SACs), setting the minimum standard for biodiversity conservation in the EU.

- To establish, in each member state, the conservation status of the habitats and species listed in the Annexes (especially priority habitats and species, as defined in Article 1), and to provide the means to monitor the further evolution of their conservation status.

- By June 1995, each member state should have proposed a list of sites indicating which of the priority natural habitat-types and species (listed in Annexes I and II) are hosted by (and native to) the proposed sites.

- By June 1998, the European Commission shall establish and adopt a draft list of 'sites of Community importance', in agreement with each member state and drawn from the member states' site-lists (mentioned above).

- As soon as possible after June 1998, and by June 2004, each member state shall designate the listed sites of Community importance as Special Areas of Conservation (SACs), establishing priority for designation in the light of the relative importance of each site for the maintenance or restoration, at a favourable conservation status, of a natural habitat-type in Annex I or a priority species in Annex II, and for the coherence of Natura 2000, and in the light of the threats of degradation or destruction to which each site is exposed.

- To avoid significant disturbance and habitat deterioration in SACs with respect to the habitats (Annex I) and species (Annex II) for which the areas have been designated.

- To assess the implications for the SAC of any plan or project likely to have a significant effect thereon, and to ensure that the competent national authorities agree to the plan or project only after having ascertained that it will not adversely affect the integrity of the site and, if appropriate, after having obtained the opinion of the general public.

- To take all compensatory measures necessary to ensure that the overall coherence of Natura 2000 is protected, if a plan or project must be carried out at or near a SAC for imperative reasons of overriding public interest, including those of a social or economic nature.

Where the site concerned hosts a priority natural habitat type and/or a priority species, the only reasons which may be considered are those relat-ing to human health or public safety, to beneficial consequences of primary importance for the environment or, further to an opinion from the Commission, to other imperative reasons of overriding public interest.

Comments The Directive has the capacity to protect substantial areas of habitat for a wide range of species. However, as with earlier environmental legislation, its success will depend on the degree of implementation by individual member states.

In the marine environment, this Directive may not provide protection for species and habitats beyond the limits of territorial waters.

Appendix 5 (p. 423) lists the 228 'natural habitat-types' of Annex I of the Directive (including 'priority' habitat-types as defined in Article 1). Annex II of the Directive lists species (of plant and animal) other than birds, the latter being covered by the Birds Directive (see above). Although the Annexes were adapted recently as a result of the accession of Austria, Sweden and Finland, further habitats of, and priority species of, the boreal zone may be added as a result of ongoing revision.

'EIA Directive': Council Directive 85/337/EEC on the assessment of the effects of certain public and private projects on the environment

Coverage All European Union member states.

Adopted June 1985.

Objectives To oblige member states to evaluate the impacts of public and private projects that are likely to have significant effects on the environment, to take account of concerns to protect human health and to maintain biodiversity and the natural resource capacity of ecosystems.

Obligations

- To carry out environmental assessments for specified classes of projects listed in the Directive, according to specified requirements. Member states should oblige developers to provide adequate information on their projects so that their impact on the environment can be assessed. Relevant authorities and the public should be consulted on the findings of such assessments. Potential transboundary effects of development projects should be communicated to other member states that may be affected.

Comments Annex I and II list those project-types which should be subject to Environmental Impact Assessment (EIA). The most important drawback of the Directive is probably that it does not imply that the final decision about the implementation of a development project should be influenced in any way by the EIA statement.

In addition, it applies only to projects, not the overall programmes, plans or policies that generate many individual projects. The European Commission plans to rectify this gap in Community law, and is working towards the adoption of a Directive on Strategic Environmental Assessment in the near future (see Box 1).

'Nitrates Directive': Council Directive 91/676/EEC protecting waters against pollution by agricultural nitrates

Coverage All European Union member states.

Adopted December 1991.

Objectives To reduce water pollution caused or introduced by nitrates from agricultural sources and to prevent such pollution.

Obligations

- To designate zones that are vulnerable to pollution from nitrates and which require special protection.

Box 1. Environmental assessment.

Strategic Environmental Assessment

Strategic Environmental Assessment (SEA) is a relatively new and still evolving means by which the environmental implications of proposed development policies, plans and programmes (PPPs) can be identified and evaluated at an early stage, before the PPPs are put into practice.

Strategic Environmental Assessment of PPPs parallels the existing process of Environmental Impact Assessment (EIA) of projects (see below). The European Commission is currently working on the development of an SEA Directive for the European Union (EU), and adopted a draft of the Directive in December 1996.

The following key issues should be taken into consideration by national governments and the EU in developing legislation and methodology for SEA.

- There is a need to consider the hierarchy of SEA of PPPs and projects. Different aspects of assessment should be considered at different stages.
- Good SEA will require adequate information on the state of the environment.
- Clear objectives, including environmental objectives, should be a key part of PPPs, so that SEA can consider not only the impact of the PPPs on the environment but their ability to achieve their own objectives as well.
- A range of alternative options should be identified, whose impacts can be assessed and compared.
- SEA requires the identification of an appropriate range of environmental indicators, and sound methods of impact prediction.
- An assessment of the cumulative impacts of a series of actions, and of interactions between separate impacts, should be a key part of an SEA.
- Once the impacts are identified, SEA should offer a wide range of opportunities for avoidance or mitigation.
- Monitoring of environmental impacts, and of the objectives of PPPs, will help to improve assessment and decision-making in the future.

SEA should be applied to a wide range of socio-economic or policy 'sectors', including agriculture, forestry, spatial development planning, transport, water-resource planning and tourism. SEAs should be consistent, transparent and accountable. SEA will have a key role in the integration of environmental policy objectives into other policy areas. SEA of PPPs will require flexible methodology, geared to the specific characteristics of the policy, plan or programme under assessment.

Environmental Impact Assessment

Environmental Impact Assessment (EIA) of a variety of types of development project, as applied through relevant national legislation and under Directive 85/337 in the EU, can provide a powerful mechanism for preventing or limiting environmental damage by such projects, if linked to planning regulations and if financial support (e.g. from the EU Structural Funds; see Box 2) is made conditional upon the results. In addition, EIA is now widely practised in Europe (e.g. UNECE 1995).

However, the scope of EIAs should be expanded to cover major forestry and agricultural projects, the assessment of which is not compulsory under current EU or most national legislation. Furthermore, standards vary, and a recent review concluded that ecological assessments are rarely given adequate consideration in EIAs (Treweek 1996). Suggested necessary improvements to EIAs include the following.

- Independent reviews of ecological statements.
- Formal monitoring of projects.
- Standardization of sampling/survey techniques.
- Quantitative impact predictions.
- Increased finance to develop national information databases.
- Research to standardize conservation evaluation procedures.
- Field-testing of impact predictions.
- Post-development monitoring.

Such improvements would require further legislation to define standards, as well as increased funding, perhaps provided by prospective developers (according to the 'polluter pays' principle).

- To establish action programmes in order to reduce nitrate pollution in these zones.

Comments Eutrophic freshwater lakes and other freshwater bodies likely to become eutrophic are addressed in this Directive, which provides criteria for identifying eutrophic waters which would be affected by nitrate pollution.

'Port State Control Directive': Council Directive 95/21/EEC concerning the enforcement, in respect of shipping using Community ports and sailing in the waters under the jurisdiction of the Member States, of international standards for ship safety, pollution prevention and shipboard living and working conditions

Coverage All European Union member states.

Adopted July 1995; implementation deadline by 30 June 1996.

Objectives To contribute to a drastic reduction of substandard shipping from the waters under the jurisdiction of member states.

Obligations

- To enhance compliance with international and relevant Community legislation on maritime safety, protection of the marine environment and living and working conditions on board ships of all flags.

- To establish common criteria for the control of ships by the State of the port and to harmonize procedures for inspection and detention, taking proper account of the commitments made by the maritime authorities of the member states with the Paris Memorandum of Understanding on Port State Control.

Council Directive 91/271/EEC on urban waste water treatment

Coverage All European Union member states.

Adopted May 1991.

Objectives

- To ensure that all urban agglomerations are provided with collecting systems for waste-water treatment, and that the collected water is sufficiently treated.

- To outline a programme which specifies the requirements for the collection, treatment and discharge of waste water from certain industrial sectors.

- To further the re-use of treated waste water, the monitoring of water subject to discharges from treatment plants, and the publication of situation reports every two years.

Obligations To designate sensitive and less sensitive areas for water discharge.

Comments Eutrophic, natural freshwater lakes are included in Annex II, where sensitive water-bodies are defined. This Directive has great potential to be influential in the improvement of the water quality in many wetlands.

Council Directive 90/313/EEC on the freedom of access to information on the environment

Coverage All European Union member states.

Adopted June 1990.

Objectives To allow any natural or legal person to receive information on the state of water, air, soil, fauna, flora, land, and natural sites held by public authorities in written, visual, oral or database form.

Obligations

- To make available information on the environment held by public authorities.

- To ensure freedom of access to, and the dissemination of, such information.

Council Directive 78/659/EEC on the conservation of fresh water for fish

Coverage All European Union member states.

Adopted 1978.

Objectives To protect or improve surface water in order to support fish life.

Obligations To designate areas and establish pollution-reduction programmes to ensure that the waters in these areas are brought into conformity with the values defined in Annex I.

Comments The Directive does not cover waters in natural or artificial fish-ponds that are used for intensive fish-farming.

Council Directive 76/464/EEC on the control of dangerous substances discharge

Coverage All European Union member states.

Adopted 1976.

Objectives Elimination or reduction of the pollution of inland, coastal and territorial waters by particularly dangerous substances.

Obligations

- To eliminate pollution by substances on List I of the Annex.

- To reduce pollution by substances on List II of the Annex.

- To establish a system of prior authorization for

the discharge of List I substances to inland surface waters, territorial waters, internal coastal
waters and groundwater, and where necessary
into sewers.

- To establish pollution-reduction programmes, with
deadlines for implementation and including prior
authorization and compliance with emission standards for discharges.

Comments Six other Directives are linked to this
one and establish emission-limit values and water-
quality objectives for List I substances.

'Agri-environment Regulation': Council Regulation 2078/92/EEC on agricultural production methods compatible with the requirements of the protection of the environment and the maintenance of the countryside

Coverage Member states of the European Union as
at the adoption date.

Adopted June 1992.

Objectives Protection of the farmed environment
through payments to farmers for farming practices which protect and manage the countryside.

Obligations

- Member states submit plans to the European
Commission for multi-annual programmes, determined at national or regional levels, covering
all territories.

- Schemes for paying farmers for extensification of
livestock or arable, maintenance of the countryside, organic farming and public access are prepared and administered by national governments.

Comments These are large (ECU 6.6 billion over
four years) and potentially valuable schemes, if
the objectives, targets, prescriptions, monitoring
and evaluation relate to conservation priorities.

ECONOMIC INSTRUMENTS

World Bank and Global Environment Facility (GEF)

The World Bank was established after the Second
World War, as part of the United Nations, to support
the reconstruction of Europe and the economic development of the world's poorer nations. It consists
of two institutions: the International Bank for Reconstruction and Development (IBRD) and the International Development Association (IDA). The former
borrows money on international capital markets and
lends it to countries for development projects. The
IDA is a fund established by the donor countries to
provide loans to the poorest countries.

The World Bank is controlled ultimately by the
Bank's member countries, i.e. practically all countries in the world. Day-to-day control is exercised by
the World Bank Executive Board, members of which
are elected or appointed by governments. The Board
has to approve all IBRD and IDA projects.

The World Bank funds a wide range of activities,
including large development projects, programmes
and sectoral adjustment projects. The World Bank
has Operational Policies (OPs) which require it to
incorporate the environment into its activities in
various ways. These include OPs on Environmental
Assessment, Natural Habitats, Forestry, Water Resources Management, and Indigenous People. The
World Bank is preparing a strategy for implementing
the Biodiversity Convention and is one of the three
implementing agencies of the Global Environment
Facility (GEF), together with the United Nations
Development Programme and the United Nations
Environmental Programme. The GEF is a fund established in 1992 to support actions to address global
environmental issues. The GEF now also acts as one
of the key funding mechanisms for implementation
of the Biodiversity Convention and Agenda 21, and
it provides grants, rather than loans, to governments.

LIFE Fund of the EU

Funds have been available since 1984 from the
European Union to support EU environmental legis-
lation, such as Birds and Habitats Directives, com-
prising such former Regulations as ACE (1872/84/
EEC and 2242/87/EEC), MEDSPA (563/91/EEC),
NORSPA, and ACNAT (3907/91/EEC).

These Regulations were superseded by the adoption in May 1992—at the same time as the Habitats
Directive—of a new all-encompassing environment
fund known as LIFE (Regulation 1973/92/EEC).
Targeting five priority fields of action, LIFE is intended to assist in the development and implementation of the Union's Environment Policy as outlined
in the Fifth Environmental Action Programme (CEC
1993).

One of the five priority fields—with an indicative
amount of 45% of the total annual LIFE budget—is
the protection of habitats and of nature, which is
achieved through part-financing projects that work
towards, and provide an incentive for, the implementation of the Habitats and Birds Directives. Other
objectives of LIFE are to support actions concerning
regional or global environmental problems that are
covered by international agreements, and to provide
technical assistance to non-EU countries from the
Mediterranean region or bordering the Baltic Sea.

The Commission's financial contribution is normally for a maximum of 50% of the total cost of the
project, but in exceptional cases where sites hold

species in danger of extinction, or habitats at risk of disappearing from the Community, the contribution may be raised to 75%. A sum of 400 million ECU was available for the first phase running from 1991 to 1995. Priority actions are decided and financed every year.

PHARE and TACIS Programmes of the EU

The PHARE Programme was established by the European Community to provide grants in support of economic and social transformation in the countries of central and eastern Europe. The TACIS programme is a similar Community initiative to support the countries of the Commonwealth of Independent States and Mongolia. There are several initiatives within PHARE and TACIS, including the PHARE Partnership Programme, the PHARE/TACIS Democracy Programme, and the PHARE/TACIS LIEN Programme.

The Partnership Programme's overall objective is to promote local economic and socio-economic development, through encouraging cross-country cooperation and the establishment and strengthening of sustainable partnerships among decentralized, non-profit organizations based in the European Community and central and eastern Europe.

The Democracy Programme aims to contribute to the consolidation of pluralist democratic procedures and practices, as well as to the rule of law, in central- and east-European countries. The LIEN (Link Inter-European NGOs) Programme supports the activities of NGOs in favour of disadvantaged groups in the same countries.

Structural Funds of the EU

The Structural Funds are the European Union's main instrument for achieving economic and social cohesion. Such cohesion is seen as a necessary prerequisite for economic and monetary union. There are four different funds under the Structural Funds: the European Social Fund (ESF), the European Agricultural Guidance Fund (EAGGF), the European Rural Development Fund (ERDF) and the Financial Instrument for Fisheries Guidance (FIFG).

The Structural Funds serve six objectives. Objective 1 covers the regions where development is lagging, defined as regions with a GDP less than 75% of the Community average. Objective 2 serves the regions of industrial decline. Objectives 3 (unemployment) and 4 (adaptation to industrial change) are not defined geographically and are served by the ESF. Objective 5a is specifically related to the adjustment of agricultural structures and 5b is concerned with wider aspects of rural development. Both 5a and 5b are served by the EAGGF. Objective

6 serves the new member states, Sweden, Finland and Austria.

For those objectives that deal with geographical regions, member states submit Regional Development Plans, including environmental information. The European Commission responds to each plan with a Community Support Framework (CSF) which establishes the parameters for Community support. After this, an Operational Programme (OP) is produced, setting out the individual projects to be supported. Decisions on projects are taken by a monitoring committee.

Nine percent of the budget for the Structural Funds is held in reserve to be spent on Community initiatives. Using this budget, the European Commission can establish specific programmes to deal with issues that have a transnational dimension or with problems which need an urgent response. Such initiatives include 'LEADER', to support specific rural developments, and 'ENVIREG' for environmental projects.

The functioning of these EU funds is unsatisfactory, as they continue to cause damage to Europe's biodiversity, and they are in need of review (Box 2).

Other EU funds

Funds are also allocated by the EU to research and technical development programmes which cover fields related to environment such as air pollution, water quality, soil protection, basic functioning of ecosystems, climate change, and so on.

BROAD POLICIES AND INITIATIVES

EU Agri-environment policy

The objectives of the Common Agricultural Policy (CAP) are as stated in Articles 38–47 of the Treaty of Rome, essentially the construction of a Single Market designed to achieve the specific objectives of Article 39. Articles 40–47 concentrate on how the CAP works. Environmental issues are not specifically mentioned. A growing recognition of the role of agriculture and the CAP in both damaging and protecting the environment has led to many policy changes. These are largely laid out in the Fifth Environmental Action Programme of the EU, and have their legal basis in Article 130r of the Maastricht Treaty.

Within the CAP, the Agri-Environment Regulation (2078/92/EEC) provides the opportunity for financial support to farmers who follow environmentally beneficial farming practices, under management agreements determined within 'zonal programmes' by member states (see 'International legislation', above).

Box 2. EU Structural Funds: the need for a review.

In 1995, BirdLife International published the report 'The Structural Funds and biodiversity conservation'. The report focuses on the Objective 1 regions in Spain, Greece and Italy, and examines what the Funds' implications are for biodiversity during the period up to 1999. It also looks critically at whether certain Fund projects yield economic benefits and represent wise use of taxpayers' money. The main recommendations of the report are as follows.

- A strategic new approach is needed in integrating biodiversity conservation with regional and rural development. A Task Force on Regional/Rural Development and Biodiversity needs to be established by the European Commission, in order to coordinate the Commission's overall strategy and to offer practical guidance to the member states on the coordination of EU funds.

- The Funds' Regulations and operational criteria should be reviewed, in order to maximize the integration of biodiversity conservation with regional/rural development. In particular, Article 3(3) of the Framework Regulation (2081/93), Article 1(f) of the ERDF Regulation (2083/93/EEC) and Article 5(i) of the EAGGF Regulation (2085/93/EEC) should explicitly allow a wide range of nature-conservation investment.

- EU environmental legislation must be upheld. The Commission should firmly enforce nature-conservation legislation (including the Birds and Habitats Directives) in Regions covered by the Structural Funds. There is a need for improved implementation of the EIA Directive.

- A creative approach is needed towards integrating environment and development within the new Community Support Frameworks and Operational Programmes. Among other aspects, this requires that environmental authorities in the Regions receive training and capacity-building, that Operational Programmes undergo environmental assessment (see Box 1), that aid rates and project-selection criteria favour environmentally desirable projects, that environmental impacts are monitored, and that Monitoring Committees have a stronger role.

- The Court of Auditors should investigate whether current Structural Fund procedures provide adequate guarantees of 'socio-economic benefits commensurate with the resources deployed'.

- Concessional loans rather than grants should be considered for certain categories of development projects, including irrigation and water supply.

- The Commission should encourage member states to use its *Guide to Cost–Benefit Analysis for Major Projects* for projects within Operational Programmes.

- An independent evaluation should be conducted of the real economic and social benefits of past EU-funded irrigation projects. The Commission and member states should investigate rural development approaches that do not depend on large-scale irrigation.

- Appraisal of ERDF 'major projects' and Cohesion Fund projects for water supply should insist on rigorous analyses of real water demand.

Pan-European Biological and Landscape Diversity Strategy

The Pan-European Biological and Landscape Diversity Strategy (PEBLanDS) is a broad framework for actions in support of the Biodiversity Convention in Europe, and was endorsed by the environment ministers of 55 European countries at the Ministerial Conference 'Environment for Europe' in October 1995 at Sofia. Two coordinating bodies (a council and a bureau) have been established in order to further develop and implement this strategy.

The Strategy recognizes the importance of conserving species and habitats on the one hand, and the close relationship between biodiversity conservation and economic activity on the other. It aims to reduce the threats to Europe's biological and landscape diversity, to strengthen the 'ecological coherence' of Europe, and to achieve full public involvement in conservation. The legal basis for implementing the actions of the Strategy is found in existing international agreements and treaties. Its priorities for the next 20 years focus on the introduction of biological and landscape conservation considerations into all

social and economic sectors. Key actors in implementing the strategy are identified in the widest possible sense, including decision-makers, the economic sector, scientists, NGOs, land-owners and the public in general.

The actions of the Strategy are to be implemented by developing a rolling work programme covering four consecutive five-year plans during the period 1996–2016 (CDPE 1995). These Action Plans will identify the fundamental actions necessary for the realization of the major objectives of the Strategy. The Action Plan for 1996–2000 contains 12 Action Themes, including the establishment of a Pan-European Ecological Network, the integration of conservation objectives into sectoral policies, growth in public involvement and awareness, and the conservation of habitats and species. The prioritization of issues concerning landscapes, ecosystems and species is based on *Europe's environment: the Dobrís Assessment* (Stanners and Bourdeau 1995) and *Parks for Life: action for protected areas in Europe* (IUCN–CNPPA 1994).

Parks for Life: action for protected areas in Europe

Parks for Life is an initiative of the IUCN World Commission of Protected Areas, in association with the Federation of Nature and National Parks of Europe (FNNPE), World Wide Fund for Nature (WWF), World Conservation Monitoring Centre (WCMC) and BirdLife International (IUCN–CNPPA 1994). It sets out a vision for Europe's protected areas in terms of policies, recommendations and actions.

The main themes of the plan include the positioning of protected areas in a wider context, addressing priorities at all levels, and better planning and management for protected areas. There are chapters addressing the main impact of sectoral policies on protected areas, actions concerning some priority sub-regions and countries, and overviewing the legal framework and financial opportunities for better protection. The plan deals with securing public support for protected areas, and outlines the implementation process.

There are 30 priority projects listed in the plan, which need urgent attention. In 1995, a coordinator was appointed to oversee the implementation of these projects. The idea is that a wide range of organizations take up a leading or active participatory role in the projects, so as to ensure effective cooperation and broad support for these activities.

Follow-up to the Earth Summit

In addition to the Biodiversity Convention, important and relevant products of, and initiatives launched at, the Earth Summit were:

- The *Rio Declaration*, a statement of principles addressing the need to integrate the protection of the environment with sustainable development.

- *Agenda 21—An Action Plan for the Next Century*, which gives political commitment to the integration of environmental concerns across a broad range of sectors, including industry, agriculture, energy, transport, education and training, recreation and tourism, land-use, and fisheries.

UNESCO Man and Biosphere Programme

The Man and Biosphere (MAB) Programme of the United Nations Educational, Scientific and Cultural Organization (UNESCO) was launched in 1970. Its aims include the development within the national and social services of a basis for the rational use and conservation of the resources of the biosphere. MAB Project 8 of the 14 projects or themes in the Programme deals with the conservation of natural areas and the genetic material they contain. The objective of this programme is to create a worldwide network of reserves called Biosphere Reserves, for the conservation of biodiversity, for sustainable human development locally, and as a logistical tool for research on global-scale phenomena. Each Reserve should qualify under one or more of the following categories.

- Representative examples of natural biomes.

- Unique communities or areas with unusual natural features of exceptional interest.

- Examples of harmonious landscapes, resulting from traditional patterns of land-use.

- Examples of modified or degraded landscapes capable of being restored to more natural conditions.

There are still relatively few coastal and marine Biosphere Reserves compared to those covering terrestrial habitats.

HABITAT RECOVERY PLANS: THE FOLLOW-UP TO THIS BOOK

Ideally, national, EU and pan-European biodiversity strategies should lead to the development of plans for the conservation of particular species and habitats of national, EU and pan-European priority. This book gives some broad indication of priority habitats, though at a national level, they might be divided and identified more finely. Habitats could be Red Listed, much as species commonly are, with priority given to those that have become scarce or are declining in extent or quality. Attention should also be given, at a national level, to those which are confined, or substantially so, to the particular country. No formal criteria for the Red Listing of habitats have been developed in the way that they have for species.

Recovery plans for habitats need to identify the resource in terms of extent and location, and in terms of defining qualities such as important or endemic species and communities of plants or animals. For clumped or rarer habitats, sites of particular importance need to be documented. Some such sites might be conserved through protected-area mechanisms. In the case of the more extensive habitats and those with multiple and competing land-uses, this will not always be appropriate.

The next stage of planning is the identification of threats to the habitat, because this is the rational route to the determination of the particular actions needed. Across a suite of plans for habitats, some actions will be specific but many will be common to several habitats and may be termed generic actions. Generic actions will usually focus on particular sectoral policies such as agriculture, land reform, forestry, transport or development.

No plausible plan can be complete without targets. Targets are necessary to indicate the scale of actions required. Conservationists have often been criticised for not being clear about what they want and, implicitly, for wanting more than is realistically achievable or acceptable to society at large. A related criticism is that conservationists tend to react late and unexpectedly when a particular new threat arises. The agreement of targets in habitat (or species) recovery plans would largely have the effect of meeting these otherwise damaging criticisms.

The setting of targets for habitat plans is not simple or straightforward. Targets need to be realistic and achievable within a stated time-frame, e.g. five to ten years. They also need to be challenging. For some of the most reduced or damaged habitats, there is simply no more room for further losses, and active restoration will be required. Targets need, as far as possible, to be quantifiable, though qualitative measures may also be required. Quantification may be in terms of area protected or actively managed. Some areas may be identified in terms of particular

sites in need of a named level of safeguard. Accompanying quality measures may be expressed in terms of species surviving or recovering to a given population level. Some examples of proposed habitat plans are provided in Housden *et al.* (1991) and Wynne *et al.* (1995), from which much of this section derives.

Finally, no planning process is complete without a monitoring programme. Monitoring serves to refine actions to ensure that targets are met. In many ways, recovery plans can be seen as experimental, with monitoring extracting new lessons which can improve understanding of efficient and effective solutions.

The adoption of formal plans is clearly a governmental responsibility, as called for in the Biodiversity Convention. Communities and NGOs have much to contribute. Indeed, such plans are most unlikely to succeed unless they are developed in a participatory process, which includes all key stakeholders. Many of the actions will need to be taken by particular communities or government departments. Some will be taken by NGOs, who might also be particularly able to help by motivating and coordinating the scientific forces needed for monitoring.

OVERVIEW OF THE STRATEGIES

FOR THE REASONS discussed in the 'Introduction' (p. 15), the results and conclusions of this study focus on bird conservation, and it is to be hoped that similar scientifically based recommendations can be made for the pan-European conservation of other organisms, as soon as the data are collated. In the meantime, it is clear that many of the broad measures that are recommended in this book will go a long way towards conserving, not just birds, but all elements of biodiversity in the wider environment in Europe.

PRIORITY HABITATS

The priority bird species for each broad habitat-type in Europe are identified in the habitat conservation strategies that form the eight main chapters of this book. As the methods for identifying these birds have been consistent throughout, comparisons can be made among the habitat-types in terms of the number of, and priority status of, bird species in order to assess the overall importance of each of the broad habitat-types. Figure 1 shows the particularly high level of importance, in terms of overall numbers of priority species, of inland wetlands, of Mediterranean forest, shrubland and rocky habitats and, to a slightly lesser extent, of arable and improved grassland.

However, interpretation of these results should take a number of important factors into account. Firstly, the analysis is highly scale-dependent. The number of priority species in the habitat-type will partly depend on the extent of the habitat-type in Europe, as species diversity normally increases as the area of habitat increases (Schoener 1976, 1986, Wiens 1989a). At the continental scale of this analysis, such increases are likely to result mainly from environmental heterogeneity: habitat-types that are widespread and extensive in Europe are likely to exhibit more variety (in terms of climate, topography, soil-type, flora, etc.) than limited and restricted ones. Structural diversity is also important, e.g. forest habitats have far more variety of physical structure than steppic habitats, and therefore have the potential to support more (specialized) species.

In addition, biogeographical and historical factors can have a significant influence on the species diversity of a habitat-type. Thus a habitat-type which spans Europe, such as inland wetlands, may incorporate species from many biogeographical regions, even if it only has a small total extent. In contrast, habitat-types which are restricted to one or only a few biogeographical regions of Europe, such as lowland Atlantic heathland, are less likely to encompass a wide variety of priority species. However, such habitat-types may be highly important for other reasons, e.g. related to the comparative rarity of the habitat itself, or of the species that it supports. For example, the European Macaronesian seas hold breeding populations of two endemic or near-endemic bird species, which are also globally threatened. Historically, lowland Atlantic heathland was much more extensive in Europe, but over the past 200 years it has been considerably reduced by conversion to other land-uses or through the abandonment of traditional management. In turn, this has resulted (in part) in its current relatively low importance for bird populations, compared to other major habitat-types dealt with in this book.

Thus, although this comparison of the relative importance of habitat-types (Figure 1) is appropriate for the prioritization of broad policies and actions in this kind of conservation strategy for the wider European environment, it should not be applied at smaller scales (regional, local or site-focused).

Secondly, the priority status (A–D) of each species reflects the relative importance of the habitat-type to the survival of that species in Europe, and should therefore be taken into account. In some habitat-types, a large proportion of species have a particularly high status (Priority A or B), e.g. steppic habitats, while in others a high proportion of species have a lower status (Priority C or D), e.g. tundra, mires and moorland.

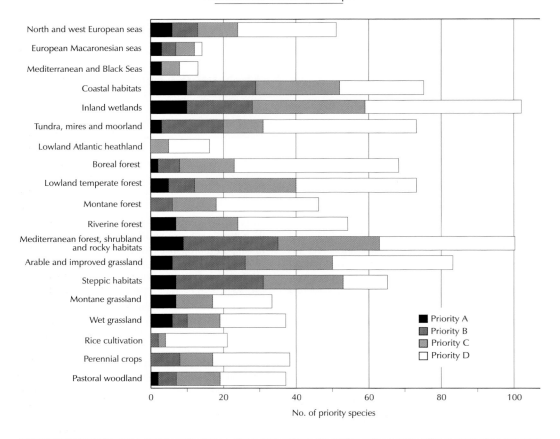

Figure1. The numbers of priority bird species in each of Europe's broad habitat-types, subdivided according to priority status (see 'Introduction', p. 15, for more details).

THE IMPORTANCE OF HABITATS IN INDIVIDUAL COUNTRIES

In order to provide some guidance on the relative importance of habitats for priority species within each country, the number of breeding species of each priority status (A–D) was obtained for each country and for each habitat, using the European Bird Database of BirdLife International/European Bird Census Council. These data are presented in full in Appendix 1 (p. 327). Table 1 summarizes the data, and takes the priority status of each species into account, by using a weighted scoring system; for each habitat and country, the weighted scores of all breeding priority species are thus summed to produce a total weighted score that is comparable between habitats (and between countries).

These data give a broad indication of the relative importance of the various habitats in each country. However, it should also be borne in mind that habitats are important for priority species outside the breeding season. Unfortunately, comprehensive data

on the presence of species on passage or in winter are not available, and it has thus not been possible to analyse habitat importance outside the breeding season. The relative importance of habitats within and between countries is thus likely to differ (from the situation presented in Table 1) during this period. In addition, the scores in Table 1 do not take account of the breeding population size of each priority species, such that a Priority A species which breeds in a particular country scores 4 whether the country holds 1% or 99% of the European population.

THREATS TO HABITATS AND PRIORITY SPECIES

The predicted threats to each habitat-type, and to the habitat features required by its priority species, are quantified within each habitat conservation strategy, using a simple scoring system (see 'Introduction', p. 15, and Appendix 4, p. 401). This helps in judging the relative importance of the different threats faced by the habitat-type, and thus in identifying priority conservation measures for that particular habitat-type.

However, it is also clearly valuable to establish the relative importance of these threats across the broad habitat-types, for the whole of Europe. It is possible to directly compare the scores for a particular threat across habitat-types, because these are derived according to the same methods. Furthermore, it is legitimate to sum these scores across habitat-types to establish a total score, because of the scoring system that has been used.

In practice, however, it has not been possible to ensure that threats are exactly matched when summing their scores across habitats. A standardized, global classification of threats is highly problematic to create, as there are several useful criteria upon which one could be based, e.g. by socio-economic sector, or by mechanism of impact. A threat in one habitat-type (e.g. intensive water management) may be more usefully discriminated as several threats in another habitat-type (e.g. drainage, impoundment of wetlands, etc.). It has therefore not been possible to derive highly accurate total scores when summing all threat scores across all habitat-types.

Nevertheless, an indicative analysis has been carried out along these lines, the results of which are summarized in Table 2 for the threats with the highest total scores (scores of 100 or greater, summed across all habitat-types). All scores in Table 2 have been converted to a three-point scale, since potentially misleading conclusions may be drawn if the numerical scores are accorded spuriously high levels of accuracy or interpreted in too exact a fashion, for the reasons described above. While not being definitive, therefore, Table 2 still gives a robust indication of the most important threats to habitats across Europe, from an ornithological perspective. In addition, it should be stressed that all of the threats dealt with in this book are important, not only those listed in Table 2.

The key prediction of Table 2 is that the major negative impacts on habitats in Europe and their priority bird populations over the next 20 years will be caused by the continuing *intensification* of human uses of Europe's environment, in such sectors as agriculture, water management, forestry, waste management, tourism, recreation and infrastructural development. The presence (or not) of scores in Table 2 indicates which chapters contain more detailed discussion and recommendations on particular threats to particular habitats, but a few of the major threats deserve to be highlighted here.

Table 2 implies that both agricultural intensification and land abandonment are major threats to birds in some habitats, which may seem contradictory at first sight. However, these threats do not affect the same areas at the same time.

Land abandonment threatens many important habitats in Europe that are managed for agriculture in a non-intensive or traditional way, such as steppes, montane grasslands, dehesas and Mediterranean shrublands. The continuation of such 'extensive' agriculture is not only essential in maintaining the biodiversity of such habitats, but also assists in the survival of rural societies. Abandonment occurs where rural people have too few financial or social incentives to manage their land. The result in the medium term is often the replacement of these open, wildlife-rich habitat mosaics by secondary, uniform, shrubby habitats of reduced conservation value, as is presently occurring in many parts of Mediterranean, central and eastern Europe.

In areas of Europe such as these, land abandonment also implies that the only way to stay competitive in agriculture is to intensify, which (as Table 2 demonstrates) is clearly harmful to priority bird species and other wildlife. One of the major challenges for decision-makers all over Europe is how to support sustainable rural development in the poorer regions.

Inappropriate water management, particularly drainage, is a most significant threat to wetland habitats but, perhaps less obviously, it also affects forest, heathland and agricultural habitats. The main socio-economic sectors that cause irreversible degradation of wetlands are agriculture (irrigation, drainage), and urban and industrial development (dams, destruction of riparian habitats, over-abstraction). The main problem is that the natural hydrological dynamics of wetlands, and the cumulative impacts of various water-use projects within the catchment, are not taken into consideration when designing and implementing water-use policies, plans and programmes.

Several aspects of forestry pose significant threats to birds in Europe. The most important threat appears to be the widespread intensification of forest management. In the most extreme cases, existing natural and semi-natural forests are being turned into uniform monoculture plantations through clear-felling and replanting, often using non-native species.

Afforestation can be a valuable step towards conserving biodiversity, if it is targeted at deforested areas whose significance for priority bird species and other wildlife has been much reduced, so long as monoculture plantations are avoided and the other principles, operational policies and habitat targets that are recommended in this book are adopted (see chapters on forest habitats). However, these recommended measures are not yet widely practised by the forestry sector, and continuing afforestation is leading to the loss of open habitat-types of high conser-

Table 1. Scoring for numbers of breeding priority species in each European country and in each habitat-type, weighted according to priority status; each Priority A species scores 4, Priority B=3, C=2, D=1 (based on breeding-bird data collected for *The EBCC atlas of European breeding birds* (Hagemeijer and Blair 1997).

	North and west European seas	European Macaronesian seas	Mediterranean and Black Seas	Coastal habitats	Inland wetlands	Tundra, mires and moorland	Lowland Atlantic heathland	Boreal forest
Albania	—	—	4	66	114	17	—	—
Andorra	—	—	—	—	7	9	—	—
Austria	—	—	—	—	92	26	—	—
Belarus	—	—	—	—	123	53	—	—
Belgium	11	—	—	50	77	27	15	—
Bulgaria	—	—	7	73	128	17	—	—
Croatia	—	—	5	44	101	18	—	—
Cyprus	—	—	4	12	17	—	—	—
Czech Republic	—	—	—	—	101	31	—	—
Denmark	25	—	—	64	95	38	17	—
Faroe Islands	47	—	—	28	17	32	—	—
Greenland	42	—	—	29	13	52	—	—
Estonia	32	—	—	66	102	64	16	72
Finland	37	—	—	66	105	107	—	92
France	42	—	17	83	118	31	22	—
Germany	28	—	—	76	121	46	20	—
Greece	—	—	17	80	120	17	—	—
Hungary	—	—	—	—	129	22	—	—
Iceland	60	—	—	41	37	58	—	10
Rep. of Ireland	52	—	—	43	53	36	7	—
Italy	—	—	18	71	108	18	—	—
Latvia	11	—	—	53	109	54	16	—
Liechtenstein	—	—	—	—	21	8	—	—
Lithuania	11	—	—	53	108	44	16	—
Luxembourg	—	—	—	—	40	8	14	—
Malta	—	—	5	2	3	—	—	—
Moldova	—	—	—	—	111	12	—	—
Netherlands	15	—	—	69	105	30	16	—
Norway	57	—	—	63	88	110	14	84
Svalbard	31	—	—	20	3	30	—	—
Poland	13	—	—	72	122	48	18	—
Portugal	12	—	—	49	84	12	17	—
Azores	—	21	—	6	2	—	—	—
Madeira	—	28	—	6	2	—	—	—
Romania	—	—	7	83	142	29	—	—
Russia	61	—	50	131	183	118	—	97
Slovakia	—	—	—	—	110	24	—	—
Slovenia	—	—	3	29	81	18	—	—
Spain	27	—	19	99	135	29	21	—
Canary Islands	—	22	—	9	8	—	—	—
Sweden	41	—	—	76	105	107	16	82
Switzerland	—	—	—	—	75	19	—	—
Turkey	—	—	13	95	156	—	—	—
Ukraine	—	—	11	95	156	42	—	—
United Kingdom	59	—	—	65	86	66	19	58
Gibraltar	—	—	—	5	4	—	—	—
Isle of Man	—	—	—	25	31	—	8	—
Guernsey	—	—	—	4	11	—	2	—
Jersey	—	—	—	5	23	—	4	—

vation value (especially steppic habitats, moorland and lowland Atlantic heathland).

In contrast, for parts of northern and eastern Europe the logging of remaining natural and semi-natural old-growth forest is the major threat. Genuinely natural habitat (of any kind) is now very rare in Europe, and these forests comprise some of the largest remaining areas.

The predicted high negative impact of human disturbance of birds' essential activities is perhaps a surprising conclusion. However, significant disturbance is caused to nesting, feeding or roosting birds of some priority species in nearly all European habitats, mainly as a result of tourism and recreation, through such activities as walking/hiking, mountain-biking, skiing, rock-climbing, watersports, off-road driving, gliding, angling and shooting. The overall threat is therefore substantial, and is particularly significant in coastal habitats and inland wetlands, as these are among the most popular areas for tourism

Lowland temperate forest	Montane forest	Riverine forest	Mediterranean forest, shrubland and rocky habitats	Arable and improved grassland	Steppic habitats	Montane grassland	Wet grassland	Rice cultivation	Perennial crops	Pastoral woodland	
107	—	69	130	106	60	36	22	—	52	51	Albania
72	47	—	80	65	—	25	—	—	35	35	Andorra
101	60	56	—	103	47	32	38	—	37	42	Austria
110	—	69	—	108	—	—	57	—	—	—	Belarus
89	—	47	—	80	—	—	41	—	27	34	Belgium
113	63	74	133	117	73	32	31	—	55	51	Bulgaria
110	59	67	127	116	67	30	32	—	52	50	Croatia
31	15	—	75	50	38	9	—	—	27	27	Cyprus
104	53	64	—	106	—	18	44	—	37	45	Czech Republic
79	—	42	—	90	—	—	44	—	26	33	Denmark
9	—	—	—	24	—	—	15	—	—	—	Faroe Islands
5	—	—	—	12	—	—	—	—	—	—	Greenland
93	—	55	—	90	—	—	47	—	—	—	Estonia
87	—	—	—	74	—	—	51	—	—	—	Finland
105	63	67	131	124	75	40	43	24	48	59	France
105	59	61	—	106	44	25	51	—	35	42	Germany
106	63	68	160	114	73	36	27	20	55	53	Greece
111	—	74	—	116	59	—	48	—	40	46	Hungary
6	—	—	—	22	—	—	18	—	—	—	Iceland
38	—	19	—	51	—	—	33	—	11	15	Rep. of Ireland
93	59	57	127	114	66	37	29	21	46	49	Italy
104	—	61	—	99	—	—	58	—	—	—	Latvia
74	—	38	—	52	—	—	15	—	20	23	Liechtenstein
94	—	56	—	97	—	—	54	—	—	—	Lithuania
79	—	42	—	71	—	—	20	—	27	35	Luxembourg
10	—	—	—	18	—	—	—	—	10	11	Malta
95	—	68	—	108	63	—	43	—	34	44	Moldova
80	—	46	—	85	—	—	44	—	28	34	Netherlands
82	—	—	—	66	—	18	46	—	—	—	Norway
—	—	—	—	6	—	—	—	—	—	—	Svalbard
114	—	67	—	106	—	—	60	—	37	49	Poland
83	48	53	120	104	76	26	21	16	49	57	Portugal
6	—	—	—	6	—	—	—	—	—	—	Azores
12	—	—	—	20	—	—	—	—	—	—	Madeira
124	65	82	—	128	82	29	46	—	46	52	Romania
122	63	84	—	139	120	27	69	—	36	48	Russia
114	64	69	—	113	56	24	44	—	40	50	Slovakia
101	62	55	—	90	37	31	26	—	38	43	Slovenia
92	55	62	145	123	93	37	38	24	49	65	Spain
16	—	—	—	20	25	—	—	—	—	—	Canary Islands
87	—	—	—	80	—	18	46	—	—	—	Sweden
95	58	49	—	94	—	33	29	—	37	43	Switzerland
105	60	71	158	126	105	48	34	21	55	55	Turkey
125	—	84	—	134	96	—	57	—	42	54	Ukraine
75	—	37	—	81	26	17	40	—	25	26	United Kingdom
15	—	—	—	25	—	—	—	—	—	—	Gibraltar
30	—	—	—	55	—	—	—	—	—	—	Isle of Man
24	—	—	—	25	—	—	—	—	—	—	Guernsey
26	—	—	—	33	—	—	—	—	—	—	Jersey

and recreation.

The short-term effects of human disturbance on birds have been well documented (e.g. Ward 1990, Hockin *et al.* 1992, Davidson and Rothwell 1993, Madsen and Fox 1995, Madsen *et al.* 1995). Such negative effects include nest-desertion or nest-trampling, reduced foraging efficiency, increased energy expenditure and even temporary avoidance of suitable habitats. However, the long-term impacts on bird populations are unknown, and further research

is therefore urgently required to quantify these (Hill *et al.* 1997). Nevertheless, the precautionary principle should be followed in the meantime, and prudent measures should be taken to reduce, alleviate or effectively mitigate disturbance impacts.

Longer-term threats
The threat analysis carried out in this book predicts the impacts on bird populations over the next 20 years. However, a longer-term view should also be

Table 2. The most important threats to habitats in Europe. Importance is judged using a simple scoring system (see main text, and Appendix 4, p. 401) which takes into account the European Threat Status of each priority bird species (Priorities A–D) and the importance of the habitat for the survival of its European population, as well as the predicted impact (Critical, High, Low), within the habitat, of each threat on the European population of each species. Scores are comparable within and between threats and habitats. Only those threats with a total score (summed across all habitats) of more than 100 are listed.

Scores: ■ >100   ● 50–100   • <50   No score indicates that the threat is not considered significant (on current knowledge) at the continental scale of this analysis (though local impacts may occur).

Threats	Pastoral woodland	Perennial crops	Rice cultivation	Wet grassland	Montane grassland	Steppic habitats	Arable & improved grassland	Med. forest, shrubland, rocky habitats	Riverine forest	Montane forest	Lowland temperate forest	Boreal forest	Lowland Atlantic heathland & moorland	Tundra, mires and bogs	Inland wetlands	Coastal habitats	Mediterranean & Black Seas	European Macaronesian seas	N & W European seas
Agriculture																			
Land abandonment	●	•	•		•			■					•	•					•
High stocking levels and/or overgrazing	•			•	•	■	■	●		•				•	•		•	•	•
Pesticide use	•	•	•	•	•	■	■	•							•	•		•	
Crop improvements	•	•	•	•	•	■	■										•	•	•
Conversion to crops or grassland	•	•	•	•	•	●	■						•				•	•	•
Elimination of marginal habitat features	•	•	•	•	•	●	■	•											
Farming operations			•	•		•	•											•	•
Loss of rotations						•	•												
Water management																			
Drainage				●	•				■				•	•	■				
Impoundment of wetlands									●				•		●	•			
Forestry																			
Intensification of forestry						■		•	●	•	■	•		•				•	•
Logging of old-growth forests					•	•	•	●	●	●	●	•		•	•	•		•	•
Pollution																			
Nutrient pollution of water																		•	•
Toxic pollution																			•
Infrastructural development																			
Urban and industrial development	•	•	•	•	•	•	•	●	•		•			•	•	■	■	•	•
Overhead cables	•			•				•						•	•			•	•
Roads and railways																		•	•
Disruption of ecological processes																			
Human disturbance of birds' activities	•		•	•	•	•	•	●	•	•	•		•	•	•	■	■	•	•
Human-induced increases in predators																•		•	•
Uncontrolled/intense/frequent fires	•							■	■										●
Introduced predators/competitors									•									•	●
Other																			
Tree disease and pests	•							•		•	•	•	•						
Habitat fragmentation by man								•	•	•			•						

taken. Of the longer-term threats, climate change driven by increasing greenhouse gases (global warming) is likely to have by far the greatest impact on habitats, birds and other wildlife, and on humanity itself (Briggs and Hossell 1995; see Box 1).

For example, as one of the most obvious symptoms of climate change, the impacts of sea-level rise over the next 20 years have been taken into account in this book (see 'Coastal Habitats', p. 93), but they are likely to be relatively small during this period. Over the next 100 years, however, much greater

impacts may be expected on coastal habitats and their bird populations.

If the climate changes as currently predicted, there will be substantial changes in the distributions of plant species in Europe (Huntley 1994), with consequent changes in biomes and in major land-uses such as agriculture and forestry (Solomon and Cramer 1993, Solomon *et al.* 1993), which will all inevitably affect the range and population size of many bird species. Although systematic and predictive modelling of the impacts on birds in Europe has

---

Box 1. Global warming.

'Greenhouse' gases form a barrier in the earth's atmosphere that prevents the sun's heat from escaping, so maintaining the atmosphere at a higher temperature than it would be otherwise. The most important greenhouse gases are carbon dioxide (CO_2), methane, nitrous oxide and chlorofluorocarbons (CFCs). By increasing the concentrations of these gases in the atmosphere, increasing amounts of heat are trapped and the global climate becomes warmer. Global atmospheric CO_2 levels have increased since pre-industrial times by 25%, and continue to grow at 1% per year (IPCC 1990, 1992). Other greenhouse gases have also increased in concentration, by an amount equivalent to the effect of an additional 25% rise in CO_2 (IPCC 1990).

These increases result from human activities, which now release over 20,000 million tonnes of CO_2 into the atmosphere each year (Stanners and Bourdeau 1995). Fossil-fuel consumption is the most important source of CO_2, contributing 80% of emissions, while another 18% come from deforestation in North America, Siberia and the tropics (IPCC 1992).

Changing agricultural practices worldwide have increased the amount of methane and nitrous oxide released into the atmosphere (Stanners and Bourdeau 1995). Industry has also played a part in the increase of these gases, and has increased the release of CFCs from virtually nothing in 1950 to their present level where they contribute 25% of the total greenhouse effect (IPCC 1990).

The Intergovernmental Panel on Climate Change (IPCC) predicts that, as a result of these man-made alterations to the atmosphere, global mean temperatures will rise to be approximately 3°C above current temperatures by the end of the twenty-first century (IPCC 1990). Temperature rises will be more pronounced at higher latitudes than in the tropics. Rates and patterns of rainfall and snowfall are likely to change, and increased temperatures will increase evaporation rates. Sea-levels may eventually rise by between 10 and 100 cm by 2100 as a result of the thermal expansion of seawater and the melting of ice-caps and glaciers.

It is likely that these changes will in turn have considerable effects on the extent and quality of habitats in Europe, although it is difficult to predict these with any certainty. Wetlands could dry out or

suffer from flooding, erosion or salinization. Mires are reliant on rainfall for their water, and would thus be particularly vulnerable to changes in the balance between rainfall and evaporation rate (Erkamo 1952). In contrast, many steppic habitats tend to occur in dry regions where even slight changes in rainfall or temperature could drastically affect these habitats and their ability to sustain wildlife (especially large herbivores) and livestock. Undoubtedly, some regions could suffer from land degradation as a result of vegetation loss.

Vegetation belts in Europe are likely to shift northwards (Liljelund 1990, Huntley 1994). Even relatively modest climate-change scenarios predict a 60% reduction in the extent of tundra and a 40% reduction in boreal forest (Solomon *et al.* 1993), with tundra being 'squeezed' from the south by boreal forest and from the north by rising sea-levels. Temperate and Mediterranean forests and shrublands, and steppic habitats, are likely to increase, at the expense of boreal forest (Solomon 1991, Huntley 1994). In Mediterranean areas, fire and drought would probably become more prevalent, thus altering the extent, structure and plant-species composition of forests and shrublands (Goudriaan *et al.* 1990).

The range and extent of agricultural habitats, moorland and heathland, being predominantly man-managed habitats, could change radically if land-use practices are altered to follow fluctuations in climatic, ecological and economic conditions. Such indirect or secondary effects of climate change are likely to have an impact just as serious as direct effects such as sea-level rise or changing rainfall patterns (Briggs and Hossell 1995).

In mountain areas, increasing temperatures would inevitably shift ecological zones upwards, and higher montane species and ecosystems may therefore be squeezed off the top of mountains, with lower montane zones experiencing a contraction in habitat extent (e.g. Hossell 1994).

Over the next few decades, human activities other than global warming are likely to have more serious impacts on bird populations. In the longer term, however, climate change will probably have a profound effect on habitats and their species across the entire continent, as well as on human uses of the environment.

apparently yet to be performed, some general predictions have been made (Huntley 1994). In particular, some ecologically restricted birds are likely to suffer reductions in range and population, whether the species is restricted to a particular plant that is likely to become less common in Europe (e.g. Parrot Crossbill *Loxia pytyopsittacus*, which feeds predominantly on Scots pine *Pinus sylvestris*), or is restricted to a particular biome or habitat likely to suffer a large reduction in range (e.g. tundra, boreal forest or montane habitats: see Box 1). Many migratory species will face increasingly long journeys as vegetation belts shift away from the equator, and they will also have to adapt to changed migration routes.

This book has not attempted to address the longer-term threats from climate change to habitats and their birds, but it is clear that longer-term conservation strategies to conserve wildlife cannot rely on the static approach of defining protected areas alone. Instead, and in addition to reducing emissions of greenhouse gases, the improvement of the ecological quality of the wider environment for birds and for biodiversity in general will be essential if the negative impacts of climate change and other longer-term threats are to be mitigated.

PRIORITY MEASURES FOR CONSERVATION OF THE WIDER ENVIRONMENT

Key policies

Some of the broadest and most influential policies in Europe, that most affect human use of the environment, are those of the European Union. There is thus a strong emphasis on EU policies in the following section.

Agriculture policy

The most important of the EU policies, affecting a large majority of the habitat-types discussed here, is the Common Agricultural Policy (CAP). The CAP is the driving force behind the EU's agricultural production and, as such, has very broad impacts on the rural environment. These occur through direct and indirect economic incentives for farmers, which strongly influence fertilizer and pesticide use, drainage, irrigation, livestock production and other important land-use issues. The CAP needs to be reformed in the coming years to include more and clearer environmental provisions, and the implementation of the EU Agri-environment Regulation should be strengthened and expanded (see 'Agricultural and Grassland Habitats', p. 267) for detailed recommendations).

Rural policy

The problems of rural unemployment, declining rural services, environmental degradation and loss of wildlife point to a failure of rural policy in Europe to address what should be its principal objectives—to protect the environment and to maintain rural communities. Too often in the past, policy-makers have assumed or hoped that these objectives can be met by pumping large sums of public money into encouraging the production of commodities such as food and timber, or through the development of built infrastructure such as motorways, dams and irrigation schemes (see 'Transport policy', below).

The continuing evolution of agriculture in Europe plays a key role in shaping habitats as well as rural economies, and the maintenance and enhancement of sustainable farming systems must remain a key objective of rural policy. At the same time, it is clear that Europe's rural areas are in need of more effective economic, social and environmental policies. Sustainable development in rural areas requires that natural resources, including biodiversity, are conserved for future generations and to meet the needs of tomorrow's rural communities. In order to achieve this, a better definition of rural policy objectives and a reform of rural support systems and incentive structures are necessary. More long-term funding to sustain rural initiatives, with nature-conservation benefits and more training in environmentally beneficial rural skills, should also be promoted.

Regional policy

Regional policy is another major factor shaping the face of Europe's landscape. Here the EU structural and cohesion policies should be highlighted, both of which come with large-scale funds attached (the Structural Funds and the Cohesion Fund). These exist to support economic development in the least-developed regions of the EU, and to promote the cohesion of member states within the Union, primarily via large-scale infrastructure development (new dams, ports, motorways, energy networks, etc.). The regions which receive the largest injections of these funds tend also to be those which support some of the highest levels of biodiversity in Europe (e.g. Spain, Greece, Portugal). It is essential that environmental objectives, including biodiversity conservation, are integrated more fully into these policies (see Box 2, p. 38, and 'Transport policy', below).

Water and fisheries policies

Other major policy instruments in the EU include the Common Fisheries Policy, which affects marine and coastal habitats, and the various Directives on water quality, which are to be brought together under a Framework Directive. Both measures should be reformed according to environmental considerations, and should incorporate more clearly the aims and principles of biodiversity conservation, as recommended in detail in this publication (see 'Marine Habitats', p. 59, and 'Inland Wetlands', p. 125).

Transport policy

All too often, roads and railways are built in peripheral regions in the hope that economic growth and new jobs will follow. There is a common belief that new or higher-volume motorways and other high-speed links will promote regional economic growth and increase employment, two justifications which are often considered of 'overriding public interest', and thus reason enough to damage areas that are internationally important for wildlife. Transport infrastructure in all its varying forms is needed, but its damaging environmental impacts are significant. Roads or railways are built through rare or vulnerable habitats, ports in estuaries destroy important feeding areas for birds, and polluting emissions from all forms of transport indirectly threaten wildlife and habitats in the longer term (see Box 1).

Such development of transport infrastructure is one of the main elements of EU cohesion policy. Substantial amounts of EU investment in transport infrastructure are directed towards peripheral regions of the EU, and most of this is allocated to roads (Bina and Briggs 1996). For example, the Trans-European Transport Networks comprise a major EU programme to create and strengthen transport links across the EU. They involve a combination of construction and upgrading of approximately 140 road schemes (including about 15,000 km of new motorways), 11 rail links, 57 combined transport projects and 26 inland waterway links throughout EU member states. They are intended to achieve several objectives within the Union, including sustainable and efficient transport, economic cohesion and environmental protection (CEC 1994b).

However, many of these schemes directly threaten important habitats (Bina et al. 1995), and the effectiveness of investing in transport infrastructure to meet economic and social objectives is increasingly doubted (Bina and Briggs 1996, Hey et al. 1996). Decisions to build new infrastructure must be based on sound and comprehensive assessment of the project's contribution towards development and employment, as well as environmental protection.

Energy policy

The direct and indirect impacts of the energy sector on habitats are significant forces in the shaping of habitats across Europe. There are currently two major forces acting in the energy sector which offer both opportunities and threats to habitats in Europe.

The first is the wide range of measures at the national and EU level, both actual and proposed, which are intended to reduce emissions of greenhouse gases. The second is the continuing liberalization of the energy markets.

Policies to deal with climate change focus primarily on increased energy efficiency, and thus reduced emissions per unit of energy supplied. These are widely regarded as the most cost-effective way of reducing all emissions, and so offer benefits to European habitats. The increased use of renewable energy sources is also being promoted by national governments and the EU. Where the infrastructure for such power generation is carefully designed and positioned, so as to avoid negative impact on vulnerable species and sites, most forms of renewable energy offer long-term advantages to birds and other wildlife.

Integration of policies across sectors

The greatest obstacle to sustainable use of the environment is the degree to which policies and practices are divided between different sectors of responsibility. Many socio-economic sectors have environmental impact and yet have no responsibility towards the environment. As a result, the costs of adverse effects are often omitted from consideration. Government departments responsible for the environment are often relatively small and command budgets that are greatly inferior to those responsible for conflicting activities.

It is wasteful to have different policies and departments either conflicting or merely failing in efficiency through lack of integration. While agriculture, transport, construction and industry might have obvious impacts, sectors such as trade or foreign affairs also influence the environment both at home and abroad. The need for integration is particularly obvious in coastal zones where a great number of potentially conflicting interests tend to concentrate along a very narrow strip of land and water. Inland wetlands also concentrate demand, and potential for conflicts in use. The integration of activities within water catchments is therefore generally required.

There is a challenge for governments to move towards the integration of environmental conservation into all policy sectors. This will require the explicit development of strategies and plans for biodiversity, which share wide public and governmental ownership and support. Biodiversity strategies at the national, EU and European level provide frameworks for such integration.

Holistic approaches to the management of large geographical areas and habitats, such as Integrated Coastal Zone Management, Catchment Management Planning or sustainable forest management, should also incorporate the restoration, where possible, of dynamic natural processes within the habitats, at the landscape level (e.g. on river flood-plains for natural control of flooding, or along coasts for natural coastal defence).

A holistic view is also invoked in this book's repeated recommendations for the development and implementation of legislation for Strategic Environmental Assessment (SEA) of sectoral policies, plans and programmes at national, transboundary and EU levels, and for a more inclusive form of Environmental Impact Assessment (EIA) of individual development projects (see Box 1, p. 34). Such early and high-level prediction of the overall environmental impacts of proposed policies, plans and programmes (from which many projects originate) is notably still missing in Europe, but could contribute greatly towards integrating policies across sectors. Through openness and consultation, different sectoral interests have the opportunity to identify and remedy potential conflicts before they occur and need costly resolution.

Such environmental assessment requires access to highly integrated information on the environment, which is still not readily available for Europe at present. The European Environment Agency and its associated Topic Centres provide the mechanisms through which such integration will need to be achieved. Inherent in environmental assessment is the identification of environmental impacts and their evaluation. Arguments for developments that appear to be economically attractive, according to mainstream cost–benefit analysis, may be found to be faulty if they have failed to internalize the costs of environmental impact.

The European Commission completed and officially adopted a draft SEA Directive in late 1996, the provisions of which should be strengthened (see Box 1, p. 34). If the SEA Directive is to be realized as national legislation in EU member states, it should include strong environmental objectives, and its jurisdiction should be broadened to cover such areas as afforestation and other large-scale forestry operations, large-scale infrastructural programmes (e.g. the proposed Trans-European Transport Networks of the EU) and agricultural and coastal development programmes and policies.

Major gaps in policy at national and EU levels

There are two notable gaps in common policy in the EU. There is still no progress on an EU-wide strategy for the coastal zone and its integrated management, and there is little initiative from the member states to deal with native forest issues on an international scale. There is, however, a drive within the Union for a more integrated rural policy, but its realization still lies some way in the future. These areas will need to be addressed rather urgently by the decision-making bodies of the EU.

Collecting comprehensive data on all major national legislation and policy areas, and comparing and analysing them across countries, has been quite beyond the scope of this book. The main recommendations for national policy-makers, however, are generally very similar to those for the EU: there is a clear need for more cross-sectoral integration and coordination of broad policies and actions, for the incorporation of clear environmental objectives in all land- and sea-use policies, and for more strategic assessment of the environmental impacts of all policies, programmes and plans.

Funding issues

From the viewpoint of biodiversity conservation, there are two main types of funding: those which cause major harm to the environment, and those which have clear conservation benefits. In this book, examples of the former include harmful agricultural subsidies which lead to intensification, overgrazing and pollution, as well as direct funding for environmentally harmful projects such as mining operations, the building of hydroelectric dams on rivers, coastal barrages and transport infrastructures. These should be subject to rigorous SEA and EIA procedures and should only go ahead if their negative environmental impacts are minimal or fully mitigated. The payment of subsidies should also be more dependent upon benefiting the environment, or minimizing damage, according to clearly defined environmental indicators. For certain development projects, concessionary loans rather than grants should be considered.

In contrast, the payments under the Agri-environment Regulation in the EU are an example of a positive environmental incentive. These should be expanded, more closely monitored and more widely implemented, not just in the EU but in other European countries as well, for example in central and eastern Europe through the EU PHARE and TACIS programmes (see 'Opportunities for Conserving the Wider Environment', p. 37). Similar schemes should also be introduced, throughout Europe, to support other non-intensive rural land-uses outside the agricultural sector, such as sustainable forest management, which also have the potential to benefit biodiversity and rural society. More detailed recommendations on such issues are contained in the BirdLife International report 'Working with nature: economies, employment and conservation in Europe' (Cuff and Rayment 1997).

More funding should also be available for pilot

projects that demonstrate to the working community of the relevant socio-economic sector (e.g. farmers) the practical nature and financial viability of the land-use changes proposed in this book, including extensification, habitat restoration, Integrated Coastal Zone Management and sustainable tourism. Advisory services should receive increased funding and promotion, so that such proposed changes in land-use policy can be put into practice more effectively, through the provision of advice and guidance to land-users and land-owners. Finally, the research programmes and projects that are outlined in the next section should also gain widespread financial support.

Research: the gaps in knowledge

Environmental conservation proceeds in spite of considerable gaps in knowledge. There is a challenge to fill these gaps, which are evident throughout this book. At the species level, there is a need to know population sizes, distributions, habitat requirements and ecological explanations of population changes, especially declines. Monitoring can pick up undesirable trends, but more detailed research is also required to properly identify present and future threats. At the habitat level, it is necessary to know about the distribution and extent of habitats and the factors adversely affecting their quality.

These gaps need to be filled, so as to sharpen the precision with which it is possible to prescribe, in detail, how to conserve biodiversity in changing environments. Still more, there are gaps in the understanding of the link between economic and policy mechanisms and their impact on individual species, or on plant and animal communities.

All of these deficiencies need to be rectified and more research efforts are required along these lines. However, even with a lack of high-quality data, it is certain that human impacts on habitats and their associated wildlife in Europe are already severe. The fact that nearly 40% of Europe's avifauna has an Unfavourable Conservation Status indicates that habitats are being degraded and lost at an alarming rate (Tucker and Heath 1994). The precautionary principle should always be uppermost in formulating land-use policies, especially those that will potentially have an impact on sensitive habitats and biological communities.

The scale of ignorance of species and habitats is sufficiently large that knowledge is unlikely to be accumulated fast enough from individual studies. There is a need to find better methods of learning from experience. Much could be done from targeted study and monitoring of particular financial or regulatory mechanisms. For instance, in spite of considerable expenditure on set-aside or Environmentally Sensitive Areas in Europe, very little is known about their large-scale effects on biodiversity. Carefully targeted monitoring of such measures could generate some very valuable information for the better formulation of future policies. From a scientific perspective, any new policy mechanism could be considered as an experiment, with considerable power to learn lessons at relatively low cost.

In addition, lessons could be better learned, through the clearer transmission and understanding of the information which does exist. In fact, far more information exists than is generally used in environmental decision-making, at all levels and scales. Better ways need to be found for organizing information and making it more available. Modern computer and communications technology provide the tools for an information revolution whose potential has barely begun to be appreciated. Agencies such as the European Environment Agency and its parts (such as the European Topic Centre for Nature Conservation) provide mechanisms for the better collation and dissemination of relevant information.

Public awareness

There is still a widely held belief in Europe that biodiversity conservation should focus on protected areas which are clearly separated from the surrounding countryside. The idea that environmental protection and biodiversity conservation should be an integral part of all uses of the environment, and of all policies of all socio-economic sectors (agriculture, forestry, tourism, etc.), is relatively new. As a result, the concept has not yet had sufficient time to become publicly known and accepted.

Even when this integration is promoted on paper there is a lot of confusion about how it should be done in practice and who should be responsible for its implementation. Clear sectoral divisions and a lack of coordination across sectoral boundaries do not help in resolving this situation. These barriers must be dissolved by new ways of thinking and by new ways of organization in governments and society.

From the message that biodiversity conservation should be integrated into all land- and sea-uses and sectoral policies, it should be clear that this task is ultimately the responsibility of all policy-makers, land-owners and land-users in Europe, at all levels. In several European countries the majority of the land is already privately owned, and the proportion of private land is also increasing in eastern Europe as a result of widespread privatization and land-redistribution initiatives. This means that an increasing proportion of people in Europe are either land-owners or land-users and are therefore involved, either positively or negatively, in the process of biodiversity conservation. However, the large majority of them are probably unaware, or do not yet acknowledge, that they play such an important collective role in determining the state of Europe's environment.

In the longer term, sustainable use of the European environment will require substantial changes in society's use of natural resources, energy and transport, and in its creation and disposal of waste material. It is therefore essential that the public is aware of the implications of the current and increasing intensity of its uses of the European environment, and of the fact that everyone has a role to play in the conservation of biodiversity in the wider environment.

MAIN CONCLUSIONS
AND RECOMMENDATIONS

Main Conclusions

1 Nearly 40% of bird species are declining across Europe. This publication clearly demonstrates that many of these widespread declines result from widespread threats to habitats. Other recent pan-European studies (e.g. Stanners and Bourdeau 1995) conclude that the same threats are likely to affect many other elements of the continent's biodiversity in similar ways.

These threats are mainly driven by the continuing intensification of man's uses of the European environment. Intensified uses often have a negative impact on birds and other species, through changing particular features of the habitat that are essential for individual species (habitat deterioration) or through destroying the habitat itself (habitat loss).

In the long term, the negative impacts of such land- and sea-uses can lead to loss of biodiversity (as defined in the Biodiversity Convention: Anon. 1994c) through extinction of species or loss of genetic diversity at a variety of geographical scales.

2 Various policies of land- and sea-use can act in concert to have a simultaneous impact upon one particular habitat-type, within and across socio-economic sectors (agriculture, fisheries, tourism, transport, etc.). These impacts are often cumulative and mutually reinforcing (e.g. intensive pesticide use and the ploughing up of marginal habitats for cultivation may combine to reduce the insect food resources of priority birds on arable land).

Furthermore, any one policy can have an impact on several habitat-types and their biodiversity (e.g. apart from severely degrading the ecological quality of grassland, intensive fertilizer use may also pollute inland wetlands and coastal and marine habitats with excess nutrients).

3 There is very little coordination of land- and sea-use policies between various sectors, at any level. Such policies are often applied to the same geographical or political region with only minimal effort towards the integration of land-use planning.

This often leads to increased stress on the environment and on natural resources. Environmental protection is seen as only another sectoral land-use policy and, as such, is often regarded as less important than the short-term economic or social interests of various other sectors.

Despite the slowly increasing recognition that environmental objectives should be more fully integrated across all socio-economic sectors and into all policies, plans and programmes, at all levels, no serious efforts have yet been made to put this into practice in Europe.

4 Nature conservation and species protection are still usually regarded as activities that focus on statutory protected areas. Biodiversity conservation in the wider environment is still a relatively new concept, the potential consequences of which have not yet gained general acceptance in society. There is also widespread confusion over who should be responsible for developing and implementing wider conservation measures in the countryside.

5 A number of international legislative instruments, notably the Biodiversity Convention and the Birds and Habitats Directives, include provisions for the conservation of habitats. However, the Biodiversity Convention and initiatives stemming from it (e.g. the Pan-European Biological and Landscape Diversity Strategy) are yet to be fully implemented, ratified or even signed by a large proportion of relevant European countries, and the Birds and Habitats Directives cover only EU member states.

6 Environmental Impact Assessment (EIA) of proposed development projects is now a widespread practice in Europe. However, project-level EIA is not mandatory for all major land-use projects (e.g. agriculture schemes), nor does it fully consider the cumulative or interacting effects of several projects in a restricted area.

Even more seriously, the major policy decisions and the large development plans or programmes that are decided upon at the regional, national or international level, which generate so many of the individual projects in Europe, are not assessed at all (a good example being the Trans-European Transport Networks proposed for the European Union).

To tackle the wide-ranging impacts of such broad initiatives, a totally new approach, Strategic Environmental Assessment (SEA), is required. However, the proper methodology and legislative background for SEA have yet to be fully developed.

7 Currently, monitoring of the impacts of various land- and sea-uses on biodiversity are inadequate, and the mechanisms behind changes or losses of biodiversity remain unclear in many cases.

Research in this area is often hindered by the complex interactions between, and cumulative effects of, various land-use factors, and there is a strong need for experiments and studies over large geographical areas and over a variety of time-scales, in order to draw meaningful conclusions that are applicable under different conditions.

It is clear from the existing data, however, that intensified land- and sea-uses are causing habitat degradation and biodiversity loss in the long term, as the widespread decline of the European avifauna shows.

8 All policy-makers, land-owners and land-users in Europe could be, and should be, significant participants in the process of conserving European biodiversity. This is not yet widely appreciated. In addition, publicly available information is often lacking on land-use measures which can promote biodiversity conservation in the wider countryside.

9 Substantial economic incentives for the destruction or degradation of natural, semi-natural or traditionally managed habitats are still widespread in Europe. Moreover, these incentives often lead to the erosion of rural societies as well. Economic measures that promote the conservation of biodiversity by users of the environment are still scarce.

The full, long-term economic and ecological costs of short-term development are not properly accounted for, partly because of the methodological shortcomings of mainstream economic cost–benefit analyses.

The 'external' or 'free' ecological services performed by habitats (e.g. water purification, coastal defence) are still not widely understood or agreed upon by society or government, and have not yet been properly integrated into mainstream economics or fully appreciated for their non-monetary values.

Main Recommendations

1 The strategic principles and recommended measures for the conservation of biodiversity in the wider environment should be clearly and widely agreed among policy-makers, land-owners and land-users at all levels, and implemented.

2 Clear environmental objectives, including targets for biodiversity conservation, should be integrated into all land- and sea-use policies at all levels (local, regional, national and international). The monitoring and evaluation of policies should be improved, and should include assessments of their effectiveness in relation to these objectives.

3 There should be more coordination and integration between sectors on policies and actions that affect land- and sea-uses. This may require the

development of innovative new structures in decision-making at the local and regional levels.

4 International legislation and major strategies and policies concerning nature conservation should be fully implemented, and their provisions wholly integrated in all policy areas by all countries in Europe. All major policy areas in the European Union (e.g. Common Agricultural Policy, Common Fisheries Policy, regional policies) should be reformed as swiftly and effectively as possible, in order to take full account of Community environmental policies. New policies on forestry and rural development should be adopted, with clear nature conservation objectives.

5 For all major policies, plans and programmes which involve more than one project and where a cumulative impact may be expected on the environment, Strategic Environmental Assessment of the whole initiative should be conducted. The proper methodology and legislative background of SEA should be fully developed and implemented at all levels in Europe.

6 New methods of economic evaluation should be developed, for the assessment of development plans, programmes, policies and projects, which take into account the full, long-term costs of any attendant environmental degradation and the loss of any natural resources or ecological processes. This requires widespread agreement by society on the value of the ecological and social services provided by habitats.

7 Economic incentives should be steered away from environmentally destructive activities towards those which actually conserve biodiversity, natural resources and natural ecological processes in the wider environment, while at the same time maintaining rural jobs and sustaining rural societies.

8 Research efforts should be widely promoted to reveal the often-complex relationships between the various human uses of the environment and any attendant losses or changes in levels of biodiversity, and to clarify the mechanisms of such loss or change. More scientific understanding of the natural dynamics of habitats at the landscape level is also needed, as

well as the development of ecological indicators to quantify the success of policy measures.

Monitoring biodiversity in the wider environment should be a component of national and international environmental monitoring programmes. International cooperation and considerably increased levels of funding for such research will be essential.

Much of this research should be funded by agricultural, industrial and other land- and sea-use sectors, to demonstrate that their activities are not damaging the environment and are sustainable. It should not be left for environmental organizations (governmental or otherwise) to demonstrate unsustainability in these sectors, before activities are examined or regulated.

9 Widespread public understanding of the services provided by habitats to society (e.g. flood control, coastal defence, water storage and purification, carbon sinks, etc.), and of the non-monetary values of all habitats (e.g. cultural or spiritual values) should be promoted, as should widespread public agreement on the importance of these services and values.

It is also essential to provide clear practical guidelines on managing habitats in an ecologically sound way, and to publicize these widely among policymakers and land-users at all levels. Eco-labelling and similar non-statutory initiatives should be developed and promoted widely, so that the public can be empowered in choosing and promoting products that are produced according to best environmental practices. The results of relevant research, restoration and demonstration projects should also be publicized.

HABITAT CONSERVATION STRATEGIES

MARINE HABITATS

compiled by

M. TASKER and L. CANOVA

from major contributions supplied by the Habitat Working Groups

North and west European seas

V. Bakken, S. Švazas, H. Skov, M. F. Leopold, G. Rolland, R. Rufino, K. H. Skarphedinsson,
M. Tasker (Coordinator), E. Stotskaia

European Macaronesian seas

D. Concepción García, F. Rodríguez Godoy, C. González, J. A. Lorenzo Gutiérrez,
A. Martín Hidalgo, J. L. Rodríguez Luengo, L. R. Monteiro, M. Nogales, P. Oliveira, I. Pereira

Mediterranean and Black Seas

L. Canova, M. Fasola

and by

E. Dunn and D. Owen

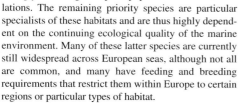

S U M M A R Y

European seas are important for the continued survival of 62 priority bird species in Europe, of which 45% have an Unfavourable Conservation Status due to their notably small or strongly declining European populations. The remaining priority species are particular specialists of these habitats and are thus highly dependent on the continuing ecological quality of the marine environment. Many of these latter species are currently still widespread across European seas, although not all are common, and many have feeding and breeding requirements that restrict them within Europe to certain regions or particular types of habitat.

Maintaining the ecological quality of marine habitats is clearly essential for the conservation of priority bird species in Europe. However, damaging changes are widespread, due mainly to the loss or deterioration of safe terrestrial nest-sites, to the pollution of seas with rubbish, oil, toxins and excessive nutrients, and to the reduction of food resources through intensification of fisheries and other industries exploiting marine habitats.

More comprehensive and stringent protection and management of birds' breeding sites is vital, particularly for many colony-nesting species. However, seabirds are dispersed when away from these colonies, and wider conservation measures for marine habitats are therefore also essential.

Governments need to ensure the integration of the conservation of marine habitats into and across all relevant sectors, programmes, policies and plans, through an appropriate balance of guidance, incentives and regulations. In particular, it is necessary to effect comprehensive and strict implementation of the various conventions and regulations aimed at controlling pollution. A substantial and permanent reduction in fishing effort is also needed in Europe, as well as intergovernmental agreement on, and implementation of, sustainable fisheries management. Substantial progress towards this can be made through non-statutory opportunities, such as those being developed by the Marine Stewardship Council.

HABITAT DESCRIPTION

Habitat definition, distribution and extent

This habitat conservation strategy covers all marine areas below low-tide mark, in the following three major geographical divisions (Figure 1):

- North and west European seas

- European Macaronesian[1] seas

- Mediterranean and Black Seas

---

[1] Macaronesia is the biogeographical region composed of the archipelagos of Madeira, the Canary Islands, the Azores and (outside Europe) the Cape Verde Islands.

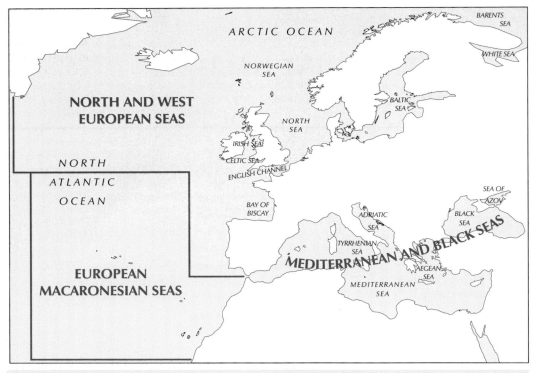

Figure 1. The seas of Europe, showing the division into marine regions used in this chapter. North Russian seas, included within the 'North and west European seas' region, extend to 90°E.

The extent of the major seas of Europe is shown in Table 1. The largest of the three divisions treated in this strategy is the 'north and west European seas', with an approximate area of some 40 million km². Of this, by far the largest sea is the European part of the North Atlantic Ocean, with an area of some 36 million km². The combined area of the north and west European seas and the Mediterranean and Black Seas thus gives a total European marine area of some 43 million km², four times the area of Europe's 10.2 million km² land mass. The European Macaronesian seas are a further 4 million km² in extent.

These figures indicate the huge scale of the habitat and its importance compared to terrestrial systems. A large proportion of these sea areas is beyond the national jurisdictions of European countries and therefore constitutes international waters, but the entire area is nevertheless influenced by European activities such as fishing, transport and waste disposal. European countries therefore have at least a partial responsibility, along with other nations using and influencing the region's seas, for their conservation.

The areas of marine habitat falling within territorial waters (normally 12 nautical miles from the coast) and within Exclusive Economic Zones for each country are indicated in Table 2. All European countries other than Andorra, Austria, Belarus, Czech Republic, Hungary, Liechtenstein, Luxembourg, former Yugoslav Republic of Macedonia, Moldova, San Marino, Switzerland, Slovakia and Vatican City have marine coastlines and therefore direct responsibility for the marine environments under their jurisdiction.

Ecological characteristics
Key abiotic factors affecting the habitat
Although at first sight marine habitats may appear uniform in character, they vary substantially in a

Table 1. The extent of the European seas (CEC 1995d).

	Surface area (km²)
North and west European seas	
Baltic Sea	412,500
Barents Sea	1,425,000
North Atlantic Ocean	36,000,000
North Sea	750,000
Norwegian Sea	1,340,000
White Sea	90,000
European Macaronesian seas	
North Atlantic Ocean	4,000,000
Mediterranean and Black Seas	
Black Sea and Sea of Azov	500,000
Mediterranean Sea	2,505,000

Table 2. Extent of marine habitats in Europe according to country (Dahl 1991, WRI 1994).

	Length of coastline (km)	Exclusive Economic Zone (km^2)
North and west European seas		
Belgium	64	2,700
Denmark	3,379	1,464,200
Faroe Islands	700	?
Greenland	39,000	?
Estonia	1,393	?[3]
Finland [1]	1,126	98,100
France [2]	3,427	3,493,100
Germany: Baltic Sea	901	9,600
Germany: North Sea	1,490	40,800
Iceland	4,988	866,900
Rep. of Ireland	1,448	380,300
Latvia	531	?[3]
Lithuania	108	?[3]
Netherlands	451	84,700
Norway [1]	5,832	2,024,800
Svalbard	2,509+	
Poland [1]	491	28,500
Portugal	1,693	1,774,200
Russia: all European seas [2]	37,653	?
Russia: Kaliningrad	c.180	?[3]
Spain [2]	4,964	1,219,400
Sweden	3,218	155,300
United Kingdom	12,429	1,785,300[4]
European Macaronesian seas		
Portugal		
Azores	780	?
Madeira	141	?
Spain		
Canary Islands	1,291	?
Mediterranean and Black Seas		
Albania	418	12,300
Bulgaria	354	32,900
Croatia	?	?
Cyprus	648	99,400
France [2]	—	—
Greece	13,676	505,100
Italy	4,996	552,100
Malta	140	66,200
Romania	225	31,900
Russia [2]	—	?
Slovenia	46	?
Spain [2]	—	—
Turkey	7,200	236,600
Ukraine	2,782	?
United Kingdom		
Gibraltar	10	?

[1] Not including islands.
[2] Figures for France, Russia and Spain under 'North and west European seas' include Mediterranean/Black Sea coast.
[3] International agreement not yet reached.
[4] Extended jurisdiction: UK does not have an Exclusive Economic Zone.

containing essential elements such as nitrogen, phosphorus and iron. Nutrients in surface waters are consumed by phytoplankton which in turn support food webs of higher animals. Deeper waters support relatively little marine life, due to lack of light and to low temperatures, but are usually rich in nutrients, due to the continuous accumulation of dead organisms sinking from above. Nutrients are thus added to surface waters by upwellings from deeper water, and also by dissolving directly from the atmosphere.

Depth, and change in depth, are important in determining amounts of nutrient reaching the surface from below. In deep oceanic waters, there is relatively little mixing between surface and deeper waters; in nearshore shallow water, turbulence caused by tides, currents and wind leads to a greater flow of nutrients to the surface. Nutrients can also be brought to the surface where oceanic currents meet the continental shelf or wash against seamounts. On a smaller scale, currents cause further mixing of nutrients as they flow around islands and headlands.

Water temperature and solar heating can also affect productivity. Shallow and coastal waters typically have lower winter and higher summer temperatures than adjacent oceanic waters. Higher water temperatures lead to higher chemical reaction rates, including photosynthesis. However, phytoplankton growth is usually restricted in warm waters by nutrient depletion at the surface. Warmer water is also less dense, and so sits above colder water, leading to layering of the water column, which further inhibits vertical nutrient flow. This occurs particularly in the summer, when solar radiation is higher and winds are lower. Where layered (or stratified) water meets well-mixed water, enhanced availability of nutrients can occur at the narrow contact zone ('front'), which is thus often particularly productive.

Formation of ice occurs at the other extreme of water temperature. Only a few scavenging birds can feed on a frozen sea, but, as soon as melted channels running into the ice ('leads') appear in spring, many more are able to find food. Such areas can be particularly productive due to the release of nutrients held in the ice and the increased light levels. The coverage of north European seas by ice varies seasonally.

Salinity may be important in determining the fauna of an area; some species prefer the relatively saline oceanic waters, others the less saline waters of estuaries and river outflows.

Sediment characteristics are important for those birds whose food occurs in or near the seabed, since they determine the distribution of prey (see 'Habitat needs of priority birds', p. 71). For example, sandy and muddy sediments support large populations of bivalve molluscs (important for seaduck Anatidae), while sandeels *Ammodytes* (very important for sev-

number of abiotic features which have a profound effect on their productivity and in turn their plant and animal communities. From the point of view of a bird, food availability at or relatively near the surface is important. This food availability relies ultimately on the inflow of nutrients, natural chemicals eral seabirds) require well-oxygenated sandy sediments. Sediment characteristics may in turn influence the optical quality of the water. Clearer water may be easier for some seabirds, which rely on eyesight, to hunt in. Generally, seabirds that dive when feeding do not appear to go deeper than about 20 m, although the auks (Alcidae) at least can far exceed this. Waters over nearshore muddy areas frequently have much higher loads of suspended sediments than those over rocks or coarse sand.

Islands and headlands are important features in their own right as breeding sites for seabirds. Offshore habitat usage can depend on the proximity of land for breeding and roosting.

North and west European seas
Tidal height is considerable along the Atlantic seaboard of Europe, typically 2–7 m, but is relatively low in the Baltic Sea. Salinities are normally near the oceanic average (33.3‰) but water in the Baltic Sea is comparatively 'fresh' (8.5–10.0‰), as most inputs come from rivers. Water temperatures here may be as high as 20°C in summer, but drop to zero in winter. To the east, much of the Barents Sea, the northern Baltic, and nearly all of the Kara and White Seas are typically covered by ice in winter.

North and west European seas generally have high levels of nutrients and of productivity. Colder northern areas and deep waters over the continental shelf have lower productivity levels, while areas such as the North Sea, with particularly shallow waters, strong currents and high nutrient inputs, may have very high productivity levels.

European Macaronesian seas
The Azores, Madeiran and Canarian Archipelagos all show similar submarine topographic features, with very narrow insular platforms that descend rapidly to depths of 2,000–3,000 m in the channels between islands. Submerged seamounts (banks) are important throughout Macaronesian seas, their combined area at 500 m depth exceeding 2,160 km². These warm-water seas are characterized by low but relatively stable levels of productivity (Raymont 1980). However, these can be substantially elevated due to local upwellings near seamounts, and to the presence of a nutrient-rich upwelling along the adjacent African coast and its resulting westerly current, which raises productivity levels in the east of the Canarian archipelago. As a result, it is thought that many seabirds in the region move to the east to feed.

Mediterranean and Black Seas
Tidal amplitudes everywhere are very low, averaging only 30 cm. Being isolated from oceanic currents and influences, water temperature and salinity in the Mediterranean Sea are more variable than in the open ocean. By contrast, the Black Sea is characterized by permanently low salinity (12.0–18.2‰). At present, the Black Sea is highly eutrophic, and as a result productivity has fallen drastically (see 'Threats to the habitat', p. 83).

The Mediterranean Sea has a very low level of biological productivity due to naturally low surface-nutrient levels. Locally, however, this pattern is altered by artificially high inputs of nutrients from some rivers, especially the Rhône, Ebro, Po and Nile, which stimulate the development of rich plankton communities, particularly where the continental shelf is wide, e.g. in the northern Adriatic Sea.

Key biotic factors and natural dynamics
Perhaps the key biotic aspect of the sea is its variability in space and time, often unpredictable, and in this it differs in a major way from terrestrial ecosystems. For reasons that are in many cases obscure, current flows can slow down, fish can fail to reproduce, and there is year-to-year variance in shellfish spatfall, for example. Variability in the quality and availability of food also changes seasonally. Fish which are about to spawn have a considerably higher energy value than spent fish. Sandeels are not available in mid-water or at the surface during winter.

The mixture and interaction of these physical and biological features leads to a patchy, heterogeneous distribution of seabird prey. The presence of concentrations of suitably-sized shellfish or finfish is particularly important for seabirds (e.g. Leopold 1993, Piatt and Nettleship 1995). There is also evidence of selection by some seabirds of the most energy-rich foods (e.g. Harris 1984b). Daytime feeding by some seabird species may be enhanced by underwater predators, such as dolphins (Delphinidae) and tuna *Thunnus*, forcing potential prey to the surface (Martín 1986). Feeding associations of some shearwaters (Procellariidae) and terns (Sternidae) with dolphins and tuna may considerably influence their breeding success, which may, therefore, in turn depend to a certain extent on healthy populations of underwater predators.

Characteristic flora and fauna
North and west European seas
The shallow seas of the boreal and temperate zone are typically less than 20 m deep, and predominantly consist of the coastal, sheltered waters of the Baltic Sea, Kattegat, White Sea, North Sea and English Channel. A variety of bivalve molluscs, shrimps and young crabs are invertebrate food of particular importance to seabirds. A wide variety of fish occur, notably flatfish, demersal (bottom-living) species

and pelagic shoaling fish. The avifauna includes seaduck, divers, cormorants (Phalacrocoracidae) and the smaller gulls (Laridae) and some terns. The seals *Phoca vitulina* and *Halichoerus grypus* are widespread. Amongst the cetaceans, only *Phocoena phocoena* and *Tursiops truncatus* use these areas regularly. Plant-life is dominated by phytoplankton, although in some shallow areas seagrass *Zostera* and *Ruppia* or beds of seaweed can dominate.

Shelf seas of the boreal and temperate zone are typically between 20 m and 200 m deep and include the seas immediately to the west of Iberia, the Bay of Biscay, the Celtic Sea, the English Channel, the Irish Sea, the North Sea, the Faroes plateau, seas around Iceland, and the west Norwegian shelf. These areas are rich in fish species (e.g. see North Sea Task Force 1993), although fewer than 20 make up over 95% of the total fish biomass. Sandeels, sprat *Sprattus sprattus* and the young of herring *Clupea harengus* and gadoid (cod-like) fish are preferred prey for many seabirds. Characteristic birds of these waters are Kittiwake *Rissa tridactyla*, Fulmar *Fulmarus glacialis*, Manx Shearwater *Puffinus puffinus*, Gannet *Sula bassana* and the larger gulls (Laridae). The most frequently encountered dolphins and whales are *Phocoena phocoena*, *Lagenorhynchus albirostris*, *Balaenoptera acutorostrata* and *Delphinus delphis*.

Seas of the Arctic zone lie between 30°W and 90°E and north of the 0°C January isotherm, and include the seas around Greenland, Jan Mayen, northern Norway and Finland, and the Kara and White Seas. Shallow seas hold seaduck, while the shelf seas are dominated by auks, *Fulmarus glacialis* and the northern large gulls. Many of these species feed on pelagic invertebrates. Those seabirds that can dive deeply may feed also on capelin *Mallotus villosus* and young Arctic gadoids. Several cetaceans occur regularly in these areas, *Orcinus orca* being one of the most characteristic. The zone was of considerable importance for the largest baleen whales before they were severely reduced in numbers by commercial whaling. There is a high diversity and abundance of seals.

The deep North Atlantic, south of the 0°C January isotherm, is relatively unproductive. Its avifauna is predominantly petrels (Hydrobatidae and Procellariidae) and shearwaters, present usually at low densities. These feed on foods occurring near the surface, such as copepods and small fish. The larger whales occur in this zone. Seabed features, such as seamounts, can concentrate food locally and can therefore be attractive both to whales and seabirds. Apart from the phytoplankton, and some drifting marine algae, no plants occur in this zone.

Deep seas in the Arctic zone, between 30°W and 90°E and north of the 0°C January isotherm, are virtually devoid of birds in winter as much of the area freezes then. In summer, the break-up of the ice leads to locally enhanced food supplies, which attract many of the same birds as are present in the shallower shelf seas of the Arctic, but at much lower densities. Both seals and cetaceans also occur at the edge of the ice-cap in summer. Phytoplankton are the only plants present.

European Macaronesian seas

The region has a rich assemblage of fish species, e.g. 600 are listed for the Canary Islands. Important food species for some seabirds include boarfish *Capros aper*, snipefish *Macrorhamphosus scolopax* and sardines *Sardina pilchardus*. Mackerel *Scomber scombrus*, bonito *Katsuwonus pelamis* and tuna are also common and can be important in facilitating the feeding of some bird species (Martín 1986).

Petrels, shearwaters, terns and gulls breed regularly, the most abundant species being Bulwer's Petrel *Bulweria bulwerii*, Madeiran Storm-petrel *Oceanodroma castro*, Cory's Shearwater *Calonectris diomedea*, Little Shearwater *Puffinus assimilis*, Roseate Tern *Sterna dougallii*, Common Tern *S. hirundo* and Yellow-legged Gull *Larus cachinnans*. Important numbers of Great Shearwater *Puffinus gravis* also occur on passage and Sandwich Tern *Sterna sandvicensis* in winter. Most of the petrels and shearwaters are typically found in offshore oceanic waters when away from their breeding grounds. In contrast, terns are characteristic of inshore shallow waters, as in the other European sea areas.

Five species of globally threatened marine turtle occur in the region, of which *Caretta caretta* is encountered frequently. About 20 species of cetacean also occur (Gordon *et al.* 1989). The globally threatened monk seal *Monachus monachus* was common in the past but is now extinct (Groombridge 1993).

Mediterranean and Black Seas

The marine fauna of the Mediterranean Sea has many endemic species and is noticeably more diverse than that of the Atlantic coasts (Stanners and Bourdeau 1995). However, many species occur at low densities due to the very low biological productivity. In contrast, the Black Sea has lower species diversity but typically a higher abundance of fauna. Deep benthic fauna are, however, largely absent from the Black Sea due to anoxic conditions.

Shallow waters extending to the edges of the continental shelf (less than 200 m deep) include a great part of the Adriatic, Black and Azov Seas, some of coastal Iberia and much of the Bay of Gibraltar, the sea surrounding the Balearic Islands, Corsica, Sardinia, Cyprus, Crete and the Gulf of Gabès (Tunisia).

Seagrass beds often cover a large part of the littoral zone and are an important component of the Mediterranean marine ecosystem. They produce organic material at the base of the food-chain, provide oxygen and are feeding and nursery areas for numerous species of fish and other marine organisms. Coral-type substrates are typical and widespread Mediterranean deposits.

Wrasses *Labrus* spp. and *Symphodus* spp., and seabream *Diplodus* spp. are common and important fish in the diet of coastal seabirds such as cormorants and gulls (Fasola *et al.* 1989). The shallow seas are typically the most important feeding areas for most seabirds of the Mediterranean and Black Seas, as this zone has the highest productivity and hence food availability. In winter, divers, grebes (Podicipedidae), *Sula bassana* and seaduck may also occur here. Two globally threatened species of marine turtle, *Caretta caretta* and *Chelonia mydas*, nest regularly and in significant numbers in the east Mediterranean. Regularly occurring cetaceans include *Delphinus delphis*, *Tursiops truncatus* and *Stenella coeruleoalba* (Cagnolaro *et al.* 1983). The globally threatened seal *Monachus monachus* was common in the past but is now exceedingly rare.

Shelf seas between 200 and 1,000 m can be important feeding habitat for shearwaters and petrels during the breeding season (Appendix 3, p. 354) and for wintering flocks. They are also used by large gulls and terns when following fishing boats to feed on discarded fish. Various cetaceans exploit this zone, including *Tursiops truncatus*, *Globicephala macrorhynchus* and *Balaenoptera physalus*. The fauna is dominated by shrimps and bottom crustaceans. Surface-fish species of importance to birds include anchovy *Engraulis encrasicholus*, sardines and species of horse mackerel *Trachurus*.

Waters over 1,000 m deep include the central Mediterranean Sea to the east of Corsica and Sardinia, central Tyrrhenian Sea and the southern Ionian Sea to the outer Lebanon coast. The surface pelagic waters are usually very low in nutrients, and biomass is mainly represented by plankton, though large shoals of clupeid fish exploit the surface and cetaceans such as *Physeter macrocephalus* and *Grampus griseus* are often present. These seas are little used by birds, except by shearwaters which may occur in large flocks.

Key relationships with other habitats

Marine habitats come into physical contact with four other habitats: adjacent sections of ocean, fresh water, coasts and the atmosphere. The oceans are the key to the world's climate, and thus to a great extent affect all other habitats. Much of this influence is related to the thermal capacity of the ocean, and the generation of oxygen (and absorption of carbon dioxide) by phytoplankton. The relationship of the atmosphere to the ocean is subject to considerable continuing study. These studies have already demonstrated substantial atmospheric inputs of pollutants to the oceans through deposition onto the ocean surface (e.g. OSPARCOM 1992a, Monteiro 1996).

Inputs to the north, west and Macaronesian European seas come through currents, most particularly the Gulf Stream. This brings both warmth and nutrients from the area off the Caribbean. Without its influence, ice would extend considerably further south off north-west Europe in winter.

Freshwater inflow is low compared to oceanic inflow, but the interface between the ocean and rivers is of great biological significance. Commonly, estuaries are formed (these are treated under 'Coastal habitats', p. 93). Rivers bring both nutrients and sediments from the land, but also bring pollution from industry, from municipal sources and from agriculture.

The oceans affect coastlands through the processes of erosion and deposition. Human changes on the coast can affect coastal currents and sediment transfer. These changes often have a greater effect on other areas of coast than on the oceans proper.

Values, roles and uses

Despite the relatively inhospitable nature of the marine environment to man, it is of considerable social and economic value. Marine habitats are widely used for various forms of food production, industry, transport and recreation.

Fisheries

The continental shelves of the north-east Atlantic are of particular importance, being one of the richest fishing areas in the world and accounting for c.10% of the world's marine fish catch (Stanners and Bourdeau 1995). Almost all fisheries are now commercial, and the economies of the Faroe Islands, Greenland and Iceland have been especially dependent on this sector. Within the north-east Atlantic, the most productive fishing area is the North Sea which, although representing only about 10% of the area, provides about one-third of the total catch. Fishing effort in the north-east Atlantic has shifted in the last 30 years towards inshore fisheries.

Most fish are caught for human consumption but over the last 40 years there has been major growth in 'industrial fishing'. The target species, sandeels, sprats, Norway pout *Gadus esmarkii*, herring and capelin are processed into feedstuffs for livestock and farmed fish. Currently, around one-third of all the fish caught in the North Sea is used for this purpose (Dunn 1996).

With the majority of north-east Atlantic stocks overfished, especially in the areas encompassed by the continental shelf, another growth sector is the deep-water fishery (trawling and longlining) on the eastern Atlantic shelf-edge.

Commercial Macaronesian fisheries comprise three main types (e.g. Krug and Silva 1990, Santos *et al.* 1995): (1) small, coastal pelagic fish such as young blue jade mackerel *Trachurus picturatus*, chub mackerel *Scomber japonicus*, *Sardina pilchardus* and bogue *Boops boops* using seine-nets, dip-nets and lift-nets; (2) offshore pelagic *Thunnus* and related large, predatory species using pole-and-line; (3) deep demersal species, using longlines and handlines. The latter two types use large quantities of small fish (not declared in fishery statistics) as bait. Stocks and landings of demersal species have decreased (e.g. Viceconsejería de Medio Ambiente 1994, Silva and Krug 1992).

In 1993 the total declared fish catch in the Mediterranean and Black Seas was 1,670,000 tonnes, of which about 80% was finfish (FAO 1993). In the Mediterranean, Italy, Greece and Spain are the main fishing countries, together accounting for about 45% of the total catch in 1993. In general, the east Mediterranean is the least productive area.

The main fish taken are *Sardina pilchardus* and *Engraulis encrasicholus*, which together accounted for about 34% of the catch in 1993. Other commercially important species include *Trachurus mediterraneus*, *Scomber japonicus*, several species of Gadiformes, Sparidae and Mullidae, and large pelagic species such as swordfish *Xiphias gladius* and tuna. Nearly all commercial finfish stocks in the Mediterranean are considered to be fully or over-exploited, with the possible exception of lower value small pelagics such as *Boops boops* and *Trachurus mediterraneus* (FAO 1993 in Stanners and Bourdeau 1995).

Until 30 years ago, the Black Sea was a major fishery, with five times the productivity of the Mediterranean, and providing a livelihood for two million people in coastal communities (Pearce 1995). However, the fisheries went into decline, and finally collapsed by the end of the 1980s, largely due to a massive increase in nutrient pollution and the accidental introduction of exotic species (Pearce 1995; see 'Threats to the habitat', p. 83).

Fish-farming

In the north and west European seas, salmon *Salmo salar* is the main species of farmed finfish, and mussels *Mytilus edulis* are the most commonly farmed shellfish. In the Mediterranean, fish- and shellfish-farming is well developed in the northern Adriatic, in the central Mediterranean and along the Iberian coast: the most important species are the gilthead *Sparus auratus*, the sea bass *Dicentrarchus labrax* and mullets of the genera *Mugil* and *Liza*.

Hunting and other harvesting activities

Seabirds are hunted for food in north European countries, and seabird eggs are harvested for food in many countries. Whaling was formerly widespread but has now stopped in most countries due to severe depletion of whale numbers. Whaling does, however, continue in the Faroes, Iceland, Greenland and Norway. Norway and Russia cull seals and use their skins, while in Greenland sealing for food and skins continues.

Transport

Shipping is an important means of commodity and passenger transport in the region. The highest density of shipping traffic outside ports is in the Dover Strait, between France and the UK, and other particularly busy shipping lanes include the Straits of Gibraltar and the Bosporus.

A huge increase in oil transportation is envisaged in the Baltic due to the expansion of the oil industry in the east. A new 'Northern Sea Route' which passes north of Norway and Russia is currently being investigated and would conceivably have major impacts in the region if it becomes operational and heavily used.

Within the European Union (EU), some member states, the European Commission and the shipping industry are currently promoting the carriage of goods by sea in preference to roads and railways, a concept termed 'short-sea shipping'.

Oil and gas extraction

The major centre for oil and gas exploration and production in Europe is the North Sea. Exploitation and exploration also occurs in the Norwegian Sea, Baltic Sea, Celtic Sea, English Channel, Irish Sea, Bay of Biscay, northern Adriatic Sea, south Tyrrhenian Sea, and Gulf of Sirte (off Libya). Large-scale oil exploitation is also anticipated in the eastern Baltic Sea in the near future. Extraction has also recently started at several sites along the north and west Black Sea coast.

Aggregate and amber extraction

Extraction of sand, gravel and shell deposits is undertaken mainly for the construction industry. For economic and technical reasons, most extraction of sediments occurs in water less than 35 m deep. To the south-west of Iceland, there is a large mining operation for shells for use in the manufacture of cement. Elsewhere in the north and west European seas, sand and gravel is extracted from the coastal waters of Belgium, Denmark, France, Netherlands and UK. A specialized extraction industry for amber exists in the Baltic Sea, for military purposes and jewellery.

Energy

Tidal energy is exploited, though only two tidal power schemes are operating in Europe: in Brittany (France) and on the Barents Sea. Wind-energy schemes are operational in Sweden, Denmark and Germany, and a pilot wave-energy plant is in operation in the Azores.

Waste disposal

Millions of tonnes of waste have been directly and intentionally disposed of into the sea, including dredged material, hazardous substances from land-based sources, oily wastes, chemical residues, garbage and sewage from shipping, drilling muds and oily cuttings from the offshore industry, and sewage from municipal sources. For example, in 1990 the UK dumped 5.4 million tonnes of sewage sludge in the North Sea (OSPARCOM 1992b cited in North Sea Task Force 1993), and Belgium, Denmark, France, Germany, Netherlands, Norway, Sweden and the UK together dumped 136 million tonnes of dredged materials in 1990. The level of dumping is, however, now decreasing as agreements under the OSPAR Convention come into force.

Dumping at sea also occurs in the Baltic, Mediterranean and Black Seas; in the western Black Sea alone, there are 16 official waste-disposal sites for dredging materials (Stanners and Bourdeau 1995). Dumping of hazardous military materials has occurred in some deep waters, e.g. dumping of nuclear reactors around Novaya Zemlya (Russia).

Military activities

Huge marine areas are used for military training exercises and weapons testing or are closed for security reasons. For example, marine areas around Novaya Zemlya (Russia) were used for experimental nuclear explosions.

Recreation

In addition to widespread coastal uses such as swimming, snorkelling, diving, sailing and windsurfing, ship-cruise tourism occurs in some seas. By far the most important recreational areas are the Mediterranean and the more popular islands of the Macaronesian seas region. The Mediterranean basin is the world's most significant tourist destination, attracting 157 million visitors in 1990 (WTO 1993), mainly to the coast, and currently some 35% of international tourists worldwide (Stanners and Bourdeau 1995), with numbers predicted to continue increasing.

The Black Sea is a major tourist destination for tourists from central and eastern Europe, attracting about 40 million people annually. However, tourism in the area is declining due to the poor economic situation and declining environmental conditions. Recreational use is also significant in parts of the north and west European seas region.

Political and socio-economic factors affecting the habitat

Factors affecting any of the uses of the region may have effects on the habitat. However, the most important current uses are fisheries, oil and gas extraction, waste disposal and transport. Political and/or socio-economic factors are therefore discussed in relation to these.

The quality of the marine environment is highly dependent on activities within water catchments, thus even landlocked countries have important roles to play in conserving marine habitats through control of pollutant inputs to river systems; hence Switzerland's participation in the North Sea Conference process. Similarly, activities such as waste disposal or fishing in one nation's marine waters may have widespread effects on other countries' or international waters. Conservation in international waters cannot therefore be left to individual countries but must be the responsibility of all users.

Fisheries

Fisheries issues have a high political profile, especially in states relying almost exclusively on them, namely the Faroe Islands, Greenland and Iceland.

Within 200 nautical miles of the coast, fisheries in the waters of EU member states are managed by the European Commission and the Council of Fisheries Ministers, under the Common Fisheries Policy (CFP). The CFP is based upon the concept of total allowable catches (TACs), which are divided up into quotas for EU member states using percentage allocations which have not changed for many years (the 'relative stability' principle). Fishermen must discard catch which is undersized or surplus to quota. The exclusive reliance on TACs, often set higher than scientific advice suggests, the inflexibility of the 'relative stability' principle, the discarding wastage and frequent lack of adherence by fishermen to TACs and quotas (in short, overfishing) all contribute to mismanagement of the fisheries (Dunn and Harrison 1995).

States outside the EU regulate their fisheries independently of the CFP, although bilateral agreements are regularly made with the EU. Part of Norway's recent decision to remain outside the EU was related to its desire to continue managing its stock independently of the CFP. Faroe Islands, Greenland,

Iceland and Norway all tend to rely more on pelagic single-species fisheries than on demersal stocks, since the former are easier to manage.

It is widely recognized that there is a need to reduce capacity in European fishing fleets to balance available fish stocks. Meanwhile, cheap imports and the proliferation of highly efficient fishing vessels are leading to declines in small fishing communities.

In central and eastern Europe, national fisheries policies were often based upon the twin aims of supplying as much fish protein as possible to the population and maximizing foreign exchange earnings (Stanners and Bourdeau 1995), and few contained measures aimed at conserving fish stocks. Some such countries, e.g. Estonia, have revised their policies but have abolished limits on netting and catches except for internationally regulated species.

In the Mediterranean, overfishing may be a problem in the Adriatic Sea as there is no agreed common policy between Italy and Croatia.

Some species are subject to international regulations, e.g. tuna and swordfish fisheries are regulated by the International Commission for the Conservation of Atlantic Tuna (ICCAT).

Oil and gas extraction

In the North Sea, reserves in the major fields are declining. However, because of the well-established infrastructure in the area, it may now be economical to exploit the minor fields. Oil and gas fields in sites outside the North Sea will not be developed unless they are clearly economical. Fields to the west of Shetland and Orkney, and in northern Norway are presently being developed. New and intensive oil-drilling projects are also underway along the Mediterranean coasts and off Ireland. Some exploration activity has started off the Faroe Islands. In the North Sea, new initiatives are driving down the amount of oil pollution from the offshore industry (ACOPS 1996).

Waste disposal

Progressive states are promoting 'clean production' and 'low-input farming' as ways to reduce the amount of waste generated by industry and agriculture respectively. Less progressive states remain unconvinced of the potential long-term financial benefits of reducing waste and continue to look for ways to manage their waste. The North Sea Conference process, as well as ministerial meetings of contracting parties to the OSPAR, Helsinki and Barcelona Conventions, provide opportunities for political initiatives in both these areas (e.g. see Esbjerg Declaration, Danish Ministry of Environment and Energy 1995).

Ship-sourced waste continues to be addressed by the International Maritime Organization (IMO), the North Sea Conference process and the European Commission, with the emphasis on better waste management rather than reducing waste. The European Commission is keen to implement its *Common Policy on Safe Seas*; a Directive on port-reception facilities is in preparation. North Sea states are adopting measures to implement the Esbjerg Declaration, including joint work taking place through IMO to achieve 'Special Area' status for the North Sea and wider under MARPOL Annex I (oily waste).

Transport

Transport by sea can result in waste disposal at sea and also introduces the risk of accidental oil and chemical spills. One means of reducing spills is to eliminate substandard ships. Given the poor record of many flag states at doing this, port state control in the region has wide political backing. Fifteen maritime states are parties to the Paris Memorandum of Understanding (MOU) on Port State Control, and in 1995 EU member states adopted a Directive on Port State Control based upon the Paris MOU.

PRIORITY BIRDS

The list of priority bird species of European marine habitats is given in Table 3. Of the three regions covered by this strategy, the north and west European seas hold by far the largest number of priority species (Tables 4 and 5). The 51 priority species in this region compare with 14 in the European Macaronesian seas and 13 in the Mediterranean and Black Seas. The high number of priority species in the north and west European seas is largely the result of the very extensive area, diverse ecology and rich seabird community of this region (see 'Ecological characteristics', p. 60, Table 1), as well as the high dependency of most of its seabirds on the habitats of the region (Table 3). Only 21 of the 51 priority species (i.e. 41%) have an Unfavourable Conservation Status in Europe (Tucker and Heath 1994), which indicates that the high number of priority species is not due to a particularly high level of threat to the habitat or its species. By comparison, 13 (93%) of the 14 Macaronesian species and 8 (61%) of the 13 Mediterranean/Black Seas' species have an Unfavourable Conservation Status. This suggests that the habitats and seabirds of these two regions are proportionately more threatened, and this is also reflected in the priority levels of the regions: 53% of the priority species of the north and west European seas have Priority D, compared to 8% and 38% in the European Macaronesian seas and Mediterranean and Black Seas respectively.

Of the six Priority A species in the north and west European seas, all of their European populations are

Table 3. The priority bird species of marine habitats in Europe.

This prioritization identifies Species of European Conservation Concern (SPECs) (see p. 17) for which the habitat is of particular importance for survival, as well as other species which depend very strongly on the habitat. It focuses on the SPEC category (which takes into account global importance and overall conservation status of the European population) and the percentage of the European population (breeding population, unless otherwise stated) that uses the habitat at any stage of the annual cycle. It *does not* take into account the threats to the species within the habitat, and therefore should not be considered as an indication of priorities for action. Indications of priorities for action for each species are given in Appendix 4 (p. 401), where the severity of habitat-specific threats is taken into account.

	SPEC category	European Threat Status	Habitat importance
NORTH AND WEST EUROPEAN SEAS			
Priority A			
Manx Shearwater *Puffinus puffinus*	2	(L)	■
Steller's Eider *Polysticta stelleri*	1	L W	■ W
Storm Petrel *Hydrobates pelagicus*	2	(L)	■
Gannet *Sula bassana*	2	L	■
Black Guillemot *Cepphus grylle*	2	D	■
Puffin *Fratercula arctica*	2	V	■
Priority B			
Red-throated Diver *Gavia stellata*	3	V	■ W
Leach's Storm-petrel *Oceanodroma leucorhoa*	3	(L)	■
Harlequin Duck *Histrionicus histrionicus*	3	V	■
Velvet Scoter *Melanitta fusca*	3 W	L W	■ W
Common Gull *Larus canus*	2	D	● W
Ivory Gull *Pagophila eburnea*	3	(E)	■
Sandwich Tern *Sterna sandvicensis*	2	D	●
Priority C			
Black-throated Diver *Gavia arctica*	3	V	● W
Cory's Shearwater *Calonectris diomedea*	2	(V)	●
Shag *Phalacrocorax aristotelis*	4	S	■
Scaup *Aythya marila*	3 W	L W	● W
Great Skua *Stercorarius skua*	4	S	■
Lesser Black-backed Gull *Larus fuscus*	4	S	■
Great Black-backed Gull *Larus marinus*	4	S	■
Caspian Tern *Sterna caspia*	3	(E)	●
Roseate Tern *Sterna dougallii*	3	E	●
Little Tern *Sterna albifrons*	3	D	●
Razorbill *Alca torda*	4	S	■
Priority D			
Great Northern Diver *Gavia immer*	—	(S)	■ W
White-billed Diver *Gavia adamsii*	—	(S)	■ W
Fulmar *Fulmarus glacialis*	—	S	■
Great Shearwater *Puffinus gravis*	—	—[1]	■ W2
Sooty Shearwater *Puffinus griseus*	—	—[1]	■ W
Yelkouan Shearwater *Puffinus yelkouan*	4	S	● W
Madeiran Storm-petrel *Oceanodroma castro*	3	V	●
Eider *Somateria mollissima*	—	S	■
King Eider *Somateria spectabilis*	—	S	■
Long-tailed Duck *Clangula hyemalis*	—	S	■
Common Scoter *Melanitta nigra*	—	S	■
Red-breasted Merganser *Mergus serrator*	—	S	■ W
Grey Phalarope *Phalaropus fulicarius*	—	(S)	■ W
Pomarine Skua *Stercorarius pomarinus*	—	(S)	■ W
Arctic Skua *Stercorarius parasiticus*	—	(S)	■ W
Long-tailed Skua *Stercorarius longicaudus*	—	(S)	■ W
Sabine's Gull *Larus sabini*	—	S	■
Herring Gull *Larus argentatus*	—	S	■
Iceland Gull *Larus glaucoides*	—	(S)	■
Glaucous Gull *Larus hyperboreus*	—	S	■
Ross's Gull *Rhodostethia rosea*	—	S	■
Kittiwake *Rissa tridactyla*	—	S	■
Arctic Tern *Sterna paradisaea*	—	S	■ W
Black Tern *Chlidonias niger*	3	D	●
Guillemot *Uria aalge*	—	S	■

cont.

Table 3. (cont.)

	SPEC category	European Threat Status	Habitat importance
Brünnich's Guillemot *Uria lomvia*	—	S	■
Little Auk *Alle alle*	—	(S)	■
EUROPEAN MACARONESIAN SEAS			
Priority A			
Fea's Petrel *Pterodroma feae*	1	E	■
Zino's Petrel *Pterodroma madeira*	1	E	■
Cory's Shearwater *Calonectris diomedea*	2	(V)	■
Priority B			
Bulwer's Petrel *Bulweria bulwerii*	3	V	■
Little Shearwater *Puffinus assimilis*	3	V	■
White-faced Storm-petrel *Pelagodroma marina*	3	L	■
Madeiran Storm-petrel *Oceanodroma castro*	3	V	■
Priority C			
Manx Shearwater *Puffinus puffinus*	2	(L)	•
Storm Petrel *Hydrobates pelagicus*	2	(L)	•
Roseate Tern *Sterna dougallii*	3	E	●
Sandwich Tern *Sterna sandvicensis*	2	D	● [W]
Puffin *Fratercula artica*	2	V	● [W]
Priority D			
Great Shearwater *Puffinus gravis*	—	—[1]	■ [W][2]
Gannet *Sula bassana* [3]	2	L	● [W]
MEDITERRANEAN AND BLACK SEAS [4]			
Priority A			
Audouin's Gull *Larus audouinii*	1	L	■
Priority B			
Cory's Shearwater *Calonectris diomedea*	2	V	●
Sandwich Tern *Sterna sandvicensis*	2	D	●
Priority C			
Yelkouan Shearwater *Puffinus yelkouan*	4	S	■
Storm Petrel *Hydrobates pelagicus*	2	(L)	•
Mediterranean Gull *Larus melanocephalus*	4	S	■
Caspian Tern *Sterna caspia*	3	E	●
Little Tern *Sterna albifrons*	3	D	●
Priority D			
Red-throated Diver *Gavia stellata*	3	V	● [W]
Black-throated Diver *Gavia arctica*	3	V	● [W]
Great Black-headed Gull *Larus ichthyaetus*	—	S	■
Slender-billed Gull *Larus genei*	—	S	■
Yellow-legged Gull *Larus cachinnans*	—	S	■

SPEC category and **European Threat Status:** see Box 2 (p. 17) for definitions.

Habitat importance for each species is assessed in terms of the maximum percentage of the European population (breeding population unless otherwise stated) that uses the habitat at any stage of the annual cycle:

    ■ >75%   ● 10–75%   • <10%

[W] Assessment relates to the European non-breeding population.

[1] Only a non-breeding visitor to Europe, thus no European Threat Status.

[2] *Puffinus gravis* moves *en masse* from the north and west European seas to the European Macaronesian seas as the non-breeding season progresses, thus more than 75% of the European population occurs in both marine regions during the annual cycle.

[3] The European Macaronesian winter population is not localized, thus the species' priority status is attributed on the basis of a Favourable Conservation Status in Europe (see Box 2, p. 17).

[4] *Aythya marila* regularly occurs in the Mediterranean in winter and has an Unfavourable Conservation Status in Europe (see Box 2, p. 17). However, it is not included as a priority species for the Mediterranean and Black Seas as its Unfavourable Conservation Status is the result of its highly localized distribution in north-west Europe, its core wintering area.

highly dependent on marine habitats in the region during some or all of the year. Only one, Steller's Eider *Polysticta stelleri*, is globally threatened with extinction, due to an unexplained 75% decline in its world population over the last 20 years (Collar *et al.* 1994). The world population (and decline) is centred on the Aleutian Islands in the north Pacific, but the species also has an Unfavourable Conservation Status in Europe because of its highly restricted winter distribution (Tucker and Heath 1994), which puts the European population under significant risk from localized events such as oil pollution.

The remaining five Priority A species have their global populations concentrated in Europe and an

Table 4. Numbers of priority bird species of marine habitats in Europe, according to their SPEC category (see p. 17) and their dependence on the habitat. Priority status of each species group is indicated by superscripts (see 'Introduction', p. 19, for definition). 'Habitat importance' is assessed in terms of the maximum percentage of each species' European population that uses the habitat at any stage of the annual cycle.

Habitat importance	SPEC category					Total no. of species
	1	2	3	4	Non-SPEC	
North and west European seas						
<10%	0 [B]	1 [C]	2 [D]	—	—	3
10–75%	0 [A]	2 [B]	5 [C]	1 [D]	—	8
>75%	1 [A]	5 [A]	5 [B]	5 [C]	24 [D]	40
Total	1	8	12	6	24	51
European Macaronesian seas						
<10%	0 [B]	4 [C]	0 [D]	—	—	4
10–75%	0 [A]	1 [B]	1 [C]	0 [D]	—	2
>75%	2 [A]	1 [A]	4 [B]	0 [C]	1 [D]	8
Total	2	6	5	0	1	14
Mediterranean and Black Seas						
<10%	0 [B]	1 [C]	2 [D]	—	—	3
10–75%	0 [A]	2 [B]	2 [C]	0 [D]	—	4
>75%	1 [A]	0 [A]	0 [B]	2 [C]	3 [D]	6
Total	1	3	4	2	3	13

Unfavourable Conservation Status in Europe. Of these, only Puffin *Fratercula arctica* has shown significant declines in its European population (Lloyd *et al.* 1991, Anker-Nilssen and Røstad 1993, Tucker and Heath 1994). The remaining species have an Unfavourable Conservation Status because they are highly localized in Europe when breeding (Tucker and Heath 1994), a condition shared by one other priority species, Leach's Storm-petrel *Oceanodroma leucorhoa* (Priority B), as well as by two others when wintering in Europe, Velvet Scoter *Melanitta fusca* (Priority B) and Scaup *Aythya marila* (Priority C).

Three species of north and west European seas are considered to be endangered in Europe (Tucker and Heath 1994): Ivory Gull *Pagophila eburnea* (Priority B), Caspian Tern *Sterna caspia* and *S. dougallii* (both Priority C), though population data on the former two are incomplete.

Although the European Macaronesian seas hold only 14 priority species, six of these do not breed anywhere else in Europe. Of these, two are globally threatened Priority A species, Fea's Petrel *Pterodroma feae* and Zino's Petrel *P. madeira* (Collar *et al.* 1994), which are also endemic to Macaronesia.

Both are endangered because of their extremely small population sizes, *P. madeira* actually being endemic solely to Madeira and having a world population of only 20–30 pairs (Tucker and Heath 1994). *Calonectris diomedea* is the only other Priority A species; its world range is almost confined to Europe, where its population is highly dependent on Macaronesian seas during the breeding season, and its European population appears to have suffered a large decline during 1970–1990 at least (Tucker and Heath 1994).

The other four species restricted in Europe to breeding in this region are *Bulweria bulwerii*, *Puffinus assimilis*, White-faced Storm-petrel *Pelagodroma marina* and *Oceanodroma castro*. One other species endangered in Europe, *Sterna dougallii*, occurs in the region, mainly on the Azores, where it has a breeding population of some 1,000 pairs, more than 60% of the European total (Tucker and Heath 1994).

Three of the priority species of the European Macaronesian seas—*Puffinus puffinus*, Storm Petrel *Hydrobates pelagicus* and *Sula bassana*—have an Unfavourable Conservation Status in Europe because they are highly localized within the north and

Table 5. Numbers of priority bird species of marine habitats in Europe, listed by priority status (see 'Introduction', p. 19). Percentages are proportions of total numbers of priority species in each marine region.

	Priority status				Total no. of species
	A	B	C	D	
North and west European seas	6 (12%)	7 (14%)	11 (21%)	27 (53%)	51
European Macaronesian seas	3 (21%)	5 (36%)	5 (36%)	1 (7%)	14
Mediterranean and Black Seas	1 (8%)	2 (15%)	5 (38%)	5 (38%)	13

west European seas when breeding. The former two species also breed within the European Macaronesian seas, though in comparatively small numbers.

The Mediterranean and Black Seas hold only 13 priority species. This low number primarily reflects the low diversity of the seabird community within the region. Many of the seabird populations here are also relatively small in size, probably because of the naturally low productivity of most of this marine region (see 'Key abiotic factors affecting the habitat', p. 60), with much larger populations elsewhere in Europe, e.g. *Hydrobates pelagicus* and Shag *Phalacrocorax aristotelis*. However, there are four species that, within Europe, are restricted to the region as breeders. These include Audouin's Gull *Larus audouinii*, a Mediterranean breeding endemic. It is also a Priority A species of global conservation concern (Tucker and Heath 1994). Although its population has increased in recent years and is now over 15,000 pairs, its global survival is considered dependent on the continuation of direct conservation measures (Collar *et al.* 1994). Mediterranean Shearwater *Puffinus yelkouan* is also confined as a breeder to the Mediterranean. Although most of their world populations breed outside Europe, within the continent Great Black-backed Gull *Larus ichthyaetus* and Slender-billed Gull *L. genei* breed only in the Mediterranean and Black Seas. Also, the global population of Mediterranean Gull *L. melanocephalus* is mostly confined to the Mediterranean.

The overall importance of individual European countries for breeding populations of priority species is indicated in Appendix 2 (p. 344). This clearly shows the importance of Russia for both the north European seas and the Black Sea. Also of considerable importance in the north and west European seas are Iceland, Norway and the UK, all of which hold breeding populations of more than half of the Priority A species and of all priority species in total. The Faroe Islands and France also hold a substantial proportion of Priority A species, but less than half of all priority species. Greenland, on the other hand, holds more than half of all the priority species, but only one third of the Priority A species.

In the European Macaronesian seas, all the island groups hold substantial and similar proportions of the priority species of the region. Madeira (Portugal), however, is noteworthy in having breeding populations of all the Priority A and B species.

In addition to Russia, other particularly important countries for breeding priority species of the Mediterranean and Black Seas are Italy, Spain, France, Greece and, to a lesser extent, Turkey and Ukraine.

Although it is not within the scope of this strategy to consider the conservation of habitats for particular subspecies, it is important to note that some subspecies of seabird have very small populations and may be threatened with extinction, e.g. the Mediterranean race *mauretanicus* of *Puffinus yelkouan*, restricted to the Balearic Islands of Spain (de Juana 1984, del Hoyo *et al.* 1992), and *Phalacrocorax aristotelis desmarestii*, endemic to the Mediterranean and Black Seas and declining in some areas (Guyot 1993; European Bird Database). Such subspecies are regionally important and merit habitat-conservation measures. Necessary measures are, however, likely to be similar to those recommended here for European priority species.

HABITAT NEEDS OF PRIORITY BIRDS

More detailed information on this subject can be found in general seabird reviews (e.g. Nettleship and Birkhead 1985, Croxall 1987, Furness 1987b, Furness and Monaghan 1987, Dunnet *et al.* 1990).

The particular requirements of priority seabirds are summarized for each region in Appendix 3 (p. 354). Comparison between and among species indicates some clear patterns in requirements.

Seabirds are often associated with a need for fish as food, and in comparatively shallow waters of subarctic areas over the continental shelf, surface-shoaling small fish are indeed very important to seabirds, especially anchovies, young herring, sardines and sprats (Clupeidae), sandeels (Ammodytidae) and capelin (Osmeridae). In the warmer and deeper waters of Europe, fish are less accessible by day and seabirds may rely on large predators, such as cetaceans, tuna or swordfish, to concentrate their prey and bring them to the surface.

However, by no means all seabirds are dependent on fish and, as Appendix 3 shows, a wide range of crustaceans (such as amphipods), cephalopods (mainly squid), other molluscs, other invertebrates, and offal and debris from man or other predators are taken. In very shallow waters, echinoderms and shellfish are important food sources, especially for seaduck. In colder polar waters in particular, fish are much less important, and priority seabirds such as Ross's Gull *Rhodostethia rosea* and Little Auk *Alle alle* may rely heavily on the huge abundance of crustacea. Carrion and offal are also important in the diet of some species, e.g. Glaucous Gull *Larus hyperboreus* and *Pagophila eburnea*.

In some cases man has apparently helped seabirds by creating additional food supplies, e.g. by discarding fish from fishing catches that are then taken by scavenging seabirds, or by catching so many large predatory fish that small fish (like sandeels) have been able to proliferate. A key feature affecting the numbers and distribution of many seabirds is the dispersion of their food supply: birds benefit when food is relatively concentrated spatially and not overdispersed, especially in the breeding season when they are constrained by attachment to their colonies. Kleptoparasitism (the deliberate stealing by one animal of food already captured by another) is common, some groups being specialists at it, e.g. skuas (Stercorariidae).

Most seabirds migrate or disperse and change diet to accommodate shifting availability of food and to optimize their foraging. These migrations may be enormous, such as those carried out by Arctic Tern *Sterna paradisaea* to the Antarctic continent and back each year. Long-distance migrations are typical of many species of shearwater, petrel, tern and skua. Many gull and seaduck species, on the other hand, tend to make shorter movements and overwinter within the European region.

Many species prefer mesotrophic or eutrophic waters, and indeed some may have benefited from the rise in productivity caused by man-made eutrophication. Others, such as species that hunt by plunge-diving, may have suffered as a result of reduced transparency of the water (Furness 1993). In several areas of the North Sea, Baltic, Mediterranean and Black Seas, eutrophication has already reached or passed its maximum sustainable level (Stanners and Bourdeau 1995), where further nutrient increases could only be detrimental. Due to the generally low productivity and scarcity of food resources on the surface of deep waters (see 'Key abiotic factors affecting the habitat', p. 60), upwellings and fronts are important feeding areas for the most pelagic species such as Sabine's Gull *Larus sabini*, *Larus audouinii*, petrels and shearwaters. River outflows and shallow coastal waters are the preferred feeding areas for most other gulls and terns.

In comparison with other types of birds, seabirds have long breeding lives, delay breeding until at least the second year of life (and usually much longer) and have low rates of reproduction (Furness and Monaghan 1987). Additional mortality in adulthood is therefore particularly likely to damage populations. Seas thus need to be free from hazards such as fishing nets or plastic rubbish that entangle seabirds, lines that hook them and pollution that poisons or kills them. Constant disturbance by human activities (e.g. boat traffic) may also make some areas energetically unprofitable for foraging (see Keller 1991).

Seabirds tend to be colonial breeders with a strong tendency to return to their colony of birth to breed (natal philopatry). Also, because seabirds are adapted to maximize adult survival rates, individual birds will readily abandon breeding if the perceived risks of death or failure are too great (Furness and Monaghan 1987). The quality of nest-sites is therefore particularly important to seabirds. Safe nest-sites are essentially those free from heavy predation, not subject to severe adverse weather (for instance flooding or overheating) and in situations that allow easy access to food. Typically, some predation is always a risk, for instance from larger seabirds or birds of prey, but mammalian predators often cause excessive pressure and lead to the abandonment of nest-sites (Burger and Gochfeld 1994).

The breeding habitats of many, particularly pelagic, species are very often offshore rocky islands, with wide boulder-slopes and cavities, or cliffs. In contrast, many coastal gulls and most terns typically breed on low-lying beaches or on islets in coastal lagoons and estuaries. In such habitats, bare ground or low vegetation are preferred and thus grazing or even occasional fires may not be detrimental if carried out sustainably. Indeed, the proper management of the vegetation, at least in some areas, may be an important factor in breeding success. A general requirement of coastal breeding species is an absence of human disturbance, but the latter is often particularly heavy during the breeding period, particularly on beaches in the Mediterranean (Canova and Fasola 1993, Fasola and Canova 1996) and parts of western Europe.

THREATS TO THE HABITAT

Reviews of threats to seabird habitats in Europe have been carried out by Dunnet *et al.* (1986), Lambertini and Leonzio (1986), Mayol Serra (1986), Tasker and Becker (1992), Aguilar *et al.* (1993) and Furness (1993). A summary of the scale and general effects of each threat is presented in Table 6. For each sea region, an assessment of the impacts of each threat on each priority species' population was also carried out (for methods, see 'Introduction', p. 21), the results of which are given in Appendix 4 (p. 401) and summarized in Figures 2 and 3. These threats are discussed below, in approximate order of importance (as shown in Figure 3).

Introduced predators

The presence of predators, especially mammals, may have a profound impact on seabird populations and distributions by precluding species from using otherwise suitable breeding sites (Appendix 3, p. 354). Thus many seabirds are restricted to mammal-free islands or steep, rocky cliffs that are inaccessible to such predators. Many of these seabird species, such as shearwaters, have also evolved nocturnal behaviour patterns to reduce predation by gulls, etc., when coming ashore.

Where seabirds and their predators do coexist under natural circumstances then seabird populations

Table 6. The most important threats to marine habitats, and to the habitat features required by their priority bird species, in Europe (compiled by the Habitat Working Groups). Threats are listed in approximate order of importance (see Figure 3), and are described in more detail in the text.

Threat	Effect on birds or habitats	Scale of threat			Regions affected, and comments
		Region	Past	Future	
1. Introduced predators (especially mammals)	Killing of eggs, young and adults, and disturbance	NW	●	●	Almost everywhere though less so in Arctic Ocean
		Mac.	●	●	
		Med.	●	●	
2. Disturbance of nesting birds	Predators taking unguarded eggs and chicks, lost incubation time and accidental falling of chicks (N.B. direct adult mortality from hunting is not considered in this book)				
a. Tourism and recreation		NW	•	•	
		Mac.	■	■	Azores and Canary Islands
		Med.	•	●	
b. Hunting; collection of eggs, feathers		NW	•	•	White, Kara and Baltic Seas, Iceland, Greenland, Faroes
		Mac.	■	●	
		Med.	•	•	
c. Military activities		NW	•	•	Very local
		Mac.	•	•	Azores only
		Med.	•	•	
3. Coastal development (inc. tourist, urban, industrial, agriculture)	Loss of nest-sites; loss of sheltered marine areas through disturbance or land-claim	NW	•	•	
		Mac.	■	■	
		Med.	●	●	Widespread future threat in north Mediterranean
4. Human-induced increases in populations of native predators	Loss of eggs, young and adults, and disturbance/competition	NW	•	•	
		Mac.	•	●	Madeira and Canary Islands
		Med.	●	•	
5. Rubbish at sea	Death through entanglement; ingestion and death or reduced feeding opportunities	NW	●	●	Everywhere
		Mac.	•	•	
		Med.	•	•	
6. Oil pollution	Death and sublethal effects; destruction of other habitats; disruption of food-chains, e.g. loss of kelp cover and associated food resources	NW	●	●	Eastern Baltic Sea and shelf seas, particularly in shipping lanes and near coasts
		Mac.	•	●	
		Med.	•	●	West Adriatic Sea
7. Overfishing of small fish:	Reduced food resources leading to decreased survival and productivity				
a. Sandeel, herring, capelin, anchovy and sardine		NW	●	■	Probably all areas
		Med.	●	■	
b. Juvenile fish for bait		Mac.	•	●	Azores and Madeira
8. Vegetation changes at nest-sites					
a. Grazing through introductions of sheep, goats and rabbits	Short-grazed vegetation provides insufficient cover for chicks; also loss of vegetation and soil	NW	•	•	Iceland and Norway; widespread problem on islands
		Mac.	●	■	Madeira archipelago
		Med.	●	?	
b. Abandonment of grazing	Growth of vegetation and loss of nesting habitat for some species	NW	?	?	
		Med.	•	●	Black Sea
c. Introduction of non-native plants	Loss of habitat through growth of dense roots and tall vegetation	Mac.	■	■	Widespread, especially Azores
9. Reduction in quantities of discards or offal from fishing vessels	Reduced food resources for discard or offal specialists; highest impact on small species	NW	—	●	Fishery areas
		Med.	—	●	Mainly discards (little offal)
10. Entanglement in fishing nets and on lines (especially gill-nets, drift-nets and longlines)	Direct mortality of diving species (and small cetaceans, sharks and marine turtles, among other wildlife)	NW	•	•	
		Mac.	•	•	Longline fishing only
		Med.	•	•	Drift-nets a widespread problem in north Tyrrhenian Sea
11. Toxic pollution (organochlorines and heavy metals)	Lethal (rare so far); reduced breeding performance; genetic impacts	NW	•	•	Everywhere, but varies with substance. Effects at population level little known
		Mac.	?	?	
		Med.	●	●	

cont.

Table 6. (cont.)

Threat	Effect on birds or habitats	Scale of threat			Regions affected, and
		Region	Past	Future	comments
12. Reduced fishing of predatory fish:	Increased populations of predatory fish compete with seabirds for small fish, leading to reduced food resources for birds				
a. Bottom-feeding predatory fish		NW	•	•?	North, Baltic and White Seas
b. Pelagic predatory fish (e.g. mackerel and tuna)		NW	•	•?	South and mid-latitudes
13. Dredging and extraction	Loss of water clarity and reduced feeding efficiency; disturbance; destruction of substrate and reduced food resources	NW	•	•	Baltic and White Seas
		Med.	•	•	Adriatic and central Tyrrhenian Seas
14. Bottom-fishing (e.g. beam-trawling and scallop-dredging)	Destruction of natural substrate causing loss of associated food resources	NW	•	•	Widespread in North Sea, also occurs in shallow waters off Norway, Irish Sea, English Channel and Atlantic coast. Potential major problem in Baltic Sea
		Med.	•	•	Adriatic Sea
15. Nutrient pollution (artificial eutrophication)	Disruption of food-chain; substrate change; growth of toxic algae, potentially leading to oxygen depletion; loss of water clarity and reduced feeding efficiency	NW	•	•	Shallow, enclosed seas with low water turnover: Baltic Sea and coastal areas of Norway, and North and White Seas
		Med.	■	?	Black Sea
16. Increasing boat traffic (transport, tourism, sports and hunting)	Disturbance of feeding birds leading to reduced feeding efficiency and disturbance when roosting	NW	•	•	Widespread. Hunting at sea in Baltic Sea, Kattegat, off Greenland, Iceland and Faroe Islands
		Mac.	•	•?	
		Med.	•	•	
17. Coastal erosion and subsidence	Loss of islands and other nesting habitats	Med.	•	•	Adriatic Sea
18. Overfishing of shellfish and echinoderms	Depletion of food resources	NW	•	•	Wadden Sea (offshore)
		Med.	•	•	

Scale of threat

NW	North and west European seas	Past	Past 20 years	■	Widespread (affects >10% of habitat)
Mac.	European Macaronesian seas	Future	Next 20 years	•	Regional (affects 1–10% of habitat)
Med.	Mediterranean and Black Seas			•	Local (affects <1% of habitat)
				?	Considerable uncertainty

are not normally harmed. However, where native or non-native predators have been introduced or have increased substantially, often due to other human activities, then local extinctions have often resulted. Seabirds may have adaptations against aerial predators but may have no defence strategy against ground predators.

Probably the single most serious seabird conservation problem in Europe concerns the predation by rats *Rattus* and cats on *Pterodroma madeira* on Madeira, which threatens to drive this species extinct in the next 20 years (Zino *et al.* 1996b). Over most of the rest of Europe, rats and American mink *Mustela vison* may cause the most serious problems (Folkestad 1982, Jönsson 1990, Burness and Morriss 1993). The latter species, a semi-aquatic small carnivore, has caused changes in distribution of Black Guillemot *Cepphus grylle* in Iceland, and in Finland

it took 70–80% of the chicks at one colony of *Sterna paradisaea* (Moors and Atkinson 1984). It has also affected seabirds in Ireland (Smal 1991), and in Scotland it is believed to have reduced breeding success of several seabird species during the 1990s (Craik 1995). For many Mediterranean seabirds, e.g. *Calonectris diomedea*, *Puffinus yelkouan* and *Hydrobates pelagicus*, rat predation may limit populations. In the Black Sea region, the non-native raccoon-dog *Nyctereutes procyonoides* has increased considerably and has become an important predator of ground-nesting seabirds and their eggs.

The introduction of predators is considered to be a problem of regional scale in all three sea regions. In the north and west European seas, 79% of breeding priority species are affected (Table 7) and 18 of these species are threatened. Although four priority A or B species may suffer relatively large population de-

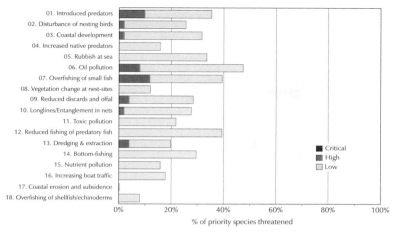

North and west European seas

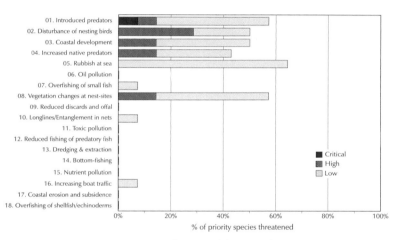

European Macaronesian seas

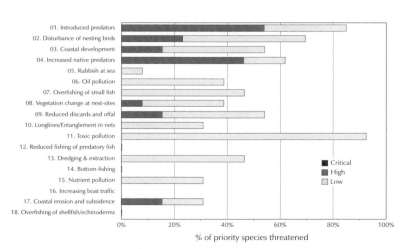

Mediterranean and Black seas

Figure 2. The predicted impact on priority bird species of threats to marine habitats, and to the species' habitat requirements, in Europe. Bars are subdivided to show the proportion of species threatened by critical, high or low impacts to their populations (see Table 6 and Appendix 4, p. 401, for more details).

Critical impact
The species is likely to go extinct in the habitat in Europe within 20 years if current trends continue

High impact
The species' population is likely to decline by more than 20% in the habitat in Europe within 20 years if current trends continue

Low impact
Only likely to have local effects, and the species' population is not likely to decline by more than 20% in the habitat in Europe within 20 years if current trends continue

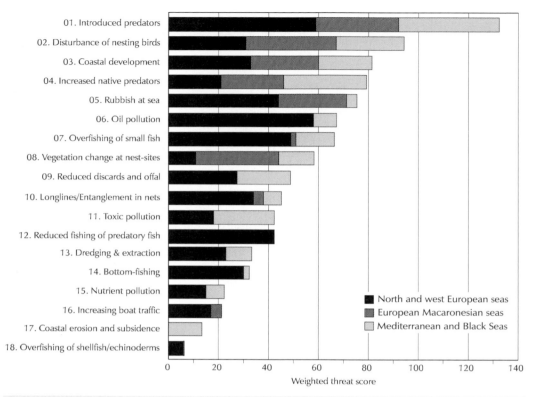

Figure 3. The most important threats to marine habitats, and to the habitat requirements of their priority bird species, in Europe. The scoring system takes into account:

- The European Threat Status of each species and the importance of the habitat for the survival of its European population (Priorities A–D); see Table 3.
- The predicted impact (Critical, High, Low), within the habitat, of each threat on the European population of each species; see Figure 2, and Appendix 4 (p. 401).

See Table 6 for more details on each threat.

clines, most other population declines are predicted to be low, primarily because many non-native predators, such as rats and *Mustela vison*, have already been introduced to many previously mammal-free seabird islands and are widespread on the mainland. Eight priority species in the European Macaronesian seas are potentially threatened, while in the Mediterranean and Black Seas all breeding priority species (11) are likely to be threatened, most of which may suffer substantial declines.

Clearly, the introduction of predators to seabird nesting areas is a major threat, particularly in the southern seas of Europe. Overall, this hazard is probably the most important threat to seabird populations in Europe.

Disturbance of nesting birds

Disturbance is undoubtedly related mainly to coastal development, especially intensive tourism and rec-

reation around the Mediterranean and Black Seas and (particularly) on the Macaronesian islands. The most popular such activities include swimming, walking, water sports and boat trips. Substantial disturbance to some seabird colonies inevitably results. In the Azores, for example, disturbance caused the complete desertion of a colony of *Sterna dougallii* with the loss of 200 clutches (Monteiro *et al.* 1996).

Some disturbance to nesting birds comes from the collection of seabird eggs for food or the hunting of adult seabirds or other species, as in the Madeiran archipelago, where disturbance from egg-collectors remains a serious threat for the globally threatened *Pterodroma feae* (Zino *et al.* 1996a). Military activities may also cause substantial disturbance. For example, low-flying aircraft cause mass panic and exodus of adults from seabird colonies, temporarily exposing eggs and chicks to predators. In the case of the globally threatened *Pterodroma feae*, competi-

Table 7. The current impact of predators on priority bird species of north and west European seas (at the birds' breeding sites). Eight priority birds of north and west European seas are all predominantly wintering or migrant species (i.e. non-breeding) in marine and coastal habitats in this region, and therefore are excluded from this table: *Puffinus gravis, Puffinus griseus, Puffinus yelkouan, Aythya marila, Polysticta stelleri, Histrionicus histrionicus, Phalaropus fulicarius* and *Chlidonias niger.*

	Gulls, skuas	Birds of prey	Red fox, Arctic fox	Rats	Stoat, mink	Crows	Hedgehog	Cats	Sheep, goats, deer	Polar bear	Otter
Gavia stellata	●	•	•	–	●	●	–	–	–	–	–
Gavia arctica	●	•	•	–	•	●	–	–	–	–	–
Gavia immer	●	–	●	–	–	–	–	–	–	–	–
Gavia adamsii	●	–	●	–	–	–	–	–	–	–	–
Fulmarus glacialis	–	–	•	–	–	–	–	–	–	–	–
Calonectris diomedea	–	–	–	–	–	–	–	●	–	–	–
Puffinus puffinus	–	●	–	●	■	–	–	●	●	–	–
Hydrobates pelagicus	■	–	–	■	■	–	–	–	–	–	–
Oceanodroma leucorhoa	■	–	–	■	■	–	–	•	–	–	–
Oceanodroma castro	■	–	–	–	–	–	–	–	–	–	–
Sula bassana	●	–	–	–	■	–	–	–	–	–	–
Phalacrocorax aristotelis	–	●	–	–	•	–	–	–	–	–	–
Somateria mollissima	■	•	■	•	■	●	●	–	–	–	●
Somateria spectabilis	–	–	–	–	–	–	–	–	–	–	–
Clangula hyemalis	■	●	●	–	–	■	–	–	–	–	–
Melanitta nigra	■	●	●	–	–	●	–	–	–	–	–
Melanitta fusca	■	●	●	–	–	●	–	–	–	–	–
Mergus serrator	■	–	–	–	–	●	–	–	–	–	–
Stercorarius pomarinus	–	–	–	–	–	–	–	–	–	–	–
Stercorarius parasiticus	–	–	–	–	–	–	–	–	–	–	–
Stercorarius longicaudus	–	–	–	–	–	–	–	–	–	–	–
Stercorarius skua	–	–	–	–	–	–	–	–	–	–	–
Larus sabini	●	–	■	–	–	–	–	–	–	–	–
Larus canus	●	–	■	●	–	•	–	●	–	–	–
Larus fuscus	●	–	●	–	–	•	–	●	–	–	–
Larus argentatus	●	–	●	–	–	•	–	●	–	–	•
Larus glaucoides	–	–	–	–	–	–	–	–	–	–	–
Larus hyperboreus	●	–	–	–	–	–	–	–	–	•	–
Larus marinus	–	–	–	–	–	–	–	–	–	–	–
Rhodostethia rosea	–	–	–	–	–	–	–	–	–	–	–
Rissa tridactyla	●	•	–	–	–	•	–	–	–	–	–
Pagophila eburnea	●	–	–	–	–	–	–	–	–	–	–
Sterna caspia	–	–	–	–	–	●	–	–	–	–	–
Sterna sandvicensis	■	–	–	–	–	–	–	●	–	–	–
Sterna dougallii	■	■	■	–	■	●	■	■	–	–	–
Sterna paradisaea	■	–	–	–	–	–	–	●	●	–	•
Sterna albifrons	■	–	–	–	●	–	–	●	–	–	–
Uria aalge	●	•	•	–	–	•	–	–	–	–	–
Uria lomvia	●	•	•	–	–	–	–	–	–	–	–
Alca torda	●	•	•	–	–	–	–	–	–	–	–
Cepphus grylle	●	•	•	–	–	–	–	–	–	–	•
Alle alle	■	•	●	–	–	–	–	–	–	•	–
Fratercula arctica	■	–	■	■	–	–	–	–	–	–	–

■ Predation has important effects on distribution or numbers
● Predation can affect distribution or numbers
• Predation recorded, but no important effects

tion for, and modification of, nest-burrows by abundant introduced rabbits *Oryctolagus cuniculus* may be a serious problem (Zino *et al.* 1996a).

Disturbance of nesting birds is considered to affect nine priority species of the Mediterranean and Black Seas region and seven priority species of the European Macaronesian seas. Impacts on most populations are likely to be low, but more serious declines are predicted in the Macaronesian islands for *Pterodroma feae* and *P. madeira* (Zino *et al.* 1996a,b), *Calonectris diomedea* and *Sterna dougallii*, and in the Mediterranean and Black Seas for *Sterna caspia, S. sandvicensis* and Little Tern *S. albifrons*. *Sterna albifrons* is also considered to be highly threatened by human disturbance during nesting in the north and west European seas region, where 12 other priority species are also thought to be threatened with small or local population declines.

Although habituation to disturbance is possible in some cases, it is clear that disturbance of nesting seabirds, particularly terns, mainly as a result of tourism and recreation, is among the most important threats to the priority marine species in Europe.

Coastal development

Coastal developments such as hotels, marinas and other tourist facilities, housing, or industrial and agricultural land-claim can threaten seabirds by destruction or degradation of nesting sites (as well as automatically increasing levels of human disturbance—see 'Disturbance of nesting birds' [above]). This is one of the important threats to *Larus audouinii* (Lambertini 1996). Some species may also be indirectly affected. For example, in the Canary Islands, large numbers of *Calonectris diomedea* are attracted at night to the lights of tourist developments and die as a result of collisions with buildings.

This threat has a widespread impact in those areas of the Mediterranean, Black and European Macaronesian seas that have existing and developing intensive tourist industries. Seven species in each of the two southerly sea regions are likely to be affected by future developments, some of which may suffer significant declines, while 16 priority species in north and west European seas are likely to be affected, mostly seaducks and terns. This threat to seabirds is among the most important in all three sea regions.

Increased native predators

Population increases in native predators, usually driven by man's activities (e.g. improper waste disposal), are likely to impact eight priority species of north and west European seas, and threaten six priority species in the European Macaronesian seas and eight in the Mediterranean and Black Seas. High impacts are predicted for several species in the latter two regions, primarily as a result of increasing populations of *Larus cachinnans*, e.g. in the Madeiran archipelago, where many of the globally threatened *Pterodroma feae* may be killed as a result (Zino and Zino 1986, Zino *et al.* 1996a). Overall, the threat is considered to be of particularly high importance in the southern two sea regions.

Rubbish at sea

Small particles of plastic used as feedstock for the plastics manufacturing industry are very common at sea, and do not degrade rapidly. These are ingested by birds (van Franeker 1985, Furness 1985, Azzarello and van Vleet 1987, Ryan 1987, 1990), but the effect of these on seabird condition and survival is unclear.

In addition, seabirds may become entangled by rubbish, which often leads to death. Entanglement in lost or discarded fishing nets ('ghost nets') is considered by some to be a substantial cause of mortality (DeGange and Newby 1980, Laist 1987; see 'Entanglement in fishing nets and on longlines', p. 81).

Rubbish at sea is thought to be of regional significance in the north and west European seas and local in the Mediterranean, Black and European Macaronesian seas. At least 25 priority species are thought to be affected by this hazard, although impacts are in all cases likely to be low. Nevertheless, its overall weighted threat score (see Figure 3 and Appendix 4, p. 401) is moderately high.

Oil pollution

Birds are often the most conspicuous victims of marine pollution, beached bird surveys frequently being the first indication of incidents (Camphuysen 1989, 1991). Several reviews of the effects of specific types of pollution on seabirds have therefore been carried out as well as comprehensive reviews by Bourne (1976) and Nisbet (1994).

Pollution can be seen as affecting seabird populations in two ways: either by causing direct mortality or through longer term, indirect effects such as reduced breeding success or adult survival. Typical of the former type is oil pollution, while the effects of some chlorinated hydrocarbons work through more indirect routes. As many seabirds have high adult survival but low reproductive rates, additional mortality in adulthood is particularly likely to damage populations.

Oil pollution represents one of the most serious threats to adult seabird survival and has consequently been the focus of many studies. Hooper *et al.* (1987) published an extensive annotated bibliography and Camphuysen and van Franeker (1992) include further references covering the north-east Atlantic. Other reviews of this specific topic have been made by Croxall (1975, 1977), Holmes and Cronshaw (1977), Clark (1982, 1984), Dunnet (1982) and Piatt *et al.* (1991).

There is no doubt that major oil spills can kill huge numbers of seabirds—careful estimates of the worst incidents on record suggest that kills of up to 500,000 birds have occurred (Mormat and Guermeur 1979, Piatt *et al.* 1990, Wiens 1995b). However, the number of birds killed varies considerably, and large numbers of casualties can result from small spills (e.g. Barrett 1979, Vaitkus *et al.* 1995).

Although accidental spills from tankers receive most media attention, most oil enters the sea from land-based sources and deliberate discharges from ships, such as when cleaning tanks (e.g. North Sea Task Force 1993). Most seabird mortality occurs as a result of oil from such chronic oil pollution rather than accidents (Clark 1984). Accidental spills or illegal operational discharges from shipping in the North Sea vary between 15,000 and 60,000 tonnes per annum—as much as 30% of the total oil input. Legal operational discharges of oil from shipping account for 1,000–2,000 tonnes per year, i.e. approximately 0.5–2% of the total oil input to the North Sea.

The size and species composition of kills depends primarily on the type and number of birds in the vicinity, but also on the type and volume of oil spilt, sea state and weather conditions (Tasker *et al.* 1987, Ritchie and O'Sullivan 1994). Seabirds are particularly vulnerable to oil because they congregate on the sea surface in large numbers to feed, moult or rest. Williams *et al.* (1995) calculated an oil vulnerability index for birds in the North Sea based upon an internationally-agreed set of criteria. A similar indication of vulnerability to oil for priority species of the north and west European seas, never previously compiled, was made by the Habitat Working Group (Table 8). Seaduck are particularly vulnerable as huge numbers can concentrate in small areas during cold winter periods (Švazas and Pareigis 1992, Durinck *et al.* 1994, Skov *et al.* 1995). Furthermore, many such concentrations are in close proximity to existing or planned oil installations, particularly in the eastern Baltic (e.g. Švazas 1992, Švazas and Vaitkus 1995, Vaitkus *et al.* 1995).

Although it has been widely documented that oil pollution from major incidents and chronic inputs kill large numbers of birds, the long-term population effects are less well understood (Dunnet 1987, Furness 1993, Nisbet 1995, Wiens 1995b). Much of the long-term impact of an oil spill may depend on the size of the non-breeding population, which will act to buffer any excess adult mortality.

Oil pollution is likely to be a regional problem in all three marine regions. Around 24 priority species of north and west European seas and five priority species of Mediterranean and Black Seas are thought to be affected by oil pollution. However, the impacts on populations are likely to be only low, except for some seaduck populations in the Baltic which are thought to be highly threatened by oil-related developments (Švazas 1992, 1995). No priority species of the European Macaronesian seas are thought likely to be affected by oil pollution because it is less frequent in the region, and because the species are less vulnerable than those of the north and west European seas. Overall, the weighted threat score (Figure 3) is high among threats considered in this chapter.

Table 8. Oil vulnerability assessment for priority birds of north and west European seas (data from the Habitat Working Group).

	Vulnerability
Gavia stellata	■
Gavia arctica	■
Gavia immer	■
Gavia adamsii	●
Fulmarus glacialis	●
Calonectris diomedea	●
Puffinus gravis	●
Puffinus griseus	●
Puffinus puffinus	■
Puffinus yelkouan	●
Hydrobates pelagicus	●
Oceanodroma leucorhoa	●
Oceanodroma castro	●
Sula bassana	●
Phalacrocorax aristotelis	■
Aythya marila	■
Somateria mollissima	■
Somateria spectabilis	■
Polysticta stelleri	■
Histrionicus histrionicus	■
Clangula hyemalis	■
Melanitta nigra	■
Melanitta fusca	■
Mergus serrator	●
Phalaropus fulicarius	■
Stercorarius pomarinus	●
Stercorarius parasiticus	●
Stercorarius longicaudus	●
Stercorarius skua	●
Larus sabini	●
Larus canus	●
Larus fuscus	●
Larus argentatus	●
Larus glaucoides	●
Larus hyperboreus	●
Larus marinus	●
Rhodostethia rosea	●
Rissa tridactyla	●
Pagophila eburnea	●
Sterna caspia	●
Sterna sandvicensis	●
Sterna dougallii	●
Sterna paradisaea	●
Sterna albifrons	●
Chlidonias niger	●
Uria aalge	■
Uria lomvia	■
Alca torda	■
Cepphus grylle	■
Alle alle	●
Fratercula arctica	■

■ High   ● Medium   • Low

Overfishing of small fish

The potential over-exploitation by man of smaller shoaling fish, especially by industrial fisheries producing high-protein meal, is of particular concern because of its capacity for direct competition for the prey taken by seabirds. Such exploitation is particularly intensive in the North Sea, where industrial fisheries have recently developed, partly in response to the prolonged paucity of stocks of larger, predatory fish due to previous overfishing. Indeed, the relative lack of such predatory pelagic fish may have led directly to a proliferation of smaller fish in the past (see 'Entanglement in fishing nets and on longlines', p. 81), thus providing further incentive for the development of industrial fisheries.

In the North Sea these fisheries, concentrating on sandeel, Norway pout and sprat, now collectively account for up to half the weight of fish landed annually (Dunn 1994b, 1995), and represent half the European industrial fish catch (Anon. 1995c). Stock management is not well regulated and probably ineffective (Bartram 1995). As a result, this has led to considerable declines in total landings (Anon. 1995c) and probably in fish stocks.

Anchovy stocks have also collapsed in the Black Sea (Pearce 1995) and declined in the Mediterranean (FAO 1993, Stanners and Bourdeau 1995). In the Black and Mediterranean Seas as a whole, a 70% decline in the landings of sprat occurred between 1988 and 1993, while over the same period sardine catches decreased by 12% (FAO 1993). In the European Macaronesian seas, overfishing of juvenile fish for bait is a problem.

Over-exploitation of small fish is potentially highly disruptive to the food-chain and to large marine ecosystems. Small fish are prey for commercially important predatory fish, such as cod, and are also the preferred prey of most seabirds (Appendix 3). Consequently where such fish stocks have collapsed in recent years, seabirds have suffered widespread breeding failures and some populations have declined. For example, the population of over 1 million pairs of *Fratercula arctica* on the Røst archipelago in Norway collapsed by two-thirds between 1979 and 1988 (Anker-Nilssen 1991) as a result of a series of disastrous breeding seasons which coincided with a severe decline in the stock of Norwegian spring-spawning herring (Harris 1984b, Anker-Nilssen 1987, 1992). Also, for six successive years between 1984 and 1989, *Sterna paradisaea* in the Shetland Islands (United Kingdom) failed to rear any young because of chick starvation, following a collapse in sandeel stocks as a result of sandeel recruitment failure (Bailey *et al.* 1991, Monaghan 1992, Monaghan *et al.* 1992).

The extent to which such collapses in stocks of short-lived fish can be attributed to fishing effort rather than natural factors (such as larval mismatch with plankton blooms, or effects of currents or sea temperatures) is the subject of much debate (Bailey *et al.* 1991, Furness 1993, 1995, Wright and Bailey 1993). There has certainly been inadequate management (e.g. of the Shetland sandeel fishery) and there

is still a considerable lack of knowledge of the ecology of these species. In view of such uncertainty, these fisheries must be considered a serious ecological hazard.

Over-exploitation of small fish is conservatively predicted to pose a threat to 20 priority species of north and west European seas and six priority species of the Mediterranean and Black Seas. No species are currently likely to be affected by industrial fishing in European Macaronesian seas, though one, *Sterna dougallii*, may be locally affected by the catching of juvenile fish for bait. Impacts on species in the Mediterranean and Black Seas are predicted to be low. Although by no means certain, high population declines may occur in six species in the north and west European seas. Despite the threat being relatively not so important in the other two regions, the weighted threat score of 66 across all regions is high. Clearly, such fisheries are potentially a significant threat to seabird populations in Europe.

Vegetation change at nest-sites

Changes in the structure and composition of vegetation can spoil seabird breeding sites and render them unsuitable for nesting. Such problems do not normally affect cliff-nesting seabirds, but species that breed on low shingle or sandy islands, among sand-dunes, or in burrows on grassy banks or plateaus can be threatened (see Appendix 3). Such habitats are often grazed by livestock or introduced, non-native species such as rabbits. Intensive grazing by such species may lead to loss of vegetation, which may have formerly provided important cover for chicks, or in severe cases to loss of the topsoil, thus destroying nesting habitat. Such overgrazing is a significant problem on the Macaronesian islands, where erosion due to goats, sheep and rabbits threatens the nesting habitat of the globally threatened *Pterodroma madeira* and *P. feae* (Zino *et al.* 1996a,b).

In contrast, the abandonment of livestock grazing can lead to the growth of vegetation that is too tall and dense for some seabirds to nest in, e.g. some terns. This is believed to be the main cause of substantial declines in *Sterna sandvicensis* in the Ukraine (V. Serebryakov and V. Siokin pers. comm. 1994).

On the Macaronesian islands, especially the Azores, the introduction of some non-native plant species, such as *Arundo donax*, *Nicotiana glauca* and *Hedychium gardeneri* has caused similar problems. These plants grow very vigorously and produce dense root growth which prevents burrowing.

Such threats are thought to have only low impacts in the north and west European seas, moderate impacts in the Mediterranean and Black Seas (five species affected), and very important impacts in

European Macaronesian seas where eight species are affected (two seriously).

Reduction of fishery discards and offal

Dumping of discards of undersized fish and of offal from commercial fishing vessels has been increasing since the last century and huge numbers of seabirds feed on this waste. As a result, rapid increases in the populations of several seabirds in the north and west European seas have been attributed to this extra food (Fisher 1952, Harris 1970, Croxall *et al.* 1984, Furness 1992, Tasker and Becker 1992).

In the North Sea alone, between 27% and 56% of all fish captures are discarded (ICES 1994), and such dumping overall comprises around one million tonnes per year of offal, fish and benthic invertebrates, supporting in turn millions of seabirds (Camphuysen *et al.* 1995). In the Mediterranean similar effects of fishing discards on seabird populations have been observed in the northern Adriatic (Fasola *et al.* 1989, Fasola and Canova 1996) and also off Spain, where the recent increase in the population of *Larus audouinii* was apparently driven by the increased availability of discards from local fisheries around this species' main colonies in the Ebro delta (Lambertini 1996), as well as due to enhanced colony protection (Beaubrun 1983, Paterson *et al.* 1992, Oró and Martínez 1994). These colonies rely largely on this food resource, and suffered very low breeding success in 1991 when a moratorium was imposed on adjacent fisheries, so it appears that there was little alternative food to be found (Paterson *et al.* 1992, Lambertini 1996).

It is possible that quantities of offal and discards may be reduced in European seas if there is a future reduction in fishing effort and increased mesh size (Furness 1992, ICES 1994, Hubold 1994 in Camphuysen *et al.* 1995). Such a reduction would adversely affect the populations of some seabirds if no alternative food sources became available.

Over most of the Mediterranean and European Macaronesian seas, fish are not usually gutted at sea and so fish offal is not normally available for birds. In the European Macaronesian seas, discarded fish are also a relatively insignificant food source and are not typically important for the region's priority birds.

However, reduced discard availability may be a moderately high threat in the Mediterranean, where seven priority seabirds may suffer population declines as a result, and it could have a particularly high impact on *Larus audouinii* (Tucker and Heath 1994, Lambertini 1996). In the north and west European seas, 12 priority species may be affected by similar declines. The overall impact of discard and offal reductions on priority seabirds in Europe is therefore moderate.

Entanglement in fishing nets and on longlines

In some parts of the world, extremely large numbers of seabirds have been entangled and killed in fishing nets. In Europe, this has been a major problem in Greenland and Norway.

Salmon drift-netting off western Greenland from 1969-1971 was estimated to have caught about 1.5 million Brünnich's Guillemots *Uria lomvia*, which probably exceeded the annual production of young. Colony sizes here showed a decline of 20–40% from the late 1950s to the late 1970s (Nettleship and Evans 1985). In northern Norway, large numbers of auks were caught by salmon drift-nets before they were banned (Barrett and Vader 1984). At the same time breeding populations of *Uria aalge* declined, and this can be accounted for by estimates of net losses (Strann *et al.* 1990 in Robins 1991). High local losses of auks have also been noted elsewhere in the north and west European seas (Whilde 1979, Oldén *et al.* 1985, Robins 1991, Stempniewicz 1993). Losses of birds to nets are highest where nets are set close to auk breeding colonies and potentially worse if nets are made of monofilament nylon, which is less visible to the swimming birds.

This kind of seabird mortality in southern European waters has been less well documented, but the overall numbers are thought to be minor or negligible (Bárcena *et al.* 1984, de Juana *et al.* 1984, Teixeira 1986, Thibault 1993). However, considerable numbers of cetaceans are known to be accidentally killed every year by gill-net fisheries off the Portuguese coast (Sequeira and Ferreira 1994).

The developing longline fishery for bottom-dwelling fish in the north-east Atlantic is known to catch *Fulmarus glacialis*, *Sula bassana* and presumably other seabirds which scavenge, but the scale of the impact is not known (Dunn 1994b, 1995). Worldwide, longlines with multiple baited hooks are killing unacceptably high numbers of albatrosses (Diomedeidae) and related birds, causing widespread reductions in populations (Brothers 1991, Bartle 1995, Prince and Croxall 1995). Drifting longlines are also used in the Mediterranean. Bird casualties from these are probably few, but these lines are known to catch large numbers of sharks and seaturtles, especially the globally threatened *Dermochelys coriacea* and *Caretta caretta*.

A widespread increase in drift-netting has occurred in the Mediterranean Sea, which now experiences more large-scale high-seas drift-netting than any other sea in the world, with highly illegal netlengths of between 10 and 12 km long often being used, e.g. in Italy (Greenpeace 1995, A. Di Natale pers. comm. 1996). Birds are rarely entangled in these types of net (A. Di Natale pers. comm. 1996), but there is a heavy bycatch, including more than 8,000 cetaceans per year by the Italian fleet alone (Greenpeace 1995).

Entanglement of seabirds in fishing nets and on longlines is likely to remain a local problem in most areas. It affects significant proportions of birds in all regions, but impacts in all cases are predicted to be only local declines. The overall threat score for all regions is therefore moderate (Figure 3).

Toxic pollution

Organochlorines and heavy metals have been recorded in eggs and tissues of many seabirds, e.g. *Larus audouinii* (Lambertini 1996). Indeed, seabirds have been widely used to monitor concentrations of persistent organochlorine compounds (such as DDE, DDT, PCBs and Dieldrin), heavy metals and trace elements (Furness 1987b, 1993). There have, however, been few recorded biological effects of these pollutants on seabirds. One well-documented case showed that a catastrophic decline in populations of *Sterna sandvicensis* and Eider *Somateria mollissima* in the western Wadden Sea had been due to their consumption of fish and shellfish that had been contaminated by effluent from a pesticide factory (Duinker and Koeman 1978). In addition, there are high natural levels of many metals in seawater, and it is likely that most seabirds are adapted to high levels of some metals. Although there is insufficient information to reliably assess the relative sensitivity of seabirds to the many chemical and metal pollutants, it is thought that nearly all priority species of the Mediterranean and Black Seas and 11 from the north and west European seas are affected by such pollutants. These impacts are all thought to be low.

Reduced fishing of predatory fish

Fishing efforts throughout the European region have increased enormously during the twentieth century and, as a result, many species have been over-exploited (Shepherd 1993), a conclusion shared by the International Council for the Exploration of the Sea (ICES) Advisory Committee on Fishery Management, which provides scientific advice for recommending quotas (Bartram 1995).

Demersal fish such as cod and plaice *Pleuronectes platessa* are the most depleted in European waters, but overfishing of predatory species such as swordfish, tuna, adult herring and mackerel is also widespread and serious. However, seabirds prefer to prey on smaller species such as sandeel, sprat, sardine and anchovy or on young herring (see Appendix 3, p. 354). Consequently, depletion of demersal fish and pelagic predators has not had detrimental effects on seabird populations. In fact, probably the reverse is true.

Models of the effects of the drastic reductions in the stock size of immature herring and mackerel in the North Sea predicted increases in sandeel stock as a result of reduced predation and competition (Andersen and Ursin 1977). Such an increase in the sandeel stock has subsequently occurred (Sherman *et al.* 1981), along with rises in populations of seabirds that can exploit sandeels (Tasker and Becker 1992). Recovery of these predator stocks, if it caused a corresponding decrease in sandeels, could therefore reduce food supply for most seabirds. The impacts of this on seabird populations are uncertain, but it is thought that 20 priority species of north and west European seas may suffer small or local declines as a result.

In southern waters, it is thought that feeding tuna can benefit seabirds by forcing small-fish prey to the surface (Martín 1986). Consequently, such seabirds may suffer from over-exploitation of large surface-feeding predatory fish, and could benefit from a reduction in fishing effort or proper control of the fishery.

In conclusion, the broader environmental principle of ensuring sustainable fisheries outweighs the probable relatively small impact on birds of reducing catches of predatory fish. Furthermore, reduced exploitation of such a major component (fish stocks) of the marine ecosystem may increase ecosystem stability, which may benefit seabirds in the longer term.

Dredging and extraction

Dredging and extractive industries may affect priority seabird populations and their habitats in several ways. A reduction in water clarity may reduce the feeding efficiency of some 14 underwater feeding species and plunge-divers. Upset or destruction of benthic communities may result in decreases in invertebrate or, indirectly, fish-food resources for priority seabirds, leading to temporary or permanent abandonment of feeding areas. Dredging and mining operations may also disturb feeding and roosting birds which may lead to increased mortality through reduced food intake and/or increased energy expenditure. If dredging and similar activities are close to nest-sites, then disturbance may increase chick losses through starvation, chilling of eggs or chicks, or increased predation.

Overall, the impacts of dredging on the populations of priority species are poorly documented but probably only local and confined to relatively few species (e.g. Furness 1993). Except where these activities take place in areas with particularly high concentrations of birds, e.g. Important Bird Areas (Grimmett and Jones 1989, Skov *et al.* 1995), they are probably of only low importance among the threats considered here.

Bottom-fishing

Scallop-dredging, beam-trawling and other fishing methods which disturb the seabed can seriously damage benthic habitats and their fauna, and decrease clarity of shallow waters (MacGarvin 1990). In turn this may disrupt food-chains and eventually seabird populations (mainly of seaducks and divers) through declines in invertebrate and fish-food resources. Many benthic communities are fragile and take years to develop and stabilize. Despite this, these practices affect up to 10% of the north and west European seas and the Black Sea (North Sea Task Force 1993, Kisiov *et al.* 1994).

Although the effects of these activities on bird populations are poorly understood, it is suggested that 15 priority species in the north and west European seas may suffer low-level impacts.

Nutrient pollution

The enrichment of natural waters by nutrients is widespread in European coastal areas as a result of the disposal of organic waste into the sea (artificial eutrophication). This occurs directly, e.g. via sewage-disposal pipelines, but most particularly via the nutrient-rich inputs from rivers which, in total, deliver thousands of tonnes of nitrogen and hundreds of tonnes of phosphorus per year into European coastal seas (Brockman *et al.* 1988, Boddeke and Hagel 1995, Stanners and Bourdeau 1995). Slight increases in nitrogen and phosphorus levels will increase productivity and may be indirectly beneficial to higher trophic levels by increasing numbers of invertebrates (van Impe 1985).

Eutrophication can also have deleterious effects, especially at higher input levels. Changes in phytoplankton communities occur, which can result in oxygen depletion of seawater (Furness 1993). Such oxygen depletion has been observed in parts of the Baltic and North Seas, resulting in mass-death of fish and benthic invertebrates (Furness 1993, North Sea Task Force 1993). Eutrophication may also lead to increased blooms of toxic algae ('red tides') which have been known to kill seabirds (e.g. Hario 1993).

However, it is in the Black Sea that the most severe effects of eutrophication have been witnessed (Pearce 1995). Between the 1950s and the 1980s, the level of nitrates and phosphates entering the sea from rivers rose five-fold, and there was large-scale dumping of dredged material from rivers into the sea, generating huge clouds of sediment. The subsequent eutrophication and loss of light caused the death of huge beds of red algae *Phyllophora* on the northwest shelf, which led to disruption of the whole ecosystem and the collapse of the Black Sea fisheries.

Surprisingly, despite these catastrophic effects on the marine ecosystem as a whole, substantial impacts on birds have not been reported. Although populations of the fish-feeding and plunge-diving *Sterna caspia* and *S. sandvicensis* are declining around the Black Sea, and this may be due in part to eutrophication effects, there is no direct evidence to support this.

In conclusion, it is very difficult to predict the indirect effects of artificial eutrophication on bird populations, due to the complexity of coastal ecosystems (Brockman *et al.* 1988, Furness 1993). However, up to 14 priority seabirds that rely on sight to catch mobile prey underwater or near the surface (see Appendix 3), such as divers and terns, are theoretically likely to suffer directly from eutrophication because of reduced water-clarity resulting from increases in phytoplankton (Furness 1993). Nevertheless, it appears that eutrophication, at least at moderate levels, is overall a relatively low priority threat to seabird populations in Europe.

Increasing boat traffic

Commercial shipping, fishing and recreation at sea can lead to disturbance of feeding and roosting seabirds. As described above, this may lead to increased mortality, which may, however, quickly decline if birds are able to habituate to relatively constant disturbance, for example close to regular ferry routes. The most likely species to be seriously disturbed at sea are divers and seaducks (e.g. Keller 1991, Mikola *et al.* 1994), though the actual effects on them are not well documented away from breeding areas.

As for dredging, impacts from this hazard are likely to be low and restricted to a small number of priority species. Except where disturbance occurs in areas with major concentrations of vulnerable species, e.g. Important Bird Areas (Grimmett and Jones 1989, Skov *et al.* 1995), the effects of this threat on priority seabird populations in Europe are likely to be of relatively low priority.

Coastal erosion and subsidence

Erosion and subsidence due to man-made changes in the environment can lead to the loss of sedimentary islands which are breeding sites for priority seabirds (gulls and terns). The problem is only known from the Mediterranean, primarily affecting the Adriatic Sea where natural erosion of soft sediment islands has been exacerbated by subsidence caused by oil-drilling operations (Fasola and Canova 1991, 1992, 1996). Extensive damming of large rivers can also reduce the sediment supply and promote erosion along the coastal zone.

On a European scale, this hazard has the second lowest importance among the threats considered here. Four priority species of the Mediterranean and

Black Seas may suffer population declines as a result, which may be large in the two tern species.

Overfishing of shellfish and echinoderms

Various seaducks, including four priority species, rely to a large extent on shellfish for food (Appendix 3). These species are therefore likely to be highly affected by overfishing of shellfish stocks. Such depletion of shellfish stocks, and apparent displacement of seaducks, has occurred in the Dutch Wadden Sea (Leopold 1993, Bartram 1995) and along the Portuguese coast (Sobral *et al.* 1989, Dias *et al.* 1994), and continues to be a threat at the former locality.

Such overfishing is, however, only likely to be local and to affect only a few priority seabirds. Overall, therefore, this hazard to seabird populations has the lowest priority among the threats considered in this strategy.

CONSERVATION OPPORTUNITIES

International legislation

A summary of the international conventions, directives and regulations affecting European seas is presented in Box 1 (and covered in more detail under 'Opportunities for Conserving the Wider Environment', p. 24). Most of these instruments contain measures for the protection both of specific sites and of the wider environment. Exceptions are the OSPAR Convention and the Port State Control Directive, where measures are not currently aimed at specific sites. The MARPOL Convention is also predominantly aimed at the wider environment, although regulations exist to allow the designation of 'Particularly Sensitive Sea Areas' and 'Areas to be Avoided' in which discharge limits are particularly strict. These special areas may, however, be large (e.g. North Sea).

Of the instruments containing measures for protection of marine sites in Europe, the Birds and Habitats Directives are particularly useful in conserving the nesting sites of priority marine birds, though much less effective at protecting valuable foraging and resting areas offshore. The Directives require EU member states to designate (by June 2004) a coherent network of protected areas in order to conserve, amongst others, 23 priority bird species of marine habitats (see Appendix 1, p. 327) and four shallow subtidal habitats of significance to seabirds (Appendix 5, p. 423). These protected areas can include marine sites to the limits of territorial waters (normally 12 nautical miles from the coast), but the Habitats Directive notably fails to address the full diversity of other habitats present within 12 nautical miles of the coast, e.g. areas of upwelling, or in the wider marine realm. These Directives also require that species be maintained at a 'favourable conservation status', which in theory requires further (unspecified) wide-scale conservation measures from member states in many cases.

The Helsinki Convention promotes the establishment of Baltic Sea Protected Areas (BSPAs) within and beyond territorial waters, as well as their management according to certain standards. Although BSPAs have been listed by contracting parties, actual management initiatives are still needed. Outside the Baltic Sea, there is currently no robust legislative framework for site protection beyond territorial waters in the area covered by this strategy. However, initiatives on this matter are currently being discussed by the Oslo and Paris Commissions (OSPARCOM) as a result of an invitation to OSPARCOM by ministers at the Fourth North Sea Conference in 1995.

North Sea Conference process

Since the first ministerial conference in 1987, the North Sea Conference process has helped to obtain ministerial commitments to improving the environmental health of the North Sea. Commitments are made by ministers of North Sea states, Switzerland and the European Commission. The conferences are valuable opportunities. Firstly, they have the potential to cover subjects that may not be covered by existing conventions or directives, such as the protection of Important Bird Areas (IBAs) beyond territorial waters. Secondly, they are public occasions, and because of this ministers may be prepared to make concessions to the environment in order for their governments to appear progressive. Following the 4th Conference in 1995, discussions on further conferences and alternatives are in progress, including preparation for the Intermediate Ministerial Meeting in Norway, March 1997, to address the integration of fisheries and environmental management.

Ministerial meetings of the Oslo and Paris Commissions (OSPARCOM)

The first ministerial meeting took place in 1992. A second meeting is due in 1997. As with the North Sea Conference process, any involvement of ministers presents opportunities for politically-binding commitments. The Programmes and Measures (PRAM) division of OSPARCOM does not currently cover activities other than reduction of land-based and sea-based pollution. However, the possibility remains for it to cover site-based protection measures for species and habitats, and the 1997 ministerial meeting provides an important opportunity for commitment to protection measures for sites beyond territorial waters.

Box 1. International environmental legislation of particular relevance to the conservation of marine habitats in Europe. See 'Opportunities for Conserving the Wider Environment' (p. 24) for details.

GLOBAL

- *Biodiversity Convention* (1992) Convention on Biological Diversity.
- *Bonn Convention* or *CMS* (1979) Convention on the Conservation of Migratory Species of Wild Animals.
- *Ramsar Convention* (1971) Convention on Wetlands of International Importance especially as Waterfowl Habitat.
- *World Heritage Convention* (1972) Convention concerning the Protection of the World Cultural and Natural Heritage.
- *MARPOL 73/78* (1973) International Convention for the Prevention of Pollution from Ships, 1973, as modified by the protocol of 1978.
- *Law of the Sea* or *UNCLOS* (1982) United Nations Convention on the Law of the Sea.

PAN-EUROPEAN

- *Bern Convention* (1979) Convention on the Conservation of European Wildlife and Natural Habitats.
- *Espoo Convention* (1991, but yet to come into force) Convention on Environmental Impact Assessment in a Transboundary Context.

REGIONAL

- *Oslo Convention* (1974) Convention for the Prevention of Marine Pollution by Dumping from Ships and Aircraft.
- *Helsinki Convention* (1974/1992) Convention for the Protection of the Marine Environment of the Baltic Sea Area.
- *Barcelona Convention* (1975) Convention for the Protection of the Mediterranean Sea against Pollution.
- *Paris Convention* (1978) Convention for the Prevention of Marine Pollution from Land-based Sources.

- *Bucharest Convention* or *Black Sea Convention* (1992) Convention on the Protection of the Black Sea against Pollution.
- *OSPAR Convention* (1992, but yet to come into force) Convention for the Protection of the Marine Environment of the North-East Atlantic (formed by the merger of the Oslo and Paris Conventions).

EUROPEAN UNION

- *Birds Directive* Council Directive on the conservation of wild birds (79/409/EEC).
- *Habitats Directive* Council Directive on the conservation of natural habitats of wild fauna and flora (92/43/EEC).
- *EIA Directive* Council Directive on the assessment of the effects of certain public and private projects on the environment (85/337/EEC).
- *Port State Control Directive* Council Directive concerning the enforcement, in respect of shipping using Community ports and sailing in the waters under the jurisdiction of the Member States, or international standards for ship safety, pollution prevention and shipboard living and working conditions (95/21/EC).
- *Nitrates Directive* Council Directive protecting waters against pollution by agricultural nitrates (91/676/EEC).
- Council Directive on urban waste water treatment (91/271/EEC).
- Council Directive on the control of dangerous substances discharge (76/464/EEC).
- Council Directive on the conservation of fresh water for fish (78/659/EEC).
- Council Directive on the freedom of access to information on the environment (90/313/EEC).
- Council Directive on the quality required of shellfish waters (79/923/EEC).

Ministerial meetings of the Helsinki Commission (HELCOM)

The first Convention on the Protection of the Marine Environment of the Baltic Sea (Helsinki Convention 1974/1992) was signed in 1974 by the coastal states of the Baltic Sea. The convention was established in order to protect the marine environment of the Baltic Sea from all sources of pollution, both from land and from ships as well as airborne. In 1992, a new revised convention was signed by countries bordering the Baltic Sea and the European Community. This new convention includes provisions for the protection of biodiversity—the fauna, flora and natural habitats—of the Baltic Sea.

The governing body of the Convention is the Helsinki Commission (HELCOM)—in full, the Baltic Marine Environment Protection Commission. The Commission meets annually and there are also meetings held at ministerial level. Decisions taken by HELCOM are regarded as being recommendations to the governments concerned, which in their turn should incorporate them into national legislation. The most important recommendations to have been made so far by the Commission in the field of nature conservation include the ones on the establishment and management of a network of Baltic Sea Protected Areas (BSPAs) and on the protection of the coastal zone. HELCOM consists of four Committees and a Programme Implementation Task Force. The last ministerial meeting to date was held in 1994, and further meetings may follow. As with the North Sea Conference process, any ministerial involvement provides opportunities for politically binding commitments.

Barcelona Convention and its related Protocols

Within the Regional Seas Programme of the United Nations Environmental Programme (UNEP), the Convention for the Protection of the Mediterranean Sea against Pollution (Barcelona Convention) came into force in 1978. There are three protocols which relate to the prevention of all types of pollution (similarly to the Helsinki Convention 1974/1992), whilst a fourth concerns the designation by contracting parties of Mediterranean Specially Protected Areas (MedSPAs) within territorial waters.

All European states bordering the Mediterranean Sea have ratified the convention and the MedSPA protocol, amongst others. In order to implement the convention and its protocols, the contracting parties adopted the 'Mediterranean Action Plan', which is coordinated by the Mediterranean Coordinating Unit in Athens. The function of Secretariat to the convention is carried out by UNEP–Europe in Geneva.

The contracting parties decided to revise the convention in 1994/1995 to take into account the results of the United Nations Conference on the Environment and Development (UNCED, or Earth Summit) held in Rio de Janeiro in 1992, and to enlarge the convention's field of competence from marine to coastal habitats as well. The convention now includes more provision for the conservation of biodiversity and the protection of species and their natural habitats in the Mediterranean Sea. All parties to the convention are now obliged to draw up a list of MedSPAs which are important for the protection of ecosystems, or which are of special interest at the scientific, aesthetic or cultural level. A total of 111 MedSPAs covering 4,602 km^2 had been designated by the nine European contracting parties as of 1993 (Stanners and Bourdeau 1995). There is also an Annex to the MedSPA protocol, containing a list of endangered or threatened species, which includes priority seabird species of the Mediterranean and Black Seas. Parties are required to take appropriate measures to protect these species and their habitats at the national and international levels.

European Program for the Mediterranean (EPM)

The European Investment Bank and the World Bank developed the EPM. Its second phase, the Mediterranean Environmental Technical Assistance Program (METAP), was launched in 1990 to identify and prepare investment projects and institutional development activities and define specific policy measures in the following areas: integrated waste-resource management, solid- and hazardous-waste management, prevention and control of marine oil and chemical pollution, and coastal zone management. The coastal zone management priority includes a biodiversity component. Its objectives include providing assistance to southern and eastern Mediterranean countries for project-preparation activities which promote the conservation of protected areas, and the organization of a network of managers of Mediterranean protected areas (MEDPAN).

Black Sea Environmental Programme (BSEP)

This intergovernmental programme of activities to halt and reverse degradation of the Black Sea (Ludikhuize 1992, Mee 1992) has been active since 1993. So far, the difficult economic conditions in the Black Sea countries have inhibited realization of the programme's objectives. In particular, there has been little success in creating new national legislation to enable proper conservation of the sea. The most active fields of endeavour have been data-gathering and scientific research, mainly on marine pollution.

UK Government response to the Donaldson Report

The report of Lord Donaldson's inquiry into the prevention of pollution from merchant shipping was published in May 1994 (Anon. 1994d). The report contained 103 recommendations to help prevent both accidental spills and deliberate discharges. The UK government published its official response in February 1995, accepting 86 of the 103 recommendations. Since many of the recommendations call for action at international levels such as the European Union, the Port State Control Committee of the Paris Memorandum of Understanding on Port State Control, and the International Maritime Organization, effective promotion and implementation could have benefits for seas elsewhere in the region.

Financial instruments

This section considers financial instruments as policy measures that achieve environmental objectives by means of incentives. Penalties levied for not adhering to legislation are not considered here as financial instruments.

Taxes are a financial instrument that could be applied to various activities affecting the seas of the region. Taxes have great potential for reducing waste disposal by acting as an incentive to minimize the amount of waste generated.

Other examples of financial instruments which might be used more widely include the Port of Rotterdam's Green Award Scheme, whereby ships with more environment-friendly attributes are charged lower port dues. The use of port waste-reception facilities may be promoted by appropriate charging systems such as the incorporation of fees for the use of reception facilities into port dues. A thorough investigation of charging systems is needed across the region in order to identify the systems that encourage better use of reception facilities.

Financial instruments for fisheries include decommissioning and tradeable quotas. Decommissioning involves the offer of money to fishermen in return for scrapping their vessels, and is one means of bringing fleet capacity into line with fish stocks. Unfortunately, the rate of decommissioning is generally lower than the increase in efficiency of the remaining fleet. A recent European Commission proposal to link early retirement incentives for fishermen to decommissioned vessels has received support from most EU member states (with the notable exception of the United Kingdom). Such an incentive scheme, if widely and generously implemented, should facilitate the decommissioning of vessels which are currently having a tangible impact on fishing effort, rather than vessels which were already of little use.

Tradeable quotas are also a financial instrument. The idea was introduced in New Zealand, as 'individual transferable quotas' (ITQs). In giving individual fishermen a stake in their local stocks, ITQs may promote stock conservation. While the ITQ system has been adopted in Iceland, it is less appropriate to the mixed demersal fisheries in the northeast Atlantic.

Policy initiatives

Common Fisheries Policy of the European Union

New Regulations and amendments to existing Regulations of the Common Fisheries Policy (CFP) are periodically proposed by the European Commission, and are considered by fisheries ministers of the EU member states. Whilst ministers may meet several times a year, there are only two formal councils each year. Though not public, these twice-yearly meetings can provide a focus for lobbying activity.

A review of the CFP is due in 2002. The political will for a fundamental reform is unlikely unless some major commercial fisheries collapse in the interim period, which, however, is possible. Many stocks are already 'below safe biological limits', meaning that recruitment is no longer adequate to maintain a sustainable spawning population. The most pressing need is for a substantial and permanent reduction of fishing effort (see 'Financial instruments', p. 86), since much of the damage to the marine ecosystem that arises from fishing is a symptom of overfishing itself.

At the Fourth North Sea Conference, ministers made a commitment to have research undertaken on the sustainability of industrial fisheries and on the effects of exploiting the target fish species on dependent predators such as seabirds and marine mammals, before any decision can be taken about better regulation of these fisheries. As of now, therefore, sandeels, though constituting around a half of the total North Sea industrial fish catch, remain unregulated by any catch limits. Adequate regulation of the sandeel fishery, whether by catch limits or by effort limitation, is thus a major challenge at and before the next review of the CFP.

'Common Policy on Safe Seas' of the European Commission

The agenda of the European Commission on shipping is set by this document (COM (93) 66 final 24-2-93), published in February 1993. It has already resulted in a Regulation on port dues for segregated ballast-tank tankers, Directives on reporting, classification societies, standards of training of seafarers, and port state control, and a proposed Directive on further reporting. The Commission is making efforts to promote other Directives proposed in the Action Programme of the document.

The will of the Commission to get things done is promising, but proposed Directives can only be accepted with the will of the EU member states, and their success or otherwise can only be judged on the adequacy of their implementation. Transport ministers of EU member states meet at least twice yearly to discuss proposals. As with the meetings of fisheries ministers, these meetings can provide a focus for lobbying activity.

Non-statutory opportunities

As with much of nature conservation, substantial progress can be made through the use of non-statutory initiatives and guidance. A recent, positive initiative has been the establishment of the Marine Stewardship Council (McGinn 1996). The MSC is a private-sector 'sustainable fishing' initiative funded by Unilever (one of the top buyers of seafood globally) and the World Wide Fund for Nature in the UK (WWF–UK), and will certify seafood as coming from sustainable fisheries. Unilever currently supplies one-quarter of the seafood market in Europe and North America, and has agreed to sell only MSC-approved seafood by the year 2005. Such a policy could exert considerable influence on fishing techniques and technology (e.g. genetic sampling of catch to verify origin, satellite-monitoring of vessels, more selective gear), and on the policies of other major seafood buyers, which are beginning to view industrial overfishing of smaller fish as a threat to the food supply of their own stocks of larger predatory fish (the main seafood-fish).

In a wider framework, local and regional initiatives for coastal and nearshore zone management, involving partnerships of government, non-government organizations and businesses which agree on the equitable use of local or regional coastal resources, may bear fruit. Such initiatives have started in UK waters and elsewhere in north-west Europe.

The hydrocarbons industry and national governments have worked together in a number of areas, outside statutory frameworks, to reduce the risk of damage to the environment. Publication of atlases of seabird vulnerability to oil pollution (e.g. Carter *et al.* 1993, Webb *et al.* 1995) from cooperatively funded projects will help sea-users to better arrange their activities to avoid areas and times of high sensitivity.

CONSERVATION RECOMMENDATIONS

Broad conservation objectives

1. To avoid the loss, or partial loss, of any internationally important marine habitat or coastal seabird breeding site (i.e. Important Bird Areas)
2. To arrest the degradation of marine habitats and seabird breeding sites, and to maintain and enhance important habitat features for their priority species
3. To improve habitat quality in degraded marine habitats and seabird sites for priority species.

Ecological targets for the habitat
In order to maintain and improve habitat quality, it is essential to retain and enhance the important habitat features required by priority bird species of marine habitats, as follows.

1. Introduced predators on the current and former nesting islands of priority species should be eradicated where feasible.
2. Disturbance of priority bird species by human activities (e.g. hunting, fishing, recreation, etc.) at or near the nesting sites should be avoided during the nesting season. Outside this season, human activities should be reduced to such levels that the species' survival rate is not significantly affected at a local or regional scale.
3. Nest-sites such as cliffs and islets should be preserved from any man-made destruction or infrastructural developments, and important associated nest-site features should be maintained, such as cliff-ledges, boulder slopes, caves, vegetation and deep soil, etc.
4. Predators on priority bird species (including eggs and young) should be maintained at historically natural population levels, and human activities that result in increases in range or population size of such predators should be avoided or reversed where these increases lead to population declines in priority species.
5. Pollutants should be reduced to levels where any sublethal or indirect effects are known or reasonably suspected not to be causing population declines in priority bird species.
6. The abundance, distribution and availability of the natural food resources of priority bird species, in particular small fish and invertebrates, should be maintained.
7. Natural, low and open vegetation at the breeding sites of priority bird species should be maintained where necessary, by, for example, appropriately managed grazing that avoids overgrazing (e.g. excessive soil erosion) or trampling of eggs and young during the breeding season, or trampling of nesting burrows at any time of year.
8. Incidental catching of priority bird species in actively used or discarded fishing nets and on baited longlines should be avoided or reduced to insignificant levels.
9. Physical damage to benthic habitats should be avoided, and their associated flora and fauna conserved.
10. High levels of nutrient pollution, e.g. where water clarity is significantly reduced or where there are plankton blooms and 'red tides', should be avoided and where necessary reduced.

Broad policies and actions
In addition to actions that should be applied to coastal and marine Important Bird Areas, as summarized in Box 1 on p. 15, the following broad actions should be implemented (see p. 22 in 'Introduction' for explanation of derivation):

Legislation

1. The Secretariats or other relevant permanent bodies of the relevant conventions (Box 1) should urgently undertake a review of the extent to which each convention is achieving conservation of marine habitats and their priority birds. Contracting parties are urged to apply by the year 2000 the relevant provisions of the conven-

tions to protect marine habitats, and to develop measures to rectify any perceived deficiencies identified in the above reviews.

The European Commission should undertake a similar review of the extent to which the relevant Directives and Regulations (Box 1) are capable of the effective protection and management of marine habitats and their priority birds, including beyond territorial waters, and should reform these instruments where necessary, with the support of national governments.

2. All states should carry out Strategic Environmental Assessments (SEAs) of any policies, plans and programmes, and Environmental Impact Assessments (EIAs) of any projects, that involve activities in the marine environment or otherwise affect priority seabirds and their habitats, including oil and gas exploration and production, short-sea shipping initiatives, shipping route programmes, coastal zone developments and fisheries programmes. These SEAs and EIAs should inform and influence decisions on award of permits, and conditions attached to any permits and any international funding.

Land-use

3. New introductions of grazing animals to priority seabird colonies should be avoided unless preceding research demonstrates that no threat will be posed to priority bird species. In areas where such problems with grazing exist, the timing and zoning of grazing should be improved, and grazing animals should be removed where they continue to threaten seabirds. The Agri-environment Regulation should be used to support grazing management of nesting habitats in EU countries (see 'Agricultural and Grassland Habitats', p. 267). Further research is, however, required to assess the impacts of grazing and to establish optimum requirements.

4. Disturbance to seabirds by human activities, especially at nesting sites, should be reduced through the regulation of access by zoning and/or timing restrictions on activities.

Policy-makers, land-owners and the general public should be informed of the problems caused by terrestrial and marine disturbance.

Further research should be carried out to assess the tolerance of priority seabird species to disturbance from different activities, and to assess the effect of these on populations.

Water quality

5. Agricultural use of fertilizers should be reduced (according to the measures outlined under 'Agricultural and Grassland Habitats', p. 267), as

should industrial inputs to seas, and waste-water treatment should be improved (possibly to be promoted via subsidies). The impact on priority bird populations of the widespread and continuing nutrient pollution of many coastal and marine ecosystems should be assessed, in order to clarify which levels of nutrient pollution are 'safe' or 'tolerable' in which situations.

6. The use of pesticides for agriculture should be reduced (according to the measures outlined under 'Agricultural and Grassland Habitats', p. 267), and industrial discharges and offshore dumping of toxic chemicals should be banned unless proper Environmental Impact Assessments forecast an ecologically favourable outcome.

Toxic levels in priority birds should be monitored locally and regionally, and the effects of chemicals, particularly of novel products, on priority bird populations should be established.

Shipping

7. States with poor records of implementing internationally agreed standards within their fleets should be urged to improve these records through action at the IMO by countries with better records.

8. States with improved records of implementing internationally agreed standards should be encouraged to participate in regional port state control agreements. This could include the membership of existing agreements (e.g. the Paris Memorandum of Understanding on Port State Control), or creation of new regional agreements.

9. Member states of the European Union which have not yet effectively implemented the Port State Control Directive, as required by 30 June 1996, should do so without further delay.

10. All MARPOL-contracting states should aim to achieve effective enforcement of the discharge limits required by the MARPOL Convention, by supplementing aerial surveillance with a regional system of waste auditing on vessels in port and stronger penalties for offenders.

11. All MARPOL-contracting states should commence work through the IMO further to amend Annexes I, II and V of the MARPOL Convention in order to achieve global zero-discharge regimes and the mandatory use of port reception facilities.

12. All MARPOL-contracting states should carry out public reviews of the adequacy of their port reception facilities. Any deficiencies in adequacy or ease of use should be rectified.

13. States should make full use of the opportunities

to reduce the risk to wildlife that are offered by the IMO concept of 'Particularly Sensitive Sea Areas' and associated IMO routeing measures (such as 'Areas to be Avoided', and 'Deep Water Routes'), and should monitor compliance of shipping at sites where routeing measures have been established.

14. Those states which have already ratified the 1990 Convention on Oil Pollution, Preparedness, Response and Cooperation (ORPC), and the 1992 Protocols to the Civil Liability Convention (1969) and the Fund Convention (1971) should urge other states to do so.

15. Further information is needed to assess the impacts of pollution on priority bird species at sea. To this end, governments should provide funding for the international coordination, analysis and dissemination of results of more frequent beached-bird surveys, particularly on coastlines with high concentrations of birds and/or high levels of shipping or industrial activity (e.g. the North Sea and Baltic Sea).

Fisheries

16. The EU Common Fisheries Policy and national fisheries policies should be reformed to ensure sustainability not just of the target species but of other dependent organisms, including marine birds. To achieve this the following actions should be taken:

   • Total allowable catches and quotas should be set for all stocks and for all regions to ensure sustainable yields. Given their special importance as prey of marine birds, industrial small-fish species (e.g. sandeel, sprats and sardine), at present subject to inadequate or minimal management, should be equally subject to catch-limits and/or regulation of fishing effort commensurate with the needs of birds and other dependent predators. The necessary research should be done to evaluate the sustainability of industrial fishing in this respect.

   • Governments should urgently reduce fishing effort by greater reduction of fleet capacity (through decommissioning) and/or by reduction of the time the fleet spends fishing.

   • Local management of fisheries should be encouraged and resourced accordingly.

   • Governments should establish areas closed to fisheries as trials for ecosystem recovery.

   • Measures should be implemented to minimize the bycatch and subsequent discarding of non-target species of fish. These policies to eliminate loss of fish resources should be pursued vigor-

ously in the interests of sustainable fisheries even though they may well act to the detriment of some scavenging seabirds.

   • Further research is required to establish the dependence of seabird populations on offal and discards as food, and the impact of changes in the availability of these resources, including indirect effects on other seabird populations as a result of any reduction in these resources.

   • Further research is needed to establish seabird diets and dietary requirements, and the extent of overlap between fisheries and seabird feeding areas. Further investigation is also needed into the biology, as it affects birds, of the important commercial and non-commercial fish species. A multi-disciplinary approach to the provision of advice for fisheries management needs to be developed by ICES which incorporates the effects of fisheries on birds (and other wildlife).

17. Areas where damage to benthic substrates from bottom-trawling is serious should be identified and these activities either more strictly regulated, or modified to reduce damage. Environmental Impact Assessments should be carried out on all proposed dredging projects.

   Research should be carried out to assess the impact of bottom-trawling on benthic communities, and on priority birds and other species dependent on them, and to develop less damaging fishing methods.

18. Existing regulations concerning catch quotas, and on exploitation of those shellfish stocks which are shared by priority marine birds and humans, should be developed and improved in order to ensure sustainability for both users, especially in Important Bird Areas. Consideration should be given to (e.g.) temporary closure of shellfish beds to fishing during critical or sensitive periods.

   Further research should be carried out into the ecology of shellfisheries and their marine bird predators.

19. Gill-net and drift-net fisheries practices, regulations and licensing should be reformed, with respect to deployment (location and timing) and net materials, in order to reduce bycatch. There should be increased enforcement of drift-netting regulations, especially in the Mediterranean.

   The impacts of such nets on birds should be assessed, vulnerable areas should be identified, and technical measures to reduce or prevent entanglement should be developed. The discarding of unwanted ('ghost') nets should be legislated against.

20. New fisheries should be subject to Strategic Environmental Assessment (SEA). Rapidly developing longline fisheries should, for example, be investigated to assess the level of threat they pose to seabirds, and mitigating measures to reduce bycatch should be introduced as necessary.

Other measures

21. Legislation should be developed and implemented to prevent deliberate introductions of predators to currently predator-free seabird colonies, and to control visits by potential carriers of such predators (e.g. anchoring and landing restrictions). Special measures should be provided at ports and landing points to prevent accidental introductions of predators (e.g. rat-proof enclosures, warning signs). Introduced predators on identified key islands, where the threats are greatest to priority bird populations, should be eradicated or controlled where feasible.

The public should be informed of the problem of introduced predators and of why measures are necessary to control them.

Further research to assess the impacts of introduced predators on seabird populations should be carried out, and key sites should be monitored for their presence or continuing absence.

22. Human activities (e.g. fish discarding, or disposal of rubbish into the sea) that have resulted in increases in predators, causing declines in priority seabird populations, should be stopped and reversed or modified accordingly. Where this is impractical or unsuccessful, culling programmes should be considered.

23. The major fish-purchasers (companies, etc.) should be encouraged to buy only from fisheries certified as sustainable. The definition of sustainable would include assessment as to whether the side-effects of the fishery are sustainable in terms of population levels of priority seabirds.

ACKNOWLEDGEMENTS

Most contributions to this chapter were made by the Habitat Working Groups during and after workshops on the north and west European seas organized by O. Hüppop and M. Tasker and hosted by the Institut für Vogelforschung 'Vogelwarte Helgoland', Helgoland, Germany (29 October–3 November 1993), and on the European Macaronesian seas arranged by A. Martín Hidalgo, C. González (Canary Islands), P. Oliveira (Madeira) and L. R. Monteiro (Azores) at the University of La Laguna, Tenerife, Canary Islands (26–28 January 1995); these contributors are listed on p. 59.

E. Dunn and D. Owen (Royal Society for the Protection of Birds) made major contributions to the sections on 'Conservation opportunities' and 'Conservation recommendations'.

Valuable information and comments were received from H. Rainey, K. Camphuysen (NIOZ, Netherlands), S. Pihl (National Environmental Research Institute, Denmark), P.

Iankov (Balgarsko Druzhestvo za Zashtita na Pticite), M. Yarar (Dogal Hayati Koruma Dernegi), D. Munteanu (Societatea Ornitologica Româna), J. Cortes (Gibraltar Ornithological and Natural History Society), E. Shaw, J. Sultana (BirdLife Malta), J. Winkelman (Vogelbescherming Nederland), A. Duncan (Ligue pour la Protection des Oiseaux), M. Strazds (Latvijas Ornitologijas Biedriba), J. Coveney (BirdWatch Ireland), G. Vaitkus (Lietuvos Ornitology Draugija) and S. Demircan (Dogal Hayati Koruma Dernegi), the participants of the *5th International Seabird Group Conference on 'Threats to seabirds'* (Glasgow, UK, 1996), and the participants of the *4th MEDMARAVIS Symposium on 'Seabird ecology and coastal zone management in the Mediterranean'* (Hammamet, Tunisia, 1995). We thank X. Monbailliu for his help in organizing the involvement of L. Canova and M. Fasola in the Habitat Working Group, and for arranging the short workshop at the MEDMARAVIS Symposium.

COASTAL HABITATS

compiled from major contributions supplied by the Habitat Working Group

R. S. Barnes, P. Bradley (Coordinator), M. Calado, F. Demirayak, P. Doody, H. Granja, N. Hecker,
R. E. Randall, C. J. Smit, A. Teixeira, J. Walmsley

and by

D. Huggett and K. Norris

SUMMARY

Europe's coastal habitats hold a very high diversity of birds, and are particularly important for the survival of 75 priority species within Europe. Nearly 70% of these priority birds have an Unfavourable Conservation Status in Europe because their European populations are small, strongly declining or highly localized. The remaining priority species are specialists which are highly dependent on the continuing existence of coastal habitats. Many of the priority species are migratory waterbirds, breeding in Arctic Canada, Greenland, northern Europe or Siberia, and using the coastal habitats of Europe as overwintering sites or as staging posts on their migration. Survival of these birds depends on the continued presence of a network of suitable sites across Europe.

Maintaining the extent and ecological quality of coastal habitats is clearly essential for the conservation of these priority bird species in Europe. However, loss and degradation of these habitats is widespread in Europe, due mainly to land-claim and infrastructural development, to water pollution, to excessive disturbance by human activities, to human-induced increases in predators, and to abandonment of traditional low intensity grazing of some habitats. Indirectly, coastal engineering and upstream changes to catchment hydrology can also cause major and far-reaching changes to coastal habitats, by affecting such processes as erosion, sedimentation and water clarity, thereby affecting habitat quality over vast areas. In the long term, sea-level rise may also lead to the loss of many important areas of intertidal coastal habitat.

The full and strict protection of networks of important coastal sites is essential, through existing relevant national and international legislation. However, more wide-ranging and holistic measures are needed to address the major threats which face coastal habitats. Implementation and strengthening of regulations to control pollution are also essential. The adoption of Integrated Coastal Zone Management at the national and European levels is a most pressing requirement, so as to integrate coastal conservation into all relevant sectors, programmes, policies and plans. Additional help would come from reform of international funds for infrastructural development, and implementation of national biodiversity plans, Catchment Management Planning, Strategic Environmental Assessments (including of coastal defences and tourist developments) and Environmental Impact Assessments.

HABITAT DESCRIPTION

Definition

This strategy covers the following coastal habitats in Europe: intertidal habitats and tidal water-bodies within estuaries and deltas, coastal lagoons, salinas (saltpans), sand and shingle beaches, rocky shores and certain terrestrial habitats that are associated with coasts, including sea cliffs (for non-seabird species), shingle spits, sand-dunes and saltmarshes. Coasts of all marine regions south of the Arctic Circle are dealt with (see map in 'Marine Habitats', p. 60), with the exception of the Caspian Sea and the coasts of European Macaronesia[1].

Coastal habitats north of the Arctic Circle are treated under 'Tundra, Mires and Moorland' (p. 159). Coastal grazing marshes (including machair) are covered under 'Agricultural and Grassland Habitats' (p. 267). Marine habitats below the low-tide line, as well as coastal habitats that are breeding sites for

[1] Macaronesia is the biogeographical region composed of the archipelagos of Madeira, the Canary Islands, the Azores and (outside Europe) the Cape Verde Islands.

priority seabirds (sea cliffs and on islands), are treated under 'Marine Habitats' (p. 59).

Distribution and extent

The distribution and extent of coastal habitats in Europe is clearly dependent on the coastal length of each country. Russia, Greece, the United Kingdom and Turkey thus contain most of Europe's coastal habitat (Table 1). Coastal habitat-types show some broad but clear geographical patterns, largely due to geology and climate. Coastal habitats of northern Europe consist mainly of rocky shores, interspersed with fjords, especially in Norway and western Scotland. Estuaries and sand-dunes are frequent but tend to be relatively small and confined to sheltered bays. Sedimentary landscapes such as estuaries, mudflats,

and sand and shingle beaches predominate in the southern North Sea and the Baltic, where saltmarshes and sand-dunes are often also present.

The Atlantic coastlines of France, Portugal and Spain incorporate a large proportion of cliffs and sand-dunes, the latter especially in France. The Mediterranean coastline is about 46,000 km long, of which nearly half is made up of islands, and rocky shores are commonplace. Coastal wetlands of the north Mediterranean coast are mainly river deltas, and estuaries are rare. As elsewhere in Europe, many coastal Mediterranean wetlands form a complex association with non-wetland habitats in the hinterland, such as sand-dunes, open grassland and shrubland. The southern Black Sea coastline includes low cliffs alternating with extensive deltaic plains, while large lagoons and deltas predominate along the northern coast.

Ecological features

Sea cliffs

Sea cliffs form where hills and ridges meet the coast and are eroded by the sea. During formation, the base of an incipient cliff is undercut by wave action. Blocks of rock collapse to form the face of the cliff, then are further eroded and carried away by waves and currents. Over the years this results in a gradual recession of the cliff. Harder rocks (e.g. basalt or granite) retreat at a slower rate and give rise to steeper cliffs than those made of softer rocks (chalk, limestone, sandstone or sand-and-clay).

Rock-type and prevailing weather are the most important physical factors determining the nature of plant and animal communities. The 'aspect' of the cliff and its degree of exposure to salt spray and wind determine the type of vegetation that will colonize the rocks and influence its zonation.

Birds are the most conspicuous fauna. In particular, the coastal cliffs of the North Atlantic and North Sea support large and very important colonies of seabirds, as described under 'Marine Habitats' (p. 59). Typical landbird species in north-west Europe include Peregrine *Falco peregrinus*, Rock Dove *Columba livia*, Rock Pipit *Anthus petrosus* and Chough *Pyrrhocorax pyrrhocorax*. In the Mediterranean, the rare Eleonora's Falcon *Falco eleonorae* is a very characteristic breeding species of this habitat.

Rocky shores

Rocky shores are intertidal accumulations of stones and boulders, usually at the base of eroding cliffs, or are intertidal areas of low-lying, hard bedrock where strong wave action prevents sediment accumulation. The distribution of rocky shores in Europe is close to that of cliffs. The only comprehensive treatment of

Table 1. Extent of coastal habitats in Europe by country (Dahl 1991, WRI 1994). No data are available for Bosnia and Herzegovina or Croatia.

	Length of coastline (km)
Albania	418
Belgium	64
Bulgaria	354
Cyprus	648
Denmark	3,379
Faroe Islands	700
Greenland [1]	39,000
Estonia	1,393
Finland [2]	1,126
France	3,427
Germany	2,389
Greece	13,676
Iceland	4,988
Rep. of Ireland	1,448
Italy	4,996
Latvia	531
Lithuania	108
Malta	140
Netherlands	451
Norway [1,2]	5,832
Svalbard [1]	2,509 +
Poland [2]	491
Portugal	1,693
Azores	780
Madeira	141
Romania	225
Russia [1]	37,653
Slovenia	46
Spain	4,964
Canary Islands	1,291
Sweden	3,218
Turkey	7,200
Ukraine	2,782
United Kingdom	12,429
Gibraltar	10

[1] Figure includes coastline north of the Arctic Circle, although such coastal habitats are covered under 'Tundra, Mires and Moorland' (p. 159).
[2] Not including islands.

the distribution of rocky shores in Europe is for Britain (Burd 1989), where such shores account for 35% of total coastal length (Ferns 1992).

Physical features affecting the ecology of rocky shore habitats include the slope of the shore, the exposure to light, sunshine and wave action, the tidal amplitude, and the hardness and texture of the rocks (Barnes and Hughes 1982). Marine plants and animals distribute themselves in relation to their capacity to withstand different degrees of stress, primarily manifested as desiccation when exposed by the tide. In addition to physical stresses, biological factors such as competition and predation are important contributors to the vertical zonation of species (see Hawkins and Jones 1992). Lichens and algae are the dominant plants, and molluscs the most obvious non-bird fauna.

Rocky shores hold few breeding birds, but some waders (Charadrii) have important wintering populations occupying European rocky shores, especially Turnstone *Arenaria interpres*, Purple Sandpiper *Calidris maritima* and Oystercatcher *Haematopus ostralegus*. Rocky coasts are the main habitat for the globally threatened monk seal *Monachus monachus* in the Mediterranean, and important feeding areas for the otter *Lutra lutra* in parts of northwest Europe.

Sand and shingle beaches

A beach is 'a body of wave-washed sediment extending along a coast between the landward limit of wave action and the outermost breakers' (Bird 1984). The growth and existence of a beach depends on a plentiful supply of sand or shingle and a favourable pattern of waves, winds and tidal currents. Shingle beaches form where wave action is stronger and prevents deposition of fine particles, while sand beaches form in more sheltered areas (Eltringham 1971). A predominant sediment-particle size of over 2 mm separates shingle from sand (King 1972). There are several categories of beach (see Chapman 1976), including spits, bars, forelands and barrier islands.

Systematic information on the location and extent of sand beaches is currently lacking for much of Europe. The European distribution of shingle beaches has only been documented for Britain (Randall 1989, Sneddon and Randall 1993) and most of western France (Géhu 1960a,b,c, 1991). Shingle beaches appear to be commonest in the United Kingdom, on the Norwegian coast, the Brest peninsula of France and the Rügen coast of Germany.

Several physical factors characterize and distinguish beaches (Eltringham 1971), most importantly sediment-particle size, but also organic content, wave action, slope, area of beach and surface stability.

Shingle beaches are highly unstable, creating an inhospitable environment for plants and animals. A sparse vegetation develops above high-water mark, ranging from extensive patches of lichen to grassland, heath and scrub in the most stable areas. Sand beaches allow establishment of more stable plant communities. Colonizing plants stabilize drifting sand and are a first stage in development of dunes.

The fauna associated with beaches largely consists of intertidal invertebrates (Philip and Maclean 1985) and birds. The former tend to be dominated by burrowing oligochaete worms or bivalve molluscs, or are closely linked to specific plant communities or shelters available on the beach itself.

Sandy and shingle beaches offer favourable nesting conditions for birds such as *Haematopus ostralegus*, Ringed Plover *Charadrius hiaticula* and various gulls (Laridae) and terns (Sternidae), on bare shingle (Fuller 1982). Kentish Plover *Charadrius alexandrinus* is a typical breeder on sandy beaches, along with terns such as Sandwich Tern *Sterna sandvicensis*, Common Tern *S. hirundo*, and especially Little Tern *S. albifrons*.

In winter, shingle beaches have a very restricted bird fauna. Various passerines such as Skylark *Alauda arvensis*, Shore Lark *Eremophila alpestris*, Linnet *Carduelis cannabina* and Twite *C. flavirostris* feed on the seeds of the various maritime plants.

In contrast, sand beaches hold larger numbers, mostly waders, in winter. The most characteristic species along north-west European coasts is Sanderling *Calidris alba*, but other waders such as *Haematopus ostralegus*, *Charadrius hiaticula*, Grey Plover *Pluvialis squatarola*, Bar-tailed Godwit *Limosa lapponica* and *Arenaria interpres* are not uncommon. Such beaches are also frequently used as roosts by gulls.

Estuaries and intertidal flats

An estuary is 'a partially enclosed body of water, open to saline water from the sea and receiving fresh water from rivers, land run-off or seepage' (Bird 1984). Estuaries are usually formed in flat areas where rivers meet the sea, although in a variety of circumstances and forms (e.g. Davidson *et al.* 1991). A characteristic of all estuaries is that, although substantial amounts of sediment are deposited from nearby eroded shores, the majority of the finest particles are brought down by rivers. Distribution of this deposited material depends on the pattern of currents created by the river flow and the tides, which in turn is highly influenced by the shape of the estuary. Typically, the finest particles are deposited on the sides of the channel, where they form extensive mudflats. In addition, sand beaches or rocky shores often occur within them.

Estuaries and intertidal flats are a scarce habitat-type in Europe, covering about 9,300 km², or less than 0.1% of the land surface (Wolff 1989). Estuaries occur throughout Europe wherever there is an abundant source of suitable sedimentary material. However, the majority of large sites occur in northern and western Europe, particularly along the southern North Sea and Baltic coasts, with the largest concentration being in Britain (Figure 1; Davidson et al. 1991). The largest intertidal flats in Europe are in the Wadden Sea, covering some 4,290 km² (de Jong et al. 1993).

The typical physical features of estuaries that most influence their ecology are their tidal character, shelter from strong wave action, pronounced water layering and mixing, steep gradients of temperature and salinity, high levels of sediment suspension and transport, and their high level and rapid exchange of nutrients (Bird 1984). The salinity gradient is a most characteristic feature (Prater 1981). This and related physical gradients, and adaptations by organisms to them, result in considerable changes in the composition of plants and benthic species over short distances.

Typically, vegetation on estuarine flats is very sparse. The abundance and distribution of intertidal invertebrates largely depends on the sediment-type and amount of available nutrients. Although species diversity may be relatively low, sediment that is rich in organic content may support exceedingly high densities—molluscs and polychaete worms can exceed 10,000 individuals/m² (Davidson et al. 1991). Typical fish include flounder *Platichthys flesus*, grey mullet *Chelon labrosus* and three-spined stickleback *Gasterosteus aculeatus*.

Estuarine mudflats are very important for many wader populations in Europe in winter and during migration, and many species feed almost exclusively on intertidal benthic invertebrates at low tide. The most abundant species include Black-tailed Godwit

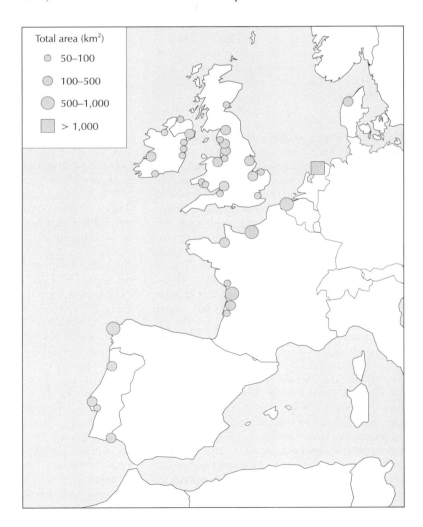

Figure 1. The distribution and extent of the major estuaries and intertidal flats of northwest Europe in 1989 (after Davidson et al. 1991). No data were available for Norway.

Total area (km²)

50–100
100–500
500–1,000
> 1,000

Limosa limosa, Curlew *Numenius arquata*, Redshank *Tringa totanus*, *Haematopus ostralegus*, Knot *Calidris canutus* and Dunlin *C. alpina*. Numbers of these birds can be huge; on the Atlantic coast of Europe, 38 estuaries regularly hold more than 20,000 waders each in winter, supporting 2.6 million birds in total (Smit and Piersma 1989).

Several species of ducks and geese (Anatidae) also winter on estuaries in large numbers. Several of these (e.g. Wigeon *Anas penelope* and Brent Goose *Branta bernicla*) rely considerably on belts of *Enteromorpha* algae and *Zostera* seagrass. Eider *Somateria mollissima* and gulls feed on estuarine mussel-beds, while the abundant fish populations are food for Cormorant *Phalacrocorax carbo*, Grey Heron *Ardea cinerea*, gulls and terns. In northern and western Europe, common seal *Phoca vitulina* is found in large numbers in this habitat-type.

Lagoons

Lagoons are shallow, virtually tideless water-bodies, partially isolated from adjacent sea by a rock or sedimentary barrier of sand or shingle. Sea water penetrates by a natural channel, by percolation or by occasional overtopping or breach of the barrier (Barnes 1980, Guélorget and Perthuisot 1992). Lagoons generally develop in sheltered areas where sediment is deposited onshore to form a spit or a barrier parallel to the coast, partially separating the area from the sea (see Guélorget and Perthuisot 1992).

In Europe, lagoons are particularly abundant around the weakly tidal shores of the Baltic, Mediterranean and Black Seas (Figure 2). They occur more rarely along the northern North Atlantic coast, where offshore deposits of shingle are to be found. They are absent from Scandinavia.

Within Europe, the location of the major coastal lagoons has been systematically documented only for some countries (Muus 1967, Comin and Parareda 1979, Sacchi 1979, Ardizzone *et al.* 1988, Barnes 1989, Perthuisot and Guélorget 1992). Chauvet (1988) maps the distribution of the principal Mediterranean lagoons.

The normally brackish water of lagoons supports submerged plants such as *Ruppia* and/or *Zostera*, and is often fringed by *Phragmites* reedbeds or by saltmarsh vegetation. Invertebrate and fish populations in lagoons are often at high density, and comprise mixtures of freshwater and marine/estuarine species capable of withstanding brackish water; characteristic invertebrates include worms (e.g. *Armandia*), shrimps (e.g. *Gammarus*) and snails (e.g. *Hydrobia*). Lagoons are consequently important feeding habitats for a wide range of birds. Breeding species include Greater Flamingo *Phoenicopterus ruber*, waders such as Avocet *Recurvirostra avosetta*

and Black-winged Stilt *Himantopus himantopus*, and various gulls and terns (e.g. Caspian Tern *Sterna caspia* and Mediterranean Gull *Larus melanocephalus*). Many species of waders, gulls and terns use coastal lagoons during winter and migration for foraging and roosting. Grebes (Podicipedidae), various ducks, and Coot *Fulica atra* are also typically present in winter, sometimes in many thousands.

Salinas

Salinas in Europe are man-made impoundments of sea water in flat coastal areas, designed to produce salt (by evaporation) for commerce and industry. A salina can vary in size from 2 ha (in Malta) to 120 km^2 (in southern France). Coastal salinas occur widely in Europe at middle to southern latitudes, but are most characteristic of the Mediterranean basin, where there are at least 53 currently operational in European countries, covering at least 520 km^2. Most are industrial operations, many others are abandoned or disused, and some 'primitive' and artisanal saltpans still exist (the former on rocky coasts) on some Mediterranean and Atlantic islands.

Salinas are probably the most inhospitable coastal habitat for plants and animals. The high salinity allows only the most salt-tolerant species to exist. Aquatic vegetation in permanently flooded ponds is dominated by submerged plants of *Ruppia*. Filamentous algae, such as *Enteromorpha*, occur in the lower salinity ponds. The majority of invertebrates occur in the 40–150 g/litre salinity range. Above 250 g/litre only the brine shrimp *Artemia* is present.

Mediterranean salinas are used as breeding habitats by several species of gulls and terns (in particular *Larus melanocephalus*, Slender-billed Gull *L. genei* and *Sterna albifrons*) and some waders, such as *Recurvirostra avosetta*, *Charadrius alexandrinus*, Little Ringed Plover *C. dubius* and *Himantopus himantopus*. It is also now one of the most important breeding habitats for *Phoenicopterus ruber*. Most of these species prefer to nest on islands within the salinas, rather than around the edge of the pans. In summer, small fish *Atherina* provide an important food for terns and gulls. During winter and on migration, numerous bird species, especially waders, congregate in significant numbers at Mediterranean salinas.

Sand-dunes

Sand-dunes are formed on coastlines where there is a suitable supply of sediment, usually within the particle size range 0.2–2.0 mm, often near estuaries and deltas. Sand-dunes in Europe can extend from a few metres to many kilometres inland, and can be up to 50 m high. The critical factor is the availability of a

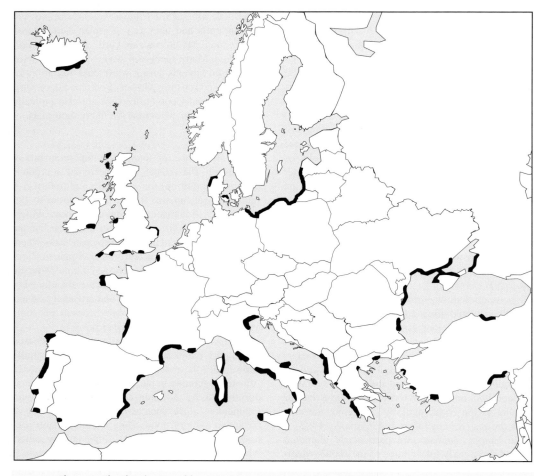

Figure 2. The distribution of former and present lagoonal coasts in Europe (after Barnes 1994).

sufficiently large intertidal sandflat which dries out at low tide and where sand grains can be blown onto the land. Vegetation, by trapping sand, plays an important role in the growth of this terrestrial habitat.

The availability of suitable sand, coupled with the effects of wind and vegetation, control the rate of growth of sand-dunes. Later, as plants gradually colonize the sand, a dune is subject to vegetational succession. There are embryo dunes located behind the strand-line, then mobile dunes, semi-fixed dunes, fixed dunes and dune-valley bottoms (slacks), finally grading into non-coastal, inland habitats.

Sand-dunes occur along most European coasts (Doody 1991). Particularly extensive sand-dunes occur in Iceland, on the eastern North Sea coast and the northern Baltic, on the west coasts of France and Portugal, on the south-west coast of Spain (e.g. Doñana National Park), and on the coastline of Greece and Turkey. However, large losses have occurred in the last 100 years (see 'Threats to the

habitat' (p. 104) and Table 6).

Sand-dunes have a highly diverse flora and invertebrate fauna (Davidson *et al.* 1991). The dune avifauna includes colonies of gulls and terns (e.g. Black-headed Gull *Larus ridibundus* and *Sterna sandvicensis*) nesting on open sand, ducks, such as *Somateria mollissima* and Shelduck *Tadorna tadorna* nesting in tall grass, and birds of prey, including harriers *Circus* and Short-eared Owl *Asio flammeus*, which also breed and hunt in the open, grassy areas. Dune-scrub may support high numbers of breeding passerines, such as Whitethroat *Sylvia communis* and *Carduelis cannabina* (e.g. Morgan 1978, Nairn and Whatmough 1978, Fuller 1982). Several species of mammals including rabbit *Oryctolagus cuniculus*, red fox *Vulpes vulpes*, bats (Chiroptera), voles and mice (Muridae), and shrews (Soricidae) occur. Reptiles and amphibians inhabiting sand-dunes include lizards *Lacerta*, tortoises *Testudo graeca* and *T. hermanni* and toads *Bufo*.

Saltmarshes

Saltmarshes can be defined as 'stands of open or closed herbaceous, salt-tolerant vegetation which develop in the intertidal zone on generally muddy substrates' (Long and Mason 1983). Saltmarshes typically develop on flat and muddy tidal plains in sheltered locations where abundant sediment is deposited and wave action restricted. These conditions favour establishment of pioneer salt-tolerant plants, which then facilitate further sedimentation by reducing wave action. As the mud surface rises, further plants colonize and the surface gradually becomes fully vegetated, except for drainage channels and isolated depressions. Completion of the process can take a few years or centuries, depending on the supply of sediment and the degree of protection that the site is afforded (Long and Mason 1983).

Saltmarshes are widely distributed around the coastline of Europe, from the Arctic to the Mediterranean (Dijkema 1984), with the highest concentrations along the northern Baltic, North Sea and south-west Atlantic coasts. They are particularly associated with the major estuaries of north-west Europe (Davidson *et al.* 1991).

The dominant factors determining the composition and distribution of saltmarsh flora are the degree of tolerance of each plant species to seawater inundation and saline sediment. As a result, the vegetation is strongly zoned (Long and Mason 1983). The lowest pioneer zone consists of the earliest colonizing plants, such as *Spartina*, *Salicornia* and *Puccinellia maritima*, which form simple communities of monospecific stands. The mid-zone is typically dominated by the shrub *Halimione portulacoides* and grass *Festuca rubra*, particularly on grazed marshes. The upper marsh has the richest plant communities. Saltmarshes present a difficult environment for colonization by invertebrates due to changes in salinity and periodic tidal immersion. The invertebrate fauna is a mixture of marine, freshwater and terrestrial species.

Breeding birds occur on the upper levels of unenclosed saltmarshes across Europe. These include ducks, waders, gulls and passerines. For example, in Britain, the five most frequent species are (in descending order) *Alauda arvensis*, *Tringa totanus*, Meadow Pipit *Anthus pratensis*, *Tadorna tadorna* and *Haematopus ostralegus* (Fuller 1982). *Larus ridibundus* and *Sterna hirundo* also occur in high numbers but less frequently. In southern Europe, other species may include Stone Curlew *Burhinus oedicnemus*, *Himantopus himantopus* and Short-toed Lark *Calandrella brachydactyla*. During the low-tide period, the majority of waders and waterfowl use estuarine mudflats as foraging areas and only a few species of wader (e.g. *Tringa totanus* and Snipe *Gallinago gallinago*) forage in saltmarshes. The majority of these birds, however, use the saltmarshes for roosting at high tide, or as supplementary feeding areas. Saltmarshes are also important foraging areas in winter for grazing ducks (e.g. *Anas penelope*) and geese (White-fronted Goose *Anser albifrons* and Barnacle Goose *Branta leucopsis*).

Deltas

Deltas are formed at larger river estuaries where the sea has a small tidal range and little wave action. In such circumstances, rivers bring suspended sediment into the sea faster than the waves carry it away. The single river channel soon becomes insufficient to transport the sediment load and subsequent blockage breaks into multiple channels, forming a delta (Bird 1984). Deltas are frequently large and often comprise a wide variety of coastal habitat-types (i.e. sand beaches, lagoons, sand-dunes, saltmarshes, etc.), as well as freshwater river channels, oxbow lakes and extensive emergent vegetation. In this strategy, references to 'delta habitats' will imply only these unique latter habitat components.

The largest concentration of deltas occurs along the Mediterranean and Black Sea coasts, although they also occur in northern Europe, especially on the Baltic shore (Table 2). The Volga delta on the north coast of the Caspian Sea is, however, the largest European delta, although not treated in this strategy (see p. 17).

Table 2. The major European deltas (Grimmett and Jones 1989).

Delta	Country	Sea	Area (km^2)
Wisla	Poland	Baltic Sea	1,700
Rhône	France	Mediterranean Sea	760
Po	Italy	Mediterranean Sea	250
Ebro	Spain	Mediterranean Sea	350
Menderes	Turkey	Mediterranean Sea	130
Göksu	Turkey	Mediterranean Sea	130
Çukurova (Ceyhan, Seyhan, Tarsus)	Turkey	Mediterranean Sea	625
Kizilirmak	Turkey	Black Sea	500
Danube	Romania/Ukraine	Black Sea	4,420
Volga	Russia	Caspian Sea	6,525

The high diversity of coastal habitat-types that is usually present means that deltas generally support an exceptional diversity of flora and fauna, especially in the more southerly latitudes of Europe. Deltas thus provide suitable habitats for a very wide variety of breeding, wintering and migrating waterbirds, including herons and egrets (Ardeidae), duck, waders, gulls and terns, as described for the various coastal habitats above.

Values, roles and uses

The use of coastal habitats by humans has a long history and can be divided into the following three main categories.

Direct use of the land for human purposes

Towns and cities have developed particularly along coasts and have enforced severe pressures upon them ever since. The building of docks, ports, housing, marinas, roads, car-parks, bridges, tunnels, barrages, leisure complexes, power stations, oil and gas installations and sea defences has been a consequence of these early developments. Clearly, such drastic uses of the land usually result in the loss of the original habitat.

Exploitation of natural resources

The most common ways by which natural resources are exploited are through sand-, gravel- and salt-extraction; fishing and aquaculture; seaweed-harvesting; and hunting. Of these, aquaculture is one of the most recent practices and is a rapidly expanding activity in lagoons and salinas, especially in the Mediterranean region. The brackish water of some coastal habitats is used for irrigation and for industrial and domestic supply. Vegetated habitats, particularly sand-dunes and saltmarshes, are used as pasture for grazing animals.

Intrinsic value

It is widely acknowledged that coastal habitats are extremely valuable for their natural beauty alone. They are attractive for their scenery and they provide public space for relaxation and for the pursuit of recreational activities, many of which are dependent upon the special geomorphology and the associated flora and fauna of these habitats. Coasts are also important for scientific research and educational activities.

Political and socio-economic factors affecting the habitat

Political and economic changes, alterations in the policies of land-use and environmental planning, population growth and increases in tourism all have an impact on the coastal environment.

Urbanization and associated infrastructural development play a major role in the destabilization of coastal ecosystems, and demographic studies predict dramatic increases in the coastal population size in some countries in the next few decades. For example, the population of the Mediterranean basin is predicted to rise from 350 million in 1985 to 530–570 million by 2025 (Stanners and Bourdeau 1995). Furthermore, many people prefer to live on the coast, thereby exacerbating urban expansion in these areas. Expansion of human activities along the coast is expected to increase the need for agriculture, industry, transport and energy developments (see 'Threats to the habitat', p. 104). This will occur especially in countries where population growth-rate is high.

Pressures on coastal habitats will also worsen through development of intensive agriculture, which requires large imputs of fresh water, agrichemicals and energy, and which produces large quantities of water-borne pollutants. Although the area of land under cultivation in northern Mediterranean countries has been stable or has decreased in recent decades, the proportion of irrigated cultivation is still expected to increase sharply in the future in southern and eastern countries.

In addition to the expansion of coastal urban populations and accompanying activities, growing international and domestic tourism will have a further impact on coasts. The pressures will be greatest in the Mediterranean basin, where it has been forecast that there will be 380 million tourist visitors by 2025 if overall economic growth is weak, and 760 million if it is strong (Stanners and Bourdeau 1995). Almost half of these tourists will be based along the coastline.

PRIORITY BIRDS

Coastal habitats are of very great importance for a large number of bird species, especially waterbirds. There are no less than 75 priority bird species which regularly use this type of habitat (Table 3), of which 52 have an Unfavourable Conservation Status in Europe (Table 4).

Priority A species number 10, an exceptionally high total, of which five are globally threatened (Collar *et al.* 1994): Dalmatian Pelican *Pelecanus crispus*, Marbled Teal *Marmaronetta angustirostris*, Ferruginous Duck *Aythya nyroca*, Slender-billed Curlew *Numenius tenuirostris* and Audouin's Gull *Larus audouinii*. The European populations of all five depend heavily on Mediterranean coastal wetlands during the breeding, migration or winter seasons, illustrating the exceptional importance of this regional habitat-type for bird conservation in Europe (Heredia *et al.* 1996).

Table 3. The priority bird species of coastal habitats in Europe.

This prioritization identifies Species of European Conservation Concern (SPECs) (see p. 17) for which the habitat is of particular importance for survival, as well as other species which depend very strongly on the habitat. It focuses on the SPEC category (which takes into account global importance and overall conservation status of the European population) and the percentage of the European population (breeding population, unless otherwise stated) that uses the habitat at any stage of the annual cycle. It *does not* take into account the threats to the species within the habitat, and therefore should not be considered as an indication of priorities for action. Indications of priorities for action for each species are given in Appendix 4 (p. 401), where the severity of habitat-specific threats is taken into account.

	SPEC category	European Threat Status	Habitat importance
Priority A			
Dalmatian Pelican *Pelecanus crispus*	1	V	●
Spoonbill *Platalea leucorodia*	2	E	■
Barnacle Goose *Branta leucopsis*	4/2 [w]	L [w]	■
Marbled Teal *Marmaronetta angustirostris*	1	E	●
Ferruginous Duck *Aythya nyroca*	1	V	●
Eleonora's Falcon *Falco eleonorae*	2	R	■
Slender-billed Curlew *Numenius tenuirostris*	1	—[1]	●
Redshank *Tringa totanus*	2	D	■
Audouin's Gull *Larus audouinii*	1	L	■
Sandwich Tern *Sterna sandvicensis*	2	D	■
Priority B			
Pygmy Cormorant *Phalacrocorax pygmeus*	2	V	●
White Pelican *Pelecanus onocrotalus*	3	R	■
Greater Flamingo *Phoenicopterus ruber*	3	L	■
Brent Goose *Branta bernicla*	3	V	■
Spanish Imperial Eagle *Aquila adalberti*	1	E	•
Purple Gallinule *Porphyrio porphyrio*	3	R	■
Avocet *Recurvirostra avosetta*	4/3 [w]	L [w]	■
Kentish Plover *Charadrius alexandrinus*	3	D	■
Spur-winged Plover *Hoplopterus spinosus*	3	(E)	■
Knot *Calidris canutus*	3 [w]	L [w]	■
Dunlin *Calidris alpina*	3 [w]	V [w]	■
Black-tailed Godwit *Limosa limosa*	2	V	●
Bar-tailed Godwit *Limosa lapponica*	3 [w]	L [w]	■
Curlew *Numenius arquata*	3 [w]	D [w]	■
Common Gull *Larus canus*	2	D	●
Caspian Tern *Sterna caspia*	3	(E)	■
Roseate Tern *Sterna dougallii*	3	E	■
Little Tern *Sterna albifrons*	3	D	■
Aquatic Warbler *Acrocephalus paludicola*	1	E	●
Priority C			
Little Bittern *Ixobrychus minutus*	3	(V)	●
Squacco Heron *Ardeola ralloides*	3	V	●
Purple Heron *Ardea purpurea*	3	V	●
Glossy Ibis *Plegadis falcinellus*	3	D	●
Bewick's Swan *Cygnus columbianus*	3 [w]	L [w]	●
Ruddy Shelduck *Tadorna ferruginea*	3	V	●
Gadwall *Anas strepera*	3	V	●
Pintail *Anas acuta*	3	V	●
Red-crested Pochard *Netta rufina*	3	D	●
Scaup *Aythya marila*	3 [w]	L [w]	●
White-tailed Eagle *Haliaeetus albicilla*	3	R	●
Hen Harrier *Circus cyaneus*	3	V	●
Osprey *Pandion haliaetus*	3	R	●
Peregrine *Falco peregrinus*	3	R	●
Collared Pratincole *Glareola pratincola*	3	E	●
Black-winged Pratincole *Glareola nordmanni*	3	R	●
Purple Sandpiper *Calidris maritima*	4	(S)	■
Jack Snipe *Lymnocryptes minimus*	3 [w]	(V) [w]	●
Whimbrel *Numenius phaeopus*	4	(S)	■
Mediterranean Gull *Larus melanocephalus*	4	S	■
Gull-billed Tern *Gelochelidon nilotica*	3	(E)	●
Whiskered Tern *Chlidonias hybridus*	3	D	●
Black Tern *Chlidonias niger*	3	D	●

cont.

Table 3. (cont.)

	SPEC category	European Threat Status	Habitat importance
Priority D			
Bittern *Botaurus stellaris*	3	(V)	•
Night Heron *Nycticorax nycticorax*	3	D	•
Whooper Swan *Cygnus cygnus*	4 [w]	S	•
Pink-footed Goose *Anser brachyrhynchus*	4	S	•
Shelduck *Tadorna tadorna*	—	S	■
Wigeon *Anas penelope*	—	S	■
Baillon's Crake *Porzana pusilla*	3	R	•
Oystercatcher *Haematopus ostralegus*	—	S	■
Ringed Plover *Charadrius hiaticula*	—	S	■
Grey Plover *Pluvialis squatarola*	—	(S)	■
Sanderling *Calidris alba*	—	S	■
Curlew Sandpiper *Calidris ferruginea*	—	—[1]	■
Greenshank *Tringa nebularia*	—	S	■
Turnstone *Arenaria interpres*	—	S	■
Great Black-headed Gull *Larus ichthyaetus*	—	S	■
Lesser Black-backed Gull *Larus fuscus*	4	S	•
Arctic Tern *Sterna paradisaea*	—	S	■
Short-eared Owl *Asio flammeus*	3	(V)	•
Kingfisher *Alcedo atthis*	3	D	•
Skylark *Alauda arvensis*	3	V	•
Tawny Pipit *Anthus campestris*	3	V	•
Rock Pipit *Anthus petrosus*	—	S	■
Chough *Pyrrhocorax pyrrhocorax*	3	V	•

SPEC category and **European Threat Status:** see Box 2 (p. 17) for definitions.

Habitat importance for each species is assessed in terms of the maximum percentage of the European population (breeding population unless otherwise stated) that uses the habitat at any stage of the annual cycle:

■ >75%   ● 10–75%   • <10%

[w] Assessment relates to the European non-breeding population.   [1] Only a passage migrant through Europe, thus no European Threat Status.

Coastal habitats are very important for a total of 35 priority bird species, since more than 75% of their European populations depend on these habitats at some stage in their annual cycle (Table 4). Estuaries hold the majority of European wintering populations of several priority wader species, including *Tringa totanus*, *Calidris canutus*, *C. alpina*, *Limosa lapponica* and *Numenius arquata*, and are also important for goose and duck species like *Branta bernicla*, *Tadorna tadorna* and *Anas penelope*. Coastal habitats are highly important for a number of gull and tern species for breeding, including *Sterna caspia*, Roseate Tern *Sterna dougallii* and Gull-billed Tern *Gelochelidon nilotica*, all of which are Endangered in Europe (Tucker and Heath 1994). The highly specialized *Phoenicopterus ruber* is almost exclusively coastal in Europe.

There are only a few birds of prey for which coastal habitats are important, notably *Falco eleonorae*, which is found solely along rocky coasts of the Mediterranean. The small Mediterranean population of Osprey *Pandion haliaetus* is also coastal, while river deltas are important for White-tailed Eagle *Haliaeetus albicilla*. Among passerine species, Aquatic Warbler *Acrocephalus paludicola* has the highest priority. A significant part of the popula-

Table 4. Numbers of priority bird species of coastal habitats in Europe, according to their SPEC category and their dependence on the habitat. Priority status of each species group is indicated by superscripts (see 'Introduction', p. 19, for definition). 'Habitat importance' is assessed in terms of the maximum percentage of each species' European population that uses the habitat at any stage of the annual cycle.

Habitat importance	SPEC category					Total no. of species
	1	2	3	4	Non-SPEC	
<10%	2 [B]	0 [C]	8 [D]	—	—	10
10–75%	4 [A]	3 [B]	20 [C]	3 [D]	—	30
>75%	1 [A]	5 [A]	14 [B]	3 [C]	12 [D]	35
Total	7	8	42	6	12	75

tion of this globally threatened species can be found in beds of sedges (*Carex*) and reeds (*Phragmites*) of estuaries and coastal lagoons during migration, and it also breeds in stunted reedbeds in some saltmarshes on the Baltic coast of Poland and Germany (Heredia 1996c).

The number of priority species breeding in each European country is given in Appendix 2 (p. 344). South European countries with particularly long coastlines—Russia, Ukraine, Turkey, Spain and France—are the only countries with breeding populations of the majority of priority coastal birds, with Russia having a particularly diverse and important avifauna of priority coastal species. It should be stressed again, though, that coastal habitats tend to support the highest numbers and diversity of priority birds outside the breeding season (see Figure 3).

HABITAT NEEDS OF PRIORITY BIRDS

The most fundamental habitat requirement is the maintenance of the physical and biological processes which underpin the ecology of these habitats. Some of the most important are the geomorphological processes which create and maintain sedimentary structures, such as lagoons, deltas, beaches, mudflats and sand-dunes. In particular, these are dependent on sediment supply, transport and deposition, so disruption of these processes can have profound effects on the habitats. Another critical habitat requirement is the maintenance of 'natural' hydrological regimes and water quality. Man-made disturbances of salinity, nutrient levels, water-level, wave exposure, sedimentation, turbidity and current regime can cause detrimental changes to the ecology of the whole ecosystem, thereby damaging flora and fauna.

At least 35 priority species (nearly 50% of the total) are highly dependent on coastal habitats outside the breeding season, as wintering quarters or as staging posts on migration (Appendix 3, p. 354). Many of these species breed on the tundra of northern Europe and western Siberia (see 'Tundra, Mires and Moorland', p. 159). Most of these species therefore require a network of suitable coastal wetlands as stopover sites in order to undertake their migrations every year (Smit and Piersma 1989). Loss or damage to these may seriously reduce the chances of waterbirds reaching their nesting grounds or breeding successfully.

Coastal lagoons support the most diverse avifauna of any of the coastal habitat-types described here, but estuaries, deltas and salinas are also very attractive to a wide variety of priority birds (Figure 3). Lagoons are especially important for species with a southerly distribution, including all of the globally threatened species (Heredia *et al.* 1996). The mosaic of different habitat-types found in deltas offers particularly good feeding and breeding opportunities for a wide range of species with diverse habitat requirements. Estuaries support huge numbers of wintering waders, ducks and geese by exposing large areas of intertidal mudflats with a very rich supply of invertebrates. Salinas provide suitable

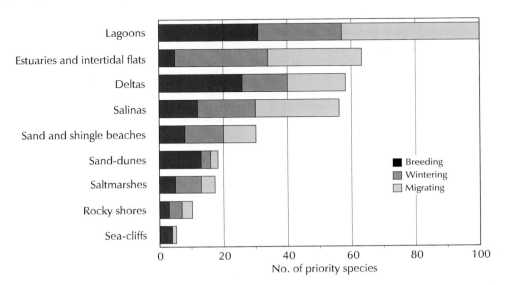

Figure 3. The number of priority bird species supported by different coastal habitat-types in Europe, while breeding, migrating or wintering (data from Habitat Working Group, see Appendix 3, p. 354). Data for sea-cliffs exclude priority breeding seabirds, which are covered under 'Marine Habitats'.

breeding sites if there are predator-free islands, and sand-dunes often also support good breeding populations of priority coastal birds, especially if they are isolated (e.g. on islands), or otherwise well protected from disturbance.

Water salinity is a feature which strongly determines the distribution of priority waterbirds in coastal habitats. There are 12 priority species tolerant of highly saline conditions, the best example being *Phoenicopterus ruber*. Other species, however, require fresh water for feeding, including most of the herons. Coastal salinity levels can change quite rapidly from highly saline to totally fresh over relatively short distances, which partly accounts for the high avian diversity of coasts.

Wetlands with shallow to moderate water-depths are preferred by most of the priority species. A large tidal range is important for 11 priority species, including *Branta bernicla*, *Anas penelope* and several waders. Clear water is an important requirement for species such as herons and terns which hunt for fish mainly by sight. The intertidal sediment-type preferred by most priority birds is mud and silt, which is richest in organic matter and in invertebrates, the main diet of several species of waders, ducks and geese. At least in Britain, *Recurvirostra avosetta* favours lagoon-like habitats, *Limosa lapponica*, *Calidris alba* and *C. canutus* prefer sandy estuaries, *C. alpina* and *Pluvialis squatarola* prefer muddy estuaries, and *C. maritima* and *Arenaria interpres* favour rocky shores (Hill *et al.* 1993).

Vegetation cover is an important habitat feature, especially at lagoons, deltas, sand-dunes and saltmarshes, but priority birds differ widely in their habitat selection in this respect. Bare ground is essential for nesting for 10 species, including *Phoenicopterus ruber* and waders. Most gull and tern species require some low vegetation cover during breeding. High cover of tall vegetation is essential during the breeding season for 12 species, including the two *Pelecanus* species, most of the herons and ducks, and Purple Gallinule *Porphyrio porphyrio*. Human intervention is sometimes required to conserve particular successional stages of vegetation in different coastal habitats, according to the needs of particular priority species. For example, maintenance of a particular structure, or of a variety of indigenous species, may require appropriate grazing pressure or eradication of invasive species (e.g. non-native cord-grass *Spartina anglica* on intertidal flats).

Food requirements of priority coastal species are also highly diverse. Nineteen species live mostly or largely on fish, including Pygmy Cormorant *Phalacrocorax pygmeus*, the two *Pelecanus* species, most of the herons, *Haliaeetus albicilla* and *Pandion haliaetus*. A similar number of species depend on terrestrial and benthic invertebrates, most being waders. Aquatic insects are also an important food source for 19 species, including herons, ducks, waders and gulls.

When migrating, birds need to accumulate fat and muscle protein for their flight and as reserves for the early stages of breeding. To achieve this they need abundant supplies of food, particularly in spring, and may need to feed unimpeded throughout periods of tidal exposure. Such constant feeding may also be essential during periods of hard weather in winter. Restriction of feeding opportunities by land-claim on upper parts of feeding grounds can jeopardize their ability to take sufficient reserves to breeding grounds to breed successfully, or even jeopardize their own survival (e.g. Davidson and Evans 1988).

Other significant landscape requirements include the need for nearby wet grasslands or other agricultural habitats (18 and 20 species respectively), which are especially important for geese, swans *Cygnus* and waders as feeding or breeding areas. Negative features or qualities of the landscape which are important in limiting the distribution and abundance of priority species along coasts include a high level of human disturbance (especially from hunting), high densities of overhead cables (which kill through collision and electrocution), and water pollution.

Further information on the habitat requirements of coastal birds is given in Hancock and Elliott (1978), Johnsgard (1978), Prater (1981) and Hill *et al.* (1993).

THREATS TO THE HABITAT

There has been no Europe-wide appraisal of the threats to coastal habitats. Smit *et al.* (1987) carried out a review, based on a previous evaluation by Wolff and Binsbergen (1985), of threats or potential threats to estuaries along the East Atlantic Flyway.

Table 5 therefore provides an assessment by the Habitat Working Group of the scale and effects of threats to all coastal habitats in Europe, and to their priority bird species. In addition, the Group predicted the impact of each threat on the European population of each priority species over the next 20 years (Appendix 4, p. 401). A summary of the results is presented in Figures 4 and 5, and the threats are discussed individually below in approximate order of importance (as shown in Figure 5).

Tourism and recreation

There are more tourist arrivals in Europe than anywhere else in the world: 280 million per year in 1990, or 60% of the world total (Stanners and Bourdeau 1995). Coastal areas are the most popular destina-

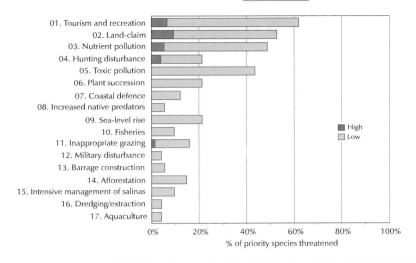

High impact
The species' population is likely to decline by more than 20% in the habitat in Europe within 20 years if current trends continue

Low impact
Only likely to have local effects, and the species' population is not likely to decline by more than 20% in the habitat in Europe within 20 years if current trends continue

Figure 4. The predicted impact on priority bird species of threats to coastal habitats, and to the species' habitat requirements, in Europe. Bars are subdivided to show the proportion of species threatened by low or high impacts to their populations (see Table 5 and Appendix 4, p. 401, for more details).

tions for tourism and general recreation within Europe, and the southern coasts are particularly affected, especially in the Mediterranean basin, which is the most popular tourist destination in the world, attracting some 100 million tourists per year.

To date, over two-thirds of the European Mediterranean and Iberian coast have been strongly affected by development for tourism, especially along the coasts of the Algarve (Portugal), the Spanish Costas, Cyprus and Turkey. Growth in tourist num-

bers has been massive, with overall numbers visiting the Mediterranean basin tripling between 1970 and 1990 (WTO 1993). In some countries growth has been even greater, with a four-fold increase in numbers in Turkey and six-fold in Greece. Numbers in the Mediterranean basin are predicted to continue increasing rapidly, reaching between 380 and 760 million arrivals per year by 2025, depending on global economic conditions (Stanners and Bourdeau 1995).

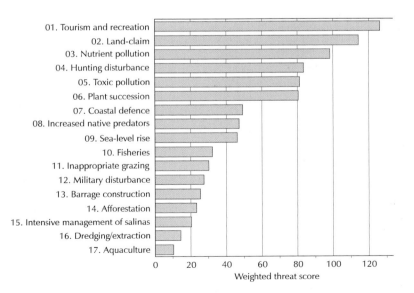

The scoring system takes into account:

• The European Threat Status of each species and the importance of the habitat for the survival of its European population (Priorities A–D); see Table 3.

• The predicted impact (Critical, High, Low), within the habitat, of each threat on the European population of each species; see Figure 4, and Appendix 4 (p. 401).

See Table 5 for more details on each threat.

Figure 5. The most important threats to coastal habitats, and to the habitat requirements of their priority bird species, in Europe.

Table 5. The most important threats to coastal habitats, and to the habitat requirements of their priority bird species, in Europe (compiled by the Habitat Working Group). Threats are listed in order of relative importance (see Figure 5).

Threat	Effect on birds or habitats	Scale of threat Habitat	Past	Future	Regions affected, and comments
1. Tourism and recreation	Disturbance and changes to vegetation from trampling; associated problems from building developments, infrastructure, pollution and increased water demand	Lg,Es, Sh,Ds, Sm	■	■	Widespread, particularly in the Mediterranean, and still increasing
2. Land-claim and coastal development	Loss and degradation of intertidal, shallow subtidal and terrestrial habitats	Lg,Es, Ds,Sl, Sm	■	●	Most countries
3. Nutrient pollution	Decreased water clarity; algal blooms and anoxic conditions; changes in vegetation structure; reduced food supply	Lg,Es, Sh,Ds, Sm	●	●	Widespread, but highest near river-mouths and industrial and urban centres
4. Hunting disturbance	Disturbance to breeding, roosting and feeding birds	Lg,Es, Sh,Ds, Sm	■	●	Widespread (no hunting within 500 m of coast in Portugal)
5. Toxic pollution	Direct mortality of species, sub-lethal effects, and indirect effects through loss of food resources	Lg,Es,	●	●	Mostly from rivers draining intensive agricultural regions and heavily industrialized areas
6. Plant succession (leading to loss of habitat)	Loss of open water; changes in vegetation composition and structure, including invasion of shrubs and trees, leading to loss of ground-nesting habitat	Lg,Ds, Sm	●	■	Widespread
7. Coastal defence	Loss of intertidal habitats, changes in hydrology, tides, sediments and salinity	Lg,Es, Sh,Ds, Sm	●	■	Widespread, especially England, Denmark, Germany and Netherlands
8. Increased native predators	Loss of eggs and young; disturbance of birds	Lg,Es, Sh,Ds, Sl,Sm	●	■	Particularly near urban and tourist areas
9. Sea-level rise	Loss of intertidal habitat, increased seawater input, erosion	Lg,Es, Sh	—	■	Everywhere
10. Fisheries	Decrease in food supply through declines in shellfish and fish stocks; changes in sediment, benthic vegetation and animal communities from bottom-trawling; disturbance of birds	Lg,Es,	■	■	
11. Inappropriate grazing	Changes in vegetation structure (initially invasion of grasses, then excessive erosion); trampling of eggs and young by livestock	Sh,Ds, Sm	•	•	UK and Republic of Ireland; mostly local elsewhere
12. Military disturbance	Changes to vegetation structure from trampling and disturbance to animal communities; associated developments (e.g. roads)	Lg,Es, Sh,Ds, Sm	●	•	Local
13. Barrage construction	Changes in salinity, sediment-type and tidal regime (normally reducing exposure of intertidal area), and increased erosion	Lg,Es	●	•	Confined to areas in north with large tidal ranges
14. Afforestation	Loss of natural and semi-natural vegetation; habitat fragmentation; changes in hydrology	Lg,Es, Ds	•	•	
15. Intensive management of salinas	Loss of nesting habitat, reduced feeding opportunities, high disturbance	Sl	•	•	Mediterranean only

cont.

Table 5. (cont.)

Threat	Effect on birds or habitats	Scale of threat Habitat	Past	Future	Regions affected, and comments
16. Dredging, and extraction of aggregate	Increases turbidity of water; destroys benthic community locally; can lead to nutrient pollution	Lg,Es, Sh,Ds	•	•	Widespread, particularly Mediterranean
17. Aquaculture	Decreased water clarity; nutrient pollution; disturbance and sediment change; conversion of natural wetlands to fish-ponds	Lg,Es, Sl	•	•	Widespread and increasing; local in Greece

Scale of threat

Cl	Cliffs	Ds	Sand-dunes	Past	Past 20 years
Rs	Rocky shores	Sl	Salinas	Future	Next 20 years
Lg	Lagoons	Sm	Saltmarsh	■	Widespread (affects >10% of habitat)
Es	Estuaries and intertidal flats	Dl	Deltas	●	Regional (affects 1–10% of habitat)
Sh	Sand and shingle beaches			•	Local (affects <1% of habitat)

In northern Europe, parts of the French and Belgian seaboard and traditional resorts in Germany, the Netherlands, UK and along the Baltic Sea coasts also attract large numbers of visitors. Tourist numbers have increased in some of these areas in recent years, including Denmark and the Wadden Sea (de Jong *et al.* 1993, Stanners and Bourdeau 1995). The Black Sea has traditionally been a major destinination for tourists from central and eastern Europe, but declining environmental conditions in the area are having a detrimental impact on the tourism industry.

In terms of direct habitat loss, about 4,000 km² of coastal habitat have already been built on, or used for, tourist accommodation and related infrastructure in Mediterranean countries, and this figure could reach 8,000 km² by 2025, inevitably causing further significant reductions in populations of fauna and flora, and in species diversity (Stanners and Bourdeau 1995). For example, the massive loss of coastal sand-dunes in Europe this century has been strongly driven by tourist and recreational development (Table 6). Tourism also results in increased demand for water and other resources, and increased environmental damage from pollution. The outdoor activi-

Table 6. Estimated area of sand-dune loss over the last 100 years in some European countries (P. Doody pers. comm.).

	Total area (km²)	Loss (%)		Causes
Albania	>20	?		Mostly reversion to scrub and woodland
Belgium	50	46		Urbanization, tourist development
Bulgaria	?	?		Tourist development
Denmark	800	35		Afforestation (300 km²), recreational development
Finland	13	?		Recreational development, reversion to scrub and woodland
France	2,500	40	(Atlantic coast)	Afforestation, tourist development
		75	(Med. coast)	Afforestation, tourist development
Germany	<120	15–20		Urbanization, recreational development
Greece	<200	40–50		Tourist development, urbanization
Iceland	1,200	?		Erosion
Rep. of Ireland	143	40–60		Recreational development, erosion
Italy	<400	80		Tourist development, urbanization, afforestation
Netherlands	450	32		Afforestation, tourist development, urbanization, water abstraction
Norway	>20	>40		Development, reversion to scrub and woodland
Poland	350	>50		Afforestation, recreational development
Portugal	1,000	45–50		Afforestation, agriculture, tourist development
Romania	?	?		Afforestation, erosion
Spain	700	30	(Atlantic coast)	Afforestation, tourist and recreational developments
		75	(Med. coast)	Afforestation, tourist and recreational developments
Sweden	>20	?		Afforestation
Turkey	360	>30		Afforestation, tourist development
Yugoslavia	?	?		Tourist development
United Kingdom [1]	560	30–40		Afforestation, urbanization, recreational development

[1] Excluding Northern Ireland.

ties of such large numbers of people have a direct impact on bird populations through disturbance. As tourism, and therefore disturbance, occurs predominantly during the spring and summer months, breeding birds of coastal habitats are among the most affected. Beach-nesting waders (e.g. *Charadrius alexandrinus*) and terns (in particular *Sterna albifrons*) are highly prone to disturbance, while their eggs and chicks are at risk from trampling.

Habitat degradation also occurs as a result of intensive trampling and the increasingly widespread use of off-road vehicles, particularly on sand and shingle-beaches and sand-dunes. Consequences include the destruction of the characteristic sparse and low vegetation, initiation of erosion and lowering of ridge heights, thus reducing nesting opportunities and making nests and vegetation more vulnerable to flooding by high spring tides (Frid and Evans 1995).

In total, the loss and degradation of coastal habitats engendered by tourism and recreation in Europe is predicted to affect 47 of the 75 priority species, more than any other threat. The European populations of most species are predicted to suffer only low impacts, but five species are likely to suffer high impacts, and this threat is thus considered to be of the highest overall importance amongst those evaluated here.

Land-claim and coastal development

Land-claim causes loss and degradation of coastal habitats which is often irreversible in practical terms. Hundreds of square kilometres of intertidal flats have been lost to agriculture and coastal-defence engineering in Europe since Roman times (e.g. Davidson *et al*. 1991), and land-claim accounts for much of the loss of coastal grazing marshes in estuaries. More recently, losses to agriculture have declined in scale, but land has been claimed for a wider variety of other purposes, especially industrial, urban and recreational developments. Since these more recent changes involve greater investment in infrastructure, they are less reversible than land-claim for agriculture or coastal defence. Land-claim can also degrade or destroy adjacent coastal and subtidal habitats indirectly, through altering tidal currents and consequent geomorphological processes, such as rates of erosion or the movement and deposition of sediments. These geomorphological effects may have profound impacts on coastal habitats in the long term, but detailed discussion or prediction of these is beyond the scope of this strategy. Some further information is given in Davidson *et al*. (1991).

Lowland areas of north-west Europe have been most affected by this threat. For example, more than 350 km² of the Wadden Sea was claimed for human use between 1963 and 1982 . The 'Delta Region' of the Rhine–Maas estuary in the Netherlands has lost 160 km² of coastal habitat to agricultural use, and 350 km² of the Zuiderzee in the Netherlands was similarly claimed by building a 34-km-long barrier across its mouth (Barnes and Hughes 1982). In Britain, 88% of estuaries have lost intertidal habitat to agricultural land-claim in the past (Davidson *et al*. 1991), and about 40 km² of saltmarshes in Britain, classified as Sites of Special Scientific Interest, have been lost to land-claim since 1950 (Long and Mason 1983).

Although current agricultural surpluses mean that habitat loss for farmland is no longer likely to pose a major threat in EU countries, further land-claim for other purposes is likely to continue. For example, in Britain during 1989 there were 135 land-claim proposals which targeted 55 estuaries, nearly all for non-agricultural purposes such as housing and car-parks, marinas, barrage schemes, transport, rubbish tips and ports (Davidson *et al*. 1991). It was thought that 50% of the internationally important estuaries in Britain would suffer further habitat losses due to land-claim, although this percentage may now be lower. Land-claim continues in the Wadden Sea, where the Küstenschutz Leybucht project alone will result in 7.4 km² of saltmarsh and mudflats being destroyed by the end of this century.

Similar habitat loss has occurred elsewhere in Europe, but is less well documented—many examples of serious damage to individual coastal sites are described in Grimmett and Jones (1989). In the Mediterranean, extensive areas of deltas (e.g. the Evros delta of Greece) and lagoons (e.g. Caorle lagoon in Italy) have been lost in recent years due to drainage for agricultural purposes. Such land-claim in Europe is likely to have a serious, population-level impact on the globally threatened *Marmaronetta angustirostris* (Green 1996). One of the major threats to the network of salinas in the Mediterranean region, many of which are now abandoned or disused, is land-claim for industrial and urban development, recreational sites, sports centres, camp sites and refuse-dumps.

Densities of feeding birds in important coastal habitats such as intertidal flats can sometimes reach their 'carrying capacity' levels, i.e. upper limits which cannot be exceeded (e.g. Zwarts 1974, Goss-Custard 1977a,b, Zwarts and Drent 1981, Meire and Kuijken 1984, Goss-Custard and Durrell 1990). The loss of such feeding grounds will force some birds to feed in less preferred areas or to leave the area altogether.

There appears seldom to be a direct relationship between the size of the intertidal area lost and the extent of impact on the bird populations dependent upon that area (Evans 1981, Evans and Pienkowski 1983, Laursen *et al.* 1984, Davidson and Evans 1986, Lambeck *et al.* 1989, Meire 1990). The populations of some intertidal species, particularly those that need to feed over especially long periods of the tidal cycle to achieve their required food intake, or those that depended most on the claimed areas, generally decline in greater proportion than would be expected from the size of the area lost alone. The full impact of such habitat loss may only become apparent during periods of severe weather, when food requirements are highest (e.g. Lambeck 1990). Hidden impacts may also occur through mortality during migration or through reduced reproductive success on breeding grounds.

Overall, this threat is thought to affect 39 of the 75 priority species, with most impacts being low, although seven species are predicted to suffer high population impacts. Land-claim and coastal development are clearly among the most important hazards to coastal habitats and their priority species in Europe.

Nutrient pollution

The rapid and widespread intensification of agriculture in Europe over the last 30 years has been associated with a massive increase in the use of inorganic fertilizers (see 'Agricultural and Grassland Habitats', p. 267). Considerable amounts of nitrogen and phosphorus in the sea originate from the riverine run-off from intensive livestock and arable farming. Nutrient pollution from sewage disposal is also a major problem, though not as great as riverine discharge of nutrients from agriculture. Considering that urban expansion is forecast to continue, pollutants directly discharged into the sea are likely to reach higher concentrations unless there are stricter controls. In many countries only primary treatment is given to sewage before it is discharged (MacGarvin 1990).

This enrichment, whilst causing general ecological deterioration of an estuary, can, however, increase the biomass of some invertebrates that provide food for estuarine waterfowl. This may permit the numbers of birds using the area to increase (e.g. van Impe 1985), although such heavy pollution can result in an overall reduction in the diversity of the estuarine system. Conversely, cutbacks in organic pollutant levels can increase the range of plants and animals able to tolerate the estuarine conditions, but may alter the overall species composition as well as reduce the size of the wintering waterfowl assemblages using the area (e.g. Campbell 1978, 1984, Furness *et al.* 1986).

Excessive amounts of nutrients entering lagoons, deltas, estuaries and coastal seas may result in blooms of algae and phytoplankton (e.g. in the 1980s, blooms of the alga *Phaeocystis* plagued beaches in northeast Europe), followed by mass release of toxins ('red tides') or oxygen depletion, which results in the death or contamination of fish and other organisms requiring aerobic conditions (Essink 1984; see 'Marine Habitats', p. 59). Anoxic conditions on intertidal flats may also make remaining invertebrates unpalatable for birds.

Overall it is estimated that 36 priority species will suffer population impacts, and that these will be high for four species. Nutrient pollution is clearly a very important threat to coastal habitats.

Hunting disturbance

The direct impact of hunting mortality on bird populations is not covered by this book (see 'Introduction', p. 15). The indirect effects of hunting on habitat quality are dealt with here.

The most widespread indirect effect of hunting is probably disturbance to nesting birds in summer and to feeding and roosting birds in winter. Hunting (and other forms of disturbance) can cause temporary disruption of the normal activities of birds (e.g. feeding and roosting), alter their daily rhythms, increase escape-flight distances, and displace birds from preferred feeding and roosting habitats (Meltofte 1982, Bell and Owen 1990, Madsen and Fox 1995, Madsen *et al.* 1995). Thus birds may starve as a result of reduced food intake and/or increased energy expenditure. Disturbance may also disrupt pair-bonds and other social structures, and during nesting periods may expose eggs and chicks to increased risk of predation or temperature stress.

Hunting is not normally legal in Europe during the breeding season, but illegal hunting is commonplace in parts of the Mediterranean (Woldhek 1980, Magnin 1991). Hunting is likely to be the most common cause of disturbance to birds in autumn and winter, when coastal habitats hold large and important populations of waterbirds (Madsen and Fox 1995). Restrictions on hunting in winter are routinely imposed during long periods of freezing conditions, but the effects of sustained disturbance in less extreme conditions may also be detrimental.

Sensitivity to disturbance varies among species. Madsen and Pihl (1993) analysed the sensitivity of coastal wildfowl to disturbance, and concluded that the most sensitive species were Bewick's Swan *Cygnus columbianus* and Whooper Swan *C. cygnus*, and all geese and ducks other than seaducks. However, these are only predictions, and while the effects of disturbance have been demonstrated locally, impacts at the population level have yet to be proved or quantified, and further research is required (Hill *et al.* 1997). Nevertheless, it is estimated that 31 priority species will suffer low impacts, and the overall importance of this threat is therefore considered moderately high.

Toxic pollution

Many coastal habitats are at risk from pollution due to the heavy concentration of hunting and of industrial activities in the coastal zone. The former can lead to lead pollution through spent gunshot, and estuaries and related intertidal habitats are often close to oil refineries and industrial centres, and hence frequently polluted by effluents or spillages of toxic chemicals. Oil is also often washed ashore from accidental or illegal spills or discharges from offshore shipping (see 'Marine Habitats', p. 59, for more details).

Lead poisoning in waterfowl occurs through the ingestion of spent lead gunshot that has fallen into wetlands where waterfowl feed. Waterfowl ingest shot along with grit and food. Shot may be retained in the muscular gizzard, eroded by the grinding action of gizzard grit, dissolved by stomach acids and absorbed into the blood stream.

Exposure is greatest where wetlands are heavily hunted over and where soils or sediments are fairly firm and contain little grit. In the Camargue (Rhône delta, southern France), shot densities of up to 2 million/ha have been recorded in the sediment (Pain 1991). More than 50% of the gizzards of some duck species were found to contain ingested shot in studies from the Camargue and elsewhere in the Mediterranean (Pain 1990, Pain and Handrinos 1990). Lead poisoning is a global problem. In the USA, before a nationwide ban on the use of lead shot for waterfowl hunting, an estimated 1.6–2.4 million waterfowl died annually from lead poisoning (Bellrose 1958, USFWS 1986).

Although waterfowl may be subject to greater exposure to lead shot than other species, the problem is not restricted to them. Raptors, especially scavengers that feed on wounded or unretrieved game, are particularly susceptible to shot ingestion and subsequent poisoning. For example, a high proportion of Marsh Harrier Circus aeruginosus in coastal areas of Charente-Maritime (France) have been found to have elevated blood-lead concentrations through shot ingestion (Pain et al. 1994, Pain and Bavoux in press).

The solution to lead poisoning lies in the replacement of lead shot with a non-toxic alternative. Although several European countries have implemented or announced legislative bans on the use of lead shot for hunting waterbirds or hunting over wetlands (Fawcett and van Vessem 1995), to date no action has been taken in the south European countries where some of the highest levels of exposure to lead have been recorded.

In addition, agricultural run-off, industrial effluents and dredging spoils are sources of toxic substances, such as heavy metals and pesticide residues. Continuous dredging in many ports and large estuaries around the North Sea region releases high concentrations of heavy metals. Over the years, heavy metals and some pesticide residues can accumulate in the sediments of these areas, where they are absorbed by plants and invertebrates, which are fed on by fish and birds, and where they accumulate in the tissues or organs (e.g. NERC 1983, Evans et al. 1987). Direct mortality of birds from toxic chemicals appears to be very infrequent and localized, although good data are lacking. One instance of high mortality attributed to heavy metal pollution was the result of poisoning by alkyl-lead compounds in the Mersey estuary, UK (Bull et al. 1983). Indirect effects may, however, be substantial, e.g. in heavily polluted estuaries where invertebrate prey are killed off by toxins or made unpalatable (Dickson et al. 1987).

The priority species most likely to be affected by toxic chemicals are those that feed mainly on larger animal prey, especially estuarine fish-eating species. Waders feeding on highly polluted mudflats may also be affected. Overall, this threat has a moderately high importance amongst those considered here, affecting 32 priority species, although all impacts are predicted to be low.

Plant succession

Succession is the natural process by which different plant species successively colonize an area, each one changing the habitat and enabling its replacement by another type (see p. 97 with respect to sand-dunes for example). Thus wetland habitats are colonized by emergent vegetation, then invaded by scrub, which typically grows into mature woodland (see Box 1, p. 131). Under natural circumstances, this ecological process would not be a threat as each stage of the succession would be transformed in some places but created elsewhere.

However, such cycles no longer occur over most of today's man-modified landscape, and habitats such as lagoons and sand-dunes, in particular, are not being created. Lagoons tend to fill with sediment and their open water is gradually encroached upon by emergent vegetation such as Phragmites reedbeds which eventually turn to scrub. Sand-dunes are also increasingly turning to scrub and woodland. Emergent vegetation in lagoons can be controlled, but ultimately little can be done to stop natural infilling short of highly disruptive and expensive management measures. Grasslands on sand-dunes and shingle beaches can be maintained by grazing and scrub removal (Doody 1989). However, in many parts of Europe the traditional extensive grazing of coastal habitats is being abandoned, particularly in eastern Europe following the collapse of centrally-planned economies. Also, the policy in some protected areas is to stop all grazing (Stock *et al.* 1994), which will be detrimental to priority waterbirds which benefit from moderate grazing of vegetation (but see 'Inappropriate grazing', p. 113).

Succession on open mudflats through the encroachment of saltmarsh plants has also been exacerbated by a recent spread of cord-grass *Spartina anglica* following deliberate planting to reinforce saltmarshes against erosion. This fertile hybrid is less restricted than the native species and has spread rapidly, so that it is now a major problem in UK estuaries (Davidson *et al.* 1991). The species promotes rapid accumulation of sediment (10 cm depth per year) at its pioneer stage and produces a level marsh with a poor diversity of plants and breeding birds. Its dense sward, low invertebrate densities, and tendency to rapidly spread down the shore reduces the intertidal feeding areas available to large numbers of waterfowl and waders at low tide (Doody 1984). Such reductions may be very serious for some species in winter, when birds need to feed for most of the tidal cycle in order to maintain their feeding rate above critical levels. Recent declines in *Calidris alpina* populations in the UK have been attributed to the encroachment of *Spartina* (Goss-Custard and Moser 1988). In recent years, natural die-back of *Spartina* has been observed at some sites, but the cause of this is poorly understood.

The priority species most likely to be adversely affected by succession include breeding herons and other species requiring scrub-free reedbeds in lagoons, breeding and wintering waders and waterfowl of saltmarshes and upper zones of muddy estuaries, terns breeding on shingle beaches and sand-dunes, and passerines that nest or feed in open habitats.

In total, 28 priority species are likely to be affected by this threat, four of which are predicted to suffer high population impacts, and its overall importance is thus moderately high compared to most other hazards.

Coastal defence

Even before the current problem of sea-level rise (see 'Sea-level rise' above), the low-lying coasts of England, Denmark, Germany and the Netherlands have been dominated by extensive coastal engineering measures which have been built to protect land and property from inundation or erosion by the sea. For example, in the Delta Region of the Netherlands, large sections were closed off from the sea by the construction of barrages, dykes and storm-surge barriers between 1950 and 1987 (Marchand and Udo de Haes 1989). This created a series of separate compartments by which a system of lakes (freshwater and saltwater) and a controlled tidal estuary were created. Elsewhere in Europe the construction of simple sea-walls is a commonplace measure to prevent flooding and encroachment by the sea. Groynes (wooden barriers placed at intervals on the beach running into the sea) are also frequently used to prevent longshore drift of sediments and the erosion of beaches.

One of the most important and wide-ranging impacts of such sea-defence measures is the disruption of natural geomorphological processes. The protection of coasts may exacerbate the problem of erosion and flood-risk, since such measures reduce the supply of sediment which helps to maintain natural barriers such as offshore banks of sand and shingle and beaches. Sea-defence barriers may also isolate estuaries and lagoons from marine or freshwater inputs (such as in the Dutch Delta Region), thereby radically reducing or restricting tidal flows and altering these habitats. In such cases, the impacts can be equivalent to that of habitat loss (see 'Landclaim and coastal development', p. 108).

In total, 21 species are predicted to suffer low impacts from this threat over the next 20 years, and in the current circumstances its overall impact is likely to be moderate. However, with the global sea-level rising, the longer term trend in most areas is likely to be the creation of more defences and the further raising of existing coastal barriers.

Increased native predators

Coastal, and particularly colonial, nesting birds are vulnerable to a wide range of native predators including mammals such as *Vulpes vulpes* and brown rat *Rattus norvegicus*, and several species of gulls. These have all shown recent population increases, probably (at least in part) due to their adaptation to feeding on organic rubbish at dumps, urban and tourist centres, etc. Consequently, these artificially increased predator populations may now be having significantly greater impacts on bird populations than under natural conditions. For example, *Larus audouinii* is suffering locally high impacts due to competition with the hugely increased populations of the non-priority Yellow-legged Gull *Larus cachinnans* (Lambertini 1996—see 'Marine Habitats', p. 59, for other examples).

Sea-level rise

One of the predicted consequences of global warming is an increase in global sea-level as a result of the thermal expansion of water and the melting of glaciers and icesheets. Accordingly, the Intergovernmental Panel on Climate Change (IPCC) has tested a number of climatic scenarios. Under the worst case 'business-as-usual' scenario (IPCC 1990), the average global mean sea-level rise over the period 1990–2100 would be about 6 cm per decade, with an uncertainty range of 3–10 cm per decade. However, any rise in sea-level would not be uniform over the globe, so the specific effects in Europe are also uncertain.

The potential effects of sea-level rise on coastal habitats include permanent inundation of low-lying land, increased frequency of temporary flooding from high tides or storm surges, and changes in rates of erosion and salinization of wetland ecosystems. Sea-level rise may be compensated for in some areas by increased sediment deposition, but this depends on a sufficient supply of sedimentary material (Jelgersma 1994).

Although the actual extent and severity of these impacts are highly uncertain, the areas at most risk include extensive parts of the southern Baltic coast, and the North Sea coasts of Denmark, Germany, Netherlands, Belgium and southern England (ECGB 1992). In southern Europe many important deltas, including the Ebro, Rhône, Po and Danube, are likely to be reduced in extent.

The effects of sea-level rise on intertidal areas will vary from place to place (Figure 6). Where there are no artificial sea-defences, there is the potential for landward encroachment of the sea where the land is low-lying. In some circumstances this may lead to an increase in intertidal area. However, if the coast is steep or if a sea-wall exists, then a shift of intertidal land will be impossible and much of the existing intertidal habitat in such areas will disappear. In response to sea-level rise, a considerable increase in the construction of sea-defences can be expected. However, the costs are likely to be so great that large areas are likely to be intentionally left to be inundated by the sea through programmes of managed retreat.

Over the next 20 years, the negative consequences of sea-level rise on bird populations are likely to be moderate, inflicting low impacts on an estimated 19 priority species. However, the effects on the habitat may well be profound in the longer term, although difficult to predict at the moment.

Fisheries
Fisheries of estuaries, lagoons, and intertidal flats may be overexploiting stocks and altering habitats

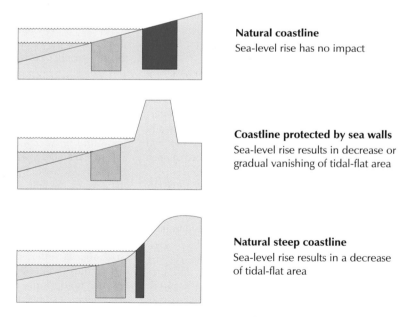

Natural coastline
Sea-level rise has no impact

Coastline protected by sea walls
Sea-level rise results in decrease or gradual vanishing of tidal-flat area

Natural steep coastline
Sea-level rise results in a decrease of tidal-flat area

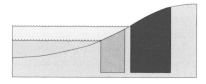

Figure 6. The effects of sea-level rise on the intertidal area of four types of coastline (after Davidson *et al.* 1991).

Natural gently rising coastline
Sea-level rise will result in an increase of tidal-flat area

for the worse, at least with regard to impacts on priority bird species. In particular, concern has been expressed over the intensity of exploitation of mussels *Mytilus edulis* and cockles *Cerastoderma edule* on the intertidal flats in the Dutch sector of the Wadden Sea (Revier 1992, Piersma *et al.* 1993).

Mechanized harvesting of cockles and mussel spat involves hydraulic suction-dredging of the top 4–10 cm of the intertidal sediment (the most productive layer). Up to 30% of the total Wadden Sea stock of cockles is taken per year, including up to 100% of dense cockle-beds in years of poor spat recruitment. Sediments are churned up on a large scale, there is mortality of non-target fauna, and components such as shell-sand, the diatom flora and the finest sediments are swept away by currents in heavily fished areas, leaving coarser, 'cleaner', less productive sediment. These wide-scale changes are thought to be altering and impoverishing the benthic animal community. The high efficiency of these systems also makes the local or regional sustainability of the fishery unlikely.

The systematic and intensive exploitation of newly established *M. edulis* beds is also thought to be unsustainable when combined with the natural die-off of old beds (as well as their premature destruction by suction-dredging for cockles). Mussel beds also have a valuable role in trapping sediments and helping to create and maintain intertidal flats.

In the early 1990s, the development of these shellfisheries coincided with mass die-offs in winter (tens of thousands of birds) and major breeding population declines in *Somateria mollissima* and *Haematopus ostralegus*, the priority species which are most dependent on these molluscs for food. Effects on other species are less easy to determine, though apparent loss of feeding habitat for the major *Calidris canutus* population has been attributed to the commercial cockle fishery (Piersma *et al.* 1993).

The depletion of small-fish stocks in European seas is also of major concern, particularly in the North Sea and north Mediterranean, and beam-trawling for demersal (bottom-dwelling) fish and scallop-dredging may also seriously damage shallow subtidal habitats. These activities threaten, or have the potential to threaten, several priority birds which nest in coastal habitats and which feed on marine fish or shellfish offshore (e.g. terns and seaducks), but this threat is treated under 'Marine Habitats' (p. 59).

Inappropriate grazing

Coastal habitats such as saltmarshes and sand-dunes are often used for grazing livestock, thus helping to prevent their loss through succession (see 'Plant succession', p. 110). However, the consequences of grazing are not straightforward. Moderate grazing (i.e. 4–6 sheep or 0.6–1 full-grown cows per ha) on upper saltmarshes can benefit breeding birds, particularly waders, whose chicks feed on invertebrates associated with livestock dung, provided that nests and eggs are not subject to excessive losses from trampling (Møller 1975, Frid and Evans 1995). Heavy grazing pressure, or grazing during the breeding season, is detrimental to such birds (and to plant diversity) but beneficial to wintering waterfowl, which prefer the short, grass-rich swards. Such grazing, particularly during spring tides, can also damage the marsh surface and initiate widespread marsh erosion.

Sand-dunes can support grazing densities of 0.5 cattle/ha or 4 sheep/ha. Higher stocking levels, particularly of unrestrained sheep, can lead to considerable damage and increased erosion rates.

Overall, problems for birds from overgrazing and other inappropriate grazing practices are likely to decrease and become more localized, since abandonment of extensive pastoral agricultural systems, such as on coastal grazing marshes, is currently widespread. The overgrazing of sand-dunes in southeast Europe is, however, likely to remain a problem. In total, 12 priority species may suffer population declines, although all but one species will suffer only low impacts, and the overall importance of this threat is therefore moderately low amongst those considered here.

Military disturbance

Some coastal sites that are used for military activities contain, or are close to, internationally important moulting, roosting or breeding areas for priority bird species. Such areas are often intensively used for target practice and training flights by helicopters and jets, which can cause considerable disturbance to birds (Platteeuw 1986; see 'Hunting disturbance', p. 109). However, such areas are usually protected from infrastructural development, and sometimes from recreational and hunting disturbance, which are all much more widespread and severe threats.

The European populations of eight priority species are predicted to suffer low impacts from this threat, which is therefore of relatively low importance amongst those considered in this strategy.

Barrage construction

Barrages are one of the most disruptive of all developments to estuarine habitats. Tidal-power barrages, designed to generate electricity, can severely alter intertidal and subtidal areas by reducing current flow, tidal amplitude and salinity, thereby influencing the movement and distribution of sediment. Vegetation, invertebrates, fish and birds are consequently affected.

Some barrages are designed to provide a lake equivalent to high-water level, simply for amenity or 'aesthetic' value (as exposed mudflats are felt by some to be unsightly). They entirely inundate and eliminate the intertidal flats and saltmarshes within them. For example, the construction of a barrage at Cardiff Bay (UK) will create a permanent lake for 'aesthetic' purposes. The barrage will inundate about 2 km² of intertidal mud habitats, which would otherwise support significant numbers of waterbirds (Hill 1994).

However, on a European scale, such barrages are currently rare, being predicted to cause low impacts to the European populations of 10 priority species, and the overall importance of this threat is therefore relatively low.

Afforestation

Afforestation to stabilize sand-dunes (after overgrazing, burning or excessive trampling) is one of the major causes of loss of open dune landscapes throughout Europe (Table 6). A recent survey found that of the 4,000–4,280 km² of sand-dunes remaining on the Atlantic coast of Europe, about 40% have been afforested. Also, about one third of the 40% loss of sand-dunes in Europe since 1900 (comprising hundreds of square kilometres: Stanners and Bourdeau 1995) has occurred during the last three decades, principally as a result of afforestation with non-native tree species.

Afforestation eliminates nesting opportunities for those priority birds which require open ground for nesting in sand-dunes (e.g. seaduck). However, such species are relatively few in Europe, and, in total, low impacts are predicted to affect 11 such species. Afforestation of coastal habitats is thus considered overall to be of relatively low importance amongst the threats considered in this strategy.

Intensive management of salinas

Natural and artificial islands (from salt-workings) were once a common sight in most industrial salinas in the Mediterranean, and were occupied by breeding colonies of gulls, terns and waders. Modern management involves levelling each salina to increase the surface area of the water, so eliminating any islands. Ground-nesting birds are thus displaced to less favourable breeding sites, e.g. on dykes bordering the salinas, where they are much more vulnerable to nest-failure, predation or human persecution. Such changes are expected to cause low impacts to seven priority species in Europe, and are therefore of low importance overall among the threats considered in this strategy.

Dredging/extraction

Considerable quantities of sand and gravel are extracted for the building industry every year from lagoons, estuaries, sand-dunes and shingle beaches. Metals and sea-salt are also commercially excavated, and regular dredging of shipping lanes occurs in very shallow estuaries, lagoons and other shallow-water environments which are undergoing active sedimentation. Such extraction and dredging increases water turbidity, causes eutrophication and results in the temporary inhibition of phytoplankton growth. The removal of the sediment itself locally destroys the benthic community and decreases the habitat quality for many invertebrates and demersal fish, thus reducing the food supply of priority birds (see also 'Marine Habitats', p. 59).

The activity also disrupts geomorphological processes. Significant parts of the Mediterranean coast are subject to unnaturally high levels of erosion due to the reduction in sediment supply caused by the high levels of sand-dredging from rivers inland. Thus, although the impacts on priority birds over the next 20 years may be low (and are currently predicted to affect only six of the priority species), longer term impacts on coastal habitats may be considerable.

Aquaculture

Lagoons and abandoned salinas in the Mediterranean are increasingly sold, leased or otherwise used for aquaculture of fish and bivalve molluscs, mainly *Mytilus edulis* and oysters *Ostrea edulis* (Ardizzone *et al.* 1988, Chauvet 1988). In north-west Europe, estuaries are commonly used for the cultivation of these bivalves. Fish-farming is particularly damaging as lagoons are often deepened, and the addition of large quantities of fish-food can lead to eutrophication of the water. The regular feeding of fish and other management activities also causes considerable disturbance to feeding birds. Shellfish farming can lead to subtle changes in the character of benthic habitats which may have impacts on birds, although in most cases these are poorly understood.

In addition, the waterbirds that frequent these habitats are often considered as fish-eating predators and are commonly persecuted.

Altogether, three priority species are thought to be threatened with low impacts (including, however, the globally threatened *Pelecanus crispus*), and thus the overall importance of this threat is low amongst those considered in this analysis.

Other threats

In addition to the above threats, a number of others may cause significant loss or degradation of coastal habitats in some parts of Europe. These include:

- Water abstraction from wetlands in sand-dunes.

- Land-subsidence, disturbance and pollution (leading to loss of breeding islands, intertidal flats and other shallow sedimentary habitats) resulting from oil and gas extraction, e.g. in the Wadden Sea (Smit *et al.* 1987) where the ground level will drop by up to 16 cm.

- Disturbance and locally high mortality of some priority birds due to wind-energy parks (Winkelman 1995, Musters *et al.* 1996) and pipeline construction (Smit *et al.* 1987).

- The effects of upstream dams on estuary and delta hydrology and geomorphology in Turkey, Greece, Bulgaria and Spain, mainly through reducing sedimentation levels (Anon. 1993b, Heredia *et al.* 1996).

CONSERVATION OPPORTUNITIES

International legislation

Management of coastal habitats for nature conservation has been encouraged in Europe by a relatively large number of international conventions, directives and regulations, as listed in Box 1. These are described in more depth in the introductory chapter on 'Opportunities for Conserving the Wider Environment' (p. 24), but information that is particularly relevant to coastal habitats is highlighted here.

An overall 'framework' strategy, as potentially provided by the Pan-European Biological and Landscape Diversity Strategy (PEBLanDS), could be very helpful in the coastal zone, which is covered by numerous legal mechanisms, and which has been subject to a multitude of initiatives, declarations and recommendations.

Action Theme 5 of PEBLanDS covers 'Coastal and marine ecosystems', and the three main objectives of the first Action Plan (1996–2000) under this theme are (1) to promote the creation of a representative and integrated coastal network of protected areas throughout Europe, (2) to promote the adoption of 'integrated coastal zone management' by authorities throughout Europe, and (3) to produce a Coastal Code of Conduct with clear recommendations on 'best practice', aimed at coastal developers (also to include the drafting of a model law for the sustainable use of coastal areas and resources). A more detailed 1996–1997 work programme for Action Theme 5 was recently proposed (UNEP–ROE 1996). At the moment it is not yet clear how far national governments will go in translating such well-meaning statements into legislation, initiatives and actions.

The strategy's first objective, that of creating an integrated network of protected areas (including coasts), has been previously promoted by numerous agencies, e.g. Anon. (1992a) with respect to coasts. The realization of such a concept is important for the conservation of priority coastal birds, many of which are highly migratory and depend on a wide and varied network of wetland sites in many countries during their annual cycle. The Bonn Convention, in particular, offers a valuable intergovernmental framework for conserving, monitoring and managing the migration systems of priority coastal waterbirds, through the implementation of its African–Eurasian Waterbird Agreement (AEWA), which is intended to come into force in 1998.

Within European Union (EU) member states, the Birds and Habitats Directives promote such a site network. Annex I of each Directive lists, respectively, 45 priority birds of coastal habitats and 41 coastal habitat-types which are considered to be threatened in the EU (see Appendix 1, p. 327, and Appendix 5, p. 423), and they require that member states designate by June 2004 a 'coherent ecological network' of protected areas ('Natura 2000'), comprising Special Areas of Conservation (SACs) and Special Protection Areas (SPAs) in sufficient numbers and locations so as to conserve these threatened habitat-types and species.

Spanning the EU and non-EU countries are the 'Regional Seas' conventions such as the Barcelona Convention and the Helsinki Convention, which address the protection of the coastal and marine environment. Protocols to these conventions require or encourage the designation of Mediterranean Specially Protected Areas (MedSPAs) and Baltic Sea Protection Areas (BSPAs) respectively. It appears likely that the Bucharest Convention, which recently came into force, will also adopt a protocol concerning the protection of the biodiversity of the Black Sea in the near future, and this may include a requirement for the designation and protection of important coastal and marine sites.

Other examples of international initiatives which aim at the creation of an integrated European network of protected areas (including coastal habitats) are the IUCN 'Action Plan for Protected Areas in Europe' (IUCN–CNPPA 1994) and the Dutch government's European Ecological Network (EECONET) concept (Bennett 1991).

The Ramsar Convention provides a good basis for achieving the second objective of the PEBLanDS Action Plan, to promote good management of the coastal zone. Contracting parties to this Convention are obliged to work towards the 'wise and sustainable use' of wetlands in general (including coastal ones), through the inventory and monitoring of all important sites and the promotion of wetland conservation in the national land-use planning system.

Box 1. International environmental legislation of particular relevance to the conservation of coastal habitats in Europe. See 'Opportunities for Conserving the Wider Environment' (p. 24) for details.

GLOBAL

- *Biodiversity Convention* (1992) Convention on Biological Diversity.
- *Bonn Convention* or *CMS* (1979) Convention on the Conservation of Migratory Species of Wild Animals.
- *Ramsar Convention* (1971) Convention on Wetlands of International Importance especially as Waterfowl Habitat.
- *World Heritage Convention* (1972) Convention concerning the Protection of the World Cultural and Natural Heritage.
- *MARPOL 73/78* (1973) International Convention for the Prevention of Pollution from Ships, 1973, as modified by the protocol of 1978.
- *Law of the Sea* or *UNCLOS* (1982) United Nations Convention on the Law of the Sea.
- Convention on the Protection and Use of Transboundary Watercourses and International Lakes (1992).

PAN-EUROPEAN

- *Bern Convention* (1979) Convention on the Conservation of European Wildlife and Natural Habitats.
- *Espoo Convention* (1991, but yet to come into force) Convention on Environmental Impact Assessment in a Transboundary Context.

REGIONAL

- *Oslo Convention* (1974) Convention for the Prevention of Marine Pollution by Dumping from Ships and Aircraft.
- *Helsinki Convention* (1974/1992) Convention for the Protection of the Marine Environment of the Baltic Sea Area.
- *Barcelona Convention* (1975) Convention for the Protection of the Mediterranean Sea against Pollution.
- *Paris Convention* (1978) Convention for the Prevention of Marine Pollution from Land-based Sources.
- *Bucharest Convention* or *Black Sea Convention* (1992) Convention on the Protection of the Black Sea against Pollution.

- *OSPAR Convention* (1992, but yet to come into force) Convention for the Protection of the Marine Environment of the North-East Atlantic (formed by the merger of the Oslo and Paris Conventions).

EUROPEAN UNION

- *Birds Directive* Council Directive on the conservation of wild birds (79/409/EEC).
- *Habitats Directive* Council Directive on the conservation of natural habitats of wild fauna and flora (92/43/EEC).
- *EIA Directive* Council Directive on the assessment of the effects of certain public and private projects on the environment (85/337/EEC).
- *Port State Control Directive* Council Directive concerning the enforcement, in respect of shipping using Community ports and sailing in the waters under the jurisdiction of the Member States, or international standards for ship safety, pollution prevention and shipboard living and working conditions (95/21/EC).
- *Nitrates Directive* Council Directive protecting waters against pollution by agricultural nitrates (91/676/EEC).
- Council Directive on urban waste water treatment (91/271/EEC).
- Council Directive on the control of dangerous substances discharge (76/464/EEC).
- Council Directive on the conservation of fresh water for fish (78/659/EEC).
- Council Directive on the freedom of access to information on the environment (90/313/EEC).
- Council Directive on the quality required of shellfish waters (79/923/EEC).
- *Agri-environment Regulation* Council Regulation on agricultural production methods compatible with the requirements of the protection of the environment and the maintenance of the countryside (2078/92/EEC).
- Council Regulations on the Structural Funds (e.g. 2081/93/EEC).

Since the 'wise use' concept is still relatively novel for many contracting parties, the encouragement that the Convention gives to practise it on the ground, at least locally, at 'showpiece' Ramsar Sites (wetlands of international importance) is very valuable. The Man and Biosphere Programme of UNESCO also encourages governments to attempt 'demonstration' or model projects involving the sustainable and integrated management of internationally important na-

ture-conservation areas (Biosphere Reserves).

Joint initiatives for the management of important and internationally shared coastal areas also exist, and provide further 'hands on' experience to governments and agencies in truly integrated coastal zone management that spans national boundaries. For example, the 1982 signing of the 'Declaration on the Protection of the Wadden Sea' by Denmark, Germany and the Netherlands has been followed up by a continuing series of trilateral governmental conferences, recommendations and actions.

Through such initiatives, experience in integrated management may be gained and later applied on a broader scale along the entire national coast, or beyond. However, most European coastlines will remain outside designated protected areas or their buffer zones, and will require other measures, legislation or otherwise, to promote the sustainability of human uses, and the integration of biodiversity conservation with all such uses.

As 'Threats to the habitat' (p. 104) demonstrates, tourism and recreation are the coastal land-uses with the single greatest negative impact on coastal habitats and their priority birds. Destruction and degradation of coastal habitats by infrastructural developments, for tourism/recreation and certain other purposes, are regulated against in EU countries to some extent by the proper implementation of the EIA Directive. However, outside the EU these negative aspects of projects are left to the consideration of national legislation and decision-making processes. These may not adequately acknowledge the international dimension of coastal habitat-use by priority migratory birds and other wildlife, nor the potentially large-scale (international) and highly indirect impacts on coastal ecosystems of national projects inland, such as dams on major rivers.

Furthermore, the application of the 'environmental assessment' concept at an earlier phase in the planning of individual development projects, i.e. at the broader scale of plans, programmes and policies (e.g. for land-uses such as tourism, forestry and agriculture) in regions or nations, does not yet feature significantly in European legislation. A Directive is in preparation that will address such Strategic Environmental Assessment (SEA) in EU countries, and it is anticipated that this will play a very important role in the integrated management of the EU coast if adopted by member states (CEC 1995c).

The development of sustainable tourism is one of the key environmental concerns of the European Commission, and various projects have been funded in the framework of *Community action plans to assist tourism*, in particular by supporting activities to increase the environmental awareness of tourists and to encourage the international exchange of experience in dealing with visitor management, etc.

Land-claim for infrastructural projects affects more priority coastal birds than almost any other threat, and some progress towards integrating the environment into infrastructural development processes has been made within the EU, although much further reform is needed (BirdLife International 1995b). The recent 'framework' Regulation for the Structural Funds requires that plans and programmes submitted under Objectives 1, 2 and 5 of the Structural Funds have to be evaluated according to certain of their environmental aspects (see CEC 1995c).

Water pollution is another major threat to coastal habitats and their priority birds, and there is a range of Directives governing water quality in the EU, aimed at reducing both nutrient and toxic pollution (see Box 1). The reform of the EU Common Agricultural Policy has also produced measures aimed at reducing farming-related water pollution amongst other things (see 'Agricultural and Grassland Habitats', p. 267). Regulation 2843/94 provides aid to farmers to invest in such environmental protection. 'Marine Habitats' (p. 84) covers current legislation and initiatives which reduce the risk of coastal and marine oil pollution.

Financial instruments

Financial instruments such as taxes and user-charges, 'the polluter pays' principle, and the allocation of property rights and incentives, are increasingly being used to manage aspects of the coast. The advantages of such instruments are that they place the burden on the polluter, allocate scarce resources, and enable better transparency and accountability. Their disadvantages include the undervaluing of environmental resources and the potential focus on short-term gains.

Failure to charge the full price for access to a resource is common. For example, throughout Europe, users are rarely charged the full price of the water they use. Excessive use of water creates problems for local authorities and has serious impacts on the coastal zone—e.g. the water-demands of tourism and intensive irrigated agriculture on the Coto Doñana (Spain) threaten one of Europe's most important coastal wetlands (Stanners and Bourdeau 1995). Incentives should be developed which not only address the excessive demand on the resource but also tackle other problems simultaneously. For example, Gibraltar has an incineration plant which not only disposes of waste, but produces electricity used in a desalination plant.

Levies or taxes for conservation might raise significant amounts of money. For example, a small 'tourist' levy on visitors to the Balearic Islands (Spain) might raise significant funds to be used not only for wildlife conservation but also to compensate local communities and landowners for financial losses resulting from restrictions on their activities within protected areas, or to subsidize traditional agricultural practices in order to make them more financially attractive to local farmers.

Open access to resources, resulting in overfishing, overgrazing and pollution, is at the root of many of the problems facing the coast. Those exploiting common resources often do not have to meet the full cost of their actions and have little incentive to manage them sustainably. The allocation or auctioning of private property rights to such resources (as has been done for offshore fishing rights in some European countries—see 'Marine Habitats', p. 87), albeit a controversial measure, might help to address these problems, particularly within protected areas. An example of this approach can be found in the Odiel Marshes National Park in Andalucía (Spain).

A range of economic instruments is available within the EU which may be used in the coastal zone to help to develop integrated management. These include the Structural Funds and Regional Development Funds (e.g. for financing water-treatment plants, or for the incorporation of environmental protection measures into port, tourist or transport developments), as well as the LIFE and PESCA funding instruments (see 'Opportunities for Conserving the Wider Environment', p. 36). Since 1986 almost 100 projects in the field of protection and management of coastal zones have been granted financial support under LIFE (Regulation 92/1973/EEC) and the now superceded ACNAT and MEDSPA budget headings (CEC 1995c).

Changes in livestock-grazing patterns which adversely affect vegetation structure at the breeding sites of priority coastal birds, either through wide-scale overgrazing or through abandonment of grazing, are important threats to some priority birds of coastal habitats. These are potentially reversible under EU financial measures such as the Agri-environment Regulation (Box 1), which can support, e.g., national schemes that promote non-intensive grazing.

Policy initiatives

Integrated Coastal Zone Management (ICZM)

As mentioned previously in this strategy, the coastal zone is often a complex of interconnected habitats affected by a variety of socio-economic pressures. To deal with this, Integrated Coastal Zone Management (ICZM) is a strategic approach to the management of coastal activities which is increasingly being introduced by countries worldwide.

The introduction of integrated policies and plans which try to balance demands for resources and to resolve conflicts of use in the coastal zone, within a national framework, is the essence of ICZM (SWCL 1993). The need for ICZM is increasingly being articulated, since in many countries the institutions responsible for coastal resources are highly fragmented. At the very least, 'integration' implies: (a) managing the coastal area as an integrated system of resources and users, and (b) coordinating manage-

ment functions among the various agencies responsible for planning and implementation with regard to policies and strategies resulting from the planning phase, tasks and functions in planning and implementation, and available management resources.

The inland extent of the 'Coastal Zone' has no internationally agreed limit or dimension, but is linked to the management plan in hand (SWCL 1993). However, the area may extend several kilometres inland of the high-water mark and as far out as, or further than, the limit of territorial waters (12 nautical miles).

The major objectives of ICZM are:

- To promote sustainable use of coastal and marine resources.
- To balance demand for coastal zone resources.
- To resolve conflicts of use.
- To promote environmentally sensitive use of the coastal zone.
- To promote strategic planning for coastal habitats.

The constraints of ICZM are that:

- Integration across the whole coastal zone can be lacking since there is often a division between the approach to planning and management of the coastal zone above and below the low-water tidal line.
- Coastal resources are often common property with open (free) access to all users. Free access often leads to excessive, unsustainable use, and to the degradation or depletion of resources.
- ICZM involves the resolution of both long- and short-term problems. Too frequently, land-use planning only considers short-term scenarios. Long-term problems include: possible climate change, sea-level rise and changing hydrological patterns; the accumulation of toxic materials in the environment and their effects on species; increasing pressure for settlement and development of coastal areas, and continued developments inland with resulting modification of the quantity, quality and time-patterns of inputs (e.g. water, nutrients, sediment, etc.) to coastal waters.
- ICZM is 'driven' by the multiple demands of society for the use of coastal resources, as well as by concerns about sea-level rise due to climate change. Rising sea-level may be negative for some uses, and positive for others, thus adding more complexity to the nature of management.

National policy on land-use planning

Most European governments plan, implement and administer sectoral programmes such as fisheries, environmental quality, national parks and wildlife. This frequently leads to numerous sectors and agencies being involved in coastal policy, with fragmentation of responsibility, duplication of effort and direct conflicts between government programmes. One of the major challenges in the development of ICZM is to relate it to existing planning programmes.

Many nations practise four types of planning that influence the shape and direction of coastal management:

- National economic planning.
- Sectoral planning.
- Coast-wide or island-wide land-use planning and regulation.
- 'Special area' or regional plans.

The European Commission recently issued a Communication on the integrated management of coastal zones (CEC 1995c), as part of its development of a Union-wide ICZM strategy at the request of the Council of Ministers (Environment) (Resolution adopted 11 February 1992), with a view to providing a coherent environmental framework for integrated and sustainable forms of coastal development. This could require member states to prepare national coastal strategies and could deliver better coordination between relevant EU and national policies.

Coast-wide or island-wide land-use planning and regulation

Land-use planning and regulation specifies the type, intensity, and rate of development and conservation for a particular area. Land-use plans may cover the entire coastal zone of the nation or state (or province). Broad goals and objectives are usually specified to direct the planning effort. Land-use plans, consisting of both maps and policies, are usually translated into guidelines and legally binding rules such as zoning ordinances.

A few countries have amended their town and country planning programmes to include sets of policies for land-use control within a delineated coastal area. For example, under the Town and Country Planning Act in Cyprus, 'there are detailed regulations governing streets, construction and alteration of buildings... in coastal areas.' In Spain, the Coastal Act (1988) establishes public property over the coastline and is setting up a range of preventive and protective measures to reestablish the integrity of the coast.

'Special area' or regional plans

These plans refer to coastal land-use or resource-use programmes for areas larger than a local jurisdiction and smaller than an entire nation. The distinguishing feature of such plans is the geographic coverage; the boundaries are usually drawn with two purposes in mind. First, they are intended to 'capture' national resource or economic development issues that cross the boundaries of state or local governments. Such issues might include watershed management, the protection of sensitive habitats, or the development of a regional transportation network. Second, the boundaries are drawn to encompass a significant natural resource, such as an embayment, river basin or estuary. 'Special area' or regional plans have a multi-sectoral perspective. They may focus on a single issue (such as tourism development), but there are interconnections with other relevant sectors, and they are usually mandated by either a legislative body or a national or state ministry.

Regional planning and analysis confers a number of advantages that are absent from local and national planning. At the regional level, it is possible to address and resolve problems faced by entire ecosystems, such as the siltation of an estuary resulting from developments in the watershed.

CONSERVATION RECOMMENDATIONS

Broad conservation objectives

1. To maintain the current overall area of coastal habitats used by priority bird species, and to avoid habitat loss, or partial loss, at any coastal Important Bird Area (IBA).

2. To arrest the degradation of coastal habitats and to compensate for any unavoidable losses by maintaining and enhancing important habitat features for their priority bird species and other wildlife.

3. To maintain and, where appropriate, to enhance the quality of artificial and degraded coastal habitats for priority bird species and other wildlife.

4. To increase the area of coastal habitats that specifically benefit priority bird species, where there is no conflict with other conservation interests or objectives.

Ecological targets for the habitat

In order to maintain and improve the quality of coastal habitats, it is essential to retain and enhance important habitat features for priority bird species. It is therefore necessary to ensure the following.

1. Networks of suitable wintering and staging sites for migratory species should be maintained intact along flyways, especially of scarce and threatened coastal habitat-types such as Mediterranean lagoons and deltas, and Atlantic estuaries.

2. There should be no interference with natural

coastal processes (e.g. hydrological or geo-morphological) where these still exist, unless this is necessary for their maintenance, or in exceptional cases to conserve other rare and important coastal habitats and features in the long term. In the latter cases, if such interference causes the unavoidable degradation or loss of other coastal habitats, all efforts should be made to reinstate such habitats elsewhere. Where natural coastal processes no longer exist, they should be reinstated whenever possible.

3. Where the hydrology of coastal wetlands is regulated, such management should aim to minimize detrimental effects on the habitats, food resources and and habitat features that are necessary for the nesting, feeding and roosting of priority birds, through maintaining natural levels and flows of water as far as possible, including natural variations and cycles over time (e.g. typical seasonal fluctuations).

4. Important habitat features such as vegetation (composition and structure) and sedimentary structures such as intertidal flats, sand and shingle beaches, lagoons and sand-dunes, should be maintained (according to the principles outlined under point 2), using appropriate management where necessary.

5. Mosaics of these habitat components and a mixture of different habitat-types should also be maintained, according to the needs of priority species using individual coastal habitat-types.

6. Disturbance of priority birds by human activities (e.g. hunting, fishing and recreation) should be avoided during the breeding season, and reduced at other times to levels that demonstrably do not have a significant impact (according to species affected and local circumstances) in feeding or roosting areas.

7. Levels of toxic pollutants in the environment should be reduced such that sublethal or indirect effects are not reasonably suspected to be causing or contributing to population declines in priority bird species.

8. Artificial eutrophication of coastal wetlands should be avoided and where necessary reversed (according to the principles outlined under point 2), especially where levels are so high as to cause major reductions in habitat quality for priority birds, e.g. through excessive loss of open water to vegetation, reduced water clarity, or declines in food resources.

9. Benthic habitats of coastal water-bodies, e.g. lagoons and estuaries, and their associated flora and fauna, should be conserved, and damage by human activities should be minimized.

10. Any human exploitation of fish, shellfish and other invertebrate populations should be managed on a sustainable basis, such that habitat extent and quality (e.g. food availability) are not reduced for priority birds, and should use methods which minimize habitat disturbance and damage.

Broad policies and actions

To meet the above objectives and habitat targets, the following broad policies must be pursued (see p. 22 in 'Introduction' for explanation of derivation):

General

1. All internationally important sites for priority birds (Important Bird Areas) and representative or priority areas of coastal habitats should be protected under appropriate national legislation, and within the European Union (EU) as Special Protection Areas (SPAs) and Special Areas for Conservation (SACs) under the Birds and Habitats Directives. Such sites should be managed according to the recommendations outlined in Box 1 (p. 15).

   Within the EU, the LIFE fund should be extensively used for the conservation and management of SPAs and of sites listed for designation as SACs that contain coastal habitats.

2. National governments and the EU should ensure that Strategic Environmental Assessments (SEAs) are made to assess the full range of coastal development policies, plans and programmes, in addition to carrying out detailed Environmental Impact Assessments (EIAs) of individual schemes (including agricultural projects and other projects which are currently often exempt).

   EIAs should be carried out to internationally recognized standard procedures (e.g. following the EIA Directive) and legislation should ensure that planning approvals for developments are dependent on an EIA report that is favourable to the conservation of the area.

3. Disbursement of international funds for coastal development programmes should be strictly governed by relevant national, EU and international environmental policies and legislation, and dependent on SEAs and EIAs that are favourable to the conservation of coastal habitats in and beyond the proposed area(s) for development.

4. Relevant recommendations of regional conventions (e.g. Helsinki, Barcelona, Bucharest, OSPAR) concerning the protection of coastal areas should be fully implemented by contracting parties. Other international conventions (e.g. Biodiversity, Bonn (AEWA), Ramsar, Bern)

should also be used to their full capacity to provide an additional layer of protection to coastal habitats and to ensure the integrated and sustainable management of coastal areas.

Integrated Coastal Zone Management (ICZM)

5. The European Commission should without delay implement the resolutions of the European Council of Ministers, the Fifth Environmental Action Programme and the recommendation of the European Environmental Agency, by developing an EU-wide coastal zone strategy. A priority area for action must be to identify the instruments to achieve the coastal zone targets within the Fifth Environmental Action Programme, and to develop an operational framework to deliver ICZM at all levels of government.

6. Features common to ICZM approaches in the USA, Australia and Canada could form the basis of an EU-wide approach to ICZM. The current coastal zone programme should be orientated in such a way as to test the best methods of delivering this.

7. The European Commission should shape and influence member-state ICZM programmes through policy and guidance, and by providing the impetus for action. To this end, the Commission should:

• establish an overall EU-wide aim for the coastal zone, develop objectives which will contribute to achieving the aim, and policies to guide the activities of member states in the coastal zone;

• establish a framework within which coordinated action at the European and member-state level can be facilitated;

• outline the overall goal, aims and objectives which should feature in individual member-state coastal zone plans and programmes.

8. The European Commission should establish the administrative framework which is needed to oversee the implementation of an EU coastal zone programme. The Commission should:

• establish a body which coordinates and integrates, at the highest level, the activities of EU institutions in the coastal zone;

• establish an agency whose specific role is to administer and oversee the day-to-day running of an EU coastal zone programme;

• establish an advisory group of ICZM specialists to assist the agency;

• ensure adequate feedback mechanisms are in place from the local and member-state level back to the EU programme.

9. To encourage member states to participate in an EU coastal programme, the Commission should:

• introduce financial incentives which encourage member states to establish coastal zone programmes;

• introduce financial assistance which aids member states in the implementation of coastal zone programmes;

• consider the introduction of a 'consistency provision' compatible with the principle of subsidiarity.

10. The Commission should define the standards which need to be met by member-state ICZM programmes in order to ensure consistency between the different plans/programmes.

11. The Commission should consider providing a legal basis for delivery of coastal zone management in Europe. Any such legislation should:

• delineate the biogeographic regions of Europe which are needed to distinguish between the different approaches to ICZM required;

• establish the administrative framework, financial arrangements and conditions for member-state participation in a European programme;

• not make participation in the European coastal zone programme statutory.

Tourism and recreation

12. Limits to the intensity of use of coastal habitats by visitors/tourists should be determined, and should be effectively enforced through the provision of adequate powers and resources to local government authorities, and appropriate revision of planning regulations where necessary, such that intensity of human pressure does not exceed the relevant limit at any time of the year.

13. Disturbance-free zones for coastal bird populations should be created in the most valuable natural habitats, after ascertaining what is sufficient size, siting, quantity, etc.

14. Coastal tourism policies that concentrate on quality improvement in existing resorts rather than on developing new ones should be adopted by the Mediterranean and Black Sea coastal states.

15. There should be a shift in national and international tourism policies from mass tourism to favour environmentally friendly, sustainable and high-quality ecotourism, particularly in coastal habitats. Pilot projects aimed at developing sustainable tourism in sensitive coastal habitats should be widely developed and their results taken into consideration when developing regional land-use policies.

16. Further research into the impact of disturbance on coastal bird populations by human activities (including hunting and water-sports) should be carried out.

Agriculture

See also the recommendations under 'Agricultural and Grassland Habitats' (p. 267) regarding the points below.

17. National governments should develop agricultural policies and support measures that maintain and promote extensive agricultural systems in water catchments, in order to reduce fertilizer use, pesticide applications and livestock densities.

18. Large-scale agricultural development schemes (such as irrigation projects) should be subject to EIAs that take into account off-site and long-term impacts on catchments and downstream habitats. Large-scale irrigation projects that reduce natural water discharge in rivers should be avoided.

19. The opportunities provided by Regulations 2078/92 and 2052/88 for the conservation and maintenance of semi-natural coastal habitats (e.g. saltmarshes and sand-dunes) should be fully utilized by EU member states.

Catchment planning

20. Large-scale projects aiming at the construction of hydroelectric dams and water reservoirs along rivers should be carefully scrutinized by SEA/EIA procedures, and their negative ecological impact on coastal wetlands mitigated. If there is sufficient evidence that such projects will seriously disrupt natural coastal dynamics, and will cause the deterioration or loss of coastal wetland habitats, these projects should not go ahead.

21. At existing hydroelectric dams and reservoirs the feasibility of creating sediment bypasses should be investigated and wherever possible these should be built.

22. Plans to increase riverine discharges to levels which are sufficient to maintain coastal wetlands should be developed and implemented. This will need substantial inter-regional and even international cooperation between relevant governments.

Coastal defences

23. Where appropriate and feasible, artificial coastal defence systems should give way to the restoration of natural coastal dynamics. Restoration activities should primarily focus on high-quality natural coastal wetland habitats.

24. National governments and the EU should set targets for progressive reductions in carbon dioxide and other greenhouse-gas emissions beyond the year 2000, in order to reduce the risk to biodiversity from climate change (especially sea-level rise and increased storminess). Targets should be based on biodiversity conservation.

25. Responses to sea-level rise by national governments and the EU should incorporate the aims of protecting and enhancing biodiversity.

26. Reforms of the EU Common Agricultural Policy should include incentive schemes which would permit the financing of specific land-management measures required for managed realignment of coastal defences and habitat creation, where appropriate.

Forestry

27. Afforestation of sand-dunes should be avoided where possible, and greater awareness should be promoted of its potentially negative consequences for priority coastal birds.

28. Where it is essential to stabilize sand-dunes, native species of trees and shrubs should be used in preference to non-native species.

29. All afforestation schemes in coastal habitats, irrespective of size, should be subject to EIAs.

Water quality

See also the recommendations under 'Inland Wetlands' (p. 125) and 'Marine Habitats' (p. 59).

30. With due regard to the 'polluter pays' and the 'precautionary' principles, governments must ensure that coastal wetlands are protected from pollution by adequate legislation and its enforcement. States should promote this by the provision of safe storage schemes for potentially harmful substances, legislating in favour of environmentally benign chemicals (where available) as alternatives to toxic chemicals (e.g. enforce Red and Grey lists of the EC Directive on dangerous substances), ensuring widespread knowledge of legislation banning pesticides, and ensuring that responsible authorities have adequate resources to enforce controls and monitor illegal discharges.

31. Coastal states should identify coastal areas or sites (including coastal Important Bird Areas) that are particularly sensitive to oil pollution or other accidental or illegal discharges from shipping, and should modify shipping routes to avoid such areas by an adequate distance wherever possible. These results should be taken into account in SEAs of land-use programmes, especially those relating to industry and agricultural development.

Fisheries

See also the recommendations listed under 'Marine Habitats' (p. 59).

32. Sustainable levels of fishing in coastal waterbodies (including river deltas, lagoons and estuaries) should be defined locally and fishery efforts should be properly regulated according to the results.

33. Large-scale and intensive mussel *Mytilus* and cockle *Cerastoderma* fisheries should be subject to Environmental Impact Assessments (EIAs), which should properly evaluate their potential impact on important prey populations for priority birds. As a result, these shellfisheries should be banned, or more closely regulated, or reduced in scale, wherever this is necessary to ensure sustainability and to minimize damage and disturbance of the intertidal habitat and its associated flora and fauna.

34. Further research should be carried out on the impact of shellfisheries on intertidal habitats and their priority birds, especially those fisheries that involve large-scale and intensive exploitation, e.g. through suction-dredging.

Aquaculture

35. Proposed aquaculture projects should be subject to planning regulations and EIAs that are favourable to the conservation of coastal habitats, in particular with regard to the threats posed by water pollution by nutrients, pesticides and other chemicals used in the industry.

36. The aquaculture industry must pay regard to the natural value of coastal habitats and should promote research into the impacts of aquaculture on the wider environment.

Public awareness

37. Decision-makers and the general public should be informed of the value of coastal habitats as natural sea-defence systems and of the importance of preserving natural coastal processes.

38. The fragility of coastal habitats and the need to lower levels of human pressure and maintain disturbance-free zones should be widely promoted among coastal communities and especially among tourists. The availability of alternative types of (sustainable) tourism should be widely promoted among potential target audiences.

Research

39. More research should be carried out to identify and quantify those biotic and abiotic factors which are most important in regulating natural coastal processes, but especially to quantify the effect of human activities in the coastal zone and elsewhere on these factors.

Other measures

40. Traditional land-uses that maintain vegetation in successional states should be promoted, e.g. digging of eelgrass *Zostera marina* for fertilizer in northern Portugal.

ACKNOWLEDGEMENTS

Most contributions to this chapter were made by the Habitat Working Group during and after a workshop organized by P. Bradley, M. Calado and P. Doody, and hosted by the Ria Formosa Natural Park, Faro, Portugal (9–11 December 1993), held in collaboration with the European Union for Coastal Conservation; these contributors are listed on p. 93. K. Norris, P. Bradley and D. Huggett (Royal Society for the Protection of Birds) also made major contributions to the sections on 'Habitat needs of priority birds' and 'Conservation opportunities' respectively. B. Kalejta-Summers compiled the first draft of the chapter.

Additionally, valuable information and comments were received from participants at the *4th MEDMARAVIS Mediterranean Seabird Symposium on 'Seabird ecology and coastal zone management in the Mediterranean'* (Hammamet, Tunisia, April 1995), J. Elts (Eesti Ornitoloogiaühing), U. Filbrandt (Naturschutzbund Deutschland), D. Hill, P. Iankov (Balgarsko Druzhestvo za Zashtita na Pticite), A. Mikityuk (Ukrainske Tovaristvo Okhoroni Ptakhiv), F. Moreira (Liga para a Protecção da Natureza), D. Munteanu (Societatea Ornitologica Româna), J. Winkelman and E. Osieck (Vogelbescherming Nederland), D. Pain (Royal Society for the Protection of Birds), I. Rusev (Natural Heritage Fund, Ukraine), M. Strazds (Latvijas Ornitologijas Biedriba), J. Winkelman (Vogelbescherming Nederland) and M. Yarar (Dogal Hayati Koruma Dernegi). D. Bloch, H. Meltofte, B. Olson, A. Michalak and S. Polak answered last-minute queries.

INLAND WETLANDS

compiled by

J. VAN VESSEM, N. HECKER AND GRAHAM M. TUCKER

from contributions supplied by the Habitat Working Group
Å. Andersson, N. Hecker (Coordinator), V. Keller, P. Musil, S. Ormerod, T. Salathé, G. Sarigül,
V. Serebryakov, V. Staneviscius, J. van Vessem, M. Wieloch, M. Yarar

SUMMARY

Due to human action, most wetlands in Europe have been drained and very few now remain in a natural condition. Despite these negative trends, inland wetlands are still important for the survival of 102 bird species within Europe. More than 55% of these priority birds have remarkably small, or strongly declining, populations in Europe. The remaining priority species have a favourable conservation status in Europe but are wetland specialists, and are thus highly dependent on the continuing existence of these habitats. Most of these latter species are currently still widespread in Europe, although many are not common and require habitat features that are only found in semi-natural or natural wetlands. A significant proportion are also naturally restricted within Europe to certain regions or to particular types of wetland.

Maintaining the extent and ecological quality of the remaining inland wetlands is clearly essential for the conservation of these 102 priority bird species in Europe. In reality, however, loss and deterioration of these habitats is still widespread, due mainly to drainage and other hydrological engineering for agriculture, flood control and land-claim, as well as to water pollution and to over-abstraction of water. Impoundments and other artificial regulation of water-levels can cause widespread damage to habitat quality downstream.

International and EU legislation, international funds and broad-based policy initiatives exist to promote, support and coordinate the required national actions to deal with pollution control and the protection of important wetland sites. However, even broader and more holistic measures are still needed to address the major threats to inland wetlands.

In particular, more integrated management of entire water catchments, through more effective land-use planning, and through the reform of international funds for infrastructural development, will help to incorporate the objectives of wetland conservation more fully into programmes, policies and plans across all relevant sectors, as will the proper enactment and implementation of national biodiversity plans, Strategic Environmental Assessments and Environmental Impact Assessments. Networks of important wetland sites should be identified and protected. The 'external' values of wetlands should be incorporated into mainstream economics, and the ecological services which are provided by these habitats need to be more widely publicized and understood.

DESCRIPTION OF THE HABITAT

Habitat definition, extent and distribution

This habitat conservation strategy covers all European areas of non-coastal wetlands (i.e. not influenced by the sea), with the exception of boreal/ombrogenous mires (i.e. raised bogs and blanket bogs) and all wetlands within the tundra zone, which are both covered under 'Tundra, Mires and Moorland Habitat' (p. 159).

Wetlands are defined by the Ramsar Convention, and in this strategy, as 'areas of marsh, fen, peatland or water, whether natural or artificial, permanent or temporary, with water that is static or flowing, fresh, brackish or salt including areas of marine water, the depth of which does not exceed 6 m.' Clearly, the predominance of water within all these habitat-types characterizes wetlands.

The movement of water through natural ecosystems forms a major part of the hydrological cycle, one of the earth's most important natural phenomena and fundamental to life. Through it, rivers, lakes, marshes and other wetlands are integrally linked to each other. For this reason, any division of wetlands is certain to be artificial as all are linked at the catchment scale. However, only inland (i.e. non-coastal) wetlands are covered by this strategy, and for the convenience of categorizing species' requirements and management issues these are divided into three main types: riverine, lake and marsh habitats (Finlayson and Moser 1991, Dugan 1993). However, the integrated catchment theme is implicit throughout the strategy as this is increasingly prominent in the management of water resources and wetlands.

Rivers

River systems are drainage systems characterized by unidirectional flow (intermittent or permanent), through which water is transferred in surface channels from well-defined catchments (often with associated flood-plains in the lower reaches) to seas and estuaries, or to land-locked lakes and marshes. Canals—which transfer water, often at slow speed, between locations in engineered channels determined largely by people—are also included in this habitat subdivision.

Lakes

Lakes are standing water-bodies with open water surfaces, of either natural or artificial origin, and with permanent or temporary water. Lakes are characterized by their large size and often deep water in which temperature changes with depth, while ponds, which are smaller and shallower, have water of more uniform temperature.

Most European lakes are part of riverine systems with a sufficient through-flow of water to flush out salts and maintain freshwater conditions. In contrast, the water of some land-locked lakes leaves only by evaporation which can result in the progressive accumulation of mineral salts and hence the formation of salt-lakes. Such salt-lakes are mainly situated in climates where evaporation exceeds precipitation throughout several months, i.e. in southern Europe and particularly in Spain and Turkey.

Marshes

Marshes are characterized by emergent, rooted vegetation such as reeds, rushes, grasses and sedges. They can be found where freshwater lake-shores are flat, or in places with a permanently or temporarily high groundwater level, with surface springs, or in the flooding areas of rivers and lakes. Where the water source is derived directly from atmospheric precipitation (i.e. ombrogenous), such acidic and nutrient-poor water (oligotrophic) leads to the development of ombrogenous mires, such as raised bogs and blanket bogs. These are typical of high rainfall areas in northern and western Europe and have a different vegetation (dominated by *Sphagnum* moss) and bird community from the other inland wetlands treated here, and they are therefore treated under 'Tundra, Mires and Moorland' (p. 159) rather than within the present chapter.

Current extent and distribution

Rivers, lakes and marshes are found throughout Europe, particularly in areas of high rainfall, but their extent is difficult to estimate as published figures often use varying definitions (see Stanners and Bourdeau 1995). Table 1 summarizes the distribution and extent of the largest rivers and their catchments. Riverine habitat is so widespread and common that all European countries have responsibilities for river conservation. North-west Russia and Fennoscandia hold three-quarters of natural lake habitat in Europe (Table 2), and therefore have special responsibilities for the conservation of biodiversity in these habitats.

National estimates of the extent of inland marshland are poorly documented and therefore not given here. Although inland marsh habitats are widespread and frequently found adjoining lake and riverine wetlands, they do predominate in low-lying, flat regions with high rainfall, particularly where drainage for agriculture, etc., has not been extensive. The countries where particularly important areas of marshland habitat remain include Belarus, Poland, Russia and Turkey.

Changes in the extent of wetland habitats in Europe are difficult to quantify, as substantial losses probably started at the onset of the agricultural revolution, 10,000 years ago or more. The rate of destruction has accelerated this century, especially during the second half. Europe is so densely populated, and has such a long history of habitat modification for agricultural, urban and industrial development, that most natural wetlands have been lost and very few now remain. Wetland loss has been least in northern Europe, particularly in tundra habitats (see 'Tundra, Mires and Moorland', p. 159). Elsewhere, wetland loss has been widespread, though perhaps greatest in southern Europe, where water resources are scarcest. Losses have been documented in various countries, though with rather inconsistent approaches, as the following examples show.

Drainage of semi-natural wetland in Belgium has taken place at the rate of 10–12 km² per year since 1960. In Britain, about 50% of the country's lowland fens and valley and basin mires have been destroyed or significantly damaged since 1949 (Baldock 1984). In France, it is estimated that 10,000–20,000 km² of wetland have been drained in total. In Portugal, 70% of the wetlands in the western Algarve have been lost to agricultural development (Dugan 1990).

Table 1. Larger European river catchments exceeding 50,000 km², ordered by area of catchment (Stanners and Bourdeau 1995).

	Country	Catchment area (10³ km²)	Mean discharge (km³/yr)	Length (km)
Volga	Russia	1,360	230	3,530
Danube	Germany, Austria, Slovakia,Hungary, Croatia, Serbia, Romania, Bulgaria, Ukraine, Switzerland*, Poland*, Italy*, Czech Republic*, Slovenia*, Bosnia-Herzegovina*, Albania*, Moldova*	817	205	2,850
Dnepr	Russia, Belarus, Ukraine	558	53	2,270
Don	Russia, Ukraine*	422	38	1,870
Severnayy Dvina	Russia	358	148	740
Pechora	Russia	322	129	1,810
Neva	Russia, Finland*, Belarus*	281	79	75
Ural	Russia, Kazakhstan	270	—	2,540
Wisla	Poland, Slovakia*, Ukraine*, Belarus*	194	31	1,050
Kura	Georgia, Turkey, Azerbaijan, Armenia*, Iran*	188	18	1,360
Rhine	Switzerland, Austria, Germany, France, Netherlands, Italy*, Luxembourg*, Belgium*	185	69	2,200
Elbe	Czech Republic, Germany, Austria*, Poland*	148	24	1,140
Oder	Czech Republic, Poland, Germany	119	16	850
Loire	France	118	32	1,010
Neman	Belarus, Lithuania, Russia, Poland*	98	22	960
Douro	Spain, Portugal	98	20	790
Rhône	Switzerland, France	96	54	810
Zapadnaya Dvina	Russia, Belarus, Latvia, Lithuania*	88	21	1,020
Garonne	France	85	21	575
Ebro	Spain	84	17	910
Tajo	Spain, Portugal	80	6	1,010
Seine	France	79	16	780
Mezen'	Russia	78	26	970
Guadiana	Spain, Portugal	72	—	800
Dnestr	Ukraine, Moldova	72	10	1,350
Po	Italy, Switzerland*	69	46	670
Yuzhnyy Bug	Ukraine	65	3	860
Kuban'	Russia	58	13	870
Onega	Russia	57	18	420
Guadalquivir	Spain	57	2	675
Kemijoki	Finland	51	17	510

* Country includes part of the catchment area but the main river does not run through it.

The loss of natural inland wetlands has been partially offset by the creation of man-made wetlands such as reservoirs, fish-ponds and gravel-pits. However, these new wetlands are very different in their physical and ecological character (e.g. they tend to be eutrophic) and are often unsuitable for many of the most specialized and threatened wetland species. The loss of complex and ancient natural wetlands cannot, therefore, be normally compensated for by the creation of artificial ones.

Origin and history of the habitat

Rivers are the primary route through which water from precipitation (snow and rainfall) travels to the sea, thus completing the hydrological cycle. Their presence and routes originate from the influence of natural factors, principally climate and the effects of geology on catchment topography. Now, however, the structure and even presence of riverine habitats is increasingly affected by man, through impoundment, abstraction, and canalization.

Most natural lakes in Europe formed after the last Ice Age in basins created by glacial action, although a minority formed in basins created by tectonic or volcanic activity (see, e.g., Björk 1994). Oxbow lakes and plunge basins may also form when a bend in a meandering river is cut off from the main stream. Man-made lakes and ponds are now commonplace and include retention basins for flood control, reservoirs (for irrigation, power production, industry and domestic consumption), fish-ponds, waste-water treatment areas and cooling ponds.

Marshes form in shallow, still or slowly moving water, especially on soft substrates which favour the growth of emergent plants. Fens may occur where the water-table is close to the surface because of

Table 2. The number and extent of lakes (larger than 1 ha except where stated) in each European country. There are no, or only a few, lakes in Belgium, Czech Republic, Luxembourg, Portugal and Slovakia. Data are not available for any other European countries missing from this table.

	No. of lakes >1 ha	Total area (km²)	% of country*	Comments
Albania	>1,000	330	1.2	
Austria	329	292	0.3	
Bulgaria	c.3,500	600	0.5	60 wetlands >100 ha
Croatia	4 (>10 ha)	3	0.0	
Denmark	c.2,000	500	1.2	74 lakes >100 ha
Estonia	1,004	379	0.8	
Finland	56,012	11,487	3.4	2,589 lakes >100 ha
France	152 (>10 ha)	37	0.0	
Georgia	892	175	0.3	
Germany	122 (>100 ha)	c.350	0.1	
Greece	18 (>1,000 ha)	270	0.2	
Hungary	1,097	1,520	1.6	
Iceland	8,843	433	0.3	
Rep. of Ireland	118 (>100 ha)	1,890	2.2	
Italy	270 (>10 ha)	730	0.3	
Latvia	3,046	418	0.7	
Lithuania	2,543	946	1.5	13 lakes >1,000 ha, 144 lakes >100 ha
Moldova	3,385	74	0.2	
Netherlands	50 (>1,000 ha)	3,650 (lakes >8 ha)	10.1	River flood-plains cover 38,000 ha
Norway	210,457	17,000	5.2	1,884 lakes >100 ha
Poland	9,296	4,280 (all lakes)	1.4	Also 50 reservoirs (47.7 km²)
Russia	476,089	37,396	0.2	Europian Russia only
Spain	1,275	164	0.0	Data include marshes
Sweden	92,400	41,000	9.3	3,533 lakes >100 ha
Switzerland	1,330	1,297 (lakes >10 ha)	3.1	c.1,500 lakes if small ones are included; also 94 reservoirs (>10 ha, total 118 km²)
Turkey	?	7,170	0.9	Lake Van alone (largest in Turkey) is 371,300 ha
Ukraine	957	345	0.1	
United Kingdom	5,500 (>4 ha)	727 (lakes >4 ha)	0.3	More than 280 km² of reservoirs in England and Wales; 300,000 ponds <1 ha; all types of standing waters total 3,300 km² (1.4% of UK)

* Care should be taken when interpreting these data as for some countries only lakes larger than a certain size have been recorded, and data for some countries may include reservoirs as well as natural lakes.

Sources. Mainly Stanners and Bourdeau (1995). Additional sources as follows: *Bulgaria* Valkanov (1966), Karapetkova *et al.* (1993), Bulgarian Society for the Protection of Birds (unpublished data); *France* Anon. (1978); *Hungary* Biró (1984), Rakonczay (1990); *Republic of Ireland* Coveney *et al.* (1993); *Italy* Gaggino *et al.* (1987); *Lithuania* Homskis (1969), Anon. (1975), Parunkstis (1975), Vasilianskiene (1975), Staneviscius (1992); *Spain* Montes (1990); *Switzerland* Anon. (1976), Dill (1990); *Turkey* Baris (1989); *United Kingdom* Owen *et al.* (1986), National Rivers Authority (1991), Anon. (1994b), D. Allen and K. Partridge (pers. comm.).

lateral drainage or groundwater inflow of alkaline to neutral water that is moderately to highly rich in nutrients (Orme 1990).

Ecological characteristics

Key abiotic factors

The ecology of wetlands is profoundly affected by their physical and chemical properties and therefore an understanding of these and fluvial geomorphology (the study of water-shaped landforms) is essential for the conservation of their habitats and species. This is beyond the scope of this book, but a useful overview of the subject for ecologists is given in Gordon *et al.* (1992).

Important sources of variation in rivers include size (from discharges of less than 1 cubic metre per second in headstreams, to many thousands per second in large lowland rivers), hydrology and flow regime (e.g. wholly seasonal, seasonally flooding, spatey, ground- or surface-fed), hydraulic characteristics (e.g. current velocity and turbulence), bedform (silt, sand, cobbles or boulders), slope, channel morphology (e.g. simple, braided, meandering and deltoid), chemistry (e.g. levels of dissolved oxygen, carbon dioxide, nitrates, phosphates, calcium carbonate and organic matter), transported materials (bed-load, suspended solids) and temperature (e.g. liability to freezing). In turn, these factors are a reflection of catchment characteristics, such as altitude, latitude, geology, precipitation and evaporation, and topography.

Abiotic factors also vary with downstream progression, between river reaches, and between mid-channel and marginal zones (Calow and Petts 1992). Thus under natural circumstances the character of a river changes from source to mouth. Headwater stages where rivers are eroding consist of small streams with turbulent flow, highly oxygenated water that is low in nutrients, little vegetation and rocky substrates. In contrast, the lower reaches, where sediments are deposited, are characterized by low turbulence and oxygen levels, high levels of dissolved nitrates, phosphates, organic matter and suspended solids, soft and silty substrates, and large meandering channels within a large flood-plain.

However, superimposed on this large-scale pattern are differences between alternating stretches of rapids and pools, with rapid flow, rocky substrates and erosion in the former, and gentle flow leading to sedimentation and soft substrates in the latter. Thus both upland and downstream stages show variation within them and consequently high physical and chemical diversity, which in turn is important for the maintenance of high biodiversity.

Different riverine habitats are therefore associated with distinct stages in this progression in physical character from upland to lowland rivers, and rapid and pool components within them. Increasingly, however, such natural characteristics are now modified by river engineering (see p. 139).

Geomorphological events (e.g. landslides, earthquakes, etc.) and inflowing sediments shape the depth and slopes of lakes, while soil chemistry and dissolved minerals influence the water quality of lakes and marshes. Compared with rivers, lakes tend to offer a relatively more stable environment. Upon entering lakes, river-transported sediment can settle out of the water so the amount of light that penetrates increases, thus allowing increased plant productivity.

Lakes often become thermally stratified in summer and again in winter. The upper, warmer waters become temporarily isolated from the colder, lower part by a thermocline zone that acts as a barrier to exchange of oxygen and nutrients. This may lead to lower productivity in the deep-water layer. Mixing, however, occurs in spring and autumn, when the two water-bodies converge in temperature, which in turn often leads to 'blooms' of phytoplankton.

Canals, in some respects, are intermediate between running and standing waters, although discharge and current velocity will vary depending on their origin and use. As such, key sources of chemical variation arise in ways similar to lakes and rivers, whereas physical characteristics will reflect slope, or channel and riparian physiography. Other important sources of variation in canals are their age, state of maintenance, and extent of use.

Like rivers, lakes and marshes are an integral part of the water cycle, depending on hydrological events in the catchment and influencing downstream hydrology. They absorb floodwater and regulate the supply of surface water and groundwater. They can absorb nutrients and retain sediments, thus purifying water supplies.

Climate has a strong influence on wetlands. The dry and hot summers of the Mediterranean basin and the Pannonian basin (lowland plains centred on Hungary) completely dry out many wetlands for part of the year. On the other hand, high salinity and greater depth prevent permanent lakes from freezing, thus providing winter habitat for waterbirds.

Key biotic factors and natural dynamics

Productivity is a key factor influencing the ecology of inland wetlands. An increase in productivity, through eutrophication (i.e. an increase in nutrient levels), can have a positive or negative effect on the natural dynamics of lakes and marshes according to its intensity. At low to moderate levels the biotic production increases, resulting in an increase of food supply (e.g. phytoplankton, invertebrates and fish). In 'natural' environments, that are virtually unaffected by modern man, this proceeds on timescales of decades to centuries. Higher rates and levels of eutrophication, rarely seen in 'natural' wetlands, usually lead to algal blooms and a decrease in submerged plants, which disrupt the normal functioning of the ecosystem and which in extreme cases may cause oxygen deficiency in the water and mass death of fish (see Box 2; Freedman 1989, Harper 1992). Such detrimental eutrophication often occurs in man-made water-bodies or in areas that are farmed intensively.

Increases in nitrogen usually increase primary production, but in general it is believed that production in most fresh waters is limited by phosphorus (Freedman 1989, Harper 1992). Additions of phosphorus will thus tend to increase productivity more than nitrogen. The acidity or otherwise of water may also have consequences for production in both lakes and rivers, and, in general, productivity is higher in more calcareous (slightly alkaline) waters (Steinberg and Wright 1994).

Characteristic flora and fauna

Globally threatened animals of wetlands in Europe are listed by Baillie and Groombridge (1996).

Rivers

In general, riverine vegetation shows a progression from upland rivers dominated by attached algae (e.g. diatoms and chlorophytes) and mosses (bryophytes), through stretches dominated by submerged and emergent flowering plants (e.g. *Ranunculus, Phalaris, Potamogeton*), to the lowest zones which are characterized by suspended phytoplankton. This longitudinal change can be thought of as downstream succession. Of particular importance to the availability of bird habitats, rivers also drive the successional change in vegetation in their immediate surrounding environment, for example during meandering cycles and flood-plain restructuring caused by flooding events.

The invertebrate communities of rivers comprise, at the microscopic level, a diverse array of protozoans, rotifers and microcrustaceans, but the most conspicuously visible forms include insects, larger crustaceans, molluscs and oligochaete worms. Invertebrate communities are highly influenced by water quality, with diversity declining rapidly with increased pollution. Indeed, invertebrate species composition is frequently used as a biological indicator of water quality (see, e.g., Hawkes 1979, Green 1989).

Fish communities change from salmonids in upland zones, to cyprinids, sometimes migratory clupeids (e.g. *Alosa*), and flounder *Platichthys flesus* closest to the sea. Amphibians and mammals associated with rivers are not diverse in Europe, but some important species occur, including eight globally threatened amphibians and three globally threatened species of aquatic mammals (*Desmana moschata, Galemys pyrenaicus* and *Mustela lutreola*), as well as the much-reduced European subspecies of otter *Lutra lutra lutra* (United Nations 1991, Baillie and Groombridge 1996). Also, commoner but declining groups such as bats (Chiroptera) feed directly on insects flying over rivers.

Riverine bird communities in Europe are not as rich as those of other wetlands and have few species restricted to them (see 'Habitat needs of priority birds', p. 133). They also differ considerably according to their position along the longitudinal progression from headwaters to lower reaches. Upstream reaches have a low diversity, but are typified by, and are important for, Goosander *Mergus merganser*, Common Sandpiper *Actitis hypoleucos*, Grey Wagtail *Motacilla cinerea* and Dipper *Cinclus cinclus*.

Lower reaches of rivers have higher diversity and population densities of birds, especially where there are abundant fish, invertebrates associated with water plants, natural riparian vegetation and banks. However, unlike upland-river species, lowland-river bird communities are not characteristic of the habitat, but are common to marshes, lakes and sometimes coastal wetlands. Typical birds of lowland rivers (and canals) include Little Grebe *Tachybaptus ruficollis*, Grey Heron *Ardea cinerea*, Mute Swan *Cygnus olor*, Moorhen *Gallinula chloropus*, Kingfisher

Alcedo atthis and Sand Martin *Riparia riparia*.

Where there is abundant marginal vegetation then Little Bittern *Ixobrychus minutus* may occur, along with more common species such as Reed Warbler *Acrocephalus scirpaceus*, Great Reed Warbler *Acrocephalus arundinaceus*, Sedge Warbler *Acrocephalus schoenobaenus* and Reed Bunting *Emberiza schoeniclus*. Where suitable banks or islands of sand or shingle are present then Little Ringed Plover *Charadrius dubius* and Little Tern *Sterna albifrons* may nest.

Lakes and marshes

The plant communities of lakes, ponds and their marginal marshland zones show many similarities in composition throughout Europe. Differences in plant communities do, however, occur as a result of geographical region and altitude, hydrological regime, nutrient status, acidity and alkalinity, oxygen content, the amount of light penetrating the water and the slope of the shore.

A zonation of vegetation-types with decreasing water depth is typical of freshwater lakes. Deep waters are characterized by attached algae and phytoplankton. The outer reaches of the littoral (shore) zone, around 2 m deep, are dominated by submerged vegetation (Characeae, *Myriophyllum*, etc.), followed by communities of floating plants (*Nuphar, Nymphea, Lemna*) with decreasing water depth. Nearer the shore, emergent vegetation appears, typically represented by reed *Phragmites*, reedmace *Typha*, common club-rush *Scirpus lacustris* or reed canary-grass *Phalaris arundinacea*, followed by sedge *Carex*. Gradually, towards the drier zones, shrubs and trees can develop into a carr woodland of alder *Alnus* and willow *Salix*. Human activities like mowing and grazing, however, have often prevented the growth of trees, and historically have replaced such habitats with wet meadows dominated by sedges and herbs.

This zonation in space can also be observed over time through ecological succession, whereby vegetation zones develop and alter the environmental conditions so that they make the habitat more suitable for the next successive community (see Box 1). Wetlands are therefore dynamic and are ultimately transformed to dry-ground ecosystems by natural development. Succession can be accelerated by human activities such as drainage, river diversion, groundwater abstraction, nutrient pollution or siltation, or slowed by regular cropping of plants or partial removal of sediments.

Marshes that are independent of lakes, on permanently wet soils or at sites with surface springs, generally possess the same vegetation as lake-side marshland, although the typical wet–dry zonation is sometimes incomplete or lower in diversity.

Box 1. Ecological succession in wetlands.

Colonizing plants of deep, open water-bodies—such as duckweeds *Lemna*, pondweeds (e.g. *Potamogeton*) and water-lilies *Nymphaea*—produce detritus which then accumulates along with other sediments washed into the water-body. This process may be slow at first, but, as the water grows shallower, emergent plants—such as reeds *Phragmites*, reed-mace *Typha* and bulrushes *Scirpus*—establish themselves, thereby speeding up the succession as they impede water flow, trap more sediment and shade out floating vegetation. As the basin gets progressively shallower the water-body may become seasonal, thereby allowing the next successional stage of terrestrial plants to establish, which in turn leads to further drying out through their high rates of evapotranspiration. Thus the water-body may disappear as the wetland becomes a marsh and, with further drying out, a fen carr and ultimately, at the successional climax, a woodland.

As the vegetation changes, the size and structure of the wetland is similarly transformed together with the animal communities associated with it. Thus birds that depend on large, open water-bodies and which feed on deep-water vegetation, fish, or other aquatic animals are replaced, as succession proceeds, by species that favour dense cover and shallow water, until eventually the wetland and its characteristic species are lost.

The littoral and benthic (bottom) zones of lakes are frequently varied in structure and typically comprise a mixture of rocks, sediment and plants, and are colonized by sessile animals and algae, together with other animals that can move freely but stay close to the interface. In the bigger lakes, these communities have lower total biomass relative to those of the open water, but the smaller and shallower a lake becomes, the more the littoral and benthic zones assume dominance. The communities of the littoral and benthic zones are relatively productive, perhaps as a result of access to nutrient sources in the sediments, and because of the structural complexity of the habitats.

Littoral communities of invertebrates are diverse. Typical of standing water are free-swimming forms, e.g. various beetles (Coleoptera) and species associated with submerged plants. On exposed stony shores, communities are essentially similar to those of running water, while on areas of bare sand most invertebrates are burrowing forms. Fish populations are important in determining the abundance and sometimes the species composition of invertebrate communities. Most of the small lake-fish species occur here, e.g. various Cyprinidae, Gasterosteidae, and other bottom-dwelling and weed-loving species. The littoral zone is also important as a spawning and nursery area for open-water fish.

The deeper benthos, where light is insufficient for rooted plants, has a lower diversity, but not necessarily low productivity. However, in thermally stratified lakes, deoxygenation of deep non-circulating water may occur so that benthic productivity may be limited or even become completely absent. Deep benthic habitats are usually of low importance for birds.

Zooplankton dominate open waters of lakes and ponds. Dependent on this primary food resource are numerous larger free-swimming species, mainly fish, which feed on the plankton community and in turn are preyed upon by larger fish and/or birds.

The bird fauna of lakes is among the richest of any habitat in Europe, particularly in large, highly productive wetlands with abundant emergent and surrounding marshland vegetation. Some widespread and particularly characteristic species include grebes (Podicipedidae), various ducks (Anatidae, including Mallard *Anas platyrhynchos*, Gadwall *A. strepera*, Red-crested Pochard *Netta rufina*, Pochard *Aythya ferina* and Tufted Duck *Aythya fuligula*), Coot *Fulica atra*, and some terns (Sternidae) including all three species of marsh-tern *Chlidonias*. Osprey *Pandion haliaetus* is a rarer bird of prey, also characteristic of this habitat.

Bird population sizes on inland lakes may also be extremely large, particularly outside the breeding season when populations of swans, geese and ducks (Anatidae) that breed in northern tundra habitats, within and outside Europe, migrate to winter on lakes, particularly in western and southern Europe (see 'Tundra, Mires and Moorland', p. 159).

In ponds, one of the main factors controlling animal communities is the quality and quantity of plant growth, which may vary from algal dominance (in highly eutrophic ponds) to abundant aquatic macrophytes or complete overgrowth by emergent vegetation. Most ponds have a rich invertebrate fauna but the avifauna is usually very poor in comparison to most wetlands, although species such as Garganey *Anas querquedula* occasionally occur in high-quality habitats.

The aquatic invertebrate fauna of swamps and marshes is dominated by groups of air-breathers, such as beetles (Coleoptera), some flies (Diptera) and pulmonate molluscs, together with specialized members of some otherwise water-breathing groups. Amphibians are often abundant, as well as other vertebrate predators such as snakes.

Marshes are often the main nesting habitat of birds which use the open water or margins of lakes and rivers. Such nesting species include grebes, Spoonbill *Platalea leucorodia*, Glossy Ibis *Plegadis falcinellus*, ducks, *Fulica atra*, and gulls (Laridae), for example Black-headed Gull *Larus ridibundus* and Little Gull *L. minutus*. A significant component of the rich marshland avifauna, however, comprises species that require dense marginal vegetation and do not use the open water; these include Bittern *Botaurus stellaris*, Purple Heron *Ardea purpurea*, Marsh Harrier *Circus aeruginosus*, Water Rail *Rallus aquaticus* and other crakes (e.g. *Porzana*), Purple Gallinule *Porphyrio porphyrio*, and most of the passerines, for example numerous reed warblers *Acrocephalus*, Cetti's Warbler *Cettia cetti*, Savi's Warbler *Locustella luscinioides*, Bearded Tit *Panurus biarmicus*, Penduline Tit *Remiz pendulinus* and *Emberiza schoeniclus*.

Relationships to other habitats

Rivers interact with many other habitats in their catchments, and are pivotal in transferring water, sediment, energy, minerals and nutrients from the atmosphere, catchment and riparian zones to their flood-plains, associated lakes and marshes, and to downstream deltas, estuaries and seas. For example, rivers transport water to lakes and marshes, and mineral- or nutrient-rich sediments to seas, estuaries and deltas. Besides these direct contacts with other aquatic and semi-aquatic habitats, rivers also provide energy and water to organisms in the surrounding riparian habitats. Influences on terrestrial vegetation through changing moisture regimes, or the provision of emergent insects, e.g. to birds and bats in the riparian zone, are important functions of rivers worldwide.

Lakes and marshes are closely linked to their surroundings, especially habitats in natural lakes where there is usually no sharp division between wet and dry habitats, but rather a transition zone characterized by changing water-levels and a gradual vegetational succession. Hydrological engineering and agricultural activities, however, often destroy this transition zone. Lakes and marshes are also recipients of nutrients from the surrounding land as well as from areas in the wider catchment.

Values, roles and uses

The management and uses of wetlands are long-standing and diverse (see Dugan 1990, Finlayson and Larsson 1991, Finlayson *et al.* 1992, Maltby *et al.* 1992, Davis 1993). The types of use can roughly be divided into three groups:

- The use of the water itself (e.g. for irrigation, industrial and domestic supply, waste disposal, navigation and transport);

- The use of the resources offered by wetlands (e.g. aquaculture, gravel or sand extraction, fishing, hunting);

- The use of the wetland as a value in itself (e.g. for recreation, environmental education, scientific research).

Although many of these uses are common and widespread in Europe, some may be only locally common, while others are found widely over Europe, but only in a small number of wetlands.

Some of these valuable uses may help maintain wetlands, but at the same time most have the potential to degrade wetlands if practised inappropriately, intensively and in a non-sustainable way (see 'Threats to the habitat', p. 137). Furthermore, to a large extent many of these uses are in conflict with each other. In addition, land-uses in catchment areas are often exercised without regard for wetlands and particularly the rivers which drain them. Terrestrial developments proceed often without recognition of their effects on river systems. For these reasons, ideal management of wetlands increasingly emphasizes integrated planning over the whole catchment basin, in order to minimize multi-use conflicts. Also, such plans should ideally be recognized and feature in strategic resource-use policy.

Clearly, where wetland uses are sustainable and not damaging to ecosystems and biodiversity, they provide an important reason for wetland conservation. Wetland conservation plans should therefore be multi-disciplinary and incorporate these uses to maximize their conservation potential through the promotion of 'wise-use' concepts (see 'Conservation recommendations', p. 153).

Political and socio-economic factors affecting the habitat

Rivers, lakes and marshes are highly interconnected, therefore factors that affect catchment use—such as policies for water supply, agriculture, industry, domestic use, land-use planning, transport, energy and tourism—can all have profound effects on wetland ecosystems. Currently, particularly important effects of this type are being caused by political change in eastern Europe, where resources such as land are being transferred increasingly from the state to the private sector, and where the tourism sector is developing very rapidly and sometimes in an uncontrolled manner. These same socio-economic factors can affect the management and use of wetlands directly, for example as marginal land is brought into, or out of, economic use. There are, in addition, special political and socio-economic pressures on wetlands which arise from:

- The intrinsic value of water as an essential resource for waste disposal, water supply (to house-holds, industry and agriculture), transport, and power generation;

- The perceived importance of river engineering and wetland drainage to protect against flooding, often associated with high investment in infra-structural and agricultural development of flood-plains;

- Restrictions on the ability to manage whole catch-ments, particularly where they straddle adminis-trative boundaries between regions or nations: e.g. in Belgium, Bulgaria, Hungary, Luxembourg, Poland, Portugal, Romania and Spain, more than 55% of the land area is in international basins (Gleik 1989);

- The effects of very widespread factors which are outside even catchment control, such as long-range air pollutants leading to acidification;

- The lack of general public awareness of the socio-economic values of wetlands.

At the same time, increasing public and political awareness of environmental problems, and better relations and cooperation between nations, provide important opportunities for wetland conservation (see 'Conservation opportunities', p. 149).

PRIORITY BIRDS

European inland wetlands are clearly of consider-able importance for birds, holding 102 priority spe-cies (Table 3), which is approximately 20% of all regularly occurring bird species in Europe, and the second-highest total of priority species for any habi-tat strategy in this book. Fifty-eight of these have an Unfavourable Conservation Status in Europe be-cause they have small, declining or highly localized populations there (Table 4). This amounts to about 30% of all bird species with an Unfavourable Con-servation Status in Europe (Tucker and Heath 1994)— only agricultural habitats hold more such species overall.

Also notable is that over 50% of the 102 priority species' European populations depend very heavily on inland wetlands at some stage in their annual cycle (Table 4). This underlines the importance of the habitat to a large proportion of Europe's avifauna, many of which have an Unfavourable Conservation Status.

Inland wetlands hold eight Priority A species, all of which are globally threatened (Collar *et al.* 1994). Of these, the European populations of Ferruginous Duck *Aythya nyroca* and White-headed Duck *Oxyura leucocephala* are very heavily dependent on inland lakes for breeding and wintering habitat, while the

European wintering population of Red-breasted Goose *Branta ruficollis* depends heavily on inland lakes for roosting (Hunter and Black 1996). Four of these species are also considered to be endangered in Europe (Tucker and Heath 1994): Dalmatian Pelican *Pelecanus crispus*, Marbled Teal *Marmaronetta angustirostris*, *Oxyura leucocephala* and Aquatic Warbler *Acrocephalus paludicola*. Since Slender-billed Curlew *Numenius tenuirostris* is a migrant through Europe, apparently without having a regu-larly breeding or wintering population on the conti-nent, it has not been ascribed a 'European Threat Status' (Tucker and Heath 1994), but it is one of the most threatened birds in the world, with an IUCN threat status of Critical (Collar *et al.* 1994, Gretton 1996).

Inland wetlands also hold 19 Priority B species and a large number of Priority C (31) and Priority D species (44). Three of the Priority B species are endangered in Europe—*Platalea leucorodia*, Greater Sand Plover *Charadrius leschenaultii* (provision-ally) and the globally threatened Greater Spotted Eagle *Aquila clanga*—and four of the Priority C species are similarly endangered.

Inland wetlands are important habitats for breed-ing, passage and wintering populations of birds. At least 97 of the 102 priority species breed in Europe. Of these, Russia, Turkey, Ukraine and Romania— countries outside the European Union—support the greatest numbers (Appendix 2). This high diversity strongly reflects the particular importance of the remaining wetlands in these countries. These wetlands are relatively extensive, diverse and intact compared to those in most other European countries (IUCN 1993), and support relatively large populations of individual species. Indeed, almost all of the priority species breed regularly in Russia.

Outside the breeding season, many species from the northern tundra or from outside the continent use European wetlands as wintering sites or staging posts on migration (see 'Tundra, Mires and Moor-land', p. 159). Indeed, inland wetlands form net-works that are vital for the conservation of whole flyway populations of priority waterbird species (e.g. Boyd and Pirot 1989, Scott and Rose 1996).

HABITAT NEEDS OF PRIORITY BIRDS

Clearly the basic and most important general re-quirement for priority wetland birds, and other wild-life, is the maintenance of the habitat as a wetland and in particular its hydrological regime. Changes in hydrology can have profound effects on the suitabil-ity of a wetland for priority birds and other wildlife, either directly, e.g. through changing water depth,

Table 3. The priority bird species of inland wetlands in Europe.

This prioritization identifies Species of European Conservation Concern (SPECs) (see p. 17) for which the habitat is of particular importance for survival, as well as other species which depend very strongly on the habitat. It focuses on the SPEC category (which takes into account global importance and overall conservation status of the European population) and the percentage of the European population (breeding population, unless otherwise stated) that uses the habitat at any stage of the annual cycle. It *does not* take into account the threats to the species within the habitat, and therefore should not be considered as an indication of priorities for action. Indications of priorities for action for each species are given in Appendix 4 (p. 401), where the severity of habitat-specific threats is taken into account.

	SPEC category	European Threat Status	Habitat importance
Priority A			
Dalmatian Pelican *Pelecanus crispus*	1	V	●
Lesser White-fronted Goose *Anser erythropus*	1	V	●
Red-breasted Goose *Branta ruficollis*	1	L [W]	■
Marbled Teal *Marmaronetta angustirostris*	1	E	●
Ferruginous Duck *Aythya nyroca*	1	V	■
White-headed Duck *Oxyura leucocephala*	1	E	■
Slender-billed Curlew *Numenius tenuirostris*	1	—[1]	●
Aquatic Warbler *Acrocephalus paludicola*	1	E	●
Priority B			
Black-throated Diver *Gavia arctica*	3	V	■
Pygmy Cormorant *Phalacrocorax pygmeus*	2	V	■
Bittern *Botaurus stellaris*	3	(V)	■
Little Bittern *Ixobrychus minutus*	3	(V)	■
Squacco Heron *Ardeola ralloides*	3	V	■
Purple Heron *Ardea purpurea*	3	V	■
Spoonbill *Platalea leucorodia*	2	E	■
Ruddy Shelduck *Tadorna ferruginea*	3	V	■
Smew *Mergus albellus*	3	V	■
Greater Spotted Eagle *Aquila clanga*	1	E	●
Osprey *Pandion haliaetus*	3	R	■
Baillon's Crake *Porzana pusilla*	3	R	■
Corncrake *Crex crex*	1	V	●
Greater Sand Plover *Charadrius leschenaultii*	3	(E)	■
Black-tailed Godwit *Limosa limosa*	2	V	●
Whiskered Tern *Chlidonias hybridus*	3	D	■
Black Tern *Chlidonias niger*	3	D	■
Kingfisher *Alcedo atthis*	3	D	■
Priority C			
Red-throated Diver *Gavia stellata*	3	V	●
White Pelican *Pelecanus onocrotalus*	3	R	●
Night Heron *Nycticorax nycticorax*	3	D	●
Black Stork *Ciconia nigra*	3	R	●
White Stork *Ciconia ciconia*	2	V	●
Glossy Ibis *Plegadis falcinellus*	3	D	●
Greater Flamingo *Phoenicopterus ruber*	3	L	●
Gadwall *Anas strepera*	3	V	●
Pintail *Anas acuta*	3	V	●
Garganey *Anas querquedula*	3	V	●
Red-crested Pochard *Netta rufina*	3	D	●
Pochard *Aythya ferina*	4	S	■
Black Kite *Milvus migrans*	3	V	●
White-tailed Eagle *Haliaeetus albicilla*	3	R	●
Spotted Crake *Porzana porzana*	4	S	■
Little Crake *Porzana parva*	4	(S)	■
Purple Gallinule *Porphyrio porphyrio*	3	R	●
Crested Coot *Fulica cristata*	3	E	●
Crane *Grus grus*	3	V	●
Avocet *Recurvirostra avosetta*	4/3 [W]	L [W]	●
Collared Pratincole *Glareola pratincola*	3	E	●
Black-winged Pratincole *Glareola nordmanni*	3	R	●
Spur-winged Plover *Hoplopterus spinosus*	3	(E)	●
Jack Snipe *Lymnocryptes minimus*	3 [W]	(V) [W]	●
Redshank *Tringa totanus*	2	D	●
Wood Sandpiper *Tringa glareola*	3	D	●

cont.

Table 3. (cont.)

	SPEC category	European Threat Status	Habitat importance
Little Gull *Larus minutus*	3	D	●
Common Gull *Larus canus*	2	D	•
Gull-billed Tern *Gelochelidon nilotica*	3	(E)	●
Sand Martin *Riparia riparia*	3	D	●
Swallow *Hirundo rustica*	3	D	●
Reed Warbler *Acrocephalus scirpaceus*	4	S	■
Savi's Warbler *Locustella luscinioides*	4	(S)	■
Priority D			
Little Grebe *Tachybaptus ruficollis*	—	S	■
Great Crested Grebe *Podiceps cristatus*	—	S	■
Red-necked Grebe *Podiceps grisegena*	—	S	■
Slavonian Grebe *Podiceps auritus*	—	(S)	■
Black-necked Grebe *Podiceps nigricollis*	—	S	■
Little Egret *Egretta garzetta*	—	S	■
Great White Egret *Egretta alba*	—	S	■
Grey Heron *Ardea cinerea*	—	S	■
Mute Swan *Cygnus olor*	—	S	■
Bewick's Swan *Cygnus columbianus*	3 W	L W	•
Mallard *Anas platyrhynchos*	—	S	■
Shoveler *Anas clypeata*	—	S	■
Tufted Duck *Aythya fuligula*	—	S	■
Goldeneye *Bucephala clangula*	—	S	■
Goosander *Mergus merganser*	—	S	■
Marsh Harrier *Circus aeruginosus*	—	S	■
Water Rail *Rallus aquaticus*	—	(S)	■
Moorhen *Gallinula chloropus*	—	S	■
Coot *Fulica atra*	—	S	■
Black-winged Stilt *Himantopus himantopus*	—	S	■
Little Ringed Plover *Charadrius dubius*	—	(S)	■
Kentish Plover *Charadrius alexandrinus*	3	D	•
Ruff *Philomachus pugnax*	4	(S)	●[1]
Marsh Sandpiper *Tringa stagnatilis*	—	(S)	■
Common Sandpiper *Actitis hypoleucos*	—	S	■
Black-headed Gull *Larus ridibundus*	—	S	■
Little Tern *Sterna albifrons*	3	D	•
White-winged Black Tern *Chlidonias leucopterus*	—	S	■
White-breasted Kingfisher *Halcyon smyrnensis*	—	(S)	■
Pied Kingfisher *Ceryle rudis*	—	(S)	■
Bee-eater *Merops apiaster*	3	D	•
Grey Wagtail *Motacilla cinerea*	—	(S)	■
Dipper *Cinclus cinclus*	—	(S)	■
Cetti's Warbler *Cettia cetti*	—	S	■
Fan-tailed Warbler *Cisticola juncidis*	—	(S)	■
Moustached Warbler *Acrocephalus melanopogon*	—	(S)	■
Sedge Warbler *Acrocephalus schoenobaenus*	4	(S)	●
Paddyfield Warbler *Acrocephalus agricola*	—	S	■
Marsh Warbler *Acrocephalus palustris*	4	S	●
Great Reed Warbler *Acrocephalus arundinaceus*	—	(S)	■
Bearded Tit *Panurus biarmicus*	—	(S)	■
Penduline Tit *Remiz pendulinus*	—	(S)	■
Reed Bunting *Emberiza schoeniclus*	—	S	■

SPEC category and **European Threat Status:** see Box 2 (p. 17) for definitions.

Habitat importance for each species is assessed in terms of the maximum percentage of the European population (breeding population unless otherwise stated) that uses the habitat at any stage of the annual cycle:

■ >75%   ● 10–75%   • <10%

W Assessment relates to the European non-breeding population.

[1] Only a passage migrant through Europe, thus no European Threat Status.

flow rate, water clarity, etc., or indirectly through changing vegetation or faunal communities upon which priority species depend.

In particular, the type and structure of wetland vegetation are important factors affecting the character of wetlands and their suitability for priority wetland birds. Vegetation affects water flow, sedimentation and light penetration, provides food (di-

Table 4. Numbers of priority bird species of inland wetlands in Europe, according to their SPEC category (see p. 17) and their dependence on the habitat. Priority status of each species group is indicated by superscripts (see 'Introduction', p. 19, for definition). 'Habitat importance' is assessed in terms of the maximum percentage of each species' European population that uses the habitat at any stage of the annual cycle.

Habitat importance	SPEC category					Total no. of species
	1	2	3	4	Non-SPEC	
<10%	2 [B]	3 [C]	4 [D]	—	—	9
10–75%	5 [A]	3 [B]	25 [C]	3 [D]	—	36
>75%	3 [A]	0 [A]	13 [B]	5 [C]	36 [D]	57
Total	10	6	42	8	36	102

rectly and indirectly) and cover (for feeding, roosting and nesting). Grazing, cutting (e.g. reed for thatch) or burning of vegetation in turn affect the flora and fauna of wetlands considerably. Such activities are not necessarily detrimental if carried out sustainably or non-intensively, indeed some priority species benefit from the varied composition and structure of such vegetation, as indicated later in this section. Thus, the management of water and vegetation are key factors influencing wetland communities.

Other important general habitat requirements include low levels of pollution and of disturbance by human activities. Although some species may actually benefit from low to moderate eutrophication, through enhanced productivity, this is detrimental at the artificially high levels now seen in many European wetlands (see Box 2).

A further important requirement affecting priority birds of inland wetlands is the maintenance of the wetlands' integrity as networks of staging posts within migratory flyways (Boyd and Pirot 1989, Scott and Rose 1996). Migratory species depend on the high productivity of wetland staging posts for the build-up of sufficient fat reserves for each step of their migration. Species that migrate by short flights ('hops') thus depend on a network of abundant wetlands, though may exhibit some flexibility in their use of these and therefore be able to cope with the loss of a few individual sites. However, species that migrate by a few long flights ('jumps') are often highly specific in their choice of wetland (Smit and Piersma 1989). Thus the loss of even one wetland staging post in a flyway could cause a net reduction of the entire population if feeding and resting areas become too far apart.

Information on the general conservation and management of wetlands for wildlife can be found in Scott (1982), Finlayson (1992), Giles (1992), Eiseltová (1994), RSPB/NRA/RSNC (1994), Andrews (1995), Burgess *et al.* (1995), Holmes and Honbury (1995) and Hawke and José (1996).

The individual habitat requirements of priority species are summarized in Appendix 3 (p. 354). These requirements fall into a few broad categories.

Firstly, factors affecting the physical and chemical nature of the water-body. Amongst these, the area of open water is important to at least 35% of priority species. More than 15 species, often large (e.g. White Pelican *Pelecanus onocrotalus* and *P. crispus*), prefer large areas of water. More species (c.25% of all priority species), however, prefer smaller areas, probably because the ratio of 'edge habitat' to open water is greater, which in turn provides more plant cover for feeding and/or nesting. This is particularly important for some herons, e.g. Squacco Heron *Ardeola ralloides* and *Ardea purpurea*, nearly all crakes and some ducks, such as *Anas querquedula* and *Marmaronetta angustirostris*.

Water depth is also an important factor for priority species, primarily through its effect on food resources and availability. Most priority wading species, including some herons, crakes and waders (e.g. Avocet *Recurvirostra avosetta*, Black-tailed Godwit *Limosa limosa* and Redshank *Tringa totanus*), require shallow to moderately deep water-bodies, or at least water-bodies with gradually sloping profiles. Similarly, species that depend on emergent vegetation require such waters to support these plant communities.

Moderately deep water that is unsuitable for emergent vegetation but supports submerged aquatic vegetation is preferred by *Cygnus olor* and some diving ducks. Species that feed on benthic invertebrates and open-water fish generally prefer deeper water, although water in excess of 2 m is preferred by few priority species. Much deeper water is unsuitable for benthic feeders and fish populations may be poor in very deep, unproductive lakes. Such deep, and particularly steep-sided, lakes therefore frequently have low bird diversity and numbers.

About 80% of priority bird species occur on still water, and most prefer this. About 35%, however, will also use habitats with slow-moving water (e.g. lowland rivers), but none of these prefer such conditions. Only three species require fast-moving water, but these are characteristic of fast-flowing rivers and indeed *Cinclus cinclus* is entirely restricted to these conditions.

The trophic status of a water-body has a profound effect on the entire wetland community and hence on priority bird species. In general, eutrophic conditions are preferred, as the high productivity in such habitats normally provides abundant food resources. Moderate artificial eutrophication of nutrient-poor water-bodies benefits some birds through the stimulation of production. However, it is also often detrimental to those that require clear water—which includes more than 20 priority species (see Appendix 3)—due to the increased proliferation of algal blooms. High levels of artificial eutrophication can lead to catastrophic declines in the food of priority birds, such as submerged plants, fish and invertebrates and, in turn, the birds themselves (see Box 2).

The second major category of habitat requirements comprises features associated with the vegetation. For most priority species, such vegetation can provide important cover for nesting, roosting, predator evasion or hunting. Thus emergent vegetation, especially reedbeds *Phragmites*, is important for herons, ducks, crakes and passerines, and is actually essential for several species. The preferred density and height of vegetation varies widely among priority species, indicating the importance of maintaining habitat mosaics in wetlands.

Vegetation is also the main food source for some species, for example several ducks, such as *Anas strepera*, feed primarily on submerged aquatic plants, whilst *Porphyrio porphyrio* is highly dependent on the vegetative parts of emergent species. Trees and bushes and other terrestrial vegetation, emergent plants and floating vegetation also support diverse and rich invertebrate and vertebrate faunas and are therefore important feeding grounds for many species, including various herons, crakes and passerines. Open, low marshland is important for a minority of species, e.g. the globally threatened Lesser White-fronted Goose *Anser erythropus* (Madsen 1996). Trees and bushes are particularly important as breeding sites for some of the larger priority birds, especially when they are on uninhabited islands. Wetland vegetation also provides spawning grounds for fish and is the source of insects fed on by aerial-feeding wetland species such as *Chlidonias* terns, *Riparia riparia* and Swallow *Hirundo rustica*. Thus although the specific requirements are complex, and cannot be described in detail here, abundant, diverse natural wetland vegetation is clearly an essential habitat requirement for most inland wetland species.

Food forms a third category of habitat requirements. Many wetland bird species are generalists and take a variety of vegetable and animal matter. However, aquatic plants, invertebrates and fish are the most frequently-taken food. Only a few priority species are herbivorous, such as *Anser erythropus*,

Branta ruficollis and *Porphyrio porphyrio*, while most ducks are omnivorous, taking predominantly aquatic vegetation and invertebrates. A few priority species specialize on aquatic invertebrates; these include include *Cinclus cinclus*, *Motacilla cinerea*, *Platalea leucorodia*, *Recurvirostra avosetta* and Greater Flamingo *Phoenicopterus ruber*, the latter being highly dependent on the brine shrimp *Artemia salina*. Most waders feed on a mixture of aquatic and soil invertebrates. Crakes, and reedbed passerines such as *Locustella luscinoides* and *Acrocephalus* warblers, primarily feed on plant invertebrates, while pratincoles *Glareola*, *Chlidonias* terns, *Riparia riparia* and *Hirundo rustica* feed on aerial invertebrates.

Fish specialists are particularly numerous, and include the two divers *Gavia*, Pygmy Cormorant *Phalacrocorax pygmeus*, the two pelicans *Pelecanus*, *Pandion haliaetus*, *Alcedo atthis* and Pied Kingfisher *Ceryle rudis*. Amphibians are commonly taken by herons, along with fish and invertebrates, while birds and mammals are the main diet of birds of prey such as Marsh Harrier *Circus aeruginosus* and *Aquila clanga*.

Clearly, abundant food resources are important for all species and these in turn are dependent on suitable habitat conditions, such as the physico-chemical characteristics of the water-body (including the absence of pollution), vegetation-type and structure, and the presence of other species in the food chain. These factors are not discussed in detail here, but the often inverse relationship between fish and invertebrate populations is of particular importance (see, e.g., 'Sport fishing and commercial fisheries', p. 147).

THREATS TO THE HABITAT

One of the most important threats comes from the direct loss of wetland habitats. Wetland loss has been reviewed by numerous authors, and has been a major subject in several international symposia (e.g. Baldock 1990b, Dugan 1990, Hollis 1992, Dugan 1993, Jones and Hughes 1993).

Common and important causes of loss of floodplain wetlands, marshes and lakes in Europe (Dugan 1993, Jones and Hughes 1993) are as follows:

- Drainage for agriculture, forestry or malaria control.
- Filling for solid-waste disposal, road construction, and commercial, residential and industrial developments.
- Construction of dykes, dams, levees and seawalls for flood control, water supply, irrigation and storm protection.

- Mining of wetland soils for peat, coal, gravel, phosphate and other materials.

- Excessive abstraction of groundwater and surface water.

- Sediment-trapping by dams, deep channels and other structures.

- Land subsidence due to extraction of groundwater, oil, gas and other minerals.

- Severe pollution with non-reversible destructive effects.

- Drought.

- Other natural biotic effects.

In addition to habitat loss, wetland species of inland areas can be threatened by changes in a wide range of physical, energetic, chemical and biological processes as a result of artificial eutrophication, acidification, alterations to the hydrological regime and disruptions to the structure and species composition of vegetation. Furthermore, because wetlands are linked dendritically to their drainage basins, threats can arise at a wide range of geographical scales from habitat alteration at a single point, up to diffuse changes in or beyond the whole catchment. For the purpose of this strategy, and because the focus for management can vary in a similar way, threats have been classified as those which occur at:

- Catchment scale: their source can be diffuse through the wetland catchment (e.g. acidification).

- Riparian scale: their sources are principally in the riparian zone around the edges of wetlands (e.g. loss of riparian vegetation).

- Aquatic scale: their sources are predominantly from changes within the perimeter of the wetland (e.g. point-source pollution).

For any individual threat, these sources are not mutually exclusive.

Reviews of anthropogenic threats to wetlands have been carried out by Dugan (1993), and for various regions, e.g. central and eastern Europe (IUCN 1993), the Mediterranean (Finlayson et al. 1992) and northern Europe (Nordic Council of Ministers 1993). An assessment by the Habitat Working Group of the causes and scale of wetland loss and degradation in Europe is summarized in Table 5.

Of the 17 main threats listed, 12 are considered to have had widespread effects on inland wetlands over the last 20 years, and furthermore all of these are predicted to continue at the same level for the next 20 years. These threats include the factors already identified as causes of wetland loss, such as drainage and water abstraction, but also factors causing habitat degradation, such as artificial eutrophication, acidi-

fication, toxic pollution, excessive sediment loads, destruction of riparian habitats, commercial fisheries and sport fishing, hunting, recreation, and regulation of water-levels.

The predicted impacts of each of the threats listed in Table 5 on European populations of priority birds of inland wetlands were assessed by the Habitat Working Group and by the BirdLife Network in Europe (see Appendix 4, p. 401). The results are discussed below (in order of predicted overall importance for priority species), and summarized in Figures 1 and 2.

Drainage and land-claim

The drainage of European wetlands, particularly for agricultural purposes, has been underway for centuries, but probably reached its peak between the early 1970s and the 1980s (Finlayson and Moser 1991). There are numerous, well-documented cases of the loss of specific wetland sites through drainage. A few examples include Lake Hornborgasjön in Sweden (Hertzman and Larsson 1991), the Straldzha marshland in Bulgaria (Anon. 1993b), in Greece the Evros delta (HOS 1992), Lake Karla (Finlayson and Moser 1991) and the Nestos delta (Vassilakis and Vassilakopoulou 1993), in Turkey the Menderes delta, Sultan marshes and Eregli marshes (Kasparek 1985, Brinkmann et al. 1991, Kiliç and Kasparek 1990), in France the Landes de Gascogne (Mermet 1984), and Romney Marsh in the UK (Baldock 1984).

The overall impacts of such projects have been considerable. For example, by the early 1980s about half the wetlands of major importance for wildfowl (Anatidae) in Europe were either affected by drainage or were at risk (Chantrel 1984). In Greece, almost all major wetlands have suffered drainage for agricultural purposes during the last 60 years (Athanasiou 1987).

There is some indication that the rate of wetland loss may be slowing. This is partly because of the diminishing number of wetlands in Europe that can be easily or economically drained. For example, in Germany and the Netherlands, most wetlands with potentially productive soils have been drained (Baldock 1990b). There is also now an increased awareness of the full economic, social, cultural and wildlife benefits of wetlands. Consequently the protection of wetlands has increased and the costs/benefits of drainage schemes are assessed more critically. Indeed, such schemes are highly expensive and often not viable in the long term, as for example seen at Lake Hornborgasjön, Sweden. The drainage of this large lake for agriculture greatly damaged its ornithological value, it having once been one of the most important bird areas in the

Table 5. Threats to inland wetlands, and to the habitat features required by their priority bird species, in Europe (compiled by the Habitat Working Group). Threats are listed in approximate order of importance (see Figure 2), and are described in more detail in the text.

Threat	Effect on birds or habitats	Scale and extent of threat				Regions affected, and comments
		Scale of effect	Habitat-types	Past	Future	
1. Drainage and land-claim	Habitat loss and fragmentation	C,R,A	R,L	■	■	Everywhere, but local in NW Europe; declining
2. Destruction of riparian habitats (e.g. for agriculture or flood control)	Physical restructuring or loss of bank-side vegetation (e.g. trees and rank vegetation, etc.) leading to loss of feeding, roosting and nesting opportunities, and potentially increased predation, trampling and disturbance	R	R,L,M	■	■	Everywhere
3. Tourism development (inc. secondary housing, etc.) and recreation activities	Change in habitat structure; disturbance affecting survival or productivity and habitat use	R,A	L,M	■	■	Everywhere; local in N Europe
4. Inappropriate management of vegetation	Loss/changes in habitat structure, e.g. reedbeds and sedge-beds and vegetation alongside rivers, thus affecting nesting and feeding requirements	R	(R),(L),M	●	●	Everywhere, particularly E Europe and Mediterranean; local in N Europe
5. Nutrient pollution	Changes in emergent and sub-merged vegetation structure, causing loss of open water; increased algal blooms; reduced oxygen content and water clarity, leading to reductions in food abundance and availability	C,A	R,L,M	■	■	Everywhere
6. Over-abstraction of water (irrigation, industrial, energy, domestic) and inter-catchment water-transfers	Lower water-table and water-level; reduced flow; change in plant communities; alteration in chemistry (reduced dilution of pollution); altered sediment dynamics	C,R,A	R,L,M	■	■	Everywhere, but particularly in S Europe
7. Toxic pollution	Addition of substances such as pesticides, heavy metals (e.g. lead-shot from hunting and angling), PCBs, radionuclides and organic materials affecting organisms and ecological processes through direct mortality and sublethal or in-direct effects, e.g. on food resources	C,R,A	R,L,M	■	■	Everywhere, but regional in N Europe
8. Regulation of water-level	Reduction or increase of water-level fluctuation affecting habitat structure, and nesting or feeding requirements of birds, e.g. loss of sediment islands, flooding of nests	C,R,A	R,L,M	■	■	Everywhere
9. Hunting	Disturbance affecting survival or productivity and habitat-use; habitat management and associated changes in vegetation structure, composition and animal communities; damage to non-target species	R,A	R,L,M	■	■	Everywhere, particularly Mediterranean
10. Impoundment of natural wetlands (dams, etc.)	Inundation leads to alteration of a wide range of ecological parameters, e.g. food resources of birds, often loss of breeding habitats, local climate change and drying out of adjacent areas	C,R,A	R,L,M	■	■	Everywhere, particularly Mediterranean, local in N Europe

cont.

Table 5. (cont.)

Threat	Effect on birds or habitats	Scale and extent of threat				Regions affected, and comments
		Scale of effect	Habitat-types	Past	Future	
11. Canalization	Affects natural dynamics of water flow, changing food resources and nesting possibilities; creates uniform habitat structure; affects wetlands in the adjacent flood-plain by facilitating drainage	C,R,A	R,(M)	●	●	Everywhere, but local in N Europe
12. Human-induced increases in predators	Disturbance and direct mortality of adults, eggs and chicks	C,R,A	R,L,M	●	●	Particularly near urban and tourist areas
13. Commercial fisheries and sport angling	Introduction and management of predators (fish) leading to competition with birds and persecution, also interference with food webs; management of vegetation; increased disturbance, mortality (from nets and fishing tackle, including lead weights); artificial releases of water to encourage fish movement, causing mortality of nesting birds and disruption of natural seasonal cycles	R,A	R,L	■	■	Everywhere
14. Acidification (from atmospheric pollution)	Loss of alkalinity and decrease in pH, mobilization of toxic metals resulting in decreases in diversity and sometimes in abundance of aquatic species (e.g. fish); also effects on ecosystem processes, e.g. decomposition is retarded	C,A	R,L,M	■	■	Particularly NW and central Europe
15. Excessive sediment loads	Mobilization of catchment soils (e.g. through soil and river erosion), leading to changed habitat structure in rivers and lakes (e.g. water depth, clarity, vegetation, etc.)	C,R,A	R,L,M	■	■	Everywhere, but regional in N Europe
16. Aquaculture	Changes in vegetation structure and trophic web (through introduction of exotic species); changes in water quality, fluctuations in water-level and destruction of habitat; spread of disease; interaction/competition with birds/humans leading to persecution; use of pesticides	R,A	R,L,M	●	●	Everywhere, but local in N Europe
17. Introduction of non-native species	Changes in food resources, habitat structure, competition and predation pressure; hybridization leading to extinction of native species	(C),R,A	R,L,M	●	●	Potential effects at catchment level

Scale and extent of threat

C	Catchment	R	Rivers	Past	Past 20 years	■	Widespread (affects >10% of habitat)
R	Riparian	L	Lakes	Future	Next 20 years	●	Regional (affects 1–10% of habitat)
A	Aquatic	M	Marshes			·	Local (affects <1% of habitat)

country. Land subsidence due to soil shrinkage eventually made the drainage scheme too expensive to maintain, and the Swedish government is now recreating the lake, at a cost of at least $10 million (Hertzman and Larsson 1991, Dugan 1993).

Nevertheless, further widespread losses over the next 20 years remain likely and may threaten some of the most important remaining wetlands. At least six globally threatened species—*Pelecanus crispus, Marmaronetta angustirostris, Aythya nyroca, Oxyura leucocephala, Aquila clanga* and *Acrocephalus paludicola*—are seriously at risk of extinction due to habitat loss through drainage, and a further five globally threatened or near-threatened species are at risk of a low population impact (Collar *et al.* 1994, Heredia *et al.* 1996). Only in northern Europe is the problem likely to be local as a result of the introduction of strict controls on new wetland drainage.

The threat of continued wetland loss through drainage and land-claim is thought to affect 73 of the 102 priority species, more than any other threat. A larger proportion of these species (38%) are also thought likely to suffer high impacts from drainage, etc., than from any other threat. Consequently, the overall importance of the hazard in Europe is considered to be extremely high and more than twice that of any other threat to inland wetlands and their birds.

Destruction of riparian habitats

As described above ('Habitat needs of priority birds', p. 133), many priority species of inland wetlands depend on the presence of natural riparian vegeta-

tion. However, such vegetation is frequently cleared either to provide access (for walkers, fishermen, boats, etc.), or during the course of river-engineering works to control floods and ease navigation. Data on the extent of this threat in Europe are unavailable but it is considered likely to be widespread in almost all regions. Indeed, the engineered isolation of many rivers from their adjacent riparian habitats and floodplains is known to have caused a severe decline in riverine forests (alluvial forests), making them now one of the most endangered habitats in Europe (Vivian 1989). These forests are covered under 'Boreal and Temperate Forests' (p. 203) but are important for the nesting of several species of priority waterbirds of inland wetlands (see Appendix 3, p. 354), and are among the most biologically diverse habitats in Europe (IUCN 1993).

The loss of bankside or tall, emergent vegetation such as *Phragmites*, *Typha* and *Scirpus*, and rank sedges and grasses, can also lead to loss of suitable

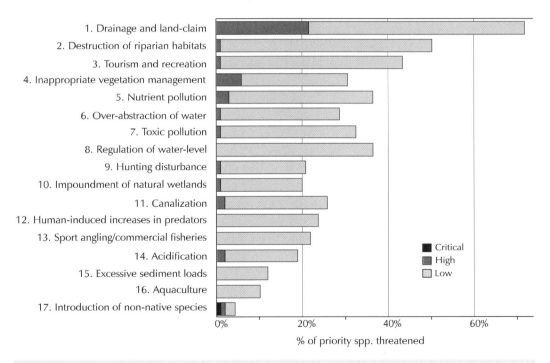

Figure 1. The predicted impact on priority bird species of threats to inland wetlands, and to the species' habitat requirements, in Europe. Bars are subdivided to show the proportion of species threatened by critical, high or low impacts to their populations (see Table 5 and Appendix 4, p. 401, for more details).

Critical impact
The species is likely to go extinct in the habitat in Europe within 20 years if current trends continue

High impact
The species' population is likely to decline by more than 20% in the habitat in Europe within 20 years if current trends continue

Low impact
Only likely to have local effects, and the species' population is not likely to decline by more than 20% in the habitat in Europe within 20 years if current trends continue

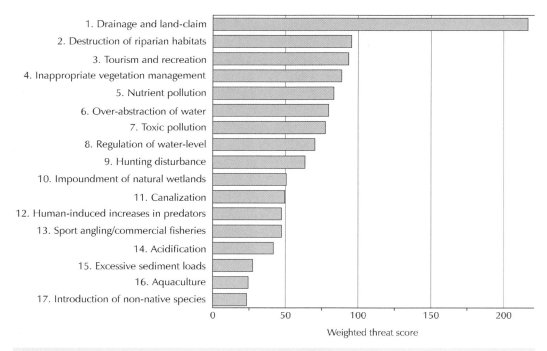

Figure 2. The most important threats to inland wetlands, and to the habitat requirements of their priority bird species, in Europe. The scoring system takes into account:

• The European Threat Status of each species and the importance of the habitat for the survival of its European population (Priorities A–D); see Table 3.

• The predicted impact (Critical, High, Low), within the habitat, of each threat on the European population of each species; see Figure 1, and Appendix 4 (p. 401).

See Table 5 for more details on each threat.

nesting, roosting and feeding habitat for a variety of priority species of ducks, rails and passerines (see 'Habitat needs of priority birds', p. 133). Mature (sometimes dead) trees are also used as nest-sites by some ducks and wetland-associated raptors.

A remarkably high total of more than 50 priority species of inland wetlands are thought to be threatened by the destruction of riparian vegetation, and this threat to the wider environment is clearly one of the most important amongst those considered here.

Tourism development and recreation

This threat to priority species of wetlands has been described in 'Coastal Habitats' (p. 104). Key features of this threat are direct habitat loss (see 'Drainage and land-claim' above) and degradation due to the building of houses, hotels and associated infrastructure, due to over-abstraction of water for consumption and irrigation, artificial eutrophication caused by discharge of untreated sewage and wastes, and major disturbance and disruption of priority birds' feeding and nesting behaviour during the breeding season, affecting survival and productivity.

Due to the concentration of international tourism there, wetlands in the Mediterranean basin are particularly at risk.

At least 43 priority species are considered to be suffering low population impacts from this widespread threat, while the globally threatened *Aquila clanga* may be suffering a high population impact. Overall, this hazard is one of the most important facing the birds of inland wetlands in Europe today.

Inappropriate management of vegetation

The presence of wetland vegetation such as reedbeds or rank grassland is often essential for the nesting, roosting or feeding of priority birds (see 'Habitat needs of priority birds', p. 133). Open areas of short grass may also be necessary as feeding habitat for species such as waders. Management is often needed to maintain such vegetation, particularly where scrub ia encroaching as a result of the advanced stages of succession (see Box 1). Birds may therefore benefit from appropriately controlled grazing, or from the cutting of vegetation for livestock fodder or thatching materials (reeds). Indeed, *Acrocephalus palu-*

dicola, a globally threatened species, is threatened in Poland as a result of the abandonment of such traditional practices (Heredia 1996c).

Burning of reedbeds may also be beneficial when high levels of dead reed have built up and scrub has started to invade. However, burning, grazing and cutting can be highly detrimental if carried out inappropriately. High grazing densities can lead to net loss of vegetation cover and excessive soil damage and erosion. Spring grazing can cause disturbance to nesting birds and trampling of their eggs and young. The varying of cutting cycles over different parts of reedbeds is best for birds and other wildlife, but annual cutting over large areas could be highly detrimental. Burning should be used with great caution and only to bring reedbeds back into normal management regimes. Further details of appropriate management regimes are given in RSPB/NRA/RSNC (1994), Sutherland and Hill (1995) and Hawke and José (1996).

This important threat is considered to affect the populations of 31 priority species, including six species which suffer a high impact (among which are three globally threatened species). Overall, inappropriate grazing, cutting and burning of vegetation pose a major threat to inland-wetland birds in Europe.

Nutrient pollution

Eutrophication is the process of nutrient enrichment in aquatic ecosystems. It occurs naturally over the geological time-scale but now often occurs artificially through human activities. In such cases the additional nutrients (primarily phosphorus in inland wetlands) usually come from sewage, agricultural fertilizers or industrial effluents (see Box 2).

There are many documented cases of wetland damage by artificial eutrophication in Europe, but a detailed and comprehensive account of this cannot be given here. An example is the eutrophication of the Norfolk Broads (UK), where nutrient pollution has greatly increased over the last 50 years due to agricultural run-off, sewage and churning of bottom-sediments by passing tourist-boats. The Broads' value to wildlife and tourism has suffered significantly, due to massive algal blooms and to widespread bank erosion caused in turn by die-off of marginal vegetation. Restoration of the Broads to their former, more touristically attractive state would now cost many millions of pounds (Burgis and Morris 1987, Phillips 1992). An assessment of river quality (based on nutrient status of the water) in Europe found that overall about 25% of river reaches are of poor or bad quality (Figure 3). However, more than 25% of river reaches were poor or bad in Bulgaria, Romania, the Czech Republic, Poland, Denmark, the Netherlands, Luxembourg, Slovenia and Italy.

Recent surveys of nutrient concentrations in European fresh waters reveal an overall widespread decline in phosphorus levels—over 25% in a third of rivers—due mainly to improved waste-water treat-

---

Box 2. Eutrophication.

The growth of algae in many natural freshwater wetlands is limited by phosphorus rather than nitrogen, carbon or other requirements. Consequently, the addition of phosphate-rich substances such as sewage (which includes high levels of phosphate-containing detergents), fertilizer run-off or industrial effluents can trigger a substantial increase in plant growth. The relationship between nutrient inputs and the effects on the ecosystem is complex and depends on many variables characterizing each waterbody, such as water clarity, temperature, depth, shading, flow-rate and turbulence.

However, eutrophication typically occurs in stages according to the level of nutrient enrichment. Low to moderate levels usually lead to an increase in submerged plants. These plants are able to obtain most of their nutrients from the substrate and at the same time produce chemicals that inhibit the growth of planktonic algae. However, as nutrient levels increase, epiphytic algae on the leaves of submerged plants and filamentous algae such as *Cladophora* increase and shade out the plants. The decrease in submerged plants then leads to a reduction in their inhibition of algae, which can then proliferate. In turn these blooms of algae further reduce light penetration which further inhibits submerged-plant growth, so that the switch to the predominance of planktonic algae becomes self-perpetuating.

In the early stages of eutrophication, the enhanced submerged-plant growth stimulates increases in bacterial activity (as the plants die) and hence promotes populations of zooplankton, and in turn other organisms in the food chain. Thus the productivity of the ecosystem increases to the benefit of some fish and many bird species. However, as eutrophication increases, the dense algal blooms reduce water clarity, which can reduce the feeding efficiency of diving birds that rely on visual detection of their animal prey (see Appendix 3, p. 354). Declines in submerged-plant populations will also reduce the suitability of the wetland to plant-feeding species such as various ducks (Anatidae).

At higher levels of eutrophication, the algal blooms cause oxygen depletion through their respiration at night and through intense bacterial activity. This may lead to the death of fish and other animals, and severe disruption of the whole wetland ecosystem, which by then will be unsuitable for most birds.
(Based on Burgis and Morris 1987, Irving 1993, Pokorný 1994.)

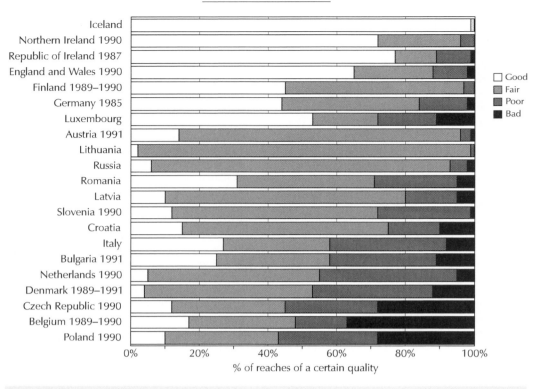

Figure 3. The quality of European river reaches (adapted from Stanners and Bourdeau 1995; sources for individual countries are listed in that work).

Good
River reaches with nutrient-poor water, low levels of organic matter; saturated with dissolved oxygen; rich invertebrate fauna; suitable spawning ground for salmonid fish.

Fair
River reaches with moderate organic pollution and nutrient content; good oxygen conditions; rich flora and fauna; large fish population.

Poor
River reaches with heavy organic pollution; oxygen concentration usually low; sediment locally anaerobic; occasional blooming of organisms insensitive to oxygen depletion; fish population small or nil; periodic fish kill.

Bad
River reaches with excessive organic pollution; prolonged periods of very low oxygen concentration or total deoxygenation; anaerobic sediment; severe toxic input; no fish.

ment (Stanners and Bourdeau 1995). In contrast to this welcome trend, however, in eastern Europe phosphorus levels rose in many rivers, largely because of very limited construction of sewage-treatment plants incorporating phosphorus removal. Nitrate levels have also tended to increase in many European rivers and lakes over the last 20–40 years (Stanners and Bourdeau 1995). This increase is largely attributable to the increased use of nitrogen fertilizers on farmland over the same period. However, increases in lake nitrate levels in areas of low human activity, for example in Norway, are mainly attributable to increased deposition of nitrogen compounds from the atmosphere, resulting from the combustion of fossil fuels and from agricultural

activities (Henriksen *et al.* 1988). Overall, nutrient pollution is widespread in almost all regions of Europe.

Impacts on birds are likely to vary according to the degree of eutrophication experienced locally. Moderate levels are likely to threaten sensitive species, such as *Cinclus cinclus*, which require healthy aquatic invertebrate populations. Predators which hunt by sight would also be affected as a result of reduced water clarity. For example, in England and Wales, *Alcedo atthis* densities have been found to be inversely correlated with high levels of nutrient pollution (Meadows 1972). High levels of nutrient pollution with severe oxygen depletion are likely to make the habitat unsuitable for most waterbirds.

Overall, this threat is considered to impact 37 priority species.

Over-abstraction of water

Abstraction of groundwater and surface water is carried out for use in agriculture (irrigation), industry, energy generation, domestic water supplies and tourism.

Over the last two decades, total water abstraction has in general increased in Europe (Stanners and Bourdeau 1995). However, there is considerable variation among countries. Increases have been particularly marked in southern Europe, but abstraction has also increased in the majority of countries in eastern and western Europe. Overexploitation occurs when groundwater abstraction exceeds the recharge rate of the aquifer, and has led to a systematic and ongoing lowering of groundwater levels in many areas of Europe. This can result in wetland loss or damage through the drying up of springs, upper river reaches, lakes and marshes, reduced river flow, and saltwater intrusion into aquifers. Furthermore, many wetlands of particular ecological importance are located in flood-plains and their hydrological systems are strongly dependent on high water-tables. Such wetlands are therefore very sensitive to minor changes in the groundwater level.

An example of such wetland damage is to be found in the Tablas de Daimiel National Park (Spain). The two main rivers feeding the wetland are fully dependent on groundwater recharge, but this source has been severely overexploited for purposes of agricultural irrigation, and at the beginning of 1993 the reserve almost completely dried out (Custodio 1991).

Such trends are predicted to continue during the next 20 years. For example, major irrigation and tourism projects in the vicinity of Doñana National Park (Spain) threaten to degrade or destroy the globally important wetlands within the park (Stanners and Bourdeau 1995), and saltwater intrusion into aquifers (which ruins them irreversibly) is a major problem in the Algarve of Portugal as a result of excessive abstraction (McCann and Appleton 1993).

Inter-catchment transfers of water—essentially, abstraction for non-local usage—have also caused region-wide wetland loss in Europe, especially in drier areas (e.g. the Mediterranean basin of Spain, Greece and Turkey) or in densely populated areas (e.g. Belgium, France, UK: McCann and Appleton 1993). A total of 89 nationally or internationally important wetlands in the UK are at risk now or in the future from abstraction, including inter-catchment transfers (English Nature 1996).

In general, birds of the marshlands and rivers of southern Europe are those most affected, and there are predicted to be population impacts on 29 priority bird species. Most impacts are likely to be low, with the exception of the globally threatened *Marmaronetta angustirostris*, where the impact could be high (Green 1996).

Toxic pollution

In general, there is insufficient information to assess the magnitude and extent of freshwater pollution by toxic substances such as heavy metals, micropollutants and radionuclides (Stanners and Bourdeau 1995). There is some evidence that the scale of most such pollution in Europe is local and/or regional rather than more extensive, as any such problems are predominantly related to point-source industrial emissions. A notable exception is the risk to priority waterbirds posed by lead poisoning following ingestion of spent or discarded lead-shot from hunting and angling at wetlands—this is a widespread problem in Europe (Pain 1992), and is described in more detail under 'Coastal Habitats' (p. 110). In addition, modelling calculations indicate that the quality of groundwater is threatened with contamination by pesticides and their residues under much of the arable land in Europe.

Although the effects of these toxic pollutants on bird populations are poorly understood, it is widely believed that virtually all predatory species are susceptible through bio-accumulation, but particularly those at the top of the food chain. For example, European populations of the globally threatened *Pelecanus crispus* are thought to be affected, at least locally, by heavy metal and pesticide pollution (Crivelli 1996). High concentrations of DDE (a residue of the pesticide DDT) were found in eggs collected at Lake Prespa (Greece) between 1984 and 1986, eggshell thickness being 12–20% less than pre-1947 values, dating from before DDT was in use (Crivelli *et al.* 1989). Later investigations in 1989, however, showed a sharp decrease in concentrations of such chlorinated hydrocarbons, although other, recent studies have also detected concentrations of heavy metals and chlorinated hydrocarbons in the species' eggs (Cook 1992, Albanis 1993).

Furthermore, sublethal effects that affect population levels through reduced breeding productivity may be more common, but this aspect has been very little studied. Indirect effects through loss of food resources are also likely to be widespread, but again little information exists to assess effects on bird populations. Some species, such as *Cinclus cinclus*, have, however, been shown to be susceptible to the effects of pollution on their invertebrate prey and they therefore make good indicators of water quality (Ormerod and Tyler 1993). It can be expected that, to some extent, other specialist invertebrate- or fish-eating species are similarly affected. Overall, 19 priority species are predicted to suffer population reductions, which could be high in the two *Gavia* species, and the threat of toxic pollution is thus considered to be moderate in importance among the threats considered here.

Inappropriate regulation of water-level

Water-levels in many rivers, lakes and reservoirs in Europe are now artificially controlled to regulate water supplies or reduce flood risks. This is a particular risk to species nesting close to the water's edge or on low-lying land or islands. Clearly, sudden increases in water-level during the breeding season can flood the nests of grebes, pelicans, ducks, waders and terns. However, rapid declines in water-level can also be harmful. Previously isolated islands may become linked to the shore, thus exposing nests to mammalian predators such as *Vulpes vulpes*. Marshy areas may also dry out, thereby reducing food availability for some species, especially waders which feed by probing soft soil for invertebrates.

There are few published reports on the extent or impacts of this problem, but it is considered to be widespread in all regions of Europe and to threaten 37 priority species with low population impacts. It is therefore of at least moderate importance among the threats considered here.

Hunting

The direct impact of hunting mortality on bird populations is not covered by this book (see 'Introduction', p. 15) but indirect effects of hunting that affect habitat quality are. Hunting is a major cause of disturbance to priority birds at wetlands in Europe, and spent lead-shot is also a major source of lead poisoning in waterbirds (Pain 1992, Fawcett and van Vessem 1995). These hazards are described in more detail under 'Coastal Habitats' (p. 109). Overall, the indirect hazards posed by hunting are considered to affect 21 priority species, particularly globally threatened species, e.g. *Branta ruficollis* is predicted to suffer a high impact from hunting disturbance (Black and Madsen 1993, Hunter and Black 1996).

Impoundment of natural wetlands

The impoundment of wetlands can result in substantial habitat degradation, either through the permanent flooding of marshland and increased water depth within impounded wetlands, or through reduction in water flow downstream of the dams.

Such damming or impoundment of wetlands for water supplies or hydroelectric power is widespread in Europe. In fact 17 internationally important wetlands (listed under the Ramsar Convention) in Europe and the rest of the Mediterranean basin are threatened by dams (Finlayson and Moser 1991). For example, in Russia, construction of the Volgograd hydropower station completed the Volga–Kama scheme of dams, producing a reservoir-controlled hydrological regime for the entire River Volga. This reduced the volume and duration of spring floods by 25%, and in dry years the dams absorb the floods completely. Large floods used to inundate the Volga delta and the prolonged high water created ideal spawning conditions for fish. Now the reduced flood volume, short flood-peak, and occasional years without floods have disrupted the ecosystem, with negative effects on fish reproduction and yields. In Spain there are almost 1,000 dams, where they affect natural river-flows more than any other factor (Ibero 1996).

Impacts on birds of inland wetlands are not well documented, but the most likely species to be affected are those of seasonally flooded wetlands and larger rivers, particularly in southern Europe. Species, such as *Actitis hypoleucos*, on upland rivers in areas suitable for hydroelectric power schemes are also likely to be at least locally threatened. In addition, priority bird species of wet grasslands are much affected by this threat (see 'Agricultural and Grassland Habitats, p. 315).

Canalization

River canalization is a widespread operation in Europe, particularly in areas of intensive agricultural production. For example, in Denmark 85–98% of the total river network has been straightened (Brookes 1987, Iversen *et al.* 1993). Canalization is normally carried out to improve flood control, navigation and drainage of surrounding land and to prevent erosion. It results in considerable physical changes. Straightening increases the slope (and hence water speed) while dredging changes the width and depth, smooth the bankside and removes vegetation. Thus it creates a homogenous system where flow and substrate become uniform, and pools and riffles are lost. Riparian vegetation and trees are often removed to allow machine-dredging and to increase drainage (see 'Drainage and land-claim', p. 138, for specific impacts). Nutrient inputs into the river may also increase after the removal of marginal natural vegetation, since the latter has a high capacity for nutrient absorption. This may lead to enhanced eutrophication, which may in turn be exacerbated by increased algal growth resulting from the increased light levels reaching the more open and unshaded water.

The overall effect is a reduction in habitat diversity, which has been shown to result in lower numbers and diversity of plant and animal species, including fish, invertebrates, small mammals and birds (Brookes 1988).

Today, the potential role of wetlands as a natural infrastructure in flood prevention and water-quality maintenance is becoming more widely appreciated, and increasing investment is being placed in conserving the remaining habitat and restoring degraded areas (e.g. on the River Rhine: Dugan 1993). Consequently, future canalization may be reduced and less severe. Furthermore, mitigation measures and more environmentally sensitive techniques are now being widely used. Such measures are described in detail in RSPB/NRA/RSNC (1994).

Human-induced increases in predators

Non-native predators that have successfully established themselves in Europe through man's agency (whether as deliberate introductions or through negligence or ignorance), e.g. semi-aquatic carnivores from North America such as *Procyon lotor* and *Mustela vison*, are spreading and may soon become regionally important predators of wetland birds, especially their eggs and young. In addition, provision of artificial food-sources (e.g. rubbish dumps) by man has led to local or regional increases in native predators such as *Vulpes vulpes*, which may be having an increasing impact on populations of priority wetland species, particularly in areas with high concentrations of ground-nesting birds. These hazards are discussed in more detail under 'Coastal Habitats' (p. 111), but overall at least 24 species may suffer local impacts on their European populations, making this a threat of moderate importance among those considered here.

Commercial fisheries and sport angling

Artificially high fish populations, through stocking for commercial or sport fisheries, have been shown to deplete populations of larger aquatic invertebrates and to be major food-competitors with waterbirds (Hill and Street 1987, Giles 1992). For example, bream *Abramis* and other species consume the bulk of the chironomid worm production on which many duckling species depend during the first few days of life, consequently affecting their survival. Fish also reduce aquatic plant growth, which affects *Anas strepera* and other priority waterbirds both directly and via the effect on invertebrate populations. Through predation of zooplankton, they may also initiate and maintain algae-dominated conditions in some water-bodies (Hrbácek 1994, Andrews 1995), which support low populations of plants and larger invertebrates and, in turn, low numbers of waterbirds.

Persecution of fish-eating species, including the globally threatened *Pelecanus crispus* and near-threatened *Phalacrocorax pygmeus*, still occurs at some important natural wetlands where there are commercial fisheries (Crivelli 1996, Crivelli *et al.* 1996). At these, and sport fisheries, other potential threats include habitat management (see 'Inappropriate management of vegetation', p. 142), increased disturbance and mortality (from entanglement in nets and fishing tackle, and poisoning from ingestion of lead), and water management to encourage fish movement (see 'Inappropriate regulation of water-level', p. 146), which can cause mortality of eggs or young and disrupt breeding cycles.

Such fisheries are a widespread threat to more than 20% of priority birds of inland wetlands, although impacts on individual species are in all cases likely to be local. This threat is therefore of low to moderate importance among the threats considered here.

Acidification

The acidification of rivers and lakes is one of the best-documented problems in the wider environment of Europe (see, e.g., Anon. 1983, Merilehto *et al.* 1988, Skjelkvåle and Wright 1990, Muniz 1991, Borodin and Kuylenstierna 1992, Stanners and Bourdeau 1995). Surface-water acidification is largely caused by the deposition of sulphur dioxide and nitrogen oxides as a result of atmospheric pollution. During the 1950s and 1960s, at the peak of sulphur deposition, the rate of acidification was several hundred times that of natural processes. Deposition of atmospheric ammonia, which originates mainly from agricultural areas with large concentrations of livestock, can also cause acidification of surface waters (van Breemen and van Dijk 1988).

Surface-water acidification generally occurs in areas where acid deposition is high, and where the catchment soil or bedrock is poor in lime or other easily weatherable materials that buffer against increased acidity. The geology of Fennoscandia renders the region particularly vulnerable to acid deposition and thousands of lakes have been acidified as a result. For example, in Finland 42% of lakes are badly acidified, with a pH below 5.0 (Bernes 1993). Most surface waters in western and central Europe are not seriously affected by acidification, despite high levels of deposition, because of adequate buffering capacity of the soils. However, surface-water acidification has been recorded in parts of Belgium, Denmark, Germany, Poland, Czech Republic, Slovakia, Austria, France, Switzerland, Italy and the United Kingdom (Merilehto *et al.* 1988).

Despite a decline in atmospheric pollution during the past two decades, and partial mitigation through liming activities, acidification continues to have a detrimental effect on lake ecosystems.

Acidification affects aquatic organisms at all levels and has a profound impact on both plant and animal communities. Aquatic organisms are influenced both directly, because of the resulting toxic conditions (in part as a result of increased leaching of aluminium ions and other metals), and indirectly, because of the loss of acid-sensitive prey. Massive loss of fish diversity and numbers in Fennoscandia has been one of the most obvious documented impacts (Overrein *et al.* 1980, Drabløs and Tollan 1980). For example, during 1940–1979, 1,750 of 5,000 lakes in southern Norway became completely devoid of fish as a result of acidification, with another 900 being seriously affected.

Impacts on birds are less well documented, and likely to be complicated by interactions between food-web components. For example, under moderate acidification, declines in fish stocks may lead to increases in suitable, alternative invertebrate prey (Eriksson 1987). However, declines in Red-throated Diver *Gavia stellata* and Black-throated Diver *G. arctica* populations in Fennoscandia may be attributable to acidification (Eriksson *et al.* 1988, Eriksson 1994). Coregonid and cyprinid fish (primarily roach *Rutilus rutilus*) are the most common prey delivered to chicks, and these are particularly sensitive to low pH. Furthermore, mercury levels in fish increase in acidified lakes (Håkanson 1980) and Eriksson *et al.* (1992) found levels in eggs sufficient for reproductive impairment to be expected. Similar effects may be expected for other fish-eating species in the region, though most populations of such species appear to be stable (Tucker and Heath 1994) and thus impacts may only be local.

Invertebrates are also affected by acidification. Common prey species such as caddis flies (Trichoptera), mayflies (Ephemeroptera), freshwater shrimps and molluscs are unable to tolerate pH ranges below 5.0. Impacts on birds relying on such species may, therefore, be expected. Indeed, local declines in populations of *Cinclus cinclus* in the United Kingdom have been attributed to acidification (Ormerod and Tyler 1987, 1993).

Excessive sediment loads

Soil erosion is a major and increasing problem in many parts of Europe, particularly in southern, central and south-eastern Europe (Blum 1990, van Lynden 1995). The resulting increased sedimentation in water-bodies, and the increase in the amount of sediment suspended in the water column, can be significant threats to a number of species. Firstly, reduced water clarity can reduce the suitability of the habitat for species which hunt by sight, such as *Pandion haliaetus* and *Alcedo atthis*. It can also reduce light penetration to levels insufficient for the growth of submerged plants, thereby changing the composition and structure of the wetland vegetation.

This can be detrimental for priority species dependent on submerged plants (see Appendix 3, p. 354) and, indirectly, for other species as well, through wider disruption of food-webs and the ecosystem.

Although at present it appears that the number of priority species affected is likely to be fairly low, there is little published research on this threat and therefore its overall impacts are uncertain.

Aquaculture

Aquaculture can provide substantial economic benefits which may help to protect wetlands. Indeed, some internationally important wetlands for birds were created for the production of fish (e.g. in the Czech Republic, Slovakia and Hungary). However, inappropriate management can lead to problems, e.g. through disturbance, pollution from the addition of excessive fish-food, and the use of chemicals such as fungicides to control fish diseases. Illegal persecution of fish-eating birds, including *Pelecanus crispus*, has been recorded at fish-farms (Crivelli 1996).

Nevertheless, the overall threat from aquaculture on a European scale is likely to be relatively low.

Introduction of non-native species

The introduction of non-native animal and plant species can, in the worst cases, disrupt entire ecosystems and profoundly alter habitats. Introduced predators are the most obvious threat to priority birds, and are treated as a distinct threat in this strategy (see 'Human-induced increases in predators', p. 147), but other, more indirectly acting threats such as the new diseases which often accompany the introduced species, or competition between similar species (native versus non-native), or habitat alteration, can also be important.

For example, the fast-growing, floating water-hyacinth *Eichhornia crassipes* can spread very rapidly over lakes, resulting in the complete loss of open water, a habitat-feature that is essential for some priority bird species. A disease introduced to Europe by North American crayfish *Procambarus clarkii* has contributed to the near-disappearance of Europe's native crayfish *Astacus astacus* and *Austrapotamobius pallipes* (Stanners and Bourdeau 1995). Numerous freshwater fish species have been introduced, such as *Gambusia affinis*, *Salmo gairdneri* and *Ameiurus nebulosus*.

The long-term impacts of these introductions on the natural biota of Europe is not well understood, and is difficult to predict, given potential interactions with global warming, eutrophication, etc. However, it is now widely realized that the introduction of Ruddy Duck *Oxyura jamaicensis* from North America poses the most serious long-term threat to the native, globally threatened *Oxyura leucocephala* (Green and Anstey 1992, Green and Hughes 1996). The former species was first introduced into Europe in the United Kingdom in the 1950s and is currently spreading across the western Palearctic, having now been recorded in 20 countries. It threatens to drive *O. leucocephala* to extinction through hybridization, which readily occurs between the two species, resulting in viable and fertile offspring.

Overall, non-native introductions are predicted to affect rather few priority bird species over the next 20 years, and at the catchment or continental scale their overall importance as a threat is currently predicted to be relatively low among those considered here. However, at the level of species or site conservation, this threat is already known to be very important, especially in island or 'non-continental' situations, and it is very difficult to predict which new species are likely to be introduced in the future, and whether they will pose a serious risk to native species. It is clearly advisable to treat this threat as a potentially serious ecological hazard.

Other threats

Other threats which may have local impacts on priority species' populations in some regions include:

- Reductions in the water-storage capacity of landscapes, through afforestation, deforestation and soil loss due to excessive erosion;

- Farming of domestic ducks and geese on natural wetlands (a common practice in eastern Europe);

- Sand and gravel extraction from existing wetlands can cause temporary habitat loss, although if the wetlands are restored correctly this can lead to wetland creation.

CONSERVATION OPPORTUNITIES

International legislation

There is no shortage of international legislative instruments to promote conservation and sustainable use of inland wetlands in Europe (Box 3). These are described in more depth in the introductory chapter on 'Opportunities for Conserving the Wider Environment' (p. 24), but information that is particularly relevant to inland wetlands is highlighted here.

Box 3. International environmental legislation of particular relevance to the conservation of inland wetlands in Europe. See 'Opportunities for Conserving the Wider Environment' (p. 24) for details.

GLOBAL

- *Biodiversity Convention* (1992) Convention on Biological Diversity.

- *Bonn Convention* or *CMS* (1979) Convention on the Conservation of Migratory Species of Wild Animals, including the Agreement on the Conservation of African–Eurasian Migratory Waterbirds (AEWA) (1995, but yet to enter into force).

- *Ramsar Convention* (1971) Convention on Wetlands of International Importance especially as Waterfowl Habitat.

- *World Heritage Convention* (1972) Convention concerning the Protection of the World Cultural and Natural Heritage.

- Convention on Long Range Transboundary Air Pollution (1979).

- Convention on the Protection and Use of Transboundary Watercourses and International Lakes (1992).

PAN-EUROPEAN

- *Bern Convention* (1979) Convention on the Conservation of European Wildlife and Natural Habitats.

- *Espoo Convention* (1991, but yet to come into force) Convention on Environmental Impact Assessment in a Transboundary Context.

EUROPEAN UNION

- *Birds Directive* Council Directive on the conservation of wild birds (79/409/EEC).

- *Habitats Directive* Council Directive on the conservation of natural habitats of wild fauna and flora (92/43/EEC).

- *EIA Directive* Council Directive on the assessment of the effects of certain public and private projects on the environment (85/337/EEC).

- *Nitrates Directive* Council Directive protecting waters against pollution by agricultural nitrates (91/676/EEC).

- Council Directive on urban waste water treatment (91/271/EEC).

- Council Directive on the control of dangerous substances discharge (76/464/EEC).

- Council Directive on the conservation of fresh water for fish (78/659/EEC).

- Council Directive on the freedom of access to information on the environment (90/313/EEC).

- *Agri-environment Regulation* Council Regulation on agricultural production methods compatible with the requirements of the protection of the environment and the maintenance of the countryside (2078/92/EEC).

- Council Regulations on the Structural Funds (e.g. 2081/93/EEC).

The EC Birds Directive and the Ramsar Convention have been at the forefront in conserving wetlands in Europe, particularly at the level of important individual sites, since one of the major threats to wetlands and their associated priority birds and other wildlife is the very widespread habitat loss and degradation (on a site-by-site basis) caused by drainage and land-claim.

The Ramsar Convention focuses wholly on the conservation of wetlands, both coastal and inland. Its contracting parties are obliged to work towards the 'wise and sustainable use' of wetlands, through the inventory and monitoring of all internationally important sites and the promotion of wetland conservation in the national land-use planning system, as well as by designating at least one internationally important wetland as a 'Ramsar site'. Such designation generally provides sites with an extra layer of protection (e.g. above national or EU-level designation), or at least heightened political recognition. In 1990, about 25% of wetland Important Bird Areas (inland and coastal) that met the Ramsar criteria in Europe were listed as Ramsar sites (Langeveld and Grimmett 1990). Since the 'wise and sustainable use' concept is still relatively novel for many contracting parties, the encouragement that the convention gives to practise it on the ground, at least locally at 'showpiece' Ramsar sites, is valuable.

The Birds Directive requires that Special Protection Areas (SPAs) be designated by EU countries for bird species listed in Annex I of the directive—species considered to be threatened in Europe. This annex includes 51 (50%) of the priority birds of inland wetlands in Europe (see Appendix 1, p. 327). The EC Habitats Directive links with the Birds Directive (Box 3), and is of considerable potential importance. It aims to combine the protection of threatened species with the wider objective of conserving habitats that are threatened in their own right, through designation of Special Areas of Conservation (SACs). Annex I of this Directive lists these threatened habitat-types, which include 20 inland wetland habitat-types (Appendix 5, p. 423). Five of these 20 are 'priority' habitat-types, which requires that they be given higher protection status.

The Birds and Habitats Directives together require that EU member states designate by June 2004 a 'coherent ecological network' of protected areas ('Natura 2000'), comprising SACs and SPAs in sufficient numbers and locations so as to conserve these threatened habitat-types and species. Implementation of Natura 2000 is moving very slowly relative to the required time-schedule, but in the long term may encourage a more integrated kind of wetland conservation, where individual sites are no longer viewed or managed in isolation of each other—at least within the European Union (EU).

The realization of Natura 2000, or better still, a pan-European network of designated protected areas, would significantly enhance the conservation of the many priority birds of inland wetlands which are highly migratory and which depend on a wide and varied network of wetlands in many countries during their annual cycle (see 'Habitat needs of priority birds', p. 133). On the pan-European (and wider) scale, the Bonn Convention offers a valuable framework for conserving, monitoring and managing the migration systems of priority waterbirds, through the recently concluded 'African–Eurasian Waterbirds Agreement' (AEWA). This agreement promotes concerted and coordinated activities among governments, conservation organizations (including the secretariats of the other relevant international conventions) and sustainable-use groups (e.g. hunting associations) to conserve and sustainably manage flyway populations of 170 waterbird species (including 59 of the 81 eligible priority species considered here) in 120 states. The obligatory activities—many of which address the conservation of the wider environment, outside protected areas—are set out in a detailed and agreed Action Plan which deals with species and habitat conservation, management of human activities, research and monitoring, education and information, and implementation. The Agreement was signed by seven states, including five European countries, on 31 December 1996, and is intended to come into force in 1998.

The EU has adopted several Directives and Regulations aimed at improving water quality (Box 3). The recently proposed framework Directive on EU water policy will be crucial in integrating this previous legislation and in making it more complete and comprehensive (CEC 1995b), with its emphasis on the need for integrated programmes which target all wetlands. However, the ultimate value of these international laws depends on how well they are implemented by each of the EU member states.

One of the land-uses with the greatest negative impact on the habitats of priority wetland birds is tourism and recreation. Destruction and degradation of inland wetlands by infrastructural developments, for some tourism/recreation projects and some other purposes (but usually not agriculture or forestry), are regulated against in EU countries to some extent by effective implementation of the Environmental Impact Assessment (EIA) Directive (Box 3). There is a proposal to amend this Directive to make more tourism and transport infrastructure projects subject to obligatory EIA procedures, and to state more clearly that wetlands are sensitive zones that particularly require EIA before any developments, and that wetlands (and other natural habitats) which are protected by official designation (e.g. as part of the Natura 2000 network) deserve particularly high regard during the evaluation of impacts.

Outside the EU, the negative impacts that such projects have on wetlands are often left to the consideration of national legislation and decision-making processes, which may not adequately acknowledge the international dimension of wetland habitat-use by priority migratory birds, nor the potentially large-scale (international) and highly indirect impacts of national projects such as dams on large river basins.

There are a number of bilateral and multilateral agreements in Europe concerning the management of such shared river basins, but most, apart from those covering the Rhine and Elbe, deal mainly with water quality and do not directly require or discuss the conservation of important aquatic species, sites and habitats, nor the maintenance of key hydrological processes associated with natural wetlands. The Espoo Convention (Box 3) requires that EIAs be carried out on large dam-building projects in shared river basins (CEC 1995b), although it has not yet come into force.

Furthermore, the application of the 'environmental assessment' concept at an earlier phase in the planning process, or on a wider scale than individual projects, i.e. towards entire policies, plans or programmes (e.g. for land-uses such as tourism or agriculture) in regions or nations, does not yet feature significantly in European legislation. An EU Directive is in preparation that will address such Strategic Environmental Assessment in EU countries, and it is anticipated that this will play a very important role in the integrated management of wetlands in the EU if adopted by member states (CEC 1995b).

Drainage and land-claim impact priority wetland birds more than any other threat, and some progress towards integrating habitat protection and infrastructural development has been made in the EU, where Council Regulation 2081/93 requires that plans and programmes submitted under Objectives 1, 2 and 5 of the Structural Funds have to be evaluated according to certain of their environmental aspects (see CEC 1995b).

Financial instruments

In the EU, some financial instruments are available for wetland conservation. For example, between 1984 and 1991 a financial instrument known as ACE (Regulations 1872/84 and 2242/87) funded 59 'biotope projects' that included a wetland conservation component, of which 56 dealt with specific wetland sites of particular importance (Mondain-Monval and Salathé 1993, CEC 1994a).

More recently, the LIFE Regulation (Regulation 1973/92), and the former Regulations of ACNAT (3907/91), MEDSPA (563/91) and NORSPA, which LIFE now replaces, have financed priority conservation actions for wetlands both within and outside the Union—5% of the budget is spent on actions in the Baltic and Mediterranean regions outside EU territory (CEC 1995b). For instance, ACNAT provided two-thirds of the funding for the initial 1992–1996 phase of MedWet, a major initiative for coordinated action to conserve wetlands throughout the Mediterranean basin. In addition, one of the main aims of the LIFE regulation is to support the implementation of the Natura 2000 network.

The EU is also a major provider of funds to wetland conservation projects in European countries outside the Union through the PHARE programme (Regulation 3906/89)—CEC (1995b) gives examples. Similar projects have also been funded by the Global Environment Facility of the World Bank and United Nations Development Programme, and by the European Bank for Reconstruction and Development, in eastern and southern Europe, some of which cover very important sites, e.g. the Danube delta (Romania) and Sultan marshes (Turkey).

Within the Common Agricultural Policy of the EU, the Agri-environment Regulation (see Box 3) could support national programmes to conserve or rehabilitate temporarily flooded pastures and wet grasslands—these habitats and opportunities are covered in more detail under 'Agricultural and Grassland Habitats' (p. 267).

More indirectly, effective promotion of such beneficial farming practices by member states, and widespread adoption by their farmers, could lead to an abatement of such important threats to wetlands as eutrophication and over-abstraction of water, by reducing the amount of land under intensive cultivation and irrigation. A current example of such a Zonal Programme is in the Mancha Occidental–Campo de Montiel area of Spain, where two wetland complexes, the Tablas de Daimiel National Park and the Lagunas de Ruidera National Park, will benefit from such measures (CEC 1995b). The Fifth Environmental Action Programme of the European Commission (Directorate-General XI) targets 15% of the arable area of the EU to be under such management contracts with farmers by the year 2000.

Funds are also allocated by the EU to research and technical development programmes which cover fields strongly related to the conservation and sustainable use of wetlands, such as reduction of air pollution, improvement of water quality, soil protection, basic research on ecosystem functioning, etc.

The EU Cohesion Fund (applicable only to Greece, Ireland, Portugal and Spain) and Structural Funds finance large infrastructural projects within the Union (roads, ports, etc.), some of which have destroyed or badly damaged wetland habitats, often indirectly through catchment-scale effects. Despite having financed facilities for water management and for the treatment of water and waste in the past (CEC 1995b), and despite some progress through Council Regulations and also a decision of the European Court (decision C-355/90, on the Santoña marshes in Spain), environmental concerns have still not been integrated into the operation of these and other international funds, and the assessment of environmental impacts of proposed projects, programmes and plans still needs major improvement, through the introduction of Strategic Environmental Assessment.

In non-EU countries of the Mediterranean basin, improvements in and provision of waste-disposal and water-treatment facilities are financed by, among others, the Mediterranean Technical Assistance Programme (METAP) of the European Investment Bank and the World Bank. The European Bank for Reconstruction and Development plays a similar role in central and eastern Europe, together with EU funds from the PHARE and TACIS programmes.

The use of financial instruments to control pollution is also a potential opportunity in these countries and elsewhere in Europe, e.g. through 'incentive charging'. This ensures that polluters (e.g. industrial companies) who pay the government for the right to pollute (e.g. to discharge waste water) are charged according to the full environmental costs of such pollution.

Policy initiatives

Increasingly in Europe, there are initiatives that can help promote wetland conservation in the wider environment as well as in protected areas. Given the physical and functional connections between wetlands themselves and their surrounding habitats, a first requirement of any such initiatives is that they should be on a catchment basis and therefore, if necessary, international in scope. At the moment, such policy is probably best developed in the European Union, where the Fifth Environmental Action Programme provides the strategic framework for the 15 member states' joint programme of policy and action on environment and sustainable development. In a recent Communication (COM(95)189 final), the European Commission concluded that the key aspect of future EU policy on the wise use and conservation of wetlands should be its integrated approach, and its four most important aims should be to achieve the following (CEC 1995b):

- Full coherence of the Natura 2000 network and respect for the obligations related to it.
- Integrated water management with respect to both quantity and quality.

- An effective land-use planning policy.
- Significant financial support of wise use and conservation of wetlands, mainly as an integral part of other policies.

Examples of how integrated water management may be achieved are the 'catchment management plans' and 'water-level management plans' that are being produced in the UK. For the former plans, the highest management authority (National Rivers Authority) will provide by 1998 a baseline overview of the current status and uses of each of the 168 river catchments in England and Wales, and set overall targets for water quality, quantity and physical features in each basin. The plans aim to provide a focus for the formation of agreements between water users about the future development of the catchment.

The 'water-level management plans' are specific to important wetland sites in England and Wales, such as SPAs, Ramsar sites, SACs, National Nature Reserves, etc., and are drawn up by the relevant flood-defence operating authority (in the Ministry of Agriculture, Food and Fisheries) following agreement between interested parties representing conservation, agriculture and flood-defence interests on that site. Each plan defines the optimal water-level management regime for that site, reflecting a balance between the needs of all interests (MAFF 1994). Management plans also define the monitoring needs of each site.

If national policies on wetlands are modified as recommended by the European Commission, it is anticipated that an increasing number of non-designated wetlands will be safeguarded from destruction or degradation as part of general policy and without the need for special designation.

The recent moves towards reducing agricultural over-production, especially in the European Union, through extensification (e.g. Environmentally Sensitive Areas), set-aside, and reductions in direct grants for 'improvements' such as drainage, all reduce direct and indirect pressure on wetlands (see 'Agricultural and Grassland Habitats' for further details on relevant policies; p. 267).

The sustainable development of tourism is another key concern of the European Commission, and various projects have been supported in the framework of *Community action plans to assist tourism*. These include supporting activities to increase the environmental awareness of tourists and suppliers of services, support and promotion of ecotourism and 'environmentally friendly' tourism initiatives, and encouragement of the transnational exchange of experience in dealing with visitor management, etc. Since 1992, the European Commission has funded 23 pilot projects in the field of tourism and the environment, three of which have a particular focus on wetlands (CEC 1995b).

Regarding wetland policy at a wider scale than the EU, the ongoing series of European environment ministers' conferences is of great strategic importance. At the most recent conference held in Sofia in 1995, the Pan-European Biological and Landscape Diversity Strategy (PEBLanDS) was endorsed by the ministers of 55 European states (see 'Opportunities for Conserving the Wider Environment', p. 24), as a broad initiative to help implement the requirements of the Biodiversity Convention (Box 3). An overall framework, as potentially provided by PEBLanDS, could be very helpful in promoting the conservation and sustainable use of inland wetlands, which are already partially covered by numerous legal mechanisms, and which have already been subject to numerous initiatives, declarations and recommendations.

Action Theme 7 of the strategy covers 'Inland wetland ecosystems', and has several main objectives under the first Action Plan (1996–2000). These include the promotion of concepts such as 'wise use' and 'integrated catchment management' among national and regional governments, sectoral planners and developers (agriculture, municipalities, transport, tourism, etc.) and businesses. It also promotes the creation of a Pan-European Ecological Network, through the designation of 'core areas', 'buffer zones' and 'corridors'/'stepping stones' for dispersal and migration of flora and fauna. This approach is of relevance to the conservation of inland wetlands, given the importance of transboundary management of international river-basins in assuring sustainable use of wetlands.

Apart from the Natura 2000 network, another current initiative which could contribute strongly to such a Pan-European Ecological Network, especially in enhancing the involvement of the non-EU countries, is the IUCN *Action Plan for Protected Areas in Europe*, which aims to ensure the development of an adequate, effective and well-managed network of protected areas in Europe by identifying the principal needs and outlining the priority actions required at regional level (IUCN–CNPPA 1994).

CONSERVATION RECOMMENDATIONS

Broad conservation objectives

1. Avoid the loss, or partial loss, of any internationally important inland wetland (IBAs) and maintain the current overall area of inland wetland habitat in Europe.

2. Stop the degradation of inland wetlands and maintain and enhance important habitat features

for their priority species.

3. Improve or restore habitat quality on artificial and degraded inland wetlands for priority species.

4. Recreate wetland habitats that specifically benefit priority species of inland wetlands where there is no conflict with other conservation interests or objectives.

Ecological targets for the habitat

In order to maintain and improve habitat quality, it is essential to retain and enhance important habitat features for priority species (see p. 133). It is therefore necessary to ensure the following conditions are met.

1. The integrity of flyway networks for migratory species should be maintained.

2. Adequate levels and flow-rates of groundwater and surface water should be maintained, or restored where there has been previous depletion of surface water or groundwater/aquifers, wherever possible.

3. Natural hydrological dynamics of wetlands should be maintained or, where appropriate and feasible, restored. Measures could include the restoration of natural flood-plain habitats, reconnecting 'dead' side-arms of rivers with main river channels, creating or recreating meandering rather than straight riverbeds, and permitting seasonal water-level fluctuations.

4. Modified river channels should incorporate 'natural' channel designs such as frequent bends, side arms, islands, pools and riffles, etc. Other important hydrological characteristics of managed wetlands (e.g. water clarity, speed of flow, temperature) should be kept at or close to natural levels and fluctuations. Water-levels and flows should be managed to minimize detrimental effects on nesting birds, food resources and feeding opportunities, according to natural variations.

5. Riparian and littoral habitats should be maintained, especially important features such as natural bankside structures and profiles (e.g. sandy cliffs, shallow muddy shores, associated wet marshes and sedimentary islets), mature native trees, dead wood, rank bankside vegetation and stands of emergent plants. Mosaics of these habitat components should also be maintained, according to the preferences of the priority species using individual wetlands.

6. Nutrient levels in wetlands should not exceed their natural carrying capacities. High levels of man-made eutrophication which cause excessive loss of open water through the growth of

plants, or which reduce water clarity or dissolved oxygen content, or which cause declines in food resources, should be avoided and reversed where necessary.

7. Toxic pollutants should be reduced below sublethal levels in all wetlands, e.g. below levels where indirect effects are known or reasonably suspected to be causing population declines in priority species.

8. High levels of water acidity that cause declines in fish and other aquatic food resources, or that cause significantly increased mobility of toxic chemicals, should be avoided. In general, the acidity levels of all wetlands should be kept within their natural range of variation.

9. Disturbance of birds by human activities (e.g. hunting, fishing and recreation) should be avoided in the vicinity of nesting sites, and reduced to acceptable levels (according to the species affected and local circumstances) when feeding or roosting, such that the carrying capacity of the disturbed area is not affected overall.

10. Benthic habitats and their associated flora and fauna should be conserved or restored.

11. Wild populations of native species of fish should be maintained, but should not be enhanced through stocking where this is detrimental to priority bird species. The introduction of non-native species of plant and animal to wetlands should be avoided, and populations of such species should be controlled or removed where possible.

Broad policies and actions

To meet the broad objectives and habitat targets given above, the following broad policies and actions must be pursued (see p. 22 in 'Introduction' for explanation of derivation).

General

1. The high economic, social and cultural value of inland wetlands, as well as their importance for birds and other wildlife, should continue to be widely promoted.

2. It is of particular importance to raise awareness among politicians and their advisers, land-managers, farmers and hydrological engineers of the need to adopt a catchment-scale, integrated approach to hydrological (wetland) management, and that individual wetlands should not be viewed or managed in isolation, or managed according to the needs of a single, local human use. Such integrated, catchment-scale plans and programmes should aim to ensure the sustainable use of land and water and the protection of biodiversity.

Legislation

3. Protection of inland wetlands should be improved through the full implementation and enforcement of existing international and national environmental legislation.

In particular, the Ramsar Convention should be used extensively to prevent further wetland loss and degradation, through the designation of more Ramsar sites and through adherence to its 'wise use' principles.

European range-states should sign and ratify the African–Eurasian Migratory Waterbird Agreement and implement the actions as listed in the Action Plan (Annex 3 to the Agreement).

In EU countries, this should include the accelerated designation (with ecologically meaningful boundaries) of all inland wetland sites that qualify as Special Protection Areas (SPAs) under the Birds Directive. As far as such SPAs also qualify as Special Areas for Conservation (SACs) under the Habitats Directive, combined SPA/SAC designation from 1998 onwards could gain time and help EU member states to catch up on their lagging progress towards the creation of Natura 2000, the 'coherent ecological network' which should be successfully established by June 2004 (as stipulated in the Habitats Directive).

The proposed framework Directive on EU water policy should be promptly adopted and implemented.

4. Establishment of new legislation by governments, and considerable improvement of existing legislation, is necessary in order to fully integrate the goal of wetland conservation and sustainable use into all aspects of national and regional policy. Such an integrated approach must be promoted as a deliberate cross-sectoral strategy.

In order to manage catchment basins in an integrated way, apart from ensuring that appropriate national and international legislation is in place to allow and promote this, it may also be necessary for governments to create new administrative structures or reorganize old ones—in either case, major institutional strengthening may also be needed. This is particularly important for transboundary water basins. All water-management agencies in Europe should have duties towards the conservation of wetland biodiversity.

5. Necessary adjustments to be made in government policy include the removal of economic incentives like tax benefits or subsidies, and of any legal encouragement to destroy wetlands.

6. Regional authorities, national governments and the European Union should ensure that Strategic Environmental Assessments (SEAs) are carried out in order to assess the full range of effects of current and proposed water-use policies, plans and programmes, in addition to carrying out properly conducted EIAs of individual schemes (including agricultural and forestry projects).

   SEAs and EIAs should be carried out to internationally recognized standard procedures (e.g. as set out in the EIA Directive; see Box 3) and legislation should ensure that planning approval for schemes that destroy or degrade wetlands (e.g. drainage) is subject to the conclusions of properly conducted SEAs/EIAs.

   Disbursement of international funds (bilateral, multilateral or EU, e.g. PHARE, Cohesion Fund and Regional Development Funds) for programmes and projects that involve wetland loss or degradation should also be strictly compliant with relevant national and EU environmental policies and legislation and subject to the conclusions of properly conducted SEAs/EIAs.

7. The authorization of water abstraction should be more closely monitored and coordinated by national and regional governments, in order to improve the management of water stocks and improve their conservation.

8. With due regard to the 'polluter pays' and the 'precautionary' principles, governments must ensure that inland wetlands are protected from pollution by adequate policy and legislation and its proper enforcement. They should achieve this by:

   • providing safe-storage schemes for potentially harmful substances;

   • legislating in favour of environmentally benign chemicals where available as alternatives to toxic chemicals (e.g. enforce Red and Grey lists of the EC Directive on dangerous substances);

   • introducing strict regulations for effective treatment of urban and industrial waste;

   • improving and enforcing legislative controls on run-off of agricultural fertilizer and waste into water-bodies;

   • ensuring widespread knowledge of any legislation banning or regulating pesticide-uses;

   • ensuring that responsible authorities have adequate regulations and resources to enforce controls on pollution, and have established procedures to widely monitor and improve water quality through integrated programmes targeting all elements of the ecological quality of water systems;

   • requiring polluters and other resource exploiters to monitor and report on their own uses and discharges, to a similarly high standard.

9. Governments must ensure that acid deposition is below critical loads throughout Europe, through international and national legislation.

10. The use of lead-shot in wetlands should be phased out in all countries and the use of non-toxic alternatives promoted.

11. The protection of the littoral/riparian strip of larger lakes and rivers should be promoted in national legislation and land-use planning regulations.

River management

12. River engineers should increase the use of natural river structures (e.g. meanders and riparian vegetation) as a means of flood protection, and conservation of these features should be incorporated into regulations governing the strategic planning of flood defence.

13. The public, and planners and water engineers in particular, should be made fully aware of the importance of riparian habitats and their potential role in bank stabilization, flood control, water purification, etc. Guidance on the management of aquatic and riparian vegetation for wildlife should also be provided to the authorities with responsibility for its management (e.g. water authorities), for example through making manuals on management techniques (e.g. RSPB/NRA/RSNC 1994) and training courses widely available.

14. Where river-channel modification is essential, then 'natural' channel designs incorporating, for example, frequent bends, side arms, islands, pools and riffles, etc., should be used wherever possible. Operations should also minimize the clearance of channel and riparian vegetation.

15. Where possible, river restoration projects should be undertaken to reinstate previously engineered channels to pre-channelized conditions. Creative use of floodwater should also be encouraged, through, for example, diversion to wet grasslands, perhaps linked to Environmentally Sensitive Area Schemes (see 'Agricultural and Grassland Habitats', p. 267).

Water supply

16. National governments and the EU must develop policies and legislative measures to, at the very least, maintain and protect current water-levels and water-sources throughout Europe. In the long term, national governments and the EU should enhance and improve water-levels and

the ecological quality of water resources where these have been degraded or over-exploited.

17. Water-resource use should be based on a long-term management planning process, which should be based on reliable data and follow the precautionary principle in all predictions of demand and supply. This process should be subject to Strategic Environmental Assessment, and should deliver cross-sectoral and holistic management of water resources.

For example, planning consents for water-dependent developments (e.g. irrigation schemes, tourist centres and certain industries) should be linked to the availability of water. Measures should be taken to ensure that future developments are more water-efficient. Water abstraction should not be allowed to exceed the recharge rate, and water-user rights should only be granted for a (renewable) limited period.

18. The concept of 'demand management' in water resources should be promoted (rather than 'supply management'), in particular through the promotion of household and industrial water-conservation measures, the reduction of leakages in water-supply systems, and the promotion of water-saving technologies.

The cost of water should reflect the environmental cost of taking that water from the environment, through application of the 'polluter pays' principle to all exploiters of water resources.

19. The concept of minimum groundwater and surface-water levels and flows necessary to avoid substantial damage to ecological processes and biodiversity should be recognized, and incorporated in national and EU legislation, with drainage and water abstraction banned where these limits are reached.

20. Existing good quality water supplies should be carefully maintained, managed and monitored, in order to avoid the need for the use of new sources because of a reduction in water quantity or quality.

21. Inter-catchment transfers of water and major abstraction schemes should be subject to the conclusions of properly conducted SEAs/EIAs, where these are not applied already.

22. All proposed impoundments of natural wetlands should be subject to SEA/EIA planning procedures which fully consider all alternatives. Where impoundment is unavoidable, mitigation measures should be mandatory.

23. Research into the effects of water-level and flow on riverine and other wetland ecosystems and their flora and fauna should be carried out, in order to develop models that predict the impacts of hydrological management regimes on priority species and other wildlife.

Water quality

24. As part of the long-term management of water resources, national governments should publish annual and long-term water quality targets for all rivers and other wetlands.

25. Pollution-sensitive areas should be identified using 'Critical Loads/Levels' methodologies, which should then be used to set water-quality targets in order to avoid further pollution, particularly in such areas, and to alleviate existing seriously affected waters.

26. Phosphate removal in sewage treatment should be improved, with maximum discharge levels set (in national and EU legislation) that ensure that critical loads are not exceeded locally, as well as comprehensive monitoring of discharges and strict enforcement of existing laws. Further action should be taken to reduce phosphate use in detergents.

27. New sewage, industrial, and other waste-water discharges into wetlands should be subject to EIAs.

28. The use of lime to treat acidified water-bodies should be carefully regulated, and avoided where further ecological damage (due to the lime) may be significant. It should also be recognized that its use can only be a short-term and local solution to the problem of acidification.

29. Further research should be undertaken into the impacts of pollutants on bird and other wildlife populations, in particular the impacts of eutrophication, of acidification, and of PCBs (polychlorinated biphenyls) and other toxic pollutants.

Agriculture and forestry

30. National governments and the EU should develop agricultural policies and support measures, in the context of catchment management planning, that maintain and promote extensive agricultural systems in water catchments, in order to reduce the need for flood defence, drainage, water abstraction, fertilizer use and pesticide applications.

31. All subsidies for agricultural drainage schemes should be removed, and such schemes should be subject to the conclusions of properly conducted EIAs before they are allowed to proceed.

32. Land tenure, when (re-)privatized from state land, should be subject to restrictions on drain-

age and/or water obstruction (building of river embankments, dams, etc.).

33. Financial support should be provided to farmers for the construction or improvement of storage facilities for silage and slurry. At the same time, inadequate storage of such materials should be made illegal in all countries.

34. The concept of 'nitrate-sensitive areas' should be expanded to include all nutrients. The location of such nutrient-sensitive areas should be determined as part of integrated catchment management.

35. Agricultural policies should be modified to reduce livestock densities in nutrient-sensitive areas and in areas susceptible to soil erosion.

36. Erosion-sensitive areas should be identified and schemes developed to alleviate risks, for example through grants for returning cultivated land to grass, or for the planting of native trees (e.g. through farm woodland grant schemes) where there is no conflict with other conservation or landscape interests. Soil-erosion protection measures should also be incorporated into land-tenure agreements where possible (e.g. in eastern Europe).

37. Afforestation projects should be avoided in acid-sensitive catchments through evaluation during SEAs and EIAs, and as part of integrated catchment management.

Other measures

38. New methods should be developed for the economic valuation of the various functions of wetland ecosystems and of net national income losses through wetland degradation and loss.

39. Pilot projects aiming at the sustainable use of wetlands involving local communities should be financially supported and the results effectively publicized.

40. The extensive use of the man-made fish-pond systems of central Europe should be promoted by the introduction of ESA-type schemes (e.g. through payments to compensate for loss of production, by reducing taxes, reducing or eliminating water charges, provision of favourable loans).

41. The effects of human disturbance on birds and wildlife, and the need for controls, should be publicized, especially to hunters, those taking part in water-sports, and anglers. Regulation of the use of water-bodies by these and other groups should be carried out where necessary through, for example, zoning, temporal or spatial restrictions, and implementation of local bye-laws. To ensure this, permission for a particular recreational use should be dependent on the development and implementation of a management plan that adequately controls the potential impacts of disturbance. The effectiveness of disturbance-regulating measures in conserving wildlife should then be monitored and reviewed where appropriate.

42. The introduction of non-native species (especially fish) into river systems and water-bodies should be legislated against, and such laws should be widely publicized and effectively implemented. Where possible, existing populations of such organisms should be eliminated, or their spread should be slowed, halted or reversed, where they are causing significant population declines or range contractions in priority birds or other native wildlife.

Research

43. Further research should be carried out into the following areas:

- wetland networks within flyways, in order to identify key breeding and wintering sites and staging posts for priority bird species;

- natural hydrological processes of undisturbed wetlands;

- natural succession dynamics;

- the mitigation of human impact on the functioning of wetland ecosystems, with a high proportion of the funding coming from those groups which cause negative impacts.

- the impacts of all forms of disturbance on wetland bird populations, and into means of reducing impacts.

Important results should be effectively communicated to the appropriate target audiences as soon as possible.

ACKNOWLEDGEMENTS

The major contributions to this chapter were made during a workshop organized by G. Sarigül and M. Yarar at Priene, Turkey (1–5 February 1994), hosted by Dogal Hayati Koruma Dernegi; these contributors are listed on p. 125.

Additional information and comments were received from A. Anselin, J.-Y. Mondain-Monval, M. Moser and C. Prentice (Wetlands International), S. Nagy (Magyar Madártani és Természetvédelmi Egyesület), V. Vinogradov,

M. Vitaloni, I. Rusev (Natural Heritage Fund, Ukraine), M. Strazds and W. Müller (Schweizer Vogelschutz), K. Norris and D. Harrison (Royal Society for the Protection of Birds), G. Rocamora (Ligue pour la Protection des Oiseaux), P. Herkenrath (Naturschutzbund Deutschland), D. Munteanu (Societatea Ornitologica Româna), J. Coveney (BirdWatch Ireland), E. Osieck and J. Winkelman (Vogelbescherming Nederland) and A. Mikityuk (Ukrainske Tovaristvo Okhoroni Ptakhiv). S. Woodward assisted with the word-processing of the text and tables. We are also grateful to Robin Standring for his valuable help while working as a volunteer at the BirdLife Secretariat.

TUNDRA, MIRES AND MOORLAND

compiled from major contributions supplied by the Habitat Working Group

I. Byrkjedal, L. Campbell, V. Galushin, J. A. Kålås, A. Mischenko, V. Morozov, L. Saari, K.-B. Strann, I. P. Tatarinkova, D. B. A. Thompson (Coordinator)

and by

M. Strazds

S U M M A R Y

Tundra (including boreal montane habitats), mires and moorland are particularly important for the survival of 73 priority bird species in Europe, where 37% have an Unfavourable Conservation Status because they have notably small, or strongly declining, populations. The remaining priority species are specialized and highly dependent on the continuing existence of these habitats. Most of these specialists are currently still widespread across their preferred habitats, although many are not common and require habitat features that are only found in natural or traditionally managed landscapes.

Maintaining the overall extent and ecological quality of tundra, mires and moorland is clearly essential. Tundra and mires are part of Europe's largest remaining wilderness, extending across much of the northernmost part of the continent as well as in a much smaller and more fragmented area in the north temperate zone. Mires and moorland are habitats which are considerably threatened by heavy grazing, afforestation, peat extraction (on mires) and, in some areas, continued drainage for agriculture. The relatively remote tundra (and boreal montane habitats) are currently less threatened, but they are being degraded in places by oil and gas extraction and by the attendant development of industries, urban centres and transport networks, as well as by the heavy grazing of domestic reindeer.

Networks of important sites of tundra, boreal montane habitats, mires and moorland should be identified and protected, particularly where these habitats have been fragmented in the north temperate zone. Outside such sites, broader conservation measures should be implemented, through the incorporation of environmental objectives into all relevant land-use policies and regulations. In particular, reform of EU and national agricultural policies is required to reduce problems caused by intensive grazing and to avoid further habitat loss.

Governments should make full use of the opportunities for coordinated conservation action that are provided by international and EU legislation, international funds, and broad policy initiatives such as the Arctic Environmental Protection Strategy. The Ramsar Convention is of particular importance for the conservation of tundra wetlands, Arctic coastal habitats and mires.

Proper enactment and implementation of Strategic Environmental Assessments and Environmental Impact Assessments are also required in order to avoid damage to these fragile habitats by industry, forestry, agriculture and other programmes and projects.

HABITAT DESCRIPTION

Habitat definition, location and extent

This strategy covers all tundra, boreal montane habitats, mires and moorland in Europe, as defined below (more detailed descriptions of the habitats are given under 'Ecological characteristics', p. 163).

All are open and usually rather nutrient-poor habitats, whose range lies totally or predominantly within northern Europe. For convenience, all inland wetlands within these habitats are also dealt with in this strategy (see 'Inland Wetlands', p. 125, for those outside these habitats), as are all coastal habitats above the low-tide line and north of the Arctic Circle (see 'Coastal Habitats', p. 93, for those south of the Arctic Circle).

Tundra

This treeless zone lies north of the boreal forest zone (taiga), and is relatively flat and low-lying over most of its range. The average July temperature is less than 10°C and the ground is affected by permafrost (permanently frozen soil or subsoil). Most tundra in Europe occurs in Russia (Figure 1), where it covers approximately 312,000 km², as well as in Greenland and Svalbard, with smaller areas (transitional with

159

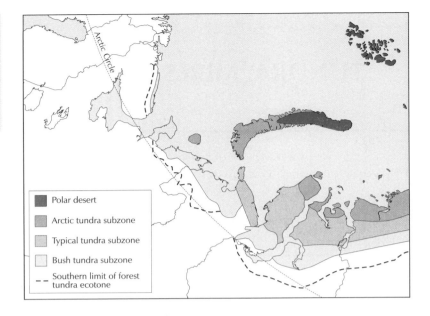

Figure 1. The distribution of tundra in Europe (from Chernov 1985), showing subzones as described in the main text. Greenland and Svalbard (not included on the map) are mainly polar desert.

Arctic Circle

- Polar desert
- Arctic tundra subzone
- Typical tundra subzone
- Bush tundra subzone
- - - Southern limit of forest tundra ecotone

boreal habitats) in the most northerly coastal parts of Norway and Finland.

Tundra remains in a mostly natural or near-natural condition, although locally some areas have changed to grassland as a result of atmospheric nitrogen pollution (nutrient enrichment) and locally heavy grazing by reindeer *Rangifer tarandus* (e.g. parts of the northern tundra plains west of the Urals in Russia).

Boreal montane habitats

These occur above the former/actual treeline, at altitudes varying from 1,200 m in central Norway to below 500 m to the north of the Arctic Circle, in the boreal zone of northern Europe (Figure 2). The environment, vegetation structure and avifauna resemble those of the tundra zone—indeed, they are often referred to as 'tundra' or 'Arctic' habitats. Montane non-forest habitats of the temperate and Mediterranean zones (e.g. in the Alps) are treated under 'Agricultural and Grassland Habitats' (p. 267).

Boreal montane habitats in Europe occur mainly in Scandinavia (along the central mountain spine), Russia (along Polyarnyy Ural, the northern spur of the Urals), Iceland and the UK. Norway, in particular, holds about 215,000 km^2 of these habitats (Dahl *et al.* 1986). They remain in a mostly natural or near-natural condition, although locally some areas have been lost to grassland as a result of atmospheric nitrogen pollution and locally heavy grazing. Some have also been locally extended by man through clearance of submontane scrub.

Mires

Mires (also called bogs) consist of waterlogged peat that is more than 1 m deep, and are found, in Europe, mainly in the boreal and north temperate zone (i.e. approximately north of 55–60°N). The vegetation is dominated by *Sphagnum* mosses, and receives water and nutrients solely from the atmosphere. Mires are consequently acidic (pH 3.5–4.5) and very nutrient-poor environments. Fens, which are treated under 'Inland Wetlands' (p. 125), differ from mires in receiving at least some of their water and nutrients from the groundwater of the surrounding landscape. As a result, they are less acidic, have a richer nutrient status, shallower peat deposits and a substantially different (and richer) vegetation and bird community. Two main morphological types of mire are distinguished here (Lindsay 1995), as follows.

- **Raised mires** are smooth, simple, shallow raised domes of peat, usually consisting of two or more contiguous domes appearing lens-shaped in cross section. The domes can be very extensive (covering many tens of square kilometres), although non-mire habitats then are usually mixed in to form a mosaic, e.g. lakes, forests and fens.

- **Blanket mires** are landscapes cloaked with peat up to 7–8 m thick. Complex patterns of pools occur on the surface. Many blanket bogs have scar-like erosion features. They develop in cool climates in wetter areas than support raised mires (i.e. more than 1,000 mm rainfall annually).

Mires originally covered large areas of northern Europe, but a considerable proportion has been destroyed by man in the south and west of this range

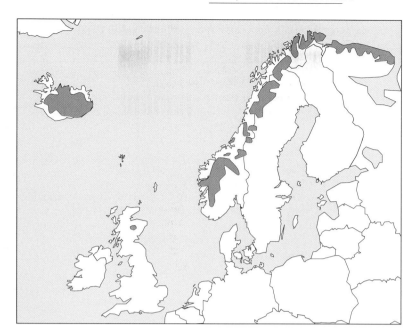

Figure 2. The distribution of boreal montane habitats in Europe (adapted from Polunin and Walters 1985).

(Goodwillie 1980, Jurkovskaya 1980, Lindsay 1995). Remaining mires are mainly in north-west Russia, Sweden and Finland (Table 1), where they total about 400,000 km² and still dominate the landscape in some regions, e.g. Ostrobothnia (Finland). Other countries that each still hold more than 1,000 km² of raised mires include Ukraine, Belarus, Poland and the Baltic states. Remaining blanket mires cover smaller areas, and are primarily found in Norway, Britain, Ireland and Iceland (Table 1, Figure 3).

However, not all surviving mires are still 'active' (laying down peat), mainly because of damage by man (Lindsay 1995). The massive overall declines in extent and condition have been caused by peat extraction or drainage for agriculture or forestry, especially of lowland mires (see 'Threats to the habitat', p. 172). Although some such changes may have been going on for more than a hundred years, they have continued at a steady or increasing rate in recent years.

Moorland
This open, man-made habitat with low and treeless vegetation is dominated by heather *Calluna*, a dwarf

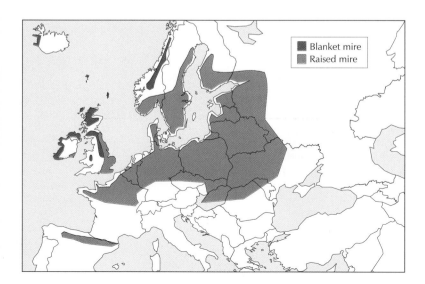

Blanket mire
Raised mire

Figure 3. The range of mires in Europe (from Lindsay 1995).

Table 1. The extent of mires in Europe (compiled mainly from Heathwaite 1993 and Lindsay 1995).

	Current extent (km²)	Notes
Belarus [1]	11,000	10,000 km² have been drained for agriculture; 3,000 km² have been cut over; 3,000 km² are protected for nature
Belgium	1	Active mire; remainder is degraded grass-mire or afforested
Denmark	25	Active mire; 30 sites but mostly at Store Vildmose and Lille Vildmose
Estonia	7,000	
Finland	104,000	
France	600–900	Almost all raised mires in lowlands destroyed, only those in Jura, Massif Central and Vosges remain in any kind of natural state
Germany	580	Maximum of near-natural mire (minimum is considered by some specialists to be much less); mostly in Bavaria, Schleswig-Holstein and Lower Saxony; originally 4,343 km², mainly in Lower Saxony
Iceland	10,000	
Rep. of Ireland	121	Intact mire; originally 3,713 km² of raised mire
Italy	?	A few small sites may occur
Latvia [2]	2,230	Intact mire; also 1,940 km² of intact wooded mire
Lithuania	5,000	
Netherlands	0	No natural areas remain; formerly 1,800 km² of mire
Norway	30,000	
Poland	13,000	
Portugal	?	A few small sites may occur
Russia	233,000	
Spain	0	Formerly a few, small raised mires in Cantabrian mountains
Sweden	70,000	
Ukraine	22,000	
United Kingdom [3]	16,779	

Sources [1] Anon. (1997), [2] VMPI (1980), [3] Gilbert and Gibbons (1996).

shrub. It develops in wet Atlantic climates (typically with more than 1,000 mm precipitation per year) on nutrient-poor soils, and is submontane in the UK and Ireland (i.e. confined to hills and lower mountains) but coastal and lowland in western Norway (see Figure 4). Fremstad and Kvenild (1993) estimate that the potential range of moorland in Norway covers about 6,000 km², although the actual extent is certainly less than this figure.

Moorland in Europe has been subject to significant losses, although not as large as suffered by mires. For example, around 20% of upland *Calluna* moorland in Great Britain has been converted to other habitats since the 1940s (Huntings Surveys 1986), mainly due to afforestation and heavy grazing (see 'Threats to the habitat', p. 172).

Origins of the habitats

Tundra, mires and boreal montane habitats are among the most natural environments remaining in Europe, while moorland is a man-made habitat of relatively recent origin.

Figure 4. The range of moorland in Europe (from Thompson *et al.* 1995b).

Tundra

Tundra was particularly widespread in Europe during the last Ice Age, perhaps reaching its maximum recent extent about 18,000 years ago when the expanding ice-sheets reached their southernmost limits. Some 13,500–5,000 years ago, as the Ice Age waned, the European climate became more temperate and the tundra contracted northwards, leaving its present-day distribution in the far north (Birks *et al*. 1988).

Boreal montane habitats

These tundra-like habitats are thought to have had a similar history and perhaps a shared origin to tundra, having been more extensive during past Ice Ages and having contracted towards their current distribution in the cold, high-altitude zone of mountains in the boreal zone.

Mires

From around 8,000–9,000 years ago onwards, as the last Ice Age came to an end, mires developed over extensive tracts of northern and north-west Europe, where the climate was wetter and tree growth was naturally limited, in what are now the boreal and north temperate zones. In some parts of Europe, human clearance of forests and subsequent changes to hydrological balances facilitated the formation of blanket mires and wooded mires; in other parts the formation of large mantles of blanket mire seems to have been associated with the natural demise of forests (under increasingly wet conditions).

Moorland

This habitat was largely created by human clearance of boreal and north temperate forests, which left the understorey of heather *Calluna* exposed. Subsequent grazing and burning combined with high rainfall to further impoverish soils and discourage tree growth, favouring the dominance of heather. Some moorlands in Europe are up to 4,000 years old, but others were created as recently as 200–400 years ago. The majority of moorlands continue to be maintained by a combination of grazing and burning (Thompson *et al*. 1995a), but where these practices are abandoned, a slow reversion to forest occurs.

Ecological characteristics

Key abiotic and biotic factors affecting the habitat

For all of these habitats except southerly mires, the climate is relatively harsh though moderated by the presence of nearby oceans (oceanic), except to the east of northern Norway, in the main tundra zone in Russia, where the climate is more extreme (continental). These are all open and generally treeless habitats, with a highly variable cover of low vegetation and generally nutrient-poor soils.

Tundra

Tundra is affected by permafrost, although in summer the surface layer thaws for a short period in all but the coldest areas. This 'active' layer of soil and rock is where most physical, chemical and biological activity occurs. The depth of the active layer varies from 30 to 100 cm. Soils are wet and shallow; only some 10% of the tundra is well-drained (Sage 1986). There is a large range of ground features created by the 'freeze–thaw' regime, ranging from mounds ('pingos') to the accumulation of surface stones into polygons. The main environment-forming plants of the tundra are mosses. They influence the summer temperature of the soils and the depth of the thawed zone above the permafrost, thereby creating and controlling habitats of herbaceous plants, shrubs and animals.

Tundra is a stressed environment, and adaptations to this amongst plants and animals are legion. Due to its inaccessibility and harsh climate, relatively little research has been done in tundra compared to more southerly habitats in Europe, and many biological mysteries remain, not least regarding the cyclic fluctuations in rodent populations in some Arctic regions, the high diversity of mating systems amongst waders (Charadrii), and fluctuations in the abundance of invertebrates. The year-to-year variations in berry production by dwarf shrubs such as *Vaccinium* and *Empetrum* are thought to have an important influence on populations of some bird species of tundra and mires.

Boreal montane habitats

Many of the abiotic and biotic factors that influence the habitat are the same as for tundra. For example, the combination of coldness, wetness (and snow-lie) and windiness determines the range of vegetation–soil–geology relationships.

Mires

Different regional climates give rise to significant variation in types of mire (Heathwaite 1993). According to Lindsay (1995), the two most important factors which together characterize a mire (bog) habitat are that:

- the living vegetation is separated from the underlying mineral soil by a layer of peat which is derived from the vegetation itself, and which is of sufficient thickness to prevent plant roots from reaching the comparatively nutrient-rich groundwater or mineral subsoil; and

- the decomposition cycle, typical of most habitats,

is inhibited by waterlogging, which leads to further accumulation of dead plant material, i.e. the active creation of peat.

A key distinction in mires is between the 'catotelm' and the 'acrotelm'. The catotelm is the base of the mire, consisting of the bulk accumulation of dead plant material (peat), and may be up to 10 m deep. Hydrological processes in this layer are slow.

The acrotelm is the thin (up to 30 cm deep) protective 'skin' of the mire that lies on top of the catotelm. This provides the surface pattern of small-scale relief and the cover of living vegetation. Hydrological processes here are rapid, with water-flow rates a thousand times greater than in the catotelm. Within the acrotelm, elevation above and below the water-table determines the composition of vegetation, to the extent that clear zones or bands of plants (and some associated invertebrates) form. These are highly dynamic habitats, with the acrotelm providing a complex environment for plants and insects.

Moorland

Moorland is found in wet, north-west European climates where the average annual precipitation is more than 1,000 mm and the soils are acidic (pH less than 6.5) and peaty, or consist of shallow peat itself (less than 10 cm deep). It occurs mainly between the upper limits of enclosed cultivation (300–400 m) and the former/present treeline. Moorland is dominated by heather *Calluna vulgaris*, a dwarf shrub, which spreads by vegetative propagation. This is a key distinction from heather-dominated heathland, where propagation is mainly by seedlings (see 'Lowland Atlantic Heathland', p. 187). Moorland shows ecological variation due to (a) latitudinal and altitudinal differences in climate, (b) dry–wet gradients in soils, (c) developmental phases of heather (occurring naturally, or after fires), and (d) man's influences, such as impacts of grazing sheep, red deer *Cervus elaphus* and other herbivores, and the pattern and intensity of burning and other management practices (MacDonald 1990, Thompson *et al.* 1995a,b).

Characteristic flora and fauna

Tundra

The absence of trees is a characteristic of most of the tundra zone, though tree species such as larch *Larix* and willow *Salix* can occur in low stunted forms, and tall trees may occur in tundra at the most southerly latitudes. This ecotone between boreal forest and tundra, the so-called 'forest tundra', is treated under 'Boreal and Temperate Forests' (p. 203).

Mosses are the dominant plants over most of the tundra, with the few other such plants consisting mainly of sedges *Carex* and cotton-grass *Eriophorum*, but there are also small bushes such as *Salix*, birch *Betula* and alder *Alnus*, as well as dwarf shrubs of various species of Ericaceae.

Within the tundra zone, there are considerable changes in vegetation composition and structure from south to north, which affect animal communities. As a result, four subzones are commonly recognized (Chernov 1985), as follows.

Bush tundra

The extensive southernmost Arctic regions cover about 175,000 km^2 of European Russia (Figure 1), and are dominated by a sporadic cover of low vegetation (15–40 cm high), consisting of sedges, grasses, herbs and bushy and stunted shrubs of *Salix*, *Betula* and *Alnus*, as well as dwarf shrubs such as *Vaccinium*, *Arctostaphylos* and *Empetrum*, with a ground cover of mosses and lichens. The tallest shrubs occur along the margins of lakes and rivers (Sage 1986).

The avifauna is relatively diverse, with more passerine species than the more northerly tundra, including Bluethroat *Luscinia svecica*, Redwing *Turdus iliacus*, Willow Warbler *Phylloscopus trochilus*, Arctic Redpoll *Carduelis hornemanni* and Little Bunting *Emberiza pusilla*. Other characteristic birds are Black-throated Diver *Gavia arctica*, Lesser White-fronted Goose *Anser erythropus*, Wigeon *Anas penelope*, Common Scoter *Melanitta nigra*, Merlin *Falco columbarius*, Gyrfalcon *F. rusticolus*, Ptarmigan *Lagopus mutus*, Golden Plover *Pluvialis apricaria*, Temminck's Stint *Calidris temminckii*, Wood Sandpiper *Tringa glareola*, Spotted Redshank *Tringa erythropus*, Jack Snipe *Lymnocryptes minumus*, Red-necked Phalarope *Phalaropus lobatus*, Meadow Pipit *Anthus pratensis* and Red-throated Pipit *A. cervinus*.

Typical tundra

This subzone covers about 32,000 km^2 of European Russia (Figure 1), and is dominated by mosses, which form a thick, unbroken carpet over the soil, 5–7 cm (or locally up to 12 cm) thick (Chernov 1985). Many moss species occur, but 10 predominate, of which three are especially typical: *Hylocomium splendens*, *Tomenthypnum nitens* and *Aulacomnium turgidum*. The herbaceous tier is dominated by sedges, including the most widespread and abundant plant species of the tundra, *Carex ensifolia*. Lichens and higher plants still occur in much greater variety than further north, as well as dwarf shrubs such as *Salix* and *Dryas*.

In low and flat areas, along the coastline and generally where drainage is impeded, a type of wet tundra or even marshland forms. Snow regularly accumulates in certain areas, e.g. sheltered hollows, which leads to the formation of snow-bed plant communities, where there is marked zonation of vegetation as a result of the gradual retreat of snow in spring.

Typical birds are Red-throated Diver *Gavia stellata*, *G. arctica*, Bean Goose *Anser fabalis*, White-fronted Goose *Anser albifrons*, various ducks (Anatidae, of which Long-tailed Duck *Clangula hyemalis* is the commonest), Rough-legged Buzzard *Buteo lagopus*, Peregrine *Falco peregrinus*, Grey Plover *Pluvialis squatarola*, Little Stint *Calidris minuta*, Dunlin *C. alpina*, Pomarine Skua *Stercorarius pomarinus*, Long-tailed Skua *S. longicaudus*, Snowy Owl *Nyctea scandiaca*, Shore Lark *Eremophila alpestris*, *Anthus cervinus* and Lapland Bunting *Calcarius lapponicus*.

The commonest mammals are Arctic fox *Alopex lagopus* and Norway lemming *Lemmus lemmus*, as well as *Rangifer tarandus*, although more than 90% of the latter are now domesticated in Europe (Jordhøy et al. 1996).

Arctic tundra

In European Russia, this habitat covers about 60,000 km^2 on Vaygach island and the southern part of Novaya Zemlya (Figure 1). Widespread patches of bare, unvegetated soil are characteristic and comprise about 50% of the ground. Vegetation is sparse, with no shrubs and a minimally developed herbaceous layer, composed mainly of grasses and herbs. The majority of plants are only 3–10 cm tall, with only a few sedges reaching 20 cm (Chernov 1985). Mosses are common, but do not form continuous carpets. The only dwarf-shrub species are *Dryas* and dwarf *Salix*. Soils are mostly free-draining and dry, except in low, boggy areas (where *Eriophorum* and sedges *Carex* prevail).

Typical birds include *Gavia stellata*, Bewick's Swan *Cygnus columbianus*, Barnacle Goose *Branta leucopsis*, Ringed Plover *Charadrius hiaticula* and Purple Sandpiper *Calidris maritima*. *Nyctea scandiaca* and Snow Bunting *Plectrophenax nivalis* also occur. Among mammals, *Alopex lagopus* and *Lemmus lemmus* are the most obvious species.

Polar desert

This is the driest and coldest zone, found in the most northerly, highest or most exposed areas, in Greenland, Svalbard, Franz-Joseph-Land and the northern part of Novaya Zemlya (Figure 1). Plant cover is very sparse, and consists mainly of lichens and mosses, with some small and very low species of grass, sedge and herb in the more sheltered positions.

The vertebrate fauna is poor and represented mostly by birds, most of which are associated with marine ecosystems. The most characteristic bird species are *Calidris maritima* and *Plectrophenax nivalis*, as well as marine birds such as Ivory Gull *Pagophila eburnea*, Glaucous Gull *Larus hyperboreus* and Little Auk *Alle alle* (which are treated under 'Marine Habitats', p. 59).

Boreal montane habitats

The vegetation and landscape are tundra-like, and can be divided into three similar subzones.

Bush montane 'tundra'

This is the lowest subzone, lying above the treeline at the ecotone with boreal woodland, and it is dominated by stunted shrubs of *Betula* and *Salix*, with plentiful wood-rush *Luzula* and dwarf shrubs *Vaccinium*, and a rich assemblage of herbs and mosses. *Calluna* occurs in the more oceanic climates, along with more mosses; however, this subzone is virtually absent from the British Isles. Breeding birds include *Buteo lagopus*, *Lagopus mutus*, Great Snipe *Gallinago media* (in Scandinavia) and *Luscinia svecica*.

Typical montane 'tundra'

Moss such as *Racomitrium* is dominant, forming fairly complete carpets in some areas. There are also sedges, grasses *Deschampsia* and dwarf shrubs (prostrate *Calluna*, *Empetrum*, *Arctostaphylos*). The areas with a more extreme climate can have a profusion of lichens *Cladonia*. Five plant communities can be distinguished within this zone (Thompson and Brown 1992), including a man-made community (tending to be grass-dominated, and prevalent in the British Isles). Breeding birds include species that are also found in the true tundra zone (bush and typical subzones), such as *Falco rusticolus* and *Calcarius lapponicus*.

Arctic montane 'tundra'

Stony deserts in the highest, most exposed montane areas, usually with permanent ice and snow. As in Arctic tundra, the vegetation cover is sparse and consists mainly of small grasses, sedges and mosses. Characteristic bird species include *Lagopus mutus*, Dotterel *Charadrius morinellus* and *Eremophila alpestris*.

Mires

Mire vegetation is complex and fairly well-researched, and is not explored in detail here. Further information is given in Osvald (1925), Kats (1948), Moore (1968, 1984), Daniels (1978), Dierssen (1982), Sjörs (1983), Rodwell (1991) and Lindsay (1995), for example. Undisturbed mire vegetation typically comprises an almost continuous carpet of various species of moss of the genus *Sphagnum*, which are adapted to the highly waterlogged, nutrient-poor and acidic conditions. Growing within this carpet are other types of plant, including sedges such as *Eriophorum* and dwarf shrubs such as *Erica*. Drier or disturbed ground may have less *Sphagnum* moss and more vascular plants. A large proportion of raised mires is naturally if sparsely wooded (often with pine *Pinus* or *Betula*), especially in the Baltic catchment. The bird communities of the two main morphological types are as follows.

Raised mires

Typical breeding bird species include Crane *Grus grus*, a wide variety of waders such as *Lymnocryptes minimus*, Curlew *Numenius arquata* and *Tringa glareola*, and passerines such as Whinchat *Saxicola rubetra* and *Luscinia svecica*. Amongst breeding raptors, Montagu's Harrier *Circus pygargus* is typical in Baltic mires, as was an unusual ground-breeding population of *Falco peregrinus* that is now starting to recover after being driven extinct in the late 1960s and early 1970s (owing to pesticide poisoning on the wintering grounds).

Blanket mires

Birds which are generally restricted to blanket mires (as opposed to raised mires) include *Calidris alpina*, *Falco columbarius*, Hen Harrier *Circus cyaneus* and Short-eared Owl *Asio flammeus*.

Moorland

Moorland can be divided into two broad types: drier heaths dominated by *Calluna* with *Vaccinium*, and wetter heaths with *Calluna*, *Eriophorum*, *Erica* and club-rush *Scirpus*. Usher and Thompson (1993) have highlighted the high diversity of the ground-dwelling invertebrates and their importance in terms of biodiversity. The species diversity of moorland butterflies and moths (Lepidoptera) is greatest on the drier *Vaccinium*-dominated heaths, and declines with increasing altitude (Coulson 1988, Coulson *et al.* 1992). Assemblages of mammals, reptiles and amphibians are low in diversity, but mountain hares *Lepus timidus* and voles *Microtus* can be abundant and are important sources of prey for raptors. In addition to the large numbers of domestic sheep, the other main mammal of note is *Cervus elaphus*. In the UK, this is managed mainly for the sporting cull of stags (Clutton-Brock and Albon 1989, Scottish Natural Heritage 1994) and occurs in large herds on open moorland throughout the year.

The bird assemblage is not particularly rich compared to other habitats in Europe, but shows a unique mixture of boreal, low and high Arctic, temperate and continental nesting species (Thompson *et al.* 1995b). The abundance of open-ground nesting birds can also be very high (Usher and Thompson 1993). Typical species include *Falco columbarius*, Willow Grouse *Lagopus lagopus*, Raven *Corvus corax*, *Anthus pratensis* and Skylark *Alauda arvensis*, and more locally Golden Eagle *Aquila chrysaetos*, *Falco peregrinus*, *Circus cyaneus*, *Numenius arquata*, *Pluvialis apricaria*, Greenshank *Tringa nebularia* and Ring Ouzel *Turdus torquatus* (Thompson *et al.* 1995b). Lapwing *Vanellus vanellus* and *Tringa nebularia* also attain high densities on mosaics of moorland and low-intensity agricultural land. Moorland-breeding populations of *Lagopus lagopus*, *Pluvialis apricaria* and Twite *Carduelis flavirostris* show subspecific differentiation from populations elsewhere in their respective ranges.

Values, roles and land-uses
Tundra and boreal montane habitats

These are amongst the wildest and most natural environments in Europe, revered for their pristine condition and rich natural heritage. By implication, they are not intensively used by man compared with other habitats in Europe. Nevertheless, the European tundra is much more developed than the Siberian tundra to the east.

Oil and gas extraction is the most obvious and widespread activity, while mining (coal and ore) and power generation (atomic and hydroelectric) occur more locally, and there are ports, industrial centres, and military bases. Major industrial centres in European Russian tundra include the lower Pechora river, Usinsk and Kolguev Island (all oil-production areas), Saremboiskye (oil/gas exploration), Vorkuta (coal mining), Apatity (mineral extraction) and Amderma (mica industry) (V. Morozov *in litt.* 1995). Coal-mining is also important on Svalbard. Associated with all of these developments on the tundra are permanent settlements, interconnected by sparse if very extensive networks of roads, railways, pylons and above-ground pipelines.

Tundra and boreal montane habitats are widely used for reindeer-grazing. The winter population size of reindeer in Europe is estimated to be 1,076,000 domestic and 77,000 wild animals, the latter including 23,000 in boreal forest and 14,000 on the islands of Novaya Zemlya and Svalbard (J. A. Kålås *in litt.* 1996).

Fishing and hunting are other important human uses of these habitats (now for commerce and sport as well as subsistence), which are usually concentrated around settlements, roads and major rivers. Tourism is becoming increasingly important, being mainly confined to the coast in the tundra zone, and involving skiing and other outdoor pursuits in the boreal montane zone.

Mires

Mires are thought to be globally important as a 'carbon sink' for atmospheric CO_2, thus inhibiting global warming (see Box 1, p. 47). In terms of overall extent affected, one of the most important uses of European mires is for extraction of peat to fuel power-stations, especially in Russia and Belarus. Neighbouring Baltic states may adopt this technology in future but, in the meantime, peat extraction for domestic heating fuel is also widespread in the eastern Baltic region.

During the past century, most lowland mires in Europe were converted to forestry or agriculture; nearly 78,000 km² of mires in European Russia had already been drained or cut over by 1957 (A. Mischenko *in litt.* 1995). Conversion to these land-uses is now less prevalent in Europe (although still significant), because fewer suitable mires are now left. Remaining lowland raised mires in western Europe are widely exploited to provide peat for the horticultural market. In western Russia and the eastern Baltic region, berry-collecting on mires is a common and widespread activity (mainly for cranberry but also various other *Vaccinium* species). A substantial proportion of officially protected Russian mires are conserved for their cranberry resource alone, and there are proposals (e.g. in Latvia) to convert mires to cranberry production on an industrial basis.

Mires have a considerable applied and historical value in serving as an archive of past changes in climate, landscape and land-use, which is interpreted by analysis of preserved plant remains in peat. In addition, an increasing proportion of remaining mires in Europe are being protected to conserve their biodiversity.

Moorland

Moorland is used for sheep-grazing, hunting of deer and grouse (Tetraonidae) and, increasingly, for tourism and recreation. It is a valuable example of a man-made, wildlife-rich habitat which can be used in a sustainable manner if managed correctly.

Tundra, boreal montane habitats and moorland are all favoured areas for military activities (training, weapons testing, etc.), due to the low human population densities, and some significant settlements in the tundra have developed largely because of their military significance. Tundra, boreal montane habitats and mires all contribute significantly to the range and richness of global biodiversity, through the high proportion of specialized species (including birds) which are confined to, or depend on, these habitats. In addition, the astonishing migrations of the waterbirds of the tundra clearly illustrate the shared environmental links between many countries in the world.

Political and socio-economic factors affecting the habitat

In Europe, there is still a strong tendency to regard tundra, mires and moorland as unproductive habitats of low economic worth. This makes them more vulnerable to destructive and unsustainable developments than most other habitats.

Tundra

The extractive industries (oil, gas, coal, ores, minerals) are a major issue in European Russia and Norway, and some (e.g. coal-mining) have continued to expand in recent years through the policies of both Norwegian and Russian governments. Large reindeer pastures have been lost to oil- and gas-field developments, with heavier grazing pressures resulting on some of the remaining tundra. Tourism is increasing—even back in 1969, 5,000 tourists per year were estimated to have visited Spitzbergen (Vaughan 1992). There are also concerns that 'subsistence' hunting may be reaching unsustainable levels, with substantial harvesting of birds (notably Brünnich's Guillemots *Uria lomvia*) in Greenland and on the north-west Russian coast.

There is a good deal of pressure, from governmental and non-governmental scientists and concerned organizations and citizens, to increase the size and number of protected areas in the tundra zone of Europe, but the socio-economic needs of the local people are also now considerable, especially compared with the less populated tundra to the east of Europe in Siberia. Large areas have already been protected, for example Norway created more than 20 nature-conservation areas in Svalbard in 1973, totalling just over 50% of the land area, and a substantial part of Greenland is also protected, including the largest national park in the world (700,000 km²), created in 1974 (Grimmett and Jones 1989). In Russia, the designation of such large-scale national parks has been more recent, but none embrace tundra in Europe as yet. Instead, there are state reserves (zapovednik) and a small number of less stringently controlled temporary reserves (zakaznik) where some economic activity is permitted. Whether there will be any future designation of tundra areas as national parks in European Russia is still a contentious issue.

Boreal montane habitats

Growing tourism and recreation increasingly affect boreal montane habitats, sometimes associated with 'invasive' infrastructural developments such as downhill-skiing centres and cable-lifts. Heavy stocking pressures from hill-sheep farming, and from red deer and reindeer husbandry, are issues in the British Isles and Norway, Scotland and northern Scandinavia respectively. For example, the national sheep flock rose from about 1.7 million to about 2.5 million in Norway between 1950 and 1994, nearly all of which are grazed in boreal montane areas between June and September.

Mires

Peat extraction to fuel power stations is a major use of mires in north-east Europe. For instance, 4–5 million tonnes of peat are extracted per year for this purpose in Belarus, comprising up to one-third of the annual total (Anon. 1997). Such extraction is increasing in importance in Russia where it is now the main cause of mire destruction, and thus possibly in the whole of Europe. There are currently proposals to build similar peat-fired power stations in Latvia and possibly in other states in the Baltic catchment.

Commercial afforestation of mires, stimulated by substantial investments made through tax 'loopholes', became a controversial public issue in the UK in the 1980s, and although such practices have now been greatly curtailed, they may be resumed in the future. Afforestation has recurred in several areas, not least to supposedly 'reforest' non-open areas of mountain and mire. There is still a considerable market for peat extracted from raised mires to supply horticulture, which is tempered by increasing public concern over this industry, and increasing use of peat 'alternatives'. Peat extraction is theoretically a sustainable activity, but regeneration of peat is a slow, decades-long process, and peat cannot be considered a renewable industrial resource in a local context.

Moorland

Sheep-farming in moorland and associated grassland habitats receives considerable government subsidies, e.g. in Norway and the UK, where sheep numbers and stocking densities have greatly increased over recent decades.

By contrast, hunting of *Lagopus lagopus* and *Cervus elaphus* in Britain are predominantly 'private' enterprises. Growing public dislike for shooting and hunting may result in the abandonment of some moorland hunting estates, where there is also evidence of dwindling numbers of hill-farms and farmers in recent years. Substantial tracts of moorland are now owned or managed by nature-conservation bodies, and here there is a movement towards encouraging the regeneration of native woodland. As there was in the 1980s, there may well be an expansion of commercial afforestation again—not least the planting of 'Christmas' trees. Poor management practices, notably in the burning of moorland, also need to be addressed.

PRIORITY BIRDS

In Europe, there are 73 priority bird species of tundra, boreal montane habitats, mires and moorland (Table 2). This total is relatively high compared to some other habitats treated in this book, especially given the relatively small area of these habitats in Europe overall. The total actually comprises a higher proportion of the entire avifauna of these habitats than is the case in the more southerly habitats of Europe.

There are only two Priority A species, *Anser erythropus* and *Gallinago media*. *Anser erythropus* is globally threatened with extinction (Collar *et al.* 1994) as the world population has greatly declined this century for reasons which are still unclear, but thought to relate to deleterious conditions on migration or while wintering in south-eastern Europe (Madsen 1996). It is totally dependent on tundra as breeding habitat, although the majority of the world population breeds outside Europe in the Siberian Arctic.

Most of the world population of *Gallinago media* breeds within Europe, where it has suffered a considerable decline during the last 150 years due to the loss of breeding habitats (especially flood-plain grasslands) in lowland temperate areas (Tucker and Heath 1994, Kålås *et al.* in press). Today, the remaining European population is considered to be very dependent on mires when breeding (Table 2), and the species is considered to be approaching globally threatened status (Collar *et al.* 1994).

The 11 Priority B species all have an Unfavourable Conservation Status in Europe (Tucker and Heath 1994). The world population of Black-tailed Godwit *Limosa limosa* breeds mainly within Europe and is significantly dependent on mires when breeding (especially the Icelandic race *L. l. islandica*). The other 10 species are very dependent on one or more of these habitats when breeding, but are not concentrated in Europe, e.g. *Falco rusticolus* and Broad-billed Sandpiper *Limicola falcinellus*.

Seven of the 12 Priority C species have an Unfavourable Conservation Status in Europe (Tucker and Heath 1994). The world populations of two of these, Redshank *Tringa totanus* and Common Gull *Larus canus*, are concentrated in Europe and are undergoing significant declines, but are only marginally dependent on the habitats treated here. The other five species with Unfavourable Conservation Status are more dependent on these habitats, but are not concentrated in Europe, e.g. Pintail *Anas acuta*, *Circus cyaneus* and *Grus grus*. The remaining five Priority C species are concentrated in Europe and are very dependent on these habitats when breeding, but are judged to have a Favourable Conservation Status in Europe (Tucker and Heath 1994), e.g. *Pluvialis*

Table 2. The priority bird species of tundra, boreal montane habitats, mires and moorland in Europe.

This prioritization identifies Species of European Conservation Concern (SPECs) (see p. 17) for which the habitat is of particular importance for survival, as well as other species which depend very strongly on the habitat. It focuses on the SPEC category (which takes into account global importance and overall conservation status of the European population) and the percentage of the European population (breeding population, unless otherwise stated) that uses the habitat at any stage of the annual cycle. It does not take into account the threats to the species within the habitat, and therefore should not be considered as an indication of priorities for action. Indications of priorities for action for each species are given in Appendix 4 (p. 401), where the severity of habitat-specific threats is taken into account.

	SPEC category	European Threat Status	Habitat importance
Priority A			
Lesser White-fronted Goose *Anser erythropus*	1	V	■
Great Snipe *Gallinago media*	2	(V)	■
Priority B			
Red-throated Diver *Gavia stellata*	3	V	■
Black-throated Diver *Gavia arctica*	3	V	■
Brent Goose *Branta bernicla*	3	V	■
Barrow's Goldeneye *Bucephala islandica*	3	E	■
Gyrfalcon *Falco rusticolus*	3	V	■
Dunlin *Calidris alpina*	3 [w]	V [w]	■
Broad-billed Sandpiper *Limicola falcinellus*	3	(V)	■
Jack Snipe *Lymnocryptes minimus*	3 [w]	(V) [w]	■
Black-tailed Godwit *Limosa limosa*	2	V	●
Wood Sandpiper *Tringa glareola*	3	D	■
Snowy Owl *Nyctea scandiaca*	3	V	■
Priority C			
Barnacle Goose *Branta leucopsis*	4/2 [w]	L [w]	■
Pintail *Anas acuta*	3	V	●
Hen Harrier *Circus cyaneus*	3	V	●
Peregrine *Falco peregrinus*	3	R	●
Crane *Grus grus*	3	V	●
Golden Plover *Pluvialis apricaria*	4	S	■
Purple Sandpiper *Calidris maritima*	4	(S)	■
Whimbrel *Numenius phaeopus*	4	(S)	■
Curlew *Numenius arquata*	3 [w]	D [w]	●
Redshank *Tringa totanus*	2	D	●
Great Skua *Stercorarius skua*	4	S	■
Common Gull *Larus canus*	2	D	●
Priority D			
Great Northern Diver *Gavia immer*	—	(S)	■
White-billed Diver *Gavia adamsii*	—	(S)	■
Bewick's Swan *Cygnus columbianus*	3 [w]	L [w]	■
Bean Goose *Anser fabilis*	—	S	■
Pink-footed Goose *Anser brachyrhynchus*	4	S	●
White-fronted Goose *Anser albifrons*	—	S	■
Scaup *Aythya marila*	3 [w]	L [w]	■
Long-tailed Duck *Clangula hyemalis*	—	S	■
Common Scoter *Melanitta nigra*	—	S	■
Velvet Scoter *Melanitta fusca*	3 [w]	L [w]	■
White-tailed Eagle *Haliaeetus albicilla*	3	R	●
Rough-legged Buzzard *Buteo lagopus*	—	S	■
Golden Eagle *Aquila chrysaetos*	3	R	●
Merlin *Falco columbarius*	—	S	■
Red/Willow Grouse *Lagopus lagopus*	—	S	■
Ptarmigan *Lagopus mutus*	—	S	■
Black Grouse *Tetrao tetrix*	3	V	●
Dotterel *Charadrius morinellus*	—	(S)	■
Grey Plover *Pluvialis squatarola*	—	(S)	■
Knot *Calidris canutus*	3 [w]	L [w]	■
Sanderling *Calidris alba*	—	S	■
Little Stint *Calidris minuta*	—	(S)	■
Temminck's Stint *Calidris temminckii*	—	(S)	■
Ruff *Philomachus pugnax*	4	(S)	●
Bar-tailed Godwit *Limosa lapponica*	3 [w]	L [w]	■
Spotted Redshank *Tringa erythropus*	—	S	■

cont.

Table 2. (cont.)

	SPEC category	European Threat Status	Habitat importance
Turnstone *Arenaria interpres*	—	S	■
Red-necked Phalarope *Phalaropus lobatus*	—	(S)	■
Grey Phalarope *Phalaropus fulicarius*	—	(S)	■
Pomarine Skua *Stercorarius pomarinus*	—	(S)	■
Arctic Skua *Stercorarius parasiticus*	—	(S)	■
Long-tailed Skua *Stercorarius longicaudus*	—	(S)	■
Sabine's Gull *Larus sabini*	—	S	■
Ross's Gull *Rhodostethia rosea*	—	S	■
Short-eared Owl *Asio flammeus*	3	(V)	•
Skylark *Alauda arvensis*	3	V	•
Shore Lark *Eremophila alpestris*	—	(S)	■
Meadow Pipit *Anthus pratensis*	4	S	●
Red-throated Pipit *Anthus cervinus*	—	(S)	■
Rock Pipit *Anthus petrosus*	—	S	■
Bluethroat *Luscinia svecica*	—	S	■
Stonechat *Saxicola torquata*	3	(D)	•
Ring Ouzel *Turdus torquatus*	4	S	●
Great Grey Shrike *Lanius excubitor*	3	D	•
Twite *Carduelis flavirostris*	—	S	■
Lapland Bunting *Calcarius lapponicus*	—	(S)	■
Snow Bunting *Plectrophenax nivalis*	—	(S)	■
Little Bunting *Emberiza pusilla*	—	(S)	■

SPEC category and **European Threat Status:** see Box 2 (p. 17) for definitions.

Habitat importance for each species is assessed in terms of the maximum percentage of the European population (breeding population unless otherwise stated) that uses the habitat at any stage of the annual cycle:

    ■ >75%   ● 10–75%   • <10%

[W] Assessment relates to the European wintering population.

apricaria, Whimbrel *Numenius phaeopus* and Great Skua *Stercorarius skua*.

In total, 27 (37%) of the 73 priority species have an Unfavourable Conservation Status in Europe (Table 3), which is a relatively low proportion compared to other habitats treated in this book. This indicates that the high number of priority species is not due to a particularly high level of threat to these habitats overall, or to their species, when compared with some other habitats in Europe. Only one of the 27 species, Barrow's Goldeneye *Bucephala islandica*, is classified as Endangered in Europe (Tucker and Heath 1994). Of the remaining 46 priority species (those having a Favourable Conservation Status), the great majority (41) are Priority D species which

are not concentrated in Europe but which are very dependent on these habitats when breeding, usually tundra (Table 3).

Nearly all of the priority species use these habitats only during the breeding season, although some waterbirds also depend on tundra for a short moulting period immediately after breeding. Most of the species are thus migratory, and this trait is especially strongly developed in tundra birds where only a few species are able to remain during the harsh and prolonged winters, e.g. *Falco rusticolus*, *Lagopus mutus* and *Corvus corax*.

Compared with other habitats covered in this book, an unusually high percentage (72%) of the priority species are very dependent on these particu-

Table 3. Numbers of priority bird species of tundra, boreal montane habitats, mires and moorland in Europe, according to their SPEC category (see p. 17) and their dependence on the habitat. Priority status of each species group is indicated by superscripts (see 'Introduction', p. 19, for definition). 'Habitat importance' is assessed in terms of the maximum percentage of each species' European population that uses the habitat at any stage of the annual cycle.

Habitat importance	SPEC category					Total no. of species
	1	2	3	4	Non-SPEC	
<10%	0 [B]	2 [C]	7 [D]	—	—	9
10–75%	0 [A]	1 [B]	5 [C]	4 [D]	—	10
>75%	1 [A]	1 [A]	10 [B]	5 [C]	37 [D]	54
Total	1	4	22	9	37	73

lar habitats (Table 3), on tundra especially——at least, during the breeding season. In particular, the manifold specializations in behaviour, physiology and anatomy that many tundra birds possess, in order to successfully exploit this extreme environment, may limit their breeding success in other habitats. Most of the priority species are either waterbirds or raptors, both with powerfully developed flying ability (to aid rapid and long-distance migration), and passerines are remarkably few compared with other habitats in Europe. This follows the pattern typical of the avifauna as a whole in these habitats.

In general, the number of breeding bird species in a given area decreases as one moves from the more southerly mires and moorland northwards into tundra (e.g. Sage 1986, Whitfield and Tomkovich 1996). The priority species of these habitats, however, follow an opposite trend, in that the majority are highly characteristic of tundra, with only a small minority being characteristic of mire or moorland, or spanning two or all three of these habitats.

Only Russia and the Fennoscandian countries support more than 75% of the breeding priority species of these habitats, and only one other country in Europe, the UK, holds 50% or more (see Appendix 2, p. 344).

HABITAT NEEDS OF PRIORITY BIRDS

The habitat requirements of the priority species have been summarized by the Habitat Working Group in Appendix 3 (p. 354). While some aspects of habitat use are relatively easy to specify (e.g. food requirements), others are more problematical (due to lack of research) yet are important. For instance, where non-uniform (mosaic) vegetation is preferred by a particular species, this is further subdivided according to whether large or small areas are preferred—a subjective judgement but nonetheless important in distinguishing between the habitat features preferred by many of the species.

Likewise, the vegetation used by these birds is classified according to its stature, ranging from scrub down to moss/lichen-dominated or bare ground. More abstract qualities of the habitat which can interact with habitat requirements are also covered in Appendix 3. Some of these factors directly influence breeding success, such as favourable weather for many of the species, cyclic fluctuations in rodent populations (for a few of the large predators and waders), and even the protection afforded by some species to others (e.g. *Buteo lagopus* to *Anser erythropus*). Similar factors influencing the survival of birds are divided into seven classes, of which disturbance by human activities, habitat loss and to a lesser extent persecution were found to be most important.

Three general points should be made regarding information in Appendix 3. First, in looking at the range of vegetation mosaics and types occupied by the 73 priority species, we need a more detailed assessment of preferred types which can be influenced by favourable land-management practices (this is particularly applicable to moorland and mires).

Second, given that many of the species spend only a short time breeding in these northern habitats, we need to distinguish between those habitat factors which render the birds susceptible to threats there, as opposed to during migration and on their wintering grounds. Appendix 3 has highlighted the importance of habitat loss, persecution, pollution, etc., and this needs to be examined more closely.

Third, more detailed research is needed on the actual 'requirements' of these priority species, because even a close association with a particular habitat should not always be regarded as a requirement for it. Some recent studies have indicated the importance of apparently 'peripheral' habitats for some birds (e.g. undisturbed old pasture fields for *Pluvialis apricaria*; forest/heath habitats for *Tringa erythropus*, and coastal/maritime areas for at least eight of the priority species). Here, associated habitats for some species, and a particular juxtaposition of habitats for others, may prove to be especially important in meeting the essential requirements of many of the birds.

About 48 priority species (67%) are regarded as obvious habitat-specialists, and 44 (60%) prefer flat to sloping land. About 18 (25%) prefer marine and 16 (22%) prefer freshwater sources of food, while preferences for different food-types are 67% for invertebrates, 46% for plants and 26% for vertebrates.

There are 17 priority species of waterfowl (Gaviidae and Anatidae), of which the great majority (14) are associated with tundra (10 species showing a marked preference for this). *Gavia stellata* needs pools or small lakes as an essential nesting requirement, but most birds feed mainly at sea. Great Northern Diver *Gavia immer* and *Clangula hyemalis* also nest in lakes or pools in the tundra zone but otherwise spend virtually all of their time at sea. Leaving aside these three species, 11 of the 14 remaining waterfowl are habitat specialists, principally of tundra. At least six species prefer vegetation mosaics within this habitat, as opposed to uniform vegetation. Not surprisingly, the main type of habitat preferred is dominated by grass/sedge or moss/sedge vegetation. Thirteen of the 17 species prefer lakes and/or pools, while only five prefer rivers.

The feeding requirements of waterfowl have been exhaustively studied. Favourable weather is considered to influence the breeding success of 11 of the waterfowl, and rodent population cycles indirectly influence the breeding success of *Anser erythropus* and *Branta leucopsis*. Natural fluctuations in the pressure of nest predation are considered to significantly alter the survival rate of at least two species—*Anser albifrons* and *A. erythropus*—and competition with *Branta leucopsis* is also thought to affect the survival rate of *A. erythropus*. Interestingly, 11 of the 17 priority species have essential requirements for particular types of water-body.

Of the seven large predators, only the two owls (*Nyctea scandiaca* and *Asio flammeus*) and *Buteo lagopus* are specialists of these habitats. They all prefer large areas of non-uniform (mosaic) habitat, with the exception of *Falco columbarius*, and only *Falco rusticolus* and *Buteo lagopus* prefer sloping ground. The productivity of five species is directly or indirectly influenced by rodent population cycles. Four species are susceptible to disturbance, five to persecution, and six to habitat loss. *Circus cyaneus* used to be more widespread in Europe and less dependent on these habitats, but through human persecution and disturbance it is now much more confined to moorland and some areas of tundra and mire. At least two species (White-tailed Eagle *Haliaeetus albicilla* and *Falco peregrinus*) are susceptible to pesticide pollution while wintering further to the south, *F. peregrinus* being particularly prone.

The gamebirds (Tetraonidae) and cranes (Gruidae) comprise four priority species, of which two are specialists of these habitats, while both *Grus grus* and Black Grouse *Tetrao tetrix* also breed in boreal forest, and the latter also in agricultural grasslands, where it shows a preference for flat areas. Habitat loss affects all species in this group, except *Lagopus mutus* which is susceptible only to natural predators. Only one essential habitat requirement is evident—medium-sized dwarf shrubs for *Lagopus lagopus*.

The 22 priority species of waders are rarely consistent in their habitat preferences. Of the six groups of birds examined here, the waders have the smallest number of essential requirements (grass/sedge vegetation as nesting habitat for *Limosa limosa*, and pools for the two species of phalaropes *Phalaropus*). Eighteen are specialists on tundra, mires and moorland, with three others having additional requirements for agricultural grasslands (*Pluvialis apricaria*), grasslands and the coast (*Tringa totanus*) and forest/heath (*T. erythropus*). About half prefer large mosaics, with *Gallinago media* having particularly complex requirements. There is a tendency among the waders to avoid rank vegetation, with only *Tringa erythropus* showing a preference for tall shrubs (0.5–1.0 m high), and four species nesting in medium-sized shrubs.

The greatest difference within the group is attributed to their preference for wet or dry ground, with nine species preferring dry ground and ten preferring wet ground, with only one species evidently requiring a stable water-table (*Limicola falcinellus*) and one species being more flexible in its requirements (Bar-tailed Godwit *Limosa lapponica*). The majority of these birds feed on terrestrial invertebrates. The breeding success of nine species is dependent on favourable weather (*Limicola falcinellus* is badly affected by heavy rain), and five species on rodent-cycles in some parts of their range. Eight species are susceptible to habitat loss, and three to persecution.

The seven species of skuas (Stercorariidae) and gulls (Laridae) include only three specialists (*Stercorarius pomarinus*, *S. longicaudus* and Ross's Gull *Rhodostethia rosea*). All seven prefer open habitat, in one case dominated by grass/sedge (*Larus canus*), and all feed largely on terrestrial invertebrates or vertebrates (skuas *Stercorarius*). *Stercorarius pomarinus*, Arctic Skua *S. parasiticus* and *S. longicaudus* are dependent on high numbers of rodents for high breeding success. *S. skua* may be susceptible to pollution, and *Rhodostethia rosea* is heavily persecuted in parts of its northern range.

The 15 priority species of passerines show diverse tendencies. Thirteen species are associated with northern or montane tundra, with Rock Pipit *Anthus petrosus* having a special preference for rocky, coastal tundra. All are habitat specialists with the exception of *Turdus torquatus* (which feeds on grassland) and *Carduelis flavirostris* (which feeds on farmland areas throughout much of its range). Seven species have essential preferences for particular vegetation-types. The preferences for small or large mosaics are difficult to quantify, yet six species prefer small mosaics, with only one possibly preferring larger patches (*Plectrophenax nivalis*). Open habitats are preferred by at least eight species, with grass–sedge communities preferred by six, and medium-sized shrub communities by five species. Only one species (*E. pusilla*) has any preference for tall shrubs. It is perhaps puzzling why so few passerines occupy moss/lichen heaths, with only *Eremophila alpestris*, *Calcarius lapponicus*, *Plectrophenax nivalis* and *Emberiza pusilla* found here. Three species benefit from favourable weather for high breeding success.

THREATS TO THE HABITAT

There has been no Europe-wide appraisal of the threats to tundra, boreal montane habitats, mires and moorland, and to their priority bird species. The Habitat Working Group therefore carried out a general review of those factors that have caused recent loss and degradation of these habitats, and of those which are anticipated as future threats (Table 4). In addition, the Group predicted the impact of each threat on the European population of each priority species over the next 20 years (full details are given in Appendix 4, p. 401). A summary of the results is presented in Figures 5–6, and the threats are discussed individually below, in order of importance (as shown in Figure 6).

Oil and gas exploitation

Although poorly documented, historical loss of tundra has undoubtedly been much less comprehensive than loss of mires and moorland, and has mainly occurred during the present century. However, the last 30 years have seen increasing efforts to exploit the natural and strategic resources of the tundra and the seas offshore, most especially through the extraction of oil and natural gas. The associated construction of buildings, roads, wells, pipelines and other infrastructure has affected the typical tundra subzone at a regional scale (especially along the Barents Sea coast of Russia), and bush tundra more locally (see Husby 1995). The threat is predicted to continue operating at the same level over the next 20 years.

The total habitat loss to these infrastructures has been relatively small (estimated at only 8 km^2 for railways and major roads in European Russia, as well as 76 km^2 for all cities, towns and industrial settlements: V. Morozov *in litt.* 1996), but the ensuing alteration of drainage patterns and fragmentation of habitat may have affected priority birds over much wider areas (see 'Afforestation', below). In addition, tundra wetlands have been widely polluted by spillages or discharges of oil, chemicals and wastes (which break down more slowly in the frigid climate), and illegal hunting and persecution of priority birds and other wildlife commonly takes place in the vicinity of roads and settlements, as does off-road driving with attendant damage to the fragile, very slow-growing tundra vegetation.

Overall, 18 out of 73 species (25%) are predicted to suffer low impacts from this threat, making it the most important among the threats considered here.

Afforestation

Forestry is not practicable in tundra or boreal montane habitats, and this threat therefore applies only to mires and moorland. Substantial areas of mire have been drained and afforested in all parts of the range, but particularly in Russia, Fennoscandia, Ireland and the UK, while moorland has been affected more locally. For example, about 18% of moorland in Scotland (UK) has been lost since the 1940s, mainly to afforestation (Scottish Natural Heritage 1995). Drainage (mainly for forestry) had reduced the extent of mires in Finland to about 44% of their original

area by 1980 (Anon. 1985). With mechanization, the rate of drainage greatly increased from 1950 but, after peaking around 1970, has been in continuous steep decline (Wahlström *et al.* 1992). Nevertheless, on a Europe-wide scale the magnitude of this threat shows no sign of diminishing over the next 20 years. In Finland (Anon. 1991), and probably in general, only the most productive mires (usually sparsely wooded, with the richest fauna and flora) are likely to be targeted for forestry, and these habitats tend not to be part of existing protected-area systems.

Moreover, there have been many examples in Europe of the loss or degradation of entire mires resulting from drainage for afforestation outside the mires themselves, as changes in the water-flow patterns in adjacent areas may have considerable lateral impacts on mire hydrology. Partly because of such effects, it is expected that 70% of Finnish mires will have lost their natural state by the year 2000 (Anon. 1985).

Fragmentation of mires and moorland by afforestation may also reduce the quality of the remaining habitat. For example, remaining patches may have become effectively of low quality for birds if they are too small, too mixed with other habitats, or too isolated from each other—even if the vegetation structure and composition remains as required. This has undoubtedly been a factor in the disappearance of some typical mire-breeding priority species, such as *Pluvialis apricaria*, from raised mires in the lowlands of the UK and some other European countries (Byrkjedal and Thompson in press.).

Overall, 17 priority species (23%) are predicted to suffer low impacts from this threat, making it one of the most important of the threats considered here.

Peat extraction

Large-scale peat extraction to fuel power-stations is the most serious threat to mires in Russia, and is the second most important threat in Belarus (Anon. 1997). In Finland, 500 km^2 of peatland are now used by the extraction industry, and of the 15 million m^3 of peat extracted per year 80% is used for power-generation, supplying 4% of national energy needs (Anon. 1991). Demand for peat appears to be rising in all of these countries—the quantity extracted in Finland is expected to rise to 20–25 million m^3 by the year 2000 (Anon. 1991), although future trends in use of energy sources is difficult to predict, since they are often highly susceptible to economic and political influences. There are proposals to adopt such methods of power generation in Latvia and possibly other mire-rich states in the Baltic catchment. Peat is normally stripped layer by layer, leaving large areas of peat 'desert' which may remain unvegetated even 10 years later. Complete recovery,

Table 4. The most important threats to tundra, boreal montane habitats, mires and moorland, and to the habitat features required by their priority bird species, in Europe (compiled by the Habitat Working Group). Threats are listed in order of importance (see Figure 6), and are described in more detail in the text.

Threat	Effect on birds or habitats	Scale of threat			Regions affected, and comments
		Habitat	Past	Future	
1. Oil and gas exploitation (and transportation)	Direct mortality from pollution; pollution leading to food loss; disturbance from operations; persecution and excessive hunting by employees; vegetation damage by off-road vehicles	Ta Tb,i	• •	• •	Russia, especially Barents Sea coast
2. Afforestation	Habitat loss and fragmentation; changes in water-levels and hydrology; flash-flooding and erosion; increased predators; water turbidity; aluminum poisoning of trees	MR ML	• •	• •	All areas, especially Russia, Fennoscandia, UK, Ireland UK, Norway
3. Peat extraction (for power generation, domestic heating, horticulture)	Long-term habitat loss and fragmentation; catchment effects on hydrology; changes in water-levels and vegetation; burning; water turbidity; aluminium poisoning of plants	MR	■	■	W European Russia, Belarus, Estonia, Latvia, Finland, Sweden, Ireland, UK
4. Overgrazing (heavy grazing)	Excessive erosion and reduction in vegetation quality; drying out of habitat through reindeer trampling/paths; increased scavengers and predators	Ta,b,c B ML MR	• • • •	• • • •	Russia, Norway, Finland Norway, Finland Norway, UK Norway, UK
5. Industrial development	Habitat loss and fragmentation; increased predators; disturbance from operations; persecution and excessive hunting by employees	Ta Ba MR	• • •	• • •	E European Russia W European Russia Finland, Russia, UK
6. Settlement (including urban development)	Habitat loss and fragmentation; increased scavengers and predators; increased disturbance, persecution, hunting, vegetation damage by off-road vehicles	T	•	•	Russia
7. Drainage for agriculture	Habitat destruction and fragmentation; drop in water-tables; excessive erosion	MR	•	•	Russia
8. Tourism	Habitat loss due to development; disturbance; increased mortality due to collisions with ski-lifts, etc.; attraction of predators; vegetation change by trampling, skiing, skidoos, etc.	B	•	•	Fennoscandia, UK
9. Transport infrastructure (roads, railways, etc.)	Habitat destruction and fragmentation; road casualties and rubbish attract predators; effects on water-table by disruption to drainage; damage by off-road vehicles to surrounding vegetation; soil erosion	T MR	• •	• •	Russia Central European Russia
10. Other agricultural improvement (other than drainage, e.g. fertilization, ploughing)	Habitat destruction and fragmentation; drop in water-tables; excessive erosion	Ba ML	• •	— —	Norway UK
11. Energy transport	Collision and electrocution (overhead cables); disturbance during construction; creates nest-sites in open areas (increased density of raptors)	Ta MR	— •	• •	Russia, Scandinavia Russia

cont.

Table 4. (cont.)

Threat	Effect on birds or habitats	Scale of threat			Regions affected, and
		Habitat	Past	Future	comments
12. Pollution (nitrogen deposition and acid rain)	Changes in composition/quality of vegetation	Ta	•	•	Russia
		B	•	•	Scandinavia, W European Russia, UK; acid rain in Finland
		MR	•	•	Central European Russia, Finland, UK
		ML	•	•	UK, Norway
13. Mining	Habitat loss; vegetation loss and damage by vehicles; disturbance from operations; persecution and excessive hunting by employees; dust pollution	Ta,b	●	●	Russia, Svalbard
		B	●	●	
		ML	—	•	UK
14. Hydropower (power stations and dams)	Habitat loss; altered hydrology; mercury in water and food-chain; change in water-table in surroundings and downstream; increased predators	Ta	•	•	Russia
		Ba,b	●	●	Scandinavia, Russia
		MR	●	●	Finland, Sweden

Scale of threat

T	Tundra zone (all types)		MR	Mires
	Ta	Bush tundra	ML	Moorland
	Tb	Typical tundra		
	Tc	Arctic tundra	Past	Past 20 years
	Td	Polar desert	Future	Next 20 years
	i	Intertidal and coastal	■	Widespread (affects >10% of habitat)
B	Boreal montane habitats (south of tundra zone, all types)		●	Regional (affects 1–10% of habitat)
	Ba	Bush montane 'tundra'	•	Local (affects <1% of habitat)
	Bb	Typical montane 'tundra'		
	Bc	Arctic montane 'tundra'		

if possible, may take a century or more. In western Europe, most peat extraction is commercial, to supply the horticultural market.

Extraction thus causes long-term habitat loss, and also degrades surrounding mires through damage to their hydrological integrity. Damage may range from localized losses of pools and surface water through to general drying out, cessation of active peat development, invasion of the mire by scrub and forest, and the eventual loss of the characteristic mire flora and fauna, including birds. This hydrological integrity may mean that the loss of a comparatively small area of mire to peat extraction may cause major changes in quality over a disproportionately large area of the remaining mire system. Small-scale but traditional extraction of peat for fuel also occurs in some European countries, and may seriously reduce mire quality locally, especially when combined with grazing and burning. Peat extraction can also cause water pollution, through increased levels of aluminium and heavy metals, and increased turbidity.

A significant proportion of mires in central European Russia were lost in this way between the 1920s and the 1970s, with attendant loss of such priority species as *Gavia arctica*, *Aquila chrysaetos*, *Lagopus lagopus* and *Lanius excubitor* in the affected areas (Bekshtrem 1927, Ptushenko and Inozemtsev 1968, Jurkovskaya 1980). Overall, 12 priority species are

predicted to suffer low impacts from this threat, making it one of the most important among those considered here.

Heavy grazing

Grazing of tundra and boreal montane habitats by reindeer is very heavy and still increasing in some European regions, e.g. Finnmark (Norway) and parts of Finland. Some such overstocking has been caused by closure of other grazing grounds due to oil and gas exploitation and to radioactive fallout from the Chernobyl disaster in the 1980s.

Boreal montane habitats and moorland are also very heavily grazed by sheep in Norway and the UK, driven by major subsidies. At least 20% of the moorland in Great Britain has been lost since the mid-1940s, with heavy grazing by sheep being the main reason in England and Wales, and the second most important reason in Scotland (where red deer are also involved). Very intense grazing converts moorland to grassland, but even below such levels widespread degradation can occur, e.g. of the remaining moorland in Great Britain, 50% is in 'poor' or 'suppressed' condition, due to heavy grazing (Thompson *et al.* 1995b). This affects the vegetation structure and results in reduced availability of suitable nest-sites and suitable food for priority species such as *Lagopus lagopus*, *Falco columbarius* and

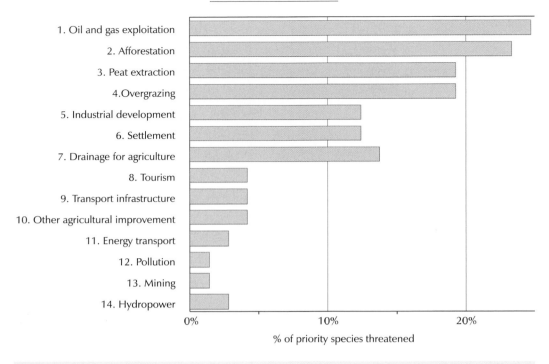

Figure 5. The predicted impact on priority bird species of threats to tundra, boreal montane habitats, mires and moorland, and to the species' habitat requirements, in Europe (as assessed by the Habitat Working Group). All threats are predicted to have a low impact on individual species, i.e. only likely to have local effects, and the species' populations are not likely to decline by more than 20% in the habitat in Europe within 20 years if current trends continue (see Table 4 for more details of the threats, and Appendix 4, p. 401, for more details of the assessment).

Circus cyaneus (Thompson *et al.* 1995b). In contrast, in some parts of Norway widespread abandonment of grazing land over the last 20 years is thought to have led to a net reduction in the national extent of moorland, due to natural regeneration of forest.

Conversely, active management, particularly through controlled burning of heather moors to maintain high populations of *L. lagopus* for shooting, has tended to retain and indeed enhance the quality of moorland, at least from the point of view of key moorland birds such as *L. lagopus* and *Circus cyaneus*, by providing a more diverse vegetation structure. Although less widespread than before (as a result of economic factors), moorland management for grouse has been estimated to be still the predominant land use on up to 22% of upland Britain (Hudson 1992). However these benefits have to some extent been offset by illegal persecution of those priority species that also feed on *L. lagopus* and compete with the shooting interest.

Overall, 14 priority species of these habitats are predicted to suffer low impacts from this threat, making it one of the most important amongst the threats considered here.

Industrial development

Apart from the direct habitat loss associated with the increasing number of industrial sites in the tundra zone (see 'Oil and gas exploitation', above), such developments locally elevate the mortality rates of priority birds through legal or illegal hunting and persecution by indigenous inhabitants or immigrant workers, and through uncontrolled disposal of refuse which inflates populations of predators such as gulls, foxes, feral dogs and corvids in the vicinity. Disturbance from industrial operations may also reduce habitat quality locally. Together, while not obviously changing the physical structure and composition of the vegetation, these processes can result in significant reductions in the abundance and productivity of some key bird species over large areas around settlements.

Overall, nine priority species are predicted to suffer low impacts from this threat, making it moderately important among the threats listed here.

Settlement

Settlement of the tundra zone has occurred locally in Russia, usually associated with oil and gas extrac-

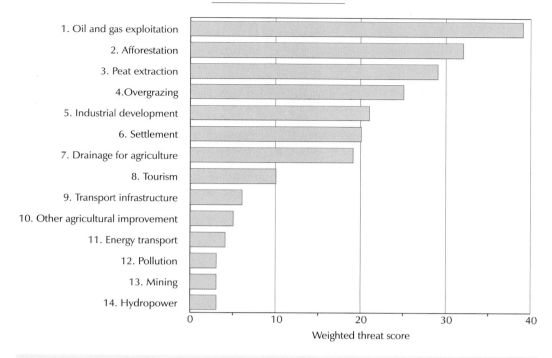

Figure 6. The most important threats to tundra, boreal montane habitats, mires and moorland, and to the habitat requirements of their priority bird species, in Europe. The scoring system (Appendix 4, p. 401) takes into account:

- The European Threat Status of each species and the importance of the habitat for the survival of its European population (Priorities A–D); see Table 2.
- The predicted impact (Critical, High, Low), within the habitat, of each threat on the European population of each species; see Figure 5, and Appendix 4 (p. 401).

See Table 4 for more details on each threat.

tion, mining and other industries, ports and military bases, and has similar effects on tundra habitats and their priority bird species as described under 'Industrial development' (above).

Overall, nine priority bird species are predicted to suffer low impacts from this threat, making it moderately important among the threats considered here.

Drainage for agriculture

Drainage for agriculture has caused widespread loss of mires in Europe, especially since the 1940s, although effects are judged to have been more localized since 1970, and have generally caused far less habitat loss and degradation than drainage for forestry. For example, almost 32% of the extent of original raised mires in Britain is now used for cultivation, while in Latvia about 15% of all peatland had been drained for agriculture by 1980 (VMPI 1980). The resulting habitat loss and degradation are as described in 'Afforestation' (above), especially in the way that drainage of land adjacent to mires can greatly damage their hydrological integrity. In the

future, conversion of mires to agriculture is unlikely in west European countries, due to agricultural overproduction, but is still likely on a local scale in Russia. In addition, tundra has been converted to agricultural use locally in European Russia, totalling about 100 km^2 to date.

Overall, 10 priority species (14%) are predicted to suffer low impacts from this threat, making it moderately important amongst those considered in this study.

Tourism and recreation

Locally, increased tourism and recreation have caused loss and degradation of boreal montane habitats in Fennoscandia and the UK. The main impacts have been through development of infrastructures (ski-resorts, cable-lifts, tourist centres, etc.), increased erosion, trampling of vegetation and disturbance of nesting birds, and locally increased populations of predators attracted to rubbish dumps. Over-intensive collection of cranberries *Vaccinium* from mires (by people in their spare time) also causes these problems locally in the Baltic region. Overall, three priority bird species are predicted to suffer low impacts from this threat, making it of relatively low importance amongst the threats set out here.

Transport infrastructure

Construction of roads, railways and pipelines has caused local habitat loss and degradation in tundra and boreal montane habitats in Europe, mainly in Russia (see 'Oil and gas exploitation', above). In addition to their habitat-fragmenting effects (see 'Afforestation', above), they also lead to the degradation of wider areas and of other habitats, e.g. through changes in the hydrology of adjacent mires, or through concentrating predators due to road casualties and rubbish, or through greatly increasing the range of off-road vehicles, which easily damage the very slow-growing vegetation. These effects may result in reduced nesting or feeding opportunities, leading to increased mortality or reduced productivity for priority birds. The significance of these threats upon bird populations in these habitats remains to be quantified in most cases, but it is predicted that three priority species will suffer local impacts as a result of this threat, making it of relatively low importance among those considered here.

Other agricultural improvements

During the last 20 years, agricultural improvement (excluding drainage—see 'Drainage for agriculture', above) has affected large areas of moorland in the UK, as well as boreal montane habitats more locally in Norway. In the main, fertilization, ploughing and re-seeding have been used to convert these habitats into grasslands, which provide more productive grazing for sheep. This habitat loss is predicted to cause local population impacts to three species during the next 20 years, although the threat overall is likely to become insignificant by the end of this period, as the overstocking in these highland habitats, which is supported by domestic subsidies, is increasingly recognized as unsustainable and undesirable.

Energy transport

The construction of electricity transport grids across mires and tundra is a local problem in Russia and Scandinavia, and is likely to continue as such in the near future. Effects on the habitats and on priority birds include mortality through collision and electrocution, temporary disturbance during construction, and increased densities of predators due to the provision of increased nest-sites (pylons). At least two species of these habitats are predicted to suffer local population impacts over the next 20 years, making this threat of low importance among the threats considered here.

Pollution

The vegetation of tundra, boreal montane habitats and mires, being dominated by slow-growing mosses and lichens, is likely to be especially sensitive to the direct effects of atmospheric pollution (e.g. nitrogen deposition and acid rain), as shown by the loss of *Sphagnum*-dominated mires in parts of the UK during the last 200 years (e.g. Pennine hills). A current example is the tundra of Finnish Lapland, which receives airborne sulphur compounds (creating acid rain, etc.) from industry on the Kola peninsula (Russia). Effects are likely to include changes in the structure and species composition of the vegetation, which are predicted to have an impact on least one priority bird species over the next 20 years.

Even more difficult to assess are the effects (on priority birds' breeding productivity and long-term survival) of pollution by, in particular, heavy metals and pesticides, as well as other pollutants such as the by-products of mining and drilling, e.g. dust and sediments. Their impact is likely to be most severe in tundra, where contaminants are slow to break down, vegetation is known to be particularly sensitive and slow to recover, and where food chains are relatively simple and short. A possible threat of unknown scale is radioactive pollution at high-Arctic sites that were formerly used for testing of nuclear weapons, or as a by-product of nuclear power generation.

Mining

Mining activity (especially coal-mining) has had regional-scale effects on tundra and boreal montane habitat (particularly in Russia and Svalbard), and will continue to do so in the near future, as well as having an impact on moorland more locally in the UK. Apart from the obvious habitat destruction caused by open-cast mining or dumping of wastes, the quality of the surrounding habitat is also reduced by various associated factors, described under 'Industrial development' (above). However, only one priority species is predicted to suffer any population impact from this activity over the next 20 years.

Hydropower

Dams for hydroelectric schemes have permanently destroyed areas of these habitats, especially mires and boreal montane habitats, and particularly in Russia and Fennoscandia. However, over the next 20 years only two priority species are predicted to suffer any population impacts as a result.

Other threats

A few other threats to these habitats were evaluated, but were not predicted to have any significant impact on the European populations of any priority species over the next 20 years, and are listed below.

- In the past, the testing of nuclear weapons destroyed and contaminated significant areas of tundra in Novaya Zemlya, Franz-Joseph-Land and the Arctic Urals (Russia), but is not likely to recur in the future.

- Windfarms can cause collision mortality to larger birds, and may cause significant avoidance behaviour in some open-country species.

- Inappropriate burning practices: the accidentally or deliberately excessive burning of moorland, associated with sheep husbandry or the management of heather for grouse, has been an important contributing factor in the decline of moorland in Great Britain since the 1940s.

- Garbage dumping on mires, due to the low economic valuation of this kind of land, is a local problem in some countries, e.g. Latvia and Ireland.

CONSERVATION OPPORTUNITIES

International legislation

Effective conservation of tundra and boreal montane habitats at a European level will depend largely on national and international measures that can be applied over very large areas of land, because much of the habitat still exists in extensive and essentially homogeneous blocks of 'natural' habitat.

However, at a regional level and particularly around the southern and western extremes of their recent historical range, mires and moorland have often been reduced to isolated fragments within extensive areas of forestry or agriculturally improved land. In such cases, biodiversity conservation may depend as much on the site-protection approach as on measures to conserve the wider environment, although the latter approach also has major benefits for sites, e.g. by preserving the hydrological integrity of mires.

The Birds and Habitats Directives are the most powerful environmental legislation in Europe with an international scope, albeit only applicable to EU member states. Annex I of the Birds Directive includes 21 (29%) of the 73 bird species considered in this book to be a priority for these habitats. Annex I of the Habitats Directive lists nine habitat-types which fall under the category of tundra, boreal montane habitats, mires and moorland, as defined in this book. Enthusiasm for the Habitats Directive is at a low ebb amongst national governments, although its effective implementation could potentially protect substantial areas of mires, boreal montane habitats and moorland in Sweden, Finland, UK and the Republic of Ireland.

The majority of priority species of tundra, mires and moorland are migratory waterbirds (see 'Priority birds', above). They are, therefore, comparatively well covered by specific international conservation initiatives, notably by the Ramsar Convention and by the Agreement on the Conservation of African–Eurasian Migratory Waterbirds (AEWA) of the Bonn Convention, which are both targeted at waterbirds and the habitats upon which they depend.

Ramsar Sites that protect the dispersed breeding populations of waterbirds have been designated in mires and tundra, e.g. Teichi Bog in Latvia, or for the globally threatened *Anser erythropus* in Sweden (Madsen 1996), and others have been explicitly proposed for such purposes and are being actively promoted (e.g. in the UK). Designation of a more extensive series of Ramsar Sites, especially in tundra areas, is both an immediate opportunity and an essential prerequisite for the development of international conservation programmes for habitats and species.

The Bonn Convention provides important additional international support, in that it encourages coordination between countries, and an appreciation that these bird populations are a shared responsibility. At least in the short to medium term, conservation of priority tundra birds in their main European strongholds in Russia may depend on international support, particularly in helping to fund the appropriate broad policies and actions.

In addition to these two core conventions, there is considerable scope for making further use of the World Heritage Convention through designation of natural World Heritage Sites (WHS), and of the Biosphere Reserve provisions under the UNESCO Man and Biosphere Programme. Designation of the Flow Country (UK) as a natural World Heritage Site is being pursued, and Russia has already designated some of its protected areas as Biosphere Reserves (e.g. the Taimyr Nature Reserve in Siberia), although it has no protected areas of tundra in Europe as yet.

Transboundary air pollution is difficult to control, and may pose a risk to tundra vegetation and ecosystems in the long term (see 'Threats to the habitat', p. 178). The Convention on Long Range Transboundary Air Pollution is one of the means by which such pollution can be combated.

Of the international agreements applying only to European countries, the Bern Convention is of most relevance since, like the Bonn Convention, it focuses attention on species and habitats whose conservation requires the cooperation of several states. Although most European countries with these habitats are contracting parties to the Ramsar, Bonn and Bern Conventions, Russia, though a contracting party to Ramsar, has only observer status on the other two, and its full participation is essential.

Box 1. **International environmental legislation of particular relevance to the conservation of tundra, boreal montane habitats, mires and moorland in Europe. See 'Opportunities for Conserving the Wider Environment' (p. 24) for details.**

GLOBAL

- *Biodiversity Convention* (1992) Convention on Biological Diversity.
- Framework Convention on Climate Change (1992)
- *Bonn Convention* or *CMS* (1979) Convention on the Conservation of Migratory Species of Wild Animals.
- *Ramsar Convention* (1971) Convention on Wetlands of International Importance especially as Waterfowl Habitat.
- *World Heritage Convention* (1972) Convention concerning the Protection of the World Cultural and Natural Heritage.
- Convention on Long Range Transboundary Air Pollution (1979).
- Convention on the Protection and Use of Transboundary Watercourses and International Lakes (1992).

PAN-EUROPEAN

- *Bern Convention* (1979) Convention on the Conservation of European Wildlife and Natural Habitats.

- *Espoo Convention* (1991, but yet to come into force) Convention on Environmental Impact Assessment in a Transboundary Context.

EUROPEAN UNION

- *Birds Directive* Council Directive on the conservation of wild birds (79/409/EEC).
- *Habitats Directive* Council Directive on the conservation of natural habitats of wild fauna and flora (92/43/EEC).
- *EIA Directive* Council Directive on the assessment of the effects of certain public and private projects on the environment (85/337/EEC).
- Council Directive on urban waste water treatment (91/271/EEC).
- Council Directive on the control of dangerous substances discharge (76/464/EEC).
- Council Directive on the conservation of fresh water for fish (78/659/EEC).
- Council Directive on the freedom of access to information on the environment (90/313/EEC).

Although these international conventions provide the essential background conservation environment, effective conservation depends on the actual way individual states are able to apply and enforce appropriate measures through their own economic and legislative procedures. At this level there are appreciable differences in the opportunities facing these habitats, which are each in turn considered in more detail below.

Economic instruments

Tundra and boreal montane habitats

Because these extreme environments are generally remote from the wide range of potential pressures from development and other human activities, their conservation might appear to be relatively straightforward, since wide-scale designation of reserves or other forms of conservation protection would have an impact on only a relatively small number of individuals or interested bodies. However, while designation by central government might be relatively easy, the very remoteness of the tundra, combined with the current social and economic pressures faced by Russia, means that actual implementation of effective conservation measures on the ground may be a difficult if not totally unrealistic prospect without considerable additional support measures.

For example, national legislation to restrict hunting pressure on wildlife around settlements in the tundra is unlikely to be effective if the hunters, or those who try to enforce the laws, have no alternative source of food or are poorly paid. As with moorland conservation in the UK, habitat and wildlife conservation in the Russian tundra must be economically attractive to those who depend on the habitat for a living.

Superimposed on these national socio-economic factors is an international dimension. Tundra and its offshore waters hold considerable mineral and hydrocarbon resources which have the potential both to enrich the countries involved and, unless carefully controlled, to damage large areas of wildlife interest. Much of the current exploitation is either funded or being carried out by major multinational companies, whose financial strength may effectively raise them above the normal constraints of national legislation. There is clearly a major opportunity for all European countries and their governments to facilitate conservation measures in the tundra by making sure that the multinationals are not allowed to operate there under lowered environmental standards.

Mires

Although mires, for example blanket bog, are sometimes grazed and could thus benefit to a degree from agri-environment measures, they are not primarily agricultural and, as indicated above, their conservation is likely to depend heavily on site-specific actions.

Strict environmental legislation in Sweden has forced the Swedish electricity generating industry to look for sources of peat in other countries, such as the UK, where controls on peat exploitation are less restrictive. Similarly, Germany is the main importer of peat from Latvia, and peat from Europe is also exported globally, e.g. to Japan.

Moorland

Within the UK, and to a lesser extent in Norway, most moorland is viewed as an agricultural habitat, and as such is subject to the social and economic pressures that apply to all farmland. Within the European Union (which includes a substantial part of Europe's moorland), the Common Agricultural Policy has the potential both to exacerbate and to reverse the negative trends in moorland area and quality. In particular, the Agri-environment Regulation (see Box 1) has various provisions aimed at providing incentives that will lead to reductions in grazing levels, and other forms of extensification, in the EU.

For example, under EC Directive 797/85 and EC Regulation 2328/91 the UK has designated several Environmentally Sensitive Areas (ESAs) which include areas of moorland, with annual payments to farmers who manage land according to specific prescriptions which reduce stocking densities and improve the habitat quality of heather. In addition there is a Moorland Scheme, which applies to land outside designated ESAs and which provides for an annual payment (per head of livestock) aimed at reducing the grazing density of sheep to specified levels. However, the success of such environmental measures may be reduced or negated if steps are not also taken to ensure that other grants that favour sheep grazing, e.g. the Hill Livestock Compensatory Allowance, include some form of cross-compliance.

Within the UK, it has been estimated that more than 20% of the uplands are still managed principally for hunting grouse and red deer. While continued management for grouse may help to ensure that good quality heather moorland is maintained, at least in some areas, there is also a need for a major change in the way that deer are managed and culled, so that grazing pressure can be substantially reduced.

Policy initiatives
Arctic Environmental Protection Strategy
Marine and coastal habitats in Europe are mostly covered by the 'regional seas' conventions, e.g. the

Helsinki Convention for the Baltic Sea, with the only major gap being the European Arctic seas. In response to this and other deficiencies in the coverage and attention paid to habitats of the far north, the Arctic Environmental Protection Strategy (AEPS) has been developed. AEPS is a broad-based policy initiative for the conservation of the entire Arctic zone (including marine and coastal habitats, as well as tundra, mires and inland wetlands), and has been adopted by governments and non-governmental organizations (NGOs) of the eight Arctic circumpolar countries, including Russia, Norway, Iceland, Greenland/Denmark and Finland. Relevant conventions may be developed under its influence in the future. Four programmes make up the strategy, as follows.

- Arctic Monitoring and Assessment Program (AMAP).
- Program for Emergency Prevention, Preparedness and Response (PEPPR).
- Program for the Protection of the Arctic Marine Environment (PPAME).
- Program for the Conservation of Arctic Flora and Flora (CAFF).

In particular, CAFF was established to address the special needs of Arctic species and their habitats. Its secretariat is in Iceland, and its main goals are as follows.

- To conserve Arctic flora and fauna, their diversity and habitats.
- To protect the Arctic ecosystem from threats.
- To improve conservation management, laws, regulations and practices for the Arctic.
- To foster collaboration in research, conservation and sustainable development.
- To ensure that Arctic interests are brought to the attention of international fora.

Initial actions towards habitat conservation in the Arctic zone have included the creation of a Circumpolar Protected Area Network (CPAN), widescale mapping of habitats and vegetation using remote sensing (e.g. in Russia), and the publication of up-to-date inventories of actual and proposed protected areas in the Arctic (CAFF 1994, 1996).

Non-statutory opportunities
Non-governmental organizations
National concerns about the destruction of mires, arising from the extraction of peat for power-generation and horticulture, have led to the formation of effective and vocal national and international consortiums (e.g. the International Mires Conservation Group) seeking to raise the profile of mire

conservation throughout Europe and to highlight the fact that mires are an international issue, both in terms of the species they hold, as under the Bonn and Bern Conventions, and of the international market pressures to which they are subjected.

TELMA Project
TELMA (Greek for mire) is an international project that has been organized by UNESCO and IUCN (The World Conservation Union) since 1967. Twenty countries participate, and one of the aims is to create a list of important mire-sites. For instance, the list for Russia includes 305 mires with a total area of about 15,000 km^2.

CONSERVATION RECOMMENDATIONS

Broad conservation objectives
1. To maintain the extent and quality of existing tundra, boreal montane habitats, mires and moorland with high conservation value, particularly mires and moorland which are restricted and threatened.

2. To improve the quality and conservation value of degraded areas of tundra, boreal montane habitats, mires and moorland which are currently of low conservation value.

3. To consider the restoration of tundra that has been damaged by pipelines or other development where this would result in net conservation benefits.

Ecological targets for the habitat
In order to maintain and improve habitat quality it is essential to retain and enhance important habitat features for priority bird species. It is therefore necessary to ensure that the following ecological targets are met.

1. Catchment hydrology should function as naturally as possible, with a minimum of man-made obstructions, drainage or chemical or physical pollution.

2. Fragmentation of existing tundra, boreal montane habitats, mires and moorland by roads, railways, pipelines and overhead cables should be minimized.

3. Large-scale and small-scale vegetation mosaics should be maintained or recreated, with a diversity of vegetational structure that is optimal for the location.

4. Grazing pressures on tundra, mires and moorland should allow the maintenance of at least the existing vegetation cover, and should ensure that scavenger/predator populations are not artifi-

cially boosted through the high livestock mortality associated with overstocking.

5. Moorland should be burned according to currently understood 'best practice' (e.g. Thompson *et al.* 1995a), and accidental fires should be minimized.

6. Pollution of land, water and air should be minimized.

7. The habitats should be free from illegal hunting, trapping, persecution, poisoning or collecting for taxidermy, and levels of legal trapping of non-target animals and of inadvertent disturbance (e.g. from industrial operations, tourism and recreation) should be minimized.

8. Powerlines should be designed not to cause bird mortality, with protective devices in place, and routed to avoid the most sensitive areas.

9. Human settlements should have efficient systems for recycling or disposing of wastes in a non-polluting manner, such that predator populations are not artificially boosted by free access to feed on refuse.

10. Use of vehicles off-road should be minimized.

Broad policies and actions
To meet the broad objectives and habitat targets listed above, in accordance with the relevant strategic principles, the following broad policies and actions should be pursued (see p. 22 in 'Introduction' for explanation of derivation).

General
1. The importance of tundra, boreal montane habitats, mires and moorland for birds and other wildlife, and the values, traditional sustainable uses and other roles of these habitats, should all be widely publicized and promoted.

   The priority threats acting against these habitats, as well as recommended actions to deal with such problems, should also be widely publicized by government agencies and NGOs among other national, regional and local government departments and branches, industries, hunters and hunting associations, other land-users, and the general public.

   In particular, the commonly held image of these habitats as inhospitable and empty wastelands, with low economic value to mankind apart from raw materials, or unless converted to other uses (e.g. forestry, agriculture), must be changed. The current weakness of planning processes in these habitats, compared with some other major habitat-types in Europe, should be rectified.

2. National and regional governments and the EU should practise and promote the 'catchment management' approach to the conservation of mires, and should encourage cross-sectoral integration of policies and cooperation on actions within water catchments. The natural hydrological regimes of mires should be maintained or restored wherever feasible.

3. Planning regulations for all land-use sectors should take into consideration the specific ecological characteristics of these habitats.

4. Potential multiple-uses of these habitats which are sustainable and non-destructive should be promoted, and should be researched further.

5. The Arctic Environmental Protection Strategy (AEPS) should be supported and developed by governments, industries and NGOs. In particular, internationally important sites for bird populations (Important Bird Areas, or IBAs) identified in the high Arctic should be included in the Circumpolar Protected Area Network of AEPS.

6. All relevant European states should contribute to (and update) the TELMA inventory of mires.

7. Those industries or agencies that are involved in destruction, exploitation or development of these habitats should also fund monitoring of populations of birds and other wildlife in the area of their activity (or selected areas), in order to better understand the effects of their activity and to determine the scale and mechanism of any impacts, according to the 'polluter pays' principle.

Legislation

8. All internationally important sites for bird populations (IBAs) and other wildlife (e.g. TELMA sites) should be protected through relevant national legislation and international conventions, e.g. as Biosphere Reserves, Natural World Heritage Sites or EMERALD sites, or as Special Protection Areas (SPAs) within the European Union (EU).

   It is especially important that the extensive areas of surviving raised mires in central and eastern Europe receive adequate legal protection that is properly enforced on the ground.

9. Within the European Union, sites containing the relevant habitat-types listed in Annex I of the Habitats Directive should be put forward as potential Special Areas of Conservation.

10. Legislation on Strategic Environmental Assessments (SEAs) and Environmental Impact Assessments (EIAs) should be further developed and properly implemented by national governments and the EU, to cover all land- and sea-uses in these habitats, but especially programmes and projects concerning industrial, forestry, agricultural and infrastructural development.

Industry

See also the recommendations under 'Marine Habitats' (p. 88) regarding offshore transport of oil.

11. Environmental safeguards in existing national laws on oil and gas extraction and other extractive industries should be properly implemented in northern Russia.

12. An intergovernmental agreement and an industry 'code of best practice' should be drawn up to regulate, monitor and advise on the extraction and transport of oil and gas, and on mining, in Arctic regions.

13. Policies, plans and programmes concerning resource exploitation in these habitats should be made subject to properly conducted SEAs, and modified accordingly. Similarly, individual projects should only proceed if subject to properly conducted EIAs that predict minimal impact on the environment, and that propose full mitigation measures. Operations should maintain certain environmental standards, and should be properly monitored throughout.

14. Governments, industry and other developers should recreate habitats or habitat features after any land-use which results in significant degradation or destruction of the habitat. The effects of restoration measures on damaged habitats, in terms of wildlife and ecosystem recovery, should be properly evaluated and monitored.

15. National governments and the EU should promote suitable grants, subsidies and taxes aimed at reducing industrial pollution. International legislation on transboundary pollution should be ratified and properly implemented.

16. National governments and the EU should reduce emissions of greenhouse gases to 1990 levels by the year 2000, (according to obligations under the Framework Convention on Climate Change), and should press for a legally binding protocol to limit emissions at the third Conference of the Parties in December 1997.

17. Pipelines for oil-transport above ground in Russian tundra should receive urgent maintenance, in order to stop and avoid further oil spillage.

Forestry

See also the recommendations under 'Boreal and Temperate Forests' (p. 234).

18. Afforestation of mires and moorland should be

avoided. Subsidies favouring the afforestation of these high-value conservation areas should be removed.

Management of large herbivores

19. Mechanisms must be developed to regulate the numbers of domestic reindeer in Russia and Fennoscandia, in order to prevent damage to habitats and species by overgrazing.

20. Stricter controls on the legal size of reindeer herds need to be developed.

21. Price-support for reindeer products from national governments or the EU should be made conditional upon low stocking densities.

22. Reindeer-herders, politicians, administrators and the general public should be made aware of the problems that overstocking causes to reindeer production.

23. The economics of reindeer herding, associated price-support and overgrazing must be further evaluated, and optimal stocking densities, timing periods and other parameters should be established for different habitat-types, and then widely promoted among herders.

24. Overgrazing by deer should be addressed in management plans which cover affected areas.

Peat extraction

25. The extent and condition of remaining 'active' mires in Europe should be identified.

26. National governments and the EU should develop their policies on peat extraction so as to promote sustainable use, e.g. as in northern Scotland (UK) where there is a 'Peatland Management Scheme'.

27. No peat extraction should be permitted on, or allowed to affect the hydrology of, IBAs/TELMA sites.

28. The adverse effects of peat extraction on global CO_2 levels, and its more local pollution of water supplies, should be widely publicized.

29. Methods should be evaluated for making peat extraction sustainable, especially from 'active' mires.

30. Research on alternatives to peat should be promoted and its results widely publicized.

31. Catchment-wide research should be promoted into the hydrological effects and impacts of peat extraction on mires and on the population dynamics of characteristic plant and animal species (for possible use as bioindicators of mire condition).

Tourism and recreation

32. Suitable areas for tourism need to be identified in tundra and boreal montane habitats. The development of sustainable tourism should be promoted by national governments and the EU, e.g. through financial assistance for pilot projects or for research into the impact of tourism on wildlife, and into the effectiveness of mitigatory measures.

Wildlife exploitation

33. Existing hunting and species-protection laws should be more strictly enforced, especially around settlements, in protected areas and in IBAs, and with regard to threatened species. Such regulations should be further assessed and developed, according to the precautionary and sustainable-use principles, and management plans for hunting should be promoted.

34. Hunters should be well aware of the importance of priority species and the need to conserve them, through incorporation of such information in existing hunting exams and permit requirements, and through effective use of media publicity.

35. Legal trappers should be informed of alternative methods of trapping quarry that do not catch non-target species, e.g. raptors.

36. Use of lead-shot in hunting and fishing should be eliminated.

Transport

37. Ecologically sensitive areas should be identified where no development of linear infrastructures (roads, railways, pipelines, overhead cables), is permitted, and should be protected through land-use planning and legislation.

38. Governments and construction companies should develop and implement methods and practices of design and construction which minimize the impact of existing and new linear infrastructure.

39. Governments and the EU should improve legislation against off-road driving in Arctic tundra, and should develop stricter legislation on snowmobile driving in winter and spring (in northern Fennoscandia especially).

40. Military activity should be banned on fragile tundra habitats.

Other measures

41. Research on man-made climate change and global warming should be continued and intensified, in order to better identify the zones and species most likely to be affected, and to predict changes more accurately in time and space.

42. The ban on test explosions of nuclear devices in the high Arctic should be permanently enforced.

43. The conversion of mire and moorland to grassland or cultivation, through re-seeding or drainage, should be halted (see 'Conservation recommendations' in 'Agricultural and Grassland Habitats', p. 321).

44. Accidental fires on moorland should be minimized, through effective publicity. A 'code of best practice' for the management of moorland by burning should be developed and widely implemented by national governments, where necessary.

ACKNOWLEDGEMENTS

Most contributions to this chapter were made by the Habitat Working Group during and after a workshop hosted by the Norsk Institutt for Naturforskning, Trondheim, Norway (22–24 November 1993) and organized by J. A. Kålås, I. J. Øien and D. Thompson; these contributors are listed on p. 159.

In addition, valuable information, comments and assistance were received from E. Lebedeva (Russian Bird Conservation Union), A. Matschke (Associació per a la Defensa de la Natura) and I. J. Øien (Norsk Ornitologisk Forening).

LOWLAND ATLANTIC HEATHLAND

compiled by

M. REBANE AND R. WYNDE

from major contributions supplied by the Habitat Working Group

W. H. Diemont, F. P. Jensen, L. Påhlsson, N. R. Webb, R. Wynde (Coordinator)

S U M M A R Y

Lowland Atlantic heathland is a traditionally managed habitat that is restricted to north-west Europe and which has been considerably reduced in recent decades, now only covering about 4,000 km². Nevertheless, it is important for the survival of 16 priority bird species in Europe, all of which have an Unfavourable Conservation Status because of their declining or small European populations. Although none of the species is a specialist of this habitat, most of the species require habitat features that are only found in similar, traditionally managed, semi-open landscapes.

The maintenance of remaining heathland and of its ecological quality is important, and recreation of heathland will also be needed in the longer term in order to reduce habitat fragmentation. Deterioration or loss of heathland continues to occur widely throughout its European range, due mainly to infrastructural development, to wildfires being too frequent or intense, to abandonment of non-intensive grazing and of other traditional land-uses, and to afforestation.

Due to the relative rarity of the habitat, it is essential that all of the remaining important areas of heathland should be strictly protected and managed appropriately. To achieve this, governments should make full use of the opportunities for coordinated action that are provided by international and European Union legislation, international funds, and other current policy initiatives. Broader measures are also needed, including the proper enactment and implementation of Strategic Environmental Assessments and Environmental Impact Assessments, in order to integrate the conservation of lowland Atlantic heathland into and across all relevant sectors, programmes, policies and plans.

HABITAT DESCRIPTION

Habitat definition, location and extent

Lowland Atlantic heathland is an open, mainly treeless habitat dominated by dwarf shrubs of the heath family (Ericaceae), usually heather *Calluna vulgaris*, or others of similar appearance. Such habitat generally occurs below 300 m altitude, within what is termed the Atlantic (bioclimatic) region of western Europe, and experiences an annual precipitation of less than 1,000 mm/year (Figure 1). Its range is centred on north-west Europe, being distributed from northern Spain (Galicia) in the south to the north-west coast of Norway in the north, and extending eastwards as far as Germany. Heathland seldom occurs more than 200 km from the coast, and its range therefore coincides with a temperate and moderate (oceanic) climate.

A distinction has been drawn between heathland (treated in this strategy) and moorland (treated under 'Tundra, Mires and Moorland', p. 159), both dwarf-shrublands usually dominated by heather, based on an altitudinal threshold of 300 m and a precipitation threshold of 1,000 mm/year. In reality, however, there is a continuum in dwarf-shrub vegetation-types in Europe, from Mediterranean shrubland through lowland heathland to moorland, mires and tundra, and rigid distinctions are not always possible between types.

A summary of the current extent of heathland in Europe is indicated in Table 1, together with estimates of former extent where known. It should, however, be noted that some of these estimates of current extent are nearly 20 years old.

Heathland in Europe probably reached a peak of at least 30,000 km² in the nineteenth century (Noirfalise and Vanesse 1976), but since that time the afforestation of heathland, and its conversion to

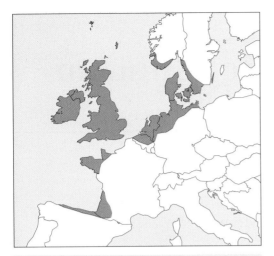

Figure 1. The potential range of lowland Atlantic heathland in Europe (after RSPB 1993). Within the area marked, the climate is suitable for the development of heathland on sandy acidic soils below 300 m elevation.

arable land, have led to an accelerating decline in extent (see 'Threats to the habitat', p. 194). In Sweden (Damman 1957) and Denmark, for instance, the area of heathland was reduced by 60–70% between 1860 and 1960 (Figure 2). In the years since 1960, its disappearance in these countries has become almost complete, apart from areas where traditional man-agement still prevails, such as in parks or nature reserves. Much the same is true of southern England (NCC 1984), the Netherlands, Belgium, northern Germany and parts of France (Table 1). However, both climate and terrain have precluded conversion in coastal regions of northern France.

Origin and history of the habitat

Over much of Atlantic Europe, where heathland is now found, forest is the vegetation climax. Large expanses of open heathland would not have originally existed, since heather *Calluna vulgaris* and other dwarf shrubs would be shaded out by trees. For the most part, dwarf shrubs would have grown only beneath the trees in gaps in the forest cover, or on particularly poor soils where the tree-cover was thin (Webb 1986). Natural climax heath was generally restricted to coastal areas where shrub and tree growth is prevented by exposure to strong prevailing winds.

The evidence for ascribing the origins of heathland very largely to human influence, throughout the greater part of the region, is now overwhelming. Instances of conversion from forest to heathland have now been found as early as the late Neolithic period (c.4,500 years ago) and, although this period of increased human impact coincides with a significant climatic change, burning and grazing played a vital part in the development of open heath, perhaps as early as the Mesolithic period and certainly by the Bronze Age (e.g. Dimbleby 1962, Tubbs 1968, Haskins 1978, Webb 1986).

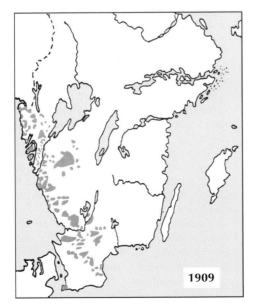

Figure 2. The decline in the extent of lowland Atlantic heathland in southern Sweden between 1909 and 1957 (after Damman 1957, in Gimingham *et al.* 1979).

Table 1. The overall maximum (nineteenth century) and recent (1970s/1980s) extent of lowland Atlantic heathland in Europe (adapted from Diemont et al. 1996), with the percentage reduction in area during this period. It is difficult to estimate the extent of heathland in Ireland, Norway and Sweden, and in Iberia, due respectively to the transition to mires and to Mediterranean shrubland.

	19th century (km²)	1970s/1980s (km²)	% reduction
Belgium	1,630	130	92
Denmark	6,580	704	89
France	2,000 [1]	790	60
Germany	10,000	550 [2]	95
Netherlands	8,000	400	95
Sweden	3,000	930	69
United Kingdom	1,450 [3]	580 [4]	60

[1] Excludes north-west France.
[2] Conservative estimate of current extent, by Habitat Working Group.
[3] Excludes Northern Ireland.
[4] England only.

As the human population grew, forest clearance gathered momentum and the resulting open country provided opportunities for increasing the scale of pastoral agriculture. With the advent of hardy breeds of cattle and sheep, there was a progressive extension northwards in Europe of grazed heathland (Gimingham et al. 1979). Shifting cultivation, burning, turf-cutting and harvesting of vegetation for fodder all contributed to the maintenance of this new and increasing habitat (Noirfalise and Vanesse 1976).

In the Netherlands, Belgium, Denmark and southern Sweden, and on the north German plain, traditional management involved a combination of grazing and turf-cutting (plaggen). This ancient agricultural system was practised for around 1,000 years until it ceased early in the twentieth century. Livestock grazed the heaths but were removed from this ground for the night to stables. Sods, which had been cut from the heath in huge amounts, were also brought to stables for bedding and as an absorbent for dung. The resulting mixture was used to fertilize the nearby arable fields (Diemont 1996). This process generated a flow of nutrients from the heath to the arable land, raised soils in the arable lands, maintained heath soils in a nutrient-poor state, and ensured the dominance of dwarf shrubs on the heath. Frequently there were breeds of sheep adapted to feeding on the heather rather than the grass (Riis-Nielsen 1995). Regular burning of heaths to ensure young, nutrient-rich fodder for the sheep did not take place—at least in recent centuries (Riis-Nielsen 1995).

Cutting of turf and peat for domestic and industrial fuel also occurred widely in past centuries (Tubbs 1968, Hopkins 1983), and the ecological consequences of extraction on this scale must have been considerable (Webb 1986). Other heathlands were used for cultivation, e.g. 'brecks' in the UK and 'driesland' in the Netherlands, where crops such as buckwheat and rye were interspersed with fallow periods of several years to restore soil fertility and to depress weeds. Again, this had the effect of depleting the soil of nutrients (Gimingham 1992).

Losses of nitrogen due to sod-cutting are estimated to be 5 kg/ha/year while losses due to sheep-grazing and heather-cutting only amount to 0.8 kg/ha/year (Riis-Nielsen 1995). Especially on phosphorus-deficient soils, sod-cutting is also much more effective in removing nutrients from heathland than prescribed burning (Diemont 1996).

Heather and gorse Ulex were also cut or mown to serve as winter fodder for sheep. In Norway, sheep grazed all year round while cattle were kept inside during the winter and fed on cut heather fodder, which accounted for about a third of their diet (Kaland 1995). Winter grazing was also an important part of heathland management in Sweden. Ulex was also used widely for fodder, fuel, composting, thatching and the production of potash, while heather was also used for thatching, as well as road construction. Other uses of heathland included the small-scale and localized extraction of sand and gravel, which created many pools.

Ecological characteristics
Key abiotic factors
Climate
Lowland Atlantic heathland experiences a temperate climate with cool moist summers and mild winters, as well as relatively long spring and autumn periods. The mean temperature of the warmest month is less than 22°C while at least four months have mean temperatures above 10°C. It is probably the summer evaporation stress which is most important in determining heathland distribution, rather than any particular temperature level (Gimingham 1972), and heathland only occurs in areas with a more strongly oceanic climate, where there is an annual precipitation of 600–1,000 mm.

A significant factor in the local occurrence of heaths beyond the defined region, at higher latitudes, higher altitudes and eastwards towards a more continental climate, is the protection afforded by snow cover to the dwarf shrubs in winter. Heathland vegetation can tolerate very considerable exposure to high winds, e.g. on ocean-facing slopes.

Soils
Lowland heathland generally develops on acidic, frequently sandy soils that are poor in exchangeable calcium or other mineral nutrients. These soils are

relatively rich in carbon and poor in nitrogen, and are almost invariably deficient in phosphorus (Gimingham 1972, Gimingham *et al.* 1979). Where soils cover calcareous strata, however, the productivity and biological diversity of the heathland is generally greater. Depending on the level of the water-table relative to the soil surface (which depends in turn upon topography and hydrology), different vegetation-types occur, ranging from dry heath through humid heath to wet heath, and finally mires (the latter being treated under 'Tundra, Mires and Moorland', p. 159).

Fire

Most heathland in Europe has a long history of use and management by man, including the use of burning to provide new nutritious growth of heather and grass for domestic stock. Fire became an important and widely used management tool for the perpetuation of heaths. The size of the area burnt will have varied from region to region. In some places small, carefully chosen patches were burnt, while elsewhere large areas were burnt rotationally, as in Jutland (Denmark) and Halland (Sweden).

Regular management by burning is no longer practised on lowland heathland in Europe, but in the recent past was more widespread, e.g. in Norway before 1940 (Rosberg *et al.* 1981).

Burning restricts heathland flora to plants with renewal buds in positions that escape the full effects of fire, and can also eliminate mosses and lichens. Burning can also give bracken *Pteridium* a competitive advantage over dwarf-shrub heathland, and thus promote its spread (e.g. Lowday *et al.* 1983).

Occasional natural fires also occurred in the past, and their effect will have depended on the age and structure of the heath, the size of the heath, and the scale and intensity of the fire. It is likely that considerable local damage was done from time to time to heaths, their vegetation and wildlife. Today, in some regions such as north-west France or southern England, extensive man-made wildfires often occur, especially in hot dry summers. As heaths are now considerably fragmented, the effects on wildlife can be disastrous, and as a consequence the creation of plans to deal with wildfires is becoming increasingly common.

Key biotic factors and natural dynamics

Plant succession

Only those lowland heathlands on peat and on exposed maritime headlands can be regarded as stable or semi-stable over time. Most other European heathlands are (or would be) transient stages in an ongoing 'succession' of vegetation-types, and are only perpetuated as heathlands by continued management using traditional land-use practices. In the absence of such turf-cutting, grazing or burning, heathland develops into grassland, scrub and woodland (Binding and Fransden 1995). The rate at which this occurs will depend on the past management history and on the structure of the dwarf-shrub community at the time of abandonment (e.g. Gimingham *et al.* 1979).

Grazing

Different livestock have varying effects on heathland, because of differences in food preferences, grazing methods, trampling pressures and dunging habits (Andrews and Rebane 1994).

Growth phases

The life span of individual plants of heather, which is often the dominant plant species in heathland, is between 30 and 40 years. During the initial growth period, plants are dense and little light reaches the ground. After 13–20 years, however, the plant's structure starts to become increasingly open (MacDonald 1990), and it is at this stage that competitors such as tree and shrub seedlings, or *Pteridium*, can become established (Watt 1955, Gimingham 1978). In unmanaged heaths, plants of all growth phases exist side-by-side, creating an uneven, patchy structure, whereas in managed heaths large patches of vegetation are in the same growth phase. In this way, the overall structure and floristic diversity of the heathland can be affected by the growth phase of the dwarf shrubs, which can thus be critical in determining the presence or absence of a number of reptiles and birds.

Heather beetle

Death of heather can result from defoliation by the heather beetle *Lochmaea saturalis* (Rodwell 1991), which can thus halt the growth cycle over wide areas, even in the absence of burning. The beetle feeds almost exclusively on heather as both larva and adult. There is some evidence from the Netherlands that outbreaks are more frequent and damaging where the nitrogen content of the heather foliage is higher than normal (Gimingham 1992), linked to atmospheric nitrogen deposition (Bobbink *et al.* 1992; see 'Threats to the habitat', below), and the beetle is also a notable influence on heathland in Denmark.

Characteristic vegetation and associated fauna

Vegetation

Heathland is characterized by plant species which grow well in nutrient-poor and acidic soils, most notably dwarf shrubs such as heather, *Erica* and *Vaccinium*, but also plants such as *Pteridium*, shrubs of *Ulex*, *Juniperus* or broom *Sarothamnus*, trees such as birch *Betula*, grasses such as *Molinia*, *Deschampsia*, *Nardus* and *Agrostis*, and also certain bryophytes, lichens and herbs. Wet heathland in the more oceanic parts of the region tends to lack trees but is important for the dwarf shrub *Erica ciliaris*, and is rich in bryophytes, while lichens tend to be characteristic of dry heaths. Overall, heathland is not floristically rich, but the species composition varies according to abiotic factors, most notably in maritime and sand-dune areas.

To the south, heathlands merge into Mediterranean shrublands, where the plant communities are generally characterized by taller shrubs with larger leaves. There is also a close relationship between heathland and grassland, which in the absence of trees may take the place of heathland under heavy grazing, or on soils of higher nutrient status, or where there is increasing continentality of climate.

Wet heathlands are often associated with mires, where dwarf shrubs cede dominance to grasses (e.g. *Molinia*, *Nardus*) or sedges (cotton-grass *Eriophorum*, deergrass *Trichophorum*) under the waterlogged conditions which encourage active peat formation.

Invertebrates

Heathland—especially where it is varied, e.g. including a mosaic of age-classes of heather, scrub and firm bare ground—generally supports large numbers and a high diversity of invertebrates, due to such features as the structural diversity of the vegetation, the presence of warm, bare, sandy soils, grass tussocks, wet heath and pools. Many species of dragonflies and damselflies (Odonata) breed in acid pools on lowland heathland, and some are almost confined to the habitat. Bees and wasps (Hymenoptera) are often abundant, there are numerous heathland moths (Lepidoptera), and beetles (Coleoptera) are also well represented with many ground-dwellers. The soil fauna of heathland is limited, owing to the generally high acidity.

Mammals, reptiles and amphibians

No mammals are characteristic of lowland Atlantic heathland in Europe, or in any way confined to it. There are few species of herbivore (e.g. brown hare *Lepus capensis* or rabbit *Oryctolagus cuniculus*), and these generally prefer grassland or grassy areas.

Among reptiles and amphibians, adder *Vipera berus* can be very plentiful, and a few species are somewhat characteristic of heathland, e.g. natterjack toad *Bufo calamita*, sand lizard *Lacerta agilis* and smooth snake *Coronella austriaca*.

Birds

As with other habitats, the occurrence and distribution of bird species on lowland heathland depends more on the vegetation structure than on the floristic composition. The carrying capacity of heathland as a bird habitat is low (Lack 1935, Glue 1973), and the species diversity is also relatively poor. Many of the typical breeding birds are also common in other open and scrubby places (Fuller 1982), e.g. Meadow Pipit *Anthus pratensis*, Wren *Troglodytes troglodytes*, Robin *Erithacus rubecula*, Willow Warbler *Phylloscopus trochilus*, Whitethroat *Sylvia communis*, Linnet *Carduelis cannabina* and Yellowhammer *Emberiza citrinella*.

There are, however, a small number of birds that are more characteristic of, or restricted to, the habitat, e.g. Dartford Warbler *Sylvia undata*, *Saxicola torquata*, Red-backed Shrike *Lanius collurio* and Great Grey Shrike *Lanius excubitor*. Where scattered trees occur, or along woodland margins, then certain species such as Green Woodpecker *Picus viridis*, Nightjar *Caprimulgus europaeus* and Woodlark *Lullula arborea* can be frequent, and more rarely Black Grouse *Tetrao tetrix*.

Stone Curlew *Burhinus oedicnemus*, Wheatear *Oenanthe oenanthe* and Skylark *Alauda arvensis* may occur on heathland where there are also large areas of short turf or bare ground, but such habitat is now rare in Europe. Some waders (Charadrii) that feed in mires, such as Curlew *Numenius arquata*, nest on adjacent dry heath.

In winter, the impoverished soils are poor in seeds and unattractive to many ground-feeding birds, there are few, if any, berry-bearing shrubs, and the invertebrate fauna of the heather itself is very poor (Bibby 1979). However, a few species make use of such open areas, especially birds of prey, with Merlin *Falco columbarius*, Hen Harrier *Circus cyaneus* and Short-eared Owl *Asio flammeus* wintering on larger heathlands. *Lanius excubitor* also favours the combination of open heath and scattered trees or bushes on which it can perch (Fuller 1982).

Values, roles and uses

Grazing is the land-use which has been practised longest and in varying degrees of intensity since the original forest cover was cleared (see 'Origin and history of the habitat', p. 188). During the last 100 years, this activity has shown widespread decline, and with the development of modern breeds of sheep and cattle, intensive agricultural systems and an increased impetus towards afforestation or conversion to arable land, grazing is no longer widely practised on remaining lowland heaths.

The close link between livestock grazing and the continued well-being of heaths has long been recognized, and attempts are being made, mainly by conservation organizations, to reverse this trend. Increasingly in Europe, hardy and sometimes rare breeds of sheep and cattle are being reintroduced, e.g. in Denmark, France, the Netherlands and the UK. A recent development has been the rearing of livestock on an non-intensive basis, with little or no artificial inputs, and marketed as 'Conservation Grade' meat.

Rabbits have been an important influence on heaths, diversifying structure and creating bare ground suitable for lichens, annual plants and invertebrates. The species can, however, become a problem on adjoining farmland (Andrews and Rebane 1994) and for a number of reasons, including frequent outbreaks of the disease myxomatosis, is infrequently farmed nowadays.

Hunting is now little practised on lowland heaths, with *Lepus capensis* and rabbits being the main quarry. The use of heaths for military training is, however, still widespread. Paradoxically, this has prevented large-scale destruction of these areas by other threats, and has ensured the retention of the heathland landscape within the countryside, albeit with some local damage.

As many heathlands in Europe are close to heavily populated areas, and access is often unrestricted, they are very popular with the general public as a focus for outdoor recreation, and also for tourism in some areas (e.g. south-west Sweden). In most European countries, many or most lowland heaths are either owned or managed by a range of organizations involved in some way with wildlife conservation, or by farmers who receive grants for non-intensive grazing or restoration of heathland.

Production of honey is important in many heathland areas in Europe, e.g. in Galicia (Spain) and southern Sweden. Plant material resulting from heathland management is sold locally as fuel or made into horticultural mulch or green compost as an alternative to peat-based products.

Political and socio-economic factors

Both the creation and the decline of heathland in Europe are driven by human factors. Large population increases during the last century, and high demands for agricultural and other products, led to widespread conversion of heathland into arable land and forestry plantations. The advent of railways and modern transportation reduced the dependence of local people on their heathland for fuel, animal forage and bedding, etc. New, intensive agricultural crop-rotation systems were introduced and manure was imported to 'improve' the heathland.

Heathland today suffers from neglect for a number of political and socio-economic reasons. One of the most important factors is modern agricultural practice. Following the development of intensive livestock production, traditional non-intensive grazing is no longer economically attractive and has consequently been widely abandoned. The fragmentation of heathland and urbanization of surrounding areas have also exacerbated the problems of maintaining livestock on heathland.

In the UK (but no longer in continental Europe), large areas of heathland (e.g. totalling 23% of the habitat in England: Evans *et al.* 1994) are still 'commons' where many people ('commoners') have rights to graze animals. This arrangement is crucial in influencing management and development threats. While it may assist in resisting developments that are detrimental to nature conservation, it can also hinder appropriate management through difficulties in obtaining approval through legislation or through a significant number of commoners (RSPB 1993).

Despite the work of a number of agencies, the lack of public awareness of heathland and its conservation importance and needs is still low. Consequently, heathland is often perceived as derelict land of little economic value. As a result, heaths have suffered disproportionate habitat loss from agriculture, afforestation schemes and industrial and housing developments during much of the twentieth century. The recent political changes in central and eastern Europe have led to cutbacks in military forces in Europe and to their redistribution, which in turn is leading to reduced use of some military training areas, particularly in the former border countries such as Germany, and intensified use of others, e.g. in the UK. Though some could be maintained as nature reserves, most such abandoned heathlands are likely to be sold and used for forestry or building developments.

Public misconceptions can also hinder conservation measures, e.g. regarding the issue of tree- and shrub-felling on heathland. Three-quarters of people living in heathland areas in the UK consider them to be natural landscapes, and more than 50% object to the cutting of trees, according to a recent survey (Atlantic Consultants 1996). Nevertheless, over 90% believed it would matter if heathlands were lost, and the most popular reason for their retention was the range of wildlife that they support.

PRIORITY BIRDS

There are 16 priority bird species of lowland Atlantic heathland in Europe (Table 2). This is a low number compared to other habitats in Europe and reflects the small extent and limited distribution of the habitat, as well as the widespread occurrence of heathland bird species in other habitats. Indeed, no birds in Europe have more than 10% of their European population

Table 2. The priority bird species of lowland Atlantic heathland in Europe.

This prioritization identifies Species of European Conservation Concern (SPECs) (see p. 17) for which the habitat is of particular importance for survival, as well as other species which depend very strongly on the habitat. It focuses on the SPEC category (which takes into account global importance and overall conservation status of the European population) and the percentage of the European population (breeding population, unless otherwise stated) that uses the habitat at any stage of the annual cycle. It does not take into account the threats to the species within the habitat, and therefore should not be considered as an indication of priorities for action. Indications of priorities for action for each species are given in Appendix 4 (p. 401), where the severity of habitat-specific threats is taken into account.

	SPEC category	European Threat Status	Habitat importance
Priority A			
No species in this category			
Priority B			
No species in this category			
Priority C			
Red-legged Partridge *Alectoris rufa*	2	V	•
Nightjar *Caprimulgus europaeus*	2	(D)	•
Green Woodpecker *Picus viridis*	2	D	•
Woodlark *Lullula arborea*	2	V	•
Dartford Warbler *Sylvia undata*	2	V	•
Priority D			
Hen Harrier *Circus cyaneus*	3	V	•
Black Grouse *Tetrao tetrix*	3	V	•
Partridge *Perdix perdix*	3	V	•
Crane *Grus grus*	3	V	•
Stone Curlew *Burhinus oedicnemus*	3	V	•
Short-eared Owl *Asio flammeus*	3	(V)	•
Skylark *Alauda arvensis*	3	V	•
Tawny Pipit *Anthus campestris*	3	V	•
Stonechat *Saxicola torquata*	3	(D)	•
Red-backed Shrike *Lanius collurio*	3	(D)	•
Great Grey Shrike *Lanius excubitor*	3	D	•

SPEC category and **European Threat Status**: see Box 2 (p. 17) for definitions.

Habitat importance for each species is assessed in terms of the maximum percentage of the European population (breeding population unless otherwise stated) that uses the habitat at any stage of the annual cycle:
■ >75%   ● 10–75%   • <10%

dependent on lowland Atlantic heathland at any time during the annual cycle (Table 2). Most of the priority birds that occur in heathland are at the fringes of their breeding ranges, in particular *Caprimulgus europaeus*, *Lullula arborea* and *Sylvia undata*. Elsewhere, the former two are widely found in open forests or along woodland edges, while the last-mentioned species is common in western Mediterranean shrubland.

Five Priority C species occur in lowland Atlantic heathland, all of which have the majority of their world population confined to Europe while breeding and/or wintering. Eleven Priority D species also occur, but are not concentrated in Europe in this way. All 16 of the priority species have an Unfavourable Conservation Status in Europe, with most being classified as Vulnerable due to their European population being notably declining and/or small (Tucker and Heath 1994). France, Spain and Germany support the highest breeding diversity of priority heathland species in Europe (see Appendix 2, p. 344).

HABITAT NEEDS OF PRIORITY BIRDS

The habitat features and attributes that are required by priority heathland species are summarized in Appendix 3 (p. 354). The species can be broadly separated according to one major factor: their need for woodland edge or scattered trees as opposed to more open and lower vegetation.

With respect to the first factor, three species are primarily found at the boundary between heathland and forest: *Caprimulgus europaeus*, *Picus viridis* and *Lullula arborea*. *C. europaeus* is a crepuscular and nocturnal feeder catching large insects on the wing over grassland, forest and wetlands (Alexander and Cresswell 1990), particularly over edge habitats as these probably hold the highest densities of suitable insect prey. Trees are also used as song-posts, and territories tend to be distributed linearly along woodland edges (Berry 1979, Cadbury 1981). Bare ground is essential for nesting and daytime roosting. *Lullula arborea* also nests on the ground and feeds in open habitats such as heathland, but requires trees as song-posts (Bowden 1990). *Picus viridis* requires trees for nesting but largely feeds on the ground on ants (Formicidae) in open, dry grassland and heathland.

In contrast, several priority species require open heathland dominated by short vegetation and an absence of trees. These include *Alauda arvensis*, *Burhinus oedicnemus* and Chough *Pyrrhocorax pyrrhocorax*. The first of these is a widespread and common generalist of heathland and many other open habitats. However, the latter two require heavily grazed turf (by rabbits or livestock) or bare ground that is rich in invertebrate food (Green 1988, Batten *et al.* 1990). Invertebrates associated with the dung of livestock are an important part of their diet, and *Burhinus oedicnemus* requires bare, stony ground for nesting.

Sylvia undata also prefers open, treeless heathland and is probably the species most restricted to dwarf-shrub habitats. In lowland Atlantic heathland the species occurs mainly in areas dominated by continuous stands of heather 60–150 cm high, though the presence of *Ulex europaeus* is also highly important as it supports very high invertebrate densities. Being a sedentary species and wholly insectivorous, the bird is very susceptible to sub-zero temperatures in winter and therefore tends to be restricted to dry heaths in maritime regions.

A further requirement of considerable importance for several heathland species is that of habitat extent and continuity (Opdam and Helmrich 1984, Stortenbeker 1987). This is particularly the case for *Tetrao tetrix* and Crane *Grus grus*, individual pairs of which require areas of heathland in excess of 100 ha. The globally threatened Great Bustard *Otis tarda* formerly occurred on heathland in Europe (e.g. in the UK and Sweden), but has been extinct in this habitat since the nineteenth century, probably due to the widespread fragmentation of heathland (see 'Threats to the habitat, below). Many other species also require heathland blocks of more than 10 ha in extent. However, as described below, the majority of remaining heaths in most countries are now smaller than this. Consequently, most heathlands are no longer suitable for many priority species, and population declines have resulted.

THREATS TO THE HABITAT

The Habitat Working Group carried out a general review of the threats to lowland Atlantic heathland and to the habitat features required by its priority bird species (Table 3). In addition, the Group predicted the impact of each threat on the European population of each priority species over the next 20

years (Appendix 4, p. 401). A summary of these results is presented in Figures 4–5, and the threats are discussed individually below in order of importance (as shown in Figure 5).

Habitat fragmentation

Moore (1968) and Webb and Hopkins (1984) demonstrated that species richness is influenced by the area of a heathland site and the variety of habitat-types within it). Therefore fragmentation may impoverish remaining heathland disproportionately to the area lost. This has implications for site management, land acquisition and policy regarding heathland restoration.

Associated with the decline in heathland in Europe (see 'Habitat definition, location and extent', p. 187) has been considerable fragmentation (e.g. Webb and Haskins 1980, Gimingham *at al.* 1981). In the Netherlands, the vast majority of remaining heathland areas are less than 10 ha in size (Figure 3), and the decrease and extinction in the habitat of certain heathland birds has been blamed on such habitat fragmentation, among other threats (Opdam and Helmrich 1984). The fauna and flora of many of the smaller patches now frequently lack characteristic heathland species (Andrews and Rebane 1994) and the distances between remaining populations have increased (Webb 1989).

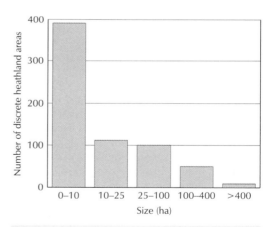

Figure 3. The size distribution of remaining areas of lowland Atlantic heathland in the Netherlands (after Anon. 1988).

Unless it is slowed, halted or reversed by broad conservation actions, the continuing fragmentation of heathland is predicted to remain widespread over the next 20 years, and will have an impact on 15 (94%) of the priority birds, including a critical impact on *Tetrao tetrix* and a high impact on *Lanius excubitor*. Fragmentation is therefore considered to

Table 3. Threats to lowland Atlantic heathland, and to the habitat features required by its priority bird species, in Europe (compiled by the Habitat Working Group). Threats are listed in order of importance (see Figure 5), and are described in more detail in the main text.

Threat	Effect on birds or habitats	Scale of threat		Regions affected, and comments
		Past	Future	
1. Habitat fragmentation (caused by many of the other threats below)	Habitat loss and deterioration reduce suitability of remaining heathland for priority birds; local extinction, less chance of recolonization	■	■	All areas
2. Abandonment of management	Increased vegetation height, especially grass sward; changed vegetation composition and structure and eventual succession to woodland	■	•	Spain; partially reversed in UK, Denmark, Netherlands, Belgium
3. Afforestation	A major historical cause of habitat loss; expensive to restore heath	●	•	UK and Denmark in past; Spain in future from eucalyptus plantations
4. Too-frequent or too-intense fires	Modifies plant succession; direct loss of wildlife	•	•	UK, France, Spain
5. Urbanization	Irreversible habitat loss	●	•	UK
6. Conversion to arable or improved grassland (including use of fertilizers and pesticides)	Habitat loss to arable or grassland, usually very difficult and expensive to reverse	●	•	UK, France and Denmark in past; Spain now and in future
7. Atmospheric nutrient pollution	Nutrient enrichment leading to change in vegetation composition, especially to grassland	●	●	All countries where soils are nitrogen-limited
8. Transport infrastructure	Irreversible habitat loss	●	•	UK and Denmark in past; UK and Netherlands in future
9. Recreation	Disturbance to wildlife and local changes to vegetation from bikes, horses, trampling, etc.	•	•	UK and Netherlands
10. Military activity	Disturbance to wildlife; damage to vegetation	•	•	Denmark, UK, Netherlands, Germany
11. Drainage and ground-water abstraction	Lower water-table; loss of open water; increased uniformity; succession to scrub woodland	●	●	All?
12. Mineral extraction (sand, gravel, clay, etc.)	Habitat loss when on large scale	•	•	UK

Scale of threat

Past	Past 20 years	■	Widespread (affects >10% of habitat)	•	Local (affects <1% of habitat)
Future	Next 20 years	●	Regional (affects 1–10% of habitat)	?	Considerable uncertainty

be one of the top two threats to heathland and its priority species in Europe, along with the neglect and abandonment of non-intensive, traditional management (see below).

Abandonment of management

A collapse in the agricultural use of remaining lowland heathland began at the start of the twentieth century and has continued to the present day. The enormous exploitation, and impoverishment of nutrients (in particular plant-available phosphorus), of heathlands in the past meant that, even in the absence of management, vegetation changes were slow. Today, however, vegetational change and succession to scrub and woodland is one of the greatest threats to heathland (RSPB 1993). Tree and shrub species,

such as oak *Quercus*, *Betula*, *Pinus*, *Juniperus* and the non-native *Rhododendron*, as well as the fern *Pteridium aquilinum*, can all invade heaths, especially when large expanses of heather become old through lack of management and open ground is available (e.g. Rosberg *et al.* 1981, Webb 1990). Once well-established, such plants can dominate and shade out heather and most other associated dwarf shrubs.

Moreover, grasses have also encroached on heathland, particularly in the Netherlands (Diemont 1996), where this has usually been attributed to reductions in the intensity of grazing, turf-cutting or other uses. However, increasing deposition of airborne nitrogen compounds may also be a cause (see 'Atmospheric nutrient pollution', below).

Scrub encroachment, due primarily to lack of

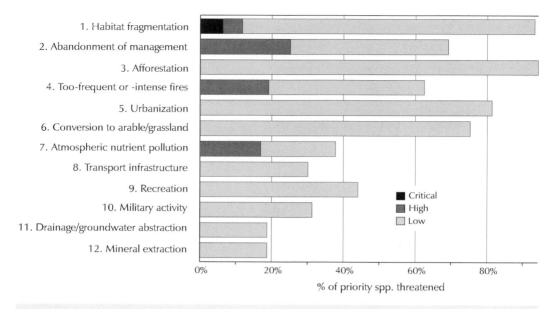

1. Habitat fragmentation
2. Abandonment of management
3. Afforestation
4. Too-frequent or -intense fires
5. Urbanization
6. Conversion to arable/grassland
7. Atmospheric nutrient pollution
8. Transport infrastructure
9. Recreation
10. Military activity
11. Drainage/groundwater abstraction
12. Mineral extraction

■ Critical
■ High
□ Low

% of priority spp. threatened

Figure 4. The predicted impact on priority bird species of threats to lowland Atlantic heathland, and to the habitat features required by those species, in Europe (as assessed by the Habitat Working Group). Bars are subdivided to show the proportion of species threatened by low, high or critical impacts to their populations (see Table 3 and Appendix 4, p. 401, for more details).

Critical impact	*High impact*	*Low impact*
The species is likely to go extinct in the habitat in Europe within 20 years if current trends continue	The species' population is likely to decline by more than 20% in the habitat in Europe within 20 years if current trends continue	Only likely to have local effects, and the species' population is not likely to decline by more than 20% in the habitat in Europe within 20 years if current trends continue

management, is now considered to be one of the most serious causes of heathland loss (Webb 1986, 1990, RSPB 1993). Introduced non-native shrub species such as *Rhododendron* can invade and completely shade out native vegetation, thus contributing to the succession of heathland towards woodland. Such vegetation is difficult and very expensive to clear. Overall, 11 (69%) of the priority birds are predicted to suffer population declines as a result of this threat (over the next 20 years), with the impact being high in four species.

Afforestation

Afforestation has probably claimed the largest area of European heathland over the last 100 years, notably in Sweden (Påhlsson 1994), Britain (Duffey 1976, Chadwick 1982) and northern Germany. Although most lowland heathland occupies land which formerly bore forest, its reforestation is sometimes difficult. However, since the introduction of deep ploughing, most plantations have been successful (Gimingham *et al.* 1981). Conifers have usually been planted, especially Scots pine *Pinus sylvestris* and Corsican

pine *P. nigra* var. *maritima* (Rodwell 1991).

In the UK, plantations established in the 1930s are now being harvested and restocked on a rotational basis, providing temporary open habitat for *Caprimulgus europaeus* and *Lullula arborea*, thus mitigating the effects of habitat loss to some extent. Restoration of heathland within plantations is now officially encouraged in several European countries, e.g. Norway (Påhlsson 1994), and in the UK tree-planting on or close to remaining heaths is officially discouraged and no longer grant-aided (Currie 1995). However, the Forestry Act in the Netherlands contains an obligation to replant after cutting woodland, including former heathland overtaken by woodland. This hampers re-establishment of heathland. Afforestation is currently not a problem in other European countries, but may yet affect Spanish heaths through the spread of *Eucalyptus* plantations.

Afforestation is one of the most important among the threats to heathland that are considered in this strategy, and is predicted to cause population declines in 15 (94%) of the priority birds, although impacts are likely to be low in all cases.

Uncontrolled, frequent or intense fires

If fires are too frequent on a site they can prevent the establishment of mature heather which is important for some of the rare heathland vertebrates including *Sylvia undata*. The scale and timing of burning is also important. The ability of heather to regenerate after fire diminishes with age, and *Ulex* regrowth is prevented by browsing on grazed sites. Intense fire can cause considerable damage to the soil if the litter layer or a peat substrate burns. The hot dry summers of 1975 and 1976 culminated in many extensive heathland fires in southern England (Cadbury 1989).

Uncontrolled fires can also precipitate serious management problems such as the spread of invasive plant species like *Pteridium* and *Betula* onto the heath (Gimingham *et al.* 1981), resulting in habitat loss to woodland and scrub.

Overall, such fires are of high relative importance among the threats considered here. They are predicted to have an impact on 10 (63%) of the priority species, including high impacts on *Burhinus oedicnemus*, *Saxicola torquata* and *Sylvia undata*.

Urbanization

The growth of urban and industrial areas represents a serious threat to heathland because of the irreversible nature of this form of habitat loss. In addition, it causes habitat fragmentation, and can also lead to increased impact on remaining heathland from recreational pressure. Development threats to heathland have been significant in the UK (Moore 1962, Webb 1990, RSPB 1993). Of the estimated 30 km² of heathland lost in the UK between 1960 and 1973, urbanization accounted for 27% (Rippey 1973), and demand for building land is likely to remain high during the next 20 years (CPRE 1995). As a result of these pressures across the European range of heathland, thirteen (81%) of the priority birds are predicted to suffer population declines, although impacts will in all cases be low.

Conversion to arable or improved grassland

Agricultural use of heaths for livestock grazing led initially to their expansion. However, improved agricultural techniques from the late eighteenth century brought about widespread conversion of heath to arable. In England, a quarter of heathland in Dorset had gone to arable by 1811, and agriculture was the greatest source of loss (38%) there between 1960 and 1973 (Rippey 1973), while the re-seeding in 1980 of Horton Common alone resulted in the loss of 5% of all *Sylvia undata* territories in Britain (Robins and Bibby 1985). In the UK, a large proportion of the heathland in eastern England was lost between 1930 and 1968 to agriculture (Chadwick 1982). Although

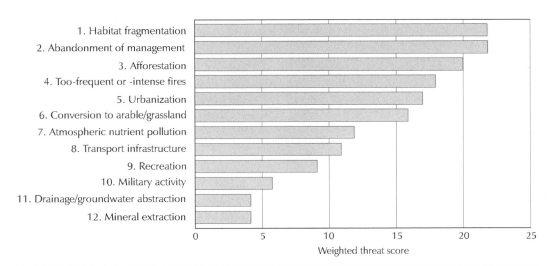

Figure 5. The most important threats to lowland Atlantic heathland, and to the habitat features required by its priority bird species, in Europe (as assessed by the Habitat Working Group). The scoring system (Appendix 4, p. 401) takes into account:

• The European Threat Status of each species and the importance of the habitat for the survival of its European population (Priorities A–D); see Table 2;

• The predicted impact (Critical, High, Low), within the habitat, of each threat on the European population of each species; see Figure 4, and Appendix 4 (p. 401).

See Table 3 for more details on each threat.

heathland soils can present considerable difficulties for cropping, i.e. lime and copper deficiencies and susceptibility to wind erosion, they have the advantage of being easy to cultivate so much heathland has been converted to arable (Hodge *et al.* 1984).

On more fertile soils, heathland can be transformed into grassland by intensive grazing or fire. Fertilizing and re-seeding can be done after burning off the existing heathland vegetation or after some form of cultivation. Conversion of heathland to grassland in one or other of these ways has taken place in many parts of the heathland region in Europe.

Following the application of intensive farming techniques, abandoned arable land on former heathland areas will not revert naturally to heathland, as either the seed-bank will have been eradicated or soil fertility will be too high. In most cases, steps will need to be taken to reduce fertility before dwarf-shrub vegetation can be re-established.

Overall, 12 (75%) of the priority birds are likely to suffer population declines as a result of this threat, although impacts will be low in all cases. While very important in the past, conversion of heath to arable or improved grassland is now considered to be of relatively moderate importance (over the next 20 years) among the threats considered here.

Atmospheric nutrient pollution

During the past few decades, large areas of heathland in Europe have become grass-dominated, and this is thought to be at least partly due to the increasing deposition of nitrogenous compounds from the atmosphere. Such compounds (mostly oxides and ammonia) are emitted mainly by vehicles, domestic heating and the power-generation industry, or result from intensive livestock rearing or high fertilizer inputs in agriculture.

High annual deposition levels of 40–80 kg/ha in the Netherlands have been associated with the loss of ferns, moss, algae and fungi from Dutch heaths and with the dramatic increase of grasses (de Smidt and van Ree 1991, Aerts and Heil 1993) which now dominate 50% of remaining Dutch heathland. The critical load of nitrogen that produces grass-dominance in heathland is estimated at 15–20 kg/ha/year (Rosberg *et al.* 1981, Power *et al.* 1995), and such levels are now widespread in some European heathland areas, e.g. the average nitrogen deposition level in Denmark is now about 20 kg/ha/year (Johansson 1994), and in southern UK 19 kg/ha/year.

However, the dominant effect of nitrogen on heathland vegetation in the Netherlands and elsewhere is challenged by some authors. The absence of regular traditional management has been considered to have much more serious effects on heathland, and the revival of traditional management practices such

as sod-cutting has been shown to restore the dominance of dwarf shrubs over grasses on Dutch heathland (Diemont 1996).

Overall, atmospheric nutrient deposition is a moderately important threat to heathland and its birds in a Europe-wide context. Unless corrective measures are taken, it is predicted to have an genuine impact on the European populations of six (38%) of the priority birds over the next 20 years, including high impacts on three species.

Transport infrastructure

Construction of roads and railways represents a moderately serious threat to heathland because of the irreversible nature of this form of habitat loss, as well as the resulting habitat fragmentation (see p. 194) and increased recreational pressure (see below) on remaining heathland. Examples are given under 'Urbanization' (above). Overall, eight (50%) of the priority species are likely to suffer low declines in their European populations as a result of habitat loss caused by the construction of transport infrastructure over the next 20 years.

Recreation

Many heaths are used for outdoor recreation and sports by the general public, e.g. walking, horse-riding, cycling, off-road driving. Where numbers of visitors are high, where the heathland area is small or where the activity is inappropriate, problems with recreation can arise. For instance, the estimated number of day-visitors to the New Forest (UK) increased from 3.5 million to at least 8 million per year between 1970 and 1987. Such increased public pressure can lead to locally excessive trampling and erosion, accidental fires, soil compaction and disturbance to livestock and wildlife (RSPB 1993). Heathland vegetation is about 10 times more susceptible to wear than adjacent acid and neutral grasslands (Harrison 1981; see Webb 1986). In particular, heather is a most sensitive plant and often dies as a result of trampling or vehicle pressure, leading to erosion and extensive bare areas (Webb 1986). The threat posed to heathland by recreation is moderately important in the Europe-wide context, and is predicted to cause low impacts to the European populations of seven (44%) of the priority bird species, over the next 20 years.

Military activity

The use of extensive areas of heathland for military training has both positive and negative implications. Due to restricted access, these training areas have been protected from infrastructural development and have not experienced recreational pressure. However, heavy military use can also lead to the destruction of vegetation and increased risk of erosion (Farrell 1993). Increasing use of some training areas heightens the need for a strategic approach to training that minimizes potential conflict with conservation interests. Overall, this and the following two threats to heathland are of relatively low importance among the threats considered here, at least on a Europe-wide basis over the next 20 years. Military activity is likely to affect five (31%) of the priority bird species, although all impacts on population size are predicted to be low.

Drainage and groundwater abstraction

Falling water-tables, due to drainage and to excessive abstraction of groundwater supplies for domestic and agricultural use, threaten some priority bird species of wet heathland, at least in the UK (Westerhoff 1992) and the Netherlands (Prins 1993). Such drainage is generally aimed at agricultural improvement of the grazing quality of wet-heath vegetation. Falling water-tables lead to an increase in nitrogen in the soil, and thus favours more vigorous plant species of drier conditions. The resulting changes in plant species composition, and thus in the overall structure of the vegetation, reduce the suitability of the habitat for priority bird species.

Mineral extraction

The extraction of sand and gravel from heathland has been undertaken for centuries but usually on a very small scale, and often with a positive, diversifying effect on the habitat. Given the current fragmented state of heathland in Europe, however, large-scale mineral extraction is now a potentially serious threat to heathland in some areas, e.g. south-west England (RSPB 1993).

CONSERVATION OPPORTUNITIES

International legislation

A summary of the international environmental legislative instruments that may be used to conserve European habitats and their birds is presented under 'Opportunities for Conserving the Wider Environment' (p. 24). Of those particularly relevant to the conservation of lowland Atlantic heathland (Box 1), the Birds and Habitats Directives provide the opportunity for member states of the European Union (EU) to protect areas of threatened habitat or sites with important populations of threatened species. Eleven of the 16 priority birds of lowland Atlantic heathland are listed on Annex I of the Birds Directive at the species or subspecies level (see Appendix 1, p. 327) and should therefore be subject to special conservation measures in EU member states. However, to date, the designation of Special Protection Areas (SPAs) under the Birds Directive has been slow in Europe (Anon. 1996). Heathland is also likely to be particularly poorly protected as this habitat is often regarded as being 'waste' or derelict land of little value, although this attitude is changing as the habitat gains more status in the planning systems of some countries.

The Habitats Directive aims to combine the pro-

Box 1. International environmental legislation of particular relevance to the conservation of lowland Atlantic heathland in Europe. See 'Opportunities for Conserving the Wider Environment' (p. 24) for details.

GLOBAL

- *Biodiversity Convention* (1992) Convention on Biological Diversity.

- *Bonn Convention* or *CMS* (1979) Convention on the Conservation of Migratory Species of Wild Animals.

- *Ramsar Convention* (1971) Convention on Wetlands of International Importance especially as Waterfowl Habitat.

- *World Heritage Convention* (1972) Convention concerning the Protection of the World Cultural and Natural Heritage.

- Convention on Long Range Transboundary Air Pollution (1979).

PAN-EUROPEAN

- *Bern Convention* (1979) Convention on the Conservation of European Wildlife and Natural Habitats.

- *Espoo Convention* (1991, but yet to come into force) Convention on Environmental Impact Assessment in a Transboundary Context.

EUROPEAN UNION

- *Birds Directive* Council Directive on the conservation of wild birds (79/409/EEC).

- *Habitats Directive* Council Directive on the conservation of natural habitats of wild fauna and flora (92/43/EEC).

- *EIA Directive* Council Directive on the assessment of the effects of certain public and private projects on the environment (85/337/EEC).

- Council Directive on fire control (86/3259/EEC).

- Council Directive on the freedom of access to information on the environment (90/313/EEC).

- Council Regulations on the Structural Funds (e.g. 2081/93/EEC).

- Council Regulation on the prevention of forest fires (2158/92/EEC).

tection of threatened species with the wider objective of conserving habitats that are of conservation concern in their own right. Annex I of the Directive lists these particular habitat-types and includes four which fall within the definition of lowland Atlantic heathland as used here (see Appendix 5, p. 423). Two of these are 'priority' habitat-types, which requires that they be given higher protection status.

All European states which hold lowland Atlantic heathland have ratified the Convention on Biological Diversity, and their environment ministers have approved the Pan-European Biological and Landscape Diversity Strategy (PEBLanDS). National biodiversity conservation strategies or action plans have also been developed by some of the relevant countries, e.g. the Natuurbeleidsplan (Nature Policy Plan) adopted by the Netherlands parliament in 1990, and these promote targeted actions to conserve lowland Atlantic heathland, e.g. 60 km² of heathland are to be re-established in the UK by 2005 (Anon. 1995b).

The EIA Directive of the EU (see Box 1) provides the opportunity for member states to prevent or limit damaging impacts to heathland from major developments, such as roads, railways, housing and industrial developments. However, forestry and agricultural activities, that pose great threats to heathland (see 'Threats to the habitat', p. 194), are not covered by the Directive. Similarly, national planning laws do not usually control such activities, although they can be effective in controlling potentially damaging building developments for roads, housing, tourism, etc.

As described in the previous section, too-frequent or too-intense fires pose a serious threat to some priority species of heathland, as well as being dangerous to human life and settlements. The EU member states are therefore implementing a Regulation (2158/92) on the prevention of wildfires in fire-prone habitats such as heathland and forest (Stanners and Bourdeau 1995); further details are given under 'Mediterranean Forest, Shrubland and Rocky Habitats' (p. 239).

Policy initiatives

In addition to legislative instruments such as site-protection measures, the conservation of remaining heathland is also highly dependent on land-use policy.

A strategy for influencing land-use across a habitat needs to consider local, regional, national and international policies for influencing these factors. What is needed is a mix of policies integrated by common and clear objectives, including those outlined in this strategy (see 'Conservation recommendations', below).

Clearly, forestry policies are of prime importance, and opportunities for influencing these are dealt with under 'Boreal and Temperate Forests'

(p. 203). Agricultural policies are also of considerable importance for the conservation of heathland, both as a result of their direct effects on livestock grazing and the demand for agricultural land, and as a result of their indirect effects through impacts from adjacent agricultural habitats (e.g. excessive water abstraction, atmospheric nutrient pollution). The following agricultural policy areas are particularly important for promoting nature conservation in lowland Atlantic heathland, and are discussed in detail under 'Agricultural and Grassland Habitats', (p. 267):

- Fundamental objectives of rural policies.
- Overall levels and forms of subsidy.
- Economic instruments (e.g. Agri-environment Regulation 2078/92).
- Information, education and advisory services.

CONSERVATION RECOMMENDATIONS
Broad conservation objectives

1. To protect and maintain all remaining lowland Atlantic heathland in Europe.

2. To arrest and reverse the deterioration of existing lowland Atlantic heathland by the introduction or improvement of management.

3. To encourage the restoration and re-establishment of lowland Atlantic heathland, particularly where this reverses the detrimental effects of habitat fragmentation.

Ecological targets for the habitat

In order to maintain and enhance the populations of priority bird species of lowland Atlantic heathland, it is important not just to maintain the extent of heathland but also to retain or enhance important habitat features according to the requirements of priority species. It is therefore necessary to ensure that the following conditions are met.

1. An optimal mixture (according to local conditions) of dwarf shrubs, bare ground, small wetlands (e.g. valley mires) and scattered scrub should be maintained in the landscape through management of natural succession processes by grazing, turf-cutting, controlled fire, and manual or mechanical removal of shrubs and trees.

2. Existing areas of lowland Atlantic heathland should be maintained and where possible re-established. In particular, the loss or fragmentation of heathland due to urban or transport development, or conversion to forestry plantations or intensive agriculture, should be avoided. Associated wetlands such as valley mires are also important and should be restored where they have been damaged by drainage.

3. The structure and composition of vegetation in heathland should be maintained according to the requirements of priority species through the appropriate management of grazing, the removal of turf, of litter-layer or of vegetation, and through the controlled use of fire. A combination of these measures may be required, depending on local circumstances.

Grazing levels should be maintained that prevent the succession of vegetation to scrub woodland, where possible creating a mosaic of low and open areas and other areas of taller and denser shrub vegetation.

Controlled fire or the physical cutting of vegetation can be used to create early, open successional stages, to increase habitat diversity and to create suitable conditions for the subsequent use of grazing animals. The use of fire and mechanical means of controlling vegetation should be carried out with great care and with due regard for other important and threatened species of flora and fauna, such as reptiles. Consequently these measures should not be implemented over discrete areas larger than 1 ha.

Consideration should be given in all management processes to ensure that nutrients do not build up in heathlands, and that they are, in most cases, actively reduced.

4. Where appropriate, planted and secondary forests in heathland should be managed so as to re-establish heathland on suitable soils. The encroachment of trees onto adjacent heaths should be controlled. The potential use of forest habitats by priority bird species (and other wildlife) that favour the forest/heathland ecotone should be maximized by appropriate forest management.

5. Disturbance of human-sensitive priority bird species should be minimized, especially during critical periods of the annual cycle (e.g. during and before the egg-laying and incubation periods).

6. Emissions of atmospheric pollutants should be reduced to below critical levels and loads, to ensure that vegetation and soils on heathlands are not affected by atmospheric deposition of nutrients and other pollutants.

7. The introduction of non-native, invasive plant and animal species should be avoided and, where this has already occurred, programmes of management to control or remove introduced species should be instigated where feasible.

Broad policies and actions

To meet the above objectives and habitat targets, in accordance with the relevant strategic principles (p. 24), the following broad measures must be pursued (see p. 22 in 'Introduction' for explanation of derivation):

General

1. The importance of lowland Atlantic heathland for nature conservation, including birds, the role of traditional management practices, and the need for continued management (particularly scrub clearance), should all be widely promoted.

2. Data on the extent and condition of existing lowland Atlantic heathland is fragmented and incomplete. Therefore, an inventory of this important resource should be completed to agreed definitions and common standards.

3. Data on the distribution and current land-use of former areas of lowland Atlantic heathland should be collated and assessed to guide any efforts to re-establish this habitat towards those areas where they would be most successful and effective.

4. The integration of heathland conservation into broader landscape management, including agriculture and forestry, should be explored.

Legislation

5. All heathland sites that are internationally important for priority birds (Important Bird Areas) and representative or priority areas of lowland Atlantic heathland should be protected under appropriate national legislation, and within the EU as Special Protection Areas (SPAs) and Special Areas for Conservation (SACs) under the Birds and Habitats Directives.

Within the EU, the LIFE fund should be extensively used for the conservation and management of SPAs and of sites listed for designation as SACs that contain lowland Atlantic heathland.

6. Strategic Environmental Assessments (SEAs) should be carried out on national and EU policies, plans and programmes for agriculture, forestry, transport, housing and mineral extraction. Environmental Impact Assessments (EIAs) should be made mandatory for major proposed agricultural, forestry or development projects under a revised EIA Directive.

7. National governments should encourage re-establishment of lowland Atlantic heath on suitable sites as a condition of after-use following mineral extraction, via the relevant planning system.

Agriculture

8. Policies for the livestock sector (notably beef and sheep) should be revised to encourage the non-intensive grazing of lowland Atlantic heathland while taking care not to encourage overgrazing.

9. Sufficient funding should be allocated to the Agri-environment Regulation so that, by the year 2000, 15% of the EU is covered by management agreements. Schemes should be targeted to regions of marginal agriculture with high importance for priority bird species (and other wildlife) of lowland Atlantic heathland, should provide support for non-intensive grazing, and should have clear environmental objectives. The habitat targets outlined above should be incorporated as appropriate, and schemes should be monitored to assess their success in achieving their objectives.

10. National and EU legislation on, e.g. land-tenure, common land, and water management, should be revised in order to implement new agri-environmental policies.

11. Further research is required to refine techniques for the re-establishment of heathland on land that has been converted to arable or intensive pasture. Information on the effectiveness of restoration measures should be widely disseminated to heathland managers and advice should be given on the most appropriate management techniques. This is particularly relevant where intensively used agricultural land is taken out of production.

12. National governments and the EU should develop mechanisms for targeted long-term diversion of land from forestry and intensive agriculture towards the re-establishment of lowland Atlantic heathland.

13. Where appropriate, heathland should be incorporated in multi-functional agricultural management.

Forestry

14. Financial mechanisms should be developed at national and EU levels to assist against scrub encroachment on lowland Atlantic heathland.

15. National and EU forestry policies should incorporate environmental objectives, including the re-establishment of lowland Atlantic heathland, among other non-forest habitats.

16. National and EU forestry policies should promote the development of integrated heathland and forestry management plans for all significant sites within the range of lowland Atlantic heathland.

Recreation and tourism

17. Publicity materials should be widely distributed informing visitors and heathland users of the wildlife importance of lowland Atlantic heathland and the potential threats to these areas. Advice should be given on how to minimize the risk of disturbance, pollution, erosion and accidental fires.

18. Heathland should be promoted as a major European cultural landscape.

19. The ecological, historical, archaeological and agricultural significance of lowland Atlantic heathland should be defined for appropriate heathland regions.

Military training

20. Integrated site-management plans, that include clear nature-conservation objectives in line with national biodiversity strategies or action plans, should be drawn up for all military training land.

Water

21. Land management within heathland regions should recognize the importance of heathland for water-filtration and the value of heathland re-establishment in the conservation of groundwater resources.

Other measures

22. Strategies for eliminating, dramatically reducing, or mitigating the effects of, atmospheric and agricultural pollution on nutrient-sensitive habitats, such as lowland Atlantic heathland, should be developed. These should include effective pollution-control measures as well as management operations to reduce the levels of key nutrients on heaths.

ACKNOWLEDGEMENTS

Most contributions to this chapter were made by the Habitat Working Group during and after a workshop organized by L. Påhlsson and hosted by Miljöenheten, Malmö, Sweden (18–20 May 1994); these contributors are listed on p. 187.

In addition, valuable information and comments were received from I. J. Øien (Norsk Ornitologisk Forening), J. A. Kålås (Norsk Institutt for Naturforskning), E. Osieck and J. Winkelman (Vogelbescherming Nederland).

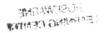

BOREAL AND TEMPERATE FORESTS

compiled by

M. REBANE, Z. WALICZKY AND R. TURNER

from major contributions supplied by the Habitat Working Group

P. Angelstam, V. Anufriev, I. P. Bainbridge, O. Biber (Coordinator), R. J. Fuller, L. Tomialojc, T. Wesolowski

and by

J. Frühauf, C. Heinrich, C. Mayr and M. Strazds

SUMMARY

The overall area of forest in Europe has been increasing since the 1920s, due to widespread afforestation of open land. Forest now covers about 30% of Europe, but most of this consists of intensively managed plantations, often monocultures of conifers or of non-native tree species, and fragmented patches of semi-natural forest. Only a few significant tracts of natural old-growth forest are left on the continent, in northern and eastern Europe.

European forests in the boreal and temperate zones hold a rich diversity of birds, of which 114 are priority species for the habitat. More than 40% of these priority birds have an Unfavourable Conservation Status in Europe, due to their European populations being notably small or declining. The remaining priority species are forest specialists which are highly dependent on the continuing existence of these habitats. Most of these species are currently still widespread across the continent, although many are not common and require habitat features that are only found in semi-natural or old-growth forests. A significant proportion are also naturally restricted within Europe to certain regions or to particular types of native forest.

In northern and eastern Europe, the remaining areas of natural old-growth forest are under considerable threat from logging, especially in Russia. Elsewhere, degradation of the ecological quality of forests is widespread, due mainly to the intensification of forestry management, expansion of plantations at the expense of semi-natural forest, overgrazing by wild herbivores, high levels of disturbance from human activities, and continued fragmentation through road-building and other developments.

All remaining areas in Europe of ancient semi-natural and, especially, natural forest must be strictly protected. In other forested areas, broad measures are required to enhance habitat quality. Currently, the most important opportunity to address these wide-ranging problems lies in the national implementation by governments of the 'Helsinki Guidelines for Sustainable Forest Management'. No international conventions are available that specifically encourage, support, coordinate and monitor national activities for the conservation of forests and their biodiversity, and there are currently no shared policies on the management and use of forests at the supranational level, e.g. across the European Union (EU).

Coordination between governments and major forestry organizations is therefore needed in order to agree on criteria for sustainable forest management, so as to promote trade in 'environmentally friendly' forest products. Governments need to balance guidance, incentives and regulations in order to promote sustainable forest management nationally, so as to change inappropriate forestry practices and to control other man-made problems such as overgrazing, wildfires and recreational pressure.

HABITAT DESCRIPTION

Habitat definition and subdivisions

This strategy covers all forests in the boreal and temperate zones of Europe, ranging from primeval old-growth to secondary plantations of native and non-native trees. The word 'forest' is used for all stages of growth and all systems of management, thus the word 'woodland' has no special meaning in this strategy, e.g. as a younger or structurally less complex version of forest. Woodlots in agricultural landscapes are also included in this habitat, but orchards, pastoral woodland (e.g. 'dehesas'), linear wooded features (e.g. hedgerows and narrow shelterbelts) and land with scattered trees (parkland) are all covered under 'Agricultural and Grassland Habitats' (p. 267). All forest vegetation within the Mediterranean zone (apart from montane forests that are dominated by boreal and temperate tree species) is covered under 'Mediterranean Forest, Shrubland and Rocky Habitats' (p. 239).

Within this broad habitat definition, the following four types of forest are recognized, all of which include plantations of native and non-native tree species (see 'Ecological characteristics' (p. 206) for a more detailed description of these forest-types). Recent plantations are differentiated from natural and semi-natural forests in the text wherever necessary and possible.

Boreal forests

Also known collectively as the taiga, these are conifer-dominated forests lying mainly between the latitudes of 55°N (hemi-boreal forest) and 70°N (forest-tundra), and extending in Europe from Scotland (UK) in the west to the Ural mountains (Russia) in the east. Only a few broadleaved deciduous tree species occur. Associated, more open, habitats in the boreal zone, such as mires and grasslands, as well as isolated patches of stunted woodland or scrub in the tundra zone, are treated under 'Tundra, Mires and Moorland' (p. 159).

Lowland temperate forests

The temperate zone lies between the boreal zone and the Mediterranean zone (see 'Mediterranean Forest, Shrubland and Rocky Habitats', p. 239), roughly between 55–60°N and 45°N. These forests lie below montane forests (approximately below 500–1,200 m), and usually on well-drained mineral soils. In general, they are dominated by broadleaved deciduous trees or are 'mixed' (broadleaved and coniferous). The proportion of broadleaved evergreen trees is low compared to Mediterranean forests (see p. 239). The transition between the boreal and temperate forest zones forms an extensive belt, the so-called hemi-boreal zone.

Montane forests

Montane forests occur in all the major mountain areas of Europe, across which they are all broadly similar despite their current separation by other intervening habitats. The lower limit of these forests occurs between 500 and 1,200 m, and the upper limit between 800 and 2,600 m, varying with latitude, aspect, exposure and climate. These forests are usually dominated by broadleaved species at the lower elevations and by conifers at higher altitudes. Forests at higher altitudes in the boreal zone are treated as boreal forests.

Riverine forests

Also called alluvial or riparian forests, these develop where there is regular natural flooding or permanent waterlogging, and include forests on alluvial floodplains along streams, rivers and lake-shores, as well as swampy groundwater forest, which develops in depressions with a permanently high water-table or where abundant springs occur. Remaining natural examples of this forest-type are now very rare in Europe, however. Riverine forests in the boreal zone are treated as boreal forests.

Distribution and extent

Currently, forests (of all types) cover about 3.1 million km² in Europe (one-third of the land), of which 1.7 million km² are located in European Russia (Stanners and Bourdeau 1995). Forest cover ranges nationally from 6% in the Republic of Ireland up to 66% in Finland (Figure 1). Forest cover in Europe has greatly increased over the 1960–1990 period, with nearly all European countries registering net growth in forest extent, although no data are available for some very important regions such as European Russia, Ukraine and former Yugoslavia (CEC 1995d). Particularly large increases occurred in Spain, France, Fennoscandia and the Baltic states, while Albania was the only country in Europe to register a strong decrease.

However, these figures conceal the almost complete replacement of ecologically high-quality virgin or old-growth forest by intensively managed forests or plantations in western Europe, and a rapid and continuing replacement in eastern Europe. The only virgin forests (i.e. forests never modified by man) left today in western Europe account for 0.24% of the current forest area (Ibero 1994). This amounts to about 0.1% of the original forest cover (before the arrival of man), believed to have been 80–90% of the region (Ibero 1994). If old-growth forests (i.e. forests that are old enough to have attained a natural variation in species and age-ranges and trees that have reached their natural lifespan) are added to the area of virgin forests, the total area covered is 1.9% of the current total forest area of western Europe, or about 0.8% of the original area (Ibero 1994). Virgin and old-growth forest becomes more extensive as one proceeds east and north across Europe, such that in the far north-east, in the Komi Republic (Russia), about 100,000 km² of ancient boreal forest still remains. In general, however, the loss of virgin and old-growth forest has been almost as great in eastern Europe as in the west for riverine, montane and lowland temperate forest, although data are, as yet, less available. Accurate data are also generally lacking throughout Europe on the extent of afforestation with monoculture stands, and on the extent and rate

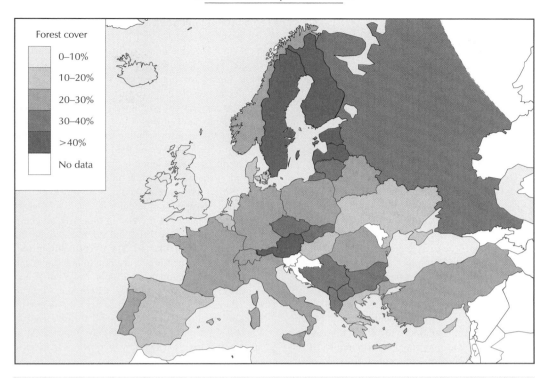

Forest cover

- 0–10%
- 10–20%
- 20–30%
- 30–40%
- >40%
- No data

Figure 1. Overall forest cover in Europe, including plantations (adapted from CEC 1995d, Stanners and Bourdeau 1995).

of planting of non-native tree species.

Comparable national data are not available on the extent of each of the four forest-types covered here. In general, boreal forests cover most of Fennoscandia (except the southern part) and northern Russia, north of 55–60°N, with an area of about 868,000 km² (excluding Russia) (Iremonger *et al.* in press).

Montane forests are most extensive in the Pyrenees, the Alps, the Carpathians, mountain ranges in the Balkans and Turkey, and in the Caucasus, but quantitative data on their extent are not readily available. As a result of their important role in protecting against erosion, flooding and avalanches, the rate of loss of montane forests is probably less than for other European forest-types.

In Europe, excluding Russia, lowland temperate forest covers about 780,000 km² (Iremonger *et al.* in press), although nearly all such forest is highly man-modified and fragmented.

Finally, riverine forest is widespread all over Europe, but is restricted to narrow strips along major rivers and tributaries, and covers about 20,000 km² (Iremonger *et al.* in press), excluding Russia. Most of this extent is plantations, since nearly all of the original riverine forest has been destroyed, especially along the major European rivers (i.e. Danube,

Po, Rhine, Rhône, Dnepr, Dnestr, Volga). Such forests once fringed all European rivers as widely as the annual inundations defined the flood-plain. However, since flood-plains have alluvial soils of high productivity, most have been converted to agriculture. For example, natural riverine forest used to cover about 2,000 km² along the Rhine. Nowadays, it is highly fragmented and covers a total of 150 km², of which less than 1.5 km² is still even semi-natural. Similarly, only several hundred km² remain along the lower Volga. Good examples of natural flood-plain forest still remain along the Mura–Drava–Sava river complex in Slovenia, Croatia and Hungary, and along the Wisla, Warta and Narew rivers in Poland and north-west Russia.

Origin and history of the habitat

The development of forests in Europe, in the face of repeated glacial periods, is complex and still not fully understood. Further information can be found in Huntley and Birks (1983) and Huntley (1990).

During the Neolithic Age in parts of Europe, areas of forest began to be cleared, probably with the aid of fire, for cultivation. Re-establishment of woodland was inhibited by livestock-grazing. Large-scale forest clearance by humans in Europe started during the Bronze Age, about 5,000 years ago (Behre 1988, Huntley 1988). Grazing and repeated burning of forest, together with leaching of exposed soils by rainfall, created extensive heathlands in north-west Europe. Certain historical periods are associated with particular surges in forest clearance, including the Roman and medieval eras. After major outbreaks of war, disease or famine, when human populations dropped significantly, overall forest area in Europe is likely to have increased temporarily at the expense of arable and grazing land.

The structure and composition of remaining forest was also radically altered by woodland pasturing, which was widespread until about the nineteenth century. Tree-species composition was largely determined by how different species were exploited. Species of oak *Quercus* were encouraged for their valued timber and to produce acorns for pigs. Beech *Fagus sylvatica* was also promoted due to its production of mast. Hornbeam *Carpinus*, elm *Ulmus* and ash *Fraxinus* were lopped, i.e. the branchlets and twigs were cut and dried for winter fodder. Grazing was a major factor in modifying forest composition and structure, and may have contributed to the decline of lime *Tilia* in European forests. Pasturing gradually advanced deeper into the interior of remaining woodland (usually on steeper slopes or poorer soils) as the human population increased.

With the development of small-scale industries, the demand for timber and fuel increased. Logging greatly reduced the abundance of *Abies* in montane forests in the Alps (Kral 1994, Mayer 1994b). Large tracts of forest were cut down or were put under coppice management for charcoal burning. Coppiced forest was also an important provider of poles and stakes for fences, orchards, vineyards and household constructions (Rackham 1990). Coppicing produced a further marked change in the species composition by favouring trees that sprout well (e.g. maple *Acer*, *Carpinus*, *Fraxinus*, *Quercus* and hazel *Corylus*) while reducing the extent of *Fagus*, spruce *Picea* and pine *Pinus*.

The consequences of deforestation were already causing concern in the Middle Ages and in some areas regulations were introduced to control the use of forests. However, this did little to prevent the misuse and loss of forest, which culminated during the eighteenth century. Towards the end of this century and during the nineteenth century, a number of changes took place to relieve the human pressure on woodland. Livestock management became more intensive and higher productivity was achievable without the use of woodland to provide acorns or pasturage. Wood was replaced by coal as a major fuel, by stone and other products for house-building, and by iron and steel for engineering works.

Reforestation began in earnest, with conifers being preferred and planted for their ability to grow more quickly and more successfully on poor degraded soils than most deciduous trees. In other areas, high-forest management was reintroduced and the decline of the coppice system began. The trend towards conifers has continued in the twentieth century but has slowed down in recent decades for a number of reasons, including an increasing public concern over land-use and the environment.

Despite these tendencies, overall forest cover in Europe continued to decline until the twentieth century. Since the Second World War, however, this trend has been reversed and during the last 30 years forest cover has been increasing in nearly all countries. However, very little indeed remains of the primary forests of Europe. The overwhelming majority of forests in Europe are now plantations, and most of the remainder are highly fragmented and degraded semi-natural forests, where very little of the original structure and diversity of vegetation remains. Historical data on overall forest extent in Europe are not available, but Figure 2 indicates, as an example, trends at a variety of locations in Britain.

Ecological characteristics

Key abiotic factors affecting the habitat

The four boreal and temperate forest-types occupy a huge range of climatic, soil and hydrological conditions, from waterlogged riverine forest on nutrient-rich alluvial soils through to montane forest on very thin and infertile bedrock-soil. Geographical location in Europe (both north–south and east–west) and altitude are important in determining forest-type, as well as local factors such as aspect and slope. The acidity (pH) and moisture content of the soil are critical in determining which tree species can grow where (Fuller 1995). Apart from these general conditions, forest structure and composition can be significantly affected by a number of processes which are summarized below.

Fire

Fire is a natural element in most forest ecosystems, and is particularly important in the natural dynamics of boreal forests as conifers burn more readily than broadleaved trees (Angelstam 1996). Under natural conditions, wild-fires ignited by lightning occur approximately once in every 50 years on dry sandy and gravel sites in boreal forest, once in every 100–150 years on moist sites, and only once per 200 or more years on the moistest sites with richest soils. Fire can act to open up the canopy for regeneration, releasing nutrients to encourage plant growth, and sometimes playing an essential role in the germination of seeds,

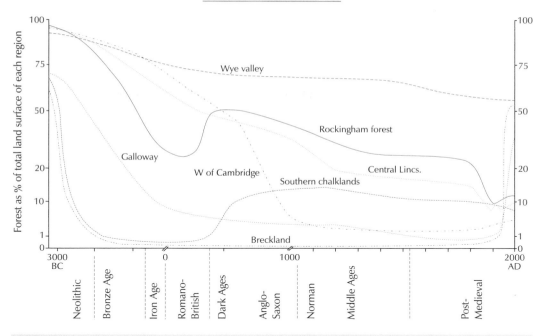

Figure 2. A reconstruction of the changes in the extent of forest cover in seven regions of Britain over the last 5,000 years (adapted from Peterken 1993), demonstrating the large differences that can exist over relatively small distances in Europe.

thereby helping to maintain the rotation between pioneer and climax stands in any one place, as well as the fertility of sites and the biodiversity of forests (Bradshaw and Hannon 1992, Dudley 1992, Kuusela 1994). On average, in a large region, 0.5–2.0% of the forest area is affected by fires annually (Bonan and Shugart 1989).

In large-scale studies along river valleys in Swedish boreal forest, the fire-disturbance intervals were 40–160 years, depending on vegetation-type and topography (Kohh 1975, Zackrisson 1977a,b). In a naturally dynamic landscape, fire frequencies depend on three major factors (Zackrisson 1977a,b). First, the intimate relationship between soil-type, soil moisture and the consequent flammability of vegetation affects the probability of fire (Schimmel 1993). Wet ground burns very rarely and may be considered as a fire refugium (Segerström *et al.* 1996). Second, fire frequencies vary according to landscape features and topography. Convex features burn more frequently than concave (Zackrisson 1977b), and south-facing slopes more than north-facing slopes (Högbom 1934). Finally, increasing air humidity reduces fire frequencies to the north (Granström 1991), and with increasing altitude (Zackrisson 1977b). Similarly, local 'continentality' (i.e. location near the centre of a large landmass) increases fire frequencies (Granström 1991).

Fire occasionally plays a role in dry temperate coniferous forests. Compared with the boreal zone, where fire is the main cause of natural turnover in forests, the natural patchiness of temperate forests is relatively fine-grained because tree-fall gaps are rarely larger than 10 ha. Fire is more of a regular occurrence in montane forests, although it is often on a small scale, and is commonest in the coniferous (higher) zones.

Flooding

An important driving factor in the natural dynamics of riverine forest is the occurrence of regular floods. On natural flood-plains, or between embankments in the case of regulated rivers, areas are inundated during periodic floods, which usually occur at least once in early spring after the snow melts (in central Europe this is usually followed by a second flood in early summer caused by heavy rains). Only a few tree species can survive longer periods of inundation, including species of poplar *Populus*, willow *Salix* and alder *Alnus*. Flooding therefore affects the development of forest and determines forest-type (i.e. riverine or otherwise), for example high-diversity *Quercus–Ulmus–Fraxinus* woodland can only develop on higher ground on flood-plains. While regular floods are therefore necessary for the maintenance of riverine forest, exceptionally large floods can cause patchy damage and disturbance to the habitat (like fires, storms, etc.).

Storms

Very strong and sustained winds can damage large areas of forest, in patches as large as 15 ha in primeval lowland forest (Tomialojc and Wesolowski 1994). Tornadoes occur very rarely in Europe, but the passing vortex can create an elongated belt of broken and uprooted trees of some 50–100 m width. The trees most vulnerable to strong winds are those on steep or exposed slopes or with shallow roots on loose or muddy soils, thus montane and riverine forests tend to be most affected.

Like fire, storm-damage tends to increase a forest's structural complexity, and is never uniform. However, storms differ from fire in that smaller areas are affected, shrubs typically survive, and more dead wood is generated, both as fallen and standing trees. The most important difference between storms and logging is that fallen trees remain on the spot after the former, while during the latter most tree and bush layers are removed. As storms effectively eliminate the high canopy and create a gap in the forest, the damage benefits mostly forest bird species which typically favour forest-edges, the shrub-layer or spatial heterogeneity (e.g. Hazel Grouse *Bonasa bonasia*, Tree Pipit *Anthus trivialis*, Robin *Erithacus rubecula*, Dunnock *Prunella modularis*, Wren *Troglodytes troglodytes*). However, the overall importance of winds and storms in shaping forests is relatively low (Angelstam 1996), and strongly inhibits forest growth only on some maritime headlands.

Snow

In high mountain areas, avalanches can destroy the upper levels of montane forests, while heavy snowfall can occasionally damage trees at all elevations, although mainly in boreal forest. Avalanches remove all trees and shrubs in their way, while the weight of heavy snowfall can break major branches, restructuring the crown and occasionally leading to the death of the tree (e.g. Scherzinger 1995). However, the overall impact of snow on the structure of European forests is relatively minor. Drought, caused by continuous winter frost, sets the upper altitudinal limit to many montane forests.

Landslides

These can occur on steep slopes after heavy rains, especially if the area has been partially deforested. Landslides only occur over a very limited area and at low frequency in Europe, and are generally of minor importance is shaping forest structure. Rockfall, through damaging and killing larger trees, also affects montane-forest distribution locally.

Key biotic factors and natural dynamics

Succession

The climax vegetation over much of boreal and temperate Europe is forest, typically broadleaved or mixed forests in lowland Europe, with conifers becoming more dominant with increasing altitude or latitude, since they are better adapted to withstand the extremes of climate. Forest was only absent from ground that was permanently waterlogged, too exposed, too dry, or naturally burnt, grazed or browsed too frequently or intensely. The most important contemporary factor that prevents the development of mature forest is human intervention.

In the absence of such intervention, all forests would go through long cycles of maturation, death and regeneration. These natural cycles occur at scales ranging from the small gaps created by the demise of individual trees up to fire-damage extending over many hectares. In such 'opened up' or disturbed habitats, initial colonization by grassland and heathland is gradually followed by the development of scrub, and finally young, then 'mature' followed by so-called 'overmature' forest (the latter being a forestry term equivalent to biologically mature forest).

The structure and tree-species composition of natural forests is thus highly variable, due to past events such as fires creating a patchwork of succession, along with continuous local variations in topography, climate, hydrology and soil. In natural climax forests there are usually many 'layers' of canopy and sub-canopy, together with a high diversity of tree species. In managed forests, one stage of forest growth gives way to another over time in a non-patchy fashion that is uniform over a much larger scale than in natural forest, and the tree-species composition remains limited and more or less constant. There is little variation in structure, and perhaps most importantly, very little biologically mature or old-growth timber.

On freshly disturbed sites, pioneer tree species such as larch *Larix*, birch *Betula*, *Pinus*, *Salix*, *Alnus*, rowan *Sorbus aucuparia* and aspen *Populus tremula* dominate in early and mid-successional stages, which are currently common in European forests as a consequence of man-made fire, logging, and the relaxation of grazing and cultivation of formerly open areas. Hence, in the boreal zone the forest landscape is often not always covered by forest in the sense that there is always a closed canopy of large trees, and on some sites the early, deciduous phase of the successional cycle may be kept almost permanent by intermittent fires (the coniferous Norway spruce *Picea abies* would normally dominate mature boreal forest).

Only plant species that are able to withstand the particular set of conditions there can grow on riverine flood-plains. They are specialized and often cannot compete with other tree species which form stable communities, e.g. *Fagus* or *Carpinus* forests. The outer limits of riverine forests are thus usually set by competition with other tree species. On their inner limits, their occurrence close to the river itself is prevented by mechanical action of flood-water and the slow but near-constant shifting of the riverbed, which impedes the development of tree communities. Thus, between the river and the forest there is usually a narrow belt of bushy riparian habitat that is kept permanently in the early stages of succession.

Browsing by wild herbivores

Regeneration of trees and shrubs is influenced by the presence and abundance of herbivores and the browsing pressures exerted by them. At naturally regulated levels, wild herbivores such as wild boar *Sus scrofa* and red deer *Cervus elaphus* may have created optimum opportunities for regeneration through ground disturbance, and created a rich mosaic of open ground and scrub within forests. However, increasing numbers of wild herbivores in many European forests are leading to changes in tree-species composition, due to increased browsing pressure on preferred food plants (Strandgaard 1982, Alverson *et al.* 1988, Pastor *et al.* 1988), and in some areas are making natural regeneration impossible (see 'Threats to the habitat', p. 223).

Disease and pests

Plant diseases and pests are widely responsible for the death of individual trees. For example, about 50% of trees in Austria suffer from root decay, due to use of unsuitable species in afforestation, poor planting techniques, and soil compression by agricultural activities. More rarely, such factors can cause major changes in tree-species composition (e.g. since 1919 Dutch elm disease has led to the almost complete disappearance of *Ulmus* species over most of Europe). Large, storm-damaged areas are often affected by disease or pests, as are monoculture plantations or monospecific stands (particularly by insects), which attack various vegetable parts (wood, leaves, roots, shoots) in various development phases. The most frequent agents include moths (Geometridae), beetles (Cerambycidae, Buprestidae and Curculionidae), and various fungi. In certain years such species can reach epidemic proportions, devastating large areas of forest plantations and causing financial damage. Air pollution and acid rain may 'weaken' trees and make them more prone to such attacks.

Management systems

Most European forests, especially in the west, are currently managed to some extent for timber production, or have been until recently. As described in 'Threats to the habitat' (p. 223), such practices can have a profound effect on the structure and tree-species composition of forest. Even the type of management several hundred years ago can strongly affect the present-day character of forest. This history, combined with the origin of the forest, leads to four main groups of forest, in terms of management (Figure 3).

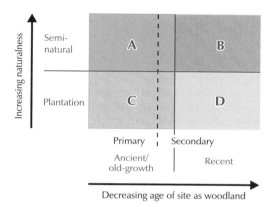

Figure 3. The four main groups of forest (A–D) with respect to forest management in Europe (adapted from Peterken 1993). Groups are arbitrarily defined in terms of two continuous variables: the age and the degree of naturalness of the forest (see arrows). There are no sharp divisions between A, B, C and D, despite the implication of the diagram. Virgin forest is excluded, since by definition it has never been modified by human activity.

Today, there is a wide variety of production systems for timber and wood in Europe, which can be placed somewhere among the following, grossly simplified continuum of 'main' systems.

- Clear-felling: large patches, usually greater than 2 ha, are harvested and replanted.

- Group-felling: a small-scale version of clear-felling, in which the patch size is below 0.5 ha and restocking is either by planting or natural regeneration.

- Shelter-wood: similar to clear-felling, except that when the crop is harvested a few scattered trees are retained to provide a seed source for regeneration of the next crop. The seed-trees may be progressively thinned over time.

- Selection systems: trees are harvested individually, thus maintaining a superficially uniform appearance in the forest.

- Coppicing: trees are cut on short rotations with natural multi-stemmed regrowth from the cut stumps; still practised widely but rarely in western Europe.

The effects of these practices on forest character and birds are described under 'Threats to the habitat' (as 'Intensification of management practices', p. 223).

In addition to these harvesting strategies, often forest management also involves thinning and removal of tree species of low economic value. This reduces the diversity of species and of forest structure. Some managed forests, particularly in the UK and Portugal, may also receive fertilizer inputs (Stanners and Bourdeau 1995).

Characteristic flora and fauna

The vegetation communities of European forests are very complex and varied and cannot be described in detail here. However, further information can be found in Polunin and Walters (1985) and Ellenberg (1988), for example.

Boreal forests

These are primarily coniferous forests, dominated by *Picea abies*, or by *Pinus sylvestris* in drier or more marginal areas, although deciduous broadleaved woods of downy birch *Betula pubescens*, *Salix* thickets and *Alnus* carr also occur. They grow on shallow soils, and the shrub and field layers are often absent or poorly developed, but in more open stands a heathland of dwarf shrubs develops, or alternatively the ground is covered in moss or lichen. The diversity of vascular plant species is generally very poor but is high for mosses and lichens (Polunin and Walters 1985).

Boreal forests still support considerable wild populations of some of the original large mammal fauna of forests, now extinct or rare elsewhere in Europe, such as elk *Alces alces*, reindeer *Rangifer tarandus*, wolverine *Gulo gulo*, wolf *Canis lupus* and beaver *Castor fiber*.

The avifauna of the central and northern taiga comprises 32 species such as Ural Owl *Strix uralensis*, Siberian Jay *Perisoreus infaustus*, Siberian Tit *Parus cinctus*, Fieldfare *Turdus pilaris* and Rustic Bunting *Emberiza rustica*. Their distribution is mainly confined to the boreal zone and most can be regarded as specialists of boreal forest itself. In contrast, typical species of the southern taiga number 11 and include more widely distributed species such as Pygmy Owl *Glaucidium passerinum*, Willow Tit *Parus montanus*, Redwing *Turdus iliacus* and Nutcracker *Nucifraga caryocatactes* (Haila and Järvinen 1990).

Lowland temperate forests

These forests are mostly dominated by *Quercus* in central and eastern Europe, and by *Fagus* in the wetter Atlantic regions, with *Pinus sylvestris* and *Picea abies* in mixed stands. The latter two conifers often predominate in natural lowland forest in southern Fennoscandia. Deciduous forests, especially the more open *Quercus petraea* and *Quercus cerris* forests of central Europe with its warmer summers, are very rich in vascular plants, e.g. in the primeval Bialowieza forest (Poland). However, original plant diversity is greatly impoverished in managed forests and plantations due to frequent, strong disturbances of the soil layer by forestry operations and heavy machinery, the invasion of non-forest and non-native plant species through increased edges and gaps, and through the use of herbicides (e.g. Angelstam and Mikusinski 1994, Halpern and Spies 1995, Christensen and Emborg 1996).

The invertebrate fauna is rich in mature forests with dead trees and fallen logs but is significantly reduced in managed forests (e.g. Pettersson *et al.* 1995). The mammal fauna of these forests is generally not diverse and large predators (e.g. brown bear *Ursus arctos, Canis lupus*) are now extinct in most such areas in Europe.

Lowland temperate forests have a rich and often abundant bird fauna, especially in primeval forest (Tomialojc 1995). Typical species of broadleaved forests include raptors such as Honey Buzzard *Pernis apivorus* and hawks *Accipiter*, Tawny Owl *Strix aluco*, most of the woodpecker (Picidae) species, and a wide variety of chats (e.g. *Erithacus rubecula, Luscinia* spp., Redstart *Phoenicurus phoenicurus*), thrushes *Turdus, Sylvia* and *Phylloscopus* warblers, flycatchers (Muscicapidae) and tits *Parus*. Natural and semi-natural lowland temperate coniferous forests have a less diverse avifauna but with several characteristic species, including Capercaillie *Tetrao urogallus*, Black Woodpecker *Dryocopus martius*, Nightjar *Caprimulgus europaeus, Anthus trivialis*, Goldcrest *Regulus regulus*, Siskin *Carduelis spinus* and crossbills *Loxia* spp.

Non-native coniferous plantations, however, contain a highly impoverished fauna, except in the early growth stages when the habitat may be used by a variety of open-ground and scrub species. Surprisingly few bird species are exclusively associated with lowland temperate forest in Europe, and there is much overlap in species composition with boreal, riverine and montane forests. Nevertheless, a small number of birds are especially dependent on this habitat-type, such as Middle Spotted Woodpecker *Dendrocopos medius*, Collared Flycatcher *Ficedula albicollis*, Nuthatch *Sitta europaea* and Hawfinch *Coccothraustes coccothraustes*. Species with important populations in lowland coniferous forests include Crested Tit *Parus cristatus*, Woodlark *Lullula arborea* and *Caprimulgus europaeus*.

Montane forests

These typically comprise the following altitudinal zones, moving from the lower elevations upwards:

1. Forests dominated by *Fagus* and *Abies* in the highest parts of the temperate zone. These forests may contain *Picea abies* or *Pinus* species in some areas.

2. The subalpine coniferous zone, often dominated by *Picea abies*, with *Abies* and *Pinus* species (*Abies* being much reduced in dominance by man during past centuries) and in the central European mountains with a distinct upper zone typified by *Pinus cembra* and *Larix decidua*. A few broadleaved species are also found at these high altitudes, particularly *Betula pubescens*, silver birch *B. pendula*, *Sorbus aucuparia* and sycamore *Acer pseudoplatanus*.

3. The 'krummholz' zone, that is, stunted and shrubby woodland at the altitudinal or exposure limits of tree-growth, usually comprising dwarf *Pinus* spp. and/or green alder *Alnus viridis*. *Salix* species and juniper *Juniperus communis* are also common in this zone, which grades into shrubby grassland at higher altitudes (see 'Montane grassland', treated under 'Agricultural and Grassland Habitats', p. 267).

At the highest elevations the flora is dominated by mosses, lichens and short vascular plants which are highly adapted to the harsh environment. The coniferous and 'krummholz' zones are similar to boreal forest in that generally there is very poor undergrowth of vascular plants, while there is a high diversity of mosses and lichens. *Fagus–Abies* forests are generally rather dark, with a closed canopy which does not favour well-developed and diverse undergrowth, and overall diversity of flora and fauna is therefore rather low. Large mammalian predators such as *Ursus arctos*, *Canis lupus* and lynx *Lynx lynx* can still be found in good numbers in the remoter parts of montane forests in Europe, however.

Generally the bird communities of montane forest contain species of both lowland temperate and boreal forest. In addition, they do support a small number of birds which are not found elsewhere, such as Black-throated Accentor *Prunella atrogularis*, the montane subspecies of Three-toed Woodpecker *Picoides tridactylus* and Citril Finch *Serinus citrinella*.

Riverine forests

These are among the richest and most complex ecosystems in Europe and support an exceptional diversity of fauna and flora (Yon and Tendron 1981, Imboden 1987, IUCN 1993). Tree species such as *Salix*, *Populus*, *Alnus* and *Fraxinus* are most common, while members of other genera, such as *Betula*,

Quercus and *Ulmus* also constitute important elements of riverine stands more locally. Typically, in southern and central Europe, flood-plains of larger rivers are covered by softwood *Salix–Populus* riverine woodland in regularly flooded parts of the valley, and by hardwood *Quercus–Fraxinus–Ulmus* woodland on higher, more stable ground (Ellenberg 1988). The flood-plains of smaller rivers, and waterlogged slopes, are mostly covered by *Fraxinus–Alnus* forests. Permanently wet places with stagnating water are covered by *Alnus* or *Salix* carrs. Forests on the flood-plains of northern Europe are composed mostly of *Betula* and *Picea*, with only small admixtures of *Salix* and *Alnus*.

Riverine forests harbour some of the densest and most diverse breeding bird assemblages found anywhere in Europe (Imboden 1987, Spitzenberger 1988, Zuna-Kratky and Frühauf 1996), which include rare solitary species (e.g. Black Stork *Ciconia nigra* and White-tailed Eagle *Haliaeetus albicilla*) and large populations of colonial breeders (e.g. Little Egret *Egretta garzetta*, Squacco Heron *Ardeola ralloides*, Night Heron *Nycticorax nycticorax*, Cormorant *Phalacrocorax carbo*). Characteristic passerine species include *Luscinia* spp., River Warbler *Locustella fluviatilis* and Penduline Tit *Remiz pendulinus*. Among mammals, otter *Lutra lutra* is still quite widespread, although increasingly threatened in Europe, while *Castor fiber* mostly occurs in northern and eastern Europe.

Relationships to other habitats

Lowland Atlantic heathland

Where grazing levels are particularly heavy or where wood-cutting and/or fires are relatively intense on poor soils, forest can revert to heathland, an open habitat dominated by dwarf shrubs (often Ericaceae, usually heather *Calluna vulgaris*). Heathlands occur from Scandinavia to north-west Spain and inland to Germany, and are treated in detail under 'Lowland Atlantic Heathland' (p. 187). The ecotone between forest and heathland is attractive to many bird species such as Black Grouse *Tetrao tetrix*, *Caprimulgus europaeus* and *Lullula arborea*, due in part to the highly varied structure of the habitat, even over short distances. Many heathlands are currently reverting to forest through lack of grazing, of other traditional uses, or of fire.

Tundra, mires and moorland

In forests where drainage is impeded, mires can occur over wide areas, following the accumulation of peat. These habitats are dominated by *Sphagnum* mosses and *Carex*, and often form an intricate mosaic of plant communities with boreal forest. Wooded bog vegetation is distinctive and covers large areas of mires in the Baltic catchment, e.g. 12% of forest in Latvia. At the northern edge of the taiga, forests grade into tundra in the forest-tundra zone. Again, the juxtaposition of these open habitats with forest is very important for birds.

Agricultural and grassland habitats

These are very important as a feeding area for many bird species which breed in forest, most notably birds of prey (see 'Habitat needs of priority birds', p. 220). Agricultural activities can also have a negative impact on neighbouring forest areas locally, e.g. through habitat fragmentation, nutrient inputs and pesticide drift. Forest loss to agriculture is no longer a major trend in Europe, and in many cases, especially in the European Union, the opposite situation can be observed in some areas. Grasslands are primarily the result of man's activities such as livestock-grazing and the cutting of grass for silage or hay. With relaxation in grazing pressure, these habitats tend to revert to forest. To the south-east, soils and climate are unsuitable for tree growth and there natural primary steppes occur (see 'Agricultural and Grassland Habitats', p. 267).

Values, roles and uses

Timber production is now the most widespread use of temperate and boreal forests by far, and most lowland forests have long been exploited for timber and wood on a variety of scales and at varying intensities. In more recent times, forests have become increasingly appreciated for their role in maintaining water-tables, as carbon sinks, and for their aesthetic qualities as landscapes, as recreational areas, and as wildlife habitats (e.g. Dudley 1992).

Timber

Europe is the world's second-highest industrial roundwood producing region. Such forestry accounts for more than 4% of industrial export earnings in Finland, Austria, Norway, Portugal and Sweden (Dudley 1992), and this proportion tends to be higher for some northern countries with economies in transition, e.g. 20% in Latvia. Total production of roundwood in Europe was about 368 million m^3 in 1989, an increase of over 18% since 1965. Some of this increased production has been achieved simply by expanding the total area of forest, but much is the result of increased productivity in most parts of Europe, through mechanization and more intensive management techniques (Stanners and Bourdeau 1995). Such capital-intensive management and wood-processing has reduced the value of mainstream forestry as a provider of employment in rural areas (Cuff and Rayment 1997).

Boreal forest in Scandinavia has been exploited and managed for timber and wood-fibre on an appreciable scale for at least 500 years. Over-exploitation of the accessible wood resources occurred during the eighteenth and nineteenth centuries, but these were gradually restored using techniques whereby clear-cutting of old-growth forest was followed by soil scarification, replacement with denser plantations of faster-growing species or varieties, 'cleaning' and thinning (destruction of young trees of unwanted species), removal of dead wood, etc. (Gamlin 1988). Hemi-boreal forest in Russia has also been exploited by selective cutting, especially along rivers, for hundreds of years. Large-scale clear-cutting has occurred since the middle of the nineteenth century in the southern part of the boreal and hemi-boreal forest, and since the 1940s in the north as well (Nilsson et al. 1992).

Historically, much lowland broadleaved woodland throughout temperate Europe was managed by coppicing or as 'wood pasture'. The latter involved treating the forest as grazing land for domestic animals as well as a source of wood and timber. Working wood pasture is now as rare as active coppice, surviving mostly in the Mediterranean zone (see 'Pastoral woodland', treated under 'Agricultural and Grassland Habitats', p. 267). Today, the only common management for semi-natural lowland temperate forest in Europe is the 'high forest' system, where trees grow from seedlings in single-stemmed form for long periods, and are felled for building timber according to a variety of patterns, such as clear-felling, group-felling, shelterwood or selection system (see p. 209).

In montane forest, timber production is an important element of management at the lower elevations. Here, substantial areas of the original forest have been modified, e.g. reforested as conifer plantations, but a significant proportion remains semi-natural in character and managed in a way which retains the original tree species diversity and composition.

The major uses of European timber are as sawn-wood and panel products for the construction and furniture industries, as paper and board for packaging, printing and writing, and as firewood or charcoal largely for household heating and cooking. Of these products, only the demands for charcoal and firewood have experienced sharp and consistent declines over recent decades (Stanners and Bourdeau 1995).

Climate

Forests moderate local climate and are increasingly seen as an important sink for carbon in mitigating the effects of global warming (MacKenzie 1994, Stanners and Bourdeau 1995). Boreal forests warm the subarctic zone by providing a dark mass capable of absorbing heat from the sun (Dudley 1992).

Recreation

The importance of forests as a source of recreation has gained increasing attention in Europe over the last two decades. Little information is available on the number of visits to forests, or even the number of forests deliberately managed for recreation. A survey by the United Nations Economic Commission for Europe (UNECE) found that most European governments viewed more than 50% of their forests as of medium to high value for public recreation (Dudley 1992). In the Nordic countries, where people have the right to visit forests regardless of who owns them, it is estimated that at least 400 million such visits are made per year (Bernes 1993, Stanners and Bourdeau 1995).

Hunting

Hunting is a popular activity in many European countries and much of it is practised in forests or on land adjacent to forests, as these areas are the breeding, roosting and feeding grounds of many of the traditional game species. Many European countries attach a great deal of cultural importance to hunting, and for many private forest-owners hunting represents an important, if not the biggest, source of income generated by their forests. Again, little data exists to quantify the degree to which this activity is followed in different regions of Europe (Stanners and Bourdeau 1995). In central and eastern Europe large forested areas were set aside in the past as state game reserves, where forest stands were managed primarily as habitats for large game (e.g. *Cervus elaphus*, roe deer *Capreolus capreolus*, fallow deer *Dama dama*, *Sus scrofa*). Timber production was generally secondary. Although these forests acted as a form of nature reserve, where human disturbance other than hunting was considerably reduced, artificially high densities of wild grazing animals exerted a very high pressure on them and greatly reduced forest regeneration (through damage to young trees and to bark). This situation continues to prevail over much central European forest (Donaubauer *et al.* 1995).

Grazing

Historically, forests have long provided pasture and fodder for a range of livestock. In all countries, however, the general trend over the last four decades has been to abandon such pasturage. Nowadays, many European countries attach only low importance to this practice (UNECE/FAO 1992b) and many have laws to regulate or prohibit grazing in forests, in order to protect them and allow natural regeneration.

However, in some countries, including Albania and Spain, livestock-grazing occurs in more than 20% of the forests (see 'Pastoral woodland' under 'Agricultural and Grassland Habitats', p. 267), and in Finland up to 80% of the state-owned forest of Lapland is grazed by reindeer (Stanners and Bourdeau 1995).

Nature conservation

The growing awareness of nature-conservation issues in European society, coupled with the realization that nearly all of Europe's natural and semi-natural forest has been destroyed, has encouraged many countries to protect part of their forests by creating national parks and forested nature reserves, or through some other form of legal protection. The proportion of forest under such protection (as defined in IUCN-CNPPA 1994) is about 8% in Europe (excluding Russia), most of which is lowland temperate forest (Iremonger *et al.* in press). However, such protection is mostly 'soft' (e.g. as protected landscape areas) which does not necessarily preclude the use of the forests for other functions such as wood production. The conservation of primeval or natural qualities in forests is often incompatible with current methods of wood production, while other functions (e.g. recreation, exploitation of non-wood products) can be more easily accommodated. Also, in at least some of the post-socialism countries (e.g. Latvia, Slovakia, Poland), high levels of protection on paper are not always realized in practice, due to lack of funds and resources.

Strict protection of forests as reserves where there is no timber exploitation is still very rare in Europe. For example, in Germany less than 1% of forested land is under protection in nature reserves or national parks, and less than 0.1% in Austria (Frank 1995). In Poland the famous Bialowieza National Park only covers about 18% of the total area of this, the largest primeval temperate forest in Europe, and remaining stands of ancient trees are threatened by logging. In proportion to their extent, boreal and riverine forests are among the least protected forest-types in Europe (Iremonger *et al.* in press).

Protection of soil and water resources

Forests, especially montane and riverine forests, play a very important role in the maintenance of water purity, the protection of fish-spawning grounds, protection against excessive soil erosion, avalanches, rockfalls and landslips, and in slowing water run-off and hence in minimizing flooding and maximizing recharge of aquifers. Many countries have introduced legislation for the protection and maintenance of montane forests as a consequence, e.g. nearly 20% of forest in Austria is 'Schutzwald' (protection forest).

Other products

Berries, pine and other nuts, fungi, game meat, honey, tannin, charcoal, resins and oils, medicinal products and Christmas trees are all gathered from forests for subsistence and for commerce. The importance and extent of the harvesting of these various renewable products in the different regions of Europe is difficult to assess because of lack of data, especially on the income derived from them. Nevertheless, the significance of this forest function should not be overlooked, especially at the local to national level (Stanners and Bourdeau 1995).

Socio-political factors affecting the habitat

Land ownership

About half (49%) of Europe's forest is privately owned, ranging from farmers with small holdings of woodland to large private companies, but there are large variations between the different regions of Europe (Kuusela 1994). In many central and east European countries, almost all forest is publicly owned, while the share of private ownership exceeds 70% in many European countries to the west, such as Austria, Denmark, France, Norway and Portugal. The division of responsibility, and the sheer number of decision-makers with differing objectives, has profound implications for any government's ability to guide the practices adopted by owners. Not all private owners make use of professional guidance and technical assistance, though they are usually under some kind of supervision by public authorities (Stanners and Bourdeau 1995).

Forestry policy

Recent European forest-management practices have largely focused on timber production, and have tended to lead to monospecific, even-aged stands, often of conifers. Society's growing awareness of the other values and functions of forests apart from timber production, particularly in northern Europe, has led to the reappraisal of many current forestry practices associated with wood production, and is beginning to influence how forests are managed (e.g. Forestry Commission 1989).

Although the principal objective remains productive forestry, there is a move towards multiple-use, with less emphasis on intensive timber production and more on non-timber values, at least by some governments and some sectors of the forestry industry. To a large extent this is driven by public pressure, an increasing demand for sustainable timber products, and discussions on the certification of forest products.

While there is no Common Forestry Policy within Europe (e.g. within the European Union), there is increasing dialogue between most European countries in attempts to undertake a coordinated approach to the sustainable use and conservation of forest biodiversity in Europe. In 1993 the second Ministerial Conference on the Protection of Forests in Europe was held in Helsinki, where the environment ministers of 45 participating countries signed up to a set of guiding principles regarding the sustainable management of Europe's forests and the conservation of their diversity (Ministerial Conference on the Protection of Forests in Europe 1993, Cosgrove and Turner 1995). Based on the 1992 recommendations of the UN Conference on Environment and Development (the 'Earth Summit'), a list of six pan-European criteria and 27 quantitative indicators for the sustainable management of forests has been recently adopted by the signatory states—see 'Conservation opportunities', p. 231, for more details.

Agricultural over-production

Current over-production of agricultural commodities in the EU, and widespread abandonment of farmland in central and east European countries (see 'Agricultural and Grassland Habitats', p. 267), is leading to afforestation schemes on low-quality agricultural land in the European Union member states (under Regulation 2080/92), and to natural regrowth of forest in the east, respectively. A drawback to these afforestation schemes is that they tend to target non-intensive cereal cultivation, which is usually of high conservation value for wildlife, as has occurred recently on the steppes of Spain, Portugal and Hungary (see 'Agricultural and Grassland Habitats', p. 267).

Central and eastern Europe

Countries in central and eastern Europe have undergone huge political upheavals since 1989, and there is an urgent need for international support in resolving the serious economic problems which have resulted. In the more northerly of such countries there is great pressure to fell (illegally) large areas of boreal forest which are now difficult to police. A major new factor affecting land-use is privatization and re-privatization (returning land to former owners) in this region. Privatization of forests is currently not widespread in most countries in central and eastern Europe, but this may change in the future. It is probable that the new owners of such forests, lacking capital and skills, will face large problems in maintaining high standards of forest management. Moreover, the expected short-term profit might well drive owners to cut down and sell their forests as quickly as possible. Facing these difficulties, it is unlikely that private forests will be managed in an ecologically sustainable way in the near future. For this reason it seems justified to keep forests under state control, or encourage the owners of small forest lots to jointly create forest cooperatives which can be managed in a more cost-effective way.

PRIORITY BIRDS

The 114 priority bird species of boreal and temperate forests in Europe are listed in Table 1. Of this rich avifauna, six species are Priority A for one or more of these forest-types (Tables 2 and 3), and two of these (Greater Spotted Eagle *Aquila clanga* and Imperial Eagle *Aquila heliaca*) are globally threatened (Collar *et al*. 1994). One of the Priority A species, Scottish Crossbill *Loxia scotica*, is endemic to UK; its European Threat Status is classed as 'Insufficiently Known' (see Appendix 1, p. 327), and Collar *et al*. (1994) describe its global status as Data Deficient. The remaining three Priority A species—Green Woodpecker *Picus viridis*, *Lullula arborea* and *Phoenicurus phoenicurus*—are widespread in Europe but all suffered substantial declines during the period 1970–1990. All but one of the Priority A species are very dependent on boreal or temperate forests (i.e. over 75% of their European population uses them) at some time during the annual cycle (Table 2).

Overall, 18 species fall within Priority B (Table 3). Of these birds, during the breeding season, Smew *Mergus albellus*, the boreal subspecies of *Picoides tridactylus*, and *Perisoreus infaustus* are confined to boreal forests, while three species (Levant Sparrowhawk *Accipiter brevipes*, Semi-collared Flycatcher *Ficedula semitorquata* and Pygmy Cormorant *Phalacrocorax pygmeus*) are restricted to eastern or south-east Europe, *P. pygmeus* being particularly associated with riverine forests. *Prunella atrogularis*, a rare and poorly known bird, is confined to montane *Picea* forest in the Ural mountains and the east coast of the White Sea. Spoonbill *Platalea leucorodia* breeds only in riverine forests while the remaining Priority B species are widespread in European forests.

Two of the Priority B species, *Platalea leucorodia* and *Ficedula semitorquata*, are considered to be endangered in Europe (see Appendix 1, p. 327), and *Phalacrocorax pygmeus*, while having formerly been classified as globally threatened (Collar and Andrew 1988), is currently listed as near-threatened (Collar *et al*. 1994).

Of the 45 Priority C species (Table 3), only Saker *Falco cherrug* is considered to be endangered in Europe (Tucker and Heath 1994), although 29 others also have an Unfavourable Conservation Status (see Appendix 1, p. 327). The remaining 15 species have

a Favourable Conservation Status but are concentrated in Europe, with Parrot Crossbill *Loxia pytyopsittacus*, *Serinus citrinella* and Red Kite *Milvus milvus* (apart from a small population in Morocco) all confined to Europe (Table 2).

In total, 48 of the 114 priority species (42%) have an Unfavourable Conservation Status (Table 2). Riverine forest has the largest proportion of priority species with an Unfavourable Conservation Status (48%), while lowland temperate forest has 45%, boreal forest 41% and montane forest 43% (Table 2).

In each forest-type a fairly large proportion of species qualify as priority species because, although they have a Favourable Conservation Status in Europe (Tucker and Heath 1994), their world populations are concentrated in Europe, ranging from up to 57% of species in montane forest down to 32% in boreal forest (Table 2). Overall, 36 species such as Tengmalm's Owl *Aegolius funereus*, Lesser Spotted Woodpecker *Dendrocopus minor* and *Coccothraustes coccothraustes* do not have their world populations concentrated in Europe but depend very heavily on forest habitat for their survival in Europe. Nearly all of these are specialists of boreal and lowland temperate forests. In these forest-types, 32% and 16% (respectively) of their priority species are heavily dependent on the forest-type (i.e. it holds more than 75% of the European population). By contrast, in both montane forest and riverine forest only about 2% of species are similarly dependent on these habitats (Table 2). This is probably because these forest-types (particularly riverine forest) are much less extensive than the former two.

As with the European avifauna in general, there is a marked increase in the species richness of priority forest birds from west to east in Europe, with far more species restricted to the boreal and temperate forests of eastern Europe than to those of western Europe. Indeed, the bird communities of the latter can be regarded as an impoverished subset of the former. This is reflected in the breeding distribution of the priority species in Europe (Appendix 2, p. 344). Ukraine, Romania and Russia are the only countries to support breeding populations of more than 90% of priority species of lowland temperate forest and of riverine forest. Sizeable countries with a relatively low diversity of such breeding species (less than 70%) lie at the western and southern extremities of the continent, i.e. Norway, Denmark, the Netherlands, UK, Portugal and Cyprus. An eastern bias is also evident for montane forest, where Romania, Slovakia and Slovenia are the countries with the highest richness of breeding priority species, only matched further to the west by France. Not surprisingly, Russia and Fennoscandia support the highest diversity of boreal forest species, with Rus-

Table 1. The priority bird species of boreal and temperate forests in Europe.

This prioritization identifies Species of European Conservation Concern (SPECs) (see p. 17) for which the habitat is of particular importance for survival, as well as other species which depend very strongly on the habitat. It focuses on the SPEC category (which takes into account global importance and overall conservation status of the European population) and the percentage of the European population (breeding population, unless otherwise stated) that uses the habitat at any stage of the annual cycle. It *does not* take into account the threats to the species within the habitat, and therefore should not be considered as an indication of priorities for action. Indications of priorities for action for each species are given in Appendix 4 (p. 401), where the severity of habitat-specific threats is taken into account.

	SPEC category	European Threat Status	Habitat importance
BOREAL FOREST			
Priority A			
Greater Spotted Eagle *Aquila clanga*	1	E	●
Scottish Crossbill *Loxia scotica*	1	Ins	■
Priority B			
Smew *Mergus albellus*	3	V	■
Nightjar *Caprimulgus europaeus*	2	(D)	●
Three-toed Woodpecker *Picoides tridactylus*	3	D	■
Woodlark *Lullula arborea*	2	V	●
Redstart *Phoenicurus phoenicurus*	2	V	●
Siberian Jay *Perisoreus infaustus*	3	(D)	■
Priority C			
White-tailed Eagle *Haliaeetus albicilla*	3	R	●
Golden Eagle *Aquila chrysaetos*	3	R	●
Osprey *Pandion haliaetus*	3	R	●
Peregrine *Falco peregrinus*	3	R	●
Black Grouse *Tetrao tetrix*	3	V	●
Turtle Dove *Streptopelia turtur*	3	D	●
Eagle Owl *Bubo bubo*	3	V	●
Wryneck *Jynx torquilla*	3	D	●
Grey-headed Woodpecker *Picus canus*	3	D	●
Stonechat *Saxicola torquata*	3	(D)	●
Redwing *Turdus iliacus*	4 W	S	■
Spotted Flycatcher *Muscicapa striata*	3	D	●
Great Grey Shrike *Lanius excubitor*	3	D	●
Parrot Crossbill *Loxia pytyopsittacus*	4	S	■
Ortolan Bunting *Emberiza hortulana*	2	(V)	●
Priority D			
Honey Buzzard *Pernis apivorus*	4	S	●
Hen Harrier *Circus cyaneus*	3	V	●
Kestrel *Falco tinnunculus*	3	D	●
Red-footed Falcon *Falco vespertinus*	3	V	●
Capercaillie *Tetrao urogallus*	—	(S)	■
Crane *Grus grus*	3	V	●
Woodcock *Scolopax rusticola*	3 W	V W	●
Wood Sandpiper *Tringa glareola*	3	D	●
Stock Dove *Columba oenas*	4	S	●
Woodpigeon *Columba palumbus*	4	S	●
Hawk Owl *Surnia ulula*	—	(S)	■
Pygmy Owl *Glaucidium passerinum*	—	(S)	■
Ural Owl *Strix uralensis*	—	(S)	■
Great Grey Owl *Strix nebulosa*	—	S	■
Tengmalm's Owl *Aegolius funereus*	—	(S)	■
Olive-backed Pipit *Anthus hodgsoni*	—	(S)	■
Waxwing *Bombycilla garrulus*	—	(S)	■
Dunnock *Prunella modularis*	4	S	●
Robin *Erithacus rubecula*	4	S	●
Thrush Nightingale *Luscinia luscinia*	4	S	●
Siberian Rubythroat *Luscinia calliope*	—	(S)	■
Red-flanked Bluetail *Tarsiger cyanurus*	—	(S)	■
White's Thrush *Zoothera dauma*	—	(S)	■
Blackbird *Turdus merula*	4	S	●
Fieldfare *Turdus pilaris*	4 W	S	●
Song Thrush *Turdus philomelos*	4	S	●

cont.

Table 1. (cont.)

	SPEC category	European Threat Status	Habitat importance
Mistle Thrush *Turdus viscivorus*	4	S	●
Icterine Warbler *Hippolais icterina*	4	S	●
Garden Warbler *Sylvia borin*	4	S	●
Blackcap *Sylvia atricapilla*	4	S	●
Arctic Warbler *Phylloscopus borealis*	—	(S)	■
Yellow-browed Warbler *Phylloscopus inornatus*	—	S	■
Wood Warbler *Phylloscopus sibilatrix*	4	(S)	●
Goldcrest *Regulus regulus*	4	(S)	●
Pied Flycatcher *Ficedula hypoleuca*	4	S	●
Siberian Tit *Parus cinctus*	—	(S)	■
Crested Tit *Parus cristatus*	4	S	●
Red-backed Shrike *Lanius collurio*	3	(D)	●
Chaffinch *Fringilla coelebs*	4	S	●
Brambling *Fringilla montifringilla*	—	S	■
Greenfinch *Carduelis chloris*	4	S	●
Siskin *Carduelis spinus*	4	S	●
Two-barred Crossbill *Loxia leucoptera*	—	(S)	■
Pine Grosbeak *Pinicola enucleator*	—	S	■
Rustic Bunting *Emberiza rustica*	—	(S)	■

LOWLAND TEMPERATE FOREST

Priority A

Greater Spotted Eagle *Aquila clanga*	1	E	●
Imperial Eagle *Aquila heliaca*	1	E	■
Green Woodpecker *Picus viridis*	2	D	■
Woodlark *Lullula arborea*	2	V	■
Redstart *Phoenicurus phoenicurus*	2	V	■

Priority B

Levant Sparrowhawk *Accipiter brevipes*	2	R	●
Lesser Spotted Eagle *Aquila pomarina*	3	R	■
Woodcock *Scolopax rusticola*	3 [w]	V [w]	■
Nightjar *Caprimulgus europaeus*	2	(D)	●
Wryneck *Jynx torquilla*	3	D	●
Grey-headed Woodpecker *Picus canus*	3	D	■
Semi-collared Flycatcher *Ficedula semitorquata*	2	(E)	●

Priority C

Black Stork *Ciconia nigra*	3	R	●
Black Kite *Milvus migrans*	3	V	●
Red Kite *Milvus milvus*	4	S	●
White-tailed Eagle *Haliaeetus albicilla*	3	R	■
Short-toed Eagle *Circaetus gallicus*	3	R	●
Golden Eagle *Aquila chrysaetos*	3	R	●
Booted Eagle *Hieraaetus pennatus*	3	R	●
Osprey *Pandion haliaetus*	3	R	●
Saker *Falco cherrug*	3	E	● [1]
Peregrine *Falco peregrinus*	3	R	●
Black Grouse *Tetrao tetrix*	3	V	●
Turtle Dove *Streptopelia turtur*	3	D	●
Scops Owl *Otus scops*	2	(D)	●
Eagle Owl *Bubo bubo*	3	V	●
Tawny Owl *Strix aluco*	4	S	■
Roller *Coracias garrulus*	2	(D)	●
Middle Spotted Woodpecker *Dendrocopos medius*	4	S	■
Three-toed Woodpecker *Picoides tridactylus*	3	D	■
Thrush Nightingale *Luscinia luscinia*	4	S	■
Nightingale *Luscinia megarhynchos*	4	(S)	■
Stonechat *Saxicola torquata*	3	(D)	●
Icterine Warbler *Hippolais icterina*	4	S	■
Garden Warbler *Sylvia borin*	4	S	■
Spotted Flycatcher *Muscicapa striata*	3	D	●
Collared Flycatcher *Ficedula albicollis*	4	S	■
Pied Flycatcher *Ficedula hypoleuca*	4	S	■
Blue Tit *Parus caeruleus*	4	S	■
Short-toed Treecreeper *Certhia brachydactyla*	4	S	■

cont.

Table 1. (cont.)

	SPEC category	European Threat Status	Habitat importance
Priority D			
Honey Buzzard *Pernis apivorus*	4	S	●
Kestrel *Falco tinnunculus*	3	D	·
Red-footed Falcon *Falco vespertinus*	3	V	·
Hazel Grouse *Bonasa bonasia*	—	S	■
Stock Dove *Columba oenas*	4	S	●
Woodpigeon *Columba palumbus*	4	S	●
Lesser Spotted Woodpecker *Dendrocopos minor*	—	S	■
Dunnock *Prunella modularis*	4	S	●
Robin *Erithacus rubecula*	4	S	●
Blackbird *Turdus merula*	4	S	●
Fieldfare *Turdus pilaris*	4 W	S	●
Song Thrush *Turdus philomelos*	4	S	●
Mistle Thrush *Turdus viscivorus*	4	S	●
Melodious Warbler *Hippolais polyglotta*	4	(S)	●
Blackcap *Sylvia atricapilla*	4	S	●
Bonelli's Warbler *Phylloscopus bonelli*	4	S	●
Wood Warbler *Phylloscopus sibilatrix*	4	(S)	●
Goldcrest *Regulus regulus*	4	(S)	●
Firecrest *Regulus ignicapillus*	4	S	●
Red-breasted Flycatcher *Ficedula parva*	—	(S)	■
Sombre Tit *Parus lugubris*	4	(S)	●
Crested Tit *Parus cristatus*	4	S	●
Nuthatch *Sitta europaea*	—	S	■
Treecreeper *Certhia familiaris*	—	S	■
Golden Oriole *Oriolus oriolus*	—	S	■
Red-backed Shrike *Lanius collurio*	3	(D)	·
Great Grey Shrike *Lanius excubitor*	3	D	●
Jackdaw *Corvus monedula*	4	(S)	●
Chaffinch *Fringilla coelebs*	4	S	●
Serin *Serinus serinus*	4	S	●
Greenfinch *Carduelis chloris*	4	S	●
Siskin *Carduelis spinus*	4	S	●
Hawfinch *Coccothraustes coccothraustes*	—	S	■
MONTANE FOREST			
Priority A			
No species in this category			
Priority B			
Imperial Eagle *Aquila heliaca*	1	E	·
Green Woodpecker *Picus viridis*	2	D	●
Woodlark *Lullula arborea*	2	V	●
Black-throated Accentor *Prunella atrogularis*	3	(V)	■
Redstart *Phoenicurus phoenicurus*	2	V	●
Semi-collared Flycatcher *Ficedula semitorquata*	2	(E)	●
Priority C			
Black Kite *Milvus migrans*	3	V	●
Short-toed Eagle *Circaetus gallicus*	3	R	●
Golden Eagle *Aquila chrysaetos*	3	R	●
Booted Eagle *Hieraaetus pennatus*	3	R	●
Black Grouse *Tetrao tetrix*	3	V	●
Eagle Owl *Bubo bubo*	3	V	●
Nightjar *Caprimulgus europaeus*	2	(D)	·
Grey-headed Woodpecker *Picus canus*	3	D	●
Three-toed Woodpecker *Picoides tridactylus*	3	D	●
Spotted Flycatcher *Muscicapa striata*	3	D	●
Citril Finch *Serinus citrinella*	4	S	■
Rock Bunting *Emberiza cia*	3	V	●
Priority D			
Black Stork *Ciconia nigra*	3	R	·
Honey Buzzard *Pernis apivorus*	4	S	●
Red Kite *Milvus milvus*	4	S	●
Woodpigeon *Columba palumbus*	4	S	●

cont.

Table 1. (cont.)

	SPEC category	European Threat Status	Habitat importance
Tawny Owl *Strix aluco*	4	S	●
Wryneck *Jynx torquilla*	3	D	●
Middle Spotted Woodpecker *Dendrocopos medius*	4	S	●
Dunnock *Prunella modularis*	4	S	●
Robin *Erithacus rubecula*	4	S	●
Ring Ouzel *Turdus torquatus*	4	S	●
Blackbird *Turdus merula*	4	S	●
Fieldfare *Turdus pilaris*	4 w	S	●
Song Thrush *Turdus philomelos*	4	S	●
Mistle Thrush *Turdus viscivorus*	4	S	●
Garden Warbler *Sylvia borin*	4	S	●
Blackcap *Sylvia atricapilla*	4	S	●
Bonelli's Warbler *Phylloscopus bonelli*	4	S	●
Wood Warbler *Phylloscopus sibilatrix*	4	(S)	●
Goldcrest *Regulus regulus*	4	(S)	●
Firecrest *Regulus ignicapillus*	4	S	●
Collared Flycatcher *Ficedula albicollis*	4	S	●
Pied Flycatcher *Ficedula hypoleuca*	4	S	●
Crested Tit *Parus cristatus*	4	S	●
Blue Tit *Parus caeruleus*	4	S	●
Red-backed Shrike *Lanius collurio*	3	(D)	●
Chaffinch *Fringilla coelebs*	4	S	●
Serin *Serinus serinus*	4	S	●
Siskin *Carduelis spinus*	4	S	●
RIVERINE FOREST			
Priority A			
Greater Spotted Eagle *Aquila clanga*	1	E	●
Priority B			
Pygmy Cormorant *Phalacrocorax pygmeus*	2	V	●
Spoonbill *Platalea leucorodia*	2	E	●
Levant Sparrowhawk *Accipiter brevipes*	2	R	●
Scops Owl *Otus scops*	2	(D)	●
Green Woodpecker *Picus viridis*	2	D	●
Semi-collared Flycatcher *Ficedula semitorquata*	2	(E)	●
Priority C			
Night Heron *Nycticorax nycticorax*	3	D	●
Squacco Heron *Ardeola ralloides*	3	V	●
Purple Heron *Ardea purpurea*	3	V	●
Black Stork *Ciconia nigra*	3	R	●
Glossy Ibis *Plegadis falcinellus*	3	D	●
Black Kite *Milvus migrans*	3	V	●
White-tailed Eagle *Haliaeetus albicilla*	3	R	●
Lesser Spotted Eagle *Aquila pomarina*	3	R	●
Osprey *Pandion haliaetus*	3	R	●
Crane *Grus grus*	3	V	●
Woodcock *Scolopax rusticola*	3 w	V w	●
Turtle Dove *Streptopelia turtur*	3	D	●
Roller *Coracias garrulus*	2	(D)	●
Wryneck *Jynx torquilla*	3	D	●
Grey-headed Woodpecker *Picus canus*	3	D	●
River Warbler *Locustella fluviatilis*	4	S	■
Spotted Flycatcher *Muscicapa striata*	3	D	●
Priority D			
Red Kite *Milvus milvus*	4	S	●
Short-toed Eagle *Circaetus gallicus*	3	R	●
Kestrel *Falco tinnunculus*	3	D	●
Red-footed Falcon *Falco vespertinus*	3	V	●
Stock Dove *Columba oenas*	4	S	●
Woodpigeon *Columba palumbus*	4	S	●
Tawny Owl *Strix aluco*	4	S	●
Middle Spotted Woodpecker *Dendrocopos medius*	4	S	●
Dunnock *Prunella modularis*	4	S	●

cont.

Table 1. (cont.)

	SPEC category	European Threat Status	Habitat importance
Robin *Erithacus rubecula*	4	S	●
Thrush Nightingale *Luscinia luscinia*	4	S	●
Nightingale *Luscinia megarhynchos*	4	(S)	●
Blackbird *Turdus merula*	4	S	●
Fieldfare *Turdus pilaris*	4[w]	S	●
Song Thrush *Turdus philomelos*	4	S	●
Marsh Warbler *Acrocephalus palustris*	4	S	●
Icterine Warbler *Hippolais icterina*	4	S	●
Melodious Warbler *Hippolais polyglotta*	4	(S)	●
Garden Warbler *Sylvia borin*	4	S	●
Blackcap *Sylvia atricapilla*	4	S	●
Collared Flycatcher *Ficedula albicollis*	4	S	●
Pied Flycatcher *Ficedula hypoleuca*	4	S	●
Blue Tit *Parus caeruleus*	4	S	●
Azure Tit *Parus cyanus*	—	(S)	■
Short-toed Treecreeper *Certhia brachydactyla*	4	S	●
Great Grey Shrike *Lanius excubitor*	3	D	·
Jackdaw *Corvus monedula*	4	(S)	●
Chaffinch *Fringilla coelebs*	4	S	●
Serin *Serinus serinus*	4	S	●
Greenfinch *Carduelis chloris*	4	S	●
Siskin *Carduelis spinus*	4	S	●

SPEC category and **European Threat Status:** see Box 2 (p. 17) for definitions.

Habitat importance for each species is assessed in terms of the maximum percentage of the European population (breeding population unless otherwise stated) that uses the habitat at any stage of the annual cycle:

　　　■ >75%　　● 10–75%　　· <10%

[w] Assessment relates to the European wintering population. [1] *Falco cherrug* depends on the habitat for nesting, although not for foraging.

HABITAT NEEDS OF PRIORITY BIRDS

The habitat requirements of the priority species are summarized in Appendix 3 (p. 354). The physical structure of forests is the major ecological factor in determining the size, diversity and species composition of bird assemblages in boreal and temperate forests. Structure is influenced by the dominant tree species, grazing pressure, age, history of silvicultural practices, and fire regime, soil-type and exposure, as described in 'Ecological characteristics' (p. 206). Forest birds often include or require several structurally different habitats within their home ranges (e.g. Raivio and Haila 1990).

Very different assemblages of birds occur in the various growth stages of both plantations and natural or semi-natural forests. As a general rule, the number of species and overall density of birds in European forests tends to increase as the trees grow (Helle and Mönkkönen 1990, Moskát and Waliczky 1992, Scherzinger 1995). An exception is found in coppice woodland, where numbers of species and overall density tend to decrease in the oldest stages (Fuller 1982). Some conifer plantations also appear to carry greater densities of breeding birds in the thicket stage than in mature forest (Constant *et al.* 1973). Some bird species are confined to the earliest stages when the vegetation is open, others to the period when the canopy is starting to close and the vegetation is very bushy (Figure 4). Other species, including most hole-nesters, select those stages with mature or dying trees. Thus, because different birds prefer different stages of growth, forests with the greatest variety of growth provide more habitats for birds, and hence support more species, than uniform woods (Fuller 1995).

Three major groups of species can be distinguished, according to their degree of dependence on forest. The first group consists of species that nest, and mainly forage, within the confines of closed-canopy, mature forest. Species which are specialists of old-growth forest include *Picoides tridactylus* and Red-breasted Flycatcher *Ficedula parva*. One of the most important habitat features of old-growth forest is the relative abundance of dead or dying wood (Virkkala *et al.* 1994). About 27 priority species—including some owls, all woodpeckers, Roller *Coracias garrulus*, tits *Parus* and *Sitta europaea*—excavate their own holes or use natural holes or abandoned holes of other species in such trees and this trait tends to be characteristic of forest-adapted species.

Table 2. Numbers of priority bird species of boreal and temperate forests in Europe, according to their SPEC category (p. 17) and their dependence on the habitat. Priority status of each species group is indicated by superscripts (see 'Introduction', p. 19, for definition). 'Habitat importance' is assessed in terms of the maximum percentage of each species' European population that uses the habitat at any stage of the annual cycle.

Habitat importance	SPEC category					Total no. of species
	1	2	3	4	Non-SPEC	
Boreal forest						
<10%	0[B]	1[C]	7[D]	—	—	8
10–75%	1[A]	3[B]	12[C]	20[D]	—	36
>75%	1[A]	0[A]	3[B]	2[C]	18[D]	24
Total	2	4	22	22	18	68
Lowland temperate forest						
<10%	0[B]	2[C]	4[D]	—	—	6
10–75%	1[A]	3[B]	15[C]	22[D]	—	41
>75%	1[A]	3[A]	4[B]	11[C]	7[D]	26
Total	2	8	23	33	7	73
Montane forest						
<10%	1[B]	1[C]	3[D]	—	—	5
10–75%	0[A]	4[B]	10[C]	25[D]	—	39
>75%	0[A]	0[A]	1[B]	1[C]	0[D]	2
Total	1	5	14	26	0	46
Riverine forest						
<10%	0[B]	1[C]	4[D]	—	—	5
10–75%	1[A]	6[B]	15[C]	26[D]	—	48
>75%	0[A]	0[A]	0[B]	1[C]	1[D]	2
Total	1	7	19	27	1	55

Unmanaged old-growth forests generally have a higher diversity of such forest-interior bird species than managed forests, as they have numerous habitat features that are rarely found in managed forest (Petty and Avery 1990, Fuller 1995) and that are preferred by priority forest species (see Appendix 3, p. 354):

- Trees are higher and more massive, with larger crowns. They produce more seed and provide better nest-sites for raptors and other species which build large nests.

- There is more dead wood and of a larger diameter, providing foraging for woodpeckers, tits, *Sitta europaea* and *Certhia*, and larger cavities as nest-sites for woodpeckers and owls (Strigidae).

- The canopy is multi-layered.

- The spacing of trees is wider and more irregular, with the development of open areas and glades which provide good foraging sites for insectivorous and frugivorous birds.

- Tree-fall gaps are more numerous. As more light reaches the ground, ground vegetation slowly develops and may provide suitable nest-sites for some ground-nesting birds, as well as more insect food and berries for ground-dwelling species, e.g. gamebirds.

- Forest structure is highly varied and 'patchy' at a landscape level, due to numerous localized disease/storm/fire-damage events in the past.

- The shrub layer is absent in many areas.

In addition to forest physiognomy, tree-species composition can also be important in determining the distribution and density patterns of some priority bird species (James and Wamer 1982, Rotenberry

Table 3. Numbers of priority bird species of boreal and temperate forests in Europe, listed by priority status (see 'Introduction', p. 19). Percentages are proportions of the total numbers of priority species in each habitat-type.

Habitat-type	Priority status				Total no. of species
	A	B	C	D	
Boreal forest	2 (3%)	6 (9%)	15 (22%)	45 (66%)	68
Lowland temperate forest	5 (7%)	7 (10%)	28 (38%)	33 (45%)	73
Montane forest	0 (0%)	6 (13%)	12 (26%)	28 (61%)	46
Riverine forest	1 (2%)	6 (11%)	17 (31%)	31 (56%)	55

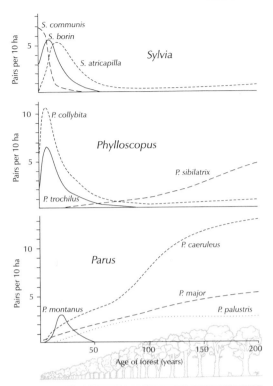

Figure 4. Changes in breeding density of birds of three genera in relation to age of oak *Quercus* forest in France (adapted from Ferry and Frochot 1970, in Fuller 1995).

1985). Some birds are mainly found in coniferous forests, including *Tetrao urogallus*, *Glaucidium passerinum*, *Picoides tridactylus* and *Parus cristatus*, and crossbills are strongly adapted to conifers, with Crossbill *Loxia curvirostra* being a generalist, *L. pytyopsittacus* and *L. scotica* favouring *Pinus*, and Two-barred Crossbill *L. leucoptera* mostly feeding on *Larix*. On the other hand, *Dendrocopos medius* is almost exclusively found in broadleaved *Quercus* woodlands and *Coccothraustes coccothraustes* shows a strong preference for feeding on *Carpinus betulus* seeds in winter. In the two closely related *Certhia* species, Treecreeper *C. familiaris* prefers tree species with smoother bark, e.g. conifers or *Fagus sylvatica*, while Short-toed Treecreeper *C. brachydactyla* favours trees with rougher bark, e.g. *Quercus cerris* and *Q. robur* (Szijj 1957).

The second major group of priority species are those that make use of the transition between the forest proper and open ground, and which require a more open forest structure incorporating mature trees, forest edges, glades and scrub. Such a structure is mimicked in forestry management by young coppice and young plantation systems, provided some mature

woodland or trees are retained. This group of birds includes species that are absent or scarce in old-growth, closed-canopy forest, e.g. *Tetrao tetrix*, *Luscinia* spp., Red-backed Shrike *Lanius collurio*, Greenfinch *Carduelis chloris* and *Emberiza rustica*. *Caprimulgus europaeus* and *Lullula arborea* also favour such transition areas where there are patches of bare ground, and have successfully adapted to the use of clear-fell areas in forests (e.g. Ranner 1990).

Forest edge, both internal and external, can differ in many ways from the interior. More light is available so that the growth of shrubs and trees is enhanced, leading to a higher primary productivity, which is likely to translate into higher insect abundances (Hansson 1983) and thus favour insectivorous birds. The open forest edge may also provide room for aerial displays and prey-catching techniques. Where scrub is missing from woodland mosaics, again it is sometimes mimicked through the forestry practice of coppicing, with *Luscinia* spp., *Prunella modularis* and Garden Warbler *Sylvia borin* being some of the species which can benefit from the creation of a mosaic of coppice underwood (Fuller 1982).

The third main group of priority forest birds are those which nest in forests but which conduct their feeding in other habitats nearby. Some raptors including *Aquila heliaca*, need tall trees, often with well-developed crowns, in which to build their large and spreading nests. Trees such as these are mainly confined to old-growth or mature forest. *Aquila clanga* may only nest a few hundred metres from the forest edge, but hunts widely over adjacent open wetland habitats such as wet meadows and marshes (Tucker and Heath 1994). *Ciconia nigra* also requires this juxtaposition of habitats, feeding inside the forest at marshland and streams but nesting in large mature trees. Other species which need trees for nesting but which feed in other habitats include Short-toed Eagle *Circaetus gallicus*, *Coracias garrulus*, and herons and egrets (Ardeidae).

The size of forests can be important to some species such as Goshawk *Accipiter gentilis*, *Bonasa bonasia*, *Dryocopus martius*, *Sitta europaea*, *Parus cinctus* and *Perisoreus infaustus*, which are more likely to occur in larger woods than in smaller ones and which are therefore susceptible to forest fragmentation (Opdam *et al.* 1985, Virkkala 1990). By contrast, *Troglodytes troglodytes*, *Erithacus rubecula* and Blackbird *Turdus merula* are found in all woods of whatever size, at least in the UK (Fuller 1982).

It should be remembered that some birds show considerable geographically-related variations in their habitat preferences. For example, in Britain *Certhia familiaris* is found in many woodland-types, but elsewhere in western Europe (where the closely related *Certhia brachydactyla* occurs) its main habitat is coniferous and mixed woodland. Fuller (1995) notes that differences between western and eastern Europe are particularly striking. In the east, Wood Warbler *Phylloscopus sibilatrix* is very common in mature coniferous forest, but it is almost exclusively a bird of broadleaved deciduous woodland in the west. Differences also extend to structural and landscape-scale factors. Woodpigeon *Columba palumbus*, *Turdus merula*, *Erithacus rubecula*, *Troglodytes troglodytes* and *Prunella modularis* are rarely found outside large tracts of forest in eastern Poland, but in western Europe they are frequently found in copses and gardens. This therefore limits the possibilities for defining, at the European scale of this study, the habitat requirements of many of the priority species at a greater level of detail than that given in Appendix 3 (p. 354).

THREATS TO THE HABITAT

The scale and impacts of the threats to boreal and temperate forests, and to the habitat features required by their priority bird species, are summarized in Table 4. The predicted impacts of each threat on each priority species over the next 20 years were also assessed for the habitat (Appendix 4, p. 401), the results being summarized in Table 5 and Figure 5. The threats are discussed below in order of importance (see Figure 5).

Inappropriate forest management

Inappropriate management of Europe's forests has drastically reduced the ecological quality of the habitat, and is the most important overall threat to boreal and temperate forests and to their priority birds in Europe, and to lowland temperate and montane forests in particular. It is predicted to have a high impact on the European populations of 12 species over the next 20 years, with low impacts on a further 40 species. Two main processes have contributed to this, as follows.

Intensification of management practices

Conventionally managed forests differ from natural forests in major ways (Harris 1984a, Hunter 1990, Angelstam 1991, 1992, Esseen *et al.* 1992, Peterken 1993). For example, stand structure and tree-species composition are altered and the frequency of (and variation within) growth-stages is changed (Van Wagner 1978). Even if large clear-cuts sometimes superficially mimic the large-scale character of forest-fires (natural or otherwise), it is important to note that burned (or storm-damaged) areas have an enormous structural diversity compared with conventional clear-cuts. For example, usually less than 10%

of the wood is consumed during a fire, leaving large volumes of standing and downed wood (Foster 1983, Payette *et al.* 1989), while the great majority of the standing wood is removed after clear-cutting.

Comparisons of forest structure in managed Swedish and in natural Russian boreal forest show that the number of dead standing trees is 33 times higher in the latter than in the former (Angelstam 1996). Currently the proportion of dead trees in most managed forest in Europe is low, e.g. less than 2% of the stock in Sweden (Kempe *et al.* 1992).

Managed temperate forests rarely contain such complex mosaics of trees as those found in natural stands, due to thinning operations and the long-term effects of past harvesting and planting systems (see p. 209). The majority of managed forests are clear-felled and consist of large-scale patches of uniform-aged trees of a small number of species, which grow through the following distinct stages: clear-fell or establishment phase, thicket phase, early pole stage and mature stage. This may favour some open-ground bird species in the early stages of re-establishment after felling, but fine-scale structural and tree-species diversity becomes low towards maturity. At a whole-forest level though, overall habitat diversity may be high, albeit as a very coarse mosaic (Fuller 1995).

By contrast, selection-felling systems promote a diverse and complex foliage profile and a high foliage volume, which may be of benefit to species needing a dense shrub layer. However, at a whole-forest scale, different growth-stage patches may be absent, which may result in a lower overall habitat diversity. Nevertheless, if thinning is used to provide an optimum balance between the density of mature trees and canopy openness, forests can be produced that are extremely rich in bird species of both mature forest and young-growth (Fuller 1990).

Two other variants on the contrasting practices of clear-felling and selection systems are group-felling and shelterwood systems. Group-felling produces a forest of small patches, typically some 50 m in diameter. This may provide more young growth than selection systems, but open areas are much smaller than under clear-fell systems and are unlikely to be large enough for many open-ground bird species (Fuller 1995). In shelterwood systems, the size of patches felled is similar to clear-fell areas, but the retention of a few mature trees leads to a far richer avifauna (Ferry and Frochot 1970, Smith 1988) and also protects soil and saplings to some extent.

In all the management systems all, or at least most, trees are typically harvested long before their biological maturity (especially in coppice systems) and dead wood and mature trees are thus rare in managed forests. A general quantitative recommen-

Table 4. Threats to boreal and temperate forests, and to the habitat features required by their priority bird species, in Europe (compiled by the Habitat Working Group). Threats are listed in order of relative importance (see Figure 5), and are described in more detail in the text.

Threat	Effect on birds or habitats	Scale of threat Habitat	Past	Future	Regions affected, and comments
1. Inappropriate forest management					
a. Intensification of management practices	Lack of old trees, shorter rotations, use of machinery, removal of dead wood, reduced tree species and structural diversity	All	■	■	All countries
b. Afforestation and reforestation (with plantations)	Creation of homogeneous stands (often of non-native species) at expense of biologically diverse old-growth forest	All	■	■	All countries
2. Logging (of natural and semi-natural old-growth forests)	Permanent habitat loss or, more frequently, temporary loss of tree-cover, often followed by intensification of forest management	B T M,R	■ ■ •	■ • •	Eastern Europe, Fennoscandia, Russia
3. Recreation and disturbance	Disturbance of birds; damage to vegetation and soils through trampling and vehicles; fire; over-collection of forest products	B T,M,R	• •	• •	All countries, but especially western Europe
4. Transport infrastructure	Habitat loss and degradation	All	•	•	All countries
5. Habitat fragmentation	Increased likelihood of disturbance; reduced forest size and therefore viability for species with large territories; inhibited dispersal of some species; increased in 'edge' and loss of interior habitats	B T,M,R	• ■	■ •	All countries
6. Clearance for agriculture	Habitat loss	B,M T R	• ■ ■	• • •	Biggest cause of forest loss since prehistory; fertile river valleys still under threat
7. Intensified water management					
a. River control	Flood protection and channelization may reduce seasonal flooding and dry out soils	B,M,T R	• ■	• ■	All countries
b. Dam construction	Flooding of forest or reduced flows and flooding episodes downstream	B,M,T R	• ■	• •	All countries
c. Drainage	Loss of wooded mires and riverine forest, etc.	B,T R	■ ■	• •	Scandinavia, Russia, alongside rivers in other central European countries
8. Overgrazing (by livestock and game)	Reduced habitat quality through changed forest structure and composition and, in severe cases, prevention of regeneration	M,T R B	• • •	■ • •	Widespread
9. Air pollution	Tree damage affecting foliage and growth; death of trees; lichen loss; acidification of forest soils; increase in rank grass	B,R T,M	• •	• •	Temperate Europe east of UK, especially southern Scandinavia and Alps
10. Severe or frequent fires	Opening of habitat, simplification of vegetation structure, and excessive erosion	B T,M R	■ • •	■ • •	Russia especially
11. Inappropriate pesticide use (in forestry)	Indirect effects, through reduction in plant and animal food-sources; direct effects through poisoning	All	•	•	

Scale of threat

B	Boreal forest	Past	Past 20 years	■	Widespread (affects >10% of habitat)	
T	Lowland temperate forest	Future	Next 20 years	●	Regional (affects 1–10% of habitat)	
M	Montane forest			•	Local (affects <1% of habitat)	
R	Riverine forest					

Table 5. The proportions of priority bird species that are adversely affected by different threats to boreal and temperate forest-types (and to habitat features required by these species) in Europe. Numbers are percentages of the total number of priority species in each forest-type. Threats are ordered and numbered as in Table 4 and Figure 5. Appendix 4 (p. 401) gives full details of the threat assessment.

Threat	Boreal forest (%)	Lowland temperate forest (%)	Montane forest (%)	Riverine forest (%)
1. Inappropriate forest management	38	53	44	35
2. Logging	44	42	33	42
3. Recreation and disturbance	12	26	21	24
4. Transport infrastructure	7	15	10	5
5. Habitat fragmentation	10	16	6	7
6. Clearance for agriculture	9	14	0	24
7. Intensified water management	25	1	0	22
8. Overgrazing	13	15	13	0
9. Air pollution	19	14	25	0
10. Severe or frequent fires	22	5	6	5
11. Inappropriate pesticide use	4	11	0	0

dation on how much dead wood is ecologically optimal in a forest is not easy to make, since the needs of different bird species for different forms of wood (tree species, standing or lying, etc.) are dissimilar or still not clear (Scherzinger 1995). Some 'moderate' recommendations, for conventional commercial forestry in central Europe, are for 50% of the

dead wood to be standing and 50% to be lying, and to allow up to 1–2% of the total stock of stems (with diameter at breast height greater than 20 cm) to remain in the forest if dead (Utschick 1991, Ammer 1994).

Logging operations with heavy machinery at least partially destroy the shrub layer, and can compact

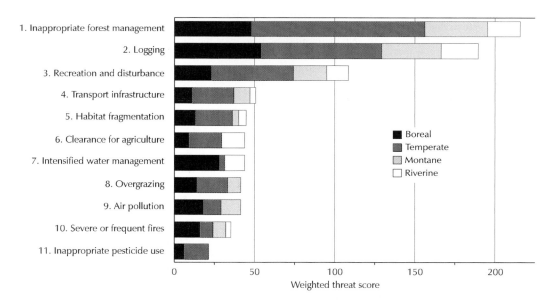

Figure 5. The most important threats to boreal and temperate forests, and to the habitat features required by their priority bird species, in Europe. The scoring system (Appendix 4, p. 401) takes into account:

- The European Threat Status of each species and the importance of the habitat for the survival of its European population (Priorities A–D); see Table 1.
- The predicted impact (Critical, High, Low), within the habitat, of each threat on the European population of each species; see Appendix 4 (p. 401).

See Table 4 for more details on each threat.

the soil such that plant regeneration is heavily reduced. Planting creates uniformly aged monospecific stands which only support a reduced diversity of wildlife. Forest is fragmented by clear-cutting, the increased density of forest roads required by intensive forestry greatly increases the amount of 'edge' habitat, so reducing the amount of 'interior' habitat.

Apart from these major impacts, intensification of management can lead to all-year-round logging and other forestry operations, which can increase disturbance, and lead to desertion or even destruction of nests of large priority birds (e.g. raptors or *Ciconia nigra*)—see 'Recreation and disturbance' (below).

With so many impacts on all of the main habitat features of forests, a large number and variety of priority bird species are likely to suffer detrimental impacts from intensive forestry management. However, the most likely to be seriously threatened are those of old-growth forest (Dudley 1992, Virkkala *et al.* 1993, Wesolowski 1995).

Afforestation and reforestation

Extensive afforestation and reforestation have resulted in the replacement of native, usually deciduous, broadleaved trees by uniform-age monocultures of a few highly productive, non-native coniferous tree species in some European countries. The modification of natural and semi-natural forests has recently become more severe with the increasing importance of pulp and fibre production. Consequently, a far wider range of tree species can be exploited, and rotation times of plantations shortened (Stanners and Bourdeau 1995).

For example, in Denmark just over half the country's total area of forest is made up of non-native species, primarily *Picea*. In Germany, two-thirds of the original broadleaved forest has been replanted with conifers (Ibero 1994). Lodgepole pine *Pinus contorta* is the main non-native tree grown in the production forests of northern Europe and accounts for about 3,300 km^2 in northern Sweden alone. Sitka spruce *Picea sitchensis* and *Pinus contorta* are the two species normally planted on mires (after drainage). On lowlands and flood-plains, extensive plantations of hybrid poplar (*Populus* ×*euramericana*) are grown. False acacia *Robinia pseudacacia* covers large areas of sandy soils and degraded hillsides in central Europe. In Portugal and Spain large plantations of *Eucalyptus* have been established.

The impact of afforestation with non-native species is particularly severe when the original structurally diverse forest is replaced by a monospecific stand, which alters not only forest structure but in many cases also the soil characteristics as well. For example, the preliminary drainage and ploughing,

and subsequent scavenging of pollutants from the air by conifer plantations all tend to increase soil acidity, while dropped *Eucalyptus* leaves contain aromatic chemicals (e.g. menthol) which strongly inhibit the development of undergrowth and makes them unpalatable for most animals. There are also great reductions in light availability in the lower layers of dense conifer plantations. Overall, these changes lead to the general impoverishment of forest fauna and flora. Typically, mature plantations hold an impoverished bird assemblage, few species of which are priority birds as identified in this book (Avery and Leslie 1990, Petty and Avery 1990, Fuller 1995).

The most detrimental properties of plantations can, however, be alleviated to some degree by, for example, maintaining unplanted areas within the forest, avoiding clear-fell systems, retaining dead wood and patches of old-growth, and incorporating patches of non-crop trees (Ratcliffe and Petty 1986, Avery and Leslie 1990, Petty and Avery 1990, Fuller 1995, Angelstam 1996).

In addition to the threat to semi-natural and natural forests, afforestation is also a significant threat to open, uncultivated habitats in Europe, in particular moorland, mires, heathland, sand-dunes and steppic habitats (see relevant chapters).

Logging

Overall, Europe's forest area has increased by over 10% since the early 1960s (Stanners and Bourdeau 1995). However, nearly all of this is due to the creation of plantations and to secondary regrowth on abandoned agricultural land. This increase has masked the continued logging of natural and semi-natural old-growth forest, particularly in Fennoscandia and eastern Europe where such forest is still much more extensive than elsewhere. Until recently, logging in the vast boreal forests of Russia was scattered and at relatively low intensity, and adequate regeneration was achieved naturally. In the Komi Republic of Russia, for example, 35% of boreal forests remain unexploited. However, logging of such forest has greatly increased since 1990, and Russian forests are now under increasing threat from internal mismanagement and from the short-term opportunism and competition of foreign timber companies (Dudley 1992) with their more intensive techniques of exploitation. In European Russia, for example, logging concessions have recently been granted to timber corporations from Japan, South Korea, USA and Finland (Gordon 1992, Grigoryev 1992).

It is feared that there is little political will or incentive for these companies to log in a sustainable manner. Although many east European countries have legal controls (of varying quality) on logging, implementation is not always strong enough to provide sufficient environmental protection, especially in the more remote areas, despite the best efforts of the authorities (Dudley 1992).

Small-scale logging is also increasing in a number of central and east European countries, due to limited law enforcement, privatization of land, and the worsening economic situation, among other factors (IUCN 1995). Such logging tends to concentrate on the most accessible areas, especially near roads, rivers and settlements.

In addition to the logging of boreal forest, the loss of the few remaining areas of semi-natural or natural riverine forest in Europe is also of special concern. For example, the only remaining riverine forest along parts of the Danube (in Hungary, Serbia and Romania) survives under a relatively natural flooding regime purely because of a historical delay in the engineering that has affected the hydrology of most other large European rivers. Between 1965 and 1992 alone, more than 40 km^2 of (mainly) riverine forest was cleared for the construction of dams for hydroelectric power generation in Austria.

This is the most serious threat to boreal and riverine forest, and is predicted to have an impact on 60 priority species overall, with the impact being high for six species in particular.

Recreation and disturbance

Nearby management operations (e.g. forestry or agriculture), or recreational use by the general public, can reduce the breeding success of certain large priority species which are particularly sensitive to human activity at or near their nest-sites, e.g. birds of prey and *Ciconia nigra*, through increased disturbance of essential behaviour (e.g. incubation) and through nest-desertion.

Locally, sheer numbers of people can also damage vegetation and increase erosion through trampling. For example, in a study in the Netherlands, local declines in a large number of bird species over 30 years were attributed to increasing recreational pressure which damaged the field layer of forest (Jansen and de Nie 1986).

The collection of berries, mushrooms, herbs and medicinal plants is partly recreational, but is increasingly of economic importance in European countries or regions where unemployment is high and/or incomes have declined. Where this occurs on a large scale, and the use of vehicles and motorbikes is widespread, then wildlife can be disturbed and soils and vegetation damaged (IUCN 1995).

Most European governments consider that forests are valuable for public recreation (see 'Values, roles and uses', p. 213). Disturbance from recreational activities is a particular problem in montane forests, since many are popular tourist areas in winter as well as in summer. For example, in Austria, disturbance from mountain-biking, off-piste skiing tours and paragliding are important and increasing problems, particularly for raptors and for gamebirds such as *Tetrao urogallus* (Georgii et al. 1994, Zeitler 1994). Overall impacts on bird populations of these habitats in Europe are poorly documented and difficult to predict—26 priority species may be affected, including two seriously—but it is likely this problem will increase in scale in the future.

Transport infrastructure

Construction of roads and railways through forests leads to direct habitat loss and indirect degradation of the surrounding forest, the indirect impacts (such as increased disturbance from recreation, hunting and forestry, or habitat fragmentation) are poorly understood. Densities of several bird species that breed in woodland appear to be substantially reduced when adjacent to busy roads (Reijen and Thissen 1987). Noise pollution and habitat fragmentation (see 'Habitat fragmentation', below) are among the causative factors proposed.

Cumulatively, such local effects may have a larger impact, since existing transport networks are already so widespread, and further expansion of such corridors can be expected throughout Europe. For instance, the Trans-European Transport Networks proposed for the European Union (EU) (CEC 1994b) involve the building of 10,000 km of new motorway and 9,000 km of new high-speed railway in the EU over 10 years. Approximately 2% of Important Bird Areas (IBAs) in the EU are located within 2 km of the proposed new infrastructures, and 12% of IBAs are within 10 km (Bina et al. 1995).

Overall though, the percentage of forest directly destroyed at a European scale is likely to be relatively small, and overall population impacts are predicted to be negligible for most species. The forest-type most threatened by such destruction is riverine forest, since it is distributed along natural lines of communication (i.e. river valleys in lowlands), and has only a small total extent in Europe. New roads continue to be built through montane forests—at a high rate in some countries (e.g. Austria). Some particularly sensitive species, such as some birds of prey, may also be indirectly affected locally, particularly in temperate forest in the more densely populated lowlands of Europe.

Habitat fragmentation

Habitat fragmentation by man is a unifying theme in the history of European forests as well as being an explanation for local and regional extinctions of forest species in the past. Fragmentation occurs when a continuous habitat is transformed into a number of smaller patches of smaller total area, isolated from each other by a matrix of habitats unlike the original (Angelstam 1996).

Continuing fragmentation of remaining patches of forest still commonly occurs in some parts of Europe, and may be a significant threat to forest biodiversity in central Europe and Scandinavia (Ibero 1994). The creation of forest roads for logging and construction projects, along with clear-felling for timber, are frequent causes of fragmentation. The very small area of forest that is destroyed is much less important than the indirect consequences of such destruction, which include the promotion of human disturbance, amplification of predation rates in surrounding forests, and inhibition of dispersal and reduction in home-range size for some of the most forest-dependent species. For example, Wilcove (1985), Andrén and Angelstam (1988), Andrén (1992) and Angelstam (1992) found that predation rates on birds were greater at the forest/farmland edge than in the forest interior.

Changes in bird communities due to fragmentation can be explained by the reduction in the total area of forest stands, and their increased isolation, increased 'edge' (perimeter length) and increased heterogeneity (Wiens 1989b). A decrease in forest area is usually paralleled by a reduction in the number of forest bird species (Moore and Hooper 1975, van Dorp and Opdam 1987), although there are exceptions (e.g. Haila *et al.* 1987). Increased edge leads to a decrease in the number of forest-interior bird species (Freemark and Merriam 1986). Opdam *et al.* (1993) demonstrated negative effects of fragmentation on some forest birds, including 'interior' species such as *Phylloscopus sibilatrix* and *Coccothraustes coccothraustes*. Less opportunistic bird species with limited dispersal appear to be more susceptible to fragmentation effects (Enoksson *et al.* 1995). Local extinction rates of forest birds in isolated patches of woodland can be quite high (up to 61%) in smaller areas of woodland (van Noorden 1986). This does not take into account recolonization, but more isolated woodland would have less chance of being recolonized since it is more difficult for birds to travel over greater distances.

Clearance for agriculture

This has been the major cause of loss of natural forest since the beginning of cultivation in prehistory. However, it has become much less significant during the twentieth century, since virtually all high-quality land is already under agricultural production. Recently, a reverse trend has developed in, for example, the European Union, where the afforestation of arable land is now subsidized by Regulation 2080/92. Abandonment of agricultural land is also widespread, particularly in eastern and southern Europe, and is leading to further increases in forest area.

However, some forest loss to agriculture is still possible in fertile river valleys. For example, the remnant patches of riverine forest in the Po valley (Italy) are under great pressure for clearance for agriculture or for plantations of *Populus*, partly as a result of the dense population of 20 million people living in the area (Ibero 1994). However, this threat is unlikely to have a major overall impact on forest and forest birds in Europe, unless changes in global trade-patterns in the future put pressure on Europe to produce significantly more crops.

Water management

Dam-building, flood protection (e.g. building of embankments) and channelization along rivers cause major ecological damage to riverine forest since they may reduce seasonal flooding, reduce natural fertilization, increase vegetation density (due to loss of mechanical stress) and dry out the soil. The last semi-natural riverine forests of the Danube and its tributaries (especially the Mura, Drava and Sava rivers) are endangered by the construction of hydro-electric power plants, although at least 360 km^2 of forest and flood-plain were recently designated as national parks (WWF 1996). Similarly, plans to build a cascade of dam lakes along the whole lower section of the Wisla in Poland also threaten the remnant riverine forests here (T. Wesolowski pers. comm.). In Spain alone there are almost 1,000 dams which, overall, are the second most important threat to riverine forest in that country (Ibero 1996). Habitat loss and damage occur not only through the loss of forest as the water-level rises but also because of the necessity to build roads through forest for access to the construction site, e.g. along the Nestos river in Greece (Ibero 1994; see 'Transport infrastructure', above).

Of a total of some 70,000 km^2 of 'wet' forest originally present in Finland at the beginning of this century, more than two-thirds have been drained (Bernes 1993, Stanners and Bourdeau 1995). Drainage (including the ploughing of open forest) affects wet lowland forest-types in both the boreal and temperate forest zone, although currently it is probably more widespread and has more impact in the former. Drainage may be preceded by logging of old-growth forest, or followed by an intensification in forest management, or the forests are converted into monospecific plantations, often of non-native species. Priority bird species which are particularly affected by drainage are those that require a mosaic of forest and small wetlands for breeding, such as *Aquila clanga*.

Overgrazing

Overgrazing as a result of the pasturage of sheep and cattle in forests was an important early cause of forest loss since the regeneration of young trees and seedlings was prevented. In recent centuries such practices have been strongly in decline. However, the cutting and/or burning of subalpine forests to increase the area of pasture still occurs in the mountains of Bosnia-Herzegovina and Serbia, although the levels of grazing, even in these countries, is decreasing. Grazing in montane forests, especially by goats, is also a problem in Bulgaria and Romania, where it limits forest regeneration and causes soil erosion, and is also probably widespread in the Balkan mountains (IUCN 1995). A new regulation in Romania forbids grazing in young and protected forests.

Persecution of wild carnivores, reductions in sport hunting in some areas, increases in commercialized hunting, introduction of non-native species (e.g. grey squirrel *Sciurus carolinensis* and muntjac deer *Muntiacus*) and apparent amelioration of winter conditions have all recently led to considerable recent increases in the populations of wild herbivores, such as *Cervus elaphus, Capreolus capreolus* and *Sus scrofa*, in many European countries (e.g. Mitchell and Kirby 1990, Jedrzejewski *et al.* 1992, 1993). In central Europe the desire of hunters to maximize populations of such game species has also led to winter-feeding. Subsequently, damage to young forestry plantations (and to adjacent agriculture) has become common, and over-browsing is also widespread in the boreal forests of Fennoscandia, where high numbers of *Alces alces* are considered to be reducing the number of *Populus tremula* trees (favoured by hole-nesting species) (Angelstam 1996), and in Scotland where *Cervus elaphus* is the main threat to remaining ancient forest (Wynde and Burrows 1991).

The effects of high grazing and browsing pressure on forests, and in turn on their priority birds, are likely to be complex and dependent on the forest-type, the grazing animals involved, the grazing period and intensity, and the species of bird present (Table 6). However, effects can be expected to include alteration of the vegetation structure, changes in plant composition, reduced plant diversity and reduced food availability (Fuller 1995). Heavy grazing may reduce or even virtually eliminate the field and shrub layer below a browse-line of 1.5–2.0 m (Putman *et al.* 1989, Mitchell and Kirby 1990). This may benefit open-forest species such as Pied Flycatcher *Ficedula hypoleuca* and *Anthus trivialis*, but is detrimental to species requiring a well-developed shrub and field layer, such as nightingales *Luscinia* or gamebirds such as *Bonasa bonasia* (Remmert 1973, 1980, IUCN 1995). High numbers of *Sus scrofa* may lead to very high nest-predation rates for some ground-nesting priority species, e.g. Eagle Owl *Bubo bubo*.

Overall, it is likely that at least a moderate number of priority species (17) will be affected, and that overgrazing will probably reduce total bird diversity in a forest but impacts are difficult to predict. Clearly, excessive grazing and browsing may block forest regeneration in the longer term, leading to forest loss which may affect a considerably larger number of species.

Air pollution

Atmospheric pollution, originating mainly as the gaseous exhaust products of industry, vehicles and power-stations, and taking the form of acid deposition and photochemical smog, is increasingly agreed to have caused extensive damage to forests in Scandinavia and central and eastern Europe, especially to forests of *Picea abies* and silver fir *Abies alba* (UNECE 1991a, Anon. 1993a, Stanners and Bourdeau 1995). Direct exposure to high emissions of sulphur dioxide can cause defoliation, reduced growth and low survival of trees (Prinz 1987, Krause 1989). Such locally intense damage has occurred in

Grazing pressure	Effects on vegetation	Possible implications for birds
High	• Very reduced shrub layer • Sparse field layer • No tree regeneration • Bare soil patches • Fewer shrub species	• Few shrub-nesting birds • Heavy predation on ground-nesters • Reduced food resources (berries, foliage invertebrates, small mammals)
Moderate or low	• Patchy shrub layer • Diverse field layer • Few bare soil patches	• Nest-sites and feeding sites available for a wide range of species
None	• Dense shrub layer • Diverse field layer • Few ground-nesters	• Many shrub-nesting birds • Few open woodland birds

Table 6. Possible implications, for birds, of different intensities of grazing pressure in forest (adapted from Fuller 1995).

industrial regions such as the Ruhr (Germany), the Czech Republic and western Poland. It is estimated that such pollution can reduce the potential timber harvest of a forest by up to 40% (Moldan and Schroor 1992, Angelstam 1996).

In addition to such direct physical damage from sulphur dioxide, close to heavily polluting industrial areas, there is also evidence to suggest that the more widely dispersed acidic deposition resulting mainly from emissions of sulphur dioxide and nitrous oxides is harmful to tree growth, making trees more susceptible to drought or to attack from insects and fungi (Stanners and Bourdeau 1995). An international co-operative programme to assess and monitor the effects of air pollution on forests among 34 European countries (excluding Russia) found that 24% of all sample trees taken from 184 million ha of coniferous and deciduous forests were damaged, that is, with defoliation greater than 25% (Anon. 1993a). Trees most affected were usually in older age classes (Stanners and Bourdeau 1995). The exact causes of this damage are still unclear but appear to relate to one or more of the following factors: multiple stress, soil acidification and resulting aluminium toxicity, ozone interactions with acid mist, magnesium deficiency, and excess nitrogen deposition (Prinz 1987, Krause 1989, Roberts et al. 1989, UNECE 1991b, Stanners and Bourdeau 1995).

In addition to impacts on trees, detrimental effects have been noted on the soil life and epiphytic and ground flora of forests (see Dudley 1992). Direct effects on animals have not been thoroughly studied, but some effects on birds have been observed. For example, various studies on tits Parus have shown that damage to conifer forests can disrupt species' feeding ecology and reduce populations of insectivorous birds (Drent and Woldendorp 1989, Gunnarsson 1990, Zang 1990, Hake 1991, Möckel 1992, Pettersson et al. 1995).

Furthermore, in parts of central Europe with pronounced forest decline, the overall breeding bird populations of forests are strongly affected by the severity of damage. In the Czech Republic, within any particular age of Picea forest, the individual density and the species diversity of breeding birds decreased with increasing severity of damage (Šťastny and Bejcek 1985, Flousek 1989). Such overall changes in density, however, conceal differences in the responses of individual species. Thus, Flousek (1989) found that, although many species declined, others remained stable, and some priority species favouring open-canopy forests (such as Anthus trivialis) increased. Other species may also benefit from increases in dead wood in damaged forests. However, such benefits may be short-lived because the most damaged stands are often clear-felled.

Overall though, severe air pollution is still relatively localized at a European scale and therefore it is predicted to cause a population decline in only a small percentage of priority birds in all forest-types. However, such prognoses are uncertain, and the longer-term impact of more widespread chronic air pollution, resulting in more widespread and severe forest degradation, could potentially affect a much larger number of species.

Severe or frequent fires

About 7,000 km² of wooded land in Europe is damaged by fire each year (Stanners and Bourdeau 1995). The number and area of forest-fires can be quantified relatively accurately in western Europe, while in eastern Europe there are fewer statistics available. Fire can be beneficial in managing some, but not all, forest sites or ecosystems (see 'Ecological characteristics', p. 206). However, long-term damage to soils, forest structure or regeneration processes will result if forest-fires are too extensive, intense or frequent. Nearly all forest-fires in Europe are caused, accidentally or deliberately, by people: UNECE–FAO statistics suggest that in 1987 natural fires made up only 3% of forest-fires in Finland and less than 0.1% in former West Germany.

Forest-fires occur naturally in all four main forest-types considered here, but are an especially important phenomenon in the conifer-dominated boreal forests, since conifers contain a high proportion of flammable substances like resin and wax and therefore burn more readily. Indeed, naturally occurring fires are an important part of the natural regeneration cycle of boreal forests, and maintain the vegetation in a varied and dynamic state. Natural fires in this forest-type consume accumulated dead material and old trees, opening up large areas of previously moss-blanketed ground for successful germination and growth of young shoots and saplings, and creating an opportunity for successional development. Not surprisingly, the threat of too-severe or too-frequent fires is considered to have more impact on priority birds of boreal forest than on those of the other three main forest-types (15 species as opposed to 3–4 elsewhere). Forest-fires are also of considerable importance in Mediterranean shrubland and forest habitats (see 'Mediterranean Forest, Shrubland and Rocky Habitats', p. 239).

Inappropriate pesticide use

Priority birds which breed in younger plantations may be affected by pesticide use within such habitats, especially those species (e.g. Caprimulgus europaeus, shrikes Lanius) which specialize in feeding on larger invertebrates, although population impacts in all cases are likely to be low on a European scale. High levels of organochlorines have been found in some priority species of forest which feed in mosaics of woodland and agricultural habitat, e.g. *Strix aluco* (Battisti *et al.* 1996), and which presumably bio-accumulate pesticide residues from contaminated prey from agricultural land. Crops such as maize or vegetables which have replaced riverine forests often require heavy use of pesticides. The chemicals can pass into the groundwater and damage surrounding areas of woodland (Yon and Tendron 1981).

CONSERVATION OPPORTUNITIES

International legislation

Box 1 summarizes the international legislation relevant to the conservation of boreal and temperate forests in the wider environment. Further information is provided in the introductory chapter on 'Opportunities for Conserving the Wider Environment' (p. 24). Conventions that provide the opportunity to protect discrete areas of threatened habitat, or sites with important populations of threatened species, include the World Heritage Convention, the Natura 2000 network of the Birds and Habitats Directives of the EU, and the EMERALD network of the Bern Convention (Box 1).

The designation of Special Protection Areas (SPAs) under the Birds Directive has been slow in Europe (Mayr 1993, CEC 1994c, Waliczky 1994, Anon. 1996). For example, in Italy 31 out of 80 SPAs designated so far are mostly forested. However, these represent less than 10% of the 'mostly forested' Important Bird Areas in Italy.

The Habitats Directive aims to combine the protection of threatened species with a wider objective of conserving habitats that are of interest in their own right. Annex I of the Directive lists 22 such habitat-types within boreal and temperate forests (as defined here), three of which are 'priority' habitats which are required to have a stronger level of protection (Appendix 5, p. 423).

Annex II of the Environmental Impact Assessment (EIA) Directive (Box 1) lists project-types which should be subject to such assessment. It does not include forestry projects, apart from initial afforestation schemes where these may lead to adverse

---

Box 1. International environmental legislation of particular relevance to the conservation of boreal and temperate forests in Europe. See 'Opportunities for Conserving the Wider Environment' (p. 24) for details.

GLOBAL

- *Biodiversity Convention* (1992) Convention on Biological Diversity.
- Framework Convention on Climate Change (1992)
- *Bonn Convention* or *CMS* (1979) Convention on the Conservation of Migratory Species of Wild Animals.
- *Ramsar Convention* (1971) Convention on Wetlands of International Importance especially as Waterfowl Habitat.
- *World Heritage Convention* (1972) Convention concerning the Protection of the World Cultural and Natural Heritage.
- Convention on Long Range Transboundary Air Pollution (1979).
- Convention on the Protection and Use of Transboundary Watercourses and International Lakes (1992).

PAN-EUROPEAN

- *Bern Convention* (1979) Convention on the Conservation of European Wildlife and Natural Habitats.
- *Espoo Convention* (1991, but yet to come into force) Convention on Environmental Impact Assessment in a Transboundary Context.

EUROPEAN UNION

- *Birds Directive* Council Directive on the conservation of wild birds (79/409/EEC).
- *Habitats Directive* Council Directive on the conservation of natural habitats of wild fauna and flora (92/43/EEC).
- *EIA Directive* Council Directive on the assessment of the effects of certain public and private projects on the environment (85/337/EEC).
- Council Directive on urban waste water treatment (91/271/EEC).
- Council Directive on the control of dangerous substances discharge (76/464/EEC).
- Council Directive on the conservation of fresh water for fish (78/659/EEC).
- Council Directive on fire control (86/3259/EEC).
- Council Directive on the freedom of access to information on the environment (90/313/EEC).
- Council Regulations on the Structural Funds (e.g. 2081/93/EEC).
- Council Regulation on the prevention of forest fires (2158/92/EEC).
- Council Regulation on afforestation of agricultural land (2080/92/EEC).
- Council Regulations on the protection of the Community's forests against atmospheric pollution (3529/ 86/EEC and 1696/87/EEC).

ecological changes. Other project-types which may have a negative impact on forests are, on the other hand, included (e.g. water management, extractive industries, infrastructure projects).

The European Union does not have a common forestry policy, but it does have many measures in place which are relevant to the forestry sector, and it offers financial assistance for a number of forestry activities, such as monitoring tree condition, preventing forest-fires, improving conditions for the marketing and processing of forestry products and woodland improvement. Regulation 2080/92, one of the package of CAP reform measures adopted in 1992, provides financial assistance for the creation of new woodlands and forests on agricultural land. The Regulation requires member states to prepare a national programme for its implementation. Forestry will also be affected by the implementation of other parts of the CAP reform package. The accession of Austria and the Scandinavian countries to the EU has markedly increased the size of the Community forestry-sector markets.

The UN Conference on Environment and Development (UNCED) in Rio de Janeiro (the Earth Summit) adopted the 'Statement of Forest Principles', the first global consensus on the management of the world's forests. Also adopted were other agreements which bear on forests, including Agenda 21, the Biodiversity Convention, and the Framework Convention on Climate Change (see 'Opportunities for Conserving the Wider Environment', p. 24). There is currently no intergovernmental legislation dealing specifically with the conservation of natural and semi-natural boreal and temperate forests. Attempts to introduce such a global forest convention at the Earth Summit were rejected. Instead a far less significant 'non-legally binding authoritative statement of principles for a global consensus on the management, conservation and sustainable development of all types of forest' was adopted. Instead of setting forests into an international framework, the statement stressed the 'sovereign and inalienable right' of states to manage forests as they judge best, although with the proviso that management should not 'cause damage to the environment of other States' (Dudley 1992, Anon. 1994a). The guiding objective of the principles was 'to contribute to the management, conservation and sustainable development of forests and to provide for their multiple and complementary functions and uses' (Anon. 1994a).

The Biodiversity Convention has the potential to act as an important framework for forest conservation. However, until it starts to be implemented, its precise scope remains unclear (Dudley 1992).

Policy initiatives

The European Ministerial Conference on the Protection of Forests, held in Helsinki in June 1993, agreed and signed four sets of guidelines. These have become known as the 'Helsinki Guidelines':

- General Guidelines for the Sustainable Management of Forests in Europe.
- General Guidelines for the Conservation of the Biodiversity of European Forests.
- Forestry Cooperation with Countries with Economies in Transition.
- Strategies for a Process of Long-term Adaptation of Forests in Europe to Climate Change.

These guidelines built upon the 'Statement of Forest Principles' agreed by governments at the Earth Summit in 1992, and they define 'sustainable forest management' for policy-makers and foresters from the perspective of European governments.

The follow-up to the Helsinki conference (the so-called 'Helsinki Process') has become one of the main ongoing political processes in international forest policy. The Helsinki Process has sought to identify measurable criteria and indicators for the evaluation of how European countries have progressed in their efforts to follow the principles of sustainable forest management and conservation of the biological diversity of European forests. A core set of criteria and indicators was adopted at a follow-up meeting in June 1994. These criteria and indicators are the tools for gathering and assessing information on how the signatory states have succeeded in implementing the general guidelines for sustainable forest management as described in the Helsinki resolutions (e.g. Cosgrove and Turner 1995). The six European criteria are:

1. Maintenance and appropriate enhancement of forest resources and their contribution to global carbon cycles.

2. Maintenance of forest-ecosystem health and vitality.

3. Maintenance and encouragement of the productive function of forests (wood and non-wood).

4. Maintenance, conservation and appropriate enhancement of biological diversity in forest ecosystems.

5. Maintenance and appropriate enhancement of protective functions in forest management (notably soil and water).

6. Maintenance of other socio-economic functions and conditions.

In response to the Biodiversity Convention, a Pan-European Biological and Landscape Diversity Strategy (PEBLanDS) was approved by virtually all of the environmental ministers in Europe at Sofia in 1995. Action Theme 9 of the Action Plan for 1996–2000 of this 20-year strategy concerns forest ecosystems. The main objectives include the conservation of all remaining virgin forests, and semi-natural and natural riverine forests, initiation of a study on the necessary adjustments to European forest-management systems, and the promotion of an action plan for biodiversity, landscape and ecological networking considerations for forest management. At the regional level it aims to initiate the restoration and regeneration of the most important fragmented forests in central and eastern Europe and the Atlantic region, to establish a programme to evaluate the impact of market conditions and privatization on forests, and to ensure greater involvement of local people in sustainable forestry.

The 'Action Plan for Protected Areas in Europe' (IUCN–CNPPA 1994) includes a section on forestry. One of the recommendations of this plan is that there should be no forestry operations in strictly protected areas (i.e. areas in IUCN categories I–III), and only limited management of moderately protected areas (i.e. in categories IV and V).

Many European countries have established national forestry policies and legal controls on forest management. At the national level, strict laws usually apply to woodland clearance, and replanting is usually required after felling. These requirements are more stringently applied in montane forests, where poor management puts at risk not only the forest but also lower-lying land which may be affected by flooding, avalanche, or siltation, etc. As for afforestation, governments commonly determine the location and type of new planting. In addition to these overall controls, more attention is being paid to the type of forestry operations conducted and the kind of forest created (see Box 2; Forestry Commission 1989). Increasing attention is being given to forest design, and now many countries operate strict measures through the local land-use planning authority before afforestation schemes may be approved.

Non-statutory opportunities

Over the last few years both governments and non-governmental organizations, such as the Forest Stewardship Council (FSC), have attempted to develop accreditation schemes for forest products (Elliott 1996). The FSC was established in 1993 to set global standards whereby those organizations certifying the sustainability of a productive forest could themselves be accredited. It has developed and registered an accreditation mark which will appear alone or alongside the logos of its accredited certifiers. So far the FSC has accredited four third-party certifiers and is working in close cooperation with governments

---

Box 2. Examples of national legislation and policies on forests in Europe.

SWEDEN There is now a consensus that forest management should mimic natural disturbance regimes of forests (Pettersson 1991, Liljelund *et al.* 1992, Angelstam *et al.* 1993), and proposed new legislation would give equal weight to the goals of producing timber and conserving biodiversity (SOU 1992).

DENMARK The National Forest and Nature Agency, part of the Ministry of the Environment, published the 'Strategy for natural forests and other forest-types of high conservation value in Denmark' in 1992. In 1989, under the Forestry Act, it was made a statutory requirement that nature conservation and protection of the environment be taken into consideration as part of forestry production.

GERMANY Conventional commercial forestry is pre-eminent, dominated by capital-intensive management techniques and even-aged monoculture plantations of non-native conifers, but this model is becoming less economically attractive due to the importation of cheaper wood products. Naturschutzbund Deutschland (NABU), the BirdLife Partner organization in Germany, is now campaigning to move the forestry sector towards management that mimics natural processes, using a pilot project at Steinbachtal as a practical demonstration that 'eco-logical' forestry can make sense both commercially (through producing high-value timber) and socio-politically (through creating permanent jobs). A significant proportion (24%) of Germany's forests are municipal property, and their management is therefore particularly open to influence by grass-roots conservation movements. The NABU campaign will target these forests via its local groups, and will also lobby regional government to change federal and regional forestry legislation more towards biodiversity-friendly management.

ROMANIA There are new policies in the new Forestry Code issued in 1996, such as ecological reconstruction of forests, conservation of biodiversity and forest landscapes, establishment of protected areas in forests, and sustainable management of forests. It is also now forbidden to reduce the extent of forest cover on state-owned land.

ITALY There is no national forestry law as such, and only a third of state-owned forests have management plans. In a report to the Ministry of Agriculture in 1987, a shift from monoculture plantations to multiple-use forests was recommended, through reforestation of forest areas damaged by fire, pollution, pests and climate change.

(e.g. Switzerland) to accredit their certification schemes. The International Standards Organization, a non-profit federation, has supported the development of 'environmental management systems' guidelines (ISO 14001). These are scheduled for implementation in 1997. However, these standards do not require any specific levels of environmental performance and do not guarantee that firms have complied with environmental regulations.

CONSERVATION RECOMMENDATIONS

Broad conservation objectives

1. To strictly protect all remaining areas of virgin forest and natural and semi-natural old-growth forest.

2. To maintain and enhance the condition of semi-natural old-growth forest by the adoption of sympathetic management procedures that mimic natural processes.

3. To restore high ecological quality to degraded areas of semi-natural forest, especially in those regions where semi-natural forest is greatly reduced or fragmented, and to restore semi-natural forest to natural old-growth condition, where feasible.

4. To recreate native and semi-natural forest in regions where existing forest cover is low, provided there is no conflict with other conservation interests or objectives, and following sound landscape conservation and other environmental principles.

5. To improve forestry plantations for wildlife by the adoption of sound conservation management practices which are compatible with sustainable forestry.

Ecological targets for the habitat

In order to maintain and enhance the habitat qualities and conservation values of European forests, it is essential to retain and enhance important habitat features for priority species, as follows.

1. A coherent and ecologically viable network of protected boreal and temperate forest areas should be created, incorporating all remaining virgin forest and a large proportion of all natural old-growth forest in Europe.

2. Man-made fragmentation of existing large forests, especially those of high conservation value, should be avoided. The restoration of priority forests, e.g. scarce, semi-natural or natural forest-types, through removal of non-native tree species and through infill planting to reduce

fragmentation, should be promoted where this does not result in harmful impacts upon other priority bird species and habitats (and other priority wildlife).

3. The presence of associated habitats of importance to forest species, such as wetlands, agricultural and grassland habitats, heathland, mires and moorland, should be maintained at the landscape scale to provide essential nesting and feeding requirements.

4. The hydrology of riverine forest should be maintained in as 'natural' a state as possible, or at least essential requirements should be fulfilled, e.g. regular flooding. Similarly, where feasible, natural fire processes should not be restricted in extensive areas of natural boreal forest, or controlled fires should be allowed that mimic essential beneficial components of these processes.

5. Forests should be dominated by tree species native to the region and habitat. Forest-management practices to be avoided include planting/restocking with non-native tree species, large-scale clear-felling, and thinning that promotes unnaturally uniform species composition.

6. Natural forest processes should be allowed to proceed unhindered in remaining virgin forests and those being returned to natural systems. Managed forests should maintain, or restore if necessary, natural features such as stands of old-growth trees, multiple layers of vegetation (field, shrub and canopy) and layered canopies, mixtures of woody species and of age-classes of tree, undamaged soil structure, and a high proportion of dead wood. Clear-felling and the creation of large areas of uniform regrowth should be avoided by adopting group-felling, shelterwood or selection systems.

7. Natural regeneration should be sufficient to ensure the long-term viability of the forest and its important species, where necessary through the control of grazing.

8. Food resources, and food-chains and webs, should be maintained through the avoidance of the use of pesticides, herbicides and fertilizers in the management of forests, wherever possible.

9. Disturbance of sensitive priority bird species, by forestry operations, recreational pursuits and other human activities, should be avoided, especially during critical periods of the life cycle (e.g. the nesting season during March–July).

Broad policies and actions

General

1. The implementation of the 'General Guidelines for the Sustainable Management of Forests in Europe' and the 'General Guidelines for the Conservation of the Biodiversity of European Forests' (i.e. the Helsinki Guidelines) should be vigorously promoted. These Guidelines offer a valuable starting point for putting biodiversity at the heart of European forestry policies and practices. Sustainable forest-management practices should:

   • protect areas of ecological fragility;

   • conserve virgin, primeval and old-growth forests, cultural heritage and the landscape;

   • safeguard the quality and quantity of water-flows and water-levels;

   • maintain and develop other protective functions of forests.

2. National governments and the EU should develop a balance between strategic guidance, regulation and economic incentives to apply to state and private forestry. Incentives will usually be funded by the public purse and, as such, should be provided only where they secure the widest public benefits.

   Additionally, European forestry policies should acknowledge that forests contribute to a variety of markets at different geographical scales: for timber and associated wood products, for energy, for leisure, for hunting and other consumable products, and for less tradable benefits such as the provision of shelter, the protection of watersheds, and the enhancement of landscapes and wildlife habitats. The values and operation of these markets influence the achievement of sustainable forest management, and therefore need to be addressed in the development and delivery of policies and mechanisms to safeguard and enhance biodiversity conservation in European forests.

3. All surviving areas of natural and semi-natural forests of high conservation value should be identified at European, national and local level. All forested Important Bird Areas should be included in this inventory.

Legislation

4. Conservation of forest biodiversity and the development of sustainable forest management should feature highly in all national biodiversity strategies and plans prepared under the framework of the Biodiversity Convention.

5. In the European Union, protection of forest habitats, and those priority birds dependent upon or affected by them, should be increased by pressing for full implementation of the Birds and Habitats Directives in by member states. All forested Important Bird Areas in these countries should be designated as Special Protection Areas under the Birds Directive.

6. Forests and associated land that are most valuable for biodiversity conservation should be protected through national, European and international systems of designation (including, where relevant, Biogenetic Reserves, the EMERALD Network of the Bern Convention, and as Biosphere Reserves under UNESCO's Man and Biosphere Programme) and through extending the EU-wide Natura 2000 network to the rest of Europe. Legal designation and protection should be supported by a consequent withdrawal of intensive, commercially orientated management, in those forests and wooded landscapes of highest nature-conservation value.

7. Governments should aim to develop ecological networks of forested habitats through the designation of forest reserves, the development of low-impact management schemes in managed natural and semi-natural forests and in plantations of native species, and through linking forested areas by woodland corridors as far as possible (and where this does not conflict with other nature conservation objectives).

8. Legislative and financial measures should be introduced to promote wood-processing techniques which generate less waste and use raw timber more effectively. Recycling of wood and paper products should also be widely encouraged by government schemes and financial incentives.

9. Strategic Environmental Assessments (SEAs) should be required for all national, international and EU forestry policies, plans, programmes and supporting delivery mechanisms. Guidelines should be introduced in all countries to inform applicants and decision-makers as to how to assess the impacts of forestry schemes upon biodiversity.

   The EIA Directive (see Box 1) should be revised to require Environmental Impact Assessments (EIAs) to be completed for all afforestation schemes over 50 ha and for those which affect designated protected areas. The scope of the Directive should also be extended to apply to large-scale tree-felling and other forestry operations and forest-uses that have significant environmental impacts.

Policy

10. Conservation objectives need to be incorporated into all national, European and international policies affecting forestry. These include national and regional policies and Operational Programmes for forestry, agriculture, transport, energy, trade, industry and regional development, as well as the EU Common Agricultural Policy, and the EU 'Community Scheme on Action in the Forestry Sector' (COM(88)255).

11. National and European forestry policies should reflect the multi-purpose nature of sustainable forest management and the cross-sectoral ownership and involvement inherent in sustainable forestry principles. To reflect this, many government authorities as well as international, national and local non-government organizations should contribute to the development of national and European forestry policies.

12. The Trans-European Networks for Transport, proposed by the EU, should be carefully evaluated using Strategic Environmental Assessment, to quantify and predict, and thus avoid, the loss and fragmentation of forest areas. Where fragmentation of forests cannot be avoided, it should be steered away from those sites of highest conservation value as represented by IBAs and the Natura 2000 network.

13. National governments and the EU should examine their policies affecting the consumption of forest products and introduce measures to reduce the levels of waste, encourage greater use of mill residues, recycled paper and trimming waste from paper and board products.

14. National governments and the EU should develop a hierarchy of strategic plans to guide future forestry and to implement sustainable forest management including the conservation of biodiversity. This would be assisted by agreement between the European Forestry Commission of the FAO, the United Nations Commission on Sustainable Development (UNCSD), the Helsinki Secretariat and the EU on a common format for developing, monitoring, reviewing and reporting upon national plans and programmes, with supporting evidence of these being promoted through local and regional measures.

15. Forestry services of national governments and the EU, and other national and international institutions responsible for developing and implementing forestry policies, should be structured and resourced to recognize the centrality of nature conservation in sustainable forest management, and should be capable of delivering such policies. Biodiversity objectives and supporting targets (quantifiable and time-limited) should be set and, together with the resources to implement them, should be transparently described in strategic and corporate plans. Government organizations should be expected to regularly report on their implementing progress in such objectives for biodiversity conservation and sustainable forest management.

Economy

16. Measures should be introduced, within the temperate and boreal zones in Europe, as well as between Europe and the rest of the world, to deter consumption of, and trade in, forest, wood and non-timber products derived from unsustainably managed forests. Conservation of biodiversity should be placed at the heart of measures to identify and promote sustainably managed forests and their products. These will include national and voluntary measures of forest and timber certification systems, such as those being developed by the Forest Stewardship Council, eco-labelling of paper, and the EU Environmental Management and Auditing Scheme. Criteria for eco-labelling should define significant populations for certain indicator wildlife species, including birds.

17. National and EU public funds for the development of timber products and their marketing should only be available for processing facilities and processes which utilize timber derived from sustainably managed forests.

18. New forestry schemes, similar to the 'Environmentally Sensitive Area' schemes created for agriculture under EU Regulations, should be introduced at national and EU level in order to promote ecologically sensible forest-management practices.

Forest management

19. Preference should be given to the use of native tree species and provenances in all afforestation and forest replanting programmes. Non-native tree species should only be used where native species have been demonstrated not to be usable, where they do not have a negative impact on the conservation of priority birds and other wildlife, when the wider environmental and social impacts of such introduced species have been properly evaluated, and when adequate refuges of native flora and fauna have been established at national, forest-type and local levels.

20. Forest-management practices which threaten priority breeding birds and other wildlife should be reduced in all natural and semi-natural forests, as follows.

- The areal extent of clear-cutting in all managed natural and semi-natural forests should be significantly reduced, and this practice should be replaced by other methods of felling which mimic natural forest processes more closely. Every European country should set clear, quantitative targets for their reduction of clear-cutting, and deadlines by which these targets should be met, after wide consultation with relevant experts in forest ecology.

- The replacement of stands of native species with monospecific stands of non-native varieties, hybrids or species should be abandoned. Wherever possible, existing stands of such varieties, hybrids or species should be replaced by local genetic varieties of native species in the next management cycle.

- Native tree and shrub species of secondary or no economic importance, but which increase both structural and floristic diversity of forests, should not be removed during forest management. In forest stands where such species have been previously removed, their replacement should be encouraged to restore the original forest diversity.

- Wherever possible, any replacement of trees in natural and semi-natural forests should be allowed to proceed through natural regeneration. If this is not possible, any artificial replanting should use local varieties of those native species which would grow in the area under natural conditions.

- Increasing the density of standing and fallen dead wood (e.g. to reach a minimum of 10–20 logs per ha) should be encouraged in all managed forests. Similarly, the proportion of old, standing trees left after felling should be increased.

- In all natural and semi-natural forests that are managed for timber, low-impact, non-intensive tree-cutting and removal practices (e.g. using workhorses to a greater extent) should gradually replace the current highly mechanized and capital-intensive methods which use heavy machinery.

- The construction within natural and semi-natural forests of tracks and roads for forestry operations should be minimized in order to reduce fragmentation, 'edge' effects, disturbance of 'sensitive' species, intrusion of non-native species, and soil erosion. In the most sensitive habitats and areas, such construction should be totally avoided.

21. National governments and the EU should encourage the Inter-Governmental Panel on Forests of the UNCSD to promote convergence of the various and sometimes conflicting principles and criteria of sustainable forest management which have been developed in different regions of the globe. This would facilitate, between such regions, global trade in forest products derived from sustainably managed sources.

Recreation

22. Recreational areas in forests should only be designated in ecologically less valuable and less sensitive areas. They should be large and well-equipped enough to cope with the projected number of visitors at peak times.

23. The impact of tourism on forests and their priority species should be monitored and regularly evaluated in order to identify threats posed by these activities.

24. Mass tourism in natural and semi-natural forests should be avoided. Recreational pursuits which are likely to have a significant negative impact on forest habitats and priority species should be subject to environmental impact assessment, and their impact reduced where necessary to ecologically acceptable levels.

Other

25. Forest expansion should be targeted away from open habitats and sites which are important for priority birds and other fauna and flora, and towards areas with reduced woodland cover and degraded land where such expansion can maintain important biodiversity values.

26. Outside extensive areas of natural boreal forest (where fire is an important component of natural forest processes), forest-fires should be strictly regulated and controlled, particularly in ancient and old-growth forests. National measures and resources to secure protection of European forests from too frequent, large or intensive fires should be improved, as should be the EC Directives on forest-fires of the EU (see Box 1).

27. Forests can be a valuable sink for greenhouse gases and act as scavengers of air pollutants. These functions offer further justification for the protection of forests. However, governments should rely upon forests as a source of such environmental benefits only where forests are part of a national programme of energy and pollution control, and provided that forest expansion does not initially contribute to such global warming as a result of afforestation of peatlands. National, European and EU programmes should reduce air and water pollution

to below the tolerance levels of forest species, while also meeting targets set by international legislation (e.g. the Framework Convention on Climate Change).

28. Population levels of domestic and wild ungulates should not be allowed to exceed the natural carrying capacity in all forest-types. Where the carrying capacity is currently being exceeded, programmes should be put in place widely to reduce ungulate numbers to ecologically acceptable levels.

ACKNOWLEDGEMENTS

Most contributions to this chapter were made by the Habitat Working Group during and after a workshop organized by L. Tomialojc and T. Wesolowski and held at Wycieczkowy PTTK, Bialowieza Forest, Poland (13–18 June 1993); these contributors are listed on p. 203. R. Turner (Royal Society for the Protection of Birds) drafted the section on 'Conservation recommendations', and C. Heinrich and C. Mayr (Naturschutzbund Deutschland), M. Strazds (Latvijas Ornitologijas Biedriba) and J. Frühauf (BirdLife-Österreich) all reviewed and commented in detail on the entire text.

In addition, valuable information and comments were received from F. Casale (Lega Italiana Protezione Uccelli) and D. Munteanu (Societatea Ornitologica Româna), J. Cuff and M. Rayment (Royal Society for the Protection of Birds).

MEDITERRANEAN FOREST, SHRUBLAND AND ROCKY HABITATS

compiled by

G. ROCAMORA

from major contributions supplied by the Habitat Working Group

J. Almeida, M. Díaz Esteban, B. Hallman, F. Petretti, R. Prodon, G. Rocamora (Coordinator)

SUMMARY

Mediterranean forest, shrubland and rocky habitats are important for the survival of 100 priority bird species in Europe, one of the highest totals for any habitat on the continent. Of these priority species, 65% have an Unfavourable Conservation Status in Europe. The rest are specialized and highly dependent on the continuing existence of such habitats. Many of these specialists require semi-open environments such as maquis, garrigue and rocky habitats which are only sustained by the continuation of traditional grazing and/or natural fire regimes. A significant proportion of the priority species are also naturally restricted within Europe.

Currently, most Mediterranean forest is intensively managed plantations; natural forest no longer exists, and remaining semi-natural forests are highly fragmented. The overall extent of Mediterranean forest and shrubland has increased (at least in southern and western Europe) during the twentieth century, at the expense of semi-open habitats, due to widespread afforestation of open land, and to natural regeneration on abandoned farmland.

The disruption of natural fire regimes, leading to frequent fires or, where controlled, rare but very intense fires, is also a widespread problem. Habitat degradation through human disturbance of priority birds, touristic and other developments, and overgrazing, is also locally severe.

The importance of these semi-natural forest, shrubland and rocky habitats needs to be recognized, and a representative network of sites should be adequately protected and managed. Elsewhere, broad measures are required and therefore governments should make full use of the opportunities for coordinated action that are provided by international and European Union (EU) legislation, international funds and other current policy initiatives such as the 'Helsinki guidelines for sustainable forest management'.

EU agri-environment measures should also be used to support traditional and environmentally beneficial pastoral farming systems, while further policy reforms are necessary in order to reduce localized problems such as overgrazing. The proper enactment and implementation of Strategic Environmental Assessments and Environmental Impact Assessments are also necessary to avoid detrimental impacts from forestry, agriculture, tourist and infrastructure-related programmes and development.

HABITAT DESCRIPTION

Habitat definition, distribution and extent

This habitat conservation strategy covers forest of 'true' Mediterranean tree species, hard-leaved shrub vegetation, including 'maquis' and 'garrigue' (also known as 'matorral', 'macchia' or 'phrygana') and rocky habitats, including inland cliffs in the Mediterranean region, and plantations dominated by Mediterranean tree species. For the purpose of this conservation strategy, the Mediterranean region is delimited by the bioclimatic boundary of Daget (1980); see Figure 1. The terms 'forest' and 'woodland' are used interchangeably in this strategy, and 'woodland' does not imply a structurally less complex form of 'forest'.

The following Mediterranean habitats are not covered in this chapter but are dealt with under 'Agricultural and Grassland Habitats' (p. 267): perennial crops (trees and shrubs, e.g. olives), pastoral woodlands (open woodlands primarily managed for livestock, e.g. 'dehesas', 'montados'), and steppic habitats that are primarily grassland, as well as all montane habitats above the treeline. Management of sea cliffs is covered under 'Marine Habitats' (p. 59) and 'Coastal Habitats' (p. 93). In addition, hard-

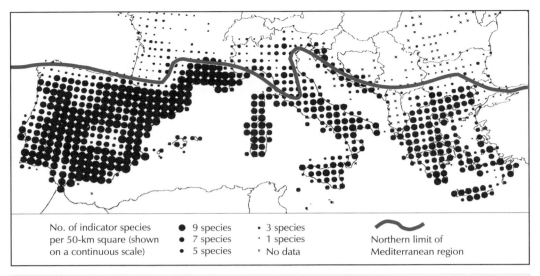

Figure 1. Map of the Mediterranean region, as defined by the bioclimatic limit of Daget (1980), showing the distribution of nine priority bird species which are indicators of Mediterranean shrubland and rocky habitats (data from Hagemeijer and Blair 1997). The indicator species are *Hieraaetus fasciatus*, *Oenanthe hispanica*, *Monticola solitarius*, *Sylvia sarda*, *S. undata*, *S. cantillans*, *S. melanocephala*, *S. hortensis* and *Petronia petronia*.

leaved forest and shrubland of Macaronesian[1] Europe is not treated in this book (see 'Introduction', p. 17).

This strategy takes into account all the successional or degraded stages of the different forest communities of Mediterranean tree species within Europe. Within this continuum, four major habitat-types can be distinguished:

- Mediterranean coniferous forest.
- Broadleaved evergreen and deciduous forest.
- Closed Mediterranean shrubland (maquis).
- Open, low shrubland (garrigue) and rocky habitats, including cliffs.

'Ecological characteristics' (p. 242) describes these four habitat-types in more detail. Maps of their current distribution are either unavailable or unsatisfactory, being based largely on the theoretical or predicted ranges of the most typical and dominant plant species (Tomaselli 1976, Polunin and Walters 1985). For Mediterranean shrublands, the recently surveyed field distribution of indicator bird species may produce a more accurate picture of the current distribution of this habitat-type (Figure 1)—indeed, this map closely fits the bioclimatic region defined by Daget (1980).

Mediterranean forest, shrubland and rocky habitats cover more than 400,000 km² in Europe

(Tomaselli 1976). The area of Mediterranean-type ecosystems (according to Müller 1982), and the extent of Mediterranean forest and shrubland habitats in each European country are summarized in Table 1. In the European Union (EU), these habitats cover 226,000 km² and represent about one third of the total Mediterranean area (Hermeline and Rey 1994).

Standardized national data on the area of the four habitat-types are difficult to obtain. Limits and definitions used to determine the Mediterranean bioclimatic area differ between studies, so figures can vary greatly. In certain countries, shrubland is described in national statistics under a general heading 'woodland', or simply not taken into account, and specific data on the extent of rocky habitats are not available (although such habitats are abundant and diverse). Rocky habitats are, however, often intermixed with garrigue and are therefore included in this habitat-type.

The extent of the different habitat-types in Europe has changed considerably during the last century. In west Mediterranean countries, total forest area has increased due to unchecked natural succession of vegetation from shrubland to woodland, and due to afforestation with non-native conifer and *Eucalyptus* plantations (see 'Threats to the habitat', p. 252). At the same time, however, the area of old-growth Mediterranean forest seems to have decreased, at least in Portugal, Italy and Greece (Rodrigues 1995). Data for the eastern Mediterranean are largely

---

[1] Macaronesia is the biogeographical region composed of the archipelagos of Madeira, the Canary Islands, the Azores and (outside Europe) the Cape Verde Islands.

Table 1. The extent of Mediterranean forest, shrubland and rocky habitats (km²) in each European country. No data are available for Andorra, Bosnia and Herzegovina, and former Yugoslav Republic of Macedonia.

	Total area with Mediterranean-type ecosystems	Habitat-type		
		Mediterranean coniferous forest	Broadleaved forest	Maquis, garrigue and rocky habitats
Albania	12,300 [1]–19,700 [2]	1,800 [3]	8,660 [3]	?
Bulgaria	?	? [10]	1,000 [3]	600 [3]
Croatia	?	4,940 [4]	12,950 [4]	?
Cyprus	9,230 [1,2]	1,380 [4]	20 [4]	?
France	32,600 [1]–65,300 [2]	—————— 22,640 [5] ——————		4,430 [5]
Greece	121,000 [1,2]	14,300 [5,6]	19,500 [6]	?
Italy	175,000 [2]–226,000 [1]	18,970 [4]	48,530 [4]	22,400 [6]
Portugal	>87,400 [1,2]	13,550 [4]	11,861 [7]	14,193 [7]
Spain	333,000 [2]–358,000 [1]	43,000 [8]	56,000 [9]	50,000 [9]
Turkey	195,000 [1]–468,000 [2]	?	?	?

Sources

Total area with Mediterranean-type ecosystems
Derived from Müller (1982) in Hobbs *et al.* (1995), according to systems of:
[1] Troll and Paffen.
[2] Köppen and Geiger.

Habitat-types
[3] P. Iankov pers. comm. (1996).
[4] CEC (1995d); data include all forests (not necessarily only true Mediterranean forests as defined here).
[5] Inventaire Forestier National (unpubl. data) and Hermeline and Rey (1994).

[6] WWF (1994).
[7] DGF (1993).
[8] Ministerio de Agricultura (1979).
[9] Ministerio de Agricultura (1980).
[10] Fragmented patches, extent unknown.

lacking, but it seems that in Turkey a considerable amount of recent forest loss has occurred through agricultural expansion.

Origin and history of the habitat

One of the main characteristics of the Mediterranean basin is the long history of human impact on natural ecosystems. Man appeared in the region at least 1,500,000 years ago (i.e. before the last glacial period), and is known to have been using fire from 400,000 years ago in southern France. In the eastern part of the basin, the first pre-neolithic villages and towns were built (e.g. in Turkey, from 6500 BC). Neolithic techniques, such as cultivation of cereals and cattle-rearing, then began to spread westwards, mainly along the coasts. This neolithic period of human development changed the whole landscape and fauna in the region. As a result, it is now difficult to know precisely what were the original and 'natural' vegetation-types and fauna of the Mediterranean.

After the neolithic period, man's impact on landscape and fauna was greatest during the Roman empire, the 12th–15th centuries, and the 18th–19th centuries. In the north-west of the Mediterranean basin, the rural population density peaked around 1880. This led to widespread overgrazing and a rapid decline in forested areas. At the end of the nineteenth century the industrial revolution began and populations began to concentrate in large towns, particularly on the coasts and plains, abandoning the more hilly areas. With the development of better transport, a large expansion of intensively cultivated fruit and vegetables took place, largely to supply export markets for early-season products and the wine industry. A rapid loss of semi-natural grazing areas therefore occurred at that time.

The early twentieth century was characterized by widespread and rapid rural depopulation. This began earliest in Mediterranean France where it was accelerated by the vine disease phylloxera and overproduction of wine. It occurred more recently in Italy, Spain, Portugal and Greece, where the phylloxera crisis also played a major role. One of the most important environmental consequences of this population drift was the abandonment of traditional non-intensive farming and livestock-grazing practices. In turn, the lack of grazing led to the development of secondary vegetation and natural reforestation over huge areas of the region (Quézel 1976, Lepart and Debussche 1992). This has also been accelerated by intensive plantation programmes often using exotic tree species. The result is an ever-increasing contrast between the urbanized or intensively cultivated open areas on plains and coasts, and the sparsely populated upland areas. In the latter, the growth of tall shrublands and forests has led to a progressive disappearance of open grassy habitats, especially in north-west Mediterranean areas.

True Mediterranean forests are now mainly restricted to rough, steep mid-altitude areas with well-drained, poor soils, since the flat and most productive areas in the region have been converted to treeless cultivations or pastoral woodlands. Secondarily abandoned areas are covered by shrubland of varying structure and species composition depending on the original forest-type.

From the sixteenth century onwards, a particularly high number of plant taxa, including trees, were introduced by Man into the Mediterranean area, mainly from Central and South America, the Middle East and Asia, South Africa and Australia. Many of these non-native plants are now widespread in the Mediterranean, for example, *Opuntia*, *Acacia*, *Citrus* and *Eucalyptus*, some of them covering large areas. This abundance of introduced taxa is a characteristic feature of many Mediterranean landscapes.

Hunting has had a strong impact on the natural fauna of the Mediterranean basin since prehistoric times. The development of the human population led to the extinction of several large mammals (e.g. leopard *Panthera pardus* and lion *P. leo*) and a reduction in the distribution and population size of many other wild species as a direct consequence of hunting (Cheylan 1991). Remaining large predators, such as the wolf *Canis lupus*, brown bear *Ursus arctos*, lynx *Lynx lynx* and the globally threatened Spanish lynx *Lynx pardina* in Iberia (Groombridge 1993) are now mainly found in mountainous refuges, and are extinct or have only small relict populations in Mediterranean habitats in most countries. These declines may have led to substantial increases in herbivore populations, which in turn have probably had a considerable impact on the vegetation and general character of Mediterranean habitats.

Ecological characteristics

Key abiotic factors

The Mediterranean region shows a high diversity of topography, geology and landscape. Mediterranean habitats are also strongly shaped by the seasonality and geographical variation of climate. This climate occurs within a relatively limited range of latitude (35° to 45°N) and is characterized by hot dry summers (average temperature 21°C) and mild winters (average temperature 6°C), typically with a winter rainfall peak. At its northern limit, Mediterranean vegetation is often restricted to south-facing slopes where higher solar radiation results in warmer and drier conditions.

Altitude is responsible for considerable climatic differences, resulting in important variations in the distribution and composition of flora and fauna, and the presence in the Mediterranean mountains of central-European-type and alpine-type habitats (which are not treated in this strategy except where they hold typical Mediterranean species).

Aspect also strongly influences vegetational composition and structure: vegetation on south-facing slopes tends to be lower and less able to recover from perturbations, such as fire, grazing or clear-cutting. The number and density of shrubs in the undercanopy rise as humidity, temperature and soil fertility increase, and fall with increasing altitude.

Fires frequently occur in the region and have a strong influence on the habitat (Prodon *et al.* 1987, Moreno and Oechel 1994, Trabaud 1994). Almost all Mediterranean vegetation-types are susceptible to fire, but those most likely to be affected are shrublands and conifer forests. Some early and middle-successional formations can even be considered to be fire-dependent.

Large areas of maquis and garrigue are subject to periodic summer fires. As a result, the vegetation is a mosaic of recently burnt areas, young secondary growth and remaining unburnt stands of 'mature' maquis or garrigue. The fires are often (illegally) set by shepherds to 'improve' pasture for their livestock. Such fires inhibit the succession of garrigue and maquis into woodland or forest.

Key biotic factors and natural dynamics

Mediterranean vegetation at its climax is normally forest, and the successional trend towards forest is active almost everywhere, except in areas with particularly poor soils or dry climates which tend to be dominated by steppe grasslands. The original forest-types of the Mediterranean basin are largely unknown (except in mountainous sectors), but there is widespread evidence of the importance of broadleaved trees before neolithic times (Pons and Quézel 1985, Vernet 1990).

Nowadays, most forests are dominated by evergreen oaks *Quercus* and conifers (pines *Pinus*, junipers *Juniperus* or cypresses *Cupressus*). The structure of the vegetation changes as succession progresses; from dry and stony grasslands to low shrublands (garrigue), high shrubland (maquis), shrublands with scattered trees, then woodland, and lastly dense forest. Some factors, either natural (e.g. seed dispersal by birds, mammals and insects) or artificial (e.g. afforestation), can accelerate these successions. Others, such as grazing/browsing, fires and wood-cutting can slow down or even stop the succession, temporarily or in the long term (Figure 2).

Due to the harsh climate, active growth and flowering tend to stop during summer to minimize water requirements. Thus, in contrast to temperate and boreal vegetation, the main growth and fruiting period of many Mediterranean plants is during autumn and even winter. Many Mediterranean trees and shrubs bear fleshy, edible fruits, the seeds of which are adapted to be dispersed by birds and mammals (Herrera 1982a,b, Jordano 1985, Debussche and Isenmann 1990). This food resource plays an important role during the autumn migration of birds, leading to the presence of relatively high numbers of wintering birds in certain Mediterranean habitats.

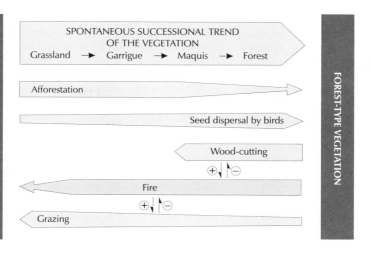

Figure 2. The most important influences on vegetation dynamics in Mediterranean forest, shrubland and rocky habitats in Europe. The arrow indicates the direction in which the factor influences plant succession. The width of each bar indicates the importance of the factor relative to the successional stage.

In addition to the effects of fires, succession may have originally been locally slowed or halted by the presence of indigenous wild herbivores, such as red deer *Cervus elaphus* and fallow deer *Dama dama*. However, these herbivores were replaced by domestic cattle, goats and sheep as early as the eighteenth century BC in the eastern part of the basin, which have subsequently been of major importance in maintaining open garrigue and grassy landscapes.

Insects and fungal diseases can also have an important impact on the structure and distribution of vegetation in some Mediterranean habitats. Such diseases have, for example, affected cork oak *Quercus suber* and other *Quercus* species, European chestnut *Castanea sativa* and even some shrub species (Brasier 1992, Cabral *et al.* 1993, Fernández and Montero 1993, Mesón and Montoya 1993). Large areas have been affected, for example, 20,000 ha in south-west Spain in 1990–1991 (ICONA 1991).

Characteristic vegetation and associated fauna

The flora of the Mediterranean basin is extremely rich and has a very high level of endemism compared with the rest of Europe. Approximately 25,000 species of flowering plants and ferns occur (i.e. including non-European areas), compared with 6,000 species in non-Mediterranean Europe (Quézel 1985). There are more than 100 common tree species in the Mediterranean basin (many of which are endemic), compared with no more than 30 in central Europe (Le Houerou 1980, Blondel and Aronson 1995). More than half of the plants of the Mediterranean basin are endemic, comprising 80% of the European total (Gomez-Campo 1985), and, of the 37 globally important centres of plant diversity and endemism that have been identified in Europe, the greatest concentration is to be found in the Mediterranean

basin (WWF/IUCN 1994).

Mediterranean vegetation is dominated by evergreen trees, shrubs and bushes that are well adapted to long dry summers. Typical and characteristic vegetation associations include evergreen oak and pine woodlands with dense shrub undergrowth, open shrublands, and dry steppe-like grasslands with scattered bushes and trees on eroded, rocky soils. In areas with higher rainfall or deeper and more humid soils, deciduous trees and shrubs become more abundant, while at higher altitudes conifers are again present.

The Mediterranean region has a high avian richness, with 345 breeding species (Blondel 1988) out of a total of 419 in the whole of Europe (Voous 1960). However, only a relatively small proportion of the breeding birds of the Mediterranean basin are endemic. Few of these endemics are forest species, the greater proportion being typical of shrublands or open-ground habitats (e.g. *Alectoris* partridges and *Sylvia* warblers), which were probably more fragmented before the neolithic age than at present (Blondel 1986, 1988). In particular, of the three globally important Endemic Bird Areas (EBAs) in Europe, one is located in Cyprus, due to the presence there of the endemic Cyprus Pied Wheatear *Oenanthe cypriaca* and Cyprus Warbler *Sylvia melanothorax*, which are both characteristic species of Mediterranean shrubland (ICBP 1992, Stattersfield *et al.* in press). Speciation in montane coniferous habitats has resulted in Corsican Nuthatch *Sitta whiteheadi*, Algerian Nuthatch *S. ledanti* and Krüper's Nuthatch *S. krueperi*.

The avifauna of Mediterranean forest, shrubland and rocky habitats is almost completely determined by the structure of the vegetation, so it is possible to some extent to predict the former from the latter (Prodon and Lebreton 1983). Thus, in a given place, the avifauna of early successional stages is completely different from that of structurally complex mature forest stages. The more structurally complex the vegetation becomes, the lower the number of bird species of Mediterranean origin (Blondel 1986, Blondel and Aronson 1995). Also, on average, those species occurring in early successional stages have a more southerly worldwide distribution than those that are found in later stages (Prodon in Rocamora 1987; Prodon 1993). There also tends to be a significant altitudinal effect on species composition, with the proportion of northern species often increasing with altitude.

Comparisons of bird communities along ecological successions (from open habitats to mature forest) in several different areas of the Mediterranean region and central Europe show a surprising convergence in composition in old, mature forests (Blondel and Aronson 1995). Each succession starts with a different set of bird species as a result of regional speciation (Blondel and Farré 1988), but the bird communities of old, mature forests are very similar wherever their location in Mediterranean Europe. This is because of the presence of 'core' forest species (Hanski 1982), i.e. bird species that are widespread in the whole Palearctic region and tend to be abundant wherever they occur. However, species with very different biogeographic affinities can co-exist in Mediterranean forests. Such faunistic mingling is characteristic of all the mountain ranges in the Mediterranean basin (Rocamora 1987, 1990).

The typical vegetation of the different habitats arising as a result of the above natural, physical and biological factors, as well as historical and current land-uses, are described below. It is beyond the scope of this account to describe the differences in these between countries or even localities. More detailed accounts of the vegetation can be found in Rikli (1943), Braun-Blanquet *et al.* (1951), Ozenda 1964, Heutz de Lemps (1970), di Castri and Mooney (1973), Quézel (1978, 1985), Daget (1980), and Polunin and Walters (1985).

Mediterranean coniferous forest

These forests are dominated by Mediterranean species of conifer, including *Pinus*, *Juniperus*, *Cupressus* and *Abies*. Recent plantations of conifers tend to be dense monocultures where shrub- and herb-layer vegetation does not grow, and are therefore of low biological interest. However, some long-established and managed plantations closely resemble natural forest.

The bird fauna of this habitat-type is, as with all other Mediterranean forests, very similar to that of most European woodlands, typically being dominated by a large number of ubiquitous forest species;

certain species—such as Woodpigeon *Columba palumbus*, Turtle Dove *Streptopelia turtur*, Firecrest *Regulus ignicapillus* and Green Woodpecker *Picus viridis*—are particularly abundant in Mediterranean coniferous forests compared with central European ones. The few species characteristic of and restricted to Mediterranean coniferous forest (in comparison with broadleaved forest) are *Sitta whiteheadi*, Crossbill *Loxia curvirostra* and *Sitta krueperi*.

The large-mammal diversity of Mediterranean coniferous forests is comparable to that of broadleaved forests, and includes *Canis lupus*, wild cat *Felis silvestris* and pine marten *Martes martes*. Chamois *Rupicapra rupicapra* and wild sheep *Ovis ammon* can be found in some high-altitude forests.

Broadleaved evergreen and deciduous forest

Evergreen oak forest (holm oak *Quercus ilex* or local equivalents) would probably be the climax vegetation over most of the Mediterranean basin. However, such forest has been widely replaced by maquis and garrigue, or pastoral woodlands, as a result of wood-cutting, livestock grazing and associated fires. In more open woodland there would have been an abundant shrub layer (e.g. of *Cistus*, *Genista* and *Erica*), and in cooler or more mountainous areas deciduous trees would have been common (e.g. *Quercus*, *Ulmus*, *Acer*, *Fraxinus*) as well as evergreen *Laurus*. This habitat-type also includes natural and semi-natural forests of *Quercus suber* which usually occur on nutrient-poor and siliceous soils. These forests tend to have a limited height and an open structure, while the shrub layer (evergreen bushes) or herb layer is generally abundant. It has, however, been removed in places to allow grazing by sheep and goats, and to accelerate the growth of *Q. suber*. Any such grazed, pastoral woodlands, from the Mediterranean and elsewhere, are treated under 'Agricultural and Grassland Habitats' (p. 267).

The avifauna is richer than that of Mediterranean coniferous forest, and includes such typical Mediterranean birds as Cinereous Vulture *Aegypius monachus*, Roller *Coracias garrulus*, Syrian Woodpecker *Dendrocopos syriacus*, Olivaceous Warbler *Hippolais pallida*, Olive-tree Warbler *H. olivetorum* and Masked Shrike *Lanius nubicus*. Species such as Woodcock *Scolopax rusticola*, Bee-eater *Merops apiaster*, Wryneck *Jynx torquilla*, Song Thrush *Turdus philomelos*, Bonelli's Warbler *Phylloscopus bonelli* and Blue Tit *Parus caeruleus* are found in both forest-types, but are much more abundant and widespread in broadleaved forest habitats. Red Kite *Milvus milvus*, Black Kite *M. migrans*, Stock Dove *Columba oenas*, Spotted Flycatcher *Muscicapa striata* and Semi-collared Flycatcher *Ficedula semitorquata* are broadleaved-forest species which do not occur in coniferous forest. Mediterranean forests can host important breeding populations of raptors, including Honey Buzzard *Pernis apivorus*, Short-toed Eagle *Circaetus gallicus*, Lesser Spotted Eagle *Aquila pomarina* and the globally threatened Spanish Imperial Eagle *Aquila adalberti*. Forests of *Q. suber* host a particularly rich avifauna (Rabaça 1983, Rocamora 1987, Pons 1991).

Large herbivores of Mediterranean broadleaved forests include *Cervus elaphus*, roe deer *Capreolus capreolus*, *Dama dama* and wild boar *Sus scrofa*, while large carnivores such as *Ursus arctos*, *Canis lupus*, *Lynx lynx* and *L. pardina* are rare.

Closed Mediterranean shrublands (maquis)

Ranging from 1 to 4 m high, and sometimes even more, maquis is a dense community of diverse, evergreen hard-leaved shrubs, which replaces evergreen oak forest after deforestation or fire. This habitat-type is well represented in the Mediterranean region. The dominant shrub genera are *Arbutus*, *Cistus*, *Erica*, *Olea*, *Phyllirea*, *Genista*, *Calycotome*, *Sarothamnus*, *Quercus*, *Ulex*, *Rhamnus*, *Pistacia* and *Myrtus*. Young or stunted evergreen oaks and pines are also often present, which may later grow to dominate the shrubs if grazing or burning are relaxed, and finally form a woodland habitat.

Maquis has a fairly characteristic bird fauna, including assemblages of Mediterranean *Sylvia* warblers (Subalpine Warbler *Sylvia cantillans*, Dartford Warbler *S. undata*, Marmora's Warbler *S. sarda*, *S. melanothorax*, Rüppell's Warbler *S. rueppelli*), buntings (e.g. Cirl Bunting *Emberiza cirlus*, Rock Bunting *E. cia* and Black-headed Bunting *E. melanocephala*) and partridges (Red-legged Partridge *Alectoris rufa*, Rock Partridge *A. graeca*, Chukar *A. chukar* and Barbary Partridge *A. barbara*). Black Francolin *Francolinus francolinus* is locally common in Cyprus and Turkey. Other warblers like Blackcap *Sylvia atricapilla*, *Hippolais pallida* or *H. olivetorum* can also be widespread and locally abundant in the maquis, as well as shrikes (Woodchat Shrike *Lanius senator*, Great Grey Shrike *L. excubitor* and Red-backed Shrike *L. collurio*) and finches (Greenfinch *Carduelis chloris*, Goldfinch *C. carduelis* and Linnet *C. cannabina*).

Maquis holds a variety of medium to large mammals, including *Cervus elaphus*, rabbits *Oryctolagus cuniculus* and *Lynx pardina* in Iberia, but small-mammal diversity is relatively low (Prodon *et al.* 1987, Blondel and Aronson 1995).

Open, low shrubland (garrigue) and rocky habitats

The term 'garrigue' ('phrygana' in the Balkans) is here used in its widest sense, to include all the different types of dwarf-shrub communities found in the Mediterranean. Garrigue has an open structure of widely spaced shrubs (e.g. *Cistus*, *Rosmarinus*, *Lavandula* and *Helichrysum*), of about 50 cm height (rarely over 1 m), interspersed with sparse grasses and, characteristically, a considerable amount of bare, stony ground. It is widespread where maquis has been degraded by livestock grazing, fires and wood-cutting, and generally occurs on rocky or sandy soils, in areas with a hot and dry climate, particularly along the coast and on islands. Despite the difficult growing conditions, garrigue is usually rich in plant species. There is a great diversity in types of garrigue, depending on soil-type and regimes of fire and grazing (Polunin and Walters 1985).

Compared to maquis, garrigue has a much richer bird fauna in terms of total number of species, and also probably in terms of total density. Its open and more diverse structure attracts a greater number of species, including many diurnal and nocturnal raptors which are attracted (along with mammalian predators) by particularly high densities of prey (Donázar *et al.* 1997). Such birds of prey include *Circaetus gallicus*, Long-legged Buzzard *Buteo rufinus*, Golden Eagle *Aquila chrysaetos*, Kestrel *Falco tinnunculus*, Lanner *F. biarmicus*, Barn Owl *Tyto alba*, Scops Owl *Otus scops*, Eagle Owl *Bubo bubo*, Little Owl *Athene noctua* and Bonelli's Eagle *Hieraaetus fasciatus*, the latter being characteristic of the habitat. Insectivorous species include Nightjar *Caprimulgus europaeus* and Red-necked Nightjar *C. ruficollis*, *Merops apiaster* and *Coracias garrulus*. The most typical and characteristic species are, however, linked to rocky or dry-grass habitats, for example, Short-toed Lark *Calandrella brachydactyla*, Crested Lark *Galerida cristata*, Thekla Lark *G. theklae*, Tawny Pipit *Anthus campestris*, Black-eared Wheatear *Oenanthe hispanica*, *O. cypriaca*, Black Wheatear *O. leucura*, Rock Thrush *Monticola saxatilis* and Spectacled Warbler *Sylvia conspicillata*. The *Alectoris* partridges are also characteristic.

Cliffs in the Mediterranean region are important as nesting sites for a number of characteristic Mediterranean birds, typically including Griffon Vulture *Gyps fulvus*, Egyptian Vulture *Neophron percnopterus*, *Hieraaetus fasciatus*, Eleonora's Falcon *Falco eleonorae*, Crag Martin *Ptyonoprogne rupestris*, Red-rumped Swallow *Hirundo daurica* and Chough *Pyrrhocorax pyrrhocorax*. They also provide, with other rocky habitats, nesting and feeding habitats for Blue Rock Thrush *Monticola solitarius* and Rock Nuthatch *Sitta neumayer*.

Mediterranean maquis and garrigue shrublands hold a rich and abundant reptile fauna, typical species including the globally threatened Hermann's tortoise *Testudo hermanni* (Groombridge 1982, 1993, Cheylan 1984), spur-thighed tortoise *T. graeca*, marginated tortoise *T. marginata*, agama *Agama stellio*, Lilford's wall lizard *Podarcis lilfordi*, large whip snake *Coluber jugularis* and western whip snake *C. viridiflavus*.

Relationships with other habitats

In montane areas, the characteristic vegetation of the Mediterranean region is replaced by temperate and boreal vegetation-types, where the characteristic summer drought of Mediterranean climates is much less apparent. These 'northern' vegetation-types include broadleaved and coniferous forest, which are covered under 'Boreal and Temperate Forests' (p. 203) where the dominant tree species are non-Mediterranean, as well as montane grasslands and associated shrublands (covered under 'Agricultural and Grassland Habitats', p. 267).

Much of the original Mediterranean vegetation in the lowlands, on the other hand, has been lost through human activities, following its initial creation by livestock grazing, fire and forestry activities. Four main types of new vegetation have arisen as a result:

- Arable cultivations and grasslands.
- Perennial crops (trees and shrubs), e.g. citrus, olives.
- Pastoral woodlands (open, extensively grazed woodland of oak).
- Plantations of non-native tree species, e.g. *Eucalyptus*.

The first three of these are covered under 'Agricultural and Grassland Habitats' (p. 267).

Values, roles and land-uses

The commonest habitat-use is undoubtedly livestock grazing, which has probably occurred in the region since the domestication of livestock by Man (see 'Origin and history of the habitat', p. 241). In the last hundred years, this has shown a widespread and rapid decline, partly due to the low productivity of the land. Soil fertility is often poor because of erosion and the removal of manure over the centuries to fertilize vegetable gardens and arable land. Huge areas of land are therefore necessary to make such pastoral farming systems viable, but these cannot compete with more intensive and productive livestock-rearing systems. Consequently, real and relative incomes from traditional pastoral farming are falling and such systems are presently threatened in countries like France and Italy. Although strongly decreasing, they are still well represented in Spain, Portugal and Greece, and in Turkey livestock grazing is still widespread and abundant.

All types of Mediterranean forest are used for timber production, but old, mature forests remain in only a very few areas, and only some are still exploited for timber. In contrast, most *Quercus ilex* woodlands and deciduous forests in the Mediterranean are managed by coppicing, which is a traditional method of managing broadleaved woodland, primarily for wood production, in which shoots are allowed to grow from the base of a felled tree. Compartments of trees are cut in rotation, commonly every 10–30 years, overall affecting about 90% of the forest. The system produces dense, uniform, shrubby growth, thereby providing a near-continuous supply of poles for fuel and building, rather than mature timber. Coppicing is a widespread activity in Mediterranean forests, often affecting 10–50% of national forest area (Kuusela 1994). As a result of this practice, forest vegetation is distributed in a mosaic of recently harvested units, young secondary growth stands (which provide good pasture for livestock), and residual stands of mature and old-growth timber with few bushes and herbs. The use of wood for charcoal production was quite a widespread activity in the past, but has now virtually disappeared from all western Mediterranean countries, with the exception of Portugal.

Quercus suber forests are mainly managed for production of cork and, secondarily, firewood. Traditionally, such forests were harvested every 8–12 years, trees living generally no longer than 120 years. This is a highly labour-intensive activity and, consequently, cork harvesting has rapidly become unprofitable in the more economically developed regions, as manpower has become considerably more costly. Cork production has been nearly abandoned in southern France, Corsica and northern Spain, and has resulted in a decrease in cork-oak forest, especially in areas where such forest had been artificially extended, to the benefit of *Quercus ilex* and maquis. Cork extraction, however, is now increasing again in these regions as some government authorities and forest owners try to revitalize this activity. In Portugal, cork production has remained economically viable and is an important industry accounting for about 4% of the country's export income.

Hunting is a traditional and still common activity in Mediterranean countries, particularly in winter when large numbers of migratory birds (e.g. *Columba palumbus*, thrushes *Turdus* and *Scolopax rusticola*) are present in forests and shrublands. Resident game such as *Alectoris* partridges are also common quarry. Hunting may help to preserve areas of semi-natural habitat, but may also lead to modification of such habitat for its own ends. Hunting has been and still is in many Mediterranean countries a major problem, involving illegal shooting of protected species and possible over-exploitation of some others. However, the direct impacts on bird populations of killing adult birds are not covered by this or other habitat conservation strategies (see 'Introduction', p. 15). Widespread introduction of non-native game species may have negatively affected native game species, through hybridization, competition or disease.

A number of other, small-scale but nevertheless important activities are also widespread, including beekeeping, resin extraction, pine-nut and chestnut harvesting. Such utilization of forest products was more diverse and widespread previous to this century. These activities provide additional sources of income that can help local farmers to maintain an economically viable activity, and are therefore currently being encouraged by national and regional agricultural policies.

These changes in land-use, with declines in pastoral farming, widespread agricultural abandonment and rural depopulation, have led to a decrease in open habitats with low vegetation, and an increase in shrubby or forested areas. This trend has recently been further encouraged by national and European policies of afforestation, or even of replacement of natural forests with non-native trees for timber and pulp production (e.g. in Portugal).

Political and socio-economic factors affecting the habitat

Most Mediterranean woodlands are privately owned (e.g. over two-thirds in the case of Spain and c.78% in Portugal), so that market profitability is necessary to ensure their persistence (Campos 1995a). Traditional use of Mediterranean forest implies a mixture of uses, which include both silviculture and livestock grazing. More recently, big-game rearing and hunting is also becoming important (see 'Values, roles and land-uses', p. 246).

Silvicultural practices (such as thinning and pruning) have been traditionally financed with the by-products of these activities, e.g. charcoal and firewood. These practices are no longer profitable in many regions due to the falling prices of the by-products, resulting from the use of more recent energy sources (e.g. electricity and gas) and increasing labour costs (Campos 1995a). The prices of other forest products such as timber, resin or cork have also fallen, whereas revenue from hunting and fishing has risen during recent decades (Campos 1992). These factors have led to the abandonment of forests and shrublands in some areas, with a consequent increase in the risk of forest-fires. Other areas of Mediterranean woodland have been replaced by non-native tree plantations (e.g. *Eucalyptus*) for industrial timber production and the pulp industry. However, these plantations have led to increases in local unemployment and a rise in forest-fires, to the

extent that the total area affected by fires can exceed that subject to afforestation schemes (Vélez 1990).

The high profitability of livestock rearing due to European Union agricultural policies (e.g. subsidy payments by head of livestock rather than, e.g., per hectare of grazed land) has in the past encouraged the transformation of grazed forests to more open pastoral woodlands. Although these maintain partial tree-cover, they increase the availability of grass for livestock. However, increasing livestock-browsing pressure, both in forests and pastoral woodlands, is leading to insufficient regeneration of trees to maintain these habitats in the long term (Elena *et al.* 1987). In some areas with better soils, subsidies and price-protection policies have promoted the total removal of trees and the transformation of woodlands to open grasslands or arable croplands (Elena *et al.* 1987, Fernández-Alés *et al.* 1992).

Under future open-market conditions, it is apparent that sustainable exploitation of Mediterranean forest, shrubland and rocky habitats is likely to be only marginally profitable, and that public intervention will be necessary to maintain them (Campos 1992, 1995a). Forestry policies have mainly promoted quick-growing plantations, whereas agricultural policies based on price-protection have led to an expansion of croplands at the expense of forest areas (Campos 1992). Recent interest in undervalued roles of forests such as soil and water conservation, and other environmental goods and services, have promoted reforestation measures (CEC 1992). These measures should, however, take into account the very low growth rates of Mediterranean woody plants in calculating subsidies to farmers and in assessments of reforestation programmes (Campos 1995b).

PRIORITY BIRDS

Mediterranean forest, shrubland and rocky habitats have a rich avifauna, including 100 priority species (Tables 2 and 3). Of these, 10 are Priority A species, one of which, *Aquila adalberti*, is classified as Endangered in Europe (Tucker and Heath 1994) and is endemic to Europe, and is consequently considered to be Globally Threatened (Collar *et al.* 1994). Four other species are also endemic breeders to the Mediterranean: *Sylvia melanothorax*, *Oenanthe cypriaca*, *Sitta whiteheadi* and *Hippolais olivetorum*. The former three are restricted to Mediterranean islands (Cyprus, Corsica). In addition, Eleonora's Falcon *Falco eleonorae* is almost endemic to the Mediterranean coast, and Cinereous Bunting *Emberiza cineracea* has a very small world range, most of which is in Greece and Turkey. Because of its small world population, *Emberiza cineracea* is considered Near-Threatened (Collar *et al.* 1994). The majority

Table 2. The priority bird species of Mediterranean forest, shrubland and rocky habitats in Europe.
  This prioritization identifies Species of European Conservation Concern (SPECs) (see p. 17) for which the habitat is of particular importance for survival, as well as other species which depend very strongly on the habitat. It focuses on the SPEC category (which takes into account global importance and overall conservation status of the European population) and the percentage of the European population (breeding population, unless otherwise stated) that uses the habitat at any stage of the annual cycle. It *does not* take into account the threats to the species within the habitat, and therefore should not be considered as an indication of priorities for action. Indications of priorities for action for each species are given in Appendix 4 (p. 401), where the severity of habitat-specific threats is taken into account.

	SPEC category	European Threat Status	Habitat importance
Priority A			
Spanish Imperial Eagle *Aquila adalberti*	1	E	■
Eleonora's Falcon *Falco eleonorae*	2	R	■
Cyprus Pied Wheatear *Oenanthe cypriaca*	2	R	■
Black-eared Wheatear *Oenanthe hispanica*	2	V	■
Olive-tree Warbler *Hippolais olivetorum*	2	(R)	■
Dartford Warbler *Sylvia undata*	2	V	■
Cyprus Warbler *Sylvia melanothorax*	2	R	■
Corsican Nuthatch *Sitta whiteheadi*	2	V	■
Masked Shrike *Lanius nubicus*	2	(V)	■
Cinereous Bunting *Emberiza cineracea*	2	(V)	■
Priority B			
Cinereous Vulture *Aegypius monachus*	3	V	■
Short-toed Eagle *Circaetus gallicus*	3	R	■
Levant Sparrowhawk *Accipiter brevipes*	2	R	•
Imperial Eagle *Aquila heliaca*	1	E	·
Booted Eagle *Hieraaetus pennatus*	3	R	■
Bonnelli's Eagle *Hieraaetus fasciatus*	3	E	■
Lanner *Falco biarmicus*	3	(E)	■
Chukar *Alectoris chukar*	3	V	■
Rock Partridge *Alectoris graeca*	2	V	•
Red-legged Partridge *Alectoris rufa*	2	V	•
Barbary Partridge *Alectoris barbara*	3	E	■
Black Francolin *Francolinus francolinus*	3	V	■
Andalusian Hemipode *Turnix sylvatica*	3	E	■
Scops Owl *Otus scops*	2	(D)	•
Nightjar *Caprimulgus europaeus*	2	(D)	•
Roller *Coracias garrulus*	2	(D)	•
Green Woodpecker *Picus viridis*	2	D	•
Thekla Lark *Galerida theklae*	3	V	■
Woodlark *Lullula arborea*	2	V	•
Black Wheatear *Oenanthe leucura*	3	E	■
Blue Rock Thrush *Monticola solitarius*	3	(V)	■
Orphean Warbler *Sylvia hortensis*	3	V	■
Semi-collared Flycatcher *Ficedula semitorquata*	2	(E)	•
Woodchat Shrike *Lanius senator*	2	V	•
Ortolan Bunting *Emberiza hortulana*	2	(V)	•
Black-headed Bunting *Emberiza melanocephala*	2	(V)	•
Priority C			
Black Stork *Ciconia nigra*	3	R	•
Black Kite *Milvus migrans*	3	V	•
Lammergeier *Gypaetus barbatus*	3	E	•
Egyptian Vulture *Neophron percnopterus*	3	E	•
Griffon Vulture *Gyps fulvus*	3	R	•
Long-legged Buzzard *Buteo rufinus*	3	E	•
Golden Eagle *Aquila chrysaetos*	3	R	•
Peregrine *Falco peregrinus*	3	R	•
Woodcock *Scolopax rusticola*	3 [w]	V [w]	•
Turtle Dove *Streptopelia turtur*	3	D	•
Eagle Owl *Bubo bubo*	3	V	•
Bee-eater *Merops apiaster*	3	D	•
Tawny Pipit *Anthus campestris*	3	V	•
Redstart *Phoenicurus phoenicurus*	2	V	·
Stonechat *Saxicola torquata*	3	(D)	•

cont.

Table 2. (cont.)

	SPEC category	European Threat Status	Habitat importance
Rock Thrush *Monticola saxatilis*	3	(D)	●
Olivaceous Warbler *Hippolais pallida*	3	(V)	●
Marmora's Warbler *Sylvia sarda*	4	(S)	■
Subalpine Warbler *Sylvia cantillans*	4	S	■
Sardinian Warbler *Sylvia melanocephala*	4	S	■
Rüppell's Warbler *Sylvia rueppelli*	4	(S)	■
Sombre Tit *Parus lugubris*	4	(S)	■
Krüper's Nuthatch *Sitta krueperi*	4	(S)	■
Rock Nuthatch *Sitta neumayer*	4	(S)	■
Chough *Pyrrhocorax pyrrhocorax*	3	V	●
Rock Bunting *Emberiza cia*	3	V	●
Cretzschmar's Bunting *Emberiza caesia*	4	(S)	■
Priority D			
Red Kite *Milvus milvus*	4	S	●
Lesser Spotted Eagle *Aquila pomarina*	3	R	•
Kestrel *Falco tinnunculus*	3	D	•
Woodpigeon *Columba palumbus*	4	S	●
Barn Owl *Tyto alba*	3	D	●
Little Owl *Athene noctua*	3	D	●
Red-necked Nightjar *Caprimulgus ruficollis*	—	S	■
White-rumped Swift *Apus caffer*	—	S	■
Little Swift *Apus affinis*	—	(S)	■
Wryneck *Jynx torquilla*	3	D	●
Syrian Woodpecker *Dendrocopos syriacus*	4	(S)	●
Short-toed Lark *Calandrella brachydactyla*	3	V	•
Crested Lark *Galerida cristata*	3	(D)	•
Crag Martin *Ptyonoprogne rupestris*	—	S	■
Red-rumped Swallow *Hirundo daurica*	—	S	■
Robin *Erithacus rubecula*	4	S	●
Nightingale *Luscinia megarhynchos*	4	(S)	●
Blackbird *Turdus merula*	4	S	●
Song Thrush *Turdus philomelos*	4	S	●
Melodious Warbler *Hippolais polyglotta*	4	(S)	●
Blackcap *Sylvia atricapilla*	4	S	●
Bonelli's Warbler *Phylloscopus bonelli*	4	S	●
Firecrest *Regulus ignicapillus*	4	S	●
Spotted Flycatcher *Muscicapa striata*	3	D	●
Crested Tit *Parus cristatus*	4	S	●
Blue Tit *Parus caeruleus*	4	S	●
Short-toed Treecreeper *Certhia brachydactyla*	4	S	●
Red-backed Shrike *Lanius collurio*	3	(D)	•
Great Grey Shrike *Lanius excubitor*	3	D	•
Spotless Starling *Sturnus unicolor*	4	S	●
Rock Sparrow *Petronia petronia*	—	S	■
Chaffinch *Fringilla coelebs*	4	S	●
Serin *Serinus serinus*	4	S	●
Citril Finch *Serinus citrinella*	4	S	●
Greenfinch *Carduelis chloris*	4	S	●
Linnet *Carduelis cannabina*	4	S	●
Cirl Bunting *Emberiza cirlus*	4	(S)	●

SPEC category and **European Threat Status:** see Box 2 (p. 17) for definitions.

Habitat importance for each species is assessed in terms of the maximum percentage of the European population (breeding population unless otherwise stated) that uses the habitat at any stage of the annual cycle:

  ■ >75%   ● 10–75%   • <10%

[W] Assessment relates to the European wintering population.

of the Priority A species are therefore of significant global importance. All of them are also heavily reliant on Mediterranean forest and shrubland habitats (Table 2).

Sylvia undata is also very dependent on Mediterranean shrubland and almost its entire world population breeds in Europe. It is also considered to have an Unfavourable Conservation Status in Europe because of declines since the early 1970s in Spain, its main European stronghold (Muntaner *et al*. 1983, Tucker and Heath 1994). However, it is an abundant and widespread species in maquis, and furthermore,

Table 3. Numbers of priority bird species of Mediterranean forest, shrubland and rocky habitats in Europe, according to their SPEC category and their dependence on the habitat. Priority status of each species group is indicated by superscripts (see 'Introduction', p. 19, for definition). 'Habitat importance' is assessed in terms of the maximum percentage of each species' European population that uses the habitat at any stage of the annual cycle.

Habitat importance	SPEC category					Total no. of species
	1	2	3	4	Non-SPEC	
<10%	1 [B]	1 [C]	10 [D]	0	0	12
10–75%	0 [A]	12 [B]	18 [C]	21 [D]	0	51
>75%	1 [A]	9 [A]	13 [B]	8 [C]	6 [D]	37
Total	2	22	41	29	6	100

there is currently some debate concerning its European population trend (R. Prodon pers. comm. 1996), especially as its preferred maquis habitat is currently increasing in area in Europe following agricultural abandonment. Nevertheless, until proven otherwise it is prudent to treat the species as a Priority A species, although its status is clearly not as critical as most other birds in the category.

Twenty-six species are classed as Priority B. Of these, two are breeders endemic to Europe, *Alectoris graeca* and *A. rufa*. In addition, Levant Sparrowhawk *Accipiter brevipes*, *Ficedula semitorquata* and *Emberiza melanocephala* have their world populations concentrated in the Mediterranean region of southeast Europe, with much smaller populations in the Middle East. These species do not, however, have more then 75% of their European populations within Mediterranean forest and shrubland habitats. One globally threatened and endangered species in Europe also occurs in this category, Imperial Eagle *Aquila heliaca*, although its use of this habitat is considered marginal. Six other species in this category are classified as Endangered in Europe: *Hieraaetus fasciatus*, *Falco biarmicus*, *Alectoris barbara*, *Turnix sylvatica*, *Oenanthe leucura* and *Ficedula semitorquata* (Tucker and Heath 1994).

Of the 27 Priority C species, three are considered Endangered in Europe (*Gypaetus barbatus*, *Neophron percnopterus* and *Buteo rufinus*), and 16 others also have an Unfavourable Conservation Status in Europe (Tucker and Heath 1994). The remaining eight species have a Favourable Conservation Status but are concentrated in Europe. Of these, *Sylvia sarda*, *Sylvia rueppelli* and *Sitta krueperi* are endemic to Mediterranean Europe.

In total, 65 out of the 100 priority species for the habitat have an Unfavourable Conservation Status in Europe. Many of the remaining species are Priority D species such as Robin *Erithacus rubecula*, Blackbird *Turdus merula*, *Sylvia atricapilla*, *Parus caeruleus* and Chaffinch *Fringilla coelebs*, which have a Favourable Conservation Status but are concentrated in Europe.

Several regions in the Mediterranean are particularly rich in priority bird species during the breeding season, including the Balkans, the coast of Croatia, parts of central Italy, the Mediterranean coast of France and much of Iberia. In national terms, Greece, Turkey and Spain hold the highest overall numbers and percentages of breeding priority species in Europe (Appendix 2. p. 344).

HABITAT NEEDS OF PRIORITY BIRDS

The preferred habitat-types and habitat requirements of the priority bird species are summarized in Figure 3 and Appendix 3 (p. 354) respectively. Of these requirements, vegetation structure is the major ecological factor determining the variability and differentiation of bird communities in Mediterranean terrestrial landscapes (Prodon and Lebreton 1981, Espeut 1984, Rocamora 1987). In general, the total number of breeding birds in a given area increases as succession progresses from garrigue and rocky habitats to maquis and then forest (Prodon 1992). However, Figure 3 indicates that this relationship does not apply to priority species, which reach their greatest diversity in garrigue and rocky habitats. Furthermore, the percentage of priority species in the upper categories A and B is slightly higher in the open habitats. In contrast, many of the forest species have a Favourable Conservation Status in Europe, and are only a priority because they are concentrated in Europe and/or the habitat.

Similar results have been noted previously. For example, in southern France, the highest proportion of rare species was found in the early open stages of succession (Prodon and Lebreton 1981, Prodon 1987). This largely results from the high proportion of forest species that are widespread over the Palearctic region (Blondel and Aronson 1995), as described in 'Habitat description' (above). Thus, in general terms, although Mediterranean forest habitats may be richer in total species, overall bird-conservation priorities are higher in open garrigue and rocky habitats.

Following the successional gradient, three main

groups of priority species can be distinguished: one linked more or less strictly to forest and woodland, one linked to shrubs and bushes, and one linked to open grassy habitats and bare rocky soils.

- In forest and woodland, many species need old, mature forest stands with dead or tall emergent trees, including Black Stork *Ciconia nigra*, *Accipiter brevipes*, *Ficedula semitorquata*, *Sitta krueperi* and *Sitta whiteheadi*. The last species is entirely dependent on the presence of sufficient tall, dead, rotten trunks to nest in. Others like Crested Tit *Parus cristatus*, Sombre Tit *P. lugubris*, *Parus caeruleus*, *Hippolais olivetorum*, *Regulus ignicapillus* and Short-toed Treecreeper *Certhia brachydactyla* can also be found in younger forests. Some like *Erithacus rubecula*, *Turdus merula*, *Turdus philomelos*, *Sylvia atricapilla*, *Phylloscopus bonelli* and *Fringilla coelebs* are actually quite ubiquitous in wooded habitats and not restricted to forests. Habitat requirements of birds linked to woodlands in the Mediterranean also vary according to the openness of the forest. Species like *Scolopax rusticola*, *Columba palumbus*, *Streptopelia turtur*, *Jynx torquilla*, *Muscicapa striata* and *Ficedula semitorquata* prefer more open forested areas compared with, for example, *Aquila pomarina*, *Parus cristatus*, *Phylloscopus bonelli* and *Regulus ignicapillus* (the last two species being abundant in Mediterranean forests).

- Shrubland species include all the Mediterranean *Sylvia* warblers (except *S. atricapilla*), as well as *Hippolais pallida* and Melodious Warbler *H. polyglotta*, and buntings *Emberiza*. These species all need bushes and shrubs, in more or less densely vegetated areas depending on the species. Some of the warblers are also present in the undergrowth of open Mediterranean forests or need the presence of a few trees (e.g. Orphean Warbler *Sylvia hortensis*). Others, like the *Alectoris* partridges, Stonechat *Saxicola torquata*, *Lanius senator*, *Lanius excubitor* and *Carduelis cannabina* also need a herbaceous layer present.

- Important species of garrigue and dry and rocky habitats include *Anthus campestris*, *Calandrella brachydactyla*, Cretzschmar's Bunting *Emberiza caesia* and *Emberiza cineracea*, which all require open, treeless habitats. *Galerida theklae*, *Oenanthe leucura* and *Oenanthe hispanica* are characteristic of this habitat too, but also occur where there are scattered bushes and trees.

One of the most important habitat requirements for a large number of priority species is the presence of cliffs, particularly as nest-sites for raptors.

Although some species are restricted to specific habitat-types, the majority of priority species actually need a mixture of woodland, shrubland and rocky habitat within their foraging area. This particularly applies to large raptors (e.g. *Aegypius monachus*) which need huge areas of such habitat mosaics (Heredia 1996a). Traditional, non-intensive agriculture and forestry often promote such landscapes by maintaining herbaceous and scrubland habitat through livestock grazing and the creation of other habitats such as cereal cultivations, vineyards, olive-groves, orchards and forest patches. Such traditional practices are therefore closely linked to many species' biological requirements and can be considered, to a certain extent, to be necessary for the maintenance of suitable habitats for them. For example, the population size of *Hieraaetus fasciatus* in Catalonia (Spain) was found to be positively correlated with the size of the rural population (Real and Mañosa 1992).

At the same time, however, larger species such as raptors are often sensitive to human disturbance.

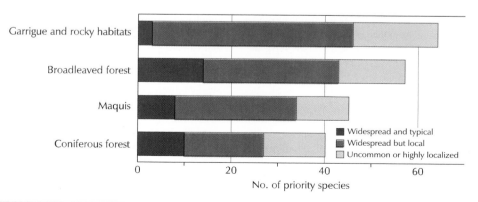

Figure 3. The occurrence of priority bird species in the main types of Mediterranean forest, shrubland and rocky habitats (compiled by the Habitat Working Group).

Regulation of human activities around nesting sites is therefore often essential to avoid nest failures.

The presence of powerlines can also cause high levels of mortality in large birds (Sériot and Rocamora 1992, Bayle 1994, Bevanger 1994) and may spoil otherwise suitable habitats for such species.

THREATS TO THE HABITAT

The scale of threats to Mediterranean forest, shrubland and rocky habitats are summarized in Table 4. An assessment of the impacts of each threat on each priority species for the habitat has also been carried out (Figures 4–5; see Appendix 4, p. 401, for the methods, scoring system and the detailed results). As most threats to these birds are unproven or unquantified, the predicted effects on populations are largely inferred and based on the majority opinions of the Habitat Working Group and other consulted experts.

Importantly, where a threat is not listed as affecting an individual species this cannot be interpreted as proof that there is no effect, but merely that no harmful effects have been detected so far. It is also not possible in the assessments to take into account interactions between threats or their additive impacts.

Land abandonment

During the last 100 years in the Mediterranean region, traditional land-uses over millions of hectares of non-intensive cultivation and pasture have been abandoned (Beaufoy et al. 1994) (see 'Origin and history of the habitat', p. 241). Without the checks to succession provided by ploughing or grazing, the vegetation has developed progressively into garrigue or maquis, and then into forest. Massive increases in the area of Mediterranean forest and shrubland have been noted in France (Lepart and Debussche 1992), Spain (Boada 1994, Escoda 1996), Portugal and Italy (WWF 1994), and have probably occurred widely elsewhere in Mediterranean Europe too. As a result, local and regional populations of formerly widespread birds of garrigue and maquis have decreased substantially (e.g. Ferrer et al. 1984, Mestre 1984). In contrast, there has been a big expansion in the distributions of many forest birds.

In addition, there is a large and rapid turnover in shrubland habitats which is often not apparent from the overall land-cover statistics. For instance, during the 1980s in France, 2–4% of the total area of maquis and garrigue was transformed into other habitat-types every year, mainly through afforestation or cultivation (Abdelli 1991). On the other hand, maquis and garrigue were also created elsewhere, during the same period, by the abandonment of agricultural land, by fire and by forest-clearance, leading to only a limited overall decrease in the area of shrubland.

The complete abandonment of grazing is highly detrimental to the many Mediterranean bird species that favour the more open garrigue and rocky habitats. It is considered to threaten 44 priority species, 12 of which are highly threatened. These include species that require patches of open, short vegetation for feeding, some large raptors that favour open habitat mosaics created by pastoral agriculture, and large scavengers such as Lammergeier *Gypaetus barbatus* and Griffon Vulture *Gyps fulvus* that are more directly affected by reduced numbers of livestock, through reduced availability of carrion.

As a result of the substantial number of species affected, the high level of threat to many of them, and their high priority status, the overall weighted threat score is 127, the highest for any threat.

Effects of fire

Fire is a natural phenomenon which has great influenced the development of Mediterranean vegetation and continues to affect the cycles of vegetation succession (see 'Habitat description', above). However, humans have severely disrupted this balance. Today, it is thought that 98% or more of fires in the Mediterranean are started by man (Vélez 1990, Trabaud and Prodon 1992), either intentionally (e.g. to clear ground) or accidentally, and it is thought that any given area now burns much more frequently than the natural rate before man, especially in conifer woodlands and *Cistus* shrublands.

The intentional starting of fires is now illegal or strictly regulated in all European Mediterranean countries. Fire-fighting techniques have also improved considerably in recent decades, and substantial financial resources have also been allocated to fight fires, ranging from 6 ECU/ha of fire-sensitive land in Greece to 45 ECU/ha in France in 1990. However, despite these actions, the incidence and total extent of fires is still great, and is increasing in many countries (e.g. Ministero dell'Ambiente 1992, Instituto Florestal 1993a,b, Boada 1994). It is estimated that 1.0–2.5% of forested Mediterranean areas in the EU burn annually (Hermeline and Rey 1994), equivalent to many thousands of square kilometres. The average fire-return interval, which has decreased dramatically this century (Blondel and Aronson 1995), may now be as little as five years in some areas (Trabaud and Prodon 1992).

Frequent, large fires are partly due to the widespread abandonment of traditional agriculture, grazing and forestry, which can lead to the growth of extensive areas of dense shrubland, and woodland with abundant dry undergrowth, that is very susceptible to fire. The increase in the human population in lowland areas and the development of tourism have

Table 4. Threats to Mediterranean forest, shrubland and rocky habitats, and to the habitat features required by their priority bird species, in Europe (compiled by the Habitat Working Group). Threats are listed in order of relative importance (see Figure 5), and are described in more detail in the text.

Threat	Effect on birds or habitats	Scale of threat			Regions affected, and comments
		Habitat	Past	Future	
1. Abandonment of grazing, or undergrazing	Increases vegetation cover; reduces vegetation diversity; increases vegetation height; loss of open areas, of feeding opportunities, and of carrion	C,B,G	●	■	Esp. in France, Spain, Italy, Portugal
2. Frequent and large fires	Opens up habitat and simplifies structure, increases susceptibility to erosion and, in extreme cases, habitat loss through 'desertification'	C B M G	■ ● ■ ■	● ● ■ ■	Generally increasing in Iberia, stable in Italy, decreasing in France
3. Afforestation with native and non-native trees, e.g. *Eucalyptus*	Habitat loss	C,B,M,G	●	·	Widespread in Italy and Iberia in the past, regional in France and Portugal in the future
4. Disturbance by human activities (especially recreation, hunting, forestry and agriculture)	Disturbance of nesting, roosting and feeding birds may lead to reduced adult survival, increased predation at nests and starvation of young	C,B,M,G	■	■	Everywhere and increasing
5. Lack of fire	Increases vegetation cover	M,G	■	■	Even less fire is predicted in the future, except Iberia
6. Wood-cutting:					
a. Coppicing	Initially opens habitat and reduces structural diversity; eventually leads to dense growth	B	■	●	Common in Italy and eastern Mediterranean
b. Selective felling	Promotes regeneration but reduces canopy and vertical structural diversity, and removes tall emergent trees; reduces food resources and nest-sites	C,B	■	■	Likely to increase; regional in Portugal
c. Clear-felling	Initially opens habitat, promotes erosion, and leads to uniform regrowth; often followed by afforestation with non-native species	C,B	●	●	
7. Reservoir creation	Habitat loss	C,B,M,G	·	·	Future regional problem in Spain and Portugal, possibly Turkey and Italy
8. Urbanization	Habitat loss	C,B M,G	· ●	· ·	Widespread in coastal areas
9. Roads, railways and other infrastructure	Habitat loss	C,B,M,G	·	·	
10. Conversion to perennial crops, e.g. vineyards, olive groves, etc.	Habitat loss	M,G	·	·	
11. Overgrazing	Promotes extremely short vegetation structure, bare ground, excessive erosion and high disturbance, and prevents regeneration of woody species	G C,B,M	■ ●	● ·	Widespread in Turkey and Greek islands, local in France, Spain, Portugal and Italy (esp. Sardinia)
12. Tree disease and pests	Change in vegetation structure, increase in dead wood	C,B	·	●	
13. Powerlines	Mortality of large bird species and, if lines at sufficiently high density, habitat avoided	C,B,M,G	●	●	

cont.

Table 4. (cont.)

Threat	Effect on birds or habitats	Scale of threat			Regions affected, and comments
		Habitat	Past	Future	
14. Habitat fragmentation	Reduces plot size, increases edge habitat, landscape diversity, habitat insularity and susceptibility to disturbance	C,B,M,G ■		●	
15. Conversion to arable agriculture	Habitat loss	B,M,G	●	●	
16. Undergrowth removal and thinning	Destroys shrub layer, promotes fast growth and uniform structure	C,B	●	●	Regional in Portugal
17. Pesticides (in forestry)	Loss of invertebrate food resources, disrupts ecosystems, and possible mortality from toxic products	C,B,M	●	●	Greece and Crete, mainly against moth *Lymandria dispar*
18. Mining and quarrying	Habitat loss	C,B,M,G	●	●	

Scale of threat

C	Coniferous forest	Past	Past 20 years	■	Widespread (affects >10% of habitat)
B	Broadleaved forest	Future	Next 20 years	●	Regional (affects 1–10% of habitat)
M	Maquis			●	Local (affects <1% of habitat)
G	Garrigue and rocky habitats			?	Considerable uncertainty

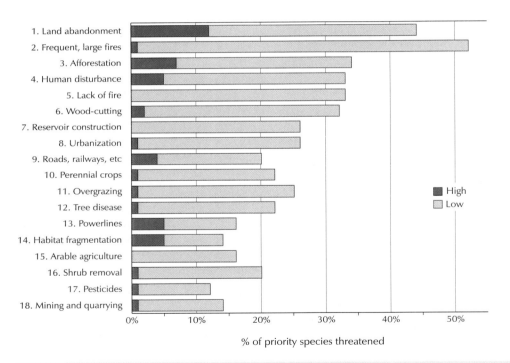

Figure 4. The predicted impact on priority bird species of threats to Mediterranean forest, shrubland and rocky habitats, and to the species' habitat requirements, in Europe. Bars are subdivided to show the proportion of species threatened by high or low impacts to their populations (none is threatened by critical impacts; see Table 4 and Appendix 4, p. 401, for more details).

High impact
The species' population is likely to decline by more than 20% in the habitat in Europe within 20 years if current trends continue

Low impact
Only likely to have local effects, and the species' population is not likely to decline by more than 20% in the habitat in Europe within 20 years if current trends continue

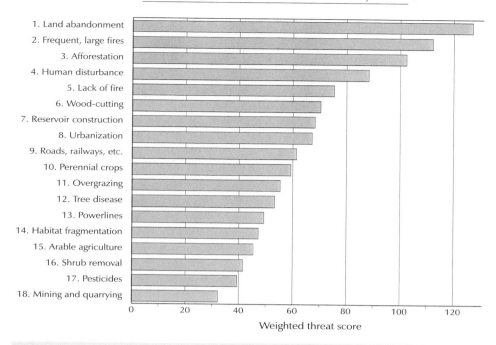

Weighted threat score

Figure 5. The most important threats to Mediterranean forest, shrubland and rocky habitats, and to the habitat requirements of their priority bird species, in Europe. The scoring system takes into account:

- The European Threat Status of each species and the importance of the habitat for the survival of its European population (Priorities A–D); see Table 2;
- The predicted impact (Critical, High, Low), within the habitat, of each threat on the European population of each species; see Figure 4, and Appendix 4 (p. 401).

See Table 4 for more details on each threat.

also considerably increased the risk of accidental fires. Illegal and often uncontrolled burning is also still used to produce fresh growth of vegetation for livestock grazing in some Mediterranean countries, even though the nutritional value of such growth is rather poor.

Bird mortality directly resulting from fire probably varies greatly depending on season, extent of fire, and bird species present. It should be negligible in carefully planned and controlled small-scale fires. The indirect effects of substantial modifications in vegetation structure and composition are very important; such effects are complex, and cannot be reviewed in detail here (see Prodon *et al.* 1984, Prodon 1987, Prodon *et al.* 1987, Faralli and Lambertini 1989). In general, open-landscape species such as *Oenanthe hispanica* increase in frequency after fire, but decline as the vegetation grows back. Bird species of maquis, such as *Sylvia undata* and Sardinian Warbler *Sylvia melanocephala*, almost disappear immediately after fire, but recover as the maquis regenerates. Forest species disappear after fire and cannot recolonize during the short to medium term.

Despite the general increase in forest resulting from widespread abandonment of land, a current feature of those parts of the Mediterranean with unnaturally frequent, large and intense fires is the widespread lack of true successional processes. Blondel and Aronson (1995) consider that many Mediterranean shrublands are 'blocked' in this way, and dominated by one or a few shrub species. Such impoverished shrublands usually have impoverished avifaunas also (Prodon 1987). Frequent fires are, therefore, a threat to forest species and also to some birds of maquis and garrigue habitats.

Some effects of fires are beneficial to bird communities of open areas, by creating landscape mosaics and increased fine-scale heterogeneity in the regenerating vegetation, and through the maintenance of bare areas (Lambertini and Faralli 1991). In this way, a lack of fire can also be a threat to some open-habitat species, especially where grazing levels are decreasing. Absence of fire over long periods leads to a build-up of woody vegetation, leaf-litter and dead wood. When fires do break out in such areas, they may be very intense and result in complete destruction of the habitat, long-term damage to the soil and severe erosion.

Some species are threatened by frequent large fires as well as by a lack of fire. This is because they require a mosaic of habitats which are threatened by both extremes of fire occurrence. For example, many raptors require tall forest trees to nest in, but open landscapes for foraging. Others, such as maquis species, require a finer-grained heterogeneity, which is favoured by less frequent, smaller fires. It should also be noted that 'anti-fire' management activities in Mediterranean forests can, if not carried out with care, cause serious disturbance and reduced breeding success in priority birds (see 'Human disturbance', below).

Frequent large fires affect more priority bird species than any other threat (Figure 4), and have almost the highest weighted threat score. Lack of fire is, however, also a very significant threat in terms of the number of priority species affected and the magnitude of the weighted threat score.

Habitat loss

Apart from 'Land abandonment' (above), the loss of Mediterranean forest, shrubland and rocky habitats also arises through deliberate conversion to other uses, principally afforestation, reservoir construction, urbanization, transport routes, agriculture, mining and quarrying. In the past 20 years, these have mostly had only local impacts, although regional-scale losses are thought to have resulted from afforestation, urbanization and conversion to agriculture. In addition, some of these changes (e.g. urbanization) are hardly reversible in the short to medium term.

Afforestation with non-native tree species (e.g. *Eucalyptus*) effectively leads to a complete loss of the existing habitat and major changes in the bird community (e.g. Sánchez and Tellería 1988, Santos and Alvarez 1990). Similarly, afforestation with indigenous species should also be considered as loss of habitat when carried out in an intensive way, with high-density tree-planting and use of artificial fertilizers and chemicals. The vegetation structure of such dense monoculture forests is very different from natural ones, and their bird communities are generally poorer in terms of species richness, especially at young stages (Rocamora 1987, Faralli 1995).

Afforestation in Iberia since the 1940s has, for example, led to the loss of thousands of square kilometres of shrubland (Ortuño 1990, Hermeline and Rey 1994). In Portugal, these schemes even led to the conversion of remaining areas of old-growth forest to non-native *Eucalyptus* plantations, until stricter legislation was implemented at the end of the 1980s. Within the EU, afforestation is now promoted as a consequence of the reform of the Common Agricultural Policy (CAP), e.g. by the Regulation concerning set-aside (1094/80/EEC), with subsidies of up to 4,000 ECU/ha. Such intensive, highly productive afforestation schemes are a threat to the regeneration of Mediterranean shrubland, rather than a means of increasing the area of natural Mediterranean forests. Other general effects of current EU agricultural policies are, however, likely to encourage land abandonment and natural reforestation, thus creating more scrubland and semi-natural forest.

Urbanization, and the construction of associated roads, railways and powerlines, has been proceeding steadily for the last 25 years in many coastal and lowland areas of the Mediterranean, especially in the vicinity of large cities. Urban growth is due to a widespread rural exodus, to the development of mass tourism and to increasing industrialization. Thousands of square kilometres of forest, maquis, garrigue and rocky coasts have been destroyed or spoiled by such growth of towns and cities in the western Mediterranean and in Greece and Turkey. The impact of these threats is not limited to the areas where the habitat has been effectively damaged, but, due to disturbance and pollution, extends considerably beyond. For example, as a consequence of the planned EU Trans-European Networks, 4% of the Important Bird Areas (IBAs) in France (see Grimmett and Jones 1989, Rocamora 1994) will be situated within 2 km of future roads or railways, and 20% within a distance of 10 km (Bina *et al.* 1995). More than 309 European IBAs (12%) are actually or potentially threatened by the implementation of the Networks.

In past decades, conversion to agriculture (e.g. arable, vineyards) of Mediterranean forest and shrubland was common but, if such conversion still occurs, it is now relatively limited in extent and largely compensated by the amount of shrubland and forest that is regenerating through land abandonment.

As a result of falling demands for agricultural land and stricter controls on *Eucalyptus* plantations, these and other threats are all likely to have only local impacts in the future. Afforestation, however, may continue on a regional scale in France, and perhaps elsewhere as a result of European Union financial incentives and attempts to control erosion.

It is predicted that habitat loss will affect a significant number of species overall. Afforestation is considered to affect 34 species, with results that are likely to be severe for several, such as gamebirds, larks (Alaudidae), *Oenanthe hispanica* and *Sylvia undata*. Consequently, it has a high overall weighted threat score. Other forms of habitat loss treated here are moderate threats overall, except impacts from mining and from conversion to arable agriculture, both of which are likely to be low.

Human disturbance

Disruption of critical aspects of the life-cycle of birds, such as nesting, feeding and roosting, due to human activities like recreation, hunting, forestry (e.g. careless anti-fire management activities) and agriculture, is widespread in all Mediterranean forest and shrubland. Some species that are known to be highly susceptible to such disturbance through reduced breeding success are large, uncommon or rare species such as *Ciconia nigra*, *Aegypius monachus*, the globally threatened *Aquila adalberti*, *Hieraaetus fasciatus* and *Bubo bubo* (González 1996, Heredia 1996a). Some colonial species such as *Merops apiaster* may also be locally affected.

Tourism is probably the main cause of disturbance to birds and other wildlife in the region. In France, for example, tourism and recreation is the second most important human threat to IBAs, mainly as a result of disturbance (Rocamora *et al.* 1995). The Mediterranean countries are the world's most popular tourist destination, with the main concentration of tourist developments being along the coasts of Spain, France, Portugal and Italy. From a 1990 level of 157 million visitors per year, numbers are predicted to increase to 380 million by 2050 if overall economic growth is weak, or up to 760 million if growth is strong (Stanners and Bourdeau 1995). Almost half of these tourists will be based along the coastline. Coastal maquis, garrigue and rocky habitats will therefore be under considerable pressure. However, other forest and shrubland habitats in rural locations will also be affected.

Hunting is also a common recreational activity that can be a major cause of disturbance. Although such pursuits may be traditional and long-standing, the creation of new mountain roads is allowing easier access to previously remote and disturbance-free forests.

It is highly likely that disturbance will increase in future as a result of further expansion of tourism in the region and continued agricultural intensification. Overall, disturbance is thought to affect a third of all priority species and has a high weighted threat score.

Wood-cutting

The coppicing of trees in forest compartments (see 'Values, roles and land-uses', p. 246) may be initially beneficial to many birds due to the opening up of the habitat, as has been observed in coppice woods in the temperate zone (Fuller and Warren 1990, Buckley 1992). However, the longer-term development of uniform and dense growth is unsuitable for many priority species. Bird-community richness is enhanced by long cutting cycles and fuller development of the shrub layer (Alvarez and Santos 1992, Tellería 1992, 1993).

Intermittent cutting also precludes the establishment of large, mature trees, thereby reducing the availability of nest-sites for large raptors, *Ciconia nigra* and hole-nesting species. This problem can, however, be overcome by leaving stands of coppice to grow to maturity.

Selective felling of Mediterranean forests is also a widespread management practice. Although less destructive than coppicing and clear-cutting, it removes the largest mature trees from forests. Thus, like coppicing, it limits nest-site availability for large raptors and hole-nesting species. It also tends to reduce food availability and bird diversity.

Clear-cutting of Mediterranean forests is less widespread than other methods of wood-harvesting. In part, this is probably to avoid soil erosion, which is frequently severe after removal of forest-cover in the Mediterranean region. Clear-felling is, however, the most destructive of wood-cutting techniques. Although some open-habitat species benefit, such as various larks and *Sylvia* warblers, the habitat is effectively destroyed for most forest species for many years, decades for some species. Felling often takes place at about the time the trees become sufficiently large for raptors, etc. to nest in. Regrowth also tends to be uniform, supporting a lower diversity of birds. However, Ferry and Frochot (1974) found that in an area of forest in Burgundy, only a few mature trees needed to remain in otherwise clear-cut compartments to maintain the presence of the majority of forest species.

With the exception of some parts of Spain and Portugal, much of Mediterranean Europe has witnessed the wide-scale abandonment of traditional forestry practices in recent years (Stanners and Bourdeau 1995), as the region cannot compete with cheap imports from the highly productive Nordic and eastern regions. Thus, although the forested area is likely to increase through land abandonment and afforestation schemes, coppicing is likely to decline. Clear-cutting is also likely to decline to a certain extent because of the high risks of erosion following such logging. In contrast, selective felling may locally increase as more trees mature and clear-felling declines.

Nearly one third of all priority Mediterranean forest and shrubland bird species are thought to be threatened by wood-cutting. However, despite the widespread nature of these activities, only local impacts are likely for most species.

Overgrazing and soil erosion

Up to the end of the nineteenth century, overgrazing was widespread in Mediterranean Europe, often resulting in severe damage to the natural vegetation and excessive soil erosion. Such erosion has been a major problem in the Mediterranean basin for millennia, especially in frequently burned, cultivated or overgrazed areas, or on steep slopes where the soil can be easily washed away by intense rainfall episodes, the latter being a characteristic of the Mediterranean climate. Excessive erosion may lead to the creation of poorly vegetated and rocky areas that can actually be beneficial to several priority birds of Mediterranean shrubland and rocky habitats. However, as the loss of soil is practically irreversible, excessive soil erosion causes serious damage to the ecosystem, especially when it affects very large areas.

Overgrazing and subsequent erosion is thought to lead to local threats to about a quarter of the priority species, including *Alectoris* spp., larks and *Sylvia* warblers of garrigue habitats. Overall, the threat is of moderate importance but, as already described, considerably less than the threats to such priority species arising from the abandonment of traditional land-uses such as grazing.

Tree disease

In recent years, tree density has also been reduced locally due to the death of *Quercus ilex*, *Q. suber* and other oak species throughout the oak-dominated woodlands of Iberia and Italy (ICONA 1991, Brasier 1992, Fernández and Montero 1993). In Portugal, average mortality rates of *Q. suber* were estimated to be about 15% in the 1980s (DGF 1990). Associated declines in maquis shrubs, such as *Cistus* and *Lavandula*, have also been recorded at some sites.

There are believed to be many causes of this phenomenon, known as 'seca', although its significance is at present unknown (Díaz *et al.* 1997) and no single primary cause has been identified. However, Brasier (1992) considers that the decline of oaks is most likely attributable to the introduced and highly aggressive root-fungus *Phytophthora cinnamomi*, and also to another fungus *Hypoxylon mediterraneum* (Hermeline and Rey 1994). These may interact with droughts, changes in land-use, insect attacks and, subsequently, secondary infections (Fernández and Montero 1993, Mesón and Montoya 1993).

P. cinnamomi is probably indigenous to the New Guinea–Sulawesi region, but has been accidentally widely distributed by man through trade. It was responsible for the recent dieback of entire *Eucalyptus* forests in western Australia and for massive deaths of native chestnuts across the south-eastern United States at the beginning of the century. It was widespread in Europe by the 1940s and caused a major epidemic in the European chestnut. Therefore, although the future spread of this pathogen is difficult to predict, it clearly has the capacity to cause massive and widespread impacts on habitats and landscapes.

The effects of tree diseases on birds are difficult to predict, but those bird species requiring mature oaks are the most likely to be affected. Impacts are therefore possible for a range of raptors and open-forest species. South-west Spain is one of the worst areas affected, coinciding with the main distribution of the globally threatened *Aquila adalberti*. Oak-tree disease may therefore be a serious threat to this species. Impacts are, however, likely to be local for other species. Overall impacts are, therefore, likely to be moderate, but as mentioned above, difficult to predict, and potentially much more serious.

Powerlines

Powerlines, telegraph wires and other overhead cables can cause high levels of mortality in large species, such as storks (Ciconiidae) and raptors, through electrocution or collisions (Ferrer *et al.* 1993, Bevanger 1994). In France, it was found that powerline mortality is mainly due to electrocution (90% of all cases), the remainder being caused by collisions, although the latter may have been under-estimated (Sériot and Rocamora 1992). Vulnerable species may consequently be absent, or avoid otherwise suitable habitats where overhead cables occur in high densities.

For example, electrocution is the commonest cause of non-natural death in the globally threatened *Aquila adalberti*, with 10–20 cases reported annually out of a total world population of some 130–160 pairs (Calderón *et al.* 1988, Ferrer and Calderón 1990, Cadenas 1992, Jiménez 1992, Oria and Caballero 1992, González 1996). In Catalonia (Spain), recent studies have also found very high mortality levels in raptors as a result of electrocution (Mañosa *et al.* 1996).

Electrocution in such large species occurs through contact with a conductor and an earthed post, or between two conductors. Due to the size of the posts, the distance between the conductors and the length of the insulators, electrocution is only frequent on powerlines with voltages of 15–45 kV. The design of the post significantly affects the probability of electrocution (González 1996). Lines near roads and tracks cause fewer deaths than those sited away from them. Mitigating measures such as non-dangerous post designs, visible markings, and layouts avoiding sensitive areas (particularly Important Bird Areas) are likely to reduce substantially this type of mortality (Negro and Ferrer 1995).

In total, 16 species are likely to be affected, five of which may be highly threatened (*Milvus milvus*, *Aquila adalberti*, *A. chrysaetos*, *Hieraaetus fasciatus* and *Bubo bubo*). Thus, although overall the number of priority species affected may be low and the weighted threat score only moderate, the majority of the affected species are highly threatened.

Habitat fragmentation

In addition to 'Habitat loss' described previously (p. 256), urban and industrial expansion, road-building, tourist developments, etc., as well as agricultural intensification, have all led to increased fragmentation of remaining Mediterranean forests, shrublands and rocky habitats, with a variety of potentially detrimental effects (Wilcove *et al.* 1986, Merriam 1988). Fragmentation reduces the size of remnant habitat patches and thus the proportion of 'edge' habitat increases. Also, boundaries often become more abrupt and patches tend to be more widely separated. For example, fragmentation of Mediterranean habitats can increase nest-predation pressure on open-ground nesters (Santos and Tellería 1991, 1992, Tellería and Santos 1992), and can increase food-competition between winter frugivores, e.g. between thrushes and rodents (Tellería *et al.* 1991).

The importance of habitat fragmentation effects is, however, a subject of considerable debate (e.g. Wiens 1989b, 1995a, Opdam *et al.* 1995, Simberloff 1995), as expectations are largely based on theoretical models and empirical evidence for the effects of fragmentation is actually rather meagre.

Indeed, although habitat fragmentation was considered to be widespread in the past and will continue to have regional impacts, it is thought unlikely to affect many priority bird species. Threats are, however, predicted to be high for five species, including the largest raptors. Nevertheless, its overall importance, as assessed by the weighted threat score, is low in comparison with other threats to Mediterranean habitats.

Undergrowth removal and thinning

The removal of undergrowth and the thinning-out of a proportion of trees from forests have been fairly common forestry practices in the region. They result in the removal of the shrub layer in forests, and they reduce tree density in order to stimulate fast and uniform growth. Dead wood and old trees which could provide nest-sites for hole-nesting species are also generally eliminated at the same time. Structural diversity is thus reduced due to the loss of an entire vegetation layer and to the increased uniformity of the trees that remain. As noted above, bird-community richness is enhanced by the development of a shrub layer and an overall high structural diversity.

Such severe forestry practices are, however, predicted to decline in the future as a result of the general decline in management for wood production described previously. Therefore, although 20 species are considered to be threatened by the practice of undergrowth removal and thinning, nearly all impacts are likely to be local. Overall, the importance of this threat is likely to be low compared with others.

Pesticide use

Pesticides have been increasingly used in forests to control pest problems. In the Mediterranean region this is a particular problem in Greece and Crete, where aerial applications of pesticides are used to control defoliating caterpillars of the moth *Lymandria dispar* in deciduous forests.

Although direct toxic effects may result from these applications, particularly to raptors, the more immediate and probable threat to birds is from a reduction in invertebrate food resources. Information on the extent and level of impact on birds is, however, very limited, but priority birds that are insectivorous (highlighted in Appendix 3, p. 354) are likely to be affected in particular.

In total, 12 species may be threatened, though nearly all impacts are likely to be local. The overall weighted threat score is therefore low. Encouraging results obtained by biological control (e.g. virus-infected clones of the chestnut bark fungus *Endothia parasitica*) may reduce the use of pesticides in forestry, and the development of less toxic and more specifically targeted pesticides may also help to reduce impacts in future.

CONSERVATION OPPORTUNITIES

International legislation

A summary of the international legislative instruments that may be used to conserve Mediterranean forest, shrubland and rocky habitats and their birds is presented in Box 1. These are described in more detail in 'Opportunities for Conserving the Wider Environment' (p. 24), but features special to these habitats are highlighted here.

To date, only a relatively small proportion of the extent of these habitats has been protected by these instruments. Annex I of the Birds Directive (Box 1) lists bird species for which Special Protection Areas (SPAs) should be designated in the EU countries, and includes 44 (44%) of the priority species of Mediterranean forest, shrubland and rocky habitats (Appendix 1, p. 327). The designation of SPAs has been slow in Europe, and particularly in the Mediterranean region. About 25,000 km^2 have been designated as SPAs in Spain (Anon. 1996), but this covers only 19% of all Important Bird Areas (IBAs) which

---

Box 1. International environmental legislation of particular relevance to the conservation of Mediterranean forest, shrubland and rocky habitats in Europe. See 'Opportunities for Conserving the Wider Environment' (p. 24) for details.

GLOBAL

- *Biodiversity Convention* (1992) Convention on Biological Diversity.
- *Bonn Convention* or *CMS* (1979) Convention on the Conservation of Migratory Species of Wild Animals.
- *World Heritage Convention* (1972) Convention concerning the Protection of the World Cultural and Natural Heritage.

PAN-EUROPEAN

- *Bern Convention* (1979) Convention on the Conservation of European Wildlife and Natural Habitats.
- *Espoo Convention* (1991, but yet to come into force) Convention on Environmental Impact Assessment in a Transboundary Context.

REGIONAL

- *Barcelona Convention* (1975) Convention for the Protection of the Mediterranean Sea against Pollution.

EUROPEAN UNION

- *Birds Directive* Council Directive on the conservation of wild birds (79/409/EEC).
- *Habitats Directive* Council Directive on the conservation of natural habitats of wild fauna and flora (92/43/EEC).
- *EIA Directive* Council Directive on the assessment of the effects of certain public and private projects on the environment (85/337/EEC).
- Council Directive on fire control (86/3259/EEC).
- Council Directive on the freedom of access to information on the environment (90/313/EEC).
- Council Regulation on the prevention of forest fires (2158/92/EEC).

---

qualify for SPA status (Grimmett and Jones 1989, Viada *et al.* 1995). Other Mediterranean countries have protected less, e.g. with France designating 18% of qualifying IBAs, Greece 13% and Italy 9% (Langeveld 1991, Waliczky 1994).

Furthermore, most existing SPAs tend to be coastal, wetland or mountain habitats, and very few are likely to hold significant amounts of Mediterranean forest, maquis or garrigue. Similarly, but more generally, these habitats are poorly represented in the protected-area systems of almost all Mediterranean countries (IUCN–CNPPA 1994)—the designation of the whole island of Menorca (Spain) as a Biosphere Reserve being one notable exception. Shrublands are particularly poorly protected as they are often widely regarded as being 'waste' or derelict land of little value. Indeed, of 22 sites listed by the European Commission (CEC 1995d) as representative of maquis or garrigue in Europe, only four are protected.

The Habitats Directive aims to combine the protection of threatened species with the protection of threatened habitats, via the designation of Special Conservation Areas (SACs), which in combination with SPAs will form the 'cohesive, integrated' Natura 2000 network of protected areas. This directive could thus play an important role in enhancing the level of protection of Mediterranean forest and especially shrubland and rocky habitats. Annex I of the Directive lists these habitat-types and includes 58 which fall within the Mediterranean forest and shrubland

habitats described here (see Appendix 5, p. 423). Eleven of these are 'priority' habitats, which requires that they be given higher protection status.

All of the European Mediterranean states have signed the Biodiversity Convention, and, as part of their obligations under this convention, at least one of these Mediterranean countries (Bulgaria) has developed a national biodiversity conservation strategy (see 'Opportunities for Conserving the Wider Environment', p. 24), which may help to conserve Mediterranean forest, shrubland and rocky habitats. The Pan-European Biological and Landscape Diversity Strategy (PEBLanDS) has been formulated by the Council of Europe as a broad initiative to address Europe's obligations under this convention. PEBLanDS was endorsed by the environment ministers of 55 European countries in October 1995.

The Environmental Impact Assessment (EIA) Directive (Box 1) provides the opportunity to prevent or limit damaging impacts to habitats from some major infrastructural projects in the EU. However, forestry and agricultural projects—that are often the most threatening to Mediterranean forest, shrubland and rocky habitats (see 'Threats to the habitat', p. 252)—are not usually covered by the Directive. Similarly, national planning laws usually do not adequately control such developments, although they can be effective in controlling potentially damaging building developments for housing and tourism, etc. However, as described previously, shrubland habitats are widely perceived as derelict land and their nature-conservation value is often unappreciated. Building developments are therefore often deliberately placed within such habitats.

As described in the previous section, frequent fires and (conversely) total lack of fires in Mediterranean ecosystems are both serious threats to many priority species and habitats. Such fires also generate serious pollution and erosion problems as well as being highly dangerous to human life and settlements. The European Union is therefore implementing a Regulation (92/2158/EEC) on the prevention of forest fires (Stanners and Bourdeau 1995). As part of this, a detailed analysis of the causes of fires and possibilities of controlling them is being undertaken in conjunction with a study of socio-economic factors prevailing in regions at risk. This investigation comprises a preliminary study of forest legislation and policies which have an impact on forest development (e.g. town-planning, environmental and agricultural policies) and which could indirectly cause fires. Attempts are also being made to evaluate the effectiveness of public awareness campaigns, and special attention is being given to economic measures that may be used to reduce the number of fires, such as town-and-country planning regulations and support for agricultural/grazing activities (CEC 1991).

Tourism is a major threat to many Mediterranean habitats through direct habitat loss to developments, as well as through human disturbance and pollution, etc. (as described in 'Threats to the habitat', p. 252). Within the European Union, various projects have been supported in the framework of *Community action plans to assist tourism*, which provide that the measures taken must be 'conducive to preserving and protecting the quality of the natural environment'. A variety of national legislative and policy measures have also been taken to control the impacts of tourism and recreation. For example, in Cyprus, land-use controls have been introduced with zoning for protected areas, controlled development areas and transferable development rights which may be used in other uncontrolled zones (Stanners and Bourdeau 1995). Controls on bed capacity have been implemented in some areas, for instance Mallorca (Spain), where 25,000 bed spaces were removed in 1993 to ease tourism pressure. Land levies for new developments, as imposed in Cyprus, can also indirectly reduce environmental pressures.

Building and urbanization is an important problem for Mediterranean forest and shrubland in the vicinity of large cities, where demand for suburban housing and new industrial sites is high. The European Commission has proposed that Strategic Environmental Assessment (SEA) of urbanization plans and programmes should be promoted and developed in these areas. In this way, the impact of many,

small-scale building projects can be better controlled and minimized.

Areas of Mediterranean forest, shrubland and rocky habitats that are managed for non-intensive livestock grazing, and areas managed for forestry, are frequently not covered by land-use and conservation-related regulations. There is a particular political reluctance to impose restrictions and additional costs on farmers, especially where this is being carried out in agriculturally marginal habitats. Furthermore, where such regulations do apply, they are often difficult to enforce. It is therefore widely recognized that the regulatory approach has limitations in forestry and especially agriculture, where there are a large number of small producers who are farming in ways that are encouraged by other policies. Policy-related measures are therefore the most likely to produce wide-scale conservation benefits to Mediterranean habitats that are managed for forestry and agriculture.

Policy initiatives

The conservation of Mediterranean forest, shrubland and rocky habitats (and their priority birds) is highly dependent on those factors that influence land-use on a wide scale, rather than merely on site-protection measures. Apart from in Spain, very few priority bird species of these habitats have substantial proportions of their populations within Important Bird Areas (IBAs). Furthermore, for those species that do, such as *Aquila adalberti*, very few of their IBAs are protected, and the habitat is still highly susceptible to changes in agricultural, forestry and other land-use policies.

A strategy for influencing land-use in selected areas, or generally across a habitat, needs to consider local, regional, national and international policies for influencing these factors. What is needed is a mix of policies integrated by common and clear objectives, including those outlined under 'Conservation recommendations' (p. 262) in this strategy.

Forestry policies are of prime importance, and opportunities within these include the *Statement of Forest Principles* made at the Earth Summit in 1992, and the subsequent *Helsinki Guidelines for Sustainable Forestry*, which seek to define and encourage sustainable management of forests globally and in Europe, respectively. Timber certification and accreditation schemes are also being created as a means of supporting sustainable and 'environmentally friendly' forestry practices (Sugal 1996). These policies and schemes are discussed in more detail under 'Boreal and Temperate Forests' (p. 203).

Agricultural policies are also of considerable importance for the conservation of Mediterranean forest, shrubland and rocky habitats because of their direct effects on livestock grazing (especially in garrigue habitats) and the demand for agricultural land. In particular, the Regulation for afforestation of surplus arable land in the EU (Box 1), one of the package of measures to reform the Common Agricultural Policy (CAP) that was adopted by the EU in 1992, provides financial assistance towards the creation of new woodlands and forests on agricultural land. The Regulation requires member states to prepare a national programme for its implementation. This requires careful planning, for although the Regulation may help to recreate lost forest in suitable areas, in some open and wildlife-rich habitats such as Mediterranean shrublands and rocky habitats, or pseudosteppes (see 'Agricultural and Grassland Habitats', p. 267), afforestation could be highly detrimental to priority bird species.

The Agri-environment Regulation (Box 1) could also have positive effects in the EU if implemented to promote appropriate stocking densities in Mediterranean shrubland and rocky habitats. The following policy areas are particularly important for achieving nature conservation in these habitats, and are discussed in detail in 'Agricultural and Grassland Habitats' (p. 267):

● Fundamental objectives of rural policies

● Overall levels and forms of subsidy

● Financial or economic instruments (e.g. EU Regulations)

● Information, education and extension services

CONSERVATION RECOMMENDATIONS

Broad conservation objectives

1. To maintain existing areas of Mediterranean forest and shrubland habitats of high conservation value, particularly open garrigue habitats and remaining ancient natural Mediterranean forests.

2. To improve the quality and conservation value of degraded areas of Mediterranean forest and shrubland habitats.

3. To consider the recreation of Mediterranean forest and shrubland habitats where this would result in net conservation benefits.

Ecological targets for the habitat

In order to maintain and improve habitat quality, it is essential to retain and enhance important habitat features for priority species. It is therefore necessary to ensure that the following conditions are met.

1. An optimal mixture (according to local condi-

tions) of forest, maquis, garrigue, grassland and rocky habitats should be maintained in the landscape, through the use and management of natural succession processes, grazing and wood-cutting.

2. Existing areas of ancient, mature Mediterranean forest should be maintained or, where possible, be allowed to re-establish. Open garrigue and rocky habitats are also particularly important and should also be maintained or recreated where these have been recently lost. In particular, the conversion of Mediterranean shrublands to intensive agriculture or forestry plantations should be avoided.

3. Fragmentation of large blocks of Mediterranean forest and shrubland habitat should be avoided.

4. The structure and composition of vegetation in shrubland and rocky habitats should be maintained for priority bird species through the appropriate management of grazing, manual or mechanical removal of shrubs and control of fire.

   Grazing levels should be maintained that avoid succession and that create areas of close-cropped grass and low shrubs, while avoiding excessive soil erosion and maintaining other areas with taller and denser shrub vegetation.

   Where appropriate, mechanical means of controlling vegetation can be used to create early, open successional stages, to increase habitat diversity, and to avoid complete succession to forest and the build-up of dry, dead vegetation that may lead to intense and extremely damaging fires. Such methods should be used with great care and with due regard for other important and threatened species of flora and fauna, such as tortoises.

   Where grazing or mechanical methods appear impractical or unfeasible, small-scale and controlled fires may be of use in managing vegetation on a local basis, with extreme care and ensuring that there are no detrimental effects on threatened flora and fauna.

5. Ancient Mediterranean forests should not be managed for forestry but allowed to revert to natural processes. Other forests should be managed on a sustainable basis and natural processes should be allowed where possible. In particular, some mature and dead trees should be maintained, and the planting of non-native tree species, clear-felling of large blocks of forest and complete removal of undergrowth should be avoided. Some areas within coppice woodlands should be allowed to become mature high forest.

6. Forest management involving pesticides that are

directly toxic to birds should be avoided, or reduced to below sublethal levels where such chemicals are known or reasonably suspected to be causing population declines in priority birds and other wildlife. In particular, these treatments should be avoided during the breeding season. Alternative methods, such as biological control, should be encouraged and developed.

7. Important food resources (vertebrates, invertebrates and plants) should not be depleted by the use of pesticides.

8. Disturbance of human-sensitive priority bird species should be minimized, especially during critical periods of the annual cycle (e.g. before and during incubation). In particular, forestry operations and access to sensitive forest roads should be restricted where these cause disturbance.

9. Mortality levels of priority species (e.g. *Aquila adalberti* and *Hieraaetus fasciatus*) from collisions with powerlines and other cables should be reduced through appropriate measures (e.g. cable re-routing, pylon design, cable marking, screening or burying), particularly at IBAs and other sites where these species occur in substantial numbers.

Broad policies and actions

To meet the broad conservation objectives and habitat targets (see above) in accordance with the strategic principles (p. 24), the broad measures listed below must be pursued (see p. 22 in 'Introduction' for explanation of derivation).

General

1. The importance of Mediterranean forest, shrubland and rocky habitats for birds and other wildlife, and for cultural values, traditional sustainable uses and other roles (e.g. erosion prevention), should be widely promoted. In particular, the view that shrubland and rocky habitats are wasteland, and therefore good locations for development, must be countered. The threats that are faced by Mediterranean forest, shrubland and rocky habitats, and their priority birds, should be publicized.

Legislation

2. Protection of Mediterranean forest, shrubland and rocky habitats should be improved through the full implementation and enforcement of existing national and international environmental legislation. This should include the designation of more protected areas of Mediterranean forest, shrubland and rocky habitats, including all Important Bird Areas (IBAs) that are mainly composed of these habitats, and their designation

under international or EU law where relevant (e.g. inclusion in the Natura 2000 network, or EMERALD network)

3. Strategic Environmental Assessments (SEAs) should be required for all national and EU policies, plans and programmes for forestry, afforestation, agriculture, transport and water resources. Guidelines should be introduced in all countries to inform applicants and decision-makers how to assess the impacts of forestry schemes on priority birds and other biodiversity.

National and EU legislation governing environmental impact assessments (e.g. the EIA Directive) should be revised so as to explicitly require EIAs to be completed for all afforestation schemes which extend over 50 ha, and for those schemes which affect designated protected areas. EIAs should be also be required to apply to felling and other forest-management operations which have significant environmental impacts in Mediterranean forest.

4. The provision of international development funds for programmes and projects should be dependent upon SEAs and EIAs that are favourable to the conservation of Mediterranean forest, shrubland and rocky habitats.

Agriculture
(See also 'General policies' in 'Agricultural and Grassland Habitats', p. 323.)

5. National and EU policies for the livestock sector (such as the Less Favoured Areas, beef, dairy, sheep and goats) should be revised so that they do not promote overgrazing or under-grazing (for example, subsidies to farmers should be made via area-payments rather than headage-payments, or policies should be modulated according to environmental criteria). Policies should support non-intensive grazing of Mediterranean shrubland and rocky habitats.

6. National legislation, e.g. on land-tenure, common land, water management and rural development, should be revised in order to implement new agri-environment policies.

7. Sufficient funding should be allocated by national governments to the EU Agri-environment Regulation and its implementation should be enforced throughout the EU. According to the Fifth Environmental Action Programme, 15% of the agricultural land of the EU should be covered by such management agreements by the year 2000. Schemes should be targeted to economically marginal areas with high nature-conservation importance, including mosaics of low intensity cultivation and grazing within Mediter-

ranean shrubland and rocky habitats, and should have clear environmental objectives. Habitat targets outlined above should be incorporated as appropriate, and the schemes should be monitored to measure their success in achieving their objectives. There should be greater coordination between the Agri-environment Regulation and incentives for afforestation of economically marginal agricultural land under Regulation 2080/92.

8. EU member-state agriculture and environment ministries should jointly determine national priorities for incentive schemes in agriculture and forestry; nature conservation authorities should develop an EU-wide network of experts in agriculture, forestry incentive schemes and related research.

9. All direct payments under the EU Common Agricultural Policy, e.g. for livestock production, should change from headage-payments to area-payments, and should be conditional on recipients protecting the environment according to certain criteria, including the avoidance of overgrazing in Mediterranean shrubland and rocky habitats. Price supports, direct payments and other subsidies should promote a lowering in the intensity of livestock grazing in overgrazed areas.

10. Environmental objectives within regional policies, such as Less Favoured Areas, Objectives 1–5 and Mediterranean policies should be improved. Allied EU policies, particularly rural development and environment policies, need to be integrated with agriculture and more fully developed. The EU Structural Funds need to promote environmentally sustainable development and should be subject to environmental assessment. The Regulations and operational criteria of the Funds (e.g. the 'framework' Regulation 2081/93) need to be revised to allow specifically for biodiversity conservation.

Fire control

11. National governments should be encouraged to develop national fire-control strategies if these do not already exist. In particular, they should ensure that old-growth Mediterranean forests are fully protected from fires, through stringent, fully enforced fire-prevention and control measures. The construction of fire-breaks and roads for fire-prevention should always be decided in consultation with local nature conservation bodies, to avoid disturbance of the most sensitive areas and habitats, and to discuss possibilities for limiting access when necessary. The EU Directive

on fire control (3259/86/EEC) should be improved according to the same criteria.

Forestry

(See also 'General policies' in 'Boreal and Temperate Forests', p. 203.)

12. National and EU forestry policies should be developed that promote sustainable forestry in the long term, and which incorporate environmental objectives specifically for Mediterranean forests.

13. Afforestation schemes with non-native tree species should be banned from Mediterranean forest, shrubland and rocky habitats. Afforestation projects with native species should be controlled through planning regulations linked to SEAs and EIAs.

14. The use of pesticides in forests should be subject to strict regulations which ensure that populations of non-target species are not detrimentally affected through direct toxic effects, or through indirect effects following bio-accumulation in the food-chain or loss of food resources.

Tourism and recreation

15. Further support should be given by national governments and the EU towards developing ecologically sustainable tourism, for example through the EU *Community action plans to assist tourism*, and towards initiatives aimed at increasing the awareness of tourists and the tourist industry of the ways that tourism can benefit or harm the natural environment. In particular, publicity materials should be widely distributed informing tourists of the wildlife importance of Mediterranean forest, shrubland and rocky habitats and of the threats that face them, as well as giving advice on how to maximize conservation benefits to them, and to minimize the risk of disturbance, pollution, excessive erosion and accidental fires.

16. The development of tourism in these habitats should be adequately planned so as to reduce the disturbance to particularly sensitive habitats and species, e.g. through limitation of visitor numbers and restriction of access to certain areas that are well-equipped to cope (e.g. with fires or refuse collection, or holiday peaks in visitors), especially during the breeding season.

Other measures

17. EIAs should be carried out on major new powerlines and other above-ground cabling projects, and in areas where there has been high mortality of priority species due to collisions with wires and electrocution. Where significant

risks are likely, which cannot be sufficiently reduced by screening or marking or by modifications to cables and pylons, cables should be re-routed or buried. Such measures should be applied first in IBAs and other sites with important numbers of priority species which are susceptible to electrocution or collisions.

18. The habitat requirements and threats faced by many Mediterranean birds are poorly understood. Further research is therefore required to provide specific, locally applicable recommendations on forestry-management and livestock-farming practices in Mediterranean forest, shrubland and rocky habitats.

ACKNOWLEDGEMENTS

Most contributions to this chapter were made during and after a workshop organized by R. Prodon at Banyuls-sur-mer, France (16–19 March 1994); these contributors are listed on p. 239.

Additionally, valuable information and comments were received from F. Casale and U. G. Orsi (Lega Italiana Protezione Uccelli), M. Charalambides (Cyprus Ornitho-logical Society), J. Cortes (Gibraltar Ornithological Society and Natural History Society), M. A. Naveso and J. Criado (Sociedad Española de Ornitología), M. Debussche, U. Faralli, P. Iankov (Balgarsko Druzhestvo za Zashtita na Pticite), A. Matschke (Associació per a la Defensa de la Natura), N. Onofre, and Z. Waliczky (BirdLife International).

AGRICULTURAL AND GRASSLAND HABITATS

Compiled by

GRAHAM M. TUCKER AND JIM DIXON

from major contributions supplied by the Habitat Working Groups

Arable and improved grassland, perennial crops, pastoral woodland

M. Díaz Esteban, J. Dixon, H. P. Kollar, E. Lebedeva, F. Markus, M. Métais, S. Nagy, D. Pain,
J. Tiainen, G. M. Tucker (Coordinator)

Wet grassland

A. J. Beintema (Coordinator), J.-J. Blanchon, R. S. K. Buisson, P. Chylarecki, H. Hötker, R. Kuresoo,
J. Plesník, K. W. Smith, O. Thorup

Steppic habitats

J. Alonso, V. Belik, J. Boutin, G. Cheylan, J. Dixon, S. Faragó, V. Galushin,
H. P. Kollar (Coordinator), C. Martínez, M. Angel Naveso, F. Suárez, S. Voisin

Montane grassland

P. Brichetti, P. F. De Franceschi, J. Frühauf, L. V. Kalabér, B. Lombatti (Coordinator), I. Ogurlu,
P. Pedrini, R. L. Potapov, U. Rehsteiner

SUMMARY

Agricultural and grassland habitats dominate Europe, covering about 5 million km^2 (c.50%) of the land surface. This extensive and diverse habitat holds a rich avifauna, including 173 priority species, more than any other major habitat-type. Nearly 70% of these priority birds have an Unfavourable Conservation Status in Europe. The remaining species, though not currently threatened, are specialists of agriculture and grassland and are thus highly dependent on these habitats; most are still widespread across the continent, although many require habitat features that are only found in traditionally or non-intensively managed habitats. A significant proportion of the priority species are also naturally restricted within Europe to certain regions or to particular types of agricultural or grassland habitat.

Maintaining the ecological quality of agricultural and grassland habitats is the main conservation priority for most of the priority bird species. However, deterioration of these habitats is widespread in Europe, due mainly to the widespread and continuing intensification of agricultural practices across the continent (e.g. crop improvement and specialization, pesticide use, elimination of marginal habitats, high stocking levels). Elsewhere, the abandonment of traditional, low-intensity farming systems is leading to the loss of ecologically rich, semi-natural habitats of particular importance for priority bird species.

The conservation of agricultural and grassland habitats and their birds is crucially dependent on broad measures. These need to be implemented principally through the incorporation of conservation objectives into all land-use policies and regulations, especially those concerning agriculture. The Agri-environment Regulation provides an excellent opportunity for supporting environmentally beneficial farming practices in member states of the EU. However, similar measures need to be developed and implemented for the rest of Europe. Further policy reforms are also required (especially of the EU Common Agricultural Policy), as well as regulations, to reduce fertilizer and pesticide use, to avoid overgrazing while maintaining extensive pastoral systems, and to maintain crop diversity and non-farmed habitats in the countryside.

Strategic Environmental Assessments and Environmental Impact Assessments also need to be applied to major forestry, agriculture, infrastructural and other programmes and projects in order to avoid detrimental impacts to these habitats. International funding for development programmes (e.g. EU Structural Funds) should be dependent on these assessments.

HABITAT DESCRIPTION

Habitat definition, distribution and extent

A wide variety of habitats are used for agriculture, i.e. the growing of crops and the rearing of domestic livestock. Such agricultural and grassland habitats in Europe are mostly found to the south of the boreal zone, which has too short a summer to allow the growing of crops and too harsh a winter for the rearing of most domestic livestock. Although low-land heathland, moorland, tundra, saltmarshes and sand-dunes may be used for grazing livestock, these habitats are covered elsewhere in this book.

Agricultural and grassland habitats have been divided into the following seven types on the basis of their bird communities and farming systems.

Arable and improved grassland

This habitat-type includes all land in Europe that is regularly ploughed and/or cultivated for the produc-

tion of food and non-food crops, with the exception of pseudosteppe (i.e. extensively farmed, rain-fed cereals grown in rotations with two or more years of dry grassland), wet rice cultivations, and perennial crops. So-called 'improved' grasslands are those that have been agriculturally improved, i.e. drained or re-seeded, with 'intensive' grassland as a subset of this category, where there are also high inputs, e.g. fertilizers, pesticides or water (irrigation). Not included are steppic, wet and montane grasslands, and those within pastoral woodland, which are all treated separately below.

Arable and improved grassland dominates the rural landscape of most countries in temperate and Mediterranean Europe, apart from high-altitude areas with harsh climates, and steep slopes or poor or heavy soils where alternative land-uses are more productive. Table 1 shows the extent of arable and permanent grassland in each European country dur-

Table 1. The extent (km²) of arable land, permanent grassland and perennial crops in each European country during 1988–1991, given as an area and as a percentage of the country's total land area (Anon. 1992b, 1994e; for Russia, Kovalenko 1994; for Estonia, arable and perennial crops, J. Elts *in litt.* 1996). No data are available for other European countries not listed here.

	Arable land (km²)	%	Permanent grassland (km²)	%	Perennial crops (km²)	%	All crops and grassland (%)
Albania	5,900	22	4,030	15	1,240	<1	41
Austria	14,400	17	19,820	24	740	<1	42
Belarus	?	?	31,000	15	?	?	44
Belgium	7,490	25	6,480	21	150	<1	47
Bulgaria	38,400	35	20,220	18	3,000	<1	56
Cyprus	1,040	11	50	<1	530	6	17
Czechoslovakia (former)	49,860	40	16,460	13	1,330	<1	54
Denmark	25,660	62	2,170	<1	40	<1	67
Estonia	11,000	24	2,690	6	150	<1	33
Finland	24,410	<1	1,250	<1	?	<1	8
France	182,670	34	117,400	22	12,800	<1	58
Germany	119,480	36	57,070	17	4,420	<1	54
Greece	28,800	22	52,550	41	10,490	<1	71
Hungary	50,500	55	12,100	13	2,370	3	70
Iceland	80	<1	22,740	22	?	<1	22
Rep. of Ireland	9,600	14	46,700	68	30	<1	82
Italy	91,040	30	49,070	17	30,450	10	58
Latvia	?	?	8,450	14	?	?	41
Lithuania	?	?	2,450	5	?	?	77
Luxembourg	550	21	400	15	10	<1	37
Malta	120	38	0	<1	10	3	41
Moldova	?	?	3,000	9	?	?	59
Netherlands	8,990	27	10,810	32	320	<1	59
Norway	8,620	<1	1,020	<1	?	<1	3
Poland	147,360	48	40,490	13	?	?	62
Portugal	20,400	22	5,300	<1	7,100	<1	36
Romania	101,000	44	44,100	19	6,000	<1	66
Russia (European)	?	?	382,010	11	?	?	35
Spain	155,770	31	103,300	21	47,910	10	61
Sweden	28,720	<1	1,680	<1	40	<1	8
Switzerland	3,910	10	16,090	41	210	<1	51
Turkey	247,500	32	86,000	11	29,800	<1	47
Ukraine	342,560	57	69,140	11	10,730	<1	70
United Kingdom	68,680	29	111,090	46	560	<1	75
Yugoslavia (former)	70,370	28	63,470	25	7,230	3	55

ing 1988–1991 and its percentage of the total agricultural land in each country. Arable and improved grassland were formerly not so widespread in the drier areas of southern and eastern Europe, but the advent of new irrigation technology in recent decades has led to a major growth in the extent of these types of farmland in these regions.

Steppic habitats

The term 'steppic' is used here to encompass three types of very open habitat with particularly low vegetation and a generally treeless aspect (Goriup 1988).

Primary or 'true' steppe

Primary steppe is mostly found on the vast Eurasian plains that stretch for 8,000 km eastwards from the Danube basin to Beijing (China). Small areas also occur elsewhere in Europe where soils are too dry, saline or nutrient-poor to support trees, e.g. some areas of Turkey. The terrain tends to be flat or gently undulating, mostly or wholly treeless in its natural state, and usually the precipitation equals or is less than evaporation. The vegetation is dominated by grasses or dwarf shrubs, maintained by a combination of natural grazing, poor soils and harsh climates.

Secondary steppe

Man-made steppes were created in Europe by the clearance of forest cover, in areas with drier climates and/or on thin or poor soils. Such steppes are often on flat or gently sloping ground, are maintained by grazing, and support vegetation similar to primary steppe (Wolkinger and Plank 1981), dominated by grasses or dwarf shrubs.

Pseudosteppe

Pseudosteppes are extensively farmed, mixed rotational systems of grassland, cereals, fodder crops and grazed fallow land in dry areas with low forest-cover. These arable landscapes mimic primary steppe in terms of their vegetation structure (Goriup 1988, Suárez et al. 1997a), and possess the habitat features needed by many bird species typical of primary and secondary steppe.

Shrublands and rocky habitats in the Mediterranean zone (e.g. garrigue, phrygana) are treated under 'Mediterranean Forest, Shrubland and Rocky Habitats' (p. 239). The approximate distribution of remaining steppic habitats in Europe has been estimated by the Habitat Working Group in Figure 1, although there appears never to have been a Europe-wide

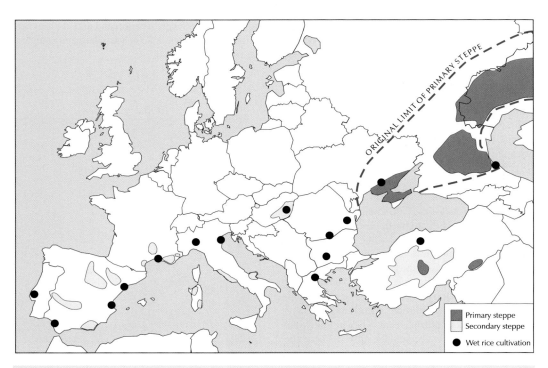

Figure 1. The distribution of primary and secondary steppe, and of the main areas of wet rice cultivation, in Europe. Pseudosteppe is not shown, due to a lack of adequate data.

inventory, and their total extent is not known. Almost no 'virgin' steppe remains in Europe as most of the primary steppe of Russia and Ukraine has been cultivated, although small patches remain in the extreme south-east. Iberia and Turkey hold substantial amounts of secondary steppe and pseudosteppe, while a significant expanse of these habitats also remains in the Pannonic region[1] (IUCN 1991). Elsewhere, only small fragments remain, mainly in southern Europe.

Montane grassland

This habitat-type includes all montane grassland and associated shrub and heath communities, above the treeline and south of the boreal zone. Although montane rocky habitats (e.g. cliffs, screes), snow-dominated habitats (the nival zone), and some montane grasslands in eastern and south-east Europe are not used for pastoral agriculture, these habitats are included for completeness.

Montane grasslands and associated habitats are widespread in the Alps, Pyrenees, and Carpathians, and in numerous smaller mountains to the south of these major ranges (e.g. in the Balkans), as well as in the Caucasus and the mountains of Turkey. Their current extent is difficult to estimate due to the lack of any systematic inventories.

Wet grassland

These are grasslands that are periodically but not perpetually flooded, or that have high water-tables, including riverine and lakeside flood-meadows, polders, upland rough grassland, machair and coastal grazed marshes.

There are no reliable summary data on the distribution and extent of wet grassland in Europe, but bird species that breed mainly in such habitats in Europe (e.g. Black-tailed Godwit *Limosa limosa*) can give a good idea of their general distribution, (Figure 3).

Rice cultivations

These are wet rice-fields in southern and eastern Europe. Wet rice cultivations in Europe are found in warm, flat areas with alluvial soils, generally at or near large wetlands due to the high demand for irrigation water, thus mainly along certain riverine flood-plains and at large river deltas of southern and east European countries (Figure 1). The total extent of rice cultivation in Europe is not known, but is small compared to other habitat-types in this book. European rice-fields account for only 0.3% of the surface area of world rice production, and in 1990 there were 3,720 km² of rice-fields within the European Union countries (Fasola and Ruíz 1997). Some non-EU countries of southern and eastern Europe formerly cultivated relatively large areas of rice, but since 1985–1990 the extent has dropped dramatically (e.g. by 80–90% in Bulgaria, Hungary and Romania).

Perennial crops

Woody plant cultivations whose fruits are harvested annually, including olive-groves, vineyards and orchards.

Table 1 gives the extent of permanent crops in each country (equivalent to this habitat-type). Such crops occur in most European countries, but particularly in the Mediterranean region.

Pastoral woodland

These are semi-natural woodlands with a variably open structure (man-made), that are managed primarily for rearing livestock. They include the 'dehesas' of Spain, 'montados' of Portugal, winter-grazed woodlands of Britain, and grazed woodland in Hungary.

The only extensive areas of pastoral woodland left in Europe are in Iberia (Figure 2), where there are still approximately 50,000 km² of dehesas in Spain and more than 5,000 km² of montados in Portugal. Information on its distribution and extent elsewhere in Europe is not available in a standard and comparable form.

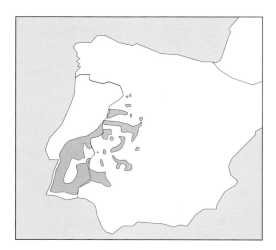

Figure 2. The distribution of pastoral woodland in Iberia. The data are not adequate to map the occurrence of this habitat elsewhere in Europe.

---

[1] Southern Romania, northern Bulgaria, north-east Yugoslavia and parts of Hungary.

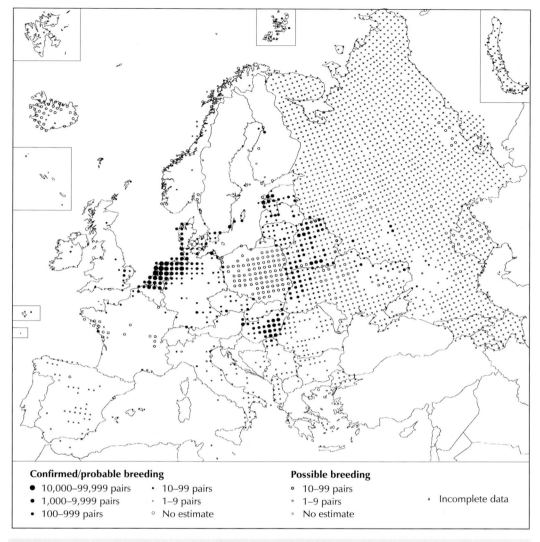

Figure 3. The distribution of wet grassland in Europe, as indicated by the breeding distribution and density of Black-tailed Godwit *Limosa limosa*, a priority bird species that is characteristic of the habitat. Mapping is on the basis of 50-km squares (data from Hagemeijer and Blair 1997). (Birds in Iceland are of the endemic race *L. l. islandica*, which breeds in tundra and moorland habitats.) No data are available for Anatolian Turkey.

Origin of the habitat

Before agriculture became widespread on the continent 7,500–4,500 years ago, forests covered most of Europe, except where tree-growth was inhibited by climate or soil conditions (Ellenberg 1986a, Goriup 1988). Natural grassland (primary steppe) would have been confined to areas of low rainfall and nutrient-poor soils, and been most extensive in south-east Europe, originally extending from the Danube basin eastwards through south-east Ukraine and southern Russia to Kazakhstan and beyond Europe to Mongolia and China. Grassland would also have occurred in mountain ranges where conditions were too exposed for trees, i.e. above the treeline (montane grassland), on saline soils (e.g. the Anatolian plateau in Turkey), and on particularly poor soils with dry conditions, e.g. the Rhine sands of the Mainz basin (Germany), calcareous grassland in the UK, and parts of the Mediterranean (Wolkinger and Plank 1981, Goriup 1988). In addition, small and patchy wet grasslands are likely to have occurred on the natural flood-plains of larger rivers, wherever floods had recently scoured away the tree-cover.

Although there is some debate over the original

271

extent of natural grassland in western and central Europe (van Dijk 1991), it is clear that apart from mires and 'true' steppe, the more-or-less treeless landscapes which now predominate in Europe are man-made, derived from the clearance of forests for agriculture.

Pastoralism has been practised in montane grassland, e.g. in the Alps, for at least 6,000 years (Bätzing 1994, Lichtenberger 1994). The opening-up of lowland woodland to provide grassland for livestock started in Neolithic times, at least in Britain (Rackham 1986), and probably gave rise initially to extensive areas of pastoral woodland. By medieval times, pastoral woodlands were widespread in Europe (Ellenberg 1986a, Rackham 1986). Further clearance for agriculture led to primarily open habitats in most areas of the continent, and now such woodlands are mainly restricted to the Iberian peninsula. Particularly in western Europe, many grasslands have subsequently been agriculturally improved.

The advent of arable agriculture in Europe also led to widespread ploughing of primary steppe and secondary grassland. Until very recently, such arable systems were usually non-intensive, with low fertilizer inputs, and incorporated crop rotations, often including several years as grass, to maintain fertility. This led to mixed habitats of grass, cereals, fodder crops and fallow land, termed 'mixed farming' in temperate areas and 'pseudosteppe' in drier, more Mediterranean climates. In northern and western Europe, in particular, such arable systems have now mostly been intensified through the use of pesticides, artificial fertilizers and irrigation water, but large areas remain in parts of the Iberian peninsula, eastern Europe and Turkey (Parr et al. 1997).

Primary steppe would originally have been grazed by an abundance of large, wild herbivores such as wild horse *Equus ferus* and saiga antelope *Saiga tatarica*. These are all now globally extinct or extinct in Europe, and have been replaced by domestic livestock, but the density and biomass of the latter are probably considerably greater than those of the original wild ungulates. Consequently, pastoral agriculture on these primary arid steppe habitats has generally caused impoverishment of the habitat, and has led to land degradation where grazing is particularly intense. In contrast, some areas have been irrigated and/or cultivated, leading to the creation of intensively managed grassland and crops.

Ecological characteristics

Key abiotic factors affecting the habitat

Although most agricultural habitats are man-made or highly modified, they remain highly influenced by abiotic factors such as climate, day length, soil topography and altitude. The relationship between climate and agriculture may be direct, acting through the effects of light, precipitation and temperature, or indirect through effects on soil. For example, the extent of the primary steppe is strongly determined by the ratio of precipitation to evaporation, while temperature and soil conditions (especially its fertility and capacity for water retention) determine the species composition of the dry grassland vegetation. The combination of climate and locality largely determines the type of agriculture that is practised in a region. However, more so than other habitats, agriculture is affected by political, social and economic factors. Crops that are sub-optimal under natural conditions may be grown if subsidized, or if technology (such as irrigation) is used to overcome limitations.

Key biotic factors and natural dynamics

As human constructs, agricultural habitats are greatly affected by their management systems (Figure 4), and some of their ecological processes differ considerably from those of natural ecosystems (Table 2). Nevertheless, natural biological processes remain important in the least man-managed agricultural habitats, which are principally the non-cultivated steppic habitats and unimproved grassland.

Table 2. A comparison between the ecological processes within field-crop ecosystems and those within natural ecosystems (adapted from Pearson 1992).

Characteristics	Field-crop ecosystem	Natural ecosystem
Control	Largely human	Largely biological
Community characteristics		
Total organic material	Low	High
Vegetation structure	Simple	Complex
Species diversity	Low	High
Food chains	Simple, short	Complex, long
Population levels	Fluctuating	Relatively constant
Response times	Relatively fast	Slow
Energetics		
Net production per unit biomass	High	Low
Standing crop of biomass	Low	High
Element cycles	Open (leaky)	Closed

Temperate regions

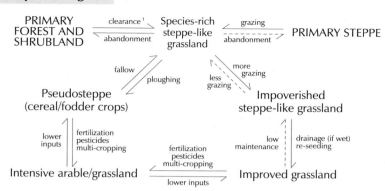

Figure 4. The processes which lead to the formation and modification of agricultural and grassland habitats in Europe (modified from Goriup 1988). Dashed lines indicate changes that are particularly slow by nature or that cannot be easily or quickly effected by man.

Semi-arid and arid regions

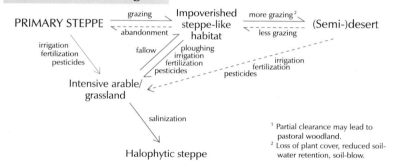

[1] Partial clearance may lead to pastoral woodland.
[2] Loss of plant cover, reduced soil-water retention, soil-blow.

Grazing and succession

Grazing animals have fundamental roles in grassland ecosystems. Such grazing influences the structure and composition of the vegetation: heavy grazing by ungulates of primary steppe, for example, leads to the disappearance of palatable herbs (often legumes), declines in *Stipa* bunch-grasses, and increases in *Festuca* grass species (feather-grasses). Intense grazing can result in the dominance of annual plants, e.g. in the Mediterranean region. Typical primary steppe, dominated by *Stipa*, can only exist under permanent but moderate grazing, since the absence of grazing results in accumulation of standing dead shoots and litter, replacement of bunch-grasses by creeping grass species, and the spread of shrubs.

Locusts (Acrididae) and small burrowing rodents from the original steppe fauna still have an important role in determining the nature of the vegetation, soil and microrelief in steppes, and 'the activities of these animals and the dynamic processes that they initiate are intrinsic features of steppes and must be regarded as indispensable to the prolonged conservation of zonal steppes in a state of dynamic equilibrium' (Kopaneva and Stebajev 1985). The vegetation

mosaics that they create may be of considerable importance in meeting the diverse habitat requirements of priority birds and other species.

Where conditions are suitable for tree growth (e.g. on secondary steppe), grazing arrests the processes of natural succession and prevents such grassland from turning into scrub and, ultimately, to forest. In the Alps, wild ungulates such as chamois *Rupicapra rupicapra* continue to survive in significant numbers, and prevent re-growth of forest trees at the treeline (e.g. Bossert 1980), contributing to the openness of bushy habitats there.

Productivity and patchiness of resources

Montane habitats are characterized by low productivity, which in drier climates (e.g the Mediterranean region) is even lower than in more humid ranges, such as the Alps or Pyrenees (Franz 1979). Productivity also varies with altitude and season. These patterns are overlaid by microclimatic conditions such as snow cover, leading to pronounced asynchrony of plant growth and therefore to an extended availability of plant resources during the yearly cycle. In the Alps, for example, Ptarmigan *Lagopus mutus* follow the altitudinal budding patterns of protein-rich willow *Salix helvetica* (Marti and Bossert 1985). Perhaps as a consequence, birds of high altitudes are adapted to a high degree of mobility (Landmann and Winding 1993); a large proportion of them are omnivorous (Franz 1979) and many of them use several different habitat-types.

Because resources are patchy and have low spatial predictability, many sedentary bird species forage or roost in groups, at least outside the breeding season, e.g. Rock Partridge *Alectoris graeca*, Black Grouse *Tetrao tetrix*, Alpine Chough *Pyrrhocorax graculus* (Glutz von Blotzheim *et al*. 1973, Lovari 1976). In winter, *Lagopus mutus* and snowcock *Tetraogallus* follow ungulates because of easier access to food covered by snow (Bossert 1980, Potapov 1985).

Management systems

Arable and improved grassland

Arable land

Management of arable land in Europe has changed profoundly during the last 150 years as a result of technical developments, particularly mechanization, water management and, over the last 50 years, the 'chemical revolution' in artificial fertilizers and pesticides. The adoption of these new technologies, though widespread, has happened at different times in European countries over different periods, and for differing reasons. Land reforms (e.g. of land-ownership rights, etc.) have had a similarly varied history and influence on agriculture in Europe. A result is that farming infrastructure and practices vary widely between European countries (Tables 1, 3, 4 and 5).

The mechanization of agriculture started with horses, but was greatly increased by the invention of steam power, the internal combustion engine and the tractor. In Britain, one of the earliest European countries to undergo this agrarian revolution, the number of tractors increased more than 20 times between 1930 and 1960, while the number of workhorses declined more than 30-fold during roughly the same period (O'Connor and Shrubb 1986). After 1960 the number of tractors remained relatively stable, but their size, power and sophistication continued to increase. As a result, farming systems became much

Table 3. Tractor use in relation to the area of cropped land (arable land and perennial crops) in Europe (data from Anon. 1992b; also Anon. 1994e, Kovalenko 1994). Data are not available for European countries not listed here.

	No. of tractors per km² of cropped land		
	1979–1981	1988	% change 1980–1988
Albania	1	2	+11
Austria	20	25	+25
Belarus	?	1 [1]	?
Belgium	14	14	0
Bulgaria	1	1	–13
Cyprus	6	10	+60
Czechoslovakia (former)	3	3	+6
Denmark	7	7	–7
Finland	8	10	+20
France	8	8	0
Germany	3	3	+18
Greece	4	5	+33
Hungary	1	1	–4
Iceland	100	100	0
Rep. of Ireland	14	14	0
Italy	8	11	+33
Latvia	?	2 [1]	?
Luxembourg	17	17	0
Malta	3	3	+7
Moldova	?	3 [1]	?
Netherlands	20	20	0
Norway	17	17	0
Poland	4	8	+85
Portugal	3	3	+8 [3]
Romania	1	2	+9
Russia (European)	?	1 [2]	?
Spain	3	3	+35
Sweden	6	6	0
Switzerland	25	25	0
Turkey	2	2	+57
Ukraine	1	1	+6
United Kingdom	7	8	+8
Yugoslavia (former)	6	14	+157

[1] 1991
[2] 1993
[3] INE (1993) records a much larger increase, of 50%, from 2.2 tractors/km² in 1979 to 3.3 tractors/km² in 1989.

less labour-intensive, with the number of farmworkers declining more than three-fold between 1950 and 1981. Similar trends are apparent over most of Europe, e.g. in Germany (George 1996, Rösler and Weins 1996), the Netherlands (Anon. 1989a), Spain (Comíns *et al.* 1994) and Norway (Statistisk Sentralbyrå 1996), and the use of tractors and other powerful machinery is now widely prevalent.

Mechanization does, however, vary geographically (Table 3). In the north-western countries of the European Union, the area of agricultural land per tractor currently lies around 10 ha, whilst in southern EU countries such as Greece, Portugal and Spain this figure is up to twice as high (or more), and up to 10 times as high (or more) in eastern Europe. In the latter region, for instance, there were still approximately one million workhorses in use on farms in Poland in 1987 (Stanners and Bourdeau 1995). Although the need for tractors varies according to land-use, in particular according to the area of tilled land, the density of tractors is a good indicator of the intensity of agricultural management of land. Thus, increases in the number of tractors in several countries over the period between 1980 and 1988 (Table 3) reflect continuing agricultural intensification in these countries.

Mechanization has brought about several widespread changes in farming practice. It is easier and more efficient, when using machinery, to work in large fields. Hence field size has been increased, often through clearing field-margin vegetation (e.g. trees, hedgerows), ditches, earth banks, and ponds. In central and eastern Europe, collectivization of private farms into state farms (each often covering square kilometres of land) earlier this century also led to increases in field size, although this process is now being reversed in some (but not all) countries as land is privatized. The high capital cost of farm machinery also encourages specialization in particular crops and amalgamation of smallholdings into larger farm enterprises.

Mechanization has also had a major impact on the practice of cyclically changing the crop grown on any particular piece of land (rotation). One of the aims of rotations was to spread the workload on farmland over the year. For example, the planting of cereal crops could be typically divided fairly evenly between spring and autumn. However, the speed of mechanized farming has removed this constraint and cereal crops are now often largely planted in autumn to extend the growing season.

The other functions of rotational farming systems were to maintain soil fertility and to reduce weeds, diseases and invertebrate pests. Traditional rotations maintained soil fertility by including nitrogen-fixing crops such as clover in the crop sequence and by the addition of animal manure from grazing animals onto temporary grass leys. These systems have largely been replaced, particularly in western Europe, by the use of artificial nitrogen, potassium and phosphate fertilizers (Table 4), and the use of pesticides: the 'chemical revolution'. Such use of chemicals is expensive and has further encouraged crop specialization, as farms have concentrated on systems that are the most economically attractive. Consequently, mixed-farming systems with livestock and a variety of arable crops have decreased considerably, particularly in western Europe (e.g. O'Connor and Shrubb 1986, Majoral 1987, Anon. 1989b, Flade 1994, George 1996, Rösler and Weins 1996, Potter 1997).

As a result of these technological advances, crop yields have increased dramatically and consistently this century. Cereal yields in western Europe are now up to 6–7 tonnes/ha (Anon. 1994e, CEC 1995d) compared with typical yields of 2 tonnes/ha at the start of the twentieth century. Cereal yields are generally highest in western countries of the European Union, and have even continued to increase in these countries in recent decades (Potter 1997). In more northern and southern EU countries, conditions for growing cereals are less suitable; nevertheless, considerable increases in yield have also been achieved. Yields in eastern Europe tend to be lower, and yields have declined since the early 1980s in several countries in this region, probably as a result of the recent transformation of their political and economic systems and consequent difficulties in obtaining artificial fertilizers, pesticides and farm machinery.

Further information on cropping systems is given by López-Bellido (1992) for the Mediterranean region, Nalborczyk and Zebor (1992) for central Europe, Gataulina (1992) for eastern Europe and Spiertz *et al.* (1992) for north-west Europe.

Agriculturally improved grassland

Rapid and drastic changes have also occurred in grassland management and livestock production systems in recent decades. Most grasslands now receive regular inputs of fertilizer, with the exception of those that are limited in productivity through dryness, flooding or extremes of climate, or where topography limits the use of machinery. Traditionally, some grasslands have received organic inputs through the use of animal manure. This has now largely been replaced by the use of artificial inorganic fertilizers, at least in western Europe (Flade 1994, George 1996, Rösler and Weins 1996, Statistisk Sentralbyrå 1996, Potter 1997).

The application of fertilizers, principally such nutrients as nitrogen and phosphorus, to semi-natural grassland generally promotes greater productiv-

Table 4. The consumption of nitrogenous fertilizer per hectare of agricultural land in Europe (CEC 1995d). No data are available for those European countries which are not listed here.

	Fertilizer use (kg/ha)		
	1970	1990	% change 1970–1990
Albania	19	66	+247
Austria	32	39	+22
Belarus	?	72	?
Belgium	104	125	+20
Bosnia-Herzegovina	?	12	?
Bulgaria	63	73	+16
Croatia	?	46	?
Cyprus	91	77	–15
Czechoslovakia (former)	59	87	+47
Denmark	97	142	+46
Estonia	?	50	?
Finland	60	81	+35
France	45	81	+80
Greece	22	47	+114
Hungary	57	55	–3
Iceland	5	5	0
Rep. of Ireland	15	66	+340
Italy	29	52	+79
Latvia	?	46	?
Lithuania	?	61	?
Malta	40	54	+35
Moldova	?	51	?
Netherlands	185	194	+5
Norway	82	113	+38
Poland	42	39	–7
Portugal	20	37	+85
Romania	25	44	+76
Russia (all)	?	20	?
Slovenia	?	25	?
Spain	18	35	+94
Sweden	60	62	+3
Switzerland	17	31	+82
Turkey	6	33	+450
Ukraine	?	44	?
United Kingdom	42	85	+102
Yugoslavia (former)	20	30	+50

ity through the dominance of a few vigorous grass species, and results in reduced floral and structural diversity, similar to that of a re-seeded grassland (Green 1990, Hopkins 1991).

Grazing and mowing also have a profound effect on the structure and floral diversity of grassland. In temperate regions, such diversity is generally highest under management that reduces the dominance of vigorous plant species (e.g. through high grazing pressure), that allows a high proportion of plants to flower and set seed (e.g. through adequate periods without grazing or mowing), and that allows plants to establish themselves (e.g. through occasional disturbance of the soil by grazing pressure or mowing) (Beaufoy *et al.* 1994). In one Belgian study of managed grassland, the highest floral diversity occurred under a low-intensity regime of one cut per year for hay, followed by late-season grazing. Similar results from such a regime have been observed in Spain (García 1992).

However, grazing and mowing can give varying results, depending on their timing and intensity, the type of grazing animal involved, and abiotic factors

such as soil, climate and altitude. For example, high grazing pressure under a rotational regime, together with winter grazing, seems to result in the highest level of floral diversity in the UK. Grassland vegetation in the Mediterranean can withstand very heavy grazing better than in other European regions because of the high proportion of annual plants that are typically present. However, moderate grazing pressure favours floral diversity and results in a more nutritious pasture with a higher proportion of nitrogen-rich leguminous plants (Fernández Alés *et al.* 1992, Malo *et al.* 1994).

As indicated above, specialization has resulted in the concentration of livestock into certain regions and major changes in production practices. Livestock are now commonly overwintered indoors and fed on silage or high-protein animal feeds. Indeed, in many areas, the practice of keeping animals in intensive rearing yards all year is now commonplace. As a result, permanent pastures in livestock-production regions have often been replaced by re-seeded and fertilized grass leys for silage production, or other fodder crops for livestock, such as maize or other cereals. The use of hay as a fodder for animals has also declined since silage is more productive and more nutritious.

There are considerable variations in livestock density among European countries (Table 5), at least partly as a result of the differences in production systems. Cattle and pig densities are highest in western European Union countries, such as Belgium, Denmark, Germany, the Netherlands and Luxembourg, largely as a result of the widespread use of intensive production systems. By comparison, cattle and pig densities in other western EU countries, like Ireland, UK and France, are comparatively low, reflecting the large areas of permanent pasture, and less intensive livestock systems. With the exception of a few, mostly smaller countries, cattle densities generally declined in Europe during the 1980s (Table 5).

Relative to the ecological carrying capacity, sheep densities are particularly high in Cyprus, Norway, Albania, Poland, Greece, Portugal, UK, Turkey and Spain (Table 5). This has harmed the environment in some areas, e.g. through excessive soil erosion. Increases in sheep densities occurred throughout most of Europe during the 1980s. Declines were, however, noted in Bulgaria, Hungary and some montane areas of Spain (see Donázar et al. 1997), and have since become widespread in eastern Europe as much agricultural land has been abandoned following the recent political and economic changes in the region (CEC 1995a), e.g. a 36% decrease in sheep numbers during 1988–1995 in Romania (National Commission for Statistics 1996). Many or most of the sheep in these countries are, however, habitually grazed on less intensively managed pastures, described in 'Steppic habitats'.

Table 5. Livestock numbers and trends in relation to area of agricultural land in Europe (Anon. 1992b; also Anon. 1994e, Kovalenko 1994). Data are not available for European countries not listed here.

	Sheep and goats		Cattle		Pigs	
	No./ha in 1988	% change 1980–1988	No./ha in 1988	% change 1980–1988	No./ha in 1988	% change 1980–1988
Albania	2.33	+31	0.66	+14	0.18	+13
Austria	0.08	+33	0.75	+7	1.13	+6
Belarus	0.10	?	0.80	?	0.60	?
Belgium	0.14	+75	1.95	−4	4.11	+18
Bosnia-Herzegovina	0.50	?	0.30	?	0.20	?
Bulgaria	1.51	−13	0.29	−6	0.65	+7
Croatia	0.30	?	0.30	?	0.50	?
Cyprus	3.11	+2	0.29	+93	1.75	+78
Czechoslovakia (former)	0.17	+21	0.75	+3	1.07	−5
Denmark	0.03	+50	0.84	−19	3.25	−3
Estonia	0.10	?	0.60	?	0.80	?
Finland	0.03	−25	0.57	−13	0.50	−5
France	0.40	0	0.72	−6	0.40	+11
Germany	0.24	+41	2.20	+4	3.92	+7
Greece	1.82	+151	0.09	−23	0.12	+21
Hungary	0.36	−20	0.27	−13	1.26	+2
Iceland	0.31	−16	0.06	+15	0.01	+9
Rep. of Ireland	0.89	+112	1.01	−4	0.17	−6
Italy	0.74	+28	0.53	+4	0.55	+9
Latvia	0.10	?	0.60	?	0.60	?
Lithuania	0.00	?	0.70	?	0.80	?
Luxembourg	0.07	+75	2.25	−2	0.80	+1
Malta	0.77	−16	1.15	+7	7.31	+695
Moldova	0.60	?	0.60	?	1.00	?
Netherlands	0.72	+71	2.40	−8	6.93	+39
Norway	2.39	+6	0.98	−9	0.77	+7
Poland	2.17	+6	0.60	−20	1.04	−3
Portugal	1.86	+18	0.42	+2	0.75	−27
Romania	1.14	+2	0.47	+1	0.95	+22
Russia (European)	0.29	?	0.11	?	0.21	?
Slovenia	0.00	?	0.60	?	0.60	?
Spain	0.89	+48	0.17	+11	0.54	+63
Sweden	0.11	0	0.50	−11	0.66	−11
Switzerland	0.22	+3	0.93	−8	0.96	−8
Turkey	1.46	+1	0.35	−20	?	?
Ukraine	0.22	+3	0.63	+3	0.46	−3
United Kingdom	1.54	+31	0.67	−8	0.44	+3
Yugoslavia (former)	0.55	+6	0.37	−13	0.59	+9

Further information on the management of agriculturally improved grassland is given for the British Isles by Green (1990), and for Spain by Montserrat and Fillat (1987).

Steppic habitats

Figure 4 illustrates how grazing, cultivation and intensification can interact to create steppic habitats. Nearly all remaining areas of primary and secondary steppe in Europe are used for livestock grazing, often free-ranging or nomadic in character, by sheep or goats, and to a lesser extent, cattle. Cereal and fodder crops (e.g. alfalfa) on pseudosteppe are typically grown with no or low use of fertilizers, pesticides or irrigation. Crops are normally grown in rotation, incorporating long periods of grass and fallow, usually three years for either stage in Iberia (Hidalgo de Trucios and Carranza Almansa 1990) but with up to 6 years of grass and 10 years of fallow in some areas. In Russia, alternation between crops and fallow (up to 3–7 years duration in arid regions) is also still common.

Traditionally, many dry grassland areas, particularly in Spain, were part of transhumance pastoral systems (see next paragraph). Livestock used the lowland dry grassland in winter and were driven along traditional droving roads to summer pastures in the mountains (Klein 1920, Ruíz and Ruíz 1986, Pain and Dunn 1995, Donázar et al. 1997). Such practices have now been mostly abandoned (Rubio and Martínez 1992).

Montane grassland

Traditional mountain farming consisted of transhumance, in which mountain pastures around and above the timberline were grazed during summer months by non-local sheep, which were moved back to pastures in the lowlands for the rest of the year. Originally the animals were moved by droving, and latterly by motorized transport. Also, hay for feeding local livestock (usually cattle) in winter was grown on the steep slopes which were unsuitable for grazing, and on rich valley-bottom soils. Trees and shrubs were cleared to increase the area of pastures and meadows, thereby considerably shaping the landscape (Franz 1979, Walter and Breckle 1986a, 1991, Mayer 1994a). Since the middle of the nineteenth century, this management system has proved more and more difficult and expensive to maintain, and has been abandoned in most mountain areas of Europe since the 1970s (Bätzing 1994, Lichtenberger 1994).

In the wealthier countries, more intensive systems have now taken over, particularly on the more productive soils. As a result, livestock densities and overall numbers have again increased in these areas, but traditional breeds have been replaced with more productive through less well-adapted animals. Damage to the vegetation through overgrazing and trampling is common (e.g. Greif and Schwackhöfer 1979, Nachurisvili 1983). In Spain, overgrazing has occurred in the most accessible areas in recent years, probably as a result of EU livestock subsidies (Donázar et al. 1997). In contrast, in less accessible montane areas, abandonment of grazing has occurred.

To increase hay production, grasslands are often levelled, re-seeded (with non-native species or new varieties) and heavily fertilized. These practices allow earlier or repeated cutting, but fundamentally change the structure and composition of natural vegetation, and can lead to soil erosion (e.g. Grabherr 1987).

Wet grassland

Nearly all wet grasslands in Europe are managed, in terms of their cropping system and hydrology, for agricultural production. Grassland may be grazed or mowed, or both. Livestock species, stocking densities, and timing of grazing/mowing influence the birdlife in wet grassland (see 'Habitat needs of priority birds', p. 299). Traditionally, wet grasslands were mown for hay, but this is quite rare in western Europe now, and they have mostly been agriculturally improved to increase stocking densities or to convert to silage production. In some parts of Europe, wet grassland is burned during dry periods in order to keep it open.

Water management on wet grassland influences most of the other management factors, and different types of wet grassland are a result of different types of water management. Most important is the absolute wetness of the soil, which may be influenced by drainage, river regulation and overall groundwater levels.

The flood-plain grasslands of eastern Europe are critically dependent for their survival on annual, natural river-flooding in late winter or early spring. This makes them particularly sensitive to hydrological management (e.g. damming or abstraction), whether of the river itself or, indirectly, elsewhere in the catchment—even hundreds of kilometres upstream.

Rice cultivation

According to Fasola and Ruíz (1996) the typical annual cycle of agricultural operations in European rice-growing is as follows. In March the soil is levelled and each field is surrounded by low dykes to retain water. Flooding to a depth of 15–25 cm and sowing takes place between early March and early May according to region. After germination and initial growth, algicides and herbicides are applied in

May–June and insecticides in July–August. The water-level is gradually lowered after flowering in July and harvesting takes place in October. During the winter most fields are drained, but in some areas, such as the Camargue (France), Ebro delta and Albufera de Valencia (Spain), water is retained to attract waterfowl for hunting (Fasola and Ruíz 1997).

Rice cultivation therefore provides a temporary aquatic ecosystem during the summer when many Mediterranean wetlands are dry. However, these rice-fields are highly dynamic, with rapid changes in water-levels and other physical and chemical properties. There have also been considerable changes in rice-growing methods over the last few decades. Traditionally, rice-fields were smaller, with uncultivated areas and higher water levels. Also, when fields were drained, some water remained in ditches which provided refuges for invertebrates and amphibians.

Today, fields are larger and often accurately levelled by tractors with laser-based equipment, which allows shallower water and results in the fields being dry for longer periods. In Italy, rice has been grown in a considerable proportion of fields with irrigation delayed until June and even using dry cultivation methods. This reduces the demand for water, which is in short supply in some regions. The timing, depth and duration of flooding and other management techniques have considerable effects on the rice-field fauna and their predators, the most notable of which are birds (Fasola and Ruíz 1996).

Perennial crops

Perennial crops in temperate Europe are mainly orchards of fruit-trees such as apples or pears, while a broader range of trees is grown in Mediterranean countries, especially olives and citrus. Most orchards and other plantations of fruit-trees in Europe are now large commercial ventures, whereas a substantial proportion of the area under olives in the Mediterranean is probably still in the form of small, family-owned groves. For example, 85–90% of olive-groves are still classified as 'traditional' in Portugal by the Ministry of Agriculture (Pain 1994a). Large, commercial olive plantations tend to be recently established in drier areas, and are often irrigated.

Production of fruit is seasonal for most if not all perennial crops in Europe. Pruning of branches is normally carried out early in the season. Periods of leaf-growth and/or fruit development are often accompanied by pesticide spraying (either prophylactic, or according to population levels of pests and diseases). Management of the grass or shrub understorey—by grazing, tilling, cutting, mowing or spraying of herbicides—is also a necessary feature in the management of most perennial crops. Fruiting

can be spread out over the season due to the use of different varieties, and is unusually extended in some species anyway (e.g. olives).

Harvesting of the fruit formerly required large numbers of part-time workers, but in recent decades picking has become increasingly mechanized. However, such machines are expensive and used only at large commercial plantations. In some depopulated rural areas of Europe, it is becoming difficult and uneconomic to recruit the seasonal workforce for picking the fruit (Pain 1994a).

Another major trend towards intensification is the increasing replacement of old crop-trees with new varieties with more favoured qualities, e.g. ones that mature faster, produce more fruit and are more amenable to mechanized harvesting (e.g. with smoother bark).

Pastoral woodland

The traditional management of pastoral woodland consists primarily of the selective removal of trees to open the canopy. In some areas (e.g. Britain) the maintenance of very old trees was encouraged (Harding and Rose 1986). In the Iberian peninsula, where most European pastoral woodlands are found, traditional use includes tree management (thinning, selection of the highest quality and most productive trees, and pruning) for the production of acorns (for livestock), charcoal, firewood, cork, tannin and timber, shifting cereal cultivation, and grazing of indigenous breeds of sheep, cattle, pigs and goats (see Díaz et al. 1997 for a review). The soil cultivation, together with scrub clearance and tree-thinning, maintain the semi-open structure of the habitat. The agricultural and forestry practices involve long (4–20 years) crop rotations, as well as slow woodland turnover, due to the poor quality of the soils (Gascó 1987, Montero 1988). It is also usual to excavate small ponds which are filled by the autumn and winter rains and which provide water for livestock during late spring and summer.

The management techniques are usually performed in such a way that shrubby, open and cultivated plots are patchily distributed within dehesas and montados (Díaz et al. 1997). Shrubby patches provide shoots and leaves for livestock and for wild herbivores, and they appear to be the main places where tree recruitment occurs, whereas winter cereals are mainly used to feed livestock during the dry summer, when grass is in short supply (Díaz et al. 1997). In the past, the decrease in food availability in pastoral woodland in the summer was also traditionally alleviated by seasonal movements (i.e. transhumance) to mountain summer pastures, where grass growth is then at its peak (Mitchell et al. 1977, Ruíz and Ruíz 1986).

In the Iberian peninsula, the increasing use of machinery, fertilizers and imported feed for livestock (under EU subsidies) have produced an increase in stock density which has caused overgrazing in some areas (Campos 1984, Díaz *et al*. 1997). Pesticide use is also increasing; for example, Malathion and other agents are used to control caterpillar pests (Robredo and Sánchez 1983). The use of local breeds of livestock has declined due to the effects of trade on sheep and cattle, or to diseases in the case of Iberian pigs (Pérez 1988). This, together with rising wages and the fall in the prices of (and demand for) pig-fat products, firewood and charcoal, has caused the abandonment of some areas or their use for hunting large game; these areas have then become overgrown by shrubs (Campos 1984, Díaz *et al*. 1997). There are also problems with woodland-ageing and lack of regeneration in some areas (Díaz *et al*. 1997).

Characteristic vegetation and associated fauna

Arable and improved grassland

Arable and agriculturally improved grassland are artificial habitats and therefore dominated by the crops within them. In Europe the most widespread crops include cultivated grasses and clover (for silage or hay), cereals such as wheat, barley, oats, ryegrass and maize, legumes such as beans, chick-peas and alfalfa, brassicas such as cabbage and oil-seed rape, root crops such as potato, carrot and sugarbeet, and sunflowers.

Due to the regular disturbance of the soil in arable farmland, non-crop plant species within the cultivated parts of fields tend to be dominated by annual and non-woody perennial species. Many of these are important food-plants for invertebrates, small mammals and seed-eating birds. However, cultivation practices and the use of herbicides often results in crops with negligible quantities of weeds.

The paucity of non-crop vegetation, combined with the use of insecticides, also results in an impoverished invertebrate ground-fauna in most intensive farmland (Litzbarski 1993, Litzbarski and Litzbarski 1996, Potts 1997, Campbell *et al*. in press). Crop pests can become abundant despite the use of pesticides, due to resistance to pesticides, loss of natural predators (e.g. through pesticides), and many pests' capacity to rapidly increase numbers after pesticide applications have worn off. Large fluctuations and spatial differences in pest numbers are typical in many intensively farmed areas (Faragó 1992). Where pesticides are not used, or specialized chemicals are used in the place of broad-spectrum products, then substantial populations of some non-pest species

may occur. These may include predatory species such as carabid and staphylinid beetles (Coleoptera), spiders and harvestmen (Arachnida), and lacewings (Neuroptera). However, in the absence of weeds as food-plants, pests which are herbivorous in the larval stage usually cannot build up numbers, e.g. moths and butterflies (Lepidoptera) and sawflies (Tenthredinidae) (Potts 1986).

Soil invertebrate populations in intensive farmland are also often impoverished in terms of diversity and abundance. Earthworms (Lumbricidae) in particular often occur at extremely low densities in arable farmland (Tucker 1992). Low numbers of soil invertebrates may be due to a lack of organic matter as food, high levels of mortality during ploughing and harrowing, or through the toxic effects of some pesticides, in particular molluscicides. High numbers of soil invertebrates can nevertheless occur in some situations in agriculturally improved grassland. For example, huge increases in tipulid larvae can result from the drainage and fertilization of rough grassland.

Open fields in intensive arable farmland are an inhospitable environment for amphibians, reptiles and small mammals, due to ploughing and other farming operations, the lack of cover early in the growing season, lack of food for species that cannot feed on the crop itself, and the use of pesticides. Few such animals therefore occur in most areas of intensive farmland, with the exception of low numbers of amphibians and small rodents, which may move into crops in the summer from refuges such as hedgerows, ponds, ditches or farmyards.

Larger mammals are, however, still common in intensive farmland, including herbivores such as rabbit *Oryctolagus cuniculus*, hare *Lepus europaeus* and even some deer (Cervidae), which feed in young crops and grassland. Red fox *Vulpes vulpes* and badger *Meles meles* are also abundant in many areas of intensive farmland.

The bird community of open fields in agricultural habitats is highly dependent on the level of intensity under which it is farmed. The most intensive arable farming systems, with autumn-sown and highly fertilized annual cereal monocultures that regularly receive pesticide applications, have a low diversity and number of birds. Such habitats will typically only support Carrion Crow *Corvus corone* and Woodpigeon *Columba palumbus* in appreciable numbers. Agriculturally improved, temporary grasslands used for silage will also have an impoverished avifauna, though these may be used (i.e. grazed) by *Columba palumbus* and even (in winter) some swans *Cygnus* and geese *Anser* and *Branta*. If soil invertebrates such as tipulid larvae are abundant then agriculturally improved grassland may also be frequented by large numbers of birds, e.g. Starling *Sturnus vulgaris*, Rook *Corvus frugilegus* and Jackdaw *Corvus monedula*.

In slightly less intensive arable systems with higher crop diversity, including spring-sown crops and fallow land (or set aside), and the presence of non-farmed marginal features such as rough grassy strips, a considerable variety of field-nesting or - foraging species may occur, such as Partridge *Perdix perdix*, Lapwing *Vanellus vanellus*, Skylark *Alauda arvensis*, Kestrel *Falco tinnunculus*, Montagu's Harrier *Circus pygargus*, Crested Lark *Galerida cristata*, Calandra Lark *Melanocorypha calandra*, Linnet *Carduelis cannabina*, Yellowhammer *Emberiza citrinella*, Ortolan Bunting *E. hortulana* and Corn Bunting *Miliaria calandra*. Less intensively farmed permanent grassland, with diverse grass swards and moderate fertilizer use and stocking densities, may hold White Stork *Ciconia ciconia*, *Alauda arvensis*, *Falco tinnunculus*, Swallow *Hirundo rustica* and, if managed as hay-meadows, Barn Owl *Tyto alba*, Quail *Coturnix coturnix* and Corncrake *Crex crex*. In winter, such grassland may also be particularly important for wintering flocks of *Vanellus vanellus*, Golden Plover *Pluvialis apricaria*, *Sturnus vulgaris* and thrushes such as Redwing *Turdus iliacus* and Fieldfare *T. pilaris*.

Farmland with high densities of hedgerows, trees and small woods, such as in parts of Ireland, UK, France, Germany and Poland, may hold a high diversity of woodland species, sometimes in substantial numbers. Particularly common species may include Wren *Troglodytes troglodytes*, Dunnock *Prunella modularis*, Robin *Erithacus rubecula*, Song Thrush *Turdus philomelos* and Chaffinch *Fringilla coelebs*. Less common species may include Tree Pipit *Anthus trivialis*, Whitethroat *Sylvia communis*, Red-backed Shrike *Lanius collurio*, *Emberiza citrinella*, *E. hortulana* and, if there is sufficient dead wood, woodpeckers such as Great Spotted Woodpecker *Dendrocopos major* and Syrian Woodpecker *D. syriacus*. Tall mature trees may provide suitable nesting sites for Tawny Owl *Strix aluco* and Scops Owl *Otus scops* and various raptors, including Red Kite *Milvus milvus*, Buzzard *Buteo buteo*, Lesser Spotted Eagle *Aquila pomarina* and *Falco tinnunculus*.

Thus, although the most intensively farmed habitats may be highly impoverished, moderate farming systems may hold reasonable numbers and diversity of birds, particularly if non-farmed habitats are also present in the landscape. In the UK, for example, O'Connor and Shrubb (1986) found that some 47 species of birds covered by the Common Birds Census were typical of farmland (i.e. were recorded in at least 50% of the sample censuses). Similarly, in northern Germany around 100 breeding bird species were found in 24 plots totalling c. 20 km^2 of open farmland with hedgerows and woodlots (Flade 1994). In fact, overall, arable and agriculturally improved grassland habitats hold the highest numbers of regularly occurring species on farmland (Tucker 1997). This is most likely due to the high diversity, large area and biogeographical range spanned by the habitats falling within this definition. Intensive agricultural habitats as defined here can be found across Europe. Species as widely separated as Black-winged Kite *Elanus caeruleus* in Iberia, Yellow-breasted Bunting *Emberiza aureola* in northern Russia, and Bimaculated Lark *Melanocorypha bimaculata* in Turkey may therefore occur. By contrast, some of the other major habitat-types defined here have geographically smaller ranges in Europe and therefore draw from a comparatively restricted pool of species.

Steppic habitats

The primary steppes of Russia and Ukraine are dominated by drought-resistant bunch-grasses *Stipa*, with species of *Festuca* also characteristic (Walter 1968, Wolkinger and Plank 1981). The diversity of herbs and their proportional contribution to biomass decreases from north to south with increasing aridity of the climate, while the proportion of dwarf shrubs (especially wormwood *Artemisia*) increases. Four zones of primary steppe successively replace one another from north to south (Walter and Breckle 1986b), as follows.

1. Wooded steppe in semi-arid climates (meadow steppe; forest steppe).

2. 'Typical' steppe in semi-arid to arid climates (*Stipa* steppe; true steppe).

3. Semi-desert steppe in arid climates (shrub steppe; semi-desert steppe). This and the next type are sometimes merged.

4. Desert steppe in hyperarid climates (shrub steppe; desert steppe).

The diversity, abundance and species composition of animal populations also change considerably from north to south, as a result of the increasing aridity (Chernov 1975). The wooded steppe has the highest abundance of animals, e.g. the invertebrate biomass in soils of this zone is four times that of typical steppe. Earthworms, in particular, decline rapidly from north to south. Of the invertebrates, grasshoppers and locusts (Acrididae) can be especially numerous and play an important role in structuring the vegetation and soil of primary steppe generally, as do small herbivorous mammals (such as steppe lemming *Lagurus lagurus*, common vole *Microtus arvalis* and suslik *Citellus pygmaeus*), which occur at high diversity and in great abundance. Huge populations of wild ungulates used to graze the primary steppes in the past, including tarpan *Equus hydruntunus*, *E. ferus* and *Saiga tatarica*, but these have been extinct in Europe for many hundreds of years.

The most typical and widespread bird species of steppic habitats in Europe, found in most (but not all) steppic regions include *Circus pygargus*, *Tetrax tetrax*, Stone Curlew *Burhinus oedicnemus*, *Melanocorypha calandra*, Short-toed Lark *Calandrella brachydactyla*, Lesser Short-toed Lark *C. rufescens*, *Galerida cristata*, *Alauda arvensis*, Tawny Pipit *Anthus campestris* and *Miliaria calandra*.

Among the birds of the primary steppes of Russia and Ukraine are a number that are largely, or entirely, restricted to this habitat, such as raptors that feed mainly on the abundant small mammals, e.g. Steppe Eagle *Aquila nipalensis*, Imperial Eagle *Aquila heliaca*, Pallid Harrier *Circus macrourus*, Saker *Falco cherrug* and, to a lesser extent, Long-legged Buzzard *Buteo rufinus*. In favourable areas, these birds of prey can reach high densities.

These eastern steppes also hold the remaining populations of Sociable Plover *Chettusia gregaria* and Caspian Plover *Charadrius asiaticus* in Europe, and the majority of Black-winged Pratincole *Glareola nordmanni* and Demoiselle Crane *Anthropoides virgo*. In addition, Black Lark *Melanocorypha yeltoniensis* and White-winged Lark *Melanocorypha leucoptera* occur nowhere else in Europe.

Much of the original area of primary steppe has now been converted to cereal cultivation, especially in Ukraine, and most of the above species of primary steppe will not use such arable habitats. However, the pseudosteppes that cover much of southern Russia are used by Great Bustard *Otis tarda*. Long double-lines of trees have often been planted in such areas to provide shelter-belts. These provide nest-sites or hunting perches for a variety of birds, including Red-footed Falcon *Falco vespertinus*, Roller *Coracias garrulus*, Lesser Grey Shrike *Lanius minor* and *Corvus frugilegus*, as well as typical woodland-edge or shrubland species (e.g. Munteanu 1990).

Extending westwards from the main Eurasian steppes are the steppes of the Pannonic region, centred on the lowland plains of Hungary. Originally, most of this area was lightly wooded steppe, and is now a mosaic of seasonally wet, saline grassland ('puszta') intermixed with areas of non-intensive arable, and smaller areas of natural sandy grassland on the wooded fringes of the lowland basin. As on the main Eurasian steppes, small herbivorous mammals such as *Citellus* are abundant and diverse.

The Pannonic steppes lack a number of the typical steppe birds found further east in the primary Eurasian steppes. Historical fragmentation of the habitat may have caused some species loss: *Tetrax tetrax* is extinct in the region, whilst *Otis tarda* has contracted considerably in range and only reduced numbers remain. Other notable breeding species of the puszta include *Falco vespertinus*, *Falco cherrug*, *Otis tarda*, *Coracias garrulus*, *Lanius minor*, and less commonly *Aquila heliaca* and Collared Pratincole *Glareola pratincola*. During migration Crane *Grus grus* and the globally threatened Lesser White-fronted Goose *Anser erythropus* depend heavily on puszta for feeding and roosting (Madsen 1996). Other important species on passage include the globally threatened Slender-billed Curlew *Numenius tenuirostris* and, especially when partly flooded, large numbers of waders (Charadrii) such as *Pluvialis apricaria*, *Vanellus vanellus*, Curlew *Numenius arquata*, Whimbrel *Numenius phaeopus*, *Limosa limosa*, Ruff *Philomachus pugnax* and Redshank *Tringa totanus*.

In Turkey, primary steppe is confined to the plains surrounding the lake of Tuz Gölü in central Anatolia, and is characterized by dwarf shrubs such as *Artemisia* and *Thymus* species, often in association with species of Chenopodiaceae (Ozenda 1979, Baris 1991). There are also fairly extensive areas of salt-steppe vegetation with *Salicornia* and *Salsola* on saline soils. Secondary steppe now covers the rest of central and east Anatolia, as a result of thousands of years of cutting, burning and grazing of the original more wooded vegetation. The current vegetation resembles primary steppe, dominated by *Bromus* grass and *Medicago* species, and at higher altitudes or where vegetation is overgrazed, by less palatable 'spine-cushion' herbs such as *Astragalus* and *Acantholimon* (Kence 1987). The steppes of southeast Anatolia are semi-desertic, with typical plants including the dwarf shrubs *Artemisia herba-alba*, *Eryngium noerium* and *Salvia spinosa* (Ertan *et al.* 1989).

Birds of these Turkish steppe habitats are dominated by larks (Alaudidae) and other typical species include *Ciconia ciconia*, *Buteo rufinus*, Lesser Kestrel *Falco naumanni*, Black-bellied Sandgrouse *Pterocles orientalis*, Pin-tailed Sandgrouse *P. alchata* and *Lanius minor*. Lanner *Falco biarmicus*, *Falco cherrug*, *Anthropoides virgo* and *Otis tarda* also occur. In areas of salt-steppe, *Glareola pratincola* and Greater Sand Plover *Charadrius leschenaultii* may be found.

The Mediterranean steppic habitats of Italy, France and Spain are dominated by grasses such as *Poa*, *Stipa*, *Lygeum spartum* and by dwarf shrubs. Accounts of their flora and fauna are given by Petretti (1988, 1995) for southern Italy and by Suárez *et al.* (1992) for Spain. Most such habitats have been replaced by pseudosteppe or irrigated farmland, but sheep pastures occur on dry and stony soils. Characteristic birds of these various Mediterranean steppic habitats include Red-legged Partridge *Alectoris rufa*, *Pterocles orientalis*, *P. alchata*, Dupont's Lark *Chersophilus duponti*, Thekla Lark *Galerida theklae*, Black-eared Wheatear *Oenanthe hispanica* and Spectacled Warbler *Sylvia conspicillata* (de Juana *et al.* 1988, Tellería *et al.* 1988, Lecomte and Voisin 1991, Suárez *et al.* 1997a). Pseudosteppe in Spain is particularly important for *Otis tarda* (Martínez and de Juana 1996), and *Glareola pratincola* also breeds.

Montane grassland

Despite some disagreement on zonation, three vegetation belts can typically be distinguished above the treeline: the shrub belt just above the treeline ('subalpine' zone), the grassland up to the line of permanent snow ('alpine' zone), and the mostly unvegetated zone up to the highest tops (nival zone) (Walter and Breckle 1986a).

The vertical extent of vegetation belts is subject to large variation, because topography influences climate and the distribution of plants. In many cases the vegetation does not separate into distinct zones, but intergrades or is patchily intermingled.

Central and west European mountain ranges (Alps, Apennines, Carpathians, and Pyrenees) are very similar in their vegetation, whereas Mediterranean mountain ranges (e.g. Sierra Nevada, Taurus), as well as the Caucasus or the Urals, differ from each other to an appreciable extent with respect to taxonomy and habits of plants (Franz 1979, Walter and Breckle 1986a, 1991).

The avifauna of European mountains has two main sources: the central Asiatic mountains and the so-called 'arcto-alpine' and 'boreo-alpine' fauna (see Berg-Schlosser 1984). The species belonging to the latter fauna are considered to have split into separate northern and montane ranges after the end of the cold Pleistocene period, resulting in separate northern and montane subspecies. Examples of such montane subspecies in the Alps are *Lagopus mutus* and Ring Ouzel *Turdus torquatus* (De Lattin 1967).

In the montane habitats of western and central Europe, there are no endemic bird species and a much smaller number of species than in central Asia, where many montane bird genera in Europe originated (e.g. *Tetraogallus*, *Montifringilla*, *Pyrrhocorax*, *Tichodroma*) (Stresemann 1920, Harrison 1982). The extent of montane habitats is smaller in Europe and hence such habitat is ecologically less diversified; moreover, the main European mountain ridges are geologically younger, and in a peripheral and isolated position on the Eurasian landmass. These factors reduce the probability of immigration or evolution of species (see Begon *et al.* 1990).

In the Caucasus, by contrast, substantial speciation has taken place. The region holds two endemic birds, Caucasian Black Grouse *Tetrao mlokosiewiczi* and Caucasian Snowcock *Tetraogallus caucasicus*, making it one of three globally important Endemic Bird Areas in Europe, and it is likewise important for other endemic taxa, e.g. at least four endemic mammals (ICBP 1992, Stattersfield *et al.* in press), and for a number of montane bird species with otherwise Asiatic ranges, including Güldenstädt's Redstart *Phoenicurus erythrogaster* and Great Rosefinch *Carpodacus rubicilla* (Tucker and Heath 1994).

Subalpine scrub, heath and grassland

The lower limit of the subalpine zone is the actual timberline. In most parts of Europe, however, the timberline has been substantially lowered (by several hundred metres) by deforestation. Where climate and soil are favourable, bushes or trees with prostrate habits occur, such as *Alnus viridis*, *Pinus mugo* and *Betula*. Montane heathland of dwarf shrubs dominates the higher ground, with *Juniperus* widespread over a large climatic range. In drier climates, e.g. in the Taurus or Sierra Nevada mountain ranges, bushy legumes become the dominant taxonomic group. In arid continental climates, a broad belt of spine-cushion vegetation can occur up to the alpine zone (Walter and Breckle 1991).

The scrub and heath of the subalpine zone are far richer in birds than alpine grassland, especially when there is a scattering of trees (e.g. Winding *et al.* 1993). Typical species include *Tetrao tetrix*, *Tetrao mlokosiewiczi* (in Turkey and the Caucasus), Whinchat *Saxicola rubetra*, Stonechat *S. torquata*, *Turdus torquatus*, Citril Finch *Serinus citrinella* and Rock Bunting *Emberiza cia*.

Where the natural scrub and heath has been cleared by man, subalpine grassland occurs. This is more productive than alpine grassland and is usually characterized by rather tall grasses and herbs. Some alpine bird species extend their range down to these habitats, e.g. Water Pipit *Anthus spinoletta* (Glutz von Blotzheim and Bauer 1985). Such meadows are also becoming increasingly important for some bird species typically inhabiting meadows, such as *Coturnix coturnix*, *Saxicola rubetra*, Woodlark *Lullula arborea* and *Emberiza hortulana*, since these habitats are rapidly diminishing in extent in the lowlands (Epple 1988).

Alpine grassland

Alpine grasslands are typically short, often with a diverse flora. Usually there is one dominant species of the genera *Carex*, *Festuca*, *Sesleria*, *Nardus* or *Juncus*. Under relatively humid conditions, mosses and lichens are numerous. As altitude or dryness increase, vegetation cover decreases and grassland gives way to grass patches and cushion vegetation.

Typical birds of the alpine zone include *Anthus spinoletta*, *Alectoris graeca*, Alpine Accentor *Prunella collaris*, *Pyrrhocorax graculus*, Chough *Pyrrhocorax pyrrhocorax*, Snowfinch *Montifringilla nivalis* and *Serinus citrinella*. The habitat is also the main foraging ground for scavenging raptors such as Lammergeier *Gypaetus barbatus*, Egyptian Vulture *Neophron percnopterus*, Griffon Vulture *Gyps fulvus* and Golden Eagle *Aquila chrysaetos*.

Rocky habitats of the nival zone

Rock, scree, bare soil, and perennial snow and ice-cover characterize the habitats of the nival zone. Such habitats are clearly not used for pastoral agriculture but are included in this strategy for completeness as they are important components of montane-habitat mosaics. Snow-fields are themselves attractive to insectivorous birds (e.g. *Montifringilla nivalis*) for two reasons: first, wind-blown insects are deposited on them (e.g. Zamora 1990, Antor 1995), and second, just after melting, insect larvae can be easily collected along their edges (Heiniger 1991). In the Caucasus, Caspian Snowcock *Tetraogallus caspius* and *T. caucasicus* are characteristic of this zone, along with *Phoenicurus erythrogaster*.

Cliffs are used as nesting sites for raptors and many others species (e.g. swifts *Apus* or choughs *Pyrrhocorax*) or roosting sites (e.g. *Montifringilla nivalis*: Heiniger 1991), and even for foraging by Wallcreeper *Tichodroma muraria*.

Wet grassland

Wet grasslands incorporate a variety of different habitat-types which are not always easy to classify. Especially in northern temperate Europe, where the habitat is most common, many areas of wet grassland fall under more than one of the categories listed below. For practical reasons, the habitat-types are sorted according to differences in their water regimes. This classification also results in some geographical separation.

Polders

Polders are flat areas that have been claimed from the sea or from rivers (e.g. in the Netherlands) and which therefore have an artificially regulated water-regime. Most are below sea-level and thus tend to have a high water-table, being mainly wet grassland on silt or peat soils. The grassland has usually been re-seeded with rye-grass mixes and other fast-growing species. Ditches may be fringed with reeds *Phragmites* and other emergent species, but other plants are largely lacking in these habitats. In re-cently claimed areas, trees are also usually absent.

The breeding avifauna of the wetter polders is rich in waders, including *Limosa limosa* and *Vanellus vanellus*, coastal species of beaches and saltmarshes such as Oystercatcher *Haematopus ostralegus* and *Tringa totanus*, and northern species such as Snipe *Gallinago gallinago* and *Philomachus pugnax*, although the latter is now rare in many areas of polders (Beintema 1988). Ducks such as Mallard *Anas platyrhynchos*, Shoveler *A. clypeata* and Garganey *A. querquedula* also breed. Passerine diversity is low, but typical species include Yellow Wagtail *Motacilla flava*, Meadow Pipit *Anthus pratensis* and *Alauda arvensis* (Beintema 1988).

Mammal populations are impoverished in such habitats, probably due to the high water-tables (which flood burrows) and the lack of cover. Consequently, mammalian predator populations tend to be low; a factor which is probably partly responsible for the high densities of breeding waders (Beintema 1988).

In winter the habitat is often of importance for geese, grazing ducks and swans (e.g. Bewick's Swan *Cygnus columbianus*).

Baltic coastal meadows

The coastal grasslands of the Baltic and some Danish coasts are temporarily flooded by brackish water. At the larger sites there are high densities of breeding ducks such as *Anas platyrhynchos*, *A. clypeata*, *A. querquedula* and Pintail *A. acuta*, and of waders such as Avocet *Recurvirostra avosetta*, *Vanellus vanellus*, Dunlin *Calidris alpina*, *Limosa limosa*, *Gallinago gallinago*, *Philomachus pugnax* and *Tringa totanus*, as well as passerines such as *Saxicola rubetra*. At the smaller sites, there are few breeding waterfowl, and the breeding species are dominated by *Haematopus ostralegus* and *Tringa totanus*, while *Calidris alpina* also occurs at some sites. *Calidris alpina* formerly bred widely inland on flood-plain meadows in the Baltic catchment, but this population has been greatly reduced by habitat destruction during the last 100 years.

When abandoned, or farmed at low intensity, this habitat is particularly important for some populations of the globally threatened Aquatic Warbler *Acrocephalus paludicola* (Sellin 1989, Bauer and Berthold 1996). The habitat is of significant importance for birds on migration, especially geese, etc. (Anatidae), *Grus grus*, *Vanellus vanellus*, several *Calidris* species, *Numenius arquata*, *Numenius phaeopus* and *Pluvialis apricaria*, but wintering bird populations are relatively low due to the harsh climate.

Coastal marshlands of western France

These are wet grasslands with systems of ditches which are filled with sea water at high tide, though the grasslands themselves are usually not flooded. The habitat has a similar breeding wader community to the Baltic coastal meadows, but lacks northern species such as *Philomachus pugnax* and *Calidris alpina*. Wintering waterfowl populations are, however, large and diverse.

Riverine and lakeside flood-plain meadows and washlands

These are wet grasslands on the flood-plains of rivers, or next to lakes and ponds, which are regularly inundated by floods in late winter or early spring. They are the most important type of wet grassland in central and eastern Europe, in terms of their extent, their abundance and diversity of birds, and their highly threatened status. Such grassland is usually grazed, and a rich and diverse flora can be present. *Limosa limosa*, Great Snipe *Gallinago media* (in eastern Europe), *Vanellus vanellus* and *Tringa totanus* may breed. Such meadows also provide good hunting habitats for *Ciconia ciconia*, Hen Harrier *Circus cyaneus*, Black Tern *Chlidonias niger*, *Tyto alba* and Short-eared Owl *Asio flammeus*. Where grazing is absent or irregular, then tall plant communities may develop, dominated by sedges (Cyperaceae). These are preferred by several passerines, including Grasshopper Warbler *Locustella naevia*, *Acrocephalus paludicola* (in parts of eastern Europe) and *Saxicola rubetra*, as well as *Anas querquedula*, *Crex crex* and Spotted Crake *Porzana porzana*.

In winter, flood-plain grasslands are internationally important for a wide variety of waterbirds, especially in more westerly or southerly parts of Europe where there is less likelihood of the habitat freezing over.

Machair

Machair is not just wet grassland, but a complex of habitat-types. It is treated here, however, as it is of particular importance for a number of birds that are typical of other wet grassland habitats. Machair is restricted to soils formed on calcareous shell-sand on the Atlantic fringes of western Scotland and Ireland. The seaward edge is usually dunes, then hollows ('slacks') which may be moist and hold different plants. This area is typically used as rough grazing. Further inland lies the dry machair, where the soil has frequently been enriched by the long-term use of seaweed as fertilizer. These areas are often cultivated on a strip rotation. At its inland edge, the shell-sand becomes a thin layer over peat and increasingly moist, with pools and wet patches in many fields. These grasslands (called the blackland zone) are used for grazing and hay.

This mosaic of habitats creates a rich community of plants and animals. Breeding waders include

Ringed Plover *Charadrius hiaticula*, *Haematopus ostralegus*, *Vanellus vanellus*, *Calidris alpina*, *Numenius arquata*, *Tringa totanus* and *Gallinago gallinago*, the densities of which can be extremely high (Nairn and Sheppard 1985, Fuller *et al.* 1986, Shepherd and Stroud 1991). Other common birds include *Alauda arvensis*, Northern Wheatear *Oenanthe oenanthe*, *Anthus pratensis* and *Miliaria calandra*, whilst some areas hold *Pyrrhocorax pyrrhocorax* and *Crex crex*.

In winter, the short machair turf is grazed by flocks of Barnacle Goose *Branta leucopsis*, whilst the wetter machair is used by White-fronted Goose *Anser albifrons flavirostris* from Greenland.

Upland wet grassland

These are grass-dominated habitats in upland areas of north-west Europe with high rainfall. They have typically become grass-dominated through the effects of high grazing pressure on heather *Calluna*-dominated moorland, but have not been agriculturally improved. Typical plants include the grasses *Nardus stricta*, *Molinia caerulea* and *Eriophorum vaginatum*, with patches of rush *Juncus* in the wetter flushes. Breeding birds commonly include *Vanellus vanellus*, *Numenius arquata*, *Tringa totanus* and *Anthus pratensis* (Fuller 1982).

Rice cultivation

Rice-growing areas are typically uniform with little or no natural vegetation (see Fasola and Ruíz 1997). However, the flooded fields can hold high densities of invertebrates, especially crustaceans (e.g. *Triops cancriformis*, *Procambarus clarkii*), and amphibians. These in turn can support large and diverse waterbird populations. Indeed rice-fields are the most important feeding grounds for herons and egrets (Ardeidae) in Europe (Fasola and Ruíz 1996, 1997). For example, rice-fields in north-west Italy are used for feeding during the breeding season by at least 20 species of waterbird, including *Ciconia ciconia*, Spoonbill *Platalea leucorodia*, Moorhen *Gallinula chloropus*, waders such as *Limosa limosa*, Black-winged Stilt *Himantopus himantopus* and *Limosa limosa*, Black-headed Gull *Larus ridibundus* and terns, including Common Tern *Sterna hirundo*, Little Tern *S. albifrons*, *Chlidonias niger* and Whiskered Tern *C. hybridus*. The fields are, however, most important for their high densities of herons, especially Night Heron *Nycticorax nycticorax*, Squacco Heron *Ardeola ralloides*, Little Egret *Egretta garzetta* and Grey Heron *Ardea cinerea*. Rice-fields in the Ebro delta (Spain) and in the Camargue (France) have a similar avifauna.

Winter-flooded rice cultivations are important for passage and wintering waterbirds.

Perennial crops

The vegetation within areas of perennial crops is dominated by the crop itself, though a grass layer is often present and in some olive-groves a shrub layer may persist. Under recent, more intensive cropping systems, many of these habitats have an impoverished avifauna, the remaining species mostly feeding off the crop itself and thus being treated as pests. Bullfinch *Pyrrhula pyrrhula*, for example, is a common pest of orchards, while thrushes such as *Turdus merula* are pests of soft fruit. Olive-groves may, however, hold large numbers and a rich diversity of birds, the density and composition of which varies according to the age and density of the trees and the management practices carried out (Pain 1994a). In Portugal, where management is often less intense than other regions, breeding species include *Alectoris rufa*, Little Owl *Athene noctua*, *Lullula arborea*, *Oenanthe hispanica*, *Galerida theklae*, *Hirundo rustica*, *Saxicola torquata*, Orphean Warbler *Sylvia hortensis*, Great Grey Shrike *Lanius excubitor*, Woodchat Shrike *L. senator*, Cirl Bunting *Emberiza cirlus* and House Sparrow *Passer domesticus* (Pina *et al.* 1990, Pain 1994a).

In Greece, the most abundant species in dense olive-groves are woodland species, such as *Fringilla coelebs*, Great Tit *Parus major* and Olive-tree Warbler *Hippolais olivetorum*, while more open groves have species such as Short-toed Treecreeper *Certhia brachydactyla* and *Lullula arborea* (Wietfeld 1981).

Mediterranean olive-groves are particularly important feeding areas for wintering and migratory birds, especially thrushes *Turdus*, warblers and finches. The ability of many omnivorous birds to increase the proportion of fruit taken allows them to overwinter in olive-groves, even in the northern Mediterranean, as olives can form an emergency food supply at the end of the winter when other food sources may be scarce or depleted (Pain 1994a). Pina *et al.* (1990) found that the most abundant species in Portuguese olive-groves in winter were Blackcap *Sylvia atricapilla*, *Parus major*, *Erithacus rubecula*, Sardinian Warbler *Sylvia melanocephala*, *Anthus pratensis*, Chiffchaff *Phylloscopus collybita*, *Fringilla coelebs* and Serin *Serinus serinus*, with large numbers of other finches and thrushes present. In Spain, *Turdus philomelos* was the commonest species in winter (comprising 30–40% of all birds) accompanied by high densities of *Turdus iliacus*, *Erithacus rubecula* and *Fringilla coelebs* (Muñoz-Cobo and Purroy 1980).

Traditionally managed orchards may support a moderate diversity of birds, especially where there are old trees with nest-holes and dead wood, and short, grazed turf. This provides favourable habitats for species such as *Otus scops*, Hoopoe *Upupa epops*, Wryneck *Jynx torquilla*, Green Woodpecker *Picus viridis*, Redstart *Phoenicurus phoenicurus*, *Lanius senator*, *L. collurio* and *Emberiza hortulana*. In winter, uncollected fruit, such as apples, can attract large numbers of *Sturnus vulgaris* and thrushes, especially *Turdus iliacus* and *T. pilaris*.

Pastoral woodland

The dehesas of Spain and montados of Portugal hold the largest diversity and numbers of birds among pastoral woodlands in Europe. They consist of stands of varying density of holm oak *Quercus ilex* and, in damper situations, cork oak *Q. suber*. As a result of management, three broad variations of habitat structure and composition occur (Díaz *et al.* 1993, 1997). Grazed plots occupy the majority of the area and are characterized by grassland with sparse shrubs and a dense cover of herbs at ground level. Ungrazed or infrequently grazed plots have high shrub cover comprising mainly *Cistus ladanifer*, *Halimium*, lavenders (e.g. *Lavandula stoechas*) and legumes (e.g. *Lygos*, *Genista*). Cultivated plots, which typically cover 10% of the farmed area, grow crops such as maize, oats or rye between March and June.

The mosaic of habitats supports a rich diversity of animals. Typical breeding birds include *Otus scops*, *Coracias garrulus*, Bee-eater *Merops apiaster*, *Upupa epops*, *Galerida theklae*, Nightingale *Luscinia megarhynchos*, Blackbird *Turdus merula*, various *Sylvia* warblers, such as *S. melanocephala* and *S. hortensis*, Blue Tit *Parus caeruleus*, *P. major*, *Lanius senator*, Azure-winged Magpie *Cyanopica cyana* and *Fringilla coelebs*. In the most open areas *Galerida cristata* and *Miliaria calandra* may occur.

Large numbers of small mammals (especially *Oryctolagus cuniculus*) and deer (red deer *Cervus elaphus* and roe deer *Capreolus capreolus*) also occur, and support high densities of carnivores, including lynx *Lynx pardinus*, wolf *Canis lupus*, genet *Genetta genetta*, red fox *Vulpes vulpes* and Egyptian mongoose *Herpestes ichneumon*, as well as numerous breeding birds of prey, including Spanish Imperial Eagle *Aquila adalberti*, Cinereous Vulture *Aegypius monachus*, *Elanus caeruleus*, Black Kite *Milvus migrans*, *Milvus milvus*, Booted Eagle *Hieraaetus pennatus* and Hobby *Falco subbuteo*.

The habitat is also very important for some birds in winter, supporting 6–7 million *Columba palumbus* and large numbers of passerines, particularly *Erithacus rubecula*, *Fringilla coelebs*, *Anthus pratensis*, Black Redstart *Phoenicurus ochruros* and *Turdus philomelos* (Díaz *et al.* 1997). Between 60,000 and 70,000 *Grus grus* also use the Iberian dehesas and montados between November and February (Alonso and Alonso 1990).

Elsewhere in Europe pastoral woodlands are

scarce and vary considerably in species composition. North-west European parklands and winter-grazed woodlands mainly consist of the oaks *Quercus robur* and *Q. petraea*, and beech *Fagus sylvatica*. Typical birds include *Buteo buteo*, Stock Dove *Columba oenas*, *C. palumbus*, *Strix aluco*, *Picus viridis*, Mistle Thrush *Turdus viscivorus* and *Sturnus vulgaris*.

In central Europe, pastoral woodlands, such as those in Hungary, can also be composed of oaks *Quercus robur* and *Q. petraea*, according to the altitude, in a mixed woodland and grassland. Bird species composition is similar to north-west European woods, but also includes *Ciconia ciconia*, *Coracias garrulus*, *Merops apiaster* and *Lanius collurio*.

Relationship with other habitats

The European landscape has been profoundly affected by human activities over the last 10,000 years. Today, large areas of truly natural habitat can only be found in parts of northern Europe and some mountainous regions. The spread and development of agriculture has been the primary force behind the reduction in the extent of natural habitats and their species. However, the creation of new semi-natural habitats such as heathland, moorland, Mediterranean shrubland, grassland and pastoral woodland, as well as entirely artificial habitats such as annually cultivated and perennial crops, may have led to a net increase in species richness in Europe overall (Hampicke 1978, Kornas 1983).

Rapid technological advances, economic development and political initiatives in the twentieth century have driven further agricultural expansion and massive intensification in Europe. Major losses of remaining natural and semi-natural habitats have resulted (Baldock 1990a).

In recent decades the spread of agriculture has mostly ceased, and overall farmland area has decreased in most European countries. In the current 15 EU countries, farmland declined by an average of 9.6% over the 30-year period 1960–1990 (CEC 1995d). Over the same period agricultural area declined in eastern Europe by a similar amount. Subsequently there has been even more widespread agricultural abandonment in the latter region (CEC 1995a). Despite this overall trend, agricultural expansion has continued locally, often at the expense of remaining natural and semi-natural habitats. Wetlands in particular have continued to be drained and converted to agriculture since the resulting farmland is often highly productive and grants for reclamation are often still available. Baldock (1990a) notes that while the scale of reclamation is now much reduced, there are still individual sites that are under threat, and documents continued losses in a number of countries.

In addition to causing the loss of habitats, the intensification of farming systems has led to widespread degradation of the remaining non-agricultural habitats. In particular, the ploughing of old grassland, application of high rates of inorganic fertilizer, and the frequent accidental pollution incidents from run-off from intensive stockyards, silage stores and manure pits, have all raised water-nutrient levels and caused widespread nutrient pollution of wetlands (Irving 1993). Nutrient pollution of terrestrial habitats, as a result of atmospheric deposition of nitrogen compounds derived from intensive stockyards in western Europe, is thought to affect vegetation (e.g. heathland) in a major way in parts of central and western Europe (Ellenberg *et al*. 1989, Bobbink *et al*. 1992, Tickle 1992). Other frequently detrimental effects of recent agricultural practices on surrounding habitats include pesticide drift (Moore 1983), and disturbance from machinery operations and aerial crop spraying. Excessive abstraction of water for agriculture is a problem in parts of Europe, particularly in the south. About 18% of agricultural land is irrigated in Italy, 16% in Portugal, 13% in Greece, and 11% in Spain (CEC 1995d). Such large demands for water have lowered water-tables in some areas and thereby have altered or even destroyed wetlands, as for example, in the Tablas de Daimiel National Park in Spain (SGOP 1983).

Agricultural expansion and intensification locally have also led to increased fragmentation of remaining natural and semi-natural habitats, such as forests, marshes and heathlands, with a variety of potentially detrimental effects (e.g. Wilcove *et al*. 1986, Merriam 1988, Bink *et al*. 1994, Kirby 1995, Opdam *et al*. 1995, Wiens 1995a).

Conversely, recent afforestation programmes have led to habitat loss and fragmentation of some important open semi-natural agricultural habitats such as upland grassland in the UK, Iberian steppic habitats (e.g. Tellería *et al*. 1994, Suárez *et al*. 1997a), and subalpine scrub and grassland (to protect from avalanches and excessive soil erosion) (Mayer 1994a).

Values, roles and land-uses

The primary use of agricultural and grassland habitats is, by definition, the production of food and other commodities through cultivation or the rearing of animals. A number of other uses are, however, also important, particularly in non-intensive farming systems.

Forestry

Cork, and therefore *Quercus suber* montados, are still a very important component of Portugal's economy, as were *Q. ilex* montados for charcoal production and timber in the past (Vieira and Eden 1995).

Hunting

As with other recreational activities, hunting is becoming more important as an economic factor in marginal agricultural regions of some European countries (Franz 1979). Hunting is a particularly important activity in the dehesas and montados of the Iberian peninsula, where game species have replaced domestic livestock in some areas (Campos 1992). Particularly important game species are *Alectoris rufa*, Turtle Dove *Streptopelia turtur* and *Coturnix coturnix*, and *Turdus* spp. and *Columba palumbus* in winter, as well as mammals such as *Oryctolagus cuniculus* and wild boar *Sus scrofa*. In more open areas of Iberia such as cereal steppe, *Alectoris rufa*, *Pluvialis apricaria* and hare *Lepus capensis* are important quarry.

On arable and lowland grassland, game shooting for rabbits, partridges, hares and Pheasant *Phasianus colchicus* can promote the preservation or creation of marginal features such as hedgerows, banks, woodlots, grassy strips and fallow fields. Similarly, wildfowl hunting may encourage the retention of winter flooding in river valleys, and therefore help to maintain wet grassland. However, potential problems are created by these land-uses, including misguided and usually illegal persecution by gamekeepers of predators, including priority birds of prey (e.g. Gamauf 1991), and lead pollution as a result of spent shot from wildfowling (see 'Inland Wetlands', p. 145). In dehesas, associated changes in land management result in increased shrub cover at the expense of open grassland, which benefits some species such as *Sylvia* warblers and *Turdus merula*, but is detrimental to those requiring open areas for foraging, such as raptors, shrikes and finches (Díaz *et al.* 1997).

Tourism and recreation

Tourism now has an important, Europe-wide impact on the rural economy (and thus, often, on agricultural and grassland habitats), through changing people's aspirations and employment opportunities. Coasts and mountains are most attractive landscapes for tourism and recreation. About 100 million people visit the Alps every year (Thaler 1994). In countries with such montane habitats (e.g. Austria, France, Switzerland, Italy), tourism is one of the most important contributors to national income and employment (e.g. Eurostat 1992).

Among tourist activities in mountains, winter sports have a prominent role because of the large scale of changes they cause at the landscape level

and the large demands on energy supply. This may lead to significant threats to montane grassland (see 'Threats to the habitat', p. 304). Elsewhere, significant transfer of employment from agriculture to tourism may lead to widespread land abandonment on a regional scale.

'Hard' developments (dams, windfarms, etc.)

Mountainous areas are often rich in water, so the growing demands on energy and water supplies lead to a great pressure on those resources. Dammed rivers in mountains provide hydroelectric power, and water for drinking and irrigation for the surrounding urban centres and lowlands. Windfarms are often placed in open agricultural and grassland habitats.

Political and socio-economic factors affecting the habitat

Agriculture in Europe

Most farms in Europe are small and independent businesses whose activities are influenced by many factors. In the countries of central and eastern Europe, the much larger state farms and collective farms were formerly influenced directly by national policies. However, land privatization is leading to policies that more closely resemble those in the rest of Europe, although it should be noted that farm structure has been slow to change, e.g. in the former East Germany in 1996, 20% of the farms still owned or utilized 80% of the agricultural area, indicating that huge farms have survived well. In the market economies of the EU and allied countries, farmers are influenced principally by the financial returns that come from the products of their farms.

In most of Europe, national or supra-national policies exist to influence the type and scale of production, through altering the financial environment by subsidies, protection and other supports. Additional factors which affect land-use decisions by farmers are market returns on crops, subsidies on crops and inputs, capital values of land and businesses, technical advice and educational background, national economic policies such as tax and interest rates, peer pressure and personal aspirations.

Reform of the EU Common Agricultural Policy

The Common Agricultural Policy (CAP) developed after the Second World War out of national policies designed to support agricultural markets while encouraging rationalization of farm structures (amalgamation, agricultural improvement, etc.). The CAP has worked principally by creating a 'single market' for produce grown within the EU by restricting imports, encouraging exports and providing payments as incentives for production (Dixon 1997, Robson 1997). To a lesser extent, aid has been provided for 'structural', farm-scale changes designed to promote production. The CAP impacts the market situation of most of Europe's farmers, particularly those growing arable crops and livestock but also fruit and vegetables, rice, tobacco and other 'minor' crops. In agricultural production terms, the CAP has been remarkably successful, although it is increasingly criticised on the grounds of economic efficiency, equitable trading, farm and rural development, and the environment.

As the European Union has been enlarged (from the original six to the current 15 countries) the CAP has been expanded and developed. It has also had its objectives and policy instruments amended by a process of incremental reform. Relatively minor reforms were made to the CAP during the late 1980s, principally by reducing the guarantees to farmers of supported prices and by introducing means of reducing supply, such as quotas on production (e.g. for milk) and on inputs such as land (e.g. for cereals). A number of concurrent crises, particularly over international trade, forced the European Council of Agriculture Ministers to adopt a package of more substantial reforms to the CAP in May 1992. Further controls on inputs of land were introduced for cereals, with a semi-compulsory set-aside being introduced in concert with a 29% reduction in prices for cereals. A package of measures was introduced accompanying these reforms to promote early retirement, forestry, and protection of the environment.

While not as radical as proposed by some, these reforms signal changed objectives for farming. The reforms were important because they considered the nature of support to farmers and not just its level. This raises important questions about the justification for making large transfers of money overtly from taxpayers to farmers. It is difficult to justify these payments on 'agricultural' grounds, given the reasons for change in the first place, and so the discussions have centred on the need to ensure that the countryside remains managed.

The 1992 CAP reforms left a legacy of new policies which are now being implemented, but there are some commodities which have been left unreformed. There are also questions over the effectiveness of the agreed reforms. These will precede and shape the longer-term agenda. Further reform is likely in:

- The wine, olive oil, fruit and vegetable, and livestock (including dairy) sectors.
- The Agri-environment Regulation (2078/92/EEC)

and other accompanying measures.

- Set-aside and arable-area aids.

The 1992 CAP reform will reduce subsidies by 20% in the period up until 2001. From 1996 to 2001 it is likely that subsidies in the cereals sector, in particular, will need to be reduced (or supply controls toughened). This will largely be an arithmetical decision about how to balance production, consumption and exports, although a political decision will have to be made on how production should be capped, if it needs to be.

A major review of the cereals regime will have to be started in 1997 to determine the effects on production of the reforms and to consider the likely response. This will be an opportunity to introduce a wider range of environmental options and schemes into arable policy, for example ways to promote more extensive and organic production methods as well as set-aside. If further set-aside is proposed, a considerable environmental input will be required to sell this to a sceptical public and to farmers.

Other sectors, particularly livestock, are likely to need policies to adjust production to the needs of markets to meet GATT requirements to the end of the century. It is possible that, even if export limits are exceeded, the EU will resist attempts to change cereals policy, save for minor adjustments to reference yields, prices, and set-aside percentages. Reform would then be delayed until the longer term (1998 onwards).

The Common Agricultural Policy in the long term

It is widely accepted that the CAP faces many issues which call into question its unchanged existence. It is worth listing these factors here.

Compatibility with GATT agreements on agriculture

While compatibility with the current 'Uruguay Round' agriculture agreement can be confirmed with some certainty, this only constrains the CAP until the year 2001. The future beyond then is unclear until the agenda for the next GATT round is set and negotiated, although it can be expected to have a similar impact to that of the Uruguay Round.

Budget and other economic costs

The CAP has been fairly resilient to criticism of its massive costs to taxpayers, to the EU budget, and to consumers and the food industry. While some member states and critics still argue that the CAP misallocates resources, that it is economically inefficient, and that it damages the interests of (poor) consumers, these objections are unlikely to be a major force for reform, particularly as the CAP has been reduced to 'only' 50% of the EU budget. Continuing economic difficulties, for instance in Germany, may lead to more overt pressure on CAP budgets, but the largest influence will be from the public concerned at the high costs of protecting farmers.

A changing rural society

Shifting demographic patterns, with farmers leaving the land, people moving into the countryside for recreation and to non-agricultural jobs, are causing a shift in policy. In some areas this will force the pace for rural development and in many it will raise demand for high environmental standards. Both of these factors question the traditional objectives and emphasis of the CAP in financially supporting a limited range of crops and farmers, but in themselves these changes are unlikely to bring about further substantial reform.

Enlargement of the EU to include central and eastern Europe

It is difficult to imagine that a high-cost, administratively complex CAP of subsidies and market regulation will survive intact into a European Union of 20 or more member states. However, the notion that the CAP will be somehow abolished and agriculture policy 'renationalized' is not realistic, whatever its merits. A substantial debate is developing about how the EU might address apparent incompatibility between the CAP and any enlargement policies. As the factor most likely to influence the future of the CAP, this is discussed in some detail below.

Beyond the immediate concerns of political and economic transition in the economies of central and eastern Europe, the future trading and political relationships between these countries and the EU will have far-reaching consequences for rural policy, both in those countries and the remainder of the EU.

All previous enlargements of the EU have had considerable impacts on the CAP, particularly the enlargements embracing the UK, the Mediterranean countries and most recently Austria, Finland and Sweden. The political, agricultural and land-use consequences of future enlargement are difficult to predict. Much greater analysis of this is necessary. However, studies of the accession of Spain indicate that opening its markets had a marked effect on its land-use, farming and the environment (Egdell 1993). The countries of central and eastern Europe most likely to join the EU (Poland, Hungary, Czech Republic, Slovakia, Bulgaria, Romania and possibly Slovenia and the Baltic states) have relatively small economies in comparison with the EU, although their agricultural sectors are large. This is particularly true of Poland whose accession is likely to prove most problematic.

The strategic options for the CAP with regard to enlargement of the Union can be characterized as lying between extremes of 'protectionism' or 'free trade'. The former extreme would be the full enlargement of the CAP to include the central and east European countries. At the other extreme, reports commissioned by Directorate-General I (for External Affairs) and Directorate-General II (for Budgets) of the European Commission calculate high financial costs for this strategy and conclude that it would be impractical, costly and damaging to the agriculture of both the EU and the central and east European countries. Instead, the reports propose a free-trade solution of reducing, to world-market levels, all CAP prices, and the abolition of all border controls, export subsidies, and supply controls (RSPB 1995).

Rural and regional policies/Structural Funds

The EU Structural Funds have supported infrastructural projects to increase agricultural efficiency, with particular reference to irrigation, water supply and rural development (dam-building, amalgamation of farms, etc.). The European Commission plans to reform the Funds, and there are serious doubts as to whether the Funds fulfil their own environmental and economic objectives consistently throughout the EU. In particular, a recent study by BirdLife International (1995b) made the following overall reform proposals regarding projects supported by the Funds:

- Ensure rigorous enforcement of EU environmental legislation.
- Establish a Commission Task Force on Regional/Rural Development and Biodiversity to coordinate a new strategic approach.
- Develop creative new ways of integrating rural development and biodiversity conservation.
- Revise the Structural Fund Regulations to allow explicitly for nature conservation.
- Investigate whether the funds are generating real economic benefits, and encourage a much more rigorous approach to economic appraisal in the Funds.
- Consider introducing loans instead of grants for certain development projects.

More information is given in BirdLife International (1995b).

Central and eastern Europe

The current political and economic situation of these transitional economies is complex. The legacy of formerly state-controlled economies, the rate of economic transformation, and the prospects for closer economic ties with western Europe all differ between countries, but strongly influence the agricultural and environmental situation.

The rapid collapse of formerly state-controlled agriculture, hyper-inflation, the collapse of markets and a lack of capital and investor confidence have all led to the wide-scale abandonment or extensification of many formerly intensively farmed areas, and also to the abandonment of extensively farmed agriculture in some areas, such as north-east Poland (CEC 1995a).

In Russia, there is now widespread poverty and a continuing lack of investment in the agricultural sector, where reversion of abandoned arable land to woodland is currently proceeding unchecked in some areas. In this context, widespread adoption of intensive farming practices is unlikely in the short term.

Elsewhere in eastern and central Europe, however, aid (especially from the EU) and some western capital is now being invested in agricultural development programmes, especially in those countries most likely to join the EU within the next decade. National legislation on water, land-use planning, forestry, pesticides and agriculture is largely being revised in these countries so as to harmonize with EU legislation. Communal and state-owned land is being privatized in nearly all countries in eastern Europe (with the notable exception of Russia), military land (often having a particularly high wildlife value) is being privatized or made communal, and restrictions on use of capital are being reduced. The ensuing land-use changes are, locally, having a major impact on the environment.

Overall, however, the general economic stagnation in central and eastern Europe continues to prevent widespread intensification of land-use. With potential EU membership, this situation could rapidly change as it did in countries such as Spain and Portugal when they became part of the Single Market and became eligible for structural assistance (the Structural Funds). BirdLife International with WWF and IUCN have begun a series of projects examining this issue, including the production of an 'Action plan to 2010 for central and eastern Europe' (Anon. 1995a).

Marginal habitats

Pastoral woodlands are common lands or owned by the Crown in Britain (Harding and Rose 1986), whereas in Spain and Portugal dehesa farms are generally large to very large private estates (average c.5 km²) which employ salaried staff (Campos 1993). Such private farms were economically profitable until the end of the 1950s, when a number of factors including rising wages, falling prices of pig-fat prod-

ucts, firewood and charcoal, and an epidemic of African swine fever, caused a major crisis in the dehesa economy. Currently, most dehesa farms have negative or only slightly positive net-profit margins, so that their continued maintenance appears to depend critically on the internalization by the market of their environmental values (Campos 1994, Díaz *et al.* 1997).

Some intensification of farming in steppic habitats has been reported in the Volgograd District of Russia, but most recent reports speak of widespread land abandonment being more common in central and eastern Europe. For example, in Hungary the proportion of arable land left unsown was three times larger during 1990–1995, on average, than in the period 1986–1990 (Central Statistical Office 1995). There has also been a widespread reduction in livestock and in extensive pastoral farming due to a collapse in demand for sheep and cattle in these countries (CEC 1995e).

PRIORITY BIRDS

In total, 173 species are priority species in one or more of the seven agricultural habitats (as listed in Table 6 and summarized in Tables 7 and 8). Arable and agriculturally improved grassland hold 81 priority species, the largest number among the habitats covered in this chapter. Steppic habitats have 16 fewer species, but they have more Priority A and B species and an almost equal number of Priority C species. This is largely because a high proportion of this habitat's species have an Unfavourable Conservation Status: 83% compared to 64% of these in arable and improved grassland. Also, 43% of species have the majority of their European populations concentrated in steppic habitats compared to only 27% of those of arable and improved grassland. This is particularly important with regard to the Priority A species. The seven Priority A species of steppic habitats are all globally threatened (Collar *et al.* 1994), and all but one has the majority of its European population within the habitat at some point in its annual cycle. In contrast, of the six Priority A species of arable and improved grassland, three are concentrated in the habitat and only one of these is globally threatened. Thus, both habitats are clearly of very high conservation importance in terms of their number of priority species.

The other agricultural habitats hold considerably fewer priority species. This is in part probably due to their smaller extent, as species richness generally decreases with decreasing area (Schoener 1976, 1986; Wiens 1989a). In addition, their small extent reduces the likelihood that the habitat will hold a high proportion of the European population of any species.

Table 6. The priority bird species of agricultural and grassland habitats in Europe.

This prioritization identifies Species of European Conservation Concern (SPECs) (see p. 17) for which the habitat is of particular importance for survival, as well as other species which depend very strongly on the habitat. It focuses on the SPEC category (which takes into account global importance and overall conservation status of the European population) and the percentage of the European population (breeding population, unless otherwise stated) that uses the habitat at any stage of the annual cycle. It does not take into account the threats to the species within the habitat, and therefore should not be considered as an indication of priorities for action. Indications of priorities for action for each species are given in Appendix 4 (p.401), where the severity of habitat-specific threats is taken into account.

	SPEC category	European Threat Status	Habitat importance
ARABLE AND IMPROVED GRASSLAND			
Priority A			
Red-breasted Goose *Branta ruficollis*	1	L ^w	■
Imperial Eagle *Aquila heliaca*	1	E	●
Red-legged Partridge *Alectoris rufa*	2	V	■
Corncrake *Crex crex*	1	V	●
Great Bustard *Otis tarda*	1	D	●
Ortolan Bunting *Emberiza hortulana*	2	(V)	■
Priority B			
White Stork *Ciconia ciconia*	2	V	●
Barnacle Goose *Branta leucopsis*	4/2 ^w	L ^w	●
Levant Sparrowhawk *Accipiter brevipes*	2	R	●
Lesser Kestrel *Falco naumanni*	1	(V)	•
Partridge *Perdix perdix*	3	V	■
Quail *Coturnix coturnix*	3	V	■
Little Bustard *Tetrax tetrax*	2	V	●
Common Gull *Larus canus*	2	D	●
Turtle Dove *Streptopelia turtur*	3	D	■
Barn Owl *Tyto alba*	3	D	■
Scops Owl *Otus scops*	2	(D)	●
Little Owl *Athene noctua*	3	D	■
Green Woodpecker *Picus viridis*	2	D	●
Crested Lark *Galerida cristata*	3	(D)	■
Skylark *Alauda arvensis*	3	V	■
Swallow *Hirundo rustica*	3	D	●
Black-eared Wheatear *Oenanthe hispanica*	2	V	●
Red-backed Shrike *Lanius collurio*	3	(D)	■
Lesser Grey Shrike *Lanius minor*	2	(D)	●
Black-headed Bunting *Emberiza melanocephala*	2	(V)	●
Priority C			
Black-winged Kite *Elanus caeruleus*	3	V	●
Black Kite *Milvus migrans*	3	V	●
Hen Harrier *Circus cyaneus*	3	V	●
Montagu's Harrier *Circus pygargus*	4	S	■
Long-legged Buzzard *Buteo rufinus*	3	(E)	●
Lesser Spotted Eagle *Aquila pomarina*	3	R	●
Kestrel *Falco tinnunculus*	3	D	●
Red-footed Falcon *Falco vespertinus*	3	V	●
Saker *Falco cherrug*	3	E	●
Crane *Grus grus*	3	V	●
Stone Curlew *Burhinus oedicnemus*	3	V	●
Collared Pratincole *Glareola pratincola*	3	E	●
Black-winged Pratincole *Glareola nordmanni*	3	R	●
Woodcock *Scolopax rusticola*	3 ^w	V ^w	●
Woodpigeon *Columba palumbus*	4	S	■
Bee-eater *Merops apiaster*	3	D	●
Roller *Coracias garrulus*	2	(D)	●
Calandra Lark *Melanocorypha calandra*	3	(D)	●
Woodlark *Lullula arborea*	2	V	•
Fieldfare *Turdus pilaris*	4 ^w	S	■
Redwing *Turdus iliacus*	4 ^w	S	■
Woodchat Shrike *Lanius senator*	2	V	●
Chough *Pyrrhocorax pyrrhocorax*	3	V	●
Corn Bunting *Miliaria calandra*	4	(S)	■

cont.

Table 6. (cont.)

	SPEC category	European Threat Status	Habitat importance
Priority D			
Cattle Egret *Bubulcus ibis*	—	S	■
Bewick's Swan *Cygnus columbianus*	3 W	L W	•
Pink-footed Goose *Anser brachyrhynchus*	4	S	•
Brent Goose *Branta bernicla*	3	V	•
Red Kite *Milvus milvus*	4	S	•
Short-toed Eagle *Circaetus gallicus*	3	R	•
Golden Plover *Pluvialis apricaria*	4	S	•
Lapwing *Vanellus vanellus*	—	(S)	■
Stock Dove *Columba oenas*	4	S	•
Tawny Owl *Strix aluco*	4	S	•
Short-eared Owl *Asio flammeus*	3	(V)	•
Tawny Pipit *Anthus campestris*	3	V	•
Meadow Pipit *Anthus pratensis*	4	S	•
Whinchat *Saxicola rubetra*	4	S	•
Blackbird *Turdus merula*	4	S	•
Song Thrush *Turdus philomelos*	4	S	•
Mistle Thrush *Turdus viscivorus*	4	S	•
Grasshopper Warbler *Locustella naevia*	4	S	•
Olivaceous Warbler *Hippolais pallida*	3	(V)	•
Barred Warbler *Sylvia nisoria*	4	(S)	•
Whitethroat *Sylvia communis*	4	S	•
Great Grey Shrike *Lanius excubitor*	3	D	•
Jackdaw *Corvus monedula*	4	(S)	•
Rook *Corvus frugilegus*	—	S	■
Starling *Sturnus vulgaris*	—	S	■
Spotless Starling *Sturnus unicolor*	4	S	■
Tree Sparrow *Passer montanus*	—	S	■
Greenfinch *Carduelis chloris*	4	S	•
Linnet *Carduelis cannabina*	4	S	•
Yellowhammer *Emberiza citrinella*	4	(S)	•
Cirl Bunting *Emberiza cirlus*	4	(S)	•
STEPPIC HABITATS			
Priority A			
Lesser White-fronted Goose *Anser erythropus*	1	V	■
Imperial Eagle *Aquila heliaca*	1	E	■
Lesser Kestrel *Falco naumanni*	1	(V)	■
Little Bustard *Tetrax tetrax*	2	V	■
Great Bustard *Otis tarda*	1	D	■
Sociable Plover *Chettusia gregaria*	1	E	■
Slender-billed Curlew *Numenius tenuirostris*	1	—	•
Priority B			
White Stork *Ciconia ciconia*	2	V	•
Pallid Harrier *Circus macrourus*	3	E	■
Long-legged Buzzard *Buteo rufinus*	3	(E)	■
Steppe Eagle *Aquila nipalensis*	3	V	■
Lanner *Falco biarmicus*	3	(E)	■
Saker *Falco cherrug*	3	E	■
Red-legged Partridge *Alectoris rufa*	2	V	•
Houbara Bustard *Chlamydotis undulata*	3	(E)	■
Stone Curlew *Burhinus oedicnemus*	3	V	■
Cream-coloured Courser *Cursorius cursor*	3	V	■
Black-winged Pratincole *Glareola nordmanni*	3	R	■
Greater Sand Plover *Charadrius leschenaultii*	3	(E)	■
Caspian Plover *Charadrius asiaticus*	3	(V)	■
Black-bellied Sandgrouse *Pterocles orientalis*	3	V	■
Pin-tailed Sandgrouse *Pterocles alchata*	3	E	■
Roller *Coracias garrulus*	2	(D)	•
Dupont's Lark *Chersophilus duponti*	3	V	■
Black Lark *Melanocorypha yeltoniensis*	3	(V)	■
Short-toed Lark *Calandrella brachydactyla*	3	V	■
Lesser Short-toed Lark *Calandrella rufescens*	3	V	■

cont.

Table 6. (cont.)

	SPEC category	European Threat Status	Habitat importance
Black-eared Wheatear *Oenanthe hispanica*	2	V	●
Lesser Grey Shrike *Lanius minor*	2	(D)	●
Trumpeter Finch *Bucanetes githagineus*	3	R	■
Priority C			
Ruddy Shelduck *Tadorna ferruginea*	3	V	●
Egyptian Vulture *Neophron percnopterus*	3	E	●
Griffon Vulture *Gyps fulvus*	3	R	●
Short-toed Eagle *Circaetus gallicus*	3	R	●
Hen Harrier *Circus cyaneus*	3	V	●
Kestrel *Falco tinnunculus*	3	D	●
Red-footed Falcon *Falco vespertinus*	3	V	●
Chukar *Alectoris chukar*	3	V	●
Partridge *Perdix perdix*	3	V	●
Quail *Coturnix coturnix*	3	V	●
Crane *Grus grus*	3	V	●
Collared Pratincole *Glareola pratincola*	3	E	●
Little Owl *Athene noctua*	3	D	●
Short-eared Owl *Asio flammeus*	3	(V)	●
Bee-eater *Merops apiaster*	3	D	●
Calandra Lark *Melanocorypha calandra*	3	(D)	●
White-winged Lark *Melanocorypha leucoptera*	4 W	(S)	■
Crested Lark *Galerida cristata*	3	(D)	●
Thekla Lark *Galerida theklae*	3	V	●
Skylark *Alauda arvensis*	3	V	●
Tawny Pipit *Anthus campestris*	3	V	●
Great Grey Shrike *Lanius excubitor*	3	D	●
Black-headed Bunting *Emberiza melanocephala*	2	(V)	•
Priority D			
Red Kite *Milvus milvus*	4	S	●
Montagu's Harrier *Circus pygargus*	4	S	●
Demoiselle Crane *Anthropoides virgo*	—	S	■
Golden Plover *Pluvialis apricaria*	4	S	●
Eagle Owl *Bubo bubo*	3	V	•
Meadow Pipit *Anthus pratensis*	4	S	●
Desert Wheatear *Oenanthe deserti*	—	(S)	■
Olivaceous Warbler *Hippolais pallida*	3	(V)	•
Jackdaw *Corvus monedula*	4	(S)	●
Spotless Starling *Sturnus unicolor*	4	S	●
Rose-coloured Starling *Sturnus roseus*	—	(S)	■
Corn Bunting *Miliaria calandra*	4	(S)	●

MONTANE GRASSLAND

Priority A

Caucasian Black Grouse *Tetrao mlokosiewiczi*	2	Ins	■
Priority B			
Lammergeier *Gypaetus barbatus*	3	E	■
Caspian Snowcock *Tetraogallus caspius*	3	Ins	■
Rock Partridge *Alectoris graeca*	2	(V)	●
Corncrake *Crex crex*	1	V	●
Radde's Accentor *Prunella ocularis*	3	(V)	■
Great Rosefinch *Carpodacus rubicilla*	3	(E)	■
Priority C			
Egyptian Vulture *Neophron percnopterus*	3	E	●
Griffon Vulture *Gyps fulvus*	3	R	●
Golden Eagle *Aquila chrysaetos*	3	R	●
Peregrine *Falco peregrinus*	3	R	●
Caucasian Snowcock *Tetraogallus caucasicus*	4	S	■
Woodlark *Lullula arborea*	2	V	•
Rock Thrush *Monticola saxatilis*	3	(D)	●
Chough *Pyrrhocorax pyrrhocorax*	3	V	●
Rock Bunting *Emberiza cia*	3	V	●
Ortolan Bunting *Emberiza hortulana*	2	(V)	•

cont.

Table 6. (cont.)

	SPEC category	European Threat Status	Habitat importance
Priority D			
Cinereous Vulture *Aegypius monachus*	3	V	●
Kestrel *Falco tinnunculus*	3	D	●
Black Grouse *Tetrao tetrix*	3	V	●
Quail *Coturnix coturnix*	3	V	●
Eagle Owl *Bubo bubo*	3	V	●
Skylark *Alauda arvensis*	3	V	●
Tawny Pipit *Anthus campestris*	3	V	●
Water Pipit *Anthus spinoletta*	—	S	■
Alpine Accentor *Prunella collaris*	—	S	■
Stonechat *Saxicola torquata*	3	(D)	●
Blue Rock Thrush *Monticola solitarius*	3	(V)	●
Alpine Chough *Pyrrhocorax graculus*	—	(S)	■
Snowfinch *Montifringilla nivalis*	—	(S)	■
Red-fronted Serin *Serinus pusillus*	—	(S)	■
Citril Finch *Serinus citrinella*	4	S	●
Crimson-winged Finch *Rhodopechys sanguinea*	—	(S)	■
WET GRASSLAND			
Priority A			
Lesser White-fronted Goose *Anser erythropus*	1	V	●
Greater Spotted Eagle *Aquila clanga*	1	E	●
Corncrake *Crex crex*	1	V	●
Great Snipe *Gallinago media*	2	(V)	■
Black-tailed Godwit *Limosa limosa*	2	V	■
Aquatic Warbler *Acrocephalus paludicola*	1	E	■
Priority B			
White Stork *Ciconia ciconia*	2	V	●
Imperial Eagle *Aquila heliaca*	1	E	●
Redshank *Tringa totanus*	2	D	●
Common Gull *Larus canus*	2	D	●
Priority C			
Bewick's Swan *Cygnus columbianus*	3 W	L W	●
Brent Goose *Branta bernicla*	3	V	●
Garganey *Anas querquedula*	3	V	●
Hen Harrier *Circus cyaneus*	3	V	●
Lesser Spotted Eagle *Aquila pomarina*	3	R	●
Curlew *Numenius arquata*	3 W	D W	●
Barn Owl *Tyto alba*	3	D	●
Short-eared Owl *Asio flammeus*	3	(V)	●
Priority D			
Purple Heron *Ardea purpurea*	3	V	●
Whooper Swan *Cygnus cygnus*	4 W	S	●
Pink-footed Goose *Anser brachyrhynchus*	4	S	●
Pintail *Anas acuta*	3	V	●
Kestrel *Falco tinnunculus*	3	D	●
Quail *Coturnix coturnix*	3	V	●
Spotted Crake *Porzana porzana*	4	S	●
Crane *Grus grus*	3	V	●
Dunlin *Calidris alpina*	3 W	V W	●
Ruff *Philomachus pugnax*	4	(S)	●
Black Tern *Chlidonias niger*	3	D	●
Little Owl *Athene noctua*	3	D	●
Skylark *Alauda arvensis*	3	V	●
Swallow *Hirundo rustica*	3	D	●
Meadow Pipit *Anthus pratensis*	4	S	●
Yellow Wagtail *Motacilla flava*	—	S	■
Whinchat *Saxicola rubetra*	4	S	●
Stonechat *Saxicola torquata*	3	(D)	●
Grasshopper Warbler *Locustella naevia*	4	S	●

cont.

Table 6. (cont.)

	SPEC category	European Threat Status	Habitat importance
RICE CULTIVATION			
Priority A			
No species in this category			
Priority B			
Night Heron *Nycticorax nycticorax*	3	D	■
Black-tailed Godwit *Limosa limosa*	2	V	●
Priority C			
Spoonbill *Platalea leucorodia*	2	E	●
Redshank *Tringa totanus*	2	D	●
Priority D			
Little Bittern *Ixobrychus minutus*	3	(V)	●
Squacco Heron *Ardeola ralloides*	3	V	●
Purple Heron *Ardea purpurea*	3	V	●
Glossy Ibis *Plegadis falcinellus*	3	D	●
Pintail *Anas acuta*	3	V	●
Garganey *Anas querquedula*	3	V	●
Red-crested Pochard *Netta rufina*	3	D	●
Black Kite *Milvus migrans*	3	V	●
Red-footed Falcon *Falco vespertinus*	3	V	●
Baillon's Crake *Porzana pusilla*	3	R	●
Collared Pratincole *Glareola pratincola*	3	E	●
Black-winged Pratincole *Glareola nordmanni*	3	R	●
Dunlin *Calidris alpina*	3 ᵂ	V ᵂ	●
Gull-billed Tern *Gelochelidon nilotica*	3	(E)	●
Black Tern *Chlidonias niger*	3	D	●
Short-eared Owl *Asio flammeus*	3	(V)	●
Swallow *Hirundo rustica*	3	D	●
PERENNIAL CROPS			
Priority A			
No species in this category			
Priority B			
Scops Owl *Otus scops*	2	(D)	●
Roller *Coracias garrulus*	2	(D)	●
Green Woodpecker *Picus viridis*	2	D	●
Woodlark *Lullula arborea*	2	V	●
Olive-tree Warbler *Hippolais olivetorum*	2	(R)	●
Woodchat Shrike *Lanius senator*	2	V	●
Masked Shrike *Lanius nubicus*	2	(V)	●
Ortolan Bunting *Emberiza hortulana*	2	(V)	●
Priority C			
Levant Sparrowhawk *Accipiter brevipes*	2	R	●
Red-legged Partridge *Alectoris rufa*	2	V	●
Bee-eater *Merops apiaster*	3	D	●
Thekla Lark *Galerida theklae*	3	V	●
Redstart *Phoenicurus phoenicurus*	2	V	●
Black-eared Wheatear *Oenanthe hispanica*	2	V	●
Olivaceous Warbler *Hippolais pallida*	3	(V)	●
Orphean Warbler *Sylvia hortensis*	3	V	●
Priority D			
Turtle Dove *Streptopelia turtur*	3	D	●
Barn Owl *Tyto alba*	3	D	●
Little Owl *Athene noctua*	3	D	●
Wryneck *Jynx torquilla*	3	D	●
Syrian Woodpecker *Dendrocopos syriacus*	4	(S)	●
Robin *Erithacus rubecula*	4	S	●
Blackbird *Turdus merula*	4	S	●
Song Thrush *Turdus philomelos*	4	S	●
Redwing *Turdus iliacus*	4 ᵂ	S	●
Mistle Thrush *Turdus viscivorus*	4	S	●
Sardinian Warbler *Sylvia melanocephala*	4	S	●
Blackcap *Sylvia atricapilla*	4	S	●

cont.

Table 6. (cont.)

	SPEC category	European Threat Status	Habitat importance
Spotted Flycatcher *Muscicapa striata*	3	D	•
Sombre Tit *Parus lugubris*	4	(S)	●
Short-toed Treecreeper *Certhia brachydactyla*	4	S	●
Red-backed Shrike *Lanius collurio*	3	(D)	•
Spotless Starling *Sturnus unicolor*	4	S	●
Chaffinch *Fringilla coelebs*	4	S	●
Serin *Serinus serinus*	4	S	●
Canary *Serinus canaria*	4	S	●
Greenfinch *Carduelis chloris*	4	S	●
PASTORAL WOODLAND			
Priority A			
Spanish Imperial Eagle *Aquila adalberti*	1	E	●
Woodchat Shrike *Lanius senator*	2	V	■
Priority B			
White Stork *Ciconia ciconia*	2	V	●
Scops Owl *Otus scops*	2	(D)	●
Roller *Coracias garrulus*	2	(D)	●
Green Woodpecker *Picus viridis*	2	D	●
Woodlark *Lullula arborea*	2	V	●
Priority C			
Black Stork *Ciconia nigra*	3	R	●
Black-winged Kite *Elanus caeruleus*	3	V	●
Black Kite *Milvus migrans*	3	V	●
Cinereous Vulture *Aegypius monachus*	3	V	●
Short-toed Eagle *Circaetus gallicus*	3	R	●
Booted Eagle *Hieraaetus pennatus*	3	R	●
Crane *Grus grus*	3	V	●
Little Owl *Athene noctua*	3	D	●
Bee-eater *Merops apiaster*	3	D	●
Thekla Lark *Galerida theklae*	3	V	●
Redstart *Phoenicurus phoenicurus*	2	V	•
Orphean Warbler *Sylvia hortensis*	3	V	●
Priority D			
Red Kite *Milvus milvus*	4	S	●
Kestrel *Falco tinnunculus*	3	D	•
Stock Dove *Columba oenas*	4	S	●
Woodpigeon *Columba palumbus*	4	S	●
Turtle Dove *Streptopelia turtur*	3	D	•
Barn Owl *Tyto alba*	3	D	•
Robin *Erithacus rubecula*	4	S	●
Blackbird *Turdus merula*	4	S	●
Song Thrush *Turdus philomelos*	4	S	●
Mistle Thrush *Turdus viscivorus*	4	S	●
Blackcap *Sylvia atricapilla*	4	S	●
Spotted Flycatcher *Muscicapa striata*	3	D	•
Short-toed Treecreeper *Certhia brachydactyla*	4	S	●
Jackdaw *Corvus monedula*	4	(S)	●
Spotless Starling *Sturnus unicolor*	4	S	●
Chaffinch *Fringilla coelebs*	4	S	●
Serin *Serinus serinus*	4	S	●
Greenfinch *Carduelis chloris*	4	S	●

SPEC category and **European Threat Status:** see Box 2 (p. 17) for definitions.

Habitat importance for each species is assessed in terms of the maximum percentage of the European population (breeding population unless otherwise stated) that uses the habitat at any stage of the annual cycle:

  ■ >75% ● 10–75% • <10%

[w] Assessment relates to the European wintering population.

Table 7. Numbers of priority bird species in agricultural and grassland habitats in Europe, according to their SPEC category (see p. 17) and their dependence on the habitat. Priority status of each species group is indicated by superscripts (see 'Introduction', p. 19, for definition). 'Habitat importance' is assessed in terms of the maximum percentage of each species' European population that uses the habitat at any stage of the annual cycle.

Habitat importance	SPEC category					Total no. of species
	1	2	3	4	Non-SPEC	
Arable and improved grassland						
<10%	1 [B]	3 [C]	7 [D]	—	—	11
10–75%	3 [A]	10 [B]	16 [C]	19 [D]	—	48
>75%	1 [A]	2 [A]	9 [B]	5 [C]	5 [D]	22
Total	5	15	32	24	5	81
Steppic habitats						
<10%	0 [B]	1 [C]	2 [D]	—	—	3
10–75%	1 [A]	5 [B]	21 [C]	7 [D]	—	34
>75%	5 [A]	1 [A]	18 [B]	1 [C]	3 [D]	28
Total	6	7	41	8	3	65
Montane grasslands						
<10%	1 [B]	2 [C]	9 [D]	—	—	12
10–75%	0 [A]	1 [B]	7 [C]	1 [D]	—	9
>75%	0 [A]	1 [A]	4 [B]	1 [C]	6 [D]	12
Total	1	4	20	2	6	33
Wet grasslands						
<10%	1 [B]	0 [C]	11 [D]	—	—	12
10–75%	3 [A]	3 [B]	8 [C]	7 [D]	—	21
>75%	1 [A]	2 [A]	0 [B]	0 [C]	1 [D]	4
Total	5	5	19	7	1	37
Rice cultivation						
<10%	0 [B]	2 [C]	17 [D]	—	—	19
10–75%	0 [A]	1 [B]	0 [C]	0 [D]	—	1
>75%	0 [A]	0 [A]	1 [B]	0 [C]	0 [D]	1
Total	0	3	18	0	0	21
Perennial crops						
<10%	0 [B]	4 [C]	6 [D]	—	—	10
10–75%	0 [A]	8 [B]	4 [C]	15 [D]	—	27
>75%	0 [A]	0 [A]	0 [B]	0 [C]	0 [D]	0
Total	0	12	10	15	0	37
Pastoral woodland						
<10%	0 [B]	1 [C]	4 [D]	—	—	5
10–75%	1 [A]	5 [B]	11 [C]	14 [D]	—	31
>75%	0 [A]	1 [A]	0 [B]	0 [C]	0 [D]	1
Total	1	7	15	14	0	37

Wet grassland, perennial crops and pastoral woodland all support 37 priority species. Six wet-grassland species are Priority A, including four globally threatened species, one of which, *Acrocephalus paludicola*, is highly dependent on the habitat (Heredia 1996c). Pastoral woodland holds two Priority A species, including the globally threatened *Aquila adalberti* (which is also endemic to Iberia), while perennial crops do not hold Priority A species or globally threatened species.

Montane grassland holds 33 priority species, of which a fairly high proportion (36%) are highly dependent on this habitat-type in Europe, in some cases having specialized habitat requirements (see 'Habitat needs of priority birds', below). Rice culti-

vations hold the fewest priority species (21), all of which, however, have an Unfavourable Conservation Status in Europe (Table 7). In particular, most of the European population of *Nycticorax nycticorax* is dependent on this habitat for foraging.

The overall diversity of priority breeding species in individual countries, in each agricultural habitat, is indicated in Appendix 2 (p. 344). As with many other habitat-types, the highest breeding diversity of priority birds of agricultural and grassland habitats (Figure 5) tends to be in the larger, more southern countries of Europe, but with less of an eastern bias compared to some other major habitat-types (e.g. lowland temperate forest). Countries with the highest diversity of breeding priority species usually have a relatively large proportion of the habitat extent within Europe, and/or habitat of relatively high ecological quality compared to other countries in Europe. It should be borne in mind, however, that many habitats are as important, or even more important, for priority birds outside the breeding season, during migration and in winter.

Russia, Ukraine, Romania, France, Spain and Turkey all have breeding populations of 75% or more of the priority species of arable and improved grassland, showing a broad east–west spread. The individual richness of country avifaunas is less for steppic habitats, where only Russia holds more than 75% of breeding priority species, while Turkey, Ukraine and Spain hold 65% or more.

Turkey holds significantly more breeding priority species of montane grassland than any other country in Europe (85%), whereas France, Italy, Spain, Greece and Albania all support 70% or more. The breeding diversity of priority species of wet grassland is highest in north temperate countries such as Russia, Poland, Latvia, Ukraine, Belarus, Germany and Lithuania.

France, Spain and Turkey hold the highest diversity of breeding priority species of rice cultivation in Europe. There is a strong south-east bias in the major countries for perennial crops, with Bulgaria, Greece, Turkey, Albania and Croatia all holding more than 80% of the breeding priority species. Finally, there is a clear south-west bias for pastoral woodland, for which Spain supports 100% of the breeding priority species, with France and Portugal both holding more than 90%.

HABITAT NEEDS OF PRIORITY SPECIES

General patterns
The particular habitat requirements of priority species of agricultural habitats are summarized in Appendix 3 (p. 354) for species of arable farmland, the various grassland habitats, and pastoral woodland.

Tables have not been produced for priority species of rice-fields and perennial crops, but key requirements are covered here and in the specific habitat accounts later in this section. Although many species are covered by this chapter, with a vast array of differing requirements, some general patterns are still apparent, especially with regard to habitat structure and food resources.

Firstly, in broad terms, a high proportion of priority species favour open habitats. Of the 130 arable and grassland species where a preference for a degree of openness is apparent, 63% occur in highly open landscapes, 55% in moderately open and only 16% in enclosed landscapes (some species occur in more than one habitat). This is probably because most agricultural habitats are fairly open and most of the species that have adapted to them originated from similarly open and treeless habitats such as natural steppe, marshes and mires, and exposed montane habitats. Such species are often ground-breeders and therefore susceptible to predators. Woodland, trees and scrub often provide breeding opportunities for predators such as *Corvus corvus*, Sparrowhawk *Accipiter nisus* and *Vulpes vulpes*. Trees, etc. also provide cover for approaching predators and can hinder escape by large, less agile species such as *Otis tarda*. Thus habitats near to such cover may be avoided by many species. Bird families or groups that are particularly adapted to ground-feeding or -nesting, such as geese, bustards (Otididae), waders, sandgrouse (Pteroclididae) and larks (Alaudidae), show particularly strong avoidance of such very tall vegetation cover. Also, some predators that require trees for nesting, such as *Falco tinnunculus* and *Buteo rufinus*, prefer isolated trees in open landscapes.

Farmland species also include those of woodland and scrub habitats, such as tits *Parus* and warblers (Sylviidae); however, in general, a relatively low proportion of such birds are priority species. This is primarily because few have an Unfavourable Conservation Status in Europe and because most are not

Table 8. Numbers of priority bird species in each agricultural and grassland habitat in Europe, in Europe, listed by priority status (see 'Introduction', p. 19). Percentages are proportions of total numbers of priority species in each habitat.

	Priority status				Total no. of priority species
	A	B	C	D	
Arable and improved grassland	6 (7%)	20 (25%)	24 (30%)	31 (38%)	81
Steppic habitats	7 (11%)	24 (37%)	22 (34%)	12 (18%)	65
Montane grasslands	1 (3%)	6 (18%)	10 (30%)	16 (48%)	33
Wet grasslands	6 (16%)	4 (11%)	9 (24%)	18 (49%)	37
Rice cultivation	0 (0%)	2 (10%)	2 (10%)	17 (81%)	21
Perennial crops	0 (0%)	8 (22%)	8 (22%)	21 (57%)	37
Pastoral woodland	2 (5%)	5 (14%)	12 (32%)	18 (49%)	37

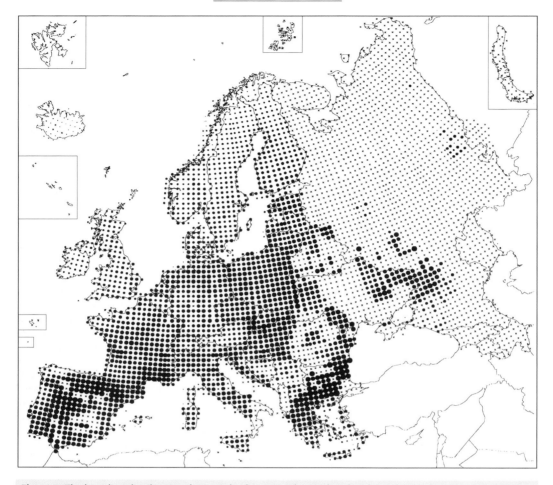

Figure 5. The breeding distribution of priority bird species of agricultural and grassland habitats in Europe during 1970–1990, mapped on the basis of 50-km squares (data from Hagemeijer and Blair 1997). Each priority species breeding in a particular square was assigned a score according to its priority status (Priority A species=4, B=3, C=2, D=1). If the species occurs in more than one of the seven main types of agricultural and grassland habitat, and its priority status differs between these habitats, the highest priority score is taken. The relative size of each dot represents the sum of all species' scores for that square. A cross × indicates incomplete data. No data are available for Anatolian Turkey.

dependent on agricultural habitats, having much larger populations in non-agricultural habitats. Those that are priority species, and that require enclosed habitats, include birds of perennial crops (e.g. orchards, olive-groves) and pastoral woodlands, such as *Otus scops*, *Coracias garrulus*, *Picus* spp. and various thrushes.

Another important landscape feature for many species is the presence of habitat mosaics. For example, several geese species feed on agricultural grassland or crops and use nearby lakes or coastal wetlands as roosting sites. Some other species, including some raptors and woodpeckers, breed in forest patches and fly to agricultural habitats to feed. By contrast, some

waders breed on agricultural grassland and move to other habitats, such as coastal mudflats, marshes or lake margins to feed. Clearly, for all these species the presence of all their required habitat-types within their home ranges is essential.

Mosaics of differing habitat-types within agricultural habitats are also essential for many species. This is often because species have different requirements for nesting and feeding. For example, in arable areas *Vanellus vanellus* often breeds in cereal fields, but prefers to feed on grassland where food resources are more plentiful (Hudson *et al.* 1994). Species' nesting and feeding requirements may also change during the course of the year. For example, individual pairs of *Alauda arvensis* can make up to three or exceptionally four nesting attempts in one year, the first pairs having eggs in early April and the late broods being fed in late July (Cramp 1988). Nesting requires vegetation of a suitable height and density. Consequently, the highest breeding densities have been found where high crop diversity provides a mosaic of habitats suitable for nesting throughout the breeding season (Schläpfer 1988, Jenny 1990).

Studies of *Otis tarda* in Spain have demonstrated how this species requires mosaics of cereal, fallow and pasture to provide varied feeding opportunities and abundant sources of invertebrates (Cramp and Simmons 1980, Martínez 1991, SEO 1992). Areas with alfalfa crops are also favoured as these are fed on in winter (Hidalgo de Trucios and Carranza Almansa 1990). In total, some 42% of the priority species require vegetation mosaics of some form within their habitats. As described in 'Threats to the habitat' (p. 304), such mosaics are becoming less frequent as a result of recently adopted, more intensive farming systems, and this presents a serious threat to many species.

The structure of vegetation is of much greater importance for most priority species than its plant-species composition. Some priority species which feed directly on vegetation, such as geese, require the presence of certain plant species, such as certain nutritious grasses, but a high floral diversity generally contributes more to habitat quality through its enhanced structural diversity than through any direct benefits from the plant species concerned.

Low and/or sparse vegetation is particularly important for species which feed on surface or soil invertebrates, such as *Vanellus vanellus*, *Burhinus oedicnemus*, *Picus viridis* and various thrushes. Similarly, ground-feeding birds such as larks that require small weed-seeds favour short vegetation and bare ground.

Taller and denser vegetation may support higher densities of invertebrates such as grasshoppers (Orthoptera) and beetles (Coleoptera), but foraging in such vegetation is more difficult. Large species, such as bustards, are able to cope with this, as can some small shrub- or tree-adapted species such as warblers, which feed on epiphytic insects. Some species have adapted to foraging in such dense ground vegetation and are indeed reliant on the presence of this cover, e.g. *Crex crex* and *Coturnix coturnix* (e.g. Flade 1991). Tall vegetation also holds higher densities of small mammals and thus certain raptors such as *Circus cyaneus* and *Falco tinnunculus* forage over such vegetation. Vegetation which is uncut allows the setting of seed, and consequently provides opportunities for some finch species which

are able to feed on the tall standing plants.

Some species have quite precise requirements for vegetation structure. For example, most shrikes require the presence of abundant perches from which to forage (Lefranc 1997).

Low and sparse vegetation generally provides insufficient cover for breeding for most ground-nesters, although some highly camouflaged species such as *Burhinus oedicnemus* are adapted to this. The presence of some moderately tall and dense vegetation is therefore necessary for most farmland species. However, vegetation can become too tall and dense to allow easy access for ground-nesting species and movement of their young. Consequently, certain particularly dense and/or tall crops such as autumn-sown cereals or heavily fertilized re-sown grassland are avoided by many priority species, e.g. *Vanellus vanellus* (Hudson *et al.* 1994). Agriculturally improved grassland also has a uniform sward which does not provide good cover for nests, and egg predation has been found to be high in such situations (e.g. Baines 1989, 1990). In this and most other grassland habitats, a high floral diversity enhances structural diversity to the benefit of priority species.

Species that do not nest on the ground require the presence of bushes or trees, either within the agricultural habitat (such as in hedgerows, or scattered throughout as in orchards and pastoral woodlands) or in other adjacent or nearby habitats, as discussed under habitat mosaics.

The availability of suitable food resources is an essential habitat component for all bird species. Many priority species of agricultural habitats are generalists, feeding on a broad range of food resources, though plant seeds, soil invertebrates (such as earthworms, tipulid and beetle larvae, etc.) and insects, especially beetles, grasshoppers and the larvae of butterflies and moths are particularly important for many species. Small mammals (e.g. voles *Microtus*) and birds and their young are also common prey for raptors. Comparatively few priority species feed directly on agricultural crops. However, some, such as geese, *Columba palumbus* and *Corvus frugilegus*, can do so in sufficient numbers to become significant pests—which may lead to conflicts between agriculture and conservation needs.

Some species do have highly specific food and feeding requirements. Some birds of prey, such as *Gypaetus barbatus* and *Aquila chrysaetos* (in certain places and times of year), depend on livestock carrion as their main food source. Some other raptors tend to rely on particular live prey, e.g. *Aquila heliaca* and *Falco cherrug* specialize on susliks *Citellus*, small herbivorous mammals which are commonest on primary steppe (e.g. Heredia 1996b).

In some cases, normally generalist species may

become highly dependent on certain abundant food sources in particular habitats. For example, olives become important for many thrushes and warblers in olive-groves during the autumn. In the pastoral woodland of Iberia, the acorn crop supports huge numbers of *Grus grus* and *Columba palumbus*.

The abundance and availability of food resources is highly dependent on farming practices, which have been highly dynamic and continue to change rapidly. As described in 'Threats to the habitat' (p. 304), these changes pose considerable threats to many priority species.

Factors affecting the mortality of adults and young can result in otherwise suitable habitats becoming unfavourable and abandoned. For example, ground-nesting species are particularly susceptible to man-induced increases in predators or to changes in agricultural practices, such as increases in stocking densities or the tendency towards rapid and earlier cutting of grassland for silage (see 'Threats to the habitat' for more details).

Some species such as raptors and the shy, globally threatened *Otis tarda* are very sensitive to human disturbance and will promptly abandon breeding as a result of obtrusive agricultural operations, or other forms of disturbance. These and other large species are also vulnerable to the presence of powerlines and other overhead cables. Collisions are a major problem, particularly where cables occur at high densities (e.g. irrigation projects), for the larger and less manoeuvrable species such as geese and swans (Anatidae) and bustards (Otidae). Powerline pylons provide highly attractive perches for raptors, but certain pylon designs cause high death-rates from electrocution, especially of the larger species (whose populations are also particularly sensitive to such heightened adult mortality).

In addition to the general habitat requirements described above, some of the habitat needs of priority species are specific to certain habitats. These are briefly described below.

Arable and improved grassland

Some species such as geese and *Columba palumbus* preferentially feed on the most nutritious and productive grasses or other field crops. Such species therefore benefit from agricultural improvements of grassland and from intensive production systems with temporary grass leys and high fertilizer use. However, the reverse is the case for the vast majority of priority species. Most species require, for example, a high crop diversity in the landscape, marginal uncultivated vegetation, semi-natural permanent grassland, abundant invertebrate and other food supplies, traditional methods, timings and frequency of harvesting, low stocking rates and, in certain land-

scapes, the presence of hedgerows, trees or scattered small woods. Thus, as described in 'Threats to the habitat' (p. 304), further intensification and the loss of these habitat features continue to threaten these arable and agriculturally improved grasslands and their priority species, as well as the other less intensively farmed semi-natural habitats described below.

Steppic habitats

Priority species are highly adapted to these open, treeless landscapes, and for most a high degree of openness is essential. Many raptors and other large species such as *Otis tarda* occur naturally at low densities and forage over wide areas. Large areas of habitat are therefore essential for such species and they are consequently highly susceptible to habitat fragmentation.

As described in 'Ecological characteristics' (p. 272), certain priority birds appear to be restricted to certain types or regions of steppic habitats. In particular, some species are highly dependent on the remaining semi-natural, uncultivated grassland steppes in eastern Europe, e.g. *Aquila nipalensis*, *Circus macrourus* and *Melanocorypha yeltoniensis*. These species do not appear to be able to tolerate agricultural improvements or the conversion of their habitat to cultivated pseudosteppe. In contrast, some species such as *Otis tarda* now appear to favour the latter habitat, so long as cultivation remains non-intensive.

Other species appear to favour shrub-steppes such as those created by historical clearance of woodland and which are now maintained by grazing and intermittent fires. Such secondary steppes in Iberia, for example, are the most important habitat of priority species such as *Chersophilus duponti* and *Galerida theklae*, and are important for other species such as *Oenanthe hispanica* and Trumpeter Finch *Bucanetes githagineus*.

Montane grassland

In winter, montane grassland is usually snow-covered and not inhabited by birds, but in summer the warmer areas with the least snow duration are important for foraging birds, especially raptors and passerines. The seasonal duration of snow-cover depends on such habitat features as slope, altitude, aspect, latitude and precipitation. Shrubs play an important role for many priority bird species in the more humid mountain ranges (e.g. Alps, Carpathians, Caucasus), where they provide a variety of important vegetable food (buds, leaves, berries) around the year (e.g. *Vaccinium*), as well as shelter and nest-sites (e.g. *Rhododendron*) (Cramp and Simmons 1980, Lang and Cocker 1991).

Carrion feeders (including all species of vultures,

and some corvids) depend for food almost wholly on large herbivores such as wild ungulates or domestic livestock. In the Alps, even *Aquila chrysaetos*, which relies on medium-sized mammals (mainly marmots *Marmota marmota* and *Lepus* spp.) in summer, feeds in winter to a high degree on ungulates (e.g. *Rupicapra rupicapra*) which have been killed by avalanches (e.g. Haller 1979).

Wet grassland

The regulation of water-levels is critical to the maintenance of suitable habitat conditions for most wet grassland species. Indeed, a high soil water-table during the breeding season (mid-March to end of June) is probably the single most important factor for breeding waders. Ideally the soil water-table should be approximately 20–30 cm below the surface during spring and summer, depending on the soil-type (Spoor and Chapman 1992). This ensures that soil invertebrate prey remain close to the surface and increases the biomass of prey accessible to waders. Surface-water features such as ponds and ditches are also important, particularly on clay sites where the soil may be impermeable.

Wet grasslands also frequently hold important numbers of priority waterfowl species in winter. For such species, large shallow winter floods provide food (by releasing seeds trapped in vegetation and flushing out terrestrial invertebrates) and security from ground predators while feeding and roosting.

Vegetation composition and structure, also important for breeding waders and wintering waterfowl, is affected by water-level and grazing regimes. For example, waders occupy different niches, from rank, unmanaged fields with tall vegetation which favour *Gallinago gallinago*, to short, grazed swards which are suitable for *Vanellus vanellus*. Grazing waterfowl, such as Wigeon *Anas penelope* and *Cygnus columbianus*, favour young grassy swards composed of softer grass species typical of heavily grazed areas.

The habitat requirements of waterfowl and waders are complex as they differ between species, seasons and sites. They cannot be discussed in detail here but further information is provided in, for example, Green (1986), Hötker (1991), Andrews and Rebane (1994), Dunn (1994a), RSPB/NRA/RSNC (1994), Beintema *et al.* (1995, 1997) and RSPB/ITE/EN (in press).

Rice cultivation

Rice cultivations in the Mediterranean region are of considerable importance for many priority species of breeding herons (Ardeidae) and passage and wintering waterfowl and waders (Pain 1994b, Fasola and Ruíz 1997). They are particularly important for breeding herons as they provide a shallow water-body during the hot and dry summer months when similar natural wetlands are often at their smallest extent. It has been found that the quality of these habitats varies considerably between regions, largely in relation to prey availability, in particular, fish, frogs (Ranidae) and aquatic invertebrates. Fasola (1986) also found that there is a critical amount of foraging habitat necessary within 5 km of a heron colony, which amounts to about 800 ha or about 10% of the area. Colony size drops rapidly when the foraging area falls below this threshold.

In addition to foraging areas, suitable nesting habitat is also essential for herons. For mixed colonies, this consists of an undisturbed site, usually protected by surrounding water, with at least 4 ha of suitable vegetation such as large bushes and tall trees.

Further information is given in Fasola and Barbieri (1978), Fasola and Ghidini (1983), Pain (1994b) and Fasola and Ruíz (1997).

Perennial crops

The priority species show considerable taxonomic and ecological variety (Table 6). Ground-feeding species such as *Alectoris rufa*, *Lullula arborea* and *Galerida theklae* require an open habitat structure with sparse trees and an open canopy, yet little or no understorey, although some taller ground vegetation is required by some ground-nesting species, e.g. *Alectoris* partridges (Pain 1994a). A fairly open habitat structure is also favoured by many of the priority arboreal species, including the shrikes *Lanius* spp., *Coracias garrulus*, *Phoenicurus phoenicurus* and *Certhia brachydactyla*. By contrast, denser habitats are preferred by more forest-adapted species such as *Hippolais olivetorum*. Some expected species, such as *Turdus merula* and *T. viscivorus*, may be absent if the crop is lacking a shrub layer.

Muñoz-Cobo (1992) found that the age of trees in olive plantations in Spain had a considerable influence on the breeding bird community. In general, the density of breeding birds increased considerably with the age of the olive-trees, although the older stages were less species-rich than the younger ones. Species such as *Oenanthe hispanica* and *Alectoris rufa* only occurred in plantations less than 19 years old, while other, more woodland-adapted species, such as *Certhia brachydactyla* and *Lullula arborea* only occurred in older groves. At least some old trees with gnarled trunks, dead wood, holes or crevices are also necessary for hole-nesting species such as *Otus scops*, *Coracias garrulus*, *Jynx torquilla*, *Dendrocopus syriacus* and *Phoenicurus phoenicurus*.

Further information on the bird communities of olive plantations is given in Muñoz-Cobo (1992),

Rey (1993, 1995), Pain (1994a) and Rabaça (1994), but the habitat requirements of birds using these and other perennial crops are complex and still not clarified in any detail.

Pastoral woodland

The most frequently shared and strongest requirements among priority species are for widely spaced, old and large trees (with suitable holes) for nesting, set in a landscape with a low level of human activity and land-use, thus with low levels of disturbance to nesting and foraging birds. There is also a strong but contradictory trend in the preference of some species for a shrubby understorey (typical forest species) and of others for an open understorey of non-intensive cereal fields (species typical of more open habitats). Clearly, the maintenance of a mosaic of both types of understorey, on a scale that satisfies the territorial or home-range requirements of both types of species, is important.

Other features or attributes of pastoral woodland which are particularly important for certain birds are the presence of fleshy fruited shrubs (important food for wintering passerines), of large rabbit populations (important food for raptors and owls), and of ponds, rivers and streams (mainly for the drinking, foraging and bathing of some rather water-dependent species). Rocky cliffs or riverine sand-cliffs are important as nesting sites for two of the priority species, in Iberia at least.

Further information on the habitat needs of priority species of pastoral woodland can be found in de Juana (1989), Almeida (1992), Alvarez and Santos (1992), Pulido and Díaz (1992), Tucker and Heath (1994), Díaz et al. (1996) and Tellería et al. (1996).

THREATS TO THE HABITAT

A comprehensive review of threats to agricultural habitats, and particularly to the habitat features required by their priority species, has not been carried out in Europe. However, further information and illustrative studies can be found in Jenkins (1984), O'Connor and Shrubb (1986), Goriup (1988), Baldock (1990a), Goriup et al. (1991), Hötker (1991), Beintema et al. (1995) and Pain and Pienkowski (1997). Information from these works, other studies, unpublished data and the expert opinions of the members of the Habitat Working Groups have all been used to review the general scale of threats to agricultural habitats in Europe and their likely impacts on priority species.

Table 9 summarizes the scale and general effects of all identified threats that are thought to be having significant impacts on priority species at the European scale. For each type of agricultural habitat, an assessment of the likely impacts of each threat on each priority species has also been made (see Appendix 4, p. 401, for methods used). The results of these assessments are provided in Appendix 4 and summarized in Table 10 and Figure 6.

The threats are discussed below in order of their importance (as shown in Figure 6), and the mechanisms by which they harm priority species (at the population level) are explained as far as is known (with especial reference to the 'Habitat needs of priority birds', above).

Crop improvements

Improvements in crop yield are mainly achieved by high fertilizer and pesticide inputs, use of new varieties, re-seeding of grassland, and irrigation. Such economically expensive measures have been widely adopted, particularly in western Europe (e.g. see Table 4), and lead to faster-growing, taller, denser crops. These usually make less suitable nesting and/or feeding habitats for priority species of arable and lowland grassland, which often need lower, more open and varied vegetation. For example, *Otis tarda* in Austria shows higher breeding success on non-irrigated than on irrigated arable land (Kollar and Wurm 1996), and the species is threatened by such crop improvements in Germany as well (Litzbarski 1993, Bauer and Berthold 1996, Litzbarski and Litzbarski 1996). As described under 'Cultivation of grassland' (p. 311), governments in Iberia and Turkey intend to convert huge areas of steppic habitats to irrigation schemes over the next few years (Suárez et al. 1997a,b).

Similarly, improvements in the agricultural productivity of wet grassland through drainage, re-seeding and chemical applications are thought to be responsible for declines in numbers of *Numenius arquata*, *Limosa limosa*, *Philomachus pugnax*, *Gallinago gallinago* and *Tringa totanus* (Baines 1988, Batten et al. 1990, Beintema 1991, Beintema et al. 1995, 1997). Such grassland improvements in turn often lead to more frequent and mechanized harvesting and/or higher stocking rates, with the resulting problems described under 'Farming operations' (p. 312) and 'High stocking levels' (p. 310).

Continued increases in crop productivity are also a major problem in already intensively farmed systems and threaten many priority species. Cereal crops are now too tall and dense for nesting in for species such as *Alauda arvensis* (Schläpfer 1988, Poulsen 1993, Wilson and Browne 1993) and *Vanellus vanellus* (Hudson et al. 1994). Declines in *Falco tinnunculus* and *Tyto alba* in Switzerland (Kaeser and Schmid 1989) and many species in Finland (Solonen 1985, Tiainen 1985, Hanski and Tiainen 1988) have also been attributed to various

Table 9. The most important threats to agricultural and grassland habitats, and to the habitat features required by their priority bird species, in Europe (compiled by the Habitat Working Groups). Threats are listed in order of importance (see Figure 6), and are described in more detail in the text.

Threat	Effect on birds or habitats	Scale of threat			Regions affected, and comments
		Habitat	Past	Future	
1. Crop improvements (including high fertilizer input, improved varieties, grass re-seeding, irrigation)	Rapid growth leading to tall dense vegetation unsuitable for nesting and/or foraging	Arb	■	■	
		Stp	●	●	
		Mon	●	●	
		Wet	■	●	
		Ric	●	●	
		Per	●	●	
2. Pesticide use	Some direct toxic effects; widespread reductions in invertebrate, vertebrate and plant food resources	Arb	■	■	
		Stp	●	●	Widespread use of rodenticide in Russia
		Wet	●	●	
		Ric	■	■	Particularly W Europe
		Per	■	■	
		Wdl	●	●	
3. Land abandonment	Plant succession leading to habitat loss through overgrowth by scrub and trees, etc.	Arb	●	●	Particularly E Europe
		Stp	●	●?	
		Mon	●	●	W Europe, may increase
		Wet	●	●	
		Ric	■	■	Particularly E Europe
		Wdl	●	●	
4. High stocking levels (or early release of livestock in spring in wet grasslands)	Overgrazed low and sparse vegetation with reduced invertebrate numbers; high risk of trampling of eggs and young (wet grasslands); acorn depletion in pastoral woodland	Arb	■	■	
		Stp	●	●	Widespread in Italy and Iberia, less in E Europe
		Mon	■	■	
		Wet	■	■	Also early spring grazing
		Wdl	●	●	
5. Loss of marginal features (e.g. trees, hedges, rank grass, scrub, etc.)	Loss of nesting, roosting and feeding opportunities	Arb	■	■	
		Stp	●	●	Widespread in Italy
		Mon	●	●	Scrub clearance, especially Turkey
		Wet	●	●	
		Ric	●	●	
		Per	●	●	
6. Afforestation	Habitat loss and fragmentation; reduced openness of habitat	Arb	●	●	
		Stp	●	●	
		Mon	●	●	
7. Cultivation of grassland (natural and semi-natural)	Lost nesting and feeding opportunities (equivalent to habitat loss for some priority species)	Stp	■	●	Now mainly Spain only
8. Farming operations (including frequency of operations and early harvesting)	Disturbance of bird behaviour; direct loss of eggs and chicks and some adults	Arb	■	■	
		Stp	●	●	
		Mon	●	●	
		Wet	■	■	
		Ric	●	●	
		Per	●	●	
9. Loss of crop diversity (loss of rotations; crop specialization; increase in field size)	Reduced crop diversity and thus reduced nesting and feeding opportunities; also increased frequency of cultivation and reduced use of organic manures leading to reduced soil invertebrate populations	Arb	■	■	
		Stp	■	●	

cont.

Table 9. (cont.)

Threat	Effect on birds or habitats	Scale of threat			Regions affected, and comments
		Region	Past	Future	
10. Recreation (including tourism, sports, hunting, etc.)	Disturbance and interference with essential bird behaviour; vegetation changes; erosion	Arb	•	•	
		Stp	•	•	
		Mon	●	■	Particularly in W Europe
		Wet	•	•	
		Ric	•	•	
		Per	●	•	
		Wdl	•	•	
11. Overhead structures (powerlines, wind-turbines, ski-lifts, tall wire fences, etc.)	Direct mortality of birds (collisions and electrocution); reduced habitat openness; disturbance; habitat loss	Arb	●	•	
		Stp	•	•	
		Mon	•	•	Particularly Alps
12. Increased predators (through human activities; e.g. crows (Corvidae), foxes *Vulpes* and dogs)	Direct mortality of eggs, chicks and adults	Arb	•	●?	
		Stp	•	•	
		Mon	•	•	
		Wet	•	•	
		Ric	•	●	
13. Urbanization and road-building	Habitat loss and fragmentation, disturbance and reduced openness	Arb	●	●	
		Stp	•	•	
		Mon	•	•	
		Wet	•	•	
14. Drainage and flood control	Loss of open water and ditches, etc.; removal of vegetation cover in ditches and cutting of steep sides; lower water-tables and reduced soil penetrability; vegetation structure and composition changes	Wet	■	●	
15. Reduction in density of oaks *Quercus* (oak disease, tree clearance, non-regeneration)	Loss of trees in pastoral woodland, and hence nesting and feeding sites for arboreal species	Wdl	●	■	
16. Loss of hay meadows (e.g. to silage, pasture or arable crops)	Loss of tall, dense grass habitat	Arb	●	●	
		Mon	•	•	Conversion to pasture, mainly in W Europe
		Wet	■	■	Conversion of Baltic coastal meadows to intensive grazing, due to massive decrease in demand for hay
17. Low stocking levels (or removal of stock over winter)	Growth of tall vegetation and changes in vegetation composition	Arb	•	●?	
		Stp	—	•	
18. Atmospheric nutrient pollution	Loss of nutrient-poor natural and semi-natural grassland and enhanced vegetation growth leading to changes in food availability	Stp	?	?	
		Mon	■	■	
		Wet	?	?	
19. Replacement of old crop-trees (with young ones and new varieties, e.g. olives)	Reduced numbers of holes for nesting; smooth trunks harbour reduced food for insectivorous species	Per	●	●	
20. Loss of old buildings	Reduced nesting opportunities for hole-nesters	Arb	•	●	
		Stp	●	•	
		Wet	•	•	
21. Reductions in livestock carcasses (e.g. due to modern disposal methods)	Reduced food supply for carrion-feeders	Stp	•	●	Particular problem in Spain and Italy
		Mon	•	●	

cont.

Table 9. (cont.)

Threat	Effect on birds or habitats	Scale of threat			Regions affected, and comments
		Region	Past	Future	
22. Autumn sowing of crops	Tall, dense vegetation in spring, unsuitable for nesting and/or feeding; also loss of winter stubble and feeding opportunities	Arb Wet Ric Wdl	● ● ● ●	● ● ● ●	Not a problem in Spain
23. Clearance of ground vegetation (e.g. in olive-groves)	Loss of feeding habitat and nesting cover	Per	■	■	
24. Conversion of pasture (to silage or hay)	Growth of tall, dense, uniform vegetation; high mortality from agricultural operations	Arb Mon Wet	■ ? ●	■ ? ●	Mainly conversion to silage
25. Dry cultivation of rice	Loss of wet feeding habitat	Ric	●	●	

Scale of threat

Arb	Arable and improved grassland	Past	Past 20 years
Stp	Steppic habitats	Future	Next 20 years
Mon	Montane grasslands	■	Widespread (affects >10% of habitat)
Wet	Wet grasslands	●	Regional (affects 1–10% of habitat)
Ric	Rice cultivation	·	Local (affects <1% of habitat)
Per	Perennial crops	?	Considerable uncertainty
Wdl	Pastoral woodland		

Table 10. The proportions of priority bird species that are adversely affected by different threats to agricultural and grassland habitat-types in Europe. Numbers are percentages of the total number of priority species in each habitat-type. Threats are ordered and numbered as in Table 9 and Figure 6. Appendix 4 (p. 401) gives full details of the threat assessment.

Threat	Arable and improved grassland	Steppic habitats	Montane grassland	Wet grassland	Rice cultivation	Perennial crops	Pastoral woodland
1. Crop improvements	57	63	30	59	0	16	0
2. Pesticide use	63	72	0	11	48	50	35
3. Land abandonment	33	71	67	27	10	21	68
4. High stocking levels	27	74	55	38	0	0	14
5. Loss of marginal features	59	38	12	32	19	16	11
6. Afforestation	22	57	70	46	0	0	0
7. Cultivation of grassland	0	83	0	0	0	0	0
8. Farming operations	13	25	9	27	10	0	0
9. Loss of crop diversity	22	32	0	0	0	0	0
10. Recreation	2	17	42	27	0	0	22
11. Overhead structures	11	18	18	24	0	0	24
12. Increased predators	8	20	24	19	0	0	0
13. Urbanization and road-building	4	25	24	22	0	0	3
14. Drainage and flood control	7	0	0	62	0	0	0
15. Reduction in oak density	2	0	0	0	0	0	70
16. Loss of hay meadows	10	0	15	22	0	0	0
17. Low stocking levels	29	0	0	0	0	0	0
18. Atmospheric nutrient pollution	2	2	52	0	0	0	0
19. Replacement of old crop-trees	0	0	0	0	0	24	0
20. Loss of old buildings	5	8	0	5	0	0	0
21. Reductions in livestock carcasses	2	5	12	0	0	0	0
22. Autumn sowing of crops	16	0	0	0	0	0	0
23. Clearance of ground vegetation	0	0	0	0	0	18	0
24. Conversion of pasture	7	0	6	3	0	0	0
25. Dry cultivation of rice	0	0	0	0	48	0	0

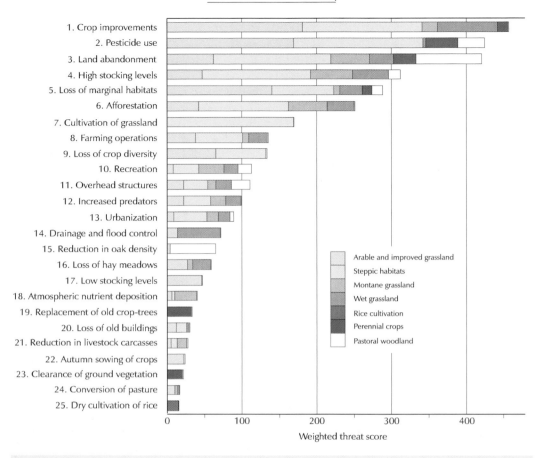

Figure 6. The most important threats to agricultural and grassland habitats, and to the habitat features required by their priority bird species, in Europe. The scoring system (Appendix 4, p. 401) takes into account:

- The European Threat Status of each species and the importance of the habitat for the survival of its European population (Priorities A–D); see Table 6.
- The predicted impact (Critical, High, Low), within the habitat, of each threat on the European population of each species; see Appendix 4 (p. 401).

See Table 9 for more details on each threat.

improvements in the productivity of agricultural systems.

Large increases in fertilizer usage have also probably contributed to the considerable reductions in grassland invertebrates. Reduced use of organic farm manures and increasing reliance on artificial fertilizers may also cause reductions in soil invertebrate availability for birds (Edwards and Lofty 1982a, Tucker 1992).

Pesticide use

Data on the use of pesticides in agriculture are complex and far from comprehensive. Nevertheless, a massive growth in pesticide use has certainly occurred across most of Europe over the last 20–30

years (Potts 1997, Potter 1997). Poisoning and other direct negative effects of some pesticides, notably organochlorines, on birds and other wildlife have been well studied and documented (e.g. Newton 1979, Potts 1986, Ratcliffe 1993) and to large extent alleviated by bans on the most toxic substances. Problems do, however, still occur. For example, in Switzerland between 1980 and 1993, 93 individuals of *Buteo buteo*, *Milvus milvus* and *Milvus migrans* were found showing symptoms of Carbofuran poisoning (Jenni-Eiermann *et al.* 1996).

Despite such incidents, the major threats from pesticides are now most likely to be through indirect effects on food resources, such as through reductions in insects and weed seeds, etc. For example, there have been documented declines in the densities of a wide range of invertebrates associated with cereal crops in the UK, although the exact causes are unknown (Aebischer 1991). Also, herbicides are now so widely used and efficient that broadleaved weeds have been enormously reduced in most intensively farmed arable land, many weed species being close to extinction in some west European countries.

However, with the exception of the well-studied *Perdix perdix* (Southwood and Cross 1969, Potts 1986, 1990), causal links between population declines in priority agricultural birds and the effects of pesticide applications on food resources are unproven (Campbell *et al.* in press). Unfortunately, most of the proposed threats from pesticides are based on associations between increased pesticide use and declines in priority birds, and on inferences from species' biological requirements. This is mainly because the effects of pesticides are superimposed on an already complex and rapidly changing system, and the large-scale and long-term studies that are necessary to unravel and prove such effects have largely not been carried out.

Studies comparing conventional farming with organic systems (with no pesticide applications) in Denmark (Braae *et al.* 1988, Petersen 1994) and the UK (Wilson 1993, Wilson and Browne 1993, Chamberlain *et al.* 1995) have found that organic farms support higher densities of many farmland birds, and show higher breeding success of *Emberiza citrinella* and *Alauda arvensis*. However, factors other than pesticides may have contributed to these observations. Nevertheless, the wealth of circumstantial evidence, as reviewed by Campbell *et al.* (in press), and the detrimental effects documented by the detailed research on *Perdix perdix*, all strongly suggest that the threats are likely to be widespread and significant.

On the basis of the opinions of the Habitat Working Group and other consulted experts (and the likely continued widespread use of pesticides), it is considered that 82 species are likely to suffer population declines. Overall, therefore the threat is considered to be among the most important of the threats considered here.

Land abandonment

Abandonment of agricultural land is no less a threat to farmland birds than many of the processes of intensification described elsewhere in this section. Semi-natural agricultural habitats are especially at risk of abandonment as their management is most likely to be financially unattractive. However, abandonment due to the failure of intensification programmes is also not unusual (Lee 1987, Egdell 1993). Such semi-natural habitats are particularly important for birds in Europe, holding a large number

of priority species (e.g. Baldock *et al.* 1993, Beaufoy *et al.* 1994). Abandonment is detrimental because rapid habitat change normally results, especially on soils that previously were heavily fertilized, through the growth of tall vegetation and scrub encroachment and the loss of the open habitat that is essential for most farmland species. Although some birds of heathland, shrubland and forest may benefit from these changes, few of them are priority species for these agricultural habitats. Indeed, most of the priority species of heathland and shrubland require semi-open habitats such as those maintained by grazing—further details are given under 'Mediterranean Forest, Shrubland and Rocky Habitats' (p. 239) and 'Lowland Atlantic Heathland' (p. 187).

Abandonment of agricultural land and conversion to other uses has been fairly widespread over Europe in recent decades. In the current 15 EU countries, farmland area has declined by an average of about 10% over the 30-year period between 1960 and 1990 (CEC 1995d). Over the same period similar declines occurred in eastern Europe. However, more recently, agricultural output has fallen considerably following the political transitions in central and eastern Europe. Output was affected by the fall in demand as consumer subsidies were removed and the general economic situation deteriorated, and by the 'price–cost' squeeze that agriculture faced (i.e. input prices rising much faster than output prices). Although total arable land remained relatively stable in these countries, the drop in output led to a considerable decline in livestock numbers between 1989 and 1994 (CEC 1995e).

Most affected were cattle and (particularly) sheep in the countries of the Central European Free Trade Association (Poland, Hungary, Czech Republic, Slovakia, Slovenia): cattle declined by 33% and sheep by 64%. In Romania and Bulgaria, sheep numbers did not decline so dramatically (e.g. by 37% in Romania), while cattle and pig numbers declined by 47% and 39% respectively. In the Baltic countries, which specialized in livestock production for the Russian market, livestock numbers (except dairy) halved during this period. Data for Russia and Ukraine are unavailable but similar trends in livestock numbers are believed to have occurred recently.

However, agricultural production in central and eastern Europe is now recovering, although there is a lot of individual variation among countries, and this general upward trend is predicted to continue, particularly for livestock, although pre-transition consumption levels are unlikely to be reached (CEC 1995e).

These recent declines in livestock farming in central and eastern Europe are reported to have caused widespread changes in the physical structure and species composition of semi-natural grasslands and in their bird communities, including those of upland grassland, pastures and lowland dry grassland. The priority birds of coastal wet grassland of the Baltic states are also highly threatened by the abandonment of traditional habitat management, as they were in Scandinavian countries in the 1960s and 1970s (Møller 1975, Soikkeli and Salo 1979). There, reintroduction of traditional farming practices in Denmark was found to lead to a gradual restoration of priority bird populations (Thorup 1991, 1992).

Rural depopulation and agricultural abandonment has also been a long-standing problem in the Mediterranean region in recent decades (see 'Mediterranean Forest, Shrubland and Rocky Habitats', p. 267). This trend is likely to continue and to pose serious threats to birds of extensively grazed grassland, pseudosteppe and pastoral woodland (Yanes 1994, Díaz et al. 1997, Donázar et al. 1997, Suárez et al. 1997b). Abandonment of perennial crops appears to be less of a threat overall, since the area of many crop species (e.g. olive) is increasing. The area under rice cultivation is also increasing in Europe.

Abandonment may at first have subtle effects on habitat quality which may nevertheless have considerable impacts on bird populations. For example, on the Russian steppes, widespread abandonment of livestock grazing has led (in the absence of natural large herbivores) to loss of *Artemisia* and dominance by tall *Stipa* bunch-grasses, which small rodents such as *Citellus* do not favour. Since the latter are the main prey of priority breeding raptors such as *Buteo rufinus*, *Aquila nipalensis* and *Aquila heliaca*, populations of these birds have also declined (e.g. Heredia 1996b).

At more advanced stages of succession, scrub encroachment and eventual woodland establishment leads to profound changes in the bird community, equivalent to that of habitat loss. Such habitat conversion may also have substantial impacts on the birds on remaining areas of agricultural habitat through the effects of fragmentation. This is particularly important for several raptor species and other large species which require extensive and continuous areas of habitat for foraging in.

Although the impacts of agricultural abandonment on priority bird populations are not well studied or documented, it can be fairly reliably anticipated that substantial impacts will result from currently predicted trends. Thus, overall, the threat is of high relative importance, particularly for steppic and montane habitats.

High stocking levels

Trampling by livestock of the eggs and young of priority breeding waders of wet grassland can be considerable (Beintema and Müskens 1987, Thorup 1992, Hudson et al. 1994). High stocking levels, or the release of livestock too early in spring, can substantially increase mortality, which may have significant impacts at population levels.

High stocking rates can also lead to excessive grazing of vegetation, resulting in reduced invertebrate numbers. This is considered to be a significant threat to priority birds in parts of southern Europe, such as Portugal and Spain, where sheep numbers have increased greatly through EU headage payments (Donázar et al. 1997). While information on the effects of such habitat degradation on bird populations is scant, impacts can be anticipated to be widespread and potentially serious through losses of suitable plant-cover for nesting and of invertebrate and other food resources. For example, the survival and abundance of many raptors in the Mediterranean zone is highest where grazed areas form varied mosaics with scrub, which supports high populations of *Oryctolagus cuniculus* (their main prey). Overgrazing would degrade and reduce the extent of such scrub, thereby threatening to impact the raptor populations.

Loss of marginal features

The elimination of non-farmed and marginal habitat features, such as trees, hedges, banks, grassy strips, scrub, ponds and ditches, is a widespread and continuing practice in arable and agriculturally improved grassland. For example, 22% (170,000 km) of the total hedgerow length in Britain was destroyed between 1947 and 1985 (Fuller et al. 1991), and in the 1980s losses continued at a rate of 28,000 km per year (Barr et al. 1993).

Several factors lead to the removal or neglect of hedges. The decline of mixed farming has meant that hedges are no longer required for stock management on many farms. Where livestock are present, the use of wire has to a great extent replaced the need to manage hedges as stock-proof structures. The latter activity is labour-intensive and is now more difficult due to widespread decline in the size of the agricultural workforce. Hedges, ditches and banks have also been removed to increase field size, partly due to the use of larger machinery and partly due to the average increase in the size of holdings and thus of herds.

Many priority birds require marginal habitat features for feeding, roosting, nesting or sheltering from predators. Hedgerows, ditches and other linear features may also play a role as 'wildlife corridors' (Rich 1996), linking fragmented habitats such as woodlands and facilitating dispersal of plants and animals (Bennett 1991). However, as yet there is little evidence to substantiate or quantify the benefits of such 'corridors'. It is important to note that some other priority bird species require some of these features to be absent. If trees and hedgerows are currently absent from any particular farmed landscape, their creation is not always appropriate and environmentally beneficial, e.g. see the threat for steppic habitats under 'Increased predators' (p. 313).

Nevertheless, the serious impacts of hedgerow- and tree-loss on priority birds and other species have been frequently observed and documented, particularly in north-west Europe (e.g. O'Connor and Shrubb 1986, Pfister *et al.* 1986, Arlettaz 1990, Baldock 1990a, Lefranc 1997). Declines in species such as *Falco tinnunculus*, *Streptopelia turtur*, *Lanius collurio* and *Emberiza hortulana* are thought to be at least partially attributable to this. However, loss of marginal habitat features is also an important threat elsewhere, and in less intensively farmed habitats. For example, in Italy much of the wildlife value of the Po valley and coastal and inland plains has been lost as a result of the destruction of hedgerows, ponds and woodlands (Petretti 1995).

The rate of intentional hedgerow destruction now appears to be slowing, at least in western Europe, due to greater appreciation of their importance for wildlife and other values, and in some areas hedge re-creation and restoration projects are underway. However, losses of hedgerows, trees and other non-farmed features are likely to continue as a result of neglect or mismanagement, and the problem is likely to remain widespread in most intensively farmed landscapes.

Afforestation

According to various sources (e.g. Lee 1987, UNECE/FAO 1992a, CEC 1995d), there has been a 10% increase in Europe's tree cover (in both area and volume) over the past 30 years. Further details are given under 'Mediterranean Forest, Shrubland and Rocky Habitats' (p. 239) and 'Boreal and Temperate Forests' (p. 203). Much of this forest expansion has been through deliberate afforestation of agricultural land, although also on other habitats (see 'Lowland Atlantic Heathland', p. 187, and 'Tundra, Mires and Moorland', p. 159). As with natural reforestation, described under 'Land abandonment' (above), deliberate afforestation effectively results in the loss of the agricultural habitat, causing profound changes in the bird community. Significant indirect effects may also result from fragmentation of the agricultural habitat.

Also as stated in 'Land abandonment' (above), non-intensively managed and economically marginal agricultural habitats are the most likely to be converted to forestry. For example, afforestation is one of the main threats to pseudosteppe in Portugal and Spain (e.g. Suárez *et al.* 1997b). Plantations there are mainly of non-native *Eucalyptus*, but those of native pine *Pinus* and *Quercus* (for the creation of pastoral woodland) also cause destruction of wildlife-rich steppic habitats, due to the non-integration of two EU Regulations (2078/92 and 2080/92) which are used for funding such forestry schemes. Also, afforestation is often promoted as a means of combating excessive soil erosion in high-risk areas, and is a widespread threat to montane grassland, again due to the economic unattractiveness of current agriculture there, as well as due to pressures to protect against avalanches.

As a result of current agricultural surpluses (at least within the EU) and recent national and EU policy changes and measures, afforestation is also increasing in the agriculturally productive lowlands of Europe, often promoted through the provision of grants. This is frequently encouraged on environmental grounds, due to a widespread perception that forest cover needs to be increased in such areas. Although such forest-creation schemes can provide environmental benefits (including for birds), especially in intensively farmed and severely deforested landscapes, they can be detrimental to many priority agricultural species. In these areas, the potential threat from afforestation schemes therefore needs to be addressed carefully, and usually in relation to local circumstances.

Cultivation of grassland

The principal cause of the loss of natural and semi-natural grassland over the last 50 years in Europe has been the advent of widespread agricultural mechanization, which has allowed vast areas of grassland to be ploughed and cultivated (Stanners and Bourdeau 1995). Today, virtually no natural grassland exists within Europe, and remaining semi-natural grasslands are mostly restricted to secondary steppes, wet grasslands or montane pastures (Goriup 1988, van Dijk 1991, Stanners and Bourdeau 1995).

Most remaining areas of natural and semi-natural grassland in central and eastern Europe are unlikely to be currently at risk of widespread destruction because of the recent drop in agricultural production over most of the region (see 'Land abandonment', above). However, this reduction has affected the livestock sector most strongly, but it is conceivable that if the crop sector shows strong growth in the future, conversion of grassland to arable could result. Nevertheless, most conversion of grassland to cereal production would be limited to areas with suitable soils and climatic conditions, and therefore some wet and montane grasslands and the remaining natural steppes might be little affected. Developments any further into the future are hard to predict, however.

In the EU, conversion of remaining semi-natural grassland to crops is minimal in most countries, due to the current policy climate against cereal over-production and to the fact that most such grasslands have already been destroyed by agricultural im-provement or cultivation. However, major irrigation projects with EU or international funding are being planned in southern Europe, threatening large areas of steppic habitats.

For example, in Spain, according to the National Hydrological Plan, it is intended to bring 6,000 km^2 under irrigation over the next 20 years. Some of this will intensify existing rain-fed cultivation on pseudosteppe, and remaining secondary steppe will undoubtedly also be brought under cultivation (Suárez et al. 1997a). Similar large-scale irrigation of steppic habitats is planned in Portugal and is ongoing in Turkey, following the completion of huge dam projects.

As well as field-crops such as vegetables, irriga-tion in these and other south European countries also involves creation of large plantations of perennial crops such as olive and almond (Rodríguez and de Juana 1991, Suárez et al. 1997a,b), which reduce the 'openness' of landscape required by many priority species of steppe. Such plantations may also support increased native predator populations, e.g. of corvids (see 'Increased predators', p. 313).

In addition, some remaining areas of secondary shrub-steppe in Spain and elsewhere are being con-verted to pseudosteppe, i.e. to rain-fed cultivation of hard wheat or barley, which threatens to impact some priority species which prefer secondary shrub-steppe to pseudosteppe, e.g. Chersophilus duponti (Suárez and Oñate in Tucker and Heath 1994).

Farming operations
Mechanized and rapid mowing methods have now largely replaced traditional hand-scything of hay crops, with greatly increased risks of egg and chick mortality. Also, agriculturally improved grasslands are often used for silage production and are conse-quently cut earlier in the year than hay, often during the main part of the breeding season of many ground-nesting species. Consequently, the mechanization of mowing and the practice of earlier cutting is a wide-spread threat to many species of grassland, including Numenius arquata (Hölzinger 1987) and the glo-bally threatened Crex crex (Myrberget 1963, Stowe et al. 1993, Øien and Folvik 1995, Crockford et al. 1996).

Ground-nesting species of arable land (e.g. Burhinus oedicnemus and Vanellus vanellus) suffer similar losses from other farming operations such as rolling, harrowing and applications of agro-chemi-cals (e.g. Larsen and Sandvik 1994, Nehls 1996).

Sensitive species may also suffer disturbance effects from agricultural operations. In particular, Otis tarda is a shy species which is vulnerable to disturbance from aerial crop-spraying, and may desert affected areas.

Loss of crop diversity
The enlargement of fields has reduced the diversity of the agricultural landscape in Europe by creating increasingly large patches of uniform habitat. This process has been further exacerbated by the ten-dency of farms to specialize in the most economi-cally attractive crop-types. This has led to a polarization of farming systems, which are increas-ingly either purely arable or purely livestock-based (O'Connor and Shrubb 1986, Potter 1997). Together with decreased use of crop rotations (see 'Manage-ment systems', p. 274), this specialization has con-siderably reduced the diversity of crops and therefore of habitat and vegetation structure in the farmed landscape, from the scale of individual farms all the way up to entire regions.

Many birds rely on crop mosaics to meet different feeding, nesting and seasonal requirements. The ef-fects of reduced crop diversity are difficult to study, but declines in Burhinus oedicnemus (Batten et al. 1990), Alauda arvensis (O'Connor and Shrubb 1986, Schläpfer 1988), Pyrrhocorax pyrrhocorax (Bignal et al. 1989) and various changes in bird-community composition in Finland (Hanski and Tiainen 1988) have been attributed to it. The globally threatened Otis tarda is threatened by such agricultural changes in Germany (Litzbarski 1993, Bauer and Berthold 1996, Litzbarski and Litzbarski 1996) and in Spain, where ongoing reductions in the extent, duration and frequency of fallow periods are one of the most important threats to the priority birds of cultivated pseudosteppe generally (Kollar 1996, Suárez et al. 1997a).

Associated with the decline in mixed farming has been the almost complete demise of the use of undersowing of ley grassland, a practice that was once carried out over a quarter of the entire European arable area (Potts 1997). Under this system, spring-sown cereals were used as a nurse for grasses and clovers so that after the cereal harvest a ley was established without further cultivation. This helped to maintain populations of overwintering inverte-brates in the soil, such as sawflies (Tenthredinidae), which are now much reduced by post-harvest culti-vations. In the past, the sawfly larvae were a particu-larly important part of the spring diet of Perdix perdix (Potts 1970), Miliaria calandra (Ward and Aebischer 1994), Alauda arvensis (Schläpfer 1988, Jenny 1990, Poulsen 1993) and probably many other species (e.g. Hill 1985). Continuous cropping with annual deep ploughing may also lead to greatly depleted soil invertebrate populations (Edwards and Lofty 1975, 1982b, Tucker 1992).

Recreation

Most agricultural habitats are not so frequently used for recreational purposes as many other habitats, or by such large numbers of people. However, walking, picnicking, horse-riding and shooting are widespread rural activities which can cause significant disturbance. In general, such effects have not been demonstrated to have a significant impact on agricultural bird populations. However, some priority species of steppe are considered to be at risk as a result of increased visitor numbers in Spain (Suárez *et al.* 1997b), and the amount of illegal hunting has increased considerably in some countries of central and eastern Europe during the 1990s (e.g. Romania), with as yet unknown effects on habitat quality.

Despite the demise of traditional agricultural activities in montane habitats in most parts of Europe, some areas (e.g. the Alps and Pyrenees) are now subject to particularly high levels of disturbance throughout the year as a result of economic development fuelled by summer tourism and winter sports. The increased disturbance has led to reductions in the area of suitable nesting habitat for some sensitive species, e.g. birds of prey such as *Aquila chrysaetos* (Pedrini 1991, Argelich *et al.* 1996). The sprawling growth of settlements, roads, powerlines and hotels, with associated facilities for refuse collection and disposal, and water supply and disposal, is often *ad hoc* and unplanned, and is a major cause of habitat loss and degradation in many montane areas (e.g. Österreichische Raumordnungskonferenz 1992).

Hillsides are graded and cleared of vegetation for the construction of special runs for skiing or tobogganing, cable-cars and ski-lifts, which are sometimes covered in man-made snow (from reservoirs, etc.) in 'dry' winters, all of which causes ecological problems such as direct destruction of soil and vegetation (e.g. Grabherr *et al.* 1987), vegetation damage and change, reduction of arthropod life, changes in local climate and hydrology, and excessive soil erosion (Cernusca 1977). Mismanaged ski-runs have been responsible for landslips and partially for flood disasters in recent decades (Aulitzky 1994).

Even the less harmful activity of summer tourism, which started in the last century, has steadily grown and is now a mass phenomenon. Hiking, camping and mountain-biking cause excessive damage to vegetation and soils by trampling, etc. Vegetation only recovers very slowly since rates of seed production and vegetative growth of even dominant species (e.g. the sedge *Carex curvula*) are very slow at high altitudes (Grabherr 1982). Disturbance of sensitive bird species (especially gamebirds in winter, e.g. Zeitler 1994) can be locally severe when people wander away from marked trails. Recently 'adventure sports' have increased notably in mountainous areas, including helicopter-skiing, gliding, parascending, motor-rallying and rock-climbing, all of which increase disturbance impacts on birds, especially breeding raptors (e.g. Slotta-Bachmayr and Werner 1992, Georgii *et al.* 1994).

Overhead structures

Overhead structures, such as cables, powerlines, wind-turbines, ski-lifts and tall wire-fences, can cause major mortality of larger priority birds through collision and electrocution (e.g. Penteriani 1994, Ruggieri *et al.* 1996). For example, collision with powerlines has been found to be the major cause of adult mortality for *Otis tarda* in some parts of Spain (Alonso *et al.* 1994), and collision with overhead cables is a major source of mortality for *Tetrao tetrix* (Miquet 1990) and Eagle Owl *Bubo bubo* (Haller 1978) in montane agricultural habitats. Electrocution is the commonest non-natural cause of death for the globally threatened *Aquila adalberti* (Calderón *et al.* 1988, Oria and Caballero 1992, González 1996).

Also, although on a European scale they cause only a small amount of direct habitat destruction, the mere presence (and operation) of such structures reduces habitat quality over much larger areas for certain priority species which prefer or need 'open' or disturbance-free habitats (see 'Habitat needs of priority birds', p. 299).

Increased predators

Local or regional increases in the population density of crows (Corvidae), gulls (Laridae), *Vulpes vulpes*, feral dogs and other native predators are often due to increased provision by man of feeding or nesting opportunities, e.g. creation of rubbish dumps, or planting of trees as shelter-belts in steppic habitats. An increased density of predators can lead to increases in direct mortality rates of eggs, chicks and adults of priority species, which may have an impact at population level if the problem occurs over a sufficiently wide area.

For example, local declines in the number of breeding waders of a wet grassland area in Denmark may have been due to increased local numbers of Common Gull *Larus canus* (Thorup 1992). In steppic habitats, high rates of such predation have been a problem for *Otis tarda*, *Burhinus oedicnemus* and *Calandrella rufescens* in Spain (Yanes 1994, Yanes and Suárez 1996, Suárez *et al.* 1997b), and for *Glareola nordmanni* and the globally threatened *Chettusia gregaria* in Russia (Molodan 1980, Gordienko 1991).

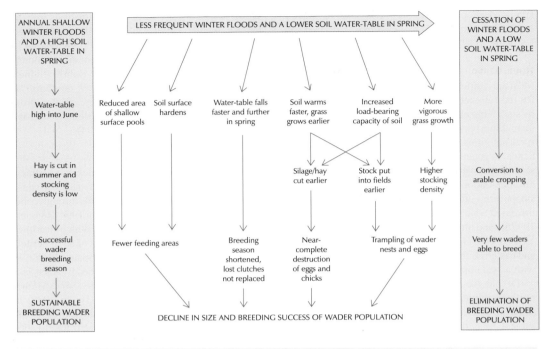

Figure 7. The effects of different kinds of water-level management on wet grassland, and on the habitat features required by priority breeding species of waders (Charadrii) (from Buisson and Williams 1991).

Increasing levels of predation may be contributing to the declines of several species on intensive farmland, e.g. in the UK (Fuller *et al.* 1995). Predation levels in the UK have probably increased overall, as there have been large increases in the national populations of several corvids, particularly *Corvus monedula*, *C. corone* and Magpie *Pica pica* (Marchant *et al.* 1990) and of *Vulpes vulpes* (Reynolds and Tapper 1995) in recent decades. On the other hand, a study of British buntings (Emberizidae) found no evidence for higher nest-predation rates since 1970 compared to earlier decades (Crick *et al.* 1994). Also, higher rates of predation do not necessarily lead to lower population densities of prey, even at a national or wider scale, since prey populations may be able to absorb such losses, either through increased nesting attempts per year, or through density-dependent increases in survival rates (e.g. a higher than usual proportion of juveniles survive, due to lessened competition for food and territories, etc.).

It is, however, possible that predation could have a significant impact on farmland bird populations locally or regionally, if such habitat is suboptimal for the species (in other ways) and results in lower than normal breeding productivity, as shown for *Vanellus vanellus* (Hudson *et al.* 1994). Predation rates for a

particular species may be higher in farmland than in other habitats as a result of poorer cover for nesting birds. Indeed, evidence of relatively high rates of predation in farmland is provided by Andrén and Angelstam (1988), Grajetzky (1993) and Suhonen *et al.* (1994). Also, although the underlying cause of the decline of *Perdix perdix* in the UK was a result of reduced food supplies, nest predation was found to be an important factor in determining the rate of decline (Potts 1986).

Urbanization and road-building

Although urbanization and road-building are widespread in Europe, direct effects on agricultural habitats are in most cases likely to be fairly restricted, and measurable impacts on bird populations will consequently be mostly local. However, in certain cases such local impacts can be very important; when, for example, developments are proposed that affect small remaining pockets of important habitats or rare and isolated populations of priority species. Threats from road-building, housing and urban development should therefore be treated seriously.

Some areas and habitats may also be subject to more concentrated pressures from these activities. Road-building is particularly frequent in mountain areas, e.g. parts of the Alps, to improve access for developments such as agriculture, forestry, winter sports, hydropower dams, etc. Luder (1993) found that the majority of bird populations in a Swiss mountain area had shown declines, probably as a result of the quadrupling of roads between 1979–1980 and 1991–1992. However, it is apparent that the roads largely acted as precursors to other changes, such as levelling of ground, drainage and increased fertilizer use, and that habitat fragmentation also resulted.

Drainage and flood control

Drainage of wetlands in Europe has been widespread this century (Baldock 1984, Williams and Bowers 1987, Williams *et al.* 1988); see 'Coastal Habitats' (p. 93) and 'Inland Wetlands' (p. 125) for examples. This has a profound effect on wet grasslands and their priority bird species. For example, lakes and marshes adjacent to wet grasslands can be drained, leading to the direct loss of foraging or safe nesting sites for priority species such as *Anas querquedula* and *A. acuta* (Rutschke 1989) or the globally threatened *Acrocephalus paludicola* (Sellin 1989). Drainage also reduces the abundance or accessibility of suitable prey, e.g. amphibians for *Ciconia ciconia* (Schulz 1994) or soil invertebrates for waders, and can lead to more rapid vegetation growth in spring, making grasslands unsuitable for nesting (Green 1986, O'Brien and Self 1994, RSPB/ITE/EN in press) Also, drainage can reduce the likelihood of winter flooding and, therefore, of feeding opportunities for a variety of waterbirds. Drainage results in a number of complex and interacting effects which cannot be discussed in detail here, but are summarized in Figure 7.

Wet, flood-plain grasslands require annual, natural river-flooding for their continued maintenance (see 'Management systems', p. 274). This means that most if not all large-scale river-development schemes, e.g. the building of river embankments and storage reservoirs for flood control, lead to partial or complete destruction of such grassland, due to reduced flooding frequency, and reduced summer water-levels and supplies. In some cases the affected areas are located hundreds of kilometres downstream of such structures (e.g. dams for irrigation, drinking water or hydroelectric power).

Most of the flood-plain grasslands of Europe have already been destroyed by such river engineering, and almost all remaining 'natural' large-river flood-plains in Europe are threatened by such developments, either ongoing or planned. Due to the connectedness of riverine wetland systems, efforts to conserve flood-plain grassland may fail if they have only a local focus—integrated conservation of the entire river catchment is the only effective ap-

proach (see 'Inland Wetlands' p. 125).

The rate of drainage of wet grassland for agricultural purposes may decline in future as a result of agricultural surpluses in the EU and the depressed livestock markets in eastern Europe. However, some further drainage for this, and due to river engineering, is still likely, especially along the larger rivers in central and eastern Europe. Increased rates of water abstraction for irrigation and drinking water may also lead to reduced water flow in rivers and to lowered water-tables in wet grassland (see 'Inland Wetlands', p. 145).

Reduction in oak density

Over large parts of Iberia, oak-tree disease (Brasier 1992) is thought to be reducing tree density in pastoral woodland, and hence leading to the loss of nesting and feeding sites for priority species of pastoral woodland. Further details of this threat are given under 'Mediterranean Forest, Shrubland and Rocky Habitats' (p. 258).

In addition, there is apparently no woodland regeneration at most dehesa farms (Montero 1988). Such lack of regeneration appears to result from the negative impact of cultivation and livestock-rearing practices on the production of acorns and the survival of seedlings and young trees (Díaz *et al.* 1997). At present, it is unclear whether this lack of regeneration is associated with recent changes in dehesa management or whether it is a problem inherent to traditional farming methods used in dehesas (Díaz *et al.* 1997).

Loss of hay meadows

The loss of hay meadows, e.g. to silage, pasture or arable crops, leads to a loss of priority birds which need the tall grassland and other features associated with this habitat (see Larsson 1976). This is a widespread threat to some priority birds of wet grassland, e.g. on the Baltic coastal meadows, where traditional joint usage for extensive grazing and for hay has been widely replaced by intensive cattle grazing, due to a massive decrease in demand for hay. Reintroduction of traditional management in Denmark led to a gradual restoration of populations of such priority species as *Calidris alpina*, *Limosa limosa* and *Tringa totanus* (Thorup 1991, 1992).

The introduction of intensive sheep-grazing is also a threat to the habitat of *Crex crex* in Scotland (Crockford *et al.* 1996).

Low stocking levels

In contrast to over-stocking, both low stocking levels and the removal of stock over the winter can also be significant threats to priority bird species that require short-grazed vegetation. Low stocking levels are not currently widespread in Europe. However, in intensive livestock production systems it is becoming more common to overwinter cattle in stockyards. The impacts of this on birds have not been documented, but it can be supposed that it may reduce food availability and foraging efficiency for species such as *Pluvialis apricaria*, *Vanellus vanellus*, *Sturnus vulgaris* and various thrushes (Turdidae) which take soil invertebrates in winter (Tucker 1992).

Atmospheric nutrient pollution

In parts of north-west Europe, particularly Belgium, the Netherlands and north-west Germany, ammonia generated by decomposition of manure slurry in the soil, or vented from sheds holding large numbers of livestock, can reach high concentrations in the atmosphere and can be transported long distances. This volatilization of ammonia, as well as of nitrogen compounds from combustion of fossil fuels, has been found to lead to atmospheric deposition rates of 10–20 kg/ha/year of nitrogen over most parts of Europe (Stanners and Bourdeau 1995). This can significantly increase soil fertility levels, which may result in the loss of nutrient-poor but wildlife-rich habitats (Ellenberg *et al.* 1989).

Detrimental impacts are likely to be restricted to semi-natural agricultural habitats that have not received fertilizer applications. Montane grasslands in central Europe are probably among the habitats most at risk due to high rainfall and proximity to the most severe pollution sources. Impacts on birds have not been documented, but it can be predicted that species which require sparse and low vegetation resulting from poor nutrient status may be at risk, e.g. *Lullula arborea* and *Anthus campestris*. Indeed, Ellenberg (1986b) and Maréchal (1993) have suggested that the triggering factor for the continent-wide decline of *Lanius collurio* is heavy inputs of nitrogen which cause vegetation to grow earlier, denser and higher in spring. High nitrogen inputs may also threaten priority birds such as *Picus viridis* which forage predominantly on ants (Formicidae) or other soil invertebrates of nutrient-poor conditions (Havelka and Rüge 1993).

Replacement of old crop-trees

The replacement of old crop-trees with young ones, or with new and more productive varieties, is currently a common practice in many commercial olive-groves (Pain 1994a) and in orchards. This reduces the amount of crevices and dead wood, which are essential for hole-nesting species such as *Otus scops*, woodpeckers (Picidae) and *Phoenicurus phoenicurus*, and also results in reduced invertebrate food availability for woodpeckers and *Certhia brachydactyla*.

Loss of old buildings

Rural depopulation is widespread in Europe, particularly in areas of marginal agriculture. However, elsewhere the renovation of old farm buildings and dwellings has become commonplace in recent decades, particularly in areas within commuting distance of towns and cities. This, together with the destruction of unsightly or unsafe derelict buildings, is thought to be reducing the availability of nest-sites for species such as *Tyto alba* and *Hirundo rustica* (Tucker and Heath 1994). Improvement of housing standards is reducing the availability of roof nest-sites for the globally threatened *Falco naumanni* with serious impacts in certain areas (Hallmann 1985, González *et al.* 1990, Biber 1996).

Reductions in animal carcasses

As mentioned previously under 'Land abandonment' (p. 309), reductions in livestock numbers in some marginal farmland areas are leading to reduced food availability for carrion-dependent birds such as vultures, and the advent of stricter disposal methods for carrion has affected these priority species more widely. This reduction in available carcasses appears to have been offset by other changes in some countries, e.g. in Spain there has been an increase in livestock numbers elsewhere and increased carrion has become available at municipal dumps (muladares) for dead or slaughtered livestock (Fernández 1988). Indeed, in recent years *Gyps fulvus* populations in Spain have increased greatly (Tucker and Heath 1994), and food availability there appears to be adequate at present (Donázar and Fernández 1990).

However, further tightening of regulations concerning the disposal of animal carcasses may eliminate 'mule tips' and any other new sources of food, and lead to further widespread reductions in carrion availability.

Autumn sowing of crops

This problem is mainly restricted to parts of north-west Europe where autumn-sown crops have become much more common on intensively farmed arable land. For example, in the UK, by the early 1980s autumn-sowing was the dominant practice in the lowlands, though the area of spring barley continued to decline up to the early 1990s (Fuller *et al.* 1995). As a consequence, winter stubbles have become much reduced. Stubbles provide important winter food supplies for seed-eating birds in the form of unharvested grain and weed seeds (Donald and Evans 1994, Evans and Smith 1994, Wilson *et al.* 1995, Evans 1997).

Also, some species such as *Vanellus vanellus* (Shrubb and Lack 1991, Hudson *et al.* 1994) and *Alauda arvensis* (Schläpfer 1988, Wilson *et al.* 1995) prefer the relatively sparsely vegetated ground associated with spring crops when breeding.

Clearance of ground vegetation

A recently adopted management practice in olive-groves is the mechanized clearance of ground vegetation between the trees (Pain 1994a). This reduces feeding habitat and invertebrate numbers for priority species such as *Oenanthe hispanica*. Although the provision of bare, open ground does benefit some bird species, even they usually require some vegetation for nesting. Thus complete clearance is also detrimental for ground-nesting species such as *Alectoris rufa*.

Conversion of pasture

In intensively farmed areas, particularly, it is becoming increasingly common for livestock to be kept in stockyards rather than being allowed to graze pastures. Pastures are then used for hay or, currently much more commonly, silage production for cattle feed. This leads to a reduction in the extent of short, grazed, permanent grassland which is the preferred foraging habitat for a wide variety of birds such as *Vanellus vanellus*, *Burhinus oedicnemus* and *Pyrrhocorax pyrrhocorax*. Silage fields often hold considerably lower soil-invertebrate populations if they are temporary re-seeded leys (Tucker 1992). Also, as mentioned elsewhere in this section, silage fields are likely to be heavily fertilized and therefore tall and dense in spring, and unsuitable for nesting and foraging for a wide range of birds.

Dry cultivation of rice

Widespread and increasing use of 'dryland' rice (fields of which do not need flooding) is a potential threat to certain waterbird species with a relatively high proportion of their European population dependent for foraging on wet rice cultivation during the breeding season, e.g. *Nycticorax nycticorax*. However, development of such new rice varieties is still under way, and current impacts are only local.

Other threats

In the grazed steppic habitats and montane grassland of some parts of Europe, livestock farmers still lay poisoned baits (e.g. strychnine) to kill *Canis lupus*, *Vulpes vulpes* and feral dogs for their perceived role in killing domestic livestock. This widespread practice causes high mortality of priority birds which are dependent on scavenging for at least part of the yearly cycle, such as vultures and *Aquila* eagles. The widespread use of poison baits in regional campaigns to control rabies also causes mortality in such species (Pedrini 1991). Since poisoning may not always be fatal, birds which recover may be inhib-

ited about foraging in such areas in the future, thus reducing the habitat quality of their home range.

CONSERVATION OPPORTUNITIES

International legislation

A variety of environmental legislative instruments and regulations apply to agricultural and grassland habitats and could contribute to their conservation (Box 1). These are described in general terms in the chapter on 'Opportunities for Conserving the Wider Environment' (p. 24), while more specific details are given here.

Many regulations exist, in western Europe in particular, to protect water, soil and air from pollution caused by agriculture. In the European Union (EU) most of these laws, such as the Nitrates Directive (see Box 1), require governments to influence farming practices so as to reduce pollution. Several countries also have legal protection for specific habitat or landscape features, such as grassland or hedges in Denmark, for sites of nature-conservation value in the UK, or for specific rare or endangered habitats in Germany. The Habitats Directive requires all important European habitat-types to be protected adequately in EU member states. However, regulations are often difficult to enforce, and there is political reluctance to impose restrictions and costs on farmers. It is widely recognized that the regulatory approach has limitations in agriculture, where there are large numbers of small producers farming mostly in ways that are encouraged by other policies.

Many potentially valuable regulations exist within the EU or are being introduced in central and east European countries. The opportunities lie in (1) finding ways of enforcing these, and (2) integrating their requirements with wider land-use policies. Priorities will be to ensure that the EU member states fulfil their obligations, e.g. through fully implementing the Habitats and Birds Directives, taking into account farming systems of high conservation value. Other legislation covering restrictions on pesticide use, protection of landscape features and controls on water pollution need to be enforced, or where necessary introduced. Planning restrictions on the conversion of agricultural land to built development need to be based on nature-conservation value and not solely on agricultural production criteria.

Policy initiatives

Clearly, the conservation of farmland birds is dependent on those factors that have a wide-ranging influence on land-use. Very few species have a significant proportion of their European population in agricultural Important Bird Areas (IBAs). Fur-

---

Box 1. International environmental legislation of particular relevance to the conservation of agricultural and grassland habitats in Europe. See 'Opportunities for Conserving the Wider Environment' (p. 24) for details.

GLOBAL

- *Biodiversity Convention* (1992) Convention on Biological Diversity.
- Framework Convention on Climate Change (1992)
- *Bonn Convention* or *CMS* (1979) Convention on the Conservation of Migratory Species of Wild Animals.
- *Ramsar Convention* (1971) Convention on Wetlands of International Importance especially as Waterfowl Habitat.
- *World Heritage Convention* (1972) Convention concerning the Protection of the World Cultural and Natural Heritage.
- Convention on Long Range Transboundary Air Pollution (1979).
- Convention on the Protection and Use of Transboundary Watercourses and International Lakes (1992).

PAN-EUROPEAN

- *Bern Convention* (1979) Convention on the Conservation of European Wildlife and Natural Habitats.
- *Espoo Convention* (1991, but yet to come into force) Convention on Environmental Impact Assessment in a Transboundary Context.

EUROPEAN UNION

- *Birds Directive* Council Directive on the conservation of wild birds (79/409/EEC).
- *Habitats Directive* Council Directive on the conservation of natural habitats of wild fauna and flora (92/43/EEC).
- *EIA Directive* Council Directive on the assessment of the effects of certain public and private projects on the environment (85/337/EEC).
- *Nitrates Directive* Council Directive protecting waters against pollution by agricultural nitrates (91/676/EEC).
- Council Directive on the freedom of access to information on the environment (90/313/EEC).
- *Agri-environment Regulation* Council Regulation on agricultural production methods compatible with the requirements of the protection of the environment and the maintenance of the countryside (2078/92/EEC).
- Council Regulation on afforestation of agricultural land (2080/92/EEC).
- Council Regulations on the Structural Funds (e.g. 2081/93/EEC).
- Council Regulation on the conservation, characterization, collection and utilization of genetic resources in agriculture (1467/94/EEC).
- Council Regulation on organic production of agricultural products and indications referring thereto on agricultural products and foodstuffs (2092/91/EEC).

---

thermore, for those species that do, such as *Otis tarda*, very few of their IBAs are protected, and the habitat is thus still highly susceptible to changes in agricultural policy.

A strategy for influencing land-use in selected farmed areas, or across farmland generally, therefore needs to consider local, regional, national and international policies for influencing these factors. What is needed is a mix of policies integrated by common and clear objectives.

Agriculture policy is highly complex, dynamic and politicized. While environmental issues are often uppermost in debates on the future of agriculture, actual opportunities for influence by nature conservation organizations are limited by competing economic, social and political concerns, as well as other environmental issues. Policies fall into three categories:

- Worldwide political and economic policies, which need to be understood but which are not easy to influence, such as multilateral policies to reduce subsidies.

- International or regional policies which have widespread effects on the environment, such as CAP reform and land privatization. Conservation interests can influence but cannot expect to direct these.

- National or regional policies with specific conservation objectives, such as the Agri-environment Regulation (2078/92/EEC) in the EU.

Within these categories the following policy areas are the most important for achieving nature conservation on farmland:

Fundamental objectives of rural policies

The legal and constitutional basis for intervention in farming often works against changing specific policies. In the EU, amendments to the Treaty of Rome under the Single European Act (Article 130r) and the Maastricht Treaty require environmental protection to be a component of all EU policies. However, this has not been matched by substantive changes to the objectives of the Common Agricultural Policy, resulting in a confused set of objectives for EU agriculture policies and the marginalization of the environment in most policies, such as set-aside. By contrast, Norway, for example, has fundamentally reappraised its objectives for rural areas to include sustainable use of inputs and protection of biodiversity.

Land-policy reforms in central and eastern Europe offer scope for incorporating environmental protection and this is happening, for example, in Russia in a series of land laws. However, due to widespread corruption and poverty, these regulations have never been enforced.

Opportunities exist for redefining the objectives of EU agriculture and rural policies so as to encompass protection of important farmed habitats, farmland biodiversity and promotion of sustainable use of resources. The objectives of environmental legislation, including international conventions and directives, should be tied legally to the objectives of rural land-use policies. Future reforms to specific CAP policies will occur periodically, with the possibility of a number of large-scale reforms to the CAP in the next ten years. EU enlargement negotiations are a specific opportunity to change the objectives of EU and future EU member states' rural policies.

The rapid transition of the national agriculture policies of central and east European countries provides an opportunity to influence the objectives of rural policy. Aid and trade arrangements with the EU could be conditional on such objectives being stated in national land-use policies and properly enforced.

Overall levels and forms of subsidy

Market prices are altered significantly in European agriculture by a number of mechanisms such as supported prices, border controls and tariffs and direct subsidies. Overall levels of subsidy have powerful influences over land-use, input use and farm incomes. For these reasons this is probably the single largest area of policy which has to be influenced to achieve conservation objectives on farmland—in particular, through the linkage of subsidies to ecological criteria for conservation and sustainable use. Agriculture subsidies in the EU account for approximately 50% of agricultural turnover, while in countries in central, eastern and south-east Europe, the 'effective' subsidy level is much lower (for example, c.8% in Hungary), although accession to the EU would in all likelihood raise this.

Attempts to reduce overall levels of subsidy have been made in the context of CAP reform, EU enlargement and the Uruguay round of the General Agreement on Tariffs and Trade (GATT). Such changes are likely to be incremental, with reductions of c.10–20% in the next ten years. Subsidies are generally shifting from mainly indirect price support to more direct payments. The agricultural economies of central and east European countries are currently in flux, with, for instance, rapid and complete removal of high subsidy levels in Poland and Hungary and the introduction of new price supports in Russia and the Czech Republic. A reduction in fertilizer use in western Europe may lead to cheaper imports of fertilizer into Russia.

Policies aimed at restricting the supply of products and hence increasing their market price and revenue to the farmer, and at reducing subsidy levels, include restrictions on outputs (quotas), land (set-aside) and inputs (extensification, taxes or bans on pesticides and fertilizers). Such policies have been introduced in different combinations in the EU; for example 72,000 km^2 of arable land was set aside in 1995 under an annual scheme. Limited livestock extensification measures including quotas and premiums for reduced stocking exist. In Austria and Sweden, prior to EU accession, nitrogen taxes operated, but these were of limited effect (on nitrogen use or on the environment) and have been abandoned.

Input restrictions are new, and have yet to be evaluated for their ability to meet policy objectives. There is much potential for deploying these for specific conservation purposes as well as for achieving general change in farmland by top-up schemes and eligibility conditions. Specific schemes to promote environmentally friendly products by enhancing market prices exist in most countries, including premiums for organic products.

Further information is required to assess the response of farmers and land-use to falling support prices. Overall trends in EU prices, and the transitional support policies of central and east European countries as they adjust to western-style policies, need to be monitored carefully. It is likely that falling support prices will have some beneficial effects, such as reduced incentives to overstock or apply fertilizers, but some low-intensity farm systems may be abandoned or change in nature.

Where it is clear that reductions in subsidy or changes in the type of subsidy (for example towards hectarage payments) will bring benefits, this should be encouraged as part of future reforms. Where direct subsidies replace indirect (price support) subsidies, environmental conditions could be attached to these (cross-compliance).

Trading arrangements

Trading patterns influence market prices and the major trading blocs have considerable control over these. The EU has, for example, encouraged trade within its borders (the 'Single Market') causing agricultural specialization, intensification and land abandonment in different regions. Restrictions or liberalization of trade policies exist with European Economic Area countries and with central and east European countries under bilateral agreements ('Association' or 'Europe' Agreements) and in the GATT. Currently there is little scope for influencing these policies, but environmental considerations are slowly being brought into trade debates.

Opportunities exist within current multilateral discussions on trade to differentiate acceptable forms of subsidy, such as management-agreement schemes, which interfere less with trade than existing subsidies on production. Also, restrictions on exports from the EU can restrict further intensification within the Union, although restrictions on the markets for produce, such as EU barriers, can harm sectors of agriculture, e.g. beef raising, that are important for the management of some important farmland habitats.

Incorporating countries with extensive farming systems, such as Spain, into a large single market, has in the past opened up regional agriculture to competition in ways that have led to the decline of traditional extensive farming systems.

Economic instruments
Numerous taxes and exemptions from taxes can apply to farmers, but so far this has not been used widely for specific environmental objectives. Examples include taxes on fertilizers or exemption from land taxes for specific purposes in France and the UK. In addition, specific payments can be made to farmers for capital items or revenue. These payments can be exclusively environmental or have mainly agricultural or social objectives. Capital grants for agricultural development have been harmful (Baldock 1990a) and in some countries, such as Denmark and the UK, these have largely been withdrawn. In areas subject to depopulation or otherwise less developed agriculturally, such as in southern Europe, capital support for land drainage, large-scale irrigation and other damaging works continue (Beaufoy et al. 1994). Payments exist in most EU countries to re-create habitats that benefit biodiversity, and there are proposals for environmental conditions to apply to many direct payments in the EU.

Management contracts paid to farmers annually can be made as compensation or as rewards for managing habitats or crop systems of value to wildlife. Schemes have existed within the EU for over 10 years, particularly in the Netherlands, Denmark, UK, France and Germany. Under the Agri-environment Regulation, agreed during the 1992 CAP reforms, total spending for all EU countries is planned to exceed ECU 6,600 million between 1993 and 1997. Outside the EU, similar proposals based on pilot areas have been made for Poland, Hungary and the Czech Republic. Funding is often limited in these

schemes, objectives often unclear and the positive benefits to wildlife rarely evaluated fully. However, this approach offers more to the bird species that are restricted to farmland than do other, less focused policies.

Specific reductions on pesticides could be achieved by taxes, however it is doubtful whether this is the right policy for controls on toxic substances. Taxes might be appropriate for limitations on fertilizer use, although for locally specific reductions to improve water quality this is unlikely to be effective. Capital grants are still given for damaging capital works in much of the EU and the remainder of Europe. Support for infrastructure development in the EU, and through aid programmes in central and eastern Europe should be subject to strategic environmental assessment. Capital grants to farmers could be made conditional on not harming sites of conservation value or could be withdrawn altogether.

Probably the area of policy with most potential is the development of specific, targeted management agreements such as Environmentally Sensitive Areas. Management contracts can be made to encourage farmers to (a) maintain existing valuable farming systems, (b) reduce the level of existing high-intensity systems and (c) create new habitats. In the EU, best use has to be made of the Agri-environment Regulation so as to ensure that schemes and prescriptions are targeted at priorities and that there is a full evaluation and monitoring programme. Pilot schemes, demonstration projects and workshops to develop technical skills among operators can also help achieve specific conservation benefits.

During 1995 and 1996, BirdLife International Partner organizations in EU countries have been reviewing the effectiveness for nature conservation of the measures under the Agri-environment Regulation, as part of a wider review of the measure. Rarely do member states devote sufficient resources to monitoring the actual environmental effects of policies. They often measure uptake, area of land in schemes, and other 'process' objectives, rather than measuring outcomes or results. Little evidence exists of the actual benefits of such policies, and much further evaluation of these policies is necessary.

Information, education, extension services
Several countries have voluntary or state-supported extension services for farmers and in some countries in the EU, such as the Netherlands, Germany and the UK, habitat management is taught in agricultural colleges.

High quality ecological and practical advice needs to be available for farmers to adjust to requirements of conservation management. Nature conservation organizations could ensure this by (1) giving limited advice on certain species, (2) advising other farm management or conservation advisers, agricultural colleges, etc., and (3) promoting the provision of an advisory service where this does not exist.

Land tenure and ownership restriction or liberalization

Land reforms are unlikely in most of western Europe, but in the transitional economies of central and eastern Europe privatization of previously state-owned or cooperative farms offers scope for ensuring that land of high ecological value is maintained in sympathetic ownership. This is a policy in its very early stages of development, and varies greatly between countries, and so cannot yet be evaluated in general terms. Privatization has progressed much further in some countries, for example in Hungary, than it has in Russia. State-ownership (e.g. of grasslands, forests, national parks and wetlands) has vested these lands with some security in the past and this is now under some threat. Alternatively, 'collective' ownership of land may have led in some cases to highly unsustainable use of water, soils and other natural resources in systems where no environmental accounting took place and the private culture of stewardship of land was not present.

Overall, however, the level of agricultural intensification was much lower in the huge collective farms of former East Germany than in the small, privately owned West German farms, and this pattern may well have held for other central and east European countries. Following unification the huge farms in eastern Germany have proved far easier to convert to organic production than the small western ones.

Banking systems, privatization processes, land-credit systems and incentives for private investment are being introduced in central and eastern Europe. There is some scope for specific nature-conservation land purchase (Mundy 1992).

Large-scale reserve creation occurs in the Netherlands and Denmark, particularly for wetlands, by the permanent or semi-permanent purchase of property rights by the government, working in collaboration with conservation agencies. Such schemes are supported in part as a means of retiring land from productive agriculture.

Afforestation and diversification schemes

Schemes to encourage the diversion of rural labour and resources away from agriculture offer positive opportunities for wildlife tourism, extensive food production and reforestation. These policies bring threats also. For example, in Germany most agricultural areas chosen to be afforested tend to be marginal, with low productivity, but of high ecological value. These areas would be better used for extensive agriculture, thus maintaining their value for rare or threatened wildlife species of agricultural habitats. National policies aimed at reducing the area of land under cultivation therefore also reduce the opportunities to convert land from intensive to extensive cultivation. For priority species of open habitats, subsequent conversion of such land to non-cultivation uses, e.g. forestry, therefore presents a problem.

CONSERVATION RECOMMENDATIONS

Broad conservation objectives

1. To maintain and enhance farming systems and habitats of high conservation value.

2. To improve the quality and conservation value of degraded agricultural habitats of currently low conservation value.

3. To consider the conversion of agricultural land of low conservation value to other habitats, where this results in net conservation benefits.

Ecological targets for the habitat

In order to maintain and improve habitat quality, it is essential to retain and enhance habitat features and attributes that are important for priority species. It is therefore necessary to ensure that the following conditions are met.

1. A high diversity of nesting sites and feeding resources should be ensured through the maintenance of mosaics of crop-types or of management regimes (e.g. different mowing, grazing or water-level regimes) within the farmland landscape. New crops should, however, not be introduced where large blocks of semi-natural habitat are essential for priority species.

2. Specific grassland and crop-types should be maintained within farm landscapes where these are highly preferred or essential requirements for priority species.

3. Important marginal habitat features such as hedgerows, trees, ponds, ditches, rank grassland and scattered rocks, etc. should be maintained wherever possible and restored where this would be of net conservation benefit.

4. The openness of habitats should be maintained where this is necessary for the majority of priority species, within steppic and wet grassland habitats.

5. Fragmentation of large blocks of semi-natural habitats of high conservation value should be avoided, especially of dry, wet and montane grasslands and pastoral woodland.

6. Suitable vegetation structure and composition for priority ground-nesting birds should be maintained, in particular through:
   a. reductions in fertilizer use;
   b. avoidance of irrigation and of re-seeding of semi-natural grasslands with fast-growing grass mixes;
   c. appropriate management of grazing and of mowing of grassland, and;
   d. spring sowing of arable crops.

7. The availability of suitable nest-sites for hole-nesting priority species in old buildings, etc., should be maintained where possible, and enhanced or replaced with artificial nest structures where nest-site availability limits populations.

8. Important vertebrate, invertebrate and plant food resources should be maintained, in particular through:
   a. the avoidance of pesticide use, or the carefully managed use of selective and rapidly biodegradable pesticides, in response to pest problems;
   b. the maintenance of arable stubble fields (especially when untreated with herbicides), fallow land and set-aside over winter (to provide food resources for seed-eating birds); and
   c. the use of organic farmyard manures in preference to artificial fertilizers (within appropriate limits that avoid pollution), and of cropping systems that include under-sowing, break crops or rotations, and the avoidance of repeated intensive cultivation techniques such as deep ploughing (to maintain soil invertebrate populations).

9. Areas of closely grazed grass and bare soil should be available for foraging where required by certain priority species which are ground-feeders.

10. The seasonal hydrological regime of wet grassland should be such that sward-heights and water-levels are appropriate and that soil invertebrates are abundant and available for priority species. For flood-plain meadows and polders, in particular, the frequency and extent of winter/early-spring flooding should be maintained at or near historically natural levels, and a high soil water-table should be maintained during the breeding season, during which man-induced floods should be avoided.

    Adequate levels and flow-rates of groundwater and surface water should be restored where there has been previous depletion of surface water or groundwater/ aquifers, wherever possible.

11. Adequate levels of livestock carrion should be maintained in the main foraging habitats of carrion-feeding priority species where feasible, e.g. through special dispensations allowing less strict disposal methods in certain areas where public health standards can be guaranteed (protected areas, military zones, etc.).

12. Old crop-trees should be retained where possible in commercial olive-groves and orchards, etc., and ground-level vegetation maintained and managed appropriately for priority species present.

13. The management of pastoral woodland must ensure the long-term maintenance of oak-tree cover by promoting tree regeneration, as well as the maintenance of a mosaic of grazed under-storey grasslands, cereal croplands and shrubby patches.

14. The re-creation of flood-plain and delta wetlands and of breeding habitat for priority species (e.g. small woodlands for heronries) should be promoted in areas of rice cultivation, especially where 'dryland' varieties of rice have replaced the need for fields to be flooded, and where former rice-fields have been abandoned altogether.

15. Disturbance of sensitive species by agricultural operations should be minimized, especially during critical periods of the annual cycle, e.g. through restrictions on the timing of agricultural operations and avoidance of particularly disturbing practices, such as aerial spraying where sensitive priority species occur.

16. Direct mortality and sublethal effects (e.g. eggshell thinning) from the use of toxic pesticides should be avoided, or at least reduced to levels which do not threaten any bird populations.

17. Mortality of adults, eggs and young of ground-nesting birds due to farm operations (especially hay- and silage-cutting) and to livestock trampling (particularly in dry and wet grasslands) should be minimized through appropriate timing and management of operations and of grazing.

18. Predation rates should be within natural, sustainable levels, and not be increased through human activities which lead to higher population densities of predators (e.g. improper disposal of rubbish.

19. Mortality levels of priority species from collisions with, and electrocution by, powerlines and other overhead structures should be low, and should where necessary be reduced through appropriate measures (e.g. marking, screening, re-routing or burying of cables, adapting pylons).

Broad policies and actions

To meet the above objectives and habitat targets in accordance with the relevant strategic principles (see p. 24), the following broad policies and actions must be pursued (see p. 22 in 'Introduction' for more explanation of their derivation).

General

1. Conservation objectives should be an integral part of policies affecting agriculture. This is vital for the Common Agricultural Policy (CAP) of the EU and other national agriculture policies, which need wider objectives including a specific requirement to protect Europe's environment.

2. Agricultural policy should move from intervention towards strategic guidance, based on targets for social, economic and environmental standards. Targets should be clearly articulated and should be 'owned' across national governments and within the EU. National and EU biodiversity strategies should be a mechanism for achieving this. Policies on trade, external relations and agriculture should be made mutually supportive, both nationally and across the EU, and should not cut across the objectives of each other at the macro-economic level.

3. A new deal should be struck between farmers, land-owners, rural communities and society at large, where a greater level of respect, understanding and awareness of rural life is developed. Key to this contract will be the recognition in public policy, that subsidies should only be paid to recompense for costs incurred by farmers as stewards of the countryside, or for defined social reasons.

    There should also be a gradual and selective reduction in overall subsidies for production, so as to remove the damaging distortions in land-use, farming practices and the economy that they cause, to a level equivalent to the Producer Subsidy Equivalent of the Organization for Economic Cooperation and Development as a percentage of total agricultural production of 25%. Countries currently below this level (such as in central and eastern Europe) should develop sustainable agricultural policies, recognizing that the environmental damage that has been caused in western Europe could be avoided.

    Special conditions should be attached to credits provided by the World Bank/International Monetary Fund to Russia, in order to prevent further environmental damage.

Incorporating environmental objectives into the Common Agricultural Policy (CAP)

4. Decision-making in the CAP should be made more accountable to parliaments and interest groups other than producers, by provision of resources and access to, e.g. the Management Committee on Agricultural Structures and Rural Development (STAR Committee), and the Advisory Committees on Agriculture of the Directorate-General for Agriculture (DG VI) of the European Commission.

5. National governments and the EU should encourage greater innovation in policy-making by strengthening 'think-tanks' that are independent of operational responsibilities, by creating more inter-departmental and national/EU working groups, and by reviewing research and development programmes so as to replace largely production-oriented research by policy research aimed at solving new policy problems.

6. Policies for the livestock sector (such as Less Favoured Areas, beef, dairy and sheep) should be revised so that they do not promote over- or under-grazing, e.g. by making payments dependent on area (e.g. of grazed land) rather than on headage (number of livestock), or by use of environmental conditions. Policies should support low-intensity, grass-based farming.

7. Sufficient funding should be allocated to the Agri-environment Regulation, and its implementation should be enforced throughout the Community; by the year 2000, 15% of the EU's agricultural land should be covered by management agreements.

8. Schemes under the Agri-environment Regulation should be targeted at areas of marginal agriculture with high nature-conservation importance, and should have clear environmental objectives. Biological targets for habitats, as outlined above, should be incorporated as appropriate, and schemes should be monitored to assess their success in achieving their objectives. Reduced intensity of arable cultivation (e.g. through conservation headlands), organic production, rare and indigenous breeds, spring cereals and crop rotations should all be supported.

9. There should be greater coordination between the Agri-environment Regulation and incentives for afforestation of agricultural land under EC Regulation 2080/92.

10. All direct payments, e.g. for arable or livestock production, under the CAP should be conditional on recipients protecting the environment. Price supports, direct payments and other subsi-

dies should promote extensification (especially reduced pesticide and fertilizer use).

11. CAP supply-control policies, such as set-aside, extensification, and production quotas should be given specific environmental objectives, e.g. the maintenance and improvement of winter food resources for seed-eating birds, and should be required to not harm the environment, in particular nesting birds. Set-aside could be used to create non-agricultural habitats and to promote break crops and extensification, but there must be restrictions on set-aside where it has harmful consequences. Quotas could be used to prevent the expansion of agricultural areas.

12. The European Commission should undertake a full audit of the implications of the CAP on the environment of existing EU, central and east European countries, and of the developing world. The EU should prepare an agri-environment action plan for the accession of countries in central and eastern Europe and the Mediterranean area.

13. Environmental objectives within regional policies (whether agricultural or not), such as Less Favoured Areas, Objectives 1–6 and Mediterranean policies, should be improved. Allied EU policies, particularly on rural development and the environment, need to be integrated with agriculture and more fully developed. The Structural Funds, expenditure under which is partly justified on agricultural grounds, need to promote environmentally sustainable development and be subject to environmental assessment. The Regulations and operational criteria of the Funds (e.g. the 'framework' Regulation 2081/93) need to be revised to allow specifically for biodiversity conservation.

14. Community environmental objectives, recognized within EU environmental law, should be integrated within agricultural policies, such as the Agri-environment Regulation. Strategic Environmental Assessment should apply to all policies, plans and programmes within the CAP. Community legislation on environmental protection, particularly relating to pollution and pesticide control, should be enforced across the EU.

EU member states and regional authorities

15. The Agri-environment Regulation should be implemented fully across member states and sufficient national funds allocated. Member-state agriculture and environment ministries should jointly determine national priorities for incentive schemes in agriculture. Nature conservation authorities should develop an EU-wide network

of experts in agriculture, incentive schemes and related policy and research.

16. National legislation, e.g. on land tenure, commonland, water management and rural development, should be revised in order to implement new agri-environment policies. Strategic Environmental Assessment (SEA) should be applied to all plans, programmes and policies, especially land-based policies. Crop diversity, farm structure and landscape diversity should be maintained (sometimes increased, sometimes decreased) through national policies.

Non-EU countries and agencies in Europe

17. Environmental protection should be integrated with transitional and new agricultural policies for countries of central, eastern and Mediterranean Europe. All investment policy, agricultural policies, trade agreements and legislation, whether international or national, should promote sustainable agriculture which maintains farmland of high nature conservation value. Farm structural adjustment in previously centrally controlled countries, particularly privatization, should recognize these environmental interests.

18. International agencies should support demonstration projects, including aid schemes and extension services to encourage environmentally sensitive farming. Technical advice on environmental incentive schemes in agriculture should be provided to government agencies under 'know-how' funds. Support should be provided to NGOs (assuming strong NGO capacity in the country) in order that they develop expertise and understanding of land-use policies.

19. NGOs should develop a catalyst role, encouraging the adoption of new environmental policies in agriculture. They should also play a watchdog role, ensuring conservation priorities are achieved by new policies. Specialist expertise in agriculture should be developed in key nature conservation NGOs, including through the appointment of staff.

20. Inorganic fertilizers, pesticides and other inputs (including land, water and energy) could be taxed according to environmental damage, so as to promote extensification, mixed farming and the maintenance of traditional, low-intensity farming systems, especially systems based on grass and organic manure. Intensively raised livestock could also be taxed. Taxes should not be used for controlling use of harmful pesticides or where fertilizers or other practices are damaging. In these situations, tighter regulatory control over use, trade, supply and approval should apply.

21. Land-use planning systems should be introduced which ensure coherent and sustainable rural development, incorporating a diversity of forests, agricultural lands, wetlands and other habitats.

Forestry policies should have environmental objectives and be subject to full cost-benefit analysis. Planning controls should exist for forestry and agricultural developments, which should be linked to national forestry strategies. (See 'Boreal and Temperate Forests', p. 203.)

Water policies should ensure that subsidies are removed and taxes introduced to reflect the true environmental costs of abstraction, impoundment, distribution and use. (See also 'Inland Wetlands', p.125.)

General research priorities

22. A classification of farming systems in relation to their conservation value should be developed and used to map important locations.

23. Research is required to provide specific, locally applicable recommendations on management practices, e.g. threatened key farmland species need their specific habitat requirements determined.

24. Large-scale and detailed research studies should be undertaken to establish and quantify reliably the impacts of pesticides on key priority bird populations. Further research should also be carried out, if impacts are substantial, to identify the most appropriate, feasible and cost-effective solutions.

ACKNOWLEDGEMENTS

Most contributions to this chapter were made by the Habitat Working Groups during and after workshops on arable and improved grassland, perennial crops and pastoral woodland organized by S. Nagy at Kecskemét, Hungary (30 August–3 September 1993), on wet grassland organized by A. Beintema and hosted by Instituut voor Bos- en Natuuronderzoek (IBN-DLO) at Terschelling, Netherlands (11–15 September 1993), on steppic habitats organized by M. A. Naveso and held at Centro de Investigación de Espacios Naturales Protegidos, Soto del Real, Spain (26 September–1 October 1993), and on montane grassland organized by B. Lombatti and held at Trento, Italy (14–18 June 1994); these contributors are listed on p. 267.

Additionally, valuable information, comments and assistance were received from A. Matschke (Associació per a la Defensa de la Natura), J. Frühauf (BirdLife Österreich–Gesellschaft für Vogelkunde), P. Iankov (Balgarsko Druzhestvo za Zashtita na Pticite), J. Elts (Eesti Ornitoloogiaühing), M. Flade and P. Herkenrath (Naturschutzbund Deutschland), S. Nagy (Magyar Madártani és Természetvédelmi Egyesület), F. Casale (Lega Italiana Protezione Uccelli), J. Winkelman, E. Osieck and T. den Boer (Vogelbescherming Nederland), I. J. Øien (Norsk Ornitologisk Forening), F. Moreira (Liga para a Protecção da Natureza), D. Munteanu (Societatea Ornitologica Romana), E. Lebedeva (Soyuz Ochrany Ptits Rossii), M. A. Naveso (Sociedad Española de Ornitología), I. Gorban, A. Mikityuk and I. Mihalevich (Ukrainske Tovaristvo Okhoroni Ptakhiv), D. Pain and P. José (Royal Society for the Protection of Birds).

Summary of the priority status of bird species in different habitats in Europe, and their international legal status

THE ORDER and nomenclature of bird species in this book follow Voous (1977) and *The birds of the western Palearctic* (Cramp *et al.* 1977–1994) respectively, except for species featured in Collar *et al.* (1994), for which that work is followed.

The following are dealt with in the table.

SPEC category

1 Species of global conservation concern, i.e. classified as Globally Threatened, Conservation Dependent or Data Deficient (Collar *et al.* 1994).

2 Concentrated in Europe and with an Unfavourable Conservation Status.

3 Not concentrated in Europe but with an Unfavourable Conservation Status.

4 Concentrated in Europe and with a Favourable Conservation Status.

For SPEC categories given as, for example, '4/2', the first number is the SPEC category relating to the breeding population, while the second number relates to the winter population.

w Category relates to winter populations.

European Threat Status

E	Endangered	L	Localized
V	Vulnerable	Ins	Insufficiently Known
R	Rare	S	Secure
D	Declining	( )	Status provisional

w Category relates to winter populations.

EC Birds Directive

For further information, see p. 32. The list below incorporates amendments to the Annexes of the Council Directive (79/409/EEC) of 6 March 1991 (91/244/EEC), 8 June 1994 (94/24/EC) and 29 August 1994 (94/C241/08).

Annex I

The Directive requires that species listed in Annex I 'shall be the subject of special conservation measures concerning their habitat in order to ensure their survival and reproduction in their area of distribution' and that 'Member States shall classify in particular the most suitable territories in number and size as special protection areas for the conservation of these species, taking into account their protection requirements in the geographical sea and land area where this Directive applies'.

In addition, 'Member States shall take similar measures for regularly occurring migratory species not listed in Annex I, bearing in mind their need for protection in the geographical sea and land area where this Directive applies, as regards their breeding, moulting and wintering areas and staging posts along their migration routes'.

Annex II

'The species referred to in Annex II/1 may be hunted in the geographical sea and land area where the Directive applies'. 'Species referred to in Annex II/2 may be hunted only in Member States in respect of which they are indicated'.

Annex III

For species listed in Annex III/1 Member States shall not prohibit 'the sale, transport for sale, keeping for sale and the offering for sale of live or dead birds and of any readily recognizable parts or derivatives of such birds', provided that the birds have been legally killed or captured or otherwise legally acquired. Member States may for the species listed in Annex III/2, allow within their territory the activities referred to above, making provision for certain restrictions, provided the birds have been legally killed or captured or otherwise legally acquired. The activities referred to above are prohibited for all other species of naturally occurring wild birds in the European territory of EU Member States.

Bern Convention

For further information, see p. 29. The list below incorporates additions made by the Standing Committee in December 1987.

Parties undertake to take appropriate and necessary measures for the conservation of the habitats of wild flora and fauna, especially those on Appendices I (plants) and II, and to give special attention to the protection of areas of importance for the migratory species on Appendices II and III, and to prohibit the deliberate damage or destruction of sites for species listed in Appendix II.

Parties undertake to regulate any exploitation of the wild fauna specified in Appendix III and prohibit blameworthy means of capture and killing.

Bonn Convention

For further information, see p. 27. The list below incorporates amendments by the Conference of the Parties up to 1994.

Appendix I: 'Species in danger of extinction throughout all or major parts of their range'

Parties to the Convention undertake to provide immediate protection to species included in Appendix I, and Range States should conserve and, where feasible and appropriate, restore those habitats of the species which are of importance in removing it from danger of extinction.

Appendix II: 'Species which would benefit from international cooperation in their conservation and management'

Parties to the Convention shall 'conclude Agreements covering the conservation and management of migratory species included in Appendix II'. Each Agreement should, where appropriate, provide for the 'maintenance of a network of suitable habitats appropriately disposed in relation to migratory routes'.

AEWA: Agreement on the Conservation of African–Eurasian Migratory Waterbirds

For information about this Agreement under the Bonn Convention, see p. 27. The list below is correct as of April 1997.

Priority status

The status (A–D) is given for each habitat; the derivation of the priority status is explained under 'Introduction' (p. 19).

Footnotes

[1] *P. y. mauritanicus* only
[2] *P. a. baroli* only
[3] *P. a. desmarestii* only
[4] *T. t. tetrix* only
[5] *T. t. britannicus* only
[6] *A. g. saxatilis* and *A. g. whitaken* only
[7] *P. p. italica* and *P. p. hispaniensis* only
[8] *C. p. azorica* only
[9] *F. c. ombriosa* only

	SPEC category	European Threat Status	Birds Directive	Bern Convention	Bonn Convention	AEWA	N&W European seas	European Macaronesian seas	Mediterranean and Black seas	Coastal habitats	Inland wetlands
Red-throated Diver *Gavia stellata*	3	V	I	II	II	•	B	–	D	–	C
Black-throated Diver *Gavia arctica*	3	V	I	II	II	•	C	–	D	–	B
Great Northern Diver *Gavia immer*	–	(S)	I	II	–	•	D	–	–	–	–
White-billed Diver *Gavia adamsii*	–	(S)	–	II	–	•	D	–	–	–	–
Little Grebe *Tachybaptus ruficollis*	–	S	–	II	–	–	–	–	–	–	D
Great Crested Grebe *Podiceps cristatus*	–	S	–	III	–	–	–	–	–	–	D
Red-necked Grebe *Podiceps grisegena*	–	S	–	II	–	•	–	–	–	–	D
Slavonian Grebe *Podiceps auritus*	–	(S)	I	II	–	•	–	–	–	–	D
Black-necked Grebe *Podiceps nigricollis*	–	S	–	II	–	–	–	–	–	–	D
Fulmar *Fulmarus glacialis*	–	S	–	III	–	–	D	–	–	–	–
Fea's Petrel *Pterodroma feae*	1	E	I	II	–	–	–	A	–	–	–
Zino's Petrel *Pterodroma madeira*	1	E	I	II	–	–	–	A	–	–	–
Bulwer's Petrel *Bulweria bulwerii*	3	V	I	II	–	–	–	B	–	–	–
Cory's Shearwater *Calonectris diomedea*	2	(V)	I	II	–	–	C	A	D	–	–
Great Shearwater *Puffinus gravis*	–	–	–	III	–	–	D	D	–	–	–
Sooty Shearwater *Puffinus griseus*	–	–	–	III	–	–	D	–	–	–	–
Manx Shearwater *Puffinus puffinus*	2	(L)	–	II	–	–	A	C	–	–	–
Yelkouan Shearwater *Puffinus yelkouan*	4	S	I[1]	II	–	–	D	–	C	–	–
Little Shearwater *Puffinus assimilis*	3	V	I	II[2]	–	–	–	B	–	–	–
White-faced Storm-petrel *Pelagodroma marina*	3	L	I	II	–	–	–	B	–	–	–
Storm Petrel *Hydrobates pelagicus*	2	(L)	I	II	–	–	A	C	C	–	–
Leach's Storm-petrel *Oceanodroma leucorhoa*	3	(L)	I	II	–	–	B	–	–	–	–
Madeiran Storm-petrel *Oceanodroma castro*	3	V	I	II	–	–	D	B	–	–	–
Gannet *Sula bassana*	2	L	–	III	–	–	A	D	–	–	–
Cormorant *Phalacrocorax carbo*	–	S	–	III	–	–	–	–	–	–	–
Shag *Phalacrocorax aristotelis*	4	S	I	III[3]	–	–	C	–	–	–	–
Pygmy Cormorant *Phalacrocorax pygmeus*	2	V	I	II	II	•	–	–	–	B	B
White Pelican *Pelecanus onocrotalus*	3	R	I	II	I/II	•	–	–	–	B	C
Dalmatian Pelican *Pelecanus crispus*	1	V	I	II	I/II	•	–	–	–	A	A
Bittern *Botaurus stellaris*	3	(V)	I	II	II	•	–	–	–	D	B
Little Bittern *Ixobrychus minutus*	3	(V)	I	II	II	•	–	–	–	C	B
Night Heron *Nycticorax nycticorax*	3	D	I	II	–	–	–	–	–	D	C
Squacco Heron *Ardeola ralloides*	3	V	I	II	–	–	–	–	–	C	B
Cattle Egret *Bubulcus ibis*	–	S	I	II	–	–	–	–	–	–	–
Little Egret *Egretta garzetta*	–	S	I	II	–	–	–	–	–	–	D
Great White Egret *Egretta alba*	–	S	I	II	–	•	–	–	–	–	D
Grey Heron *Ardea cinerea*	–	S	–	III	–	–	–	–	–	–	D
Purple Heron *Ardea purpurea*	3	V	I	II	II	•	–	–	–	C	B
Black Stork *Ciconia nigra*	3	R	I	II	II	•	–	–	–	–	C
White Stork *Ciconia ciconia*	2	V	I	II	II	•	–	–	–	–	C
Glossy Ibis *Plegadis falcinellus*	3	D	I	II	II	•	–	–	–	C	C
Spoonbill *Platalea leucorodia*	2	E	I	II	II	•	–	–	–	A	C
Greater Flamingo *Phoenicopterus ruber*	3	L	I	II	II	•	–	–	–	B	C
Mute Swan *Cygnus olor*	–	S	–	III	II	•	–	–	–	–	D
Bewick's Swan *Cygnus columbianus*	3[W]	L[W]	I	II	II	•	–	–	–	C	D
Whooper Swan *Cygnus cygnus*	4[W]	S	I	II	II	•	–	–	–	D	–
Bean Goose *Anser fabalis*	–	S	–	III	II	•	–	–	–	–	–
Pink-footed Goose *Anser brachyrhynchus*	4	S	II	III	II	•	–	–	–	D	–
White-fronted Goose *Anser albifrons*	–	S	–	III	II	•	–	–	–	–	–
Lesser White-fronted Goose *Anser erythropus*	1	V	I	II	II	•	–	–	–	–	A
Greylag Goose *Anser anser*	–	S	–	III	II	•	–	–	–	–	–
Barnacle Goose *Branta leucopsis*	4/2	L[W]	I	II	II	•	–	–	–	A	–
Brent Goose *Branta bernicla*	3	V	–	III	II	•	–	–	–	B	–
Red-breasted Goose *Branta ruficollis*	1	L[W]	I	II	II	•	–	–	–	–	A
Ruddy Shelduck *Tadorna ferruginea*	3	V	I	II	II	•	–	–	–	C	B
Shelduck *Tadorna tadorna*	–	S	–	II	II	•	–	–	–	D	–
Wigeon *Anas penelope*	–	S	–	III	II	•	–	–	–	D	–
Gadwall *Anas strepera*	3	V	II	III	II	•	–	–	–	C	C
Teal *Anas crecca*	–	S	–	III	II	•	–	–	–	–	–
Mallard *Anas platyrhynchos*	–	S	–	III	II	•	–	–	–	–	D
Pintail *Anas acuta*	3	V	II/III	III	II	•	–	–	–	C	C
Garganey *Anas querquedula*	3	V	–	III	II	•	–	–	–	–	C
Shoveler *Anas clypeata*	–	S	–	III	II	•	–	–	–	–	D
Marbled Teal *Marmaronetta angustirostris*	1	E	I	II	II	•	–	–	–	A	A
Red-crested Pochard *Netta rufina*	3	D	–	III	II	•	–	–	–	C	C

Appendix 1: Birds' Priority Status in European Habitats, and International Legal Status

Tundra, mires, moorland	Lowland Atlantic heathland	Boreal forest	Lowland temperate forest	Montane forest	Riverine forest	Mediterranean habitats	Arable and improved grassland	Steppic habitats	Montane grassland	Wet grassland	Rice cultivation	Perennial crops	Pastoral woodland	
B	–	–	–	–	–	–	–	–	–	–	–	–	–	*Gavia stellata*
B	–	–	–	–	–	–	–	–	–	–	–	–	–	*Gavia arctica*
D	–	–	–	–	–	–	–	–	–	–	–	–	–	*Gavia immer*
D	–	–	–	–	–	–	–	–	–	–	–	–	–	*Gavia adamsii*
–	–	–	–	–	–	–	–	–	–	–	–	–	–	*Tachybaptus ruficollis*
–	–	–	–	–	–	–	–	–	–	–	–	–	–	*Podiceps cristatus*
–	–	–	–	–	–	–	–	–	–	–	–	–	–	*Podiceps grisegena*
–	–	–	–	–	–	–	–	–	–	–	–	–	–	*Podiceps auritus*
–	–	–	–	–	–	–	–	–	–	–	–	–	–	*Podiceps nigricollis*
–	–	–	–	–	–	–	–	–	–	–	–	–	–	*Fulmarus glacialis*
–	–	–	–	–	–	–	–	–	–	–	–	–	–	*Pterodroma feae*
–	–	–	–	–	–	–	–	–	–	–	–	–	–	*Pterodroma madeira*
–	–	–	–	–	–	–	–	–	–	–	–	–	–	*Bulweria bulwerii*
–	–	–	–	–	–	–	–	–	–	–	–	–	–	*Calonectris diomedea*
–	–	–	–	–	–	–	–	–	–	–	–	–	–	*Puffinus gravis*
–	–	–	–	–	–	–	–	–	–	–	–	–	–	*Puffinus griseus*
–	–	–	–	–	–	–	–	–	–	–	–	–	–	*Puffinus puffinus*
–	–	–	–	–	–	–	–	–	–	–	–	–	–	*Puffinus yelkouan*
–	–	–	–	–	–	–	–	–	–	–	–	–	–	*Puffinus assimilis*
–	–	–	–	–	–	–	–	–	–	–	–	–	–	*Pelagodroma marina*
–	–	–	–	–	–	–	–	–	–	–	–	–	–	*Hydrobates pelagicus*
–	–	–	–	–	–	–	–	–	–	–	–	–	–	*Oceanodroma leucorhoa*
–	–	–	–	–	–	–	–	–	–	–	–	–	–	*Oceanodroma castro*
–	–	–	–	–	–	–	–	–	–	–	–	–	–	*Sula bassana*
–	–	–	–	–	–	–	–	–	–	–	–	–	–	*Phalacrocorax carbo*
–	–	–	–	–	–	–	–	–	–	–	–	–	–	*Phalacrocorax aristotelis*
–	–	–	–	–	B	–	–	–	–	–	–	–	–	*Phalacrocorax pygmeus*
–	–	–	–	–	–	–	–	–	–	–	–	–	–	*Pelecanus onocrotalus*
–	–	–	–	–	–	–	–	–	–	–	–	–	–	*Pelecanus crispus*
–	–	–	–	–	–	–	–	–	–	–	–	–	–	*Botaurus stellaris*
–	–	–	–	–	–	–	–	–	–	–	D	–	–	*Ixobrychus minutus*
–	–	–	–	–	C	–	–	–	–	B	–	–	–	*Nycticorax nycticorax*
–	–	–	–	–	C	–	–	–	–	D	–	–	–	*Ardeola ralloides*
–	–	–	–	–	–	D	–	–	–	–	–	–	–	*Bubulcus ibis*
–	–	–	–	–	–	–	–	–	–	–	–	–	–	*Egretta garzetta*
–	–	–	–	–	–	–	–	–	–	–	–	–	–	*Egretta alba*
–	–	–	–	–	–	–	–	–	–	–	–	–	–	*Ardea cinerea*
–	–	–	–	–	–	–	–	–	–	D	D	–	–	*Ardea purpurea*
–	–	C	D	C	C	–	–	–	–	–	–	–	C	*Ciconia nigra*
–	–	–	–	–	–	B	B	–	B	–	–	–	B	*Ciconia ciconia*
–	–	–	–	–	C	–	–	–	–	–	D	–	–	*Plegadis falcinellus*
–	–	–	–	–	B	–	–	–	–	–	C	–	–	*Platalea leucorodia*
–	–	–	–	–	–	–	–	–	–	–	–	–	–	*Phoenicopterus ruber*
–	–	–	–	–	–	–	–	–	–	–	–	–	–	*Cygnus olor*
D	–	–	–	–	–	–	D	–	–	C	–	–	–	*Cygnus columbianus*
–	–	–	–	–	–	–	–	–	–	D	–	–	–	*Cygnus cygnus*
D	–	–	–	–	–	–	–	–	–	–	–	–	–	*Anser fabalis*
D	–	–	–	–	–	–	D	–	–	D	–	–	–	*Anser brachyrhynchus*
D	–	–	–	–	–	–	–	–	–	–	–	–	–	*Anser albifrons*
A	–	–	–	–	–	–	–	A	–	A	–	–	–	*Anser erythropus*
–	–	–	–	–	–	–	–	–	–	–	–	–	–	*Anser anser*
C	–	–	–	–	–	–	B	–	–	–	–	–	–	*Branta leucopsis*
B	–	–	–	–	–	–	D	–	–	C	–	–	–	*Branta bernicla*
–	–	–	–	–	–	–	A	–	–	–	–	–	–	*Branta ruficollis*
–	–	–	–	–	–	–	–	C	–	–	–	–	–	*Tadorna ferruginea*
–	–	–	–	–	–	–	–	–	–	–	–	–	–	*Tadorna tadorna*
–	–	–	–	–	–	–	–	–	–	–	–	–	–	*Anas penelope*
–	–	–	–	–	–	–	–	–	–	–	–	–	–	*Anas strepera*
–	–	–	–	–	–	–	–	–	–	–	–	–	–	*Anas crecca*
–	–	–	–	–	–	–	–	–	–	–	–	–	–	*Anas platyrhynchos*
C	–	–	–	–	–	–	–	–	–	D	D	–	–	*Anas acuta*
–	–	–	–	–	–	–	–	–	–	C	D	–	–	*Anas querquedula*
–	–	–	–	–	–	–	–	–	–	–	–	–	–	*Anas clypeata*
–	–	–	–	–	–	–	–	–	–	–	–	–	–	*Marmaronetta angustirostris*
–	–	–	–	–	–	–	–	–	–	–	D	–	–	*Netta rufina*

	SPEC category	European Threat Status	Birds Directive	Bern Convention	Bonn Convention	AEWA	N&W European seas	European Macaronesian seas	Mediterranean and Black seas	Coastal habitats	Inland wetlands
Pochard *Aythya ferina*	4	S	II/III	III	II	•	–	–	–	–	C
Ferruginous Duck *Aythya nyroca*	1	V	I	III	II	•	–	–	–	A	A
Tufted Duck *Aythya fuligula*	–	S	–	III	II	•	–	–	–	–	D
Scaup *Aythya marila*	3[W]	L[W]	II/III	III	II	•	C	–	–	C	–
Eider *Somateria mollissima*	–	S	–	III	II	•	D	–	–	–	–
King Eider *Somateria spectabilis*	–	S	–	II	II	•	D	–	–	–	–
Steller's Eider *Polysticta stelleri*	1	L[W]	–	II	II	•	A	–	–	–	–
Harlequin Duck *Histrionicus histrionicus*	3	V	–	II	II	–	B	–	–	–	–
Long-tailed Duck *Clangula hyemalis*	–	S	–	III	II	•	D	–	–	–	–
Common Scoter *Melanitta nigra*	–	S	–	III	II	•	D	–	–	–	–
Velvet Scoter *Melanitta fusca*	3[W]	L[W]	II	III	II	•	B	–	–	–	–
Barrow's Goldeneye *Bucephala islandica*	3	E	–	II	II	–	–	–	–	–	–
Goldeneye *Bucephala clangula*	–	S	–	III	II	•	–	–	–	–	D
Smew *Mergus albellus*	3	V	–	II	II	•	–	–	–	–	D
Red-breasted Merganser *Mergus serrator*	–	S	–	III	II	•	D	–	–	–	–
Goosander *Mergus merganser*	–	S	–	III	II	•	–	–	–	–	D
White-headed Duck *Oxyura leucocephala*	1	E	I	II	I/II	•	–	–	–	–	A
Honey Buzzard *Pernis apivorus*	4	S	I	II	II	–	–	–	–	–	–
Black-winged Kite *Elanus caeruleus*	3	V	I	II	II	–	–	–	–	–	–
Black Kite *Milvus migrans*	3	V	I	II	II	–	–	–	–	–	C
Red Kite *Milvus milvus*	4	S	I	II	II	–	–	–	–	–	–
White-tailed Eagle *Haliaeetus albicilla*	3	R	I	II	I	–	–	–	–	C	C
Lammergeier *Gypaetus barbatus*	3	E	I	II	II	–	–	–	–	–	–
Egyptian Vulture *Neophron percnopterus*	3	E	I	II	II	–	–	–	–	–	–
Griffon Vulture *Gyps fulvus*	3	R	I	II	II	–	–	–	–	–	–
Cinereous Vulture *Aegypius monachus*	3	V	I	II	II	–	–	–	–	–	–
Short-toed Eagle *Circaetus gallicus*	3	R	I	II	II	–	–	–	–	–	–
Marsh Harrier *Circus aeruginosus*	–	S	I	II	II	–	–	–	–	–	D
Hen Harrier *Circus cyaneus*	3	V	I	II	II	–	–	–	–	C	–
Pallid Harrier *Circus macrourus*	3	E	I	II	II	–	–	–	–	–	–
Montagu's Harrier *Circus pygargus*	4	S	I	II	II	–	–	–	–	–	–
Goshawk *Accipiter gentilis*	–	S	–	II	II	–	–	–	–	–	–
Sparrowhawk *Accipiter nisus*	–	S	–	II	II	–	–	–	–	–	–
Levant Sparrowhawk *Accipiter brevipes*	2	R	I	II	II	–	–	–	–	–	–
Buzzard *Buteo buteo*	–	S	–	II	II	–	–	–	–	–	–
Long-legged Buzzard *Buteo rufinus*	3	(E)	I	II	II	–	–	–	–	–	–
Rough-legged Buzzard *Buteo lagopus*	–	S	–	II	II	–	–	–	–	–	–
Lesser Spotted Eagle *Aquila pomarina*	3	R	I	II	II	–	–	–	–	–	–
Greater Spotted Eagle *Aquila clanga*	1	E	I	II	II	–	–	–	–	–	A
Steppe Eagle *Aquila nipalensis*	3	V	–	II	II	–	–	–	–	–	–
Imperial Eagle *Aquila heliaca*	1	E	I	II	II	–	–	–	–	–	–
Spanish Imperial Eagle *Aquila adalberti*	1	E	I	II	II	–	–	–	–	B	–
Golden Eagle *Aquila chrysaetos*	3	R	I	II	II	–	–	–	–	–	–
Booted Eagle *Hieraaetus pennatus*	3	R	I	II	II	–	–	–	–	–	–
Bonelli's Eagle *Hieraaetus fasciatus*	3	E	I	II	II	–	–	–	–	–	–
Osprey *Pandion haliaetus*	3	R	I	II	II	–	–	–	–	C	A
Lesser Kestrel *Falco naumanni*	1	(V)	I	II	II	–	–	–	–	–	–
Kestrel *Falco tinnunculus*	3	D	–	II	II	–	–	–	–	–	–
Red-footed Falcon *Falco vespertinus*	3	V	–	II	II	–	–	–	–	–	–
Merlin *Falco columbarius*	–	S	I	II	II	–	–	–	–	–	–
Hobby *Falco subbuteo*	–	S	–	II	II	–	–	–	–	–	–
Eleonora's Falcon *Falco eleonorae*	2	R	I	II	II	–	–	–	–	A	–
Lanner *Falco biarmicus*	3	(E)	I	III	II	–	–	–	–	–	–
Saker *Falco cherrug*	3	E	–	II	II	–	–	–	–	–	–
Gyrfalcon *Falco rusticolus*	3	V	–	II	II	–	–	–	–	–	–
Peregrine *Falco peregrinus*	3	R	I	II	II	–	–	–	–	C	–
Barbary Falcon *Falco pelegrinoides*	–	S	–	II	II	–	–	–	–	–	–
Hazel Grouse *Bonasa bonasia*	–	S	I	III	–	–	–	–	–	–	–
Red/Willow Grouse *Lagopus lagopus*	–	S	–	III	–	–	–	–	–	–	–
Ptarmigan *Lagopus mutus*	–	S	–	III	–	–	–	–	–	–	–
Black Grouse *Tetrao tetrix*	3	V	I[4]/II/III[5]	III	–	–	–	–	–	–	–
Caucasian Black Grouse *Tetrao mlokosiewiczi*	2	Ins	–	III	–	–	–	–	–	–	–
Capercaillie *Tetrao urogallus*	–	(S)	I	II	–	–	–	–	–	–	–
Caucasian Snowcock *Tetraogallus caucasicus*	4	S	–	III	–	–	–	–	–	–	–
Caspian Snowcock *Tetraogallus caspius*	3	Ins	–	III	–	–	–	–	–	–	–

	Tundra, mires, moorland	Lowland Atlantic heathland	Boreal forest	Lowland temperate forest	Montane forest	Riverine forest	Mediterranean habitats	Arable and improved grassland	Steppic habitats	Montane grassland	Wet grassland	Rice cultivation	Perennial crops	Pastoral woodland	
–	–	–	–	–	–	–	–	–	–	–	–	–	–	–	*Aythya ferina*
–	–	–	–	–	–	–	–	–	–	–	–	–	–	–	*Aythya nyroca*
–	–	–	–	–	–	–	–	–	–	–	–	–	–	–	*Aythya fuligula*
D	–	–	–	–	–	–	–	–	–	–	–	–	–	–	*Aythya marila*
–	–	–	–	–	–	–	–	–	–	–	–	–	–	–	*Somateria mollissima*
–	–	–	–	–	–	–	–	–	–	–	–	–	–	–	*Somateria spectabilis*
–	–	–	–	–	–	–	–	–	–	–	–	–	–	–	*Polysticta stelleri*
–	–	–	–	–	–	–	–	–	–	–	–	–	–	–	*Histrionicus histrionicus*
D	–	–	–	–	–	–	–	–	–	–	–	–	–	–	*Clangula hyemalis*
D	–	–	–	–	–	–	–	–	–	–	–	–	–	–	*Melanitta nigra*
B	–	–	–	–	–	–	–	–	–	–	–	–	–	–	*Melanitta fusca*
–	–	–	–	–	–	–	–	–	–	–	–	–	–	–	*Bucephala islandica*
–	–	–	–	–	–	–	–	–	–	–	–	–	–	–	*Bucephala clangula*
–	–	B	–	–	–	–	–	–	–	–	–	–	–	–	*Mergus albellus*
–	–	–	–	–	–	–	–	–	–	–	–	–	–	–	*Mergus serrator*
–	–	–	–	–	–	–	–	–	–	–	–	–	–	–	*Mergus merganser*
–	–	–	–	–	–	–	–	–	–	–	–	–	–	–	*Oxyura leucocephala*
–	–	D	D	D	–	–	–	–	–	–	–	–	–	–	*Pernis apivorus*
–	–	–	–	–	–	–	C	–	–	–	–	–	–	C	*Elanus caeruleus*
–	–	–	C	C	C	C	C	–	–	–	–	D	–	C	*Milvus migrans*
–	–	–	C	D	D	D	D	D	–	–	–	–	–	D	*Milvus milvus*
D	–	C	C	–	C	–	–	–	–	–	–	–	–	–	*Haliaeetus albicilla*
–	–	–	–	–	–	C	–	–	B	–	–	–	–	–	*Gypaetus barbatus*
–	–	–	–	–	–	C	–	C	C	–	–	–	–	–	*Neophron percnopterus*
–	–	–	–	–	–	C	–	C	C	–	–	–	–	–	*Gyps fulvus*
–	–	–	–	–	–	B	–	–	D	–	–	–	–	C	*Aegypius monachus*
–	–	–	C	C	D	B	D	C	–	–	–	–	–	C	*Circaetus gallicus*
–	–	–	–	–	–	–	–	–	–	–	–	–	–	–	*Circus aeruginosus*
C	D	D	–	–	–	–	C	C	–	C	–	–	–	–	*Circus cyaneus*
–	–	–	–	–	–	–	–	B	–	–	–	–	–	–	*Circus macrourus*
–	–	–	–	–	–	–	C	D	–	–	–	–	–	–	*Circus pygargus*
–	–	–	–	–	–	–	–	–	–	–	–	–	–	–	*Accipiter gentilis*
–	–	–	–	–	–	–	–	–	–	–	–	–	–	–	*Accipiter nisus*
–	–	B	–	B	B	B	–	–	–	–	C	–	–	–	*Accipiter brevipes*
–	–	–	–	–	–	–	–	–	–	–	–	–	–	–	*Buteo buteo*
–	–	–	–	–	C	C	B	–	–	–	–	–	–	–	*Buteo rufinus*
D	–	–	–	–	–	–	–	–	–	–	–	–	–	–	*Buteo lagopus*
–	–	B	–	C	D	C	–	–	C	–	–	–	–	–	*Aquila pomarina*
–	A	A	–	A	–	–	–	–	A	–	–	–	–	–	*Aquila clanga*
–	–	–	–	–	–	B	–	–	–	–	–	–	–	–	*Aquila nipalensis*
–	–	A	B	–	B	A	A	–	B	–	–	–	–	–	*Aquila heliaca*
–	–	–	–	–	A	–	–	–	–	–	–	–	A	–	*Aquila adalberti*
D	–	C	C	C	–	C	–	C	–	–	–	–	–	–	*Aquila chrysaetos*
–	–	C	C	–	B	–	–	–	–	–	–	–	C	–	*Hieraaetus pennatus*
–	–	–	–	–	B	–	–	–	–	–	–	–	–	–	*Hieraaetus fasciatus*
–	C	C	–	C	–	–	–	–	–	–	–	–	–	–	*Pandion haliaetus*
–	–	–	–	–	–	B	A	–	–	–	–	–	–	–	*Falco naumanni*
–	D	D	–	D	D	C	C	D	D	–	–	–	D	–	*Falco tinnunculus*
–	D	D	–	D	–	C	C	–	–	D	–	–	–	–	*Falco vespertinus*
D	–	–	–	–	–	–	–	–	–	–	–	–	–	–	*Falco columbarius*
–	–	–	–	–	–	–	–	–	–	–	–	–	–	–	*Falco subbuteo*
–	–	–	–	A	–	–	–	–	–	–	–	–	–	–	*Falco eleonorae*
–	–	–	–	B	–	B	–	–	–	–	–	–	–	–	*Falco biarmicus*
–	–	C	–	–	C	B	–	–	–	–	–	–	–	–	*Falco cherrug*
B	–	–	–	–	–	–	–	–	–	–	–	–	–	–	*Falco rusticolus*
C	–	C	C	–	C	–	C	–	–	–	–	–	–	–	*Falco peregrinus*
–	–	–	–	–	–	–	–	–	–	–	–	–	–	–	*Falco pelegrinoides*
–	–	D	–	–	–	–	–	–	–	–	–	–	–	–	*Bonasa bonasia*
D	–	–	–	–	–	–	–	–	–	–	–	–	–	–	*Lagopus lagopus*
D	–	–	–	–	–	–	–	–	–	–	–	–	–	–	*Lagopus mutus*
D	D	C	C	C	–	–	–	–	D	–	–	–	–	–	*Tetrao tetrix*
–	–	–	–	–	–	–	–	A	–	–	–	–	–	–	*Tetrao mlokosiewiczi*
–	–	D	–	–	–	–	–	–	–	–	–	–	–	–	*Tetrao urogallus*
–	–	–	–	–	–	–	–	C	–	–	–	–	–	–	*Tetraogallus caucasicus*
–	–	–	–	–	–	–	–	B	–	–	–	–	–	–	*Tetraogallus caspius*

	SPEC category	European Threat Status	Birds Directive	Bern Convention	Bonn Convention	AEWA	N&W European seas	European Macaronesian seas	Mediterranean and Black seas	Coastal habitats	Inland wetlands
Chukar *Alectoris chukar*	3	V	–	III	–	–	–	–	–	–	–
Rock Partridge *Alectoris graeca*	2	(V)	I[6]/II	III	–	–	–	–	–	–	–
Red-legged Partridge *Alectoris rufa*	2	V	II	III	–	–	–	–	–	–	–
Barbary Partridge *Alectoris barbara*	3	(E)	I/II/III	III	–	–	–	–	–	–	–
See-see *Ammoperdix griseogularis*	–	(S)	–	III	–	–	–	–	–	–	–
Black Francolin *Francolinus francolinus*	3	V	–	III	–	–	–	–	–	–	–
Partridge *Perdix perdix*	3	V	I[7]/II/III	III	–	–	–	–	–	–	–
Quail *Coturnix coturnix*	3	V	–	III	II	–	–	–	–	–	–
Pheasant *Phasianus colchicus*	–	S	–	III	–	–	–	–	–	–	–
Andalusian Hemipode *Turnix sylvatica*	3	E	I	II	–	–	–	–	–	–	D
Water Rail *Rallus aquaticus*	–	(S)	–	III	–	–	–	–	–	–	D
Spotted Crake *Porzana porzana*	4	S	I	II	II	•	–	–	–	–	C
Little Crake *Porzana parva*	4	(S)	I	II	II	•	–	–	–	–	C
Baillon's Crake *Porzana pusilla*	3	R	I	II	II	•	–	–	–	D	B
Corncrake *Crex crex*	1	V	I	II	–	–	–	–	–	–	B
Moorhen *Gallinula chloropus*	–	S	–	III	–	–	–	–	–	–	D
Purple Gallinule *Porphyrio porphyrio*	3	R	I	II	–	–	–	–	–	B	C
Coot *Fulica atra*	–	S	–	III	–	•	–	–	–	–	D
Crested Coot *Fulica cristata*	3	E	I	I	–	–	–	–	–	–	C
Crane *Grus grus*	3	V	I	II	II	•	–	–	–	–	C
Demoiselle Crane *Anthropoides virgo*	–	S	–	II	II	•	–	–	–	–	–
Little Bustard *Tetrax tetrax*	2	V	I	II	–	–	–	–	–	–	–
Houbara Bustard *Chlamydotis undulata*	3	(E)	I	II	I	–	–	–	–	–	–
Great Bustard *Otis tarda*	1	D	I	II	I/II	–	–	–	–	–	–
Oystercatcher *Haematopus ostralegus*	–	S	–	III	II	•	–	–	–	D	–
Black-winged Stilt *Himantopus himantopus*	–	S	I	II	II	•	–	–	–	–	D
Avocet *Recurvirostra avosetta*	4/3[W]	L[W]	I	II	II	•	–	–	–	B	C
Stone Curlew *Burhinus oedicnemus*	3	V	I	II	II	•	–	–	–	–	–
Cream-coloured Courser *Cursorius cursor*	3	V	I	II	–	–	–	–	–	–	–
Collared Pratincole *Glareola pratincola*	3	E	I	II	II	•	–	–	–	C	C
Black-winged Pratincole *Glareola nordmanni*	3	R	–	II	II	•	–	–	–	C	C
Little Ringed Plover *Charadrius dubius*	–	(S)	–	II	II	•	–	–	–	–	D
Ringed Plover *Charadrius hiaticula*	–	S	–	II	II	•	–	–	–	D	–
Kentish Plover *Charadrius alexandrinus*	3	D	–	II	II	•	–	–	–	B	D
Greater Sand Plover *Charadrius leschenaultii*	3	(E)	–	II	II	•	–	–	–	–	B
Caspian Plover *Charadrius asiaticus*	3	(V)	–	III	II	•	–	–	–	–	–
Dotterel *Charadrius morinellus*	–	(S)	I	II	II	•	–	–	–	–	–
Golden Plover *Pluvialis apricaria*	4	S	I/II/III	III	II	•	–	–	–	–	–
Grey Plover *Pluvialis squatarola*	–	(S)	–	III	II	•	–	–	–	D	–
Spur-winged Plover *Hoplopterus spinosus*	3	(E)	I	II	II	•	–	–	–	B	C
Red-wattled Plover *Hoplopterus indicus*	–	S	–	III	II	–	–	–	–	–	–
Sociable Plover *Chettusia gregaria*	1	E	–	III	II	•	–	–	–	–	–
White-tailed Plover *Chettusia leucura*	–	(S)	–	III	–	–	–	–	–	–	–
Lapwing *Vanellus vanellus*	–	(S)	–	III	II	•	–	–	–	–	–
Knot *Calidris canutus*	3[W]	L[W]	II	III	–	•	–	–	–	B	–
Sanderling *Calidris alba*	–	S	–	II	–	•	–	–	–	D	–
Little Stint *Calidris minuta*	–	(S)	–	II	–	•	–	–	–	–	–
Temminck's Stint *Calidris temminckii*	–	(S)	–	II	–	•	–	–	–	–	–
Curlew Sandpiper *Calidris ferruginea*	–	–	–	II	–	•	–	–	–	D	–
Purple Sandpiper *Calidris maritima*	4	(S)	–	II	–	•	–	–	–	C	–
Dunlin *Calidris alpina*	3[W]	V[W]	–	II	–	•	–	–	–	B	–
Broad-billed Sandpiper *Limicola falcinellus*	3	(V)	–	II	II	•	–	–	–	–	–
Ruff *Philomachus pugnax*	4	(S)	I/II	III	II	•	–	–	–	–	D
Jack Snipe *Lymnocryptes minimus*	3[W]	(V)[W]	II/III	III	–	•	–	–	–	C	C
Snipe *Gallinago gallinago*	–	(S)	–	III	–	•	–	–	–	–	–
Great Snipe *Gallinago media*	2	(V)	I	II	II	•	–	–	–	–	–
Pintail Snipe *Gallinago stenura*	–	S	–	III	–	–	–	–	–	–	–
Woodcock *Scolopax rusticola*	3[W]	V[W]	II/III	III	–	•	–	–	–	–	–
Black-tailed Godwit *Limosa limosa*	2	V	II	III	II	•	–	–	–	B	B
Bar-tailed Godwit *Limosa lapponica*	3[W]	L[W]	II	III	II	•	–	–	–	B	–
Whimbrel *Numenius phaeopus*	4	(S)	II	III	II	•	–	–	–	C	–
Slender-billed Curlew *Numenius tenuirostris*	1	–	I	II	I/II	•	–	–	–	A	A
Curlew *Numenius arquata*	3[W]	D[W]	II	III	II	•	–	–	–	B	–
Spotted Redshank *Tringa erythropus*	–	S	–	III	II	•	–	–	–	–	–
Redshank *Tringa totanus*	2	D	II	III	II	•	–	–	–	A	C

Appendix 1: Birds' Priority Status in European Habitats, and International Legal Status

Tundra, mires, moorland	Lowland Atlantic heathland	Boreal forest	Lowland temperate forest	Montane forest	Riverine forest	Mediterranean habitats	Arable and improved grassland	Steppic habitats	Montane grassland	Wet grassland	Rice cultivation	Perennial crops	Pastoral woodland	Species
–	–	–	–	–	–	B	–	C	–	–	–	–	–	*Alectoris chukar*
–	–	–	–	–	–	B	–	–	B	–	–	–	–	*Alectoris graeca*
–	C	–	–	–	–	B	A	B	–	–	–	C	–	*Alectoris rufa*
–	–	–	–	–	–	B	–	–	–	–	–	–	–	*Alectoris barbara*
–	–	–	–	–	–	–	–	–	–	–	–	–	–	*Ammoperdix griseogularis*
–	–	–	–	–	–	B	–	–	–	–	–	–	–	*Francolinus francolinus*
–	D	–	–	–	–	–	B	C	–	–	–	–	–	*Perdix perdix*
–	–	–	–	–	–	–	B	C	D	D	–	–	–	*Coturnix coturnix*
–	–	–	–	–	–	–	–	–	–	–	–	–	–	*Phasianus colchicus*
–	–	–	–	–	–	B	–	–	–	–	–	–	–	*Turnix sylvatica*
–	–	–	–	–	–	–	–	–	–	–	–	–	–	*Rallus aquaticus*
–	–	–	–	–	–	–	–	–	–	D	–	–	–	*Porzana porzana*
–	–	–	–	–	–	–	–	–	–	–	–	–	–	*Porzana parva*
–	–	–	–	–	–	–	–	–	–	–	D	–	–	*Porzana pusilla*
–	–	–	–	–	–	–	A	–	B	A	–	–	–	*Crex crex*
–	–	–	–	–	–	–	–	–	–	–	–	–	–	*Gallinula chloropus*
–	–	–	–	–	–	–	–	–	–	–	–	–	–	*Porphyrio porphyrio*
–	–	–	–	–	–	–	–	–	–	–	–	–	–	*Fulica atra*
–	–	–	–	–	–	–	–	–	–	–	–	–	–	*Fulica cristata*
C	D	D	–	–	C	–	C	C	–	D	–	–	C	*Grus grus*
–	–	–	–	–	–	–	–	D	–	–	–	–	–	*Anthropoides virgo*
–	–	–	–	–	–	–	B	A	–	–	–	–	–	*Tetrax tetrax*
–	–	–	–	–	–	–	–	B	–	–	–	–	–	*Chlamydotis undulata*
–	–	–	–	–	–	–	A	A	–	–	–	–	–	*Otis tarda*
–	–	–	–	–	–	–	–	–	–	–	–	–	–	*Haematopus ostralegus*
–	–	–	–	–	–	–	–	–	–	–	–	–	–	*Himantopus himantopus*
–	–	–	–	–	–	–	–	–	–	–	–	–	–	*Recurvirostra avosetta*
–	D	–	–	–	–	–	C	B	–	–	–	–	–	*Burhinus oedicnemus*
–	–	–	–	–	–	–	–	B	–	–	–	–	–	*Cursorius cursor*
–	–	–	–	–	–	–	C	C	–	–	D	–	–	*Glareola pratincola*
–	–	–	–	–	–	–	C	B	–	–	D	–	–	*Glareola nordmanni*
–	–	–	–	–	–	–	–	–	–	–	–	–	–	*Charadrius dubius*
–	–	–	–	–	–	–	–	–	–	–	–	–	–	*Charadrius hiaticula*
–	–	–	–	–	–	–	–	–	–	–	–	–	–	*Charadrius alexandrinus*
–	–	–	–	–	–	–	–	B	–	–	–	–	–	*Charadrius leschenaultii*
–	–	–	–	–	–	–	–	B	–	–	–	–	–	*Charadrius asiaticus*
D	–	–	–	–	–	–	–	–	–	–	–	–	–	*Charadrius morinellus*
C	–	–	–	–	–	–	D	D	–	–	–	–	–	*Pluvialis apricaria*
D	–	–	–	–	–	–	–	–	–	–	–	–	–	*Pluvialis squatarola*
–	–	–	–	–	–	–	–	–	–	–	–	–	–	*Hoplopterus spinosus*
–	–	–	–	–	–	–	–	–	–	–	–	–	–	*Hoplopterus indicus*
–	–	–	–	–	–	–	–	A	–	–	–	–	–	*Chettusia gregaria*
–	–	–	–	–	–	–	–	–	–	–	–	–	–	*Chettusia leucura*
–	–	–	–	–	–	–	D	–	–	–	–	–	–	*Vanellus vanellus*
D	–	–	–	–	–	–	–	–	–	–	–	–	–	*Calidris canutus*
D	–	–	–	–	–	–	–	–	–	–	–	–	–	*Calidris alba*
D	–	–	–	–	–	–	–	–	–	–	–	–	–	*Calidris minuta*
D	–	–	–	–	–	–	–	–	–	–	–	–	–	*Calidris temminckii*
–	–	–	–	–	–	–	–	–	–	–	–	–	–	*Calidris ferruginea*
C	–	–	–	–	–	–	–	–	–	–	–	–	–	*Calidris maritima*
B	–	–	–	–	–	–	–	–	–	D	D	–	–	*Calidris alpina*
B	–	–	–	–	–	–	–	–	–	–	–	–	–	*Limicola falcinellus*
D	–	–	–	–	–	–	–	–	–	D	–	–	–	*Philomachus pugnax*
B	–	–	–	–	–	–	–	–	–	–	–	–	–	*Lymnocryptes minimus*
–	–	–	–	–	–	–	–	–	–	–	–	–	–	*Gallinago gallinago*
A	–	–	–	–	–	–	–	–	–	A	–	–	–	*Gallinago media*
–	–	–	–	–	–	–	–	–	–	–	–	–	–	*Gallinago stenura*
–	–	D	B	–	C	C	C	–	–	–	–	–	–	*Scolopax rusticola*
B	–	–	–	–	–	–	–	–	–	A	B	–	–	*Limosa limosa*
D	–	–	–	–	–	–	–	–	–	–	–	–	–	*Limosa lapponica*
C	–	–	–	–	–	–	–	–	–	–	–	–	–	*Numenius phaeopus*
–	–	–	–	–	–	–	–	A	–	–	–	–	–	*Numenius tenuirostris*
C	–	–	–	–	–	–	–	–	–	C	–	–	–	*Numenius arquata*
D	–	–	–	–	–	–	–	–	–	–	–	–	–	*Tringa erythropus*
C	–	–	–	–	–	–	–	–	–	B	C	–	–	*Tringa totanus*

Species	SPEC category	European Threat Status	Birds Directive	Bern Convention	Bonn Convention	AEWA	N&W European seas	European Macaronesian seas	Mediterranean and Black seas	Coastal habitats	Inland wetlands
Marsh Sandpiper *Tringa stagnatilis*	–	(S)	–	II	II	•	–	–	–	–	D
Greenshank *Tringa nebularia*	–	S	–	III	II	•	–	–	–	D	–
Green Sandpiper *Tringa ochropus*	–	(S)	–	II	II	•	–	–	–	–	–
Wood Sandpiper *Tringa glareola*	3	D	I	II	II	•	–	–	–	–	C
Terek Sandpiper *Xenus cinereus*	–	(S)	–	II	II	•	–	–	–	–	–
Common Sandpiper *Actitis hypoleucos*	–	S	–	II	II	•	–	–	–	–	D
Turnstone *Arenaria interpres*	–	S	–	II	–	•	–	–	–	D	–
Red-necked Phalarope *Phalaropus lobatus*	–	(S)	I	III	II	•	–	–	–	–	–
Grey Phalarope *Phalaropus fulicarius*	–	(S)	–	III	II	•	D	–	–	–	–
Pomarine Skua *Stercorarius pomarinus*	–	(S)	–	III	–	–	D	–	–	–	–
Arctic Skua *Stercorarius parasiticus*	–	(S)	–	III	–	–	D	–	–	–	–
Long-tailed Skua *Stercorarius longicaudus*	–	(S)	–	III	–	–	D	–	–	–	–
Great Skua *Stercorarius skua*	4	S	–	III	–	–	C	–	–	–	–
Great Black-headed Gull *Larus ichthyaetus*	–	S	–	III	–	•	–	–	D	D	–
Mediterranean Gull *Larus melanocephalus*	4	S	I	II	II	•	–	–	C	C	–
Little Gull *Larus minutus*	3	D	–	II	–	–	–	–	–	–	C
Sabine's Gull *Larus sabini*	–	S	–	II	–	•	D	–	–	–	–
Black-headed Gull *Larus ridibundus*	–	S	–	III	–	–	–	–	–	–	D
Slender-billed Gull *Larus genei*	–	(S)	–	II	–	•	–	–	D	–	–
Audouin's Gull *Larus audouinii*	1	L	I	II	I/II	•	–	–	A	A	–
Common Gull *Larus canus*	2	D	II	III	–	–	B	–	–	B	C
Lesser Black-backed Gull *Larus fuscus*	4	S	II	X	–	–	C	–	–	D	–
Herring Gull *Larus argentatus*	–	S	–	III	–	–	D	–	–	–	–
Yellow-legged Gull *Larus cachinnans*	–	(S)	–	III	–	–	–	–	D	–	–
Armenian Gull *Larus armenicus*	–	(S)	–	III	–	•	–	–	–	–	–
Iceland Gull *Larus glaucoides*	–	(S)	–	III	–	–	D	–	–	–	–
Glaucous Gull *Larus hyperboreus*	–	S	–	III	–	–	D	–	–	–	–
Great Black-backed Gull *Larus marinus*	4	S	II	X	–	–	C	–	–	–	–
Ross's Gull *Rhodostethia rosea*	–	S	–	III	–	–	D	–	–	–	–
Kittiwake *Rissa tridactyla*	–	S	–	III	–	–	D	–	–	–	–
Ivory Gull *Pagophila eburnea*	3	(E)	–	II	–	–	B	–	–	–	–
Gull-billed Tern *Gelochelidon nilotica*	3	(E)	I	II	–	•	–	–	–	C	C
Caspian Tern *Sterna caspia*	3	(E)	I	II	II	•	C	–	C	B	–
Sandwich Tern *Sterna sandvicensis*	2	D	I	II	II	•	B	C	B	A	–
Roseate Tern *Sterna dougallii*	3	E	I	II	II	•	C	C	–	B	–
Common Tern *Sterna hirundo*	–	S	I	II	–	•	–	–	–	–	–
Arctic Tern *Sterna paradisaea*	–	S	I	II	–	•	D	–	–	D	–
Little Tern *Sterna albifrons*	3	D	I	II	II	•	C	–	C	B	D
Whiskered Tern *Chlidonias hybridus*	3	D	I	II	–	–	–	–	–	C	B
Black Tern *Chlidonias niger*	3	D	I	II	II	•	D	–	–	C	B
White-winged Black Tern *Chlidonias leucopterus*	–	S	–	II	–	•	–	–	–	–	D
Guillemot *Uria aalge*	–	S	–	III	–	–	D	–	–	–	–
Brünnich's Guillemot *Uria lomvia*	–	S	–	III	–	–	D	–	–	–	–
Razorbill *Alca torda*	4	S	–	III	–	–	C	–	–	–	–
Black Guillemot *Cepphus grylle*	2	D	–	III	–	–	A	–	–	–	–
Little Auk *Alle alle*	–	(S)	–	III	–	–	D	–	–	–	–
Puffin *Fratercula arctica*	2	V	–	III	–	–	A	C	–	–	–
Black-bellied Sandgrouse *Pterocles orientalis*	3	V	I	II	–	–	–	–	–	–	–
Pin-tailed Sandgrouse *Pterocles alchata*	3	E	I	II	–	–	–	–	–	–	–
Rock Dove *Columba livia*	–	S	–	III	–	–	–	–	–	–	–
Stock Dove *Columba oenas*	4	S	II	III	–	–	–	–	–	–	–
Woodpigeon *Columba palumbus*	4	S	I[8]/II/III	III	–	–	–	–	–	–	–
Long-toed Pigeon *Columba trocaz*	1	V	I	II	–	–	–	–	–	–	–
Dark-tailed Laurel Pigeon *Columba bollii*	1	V	I	II	–	–	–	–	–	–	–
White-tailed Laurel Pigeon *Columba junoniae*	1	V	I	II	–	–	–	–	–	–	–
Collared Dove *Streptopelia decaocto*	–	(S)	–	III	–	–	–	–	–	–	–
Turtle Dove *Streptopelia turtur*	3	D	II	III	–	–	–	–	–	–	–
Rufous Turtle Dove *Streptopelia orientalis*	–	(S)	–	III	–	–	–	–	–	–	–
Laughing Dove *Streptopelia senegalensis*	–	(S)	–	III	–	–	–	–	–	–	–
Great Spotted Cuckoo *Clamator glandarius*	–	S	–	II	–	–	–	–	–	–	–
Cuckoo *Cuculus canorus*	–	S	–	III	–	–	–	–	–	–	–
Oriental Cuckoo *Cuculus saturatus*	–	(S)	–	III	–	–	–	–	–	–	–
Barn Owl *Tyto alba*	3	D	–	II	–	–	–	–	–	–	–
Striated Scops Owl *Otus brucei*	–	(S)	–	II	–	–	–	–	–	–	–
Scops Owl *Otus scops*	2	(D)	–	II	–	–	–	–	–	–	–

Tundra, mires, moorland	Lowland Atlantic heathland	Boreal forest	Lowland temperate forest	Montane forest	Riverine forest	Mediterranean habitats	Arable and improved grassland	Steppic habitats	Montane grassland	Wet grassland	Rice cultivation	Perennial crops	Pastoral woodland	Species
–	–	–	–	–	–	–	–	–	–	–	–	–	–	*Tringa stagnatilis*
–	–	–	–	–	–	–	–	–	–	–	–	–	–	*Tringa nebularia*
–	–	–	–	–	–	–	–	–	–	–	–	–	–	*Tringa ochropus*
B	–	D	–	–	–	–	–	–	–	–	–	–	–	*Tringa glareola*
–	–	–	–	–	–	–	–	–	–	–	–	–	–	*Xenus cinereus*
–	–	–	–	–	–	–	–	–	–	–	–	–	–	*Actitis hypoleucos*
D	–	–	–	–	–	–	–	–	–	–	–	–	–	*Arenaria interpres*
D	–	–	–	–	–	–	–	–	–	–	–	–	–	*Phalaropus lobatus*
D	–	–	–	–	–	–	–	–	–	–	–	–	–	*Phalaropus fulicarius*
D	–	–	–	–	–	–	–	–	–	–	–	–	–	*Stercorarius pomarinus*
D	–	–	–	–	–	–	–	–	–	–	–	–	–	*Stercorarius parasiticus*
D	–	–	–	–	–	–	–	–	–	–	–	–	–	*Stercorarius longicaudus*
C	–	–	–	–	–	–	–	–	–	–	–	–	–	*Stercorarius skua*
–	–	–	–	–	–	–	–	–	–	–	–	–	–	*Larus ichthyaetus*
–	–	–	–	–	–	–	–	–	–	–	–	–	–	*Larus melanocephalus*
–	–	–	–	–	–	–	–	–	–	–	–	–	–	*Larus minutus*
D	–	–	–	–	–	–	–	–	–	–	–	–	–	*Larus sabini*
–	–	–	–	–	–	–	–	–	–	–	–	–	–	*Larus ridibundus*
–	–	–	–	–	–	–	–	–	–	–	–	–	–	*Larus genei*
–	–	–	–	–	–	–	–	–	–	–	–	–	–	*Larus audouinii*
C	–	–	–	–	–	B	–	–	–	B	–	–	–	*Larus canus*
–	–	–	–	–	–	–	–	–	–	–	–	–	–	*Larus fuscus*
–	–	–	–	–	–	–	–	–	–	–	–	–	–	*Larus argentatus*
–	–	–	–	–	–	–	–	–	–	–	–	–	–	*Larus cachinnans*
–	–	–	–	–	–	–	–	–	–	–	–	–	–	*Larus armenicus*
–	–	–	–	–	–	–	–	–	–	–	–	–	–	*Larus glaucoides*
–	–	–	–	–	–	–	–	–	–	–	–	–	–	*Larus hyperboreus*
–	–	–	–	–	–	–	–	–	–	–	–	–	–	*Larus marinus*
D	–	–	–	–	–	–	–	–	–	–	–	–	–	*Rhodostethia rosea*
–	–	–	–	–	–	–	–	–	–	–	–	–	–	*Rissa tridactyla*
–	–	–	–	–	–	–	–	–	–	–	–	–	–	*Pagophila eburnea*
–	–	–	–	–	–	–	–	–	–	–	D	–	–	*Gelochelidon nilotica*
–	–	–	–	–	–	–	–	–	–	–	–	–	–	*Sterna caspia*
–	–	–	–	–	–	–	–	–	–	–	–	–	–	*Sterna sandvicensis*
–	–	–	–	–	–	–	–	–	–	–	–	–	–	*Sterna dougallii*
–	–	–	–	–	–	–	–	–	–	–	–	–	–	*Sterna hirundo*
–	–	–	–	–	–	–	–	–	–	–	–	–	–	*Sterna paradisaea*
–	–	–	–	–	–	–	–	–	–	–	–	–	–	*Sterna albifrons*
–	–	–	–	–	–	–	–	–	–	–	–	–	–	*Chlidonias hybridus*
–	–	–	–	–	–	–	–	–	–	D	D	–	–	*Chlidonias niger*
–	–	–	–	–	–	–	–	–	–	–	–	–	–	*Chlidonias leucopterus*
–	–	–	–	–	–	–	–	–	–	–	–	–	–	*Uria aalge*
–	–	–	–	–	–	–	–	–	–	–	–	–	–	*Uria lomvia*
–	–	–	–	–	–	–	–	–	–	–	–	–	–	*Alca torda*
–	–	–	–	–	–	–	–	–	–	–	–	–	–	*Cepphus grylle*
–	–	–	–	–	–	–	–	–	–	–	–	–	–	*Alle alle*
–	–	–	–	–	–	–	–	–	–	–	–	–	–	*Fratercula arctica*
–	–	–	–	–	–	–	B	–	–	–	–	–	–	*Pterocles orientalis*
–	–	–	–	–	–	–	B	–	–	–	–	–	–	*Pterocles alchata*
–	–	–	–	–	–	–	–	–	–	–	–	–	–	*Columba livia*
–	–	D	D	–	D	–	D	–	–	–	–	–	D	*Columba oenas*
–	–	D	D	D	D	D	C	–	–	–	–	–	D	*Columba palumbus*
–	–	–	–	–	–	–	–	–	–	–	–	–	–	*Columba trocaz*
–	–	–	–	–	–	–	–	–	–	–	–	–	–	*Columba bollii*
–	–	–	–	–	–	–	–	–	–	–	–	–	–	*Columba junoniae*
–	–	–	–	–	–	–	–	–	–	–	–	–	–	*Streptopelia decaocto*
–	–	C	C	–	C	C	B	–	–	–	–	D	D	*Streptopelia turtur*
–	–	–	–	–	–	–	–	–	–	–	–	–	–	*Streptopelia orientalis*
–	–	–	–	–	–	–	–	–	–	–	–	–	–	*Streptopelia senegalensis*
–	–	–	–	–	–	–	–	–	–	–	–	–	–	*Clamator glandarius*
–	–	–	–	–	–	–	–	–	–	–	–	–	–	*Cuculus canorus*
–	–	–	–	–	–	–	–	–	–	–	–	–	–	*Cuculus saturatus*
–	–	–	–	–	D	B	–	–	–	C	–	D	D	*Tyto alba*
–	–	–	–	–	–	–	–	–	–	–	–	–	–	*Otus brucei*
–	–	C	–	B	B	B	–	–	–	–	–	B	B	*Otus scops*

	SPEC category	European Threat Status	Birds Directive	Bern Convention	Bonn Convention	AEWA	N&W European seas	European Macaronesian seas	Mediterranean and Black seas	Coastal habitats	Inland wetlands
Eagle Owl *Bubo bubo*	3	V	I	II	–	–	–	–	–	–	–
Brown Fish Owl *Ketupa zeylonensis*	–	(S)	–	II	–	–	–	–	–	–	–
Snowy Owl *Nyctea scandiaca*	3	V	I	II	–	–	–	–	–	–	–
Hawk Owl *Surnia ulula*	–	(S)	–	II	–	–	–	–	–	–	–
Pygmy Owl *Glaucidium passerinum*	–	(S)	I	II	–	–	–	–	–	–	–
Little Owl *Athene noctua*	3	D	–	II	–	–	–	–	–	–	–
Tawny Owl *Strix aluco*	4	S	–	II	–	–	–	–	–	–	–
Ural Owl *Strix uralensis*	–	(S)	–	II	–	–	–	–	–	–	–
Great Grey Owl *Strix nebulosa*	–	S	–	III	–	–	–	–	–	–	–
Long-eared Owl *Asio otus*	–	S	–	II	–	–	–	–	–	–	–
Short-eared Owl *Asio flammeus*	3	(V)	I	II	–	–	–	–	–	D	–
Tengmalm's Owl *Aegolius funereus*	–	(S)	I	II	–	–	–	–	–	–	–
Nightjar *Caprimulgus europaeus*	2	(D)	I	II	–	–	–	–	–	–	–
Red-necked Nightjar *Caprimulgus ruficollis*	–	S	–	III	–	–	–	–	–	–	–
Plain Swift *Apus unicolor*	4	S	–	II	–	–	–	–	–	–	–
Swift *Apus apus*	–	S	–	III	–	–	–	–	–	–	–
Pallid Swift *Apus pallidus*	–	(S)	–	II	–	–	–	–	–	–	–
Alpine Swift *Apus melba*	–	(S)	–	II	–	–	–	–	–	–	–
White-rumped Swift *Apus caffer*	–	S	–	II	–	–	–	–	–	–	–
Little Swift *Apus affinis*	–	(S)	–	III	–	–	–	–	–	–	–
White-breasted Kingfisher *Halcyon smyrnensis*	–	(S)	–	II	–	–	–	–	–	–	D
Kingfisher *Alcedo atthis*	3	D	I	II	–	–	–	–	–	D	B
Pied Kingfisher *Ceryle rudis*	–	(S)	–	II	–	–	–	–	–	–	D
Blue-cheeked Bee-eater *Merops superciliosus*	–	(S)	–	III	–	–	–	–	–	–	–
Bee-eater *Merops apiaster*	3	D	–	II	II	–	–	–	–	–	D
Roller *Coracias garrulus*	2	(D)	I	II	II	–	–	–	–	–	–
Hoopoe *Upupa epops*	–	S	–	II	–	–	–	–	–	–	–
Wryneck *Jynx torquilla*	3	D	–	II	–	–	–	–	–	–	–
Grey-headed Woodpecker *Picus canus*	3	D	I	II	–	–	–	–	–	–	–
Green Woodpecker *Picus viridis*	2	D	–	II	–	–	–	–	–	–	–
Black Woodpecker *Dryocopus martius*	–	S	I	II	–	–	–	–	–	–	–
Great Spotted Woodpecker *Dendrocopos major*	–	S	–	II	–	–	–	–	–	–	–
Syrian Woodpecker *Dendrocopos syriacus*	4	(S)	I	II	–	–	–	–	–	–	–
Middle Spotted Woodpecker *Dendrocopos medius*	4	S	I	II	–	–	–	–	–	–	–
White-backed Woodpecker *Dendrocopos leucotos*	–	S	I	II	–	–	–	–	–	–	–
Lesser Spotted Woodpecker *Dendrocopos minor*	–	S	–	II	–	–	–	–	–	–	–
Three-toed Woodpecker *Picoides tridactylus*	3	D	I	II	–	–	–	–	–	–	–
Desert Lark *Ammomanes deserti*	–	(S)	–	III	–	–	–	–	–	–	–
Dupont's Lark *Chersophilus duponti*	3	V	I	II	–	–	–	–	–	–	–
Calandra Lark *Melanocorypha calandra*	3	(D)	I	II	–	–	–	–	–	–	–
Bimaculated Lark *Melanocorypha bimaculata*	–	(S)	–	II	–	–	–	–	–	–	–
White-winged Lark *Melanocorypha leucoptera*	4[w]	(S)	–	II	–	–	–	–	–	–	–
Black Lark *Melanocorypha yeltoniensis*	3	(V)	–	II	–	–	–	–	–	–	–
Short-toed Lark *Calandrella brachydactyla*	3	V	I	II	–	–	–	–	–	–	–
Lesser Short-toed Lark *Calandrella rufescens*	3	V	–	II	–	–	–	–	–	–	–
Crested Lark *Galerida cristata*	3	(D)	–	III	–	–	–	–	–	–	–
Thekla Lark *Galerida theklae*	3	V	I	II	–	–	–	–	–	–	–
Woodlark *Lullula arborea*	2	V	I	III	–	–	–	–	–	–	–
Skylark *Alauda arvensis*	3	V	II	III	–	–	–	–	–	D	–
Shore Lark *Eremophila alpestris*	–	(S)	–	II	–	–	–	–	–	–	–
Sand Martin *Riparia riparia*	3	D	–	II	–	–	–	–	–	–	C
Crag Martin *Ptyonoprogne rupestris*	–	S	–	II	–	–	–	–	–	–	–
Swallow *Hirundo rustica*	3	D	–	II	–	–	–	–	–	–	C
Red-rumped Swallow *Hirundo daurica*	–	S	–	II	–	–	–	–	–	–	–
House Martin *Delichon urbica*	–	S	–	II	–	–	–	–	–	–	–
Tawny Pipit *Anthus campestris*	3	V	I	II	–	–	–	–	–	D	–
Berthelot's Pipit *Anthus berthelotii*	4	S	–	II	–	–	–	–	–	–	–
Olive-backed Pipit *Anthus hodgsoni*	–	(S)	–	II	–	–	–	–	–	–	–
Tree Pipit *Anthus trivialis*	–	S	–	II	–	–	–	–	–	–	–
Pechora Pipit *Anthus gustavi*	–	(S)	–	II	–	–	–	–	–	–	–
Meadow Pipit *Anthus pratensis*	4	S	–	II	–	–	–	–	–	–	–
Red-throated Pipit *Anthus cervinus*	–	(S)	–	II	–	–	–	–	–	–	–
Rock Pipit *Anthus petrosus*	–	S	–	III	–	–	–	–	–	D	–
Water Pipit *Anthus spinoletta*	–	S	–	II	–	–	–	–	–	–	–
Yellow Wagtail *Motacilla flava*	–	S	–	II	–	–	–	–	–	–	–

Tundra, mires, moorland	Lowland Atlantic heathland	Boreal forest	Lowland temperate forest	Montane forest	Riverine forest	Mediterranean habitats	Arable and improved grassland	Steppic habitats	Montane grassland	Wet grassland	Rice cultivation	Perennial crops	Pastoral woodland	Species
–	–	C	C	C	–	C	–	D	D	–	–	–	–	*Bubo bubo*
–	–	–	–	–	–	–	–	–	–	–	–	–	–	*Ketupa zeylonensis*
B	–	–	–	–	–	–	–	–	–	–	–	–	–	*Nyctea scandiaca*
–	–	D	–	–	–	–	–	–	–	–	–	–	–	*Surnia ulula*
–	–	D	–	–	–	–	–	–	–	–	–	–	–	*Glaucidium passerinum*
–	–	–	–	–	–	D	B	C	–	D	–	D	C	*Athene noctua*
–	–	–	C	D	D	D	–	–	–	–	–	–	–	*Strix aluco*
–	–	D	–	–	–	–	–	–	–	–	–	–	–	*Strix uralensis*
–	–	D	–	–	–	–	–	–	–	–	–	–	–	*Strix nebulosa*
–	–	–	–	–	–	–	–	–	–	–	–	–	–	*Asio otus*
D	D	–	–	–	–	–	D	C	–	C	D	–	–	*Asio flammeus*
–	–	D	–	–	–	–	–	–	–	–	–	–	–	*Aegolius funereus*
–	C	B	B	C	–	B	–	–	–	–	–	–	–	*Caprimulgus europaeus*
–	–	–	–	–	–	D	–	–	–	–	–	–	–	*Caprimulgus ruficollis*
–	–	–	–	–	–	–	–	–	–	–	–	–	–	*Apus unicolor*
–	–	–	–	–	–	–	–	–	–	–	–	–	–	*Apus apus*
–	–	–	–	–	–	–	–	–	–	–	–	–	–	*Apus pallidus*
–	–	–	–	–	–	–	–	–	–	–	–	–	–	*Apus melba*
–	–	–	–	–	D	–	–	–	–	–	–	–	–	*Apus caffer*
–	–	–	–	–	D	–	–	–	–	–	–	–	–	*Apus affinis*
–	–	–	–	–	–	–	–	–	–	–	–	–	–	*Halcyon smyrnensis*
–	–	–	–	–	–	–	–	–	–	–	–	–	–	*Alcedo atthis*
–	–	–	–	–	–	–	–	–	–	–	–	–	–	*Ceryle rudis*
–	–	–	–	–	–	–	–	–	–	–	–	–	–	*Merops superciliosus*
–	–	–	–	–	–	C	C	C	–	–	–	C	C	*Merops apiaster*
–	–	–	C	–	C	B	C	B	–	–	–	B	B	*Coracias garrulus*
–	–	–	–	–	–	–	–	–	–	–	–	–	–	*Upupa epops*
–	–	C	B	D	C	D	–	–	–	–	–	D	–	*Jynx torquilla*
–	–	C	B	C	C	–	–	–	–	–	–	–	–	*Picus canus*
–	C	–	A	B	B	B	B	–	–	–	–	B	B	*Picus viridis*
–	–	–	–	–	–	–	–	–	–	–	–	–	–	*Dryocopus martius*
–	–	–	–	–	–	–	–	–	–	–	–	–	–	*Dendrocopos major*
–	–	–	–	–	–	D	–	–	–	–	–	D	–	*Dendrocopos syriacus*
–	–	–	C	D	D	–	–	–	–	–	–	–	–	*Dendrocopos medius*
–	–	–	–	–	–	–	–	–	–	–	–	–	–	*Dendrocopos leucotos*
–	–	–	D	–	–	–	–	–	–	–	–	–	–	*Dendrocopos minor*
–	–	B	C	C	–	–	–	–	–	–	–	–	–	*Picoides tridactylus*
–	–	–	–	–	–	–	–	–	–	–	–	–	–	*Ammomanes deserti*
–	–	–	–	–	–	–	B	–	–	–	–	–	–	*Chersophilus duponti*
–	–	–	–	–	–	–	C	C	–	–	–	–	–	*Melanocorypha calandra*
–	–	–	–	–	–	–	–	–	–	–	–	–	–	*Melanocorypha bimaculata*
–	–	–	–	–	–	–	–	C	–	–	–	–	–	*Melanocorypha leucoptera*
–	–	–	–	–	–	–	–	B	–	–	–	–	–	*Melanocorypha yeltoniensis*
–	–	–	–	–	–	D	–	B	–	–	–	–	–	*Calandrella brachydactyla*
–	–	–	–	–	–	–	–	B	–	–	–	–	–	*Calandrella rufescens*
–	–	–	–	–	–	D	B	C	–	–	–	–	–	*Galerida cristata*
–	–	–	–	–	–	B	–	C	–	–	–	C	C	*Galerida theklae*
–	C	B	A	B	–	B	B	–	C	–	–	B	B	*Lullula arborea*
D	D	–	–	–	–	–	B	C	D	D	–	–	–	*Alauda arvensis*
D	D	–	–	–	–	–	–	–	–	–	–	–	–	*Eremophila alpestris*
–	–	–	–	–	–	–	–	–	–	–	–	–	–	*Riparia riparia*
–	–	–	–	–	D	–	–	–	–	–	–	–	–	*Ptyonoprogne rupestris*
–	–	–	–	–	–	B	–	–	–	D	D	–	–	*Hirundo rustica*
–	–	–	–	–	D	–	–	–	–	–	–	–	–	*Hirundo daurica*
–	–	–	–	–	–	–	–	–	–	–	–	–	–	*Delichon urbica*
–	D	–	–	–	–	C	D	C	D	–	–	–	–	*Anthus campestris*
–	–	–	–	–	–	–	–	–	–	–	–	–	–	*Anthus berthelotii*
–	–	D	–	–	–	–	–	–	–	–	–	–	–	*Anthus hodgsoni*
–	–	–	–	–	–	–	–	–	–	–	–	–	–	*Anthus trivialis*
–	–	–	–	–	–	–	–	–	–	–	–	–	–	*Anthus gustavi*
D	–	–	–	–	–	D	D	–	D	–	–	–	–	*Anthus pratensis*
D	–	–	–	–	–	–	–	–	–	–	–	–	–	*Anthus cervinus*
D	–	–	–	–	–	–	–	–	–	–	–	–	–	*Anthus petrosus*
–	–	–	–	–	–	–	–	D	–	–	–	–	–	*Anthus spinoletta*
–	–	–	–	–	–	–	–	–	–	D	–	–	–	*Motacilla flava*

	SPEC category	European Threat Status	Birds Directive	Bern Convention	Bonn Convention	AEWA	N&W European seas	European Macaronesian seas	Mediterranean and Black seas	Coastal habitats	Inland wetlands
Citrine Wagtail *Motacilla citreola*	–	(S)	–	II	–	–	–	–	–	–	–
Grey Wagtail *Motacilla cinerea*	–	(S)	–	II	–	–	–	–	–	–	D
Pied Wagtail *Motacilla alba*	–	S	–	II	–	–	–	–	–	–	–
Yellow-vented Bulbul *Pycnonotus xanthopygos*	–	(S)	–	II	–	–	–	–	–	–	–
Waxwing *Bombycilla garrulus*	–	(S)	–	II	–	–	–	–	–	–	–
Dipper *Cinclus cinclus*	–	(S)	–	II	–	–	–	–	–	–	D
Wren *Troglodytes troglodytes*	–	S	–	III	–	–	–	–	–	–	–
Dunnock *Prunella modularis*	4	S	–	II	–	–	–	–	–	–	–
Siberian Accentor *Prunella montanella*	–	(S)	–	II	–	–	–	–	–	–	–
Radde's Accentor *Prunella ocularis*	3	(V)	–	II	–	–	–	–	–	–	–
Black-throated Accentor *Prunella atrogularis*	3	(V)	–	II	–	–	–	–	–	–	–
Alpine Accentor *Prunella collaris*	–	S	–	II	–	–	–	–	–	–	–
Rufous Bush Robin *Cercotrichas galactotes*	–	S	–	II	II	–	–	–	–	–	–
Robin *Erithacus rubecula*	4	S	–	II	II	–	–	–	–	–	–
Thrush Nightingale *Luscinia luscinia*	4	S	–	II	II	–	–	–	–	–	–
Nightingale *Luscinia megarhynchos*	4	(S)	–	II	II	–	–	–	–	–	–
Siberian Rubythroat *Luscinia calliope*	–	(S)	–	III	II	–	–	–	–	–	–
Bluethroat *Luscinia svecica*	–	S	I	II	II	–	–	–	–	–	–
Red-flanked Bluetail *Tarsiger cyanurus*	–	(S)	–	II	II	–	–	–	–	–	–
White-throated Robin *Irania gutturalis*	–	(S)	–	II	II	–	–	–	–	–	–
Black Redstart *Phoenicurus ochruros*	–	S	–	II	II	–	–	–	–	–	–
Redstart *Phoenicurus phoenicurus*	2	V	–	II	II	–	–	–	–	–	–
Güldenstädt's Redstart *Phoenicurus erythrogaster*	3	Ins	–	III	II	–	–	–	–	–	–
Whinchat *Saxicola rubetra*	4	S	–	II	II	–	–	–	–	–	–
Fuerteventura Chat *Saxicola dacotiae*	2	V	I	II	II	–	–	–	–	–	–
Stonechat *Saxicola torquata*	3	(D)	–	II	II	–	–	–	–	–	–
Isabelline Wheatear *Oenanthe isabellina*	–	(S)	–	II	II	–	–	–	–	–	–
Wheatear *Oenanthe oenanthe*	–	S	–	II	II	–	–	–	–	–	–
Pied Wheatear *Oenanthe pleschanka*	–	(S)	–	II	II	–	–	–	–	–	–
Cyprus Pied Wheatear *Oenanthe cypriaca*	2	R	–	III	–	–	–	–	–	–	–
Black-eared Wheatear *Oenanthe hispanica*	2	V	–	II	II	–	–	–	–	–	–
Desert Wheatear *Oenanthe deserti*	–	(S)	–	III	II	–	–	–	–	–	–
Finsch's Wheatear *Oenanthe finschii*	–	(S)	–	II	II	–	–	–	–	–	–
Red-tailed Wheatear *Oenanthe xanthoprymna*	–	(S)	–	III	II	–	–	–	–	–	–
Black Wheatear *Oenanthe leucura*	3	E	I	II	II	–	–	–	–	–	–
Rock Thrush *Monticola saxatilis*	3	(D)	–	II	II	–	–	–	–	–	–
Blue Rock Thrush *Monticola solitarius*	3	(V)	–	II	II	–	–	–	–	–	–
White's Thrush *Zoothera dauma*	–	(S)	–	III	II	–	–	–	–	–	–
Ring Ouzel *Turdus torquatus*	4	S	–	II	II	–	–	–	–	–	–
Blackbird *Turdus merula*	4	S	II	III	II	–	–	–	–	–	–
Black-throated Thrush *Turdus ruficollis*	–	(S)	–	III	II	–	–	–	–	–	–
Fieldfare *Turdus pilaris*	4[w]	S	II	III	II	–	–	–	–	–	–
Song Thrush *Turdus philomelos*	4	S	II	III	II	–	–	–	–	–	–
Redwing *Turdus iliacus*	4[w]	S	II	III	II	–	–	–	–	–	–
Mistle Thrush *Turdus viscivorus*	4	S	II	III	II	–	–	–	–	–	–
Cetti's Warbler *Cettia cetti*	–	S	–	II	II	–	–	–	–	–	D
Fan-tailed Warbler *Cisticola juncidis*	–	(S)	–	II	II	–	–	–	–	–	D
Graceful Warbler *Prinia gracilis*	–	(S)	–	II	II	–	–	–	–	–	–
Lanceolated Warbler *Locustella lanceolata*	–	(S)	–	II	II	–	–	–	–	–	–
Grasshopper Warbler *Locustella naevia*	4	S	–	II	II	–	–	–	–	–	–
River Warbler *Locustella fluviatilis*	4	S	–	II	II	–	–	–	–	–	–
Savi's Warbler *Locustella luscinioides*	4	(S)	–	II	II	–	–	–	–	–	C
Moustached Warbler *Acrocephalus melanopogon*	–	(S)	I	II	II	–	–	–	–	–	D
Aquatic Warbler *Acrocephalus paludicola*	1	E	I	II	II	–	–	–	–	B	A
Sedge Warbler *Acrocephalus schoenobaenus*	4	(S)	–	II	II	–	–	–	–	–	D
Paddyfield Warbler *Acrocephalus agricola*	–	S	–	II	II	–	–	–	–	–	D
Blyth's Reed Warbler *Acrocephalus dumetorum*	–	(S)	–	II	II	–	–	–	–	–	–
Marsh Warbler *Acrocephalus palustris*	4	S	–	II	II	–	–	–	–	–	D
Reed Warbler *Acrocephalus scirpaceus*	4	S	–	II	II	–	–	–	–	–	C
Great Reed Warbler *Acrocephalus arundinaceus*	–	(S)	–	II	II	–	–	–	–	–	D
Olivaceous Warbler *Hippolais pallida*	3	(V)	–	II	II	–	–	–	–	–	–
Booted Warbler *Hippolais caligata*	–	(S)	–	II	II	–	–	–	–	–	–
Upcher's Warbler *Hippolais languida*	–	(S)	–	II	II	–	–	–	–	–	–
Olive-tree Warbler *Hippolais olivetorum*	2	(R)	I	II	II	–	–	–	–	–	–
Icterine Warbler *Hippolais icterina*	4	S	–	II	II	–	–	–	–	–	–

Appendix 1: Birds' Priority Status in European Habitats, and International Legal Status

Tundra, mires, moorland	Lowland Atlantic heathland	Boreal forest	Lowland temperate forest	Montane forest	Riverine forest	Mediterranean habitats	Arable and improved grassland	Steppic habitats	Montane grassland	Wet grassland	Rice cultivation	Perennial crops	Pastoral woodland	Species
–	–	–	–	–	–	–	–	–	–	–	–	–	–	*Motacilla citreola*
–	–	–	–	–	–	–	–	–	–	–	–	–	–	*Motacilla cinerea*
–	–	–	–	–	–	–	–	–	–	–	–	–	–	*Motacilla alba*
–	–	–	–	–	–	–	–	–	–	–	–	–	–	*Pycnonotus xanthopygos*
–	–	D	–	–	–	–	–	–	–	–	–	–	–	*Bombycilla garrulus*
–	–	–	–	–	–	–	–	–	–	–	–	–	–	*Cinclus cinclus*
–	–	–	–	–	–	–	–	–	–	–	–	–	–	*Troglodytes troglodytes*
–	–	D	D	D	D	–	–	–	–	–	–	–	–	*Prunella modularis*
–	–	–	–	–	–	–	–	–	–	–	–	–	–	*Prunella montanella*
–	–	–	–	–	–	–	–	B	–	–	–	–	–	*Prunella ocularis*
–	–	–	–	B	–	–	–	–	–	–	–	–	–	*Prunella atrogularis*
–	–	–	–	–	–	–	–	–	D	–	–	–	–	*Prunella collaris*
–	–	–	–	–	–	–	–	–	–	–	–	–	–	*Cercotrichas galactotes*
–	–	D	D	D	D	D	–	–	–	–	–	D	D	*Erithacus rubecula*
–	–	D	C	–	D	–	–	–	–	–	–	–	–	*Luscinia luscinia*
–	–	–	C	–	D	D	–	–	–	–	–	–	–	*Luscinia megarhynchos*
–	–	D	–	–	–	–	–	–	–	–	–	–	–	*Luscinia calliope*
D	–	–	–	–	–	–	–	–	–	–	–	–	–	*Luscinia svecica*
–	–	D	–	–	–	–	–	–	–	–	–	–	–	*Tarsiger cyanurus*
–	–	–	–	–	–	–	–	–	–	–	–	–	–	*Irania gutturalis*
–	–	–	–	–	–	–	–	–	–	–	–	–	–	*Phoenicurus ochruros*
–	–	B	A	B	–	C	–	–	–	–	–	C	C	*Phoenicurus phoenicurus*
–	–	–	–	–	–	–	–	–	–	–	–	–	–	*Phoenicurus erythrogaster*
–	–	–	–	–	–	D	–	–	D	–	–	–	–	*Saxicola rubetra*
–	–	–	–	–	–	–	–	–	–	–	–	–	–	*Saxicola dacotiae*
D	D	C	C	–	–	C	–	–	D	D	–	–	–	*Saxicola torquata*
–	–	–	–	–	–	–	–	–	–	–	–	–	–	*Oenanthe isabellina*
–	–	–	–	–	–	–	–	–	–	–	–	–	–	*Oenanthe oenanthe*
–	–	–	–	–	–	–	–	–	–	–	–	–	–	*Oenanthe pleschanka*
–	–	–	–	–	–	A	–	–	–	–	–	–	–	*Oenanthe cypriaca*
–	–	–	–	–	–	A	B	B	–	–	–	C	–	*Oenanthe hispanica*
–	–	–	–	–	–	–	–	D	–	–	–	–	–	*Oenanthe deserti*
–	–	–	–	–	–	–	–	–	–	–	–	–	–	*Oenanthe finschii*
–	–	–	–	–	–	–	–	–	–	–	–	–	–	*Oenanthe xanthoprymna*
–	–	–	–	–	–	B	–	–	–	–	–	–	–	*Oenanthe leucura*
–	–	–	–	–	–	C	–	C	–	–	–	–	–	*Monticola saxatilis*
–	–	–	–	–	–	B	–	D	–	–	–	–	–	*Monticola solitarius*
–	–	D	–	–	–	–	–	–	–	–	–	–	–	*Zoothera dauma*
D	–	–	D	–	–	–	–	–	–	–	–	–	–	*Turdus torquatus*
–	–	D	D	D	D	D	D	–	–	–	–	D	D	*Turdus merula*
–	–	–	–	–	–	–	–	–	–	–	–	–	–	*Turdus ruficollis*
–	–	D	D	D	D	–	C	–	–	–	–	–	–	*Turdus pilaris*
–	–	D	D	D	D	D	D	–	–	–	–	D	D	*Turdus philomelos*
–	–	C	–	–	–	–	C	–	–	–	–	D	–	*Turdus iliacus*
–	–	D	D	D	–	–	D	–	–	–	–	D	D	*Turdus viscivorus*
–	–	–	–	–	–	–	–	–	–	–	–	–	–	*Cettia cetti*
–	–	–	–	–	–	–	–	–	–	–	–	–	–	*Cisticola juncidis*
–	–	–	–	–	–	–	–	–	–	–	–	–	–	*Prinia gracilis*
–	–	–	–	–	–	–	–	–	–	–	–	–	–	*Locustella lanceolata*
–	–	–	–	–	–	–	D	–	D	–	–	–	–	*Locustella naevia*
–	–	–	–	–	C	–	–	–	–	–	–	–	–	*Locustella fluviatilis*
–	–	–	–	–	–	–	–	–	–	–	–	–	–	*Locustella luscinioides*
–	–	–	–	–	–	–	–	–	–	–	–	–	–	*Acrocephalus melanopogon*
–	–	–	–	–	–	–	–	–	A	–	–	–	–	*Acrocephalus paludicola*
–	–	–	–	–	–	–	–	–	–	–	–	–	–	*Acrocephalus schoenobaenus*
–	–	–	–	–	–	–	–	–	–	–	–	–	–	*Acrocephalus agricola*
–	–	–	–	–	–	–	–	–	–	–	–	–	–	*Acrocephalus dumetorum*
–	–	–	–	–	D	–	–	–	–	–	–	–	–	*Acrocephalus palustris*
–	–	–	–	–	–	–	–	–	–	–	–	–	–	*Acrocephalus scirpaceus*
–	–	–	–	–	–	–	–	–	–	–	–	–	–	*Acrocephalus arundinaceus*
–	–	–	–	–	–	C	D	D	–	–	–	C	–	*Hippolais pallida*
–	–	–	–	–	–	–	–	–	–	–	–	–	–	*Hippolais caligata*
–	–	–	–	–	–	–	–	–	–	–	–	–	–	*Hippolais languida*
–	–	–	–	–	–	A	–	–	–	–	–	B	–	*Hippolais olivetorum*
–	–	D	C	–	D	–	–	–	–	–	–	–	–	*Hippolais icterina*

	SPEC category	European Threat Status	Birds Directive	Bern Convention	Bonn Convention	AEWA	N&W European seas	European Macaronesian seas	Mediterranean and Black seas	Coastal habitats	Inland wetlands
Melodious Warbler *Hippolais polyglotta*	4	(S)	–	II	II	–	–	–	–	–	–
Marmora's Warbler *Sylvia sarda*	4	(S)	I	II	II	–	–	–	–	–	–
Dartford Warbler *Sylvia undata*	2	V	I	II	II	–	–	–	–	–	–
Spectacled Warbler *Sylvia conspicillata*	–	(S)	–	II	II	–	–	–	–	–	–
Subalpine Warbler *Sylvia cantillans*	4	S	–	II	II	–	–	–	–	–	–
Ménétries's Warbler *Sylvia mystacea*	–	(S)	–	II	II	–	–	–	–	–	–
Sardinian Warbler *Sylvia melanocephala*	4	S	–	II	II	–	–	–	–	–	–
Cyprus Warbler *Sylvia melanothorax*	2	R	–	II	II	–	–	–	–	–	–
Rüppell's Warbler *Sylvia rueppelli*	4	(S)	I	II	II	–	–	–	–	–	–
Desert Warbler *Sylvia nana*	–	(S)	–	II	II	–	–	–	–	–	–
Orphean Warbler *Sylvia hortensis*	3	V	–	II	II	–	–	–	–	–	–
Barred Warbler *Sylvia nisoria*	4	(S)	I	II	II	–	–	–	–	–	–
Lesser Whitethroat *Sylvia curruca*	–	S	–	II	II	–	–	–	–	–	–
Whitethroat *Sylvia communis*	4	S	–	II	II	–	–	–	–	–	–
Garden Warbler *Sylvia borin*	4	S	–	II	II	–	–	–	–	–	–
Blackcap *Sylvia atricapilla*	4	S	–	II	II	–	–	–	–	–	–
Green Warbler *Phylloscopus nitidus*	–	(S)	–	II	II	–	–	–	–	–	–
Greenish Warbler *Phylloscopus trochiloides*	–	(S)	–	II	II	–	–	–	–	–	–
Arctic Warbler *Phylloscopus borealis*	–	(S)	–	II	II	–	–	–	–	–	–
Yellow-browed Warbler *Phylloscopus inornatus*	–	S	–	III	–	–	–	–	–	–	–
Bonelli's Warbler *Phylloscopus bonelli*	4	S	–	II	II	–	–	–	–	–	–
Wood Warbler *Phylloscopus sibilatrix*	4	(S)	–	II	II	–	–	–	–	–	–
Mountain Chiffchaff *Phylloscopus sindianus*	–	(S)	–	II	II	–	–	–	–	–	–
Chiffchaff *Phylloscopus collybita*	–	(S)	–	II	II	–	–	–	–	–	–
Willow Warbler *Phylloscopus trochilus*	–	S	–	II	II	–	–	–	–	–	–
Goldcrest *Regulus regulus*	4	(S)	–	II	II	–	–	–	–	–	–
Tenerife Goldcrest *Regulus teneriffae*	4	S	–	II	–	–	–	–	–	–	–
Firecrest *Regulus ignicapillus*	4	S	–	II	II	–	–	–	–	–	–
Spotted Flycatcher *Muscicapa striata*	3	D	–	II	II	–	–	–	–	–	–
Red-breasted Flycatcher *Ficedula parva*	–	(S)	I	II	II	–	–	–	–	–	–
Semi-collared Flycatcher *Ficedula semitorquata*	2	(E)	I	II	II	–	–	–	–	–	–
Collared Flycatcher *Ficedula albicollis*	4	S	I	II	II	–	–	–	–	–	–
Pied Flycatcher *Ficedula hypoleuca*	4	S	–	II	II	–	–	–	–	–	–
Bearded Tit *Panurus biarmicus*	–	(S)	–	II	–	–	–	–	–	–	D
Long-tailed Tit *Aegithalos caudatus*	–	S	–	III	–	–	–	–	–	–	–
Marsh Tit *Parus palustris*	–	S	–	II	–	–	–	–	–	–	–
Sombre Tit *Parus lugubris*	4	(S)	–	II	–	–	–	–	–	–	–
Willow Tit *Parus montanus*	–	(S)	–	II	–	–	–	–	–	–	–
Siberian Tit *Parus cinctus*	–	(S)	–	II	–	–	–	–	–	–	–
Crested Tit *Parus cristatus*	4	S	–	II	–	–	–	–	–	–	–
Coal Tit *Parus ater*	–	S	–	II	–	–	–	–	–	–	–
Blue Tit *Parus caeruleus*	4	S	–	II	–	–	–	–	–	–	–
Azure Tit *Parus cyanus*	–	(S)	–	II	–	–	–	–	–	–	–
Great Tit *Parus major*	–	S	–	II	–	–	–	–	–	–	–
Krüper's Nuthatch *Sitta krueperi*	4	(S)	I	II	–	–	–	–	–	–	–
Corsican Nuthatch *Sitta whiteheadi*	2	V	I	II	–	–	–	–	–	–	–
Nuthatch *Sitta europaea*	–	S	–	II	–	–	–	–	–	–	–
Eastern Rock Nuthatch *Sitta tephronota*	–	(S)	–	II	–	–	–	–	–	–	–
Rock Nuthatch *Sitta neumayer*	4	(S)	–	II	–	–	–	–	–	–	–
Wallcreeper *Tichodroma muraria*	–	(S)	–	II	–	–	–	–	–	–	–
Treecreeper *Certhia familiaris*	–	S	–	II	–	–	–	–	–	–	–
Short-toed Treecreeper *Certhia brachydactyla*	4	S	–	II	–	–	–	–	–	–	–
Penduline Tit *Remiz pendulinus*	–	(S)	–	III	–	–	–	–	–	–	D
Golden Oriole *Oriolus oriolus*	–	S	–	II	–	–	–	–	–	–	–
Red-backed Shrike *Lanius collurio*	3	(D)	I	II	–	–	–	–	–	–	–
Lesser Grey Shrike *Lanius minor*	2	(D)	I	II	–	–	–	–	–	–	–
Great Grey Shrike *Lanius excubitor*	3	D	–	II	–	–	–	–	–	–	–
Woodchat Shrike *Lanius senator*	2	V	–	II	–	–	–	–	–	–	–
Masked Shrike *Lanius nubicus*	2	(V)	–	II	–	–	–	–	–	–	–
Jay *Garrulus glandarius*	–	(S)	–	III	–	–	–	–	–	–	–
Siberian Jay *Perisoreus infaustus*	3	(D)	–	II	–	–	–	–	–	–	–
Azure-winged Magpie *Cyanopica cyanus*	–	S	–	II	–	–	–	–	–	–	–
Magpie *Pica pica*	–	S	–	III	–	–	–	–	–	–	–
Nutcracker *Nucifraga caryocatactes*	–	(S)	–	II	–	–	–	–	–	–	–
Alpine Chough *Pyrrhocorax graculus*	–	(S)	–	II	–	–	–	–	–	–	–

Appendix 1: Birds' Priority Status in European Habitats, and International Legal Status

Tundra, mires, moorland	Lowland Atlantic heathland	Boreal forest	Lowland temperate forest	Montane forest	Riverine forest	Mediterranean habitats	Arable and improved grassland	Steppic habitats	Montane grassland	Wet grassland	Rice cultivation	Perennial crops	Pastoral woodland	Species
–	–	–	D	–	D	D	–	–	–	–	–	–	–	*Hippolais polyglotta*
–	–	–	–	–	–	C	–	–	–	–	–	–	–	*Sylvia sarda*
–	C	–	–	–	–	A	–	–	–	–	–	–	–	*Sylvia undata*
–	–	–	–	–	–	–	–	–	–	–	–	–	–	*Sylvia conspicillata*
–	–	–	–	–	–	C	–	–	–	–	–	–	–	*Sylvia cantillans*
–	–	–	–	–	–	–	–	–	–	–	–	–	–	*Sylvia mystacea*
–	–	–	–	–	–	C	–	–	–	–	–	D	–	*Sylvia melanocephala*
–	–	–	–	–	–	A	–	–	–	–	–	–	–	*Sylvia melanothorax*
–	–	–	–	–	–	C	–	–	–	–	–	–	–	*Sylvia rueppelli*
–	–	–	–	–	–	–	–	–	–	–	–	–	–	*Sylvia nana*
–	–	–	–	–	–	B	–	–	–	–	–	C	C	*Sylvia hortensis*
–	–	–	–	–	–	–	D	–	–	–	–	–	–	*Sylvia nisoria*
–	–	–	–	–	–	–	–	–	–	–	–	–	–	*Sylvia curruca*
–	–	–	–	–	–	–	D	–	–	–	–	–	–	*Sylvia communis*
–	–	D	C	D	D	–	–	–	–	–	–	–	–	*Sylvia borin*
–	–	D	D	D	D	D	–	–	–	–	–	D	D	*Sylvia atricapilla*
–	–	–	–	–	–	–	–	–	–	–	–	–	–	*Phylloscopus nitidus*
–	–	–	–	–	–	–	–	–	–	–	–	–	–	*Phylloscopus trochiloides*
–	–	D	–	–	–	–	–	–	–	–	–	–	–	*Phylloscopus borealis*
–	–	D	–	–	–	–	–	–	–	–	–	–	–	*Phylloscopus inornatus*
–	–	–	D	D	–	D	–	–	–	–	–	–	–	*Phylloscopus bonelli*
–	–	D	D	D	–	–	–	–	–	–	–	–	–	*Phylloscopus sibilatrix*
–	–	–	–	–	–	–	–	–	–	–	–	–	–	*Phylloscopus sindianus*
–	–	–	–	–	–	–	–	–	–	–	–	–	–	*Phylloscopus collybita*
–	–	–	–	–	–	–	–	–	–	–	–	–	–	*Phylloscopus trochilus*
–	–	D	D	D	–	–	–	–	–	–	–	–	–	*Regulus regulus*
–	–	–	–	–	–	–	–	–	–	–	–	–	–	*Regulus teneriffae*
–	–	–	D	D	–	D	–	–	–	–	–	–	–	*Regulus ignicapillus*
–	–	C	C	C	C	D	–	–	–	–	–	D	D	*Muscicapa striata*
–	–	–	D	–	–	–	–	–	–	–	–	–	–	*Ficedula parva*
–	–	B	B	B	B	–	–	–	–	–	–	–	–	*Ficedula semitorquata*
–	–	C	D	D	–	–	–	–	–	–	–	–	–	*Ficedula albicollis*
–	–	D	C	D	D	–	–	–	–	–	–	–	–	*Ficedula hypoleuca*
–	–	–	–	–	–	–	–	–	–	–	–	–	–	*Panurus biarmicus*
–	–	–	–	–	–	–	–	–	–	–	–	–	–	*Aegithalos caudatus*
–	–	–	–	–	–	–	–	–	–	–	–	–	–	*Parus palustris*
–	–	–	D	–	–	C	–	–	–	–	–	D	–	*Parus lugubris*
–	–	–	–	–	–	–	–	–	–	–	–	–	–	*Parus montanus*
–	–	D	–	–	–	–	–	–	–	–	–	–	–	*Parus cinctus*
–	–	D	D	D	–	D	–	–	–	–	–	–	–	*Parus cristatus*
–	–	–	–	–	–	–	–	–	–	–	–	–	–	*Parus ater*
–	–	C	D	D	D	D	–	–	–	–	–	–	–	*Parus caeruleus*
–	–	–	–	–	D	–	–	–	–	–	–	–	–	*Parus cyanus*
–	–	–	–	–	–	–	–	–	–	–	–	–	–	*Parus major*
–	–	–	–	–	–	C	–	–	–	–	–	–	–	*Sitta krueperi*
–	–	–	–	–	–	A	–	–	–	–	–	–	–	*Sitta whiteheadi*
–	–	D	–	–	–	–	–	–	–	–	–	–	–	*Sitta europaea*
–	–	–	–	–	–	–	–	–	–	–	–	–	–	*Sitta tephronota*
–	–	–	–	–	–	C	–	–	–	–	–	–	–	*Sitta neumayer*
–	–	–	–	–	–	–	–	–	–	–	–	–	–	*Tichodroma muraria*
–	–	D	–	–	–	–	–	–	–	–	–	–	–	*Certhia familiaris*
–	–	C	–	D	D	–	–	–	–	–	–	D	D	*Certhia brachydactyla*
–	–	D	–	–	–	–	–	–	–	–	–	–	–	*Remiz pendulinus*
–	–	D	–	–	–	–	–	–	–	–	–	–	–	*Oriolus oriolus*
D	D	D	D	–	D	D	B	–	–	–	D	–	–	*Lanius collurio*
–	–	–	–	–	–	–	B	B	–	–	–	–	–	*Lanius minor*
D	D	C	D	–	D	D	D	C	–	–	–	–	–	*Lanius excubitor*
–	–	–	–	–	–	B	C	–	–	–	B	–	A	*Lanius senator*
–	–	–	–	–	–	A	–	–	–	–	B	–	–	*Lanius nubicus*
–	–	–	–	–	–	–	–	–	–	–	–	–	–	*Garrulus glandarius*
–	–	B	–	–	–	–	–	–	–	–	–	–	–	*Perisoreus infaustus*
–	–	–	–	–	–	–	–	–	–	–	–	–	–	*Cyanopica cyanus*
–	–	–	–	–	–	–	–	–	–	–	–	–	–	*Pica pica*
–	–	–	–	–	–	–	–	–	–	–	–	–	–	*Nucifraga caryocatactes*
–	–	–	–	–	–	–	–	–	D	–	–	–	–	*Pyrrhocorax graculus*

	SPEC category	European Threat Status	Birds Directive	Bern Convention	Bonn Convention	AEWA	N&W European seas	European Macaronesian seas	Mediterranean and Black seas	Coastal habitats	Inland wetlands
Chough *Pyrrhocorax pyrrhocorax*	3	V	I	II	–	–	–	–	–	D	–
Jackdaw *Corvus monedula*	4	(S)	–	X	–	–	–	–	–	–	–
Rook *Corvus frugilegus*	–	S	–	III	–	–	–	–	–	–	–
Carrion Crow *Corvus corone*	–	S	–	III	–	–	–	–	–	–	–
Raven *Corvus corax*	–	(S)	–	III	–	–	–	–	–	–	–
Starling *Sturnus vulgaris*	–	S	–	III	–	–	–	–	–	–	–
Spotless Starling *Sturnus unicolor*	4	S	–	II	–	–	–	–	–	–	–
Rose-coloured Starling *Sturnus roseus*	–	(S)	–	II	–	–	–	–	–	–	–
House Sparrow *Passer domesticus*	–	S	–	III	–	–	–	–	–	–	–
Spanish Sparrow *Passer hispaniolensis*	–	(S)	–	III	–	–	–	–	–	–	–
Dead Sea Sparrow *Passer moabiticus*	–	(S)	–	III	–	–	–	–	–	–	–
Tree Sparrow *Passer montanus*	–	S	–	III	–	–	–	–	–	–	–
Pale Rock Sparrow *Carpospiza brachydactyla*	–	(S)	–	III	–	–	–	–	–	–	–
Yellow-throated Sparrow *Petronia xanthocollis*	–	(S)	–	III	–	–	–	–	–	–	–
Rock Sparrow *Petronia petronia*	–	S	–	II	–	–	–	–	–	–	–
Snowfinch *Montifringilla nivalis*	–	(S)	–	II	–	–	–	–	–	–	–
Chaffinch *Fringilla coelebs*	4	S	I[9]	III	–	–	–	–	–	–	–
Blue Chaffinch *Fringilla teydea*	1	V	I	II	–	–	–	–	–	–	–
Brambling *Fringilla montifringilla*	–	S	–	III	–	–	–	–	–	–	–
Red-fronted Serin *Serinus pusillus*	–	(S)	–	II	–	–	–	–	–	–	–
Serin *Serinus serinus*	4	S	–	II	–	–	–	–	–	–	–
Canary *Serinus canaria*	4	S	–	III	–	–	–	–	–	–	–
Citril Finch *Serinus citrinella*	4	S	–	II	–	–	–	–	–	–	–
Greenfinch *Carduelis chloris*	4	S	–	II	–	–	–	–	–	–	–
Goldfinch *Carduelis carduelis*	–	(S)	–	II	–	–	–	–	–	–	–
Siskin *Carduelis spinus*	4	S	–	II	–	–	–	–	–	–	–
Linnet *Carduelis cannabina*	4	S	–	II	–	–	–	–	–	–	–
Twite *Carduelis flavirostris*	–	S	–	II	–	–	–	–	–	–	–
Redpoll *Carduelis flammea*	–	(S)	–	II	–	–	–	–	–	–	–
Arctic Redpoll *Carduelis hornemanni*	–	(S)	–	II	–	–	–	–	–	–	–
Two-barred Crossbill *Loxia leucoptera*	–	(S)	–	II	–	–	–	–	–	–	–
Crossbill *Loxia curvirostra*	–	S	–	II	–	–	–	–	–	–	–
Scottish Crossbill *Loxia scotica*	1	Ins	I	II	–	–	–	–	–	–	–
Parrot Crossbill *Loxia pytyopsittacus*	4	S	–	II	–	–	–	–	–	–	–
Crimson-winged Finch *Rhodopechys sanguinea*	–	(S)	–	?	–	–	–	–	–	–	–
Desert Finch *Rhodospiza obsoleta*	–	(S)	–	III	–	–	–	–	–	–	–
Mongolian Trumpeter Finch *Bucanetes mongolicus*	–	(S)	–	–	–	–	–	–	–	–	–
Trumpeter Finch *Bucanetes githagineus*	3	R	I	II	–	–	–	–	–	–	–
Scarlet Rosefinch *Carpodacus erythrinus*	–	(S)	–	II	–	–	–	–	–	–	–
Great Rosefinch *Carpodacus rubicilla*	3	(E)	–	III	–	–	–	–	–	–	–
Pine Grosbeak *Pinicola enucleator*	–	S	–	II	–	–	–	–	–	–	–
Bullfinch *Pyrrhula pyrrhula*	–	S	–	III	–	–	–	–	–	–	–
Hawfinch *Coccothraustes coccothraustes*	–	S	–	II	–	–	–	–	–	–	–
Lapland Bunting *Calcarius lapponicus*	–	(S)	–	II	–	–	–	–	–	–	–
Snow Bunting *Plectrophenax nivalis*	–	(S)	–	II	–	–	–	–	–	–	–
Yellowhammer *Emberiza citrinella*	4	(S)	–	II	–	–	–	–	–	–	–
Cirl Bunting *Emberiza cirlus*	4	(S)	–	II	–	–	–	–	–	–	–
Rock Bunting *Emberiza cia*	3	V	–	II	–	–	–	–	–	–	–
Cinereous Bunting *Emberiza cineracea*	2	(V)	I	II	–	–	–	–	–	–	–
Ortolan Bunting *Emberiza hortulana*	2	(V)	I	III	–	–	–	–	–	–	–
Grey-necked Bunting *Emberiza buchanani*	–	(S)	–	III	–	–	–	–	–	–	–
Cretzschmar's Bunting *Emberiza caesia*	4	(S)	I	II	–	–	–	–	–	–	–
Rustic Bunting *Emberiza rustica*	–	(S)	–	II	–	–	–	–	–	–	–
Little Bunting *Emberiza pusilla*	–	(S)	–	II	–	–	–	–	–	–	–
Yellow-breasted Bunting *Emberiza aureola*	–	(S)	–	II	–	–	–	–	–	–	–
Reed Bunting *Emberiza schoeniclus*	–	S	–	II	–	–	–	–	–	–	D
Red-headed Bunting *Emberiza bruniceps*	–	S	–	III	–	–	–	–	–	–	–
Black-headed Bunting *Emberiza melanocephala*	2	(V)	–	II	–	–	–	–	–	–	–
Corn Bunting *Miliaria calandra*	4	(S)	–	III	–	–	–	–	–	–	–

Tundra, mires, moorland	Lowland Atlantic heathland	Boreal forest	Lowland temperate forest	Montane forest	Riverine forest	Mediterranean habitats	Arable and improved grassland	Steppic habitats	Montane grassland	Wet grassland	Rice cultivation	Perennial crops	Pastoral woodland	
–	–	–	–	–	–	C	C	–	C	–	–	–	–	*Pyrrhocorax pyrrhocorax*
–	–	–	D	–	D	–	D	D	–	–	–	–	D	*Corvus monedula*
–	–	–	–	–	–	–	D	–	–	–	–	–	–	*Corvus frugilegus*
–	–	–	–	–	–	–	–	–	–	–	–	–	–	*Corvus corone*
–	–	–	–	–	–	–	–	–	–	–	–	–	–	*Corvus corax*
–	–	–	–	–	–	–	D	–	–	–	–	–	–	*Sturnus vulgaris*
–	–	–	–	–	–	D	D	D	–	–	–	D	D	*Sturnus unicolor*
–	–	–	–	–	–	–	–	D	–	–	–	–	–	*Sturnus roseus*
–	–	–	–	–	–	–	–	–	–	–	–	–	–	*Passer domesticus*
–	–	–	–	–	–	–	–	–	–	–	–	–	–	*Passer hispaniolensis*
–	–	–	–	–	–	–	–	–	–	–	–	–	–	*Passer moabiticus*
–	–	–	–	–	–	–	D	–	–	–	–	–	–	*Passer montanus*
–	–	–	–	–	–	–	–	–	–	–	–	–	–	*Carpospiza brachydactyla*
–	–	–	–	–	–	–	–	–	–	–	–	–	–	*Petronia xanthocollis*
–	–	–	–	–	–	D	–	–	–	–	–	–	–	*Petronia petronia*
–	–	–	–	–	–	–	–	–	D	–	–	–	–	*Montifringilla nivalis*
–	–	D	D	D	D	D	–	–	–	–	–	D	D	*Fringilla coelebs*
–	–	–	–	–	–	–	–	–	–	–	–	–	–	*Fringilla teydea*
–	–	D	–	–	–	–	–	–	–	–	–	–	–	*Fringilla montifringilla*
–	–	–	–	–	–	–	–	–	D	–	–	–	–	*Serinus pusillus*
–	–	–	D	D	D	D	–	–	–	–	–	D	D	*Serinus serinus*
–	–	–	–	–	–	–	–	–	–	–	–	D	–	*Serinus canaria*
–	–	–	C	–	D	–	–	–	D	–	–	–	–	*Serinus citrinella*
–	–	D	D	–	D	D	D	–	–	–	–	D	D	*Carduelis chloris*
–	–	–	–	–	–	–	–	–	–	–	–	–	–	*Carduelis carduelis*
–	–	D	D	D	D	–	–	–	–	–	–	–	–	*Carduelis spinus*
–	–	–	–	–	–	D	D	–	–	–	–	–	–	*Carduelis cannabina*
D	–	–	–	–	–	–	–	–	–	–	–	–	–	*Carduelis flavirostris*
–	–	–	–	–	–	–	–	–	–	–	–	–	–	*Carduelis flammea*
–	–	–	–	–	–	–	–	–	–	–	–	–	–	*Carduelis hornemanni*
–	–	D	–	–	–	–	–	–	–	–	–	–	–	*Loxia leucoptera*
–	–	–	–	–	–	–	–	–	–	–	–	–	–	*Loxia curvirostra*
–	–	A	–	–	–	–	–	–	–	–	–	–	–	*Loxia scotica*
–	–	C	–	–	–	–	–	–	–	–	–	–	–	*Loxia pytyopsittacus*
–	–	–	–	–	–	–	–	–	D	–	–	–	–	*Rhodopechys sanguinea*
–	–	–	–	–	–	–	–	–	–	–	–	–	–	*Rhodospiza obsoleta*
–	–	–	–	–	–	–	–	–	–	–	–	–	–	*Bucanetes mongolicus*
–	–	–	–	–	–	–	–	B	–	–	–	–	–	*Bucanetes githagineus*
–	–	–	–	–	–	–	–	–	–	–	–	–	–	*Carpodacus erythrinus*
–	–	–	–	–	–	–	–	–	B	–	–	–	–	*Carpodacus rubicilla*
–	–	D	–	–	–	–	–	–	–	–	–	–	–	*Pinicola enucleator*
–	–	–	–	–	–	–	–	–	–	–	–	–	–	*Pyrrhula pyrrhula*
–	–	–	D	–	–	–	–	–	–	–	–	–	–	*Coccothraustes coccothraustes*
D	–	–	–	–	–	–	–	–	–	–	–	–	–	*Calcarius lapponicus*
D	–	–	–	–	–	–	–	–	–	–	–	–	–	*Plectrophenax nivalis*
–	–	–	–	–	–	–	D	–	–	–	–	–	–	*Emberiza citrinella*
–	–	–	–	–	–	D	D	–	–	–	–	–	–	*Emberiza cirlus*
–	–	–	C	–	C	–	–	–	C	–	–	–	–	*Emberiza cia*
–	–	–	–	–	–	A	–	–	–	–	–	–	–	*Emberiza cineracea*
–	–	C	–	–	–	B	A	–	C	–	B	–	–	*Emberiza hortulana*
–	–	–	–	–	–	–	–	–	–	–	–	–	–	*Emberiza buchanani*
–	–	–	–	–	–	C	–	–	–	–	–	–	–	*Emberiza caesia*
–	–	D	–	–	–	–	–	–	–	–	–	–	–	*Emberiza rustica*
D	–	–	–	–	–	–	–	–	–	–	–	–	–	*Emberiza pusilla*
–	–	–	–	–	–	–	–	–	–	–	–	–	–	*Emberiza aureola*
–	–	–	–	–	–	–	–	–	–	–	–	–	–	*Emberiza schoeniclus*
–	–	–	–	–	–	–	–	–	–	–	–	–	–	*Emberiza bruniceps*
–	–	–	–	–	–	B	B	C	–	–	–	–	–	*Emberiza melanocephala*
–	–	–	–	–	–	–	C	D	–	–	–	–	–	*Miliaria calandra*

THE FOLLOWING tables show the total number of priority species which breed in each European country, according to the species' priority status (A–D). The results are summarized under 'Overview of the Strategies' (Table 1, p. 44). Data are derived from Tucker and Heath (1994) and the joint European Bird Database of the European Bird Census Council and BirdLife International (see Hagemeijer and Blair 1997). Percentages are proportions of the total number of species in the priority category that breed in the habitat in Europe. Data were not available for Bosnia and Herzegovina or for the former Yugoslav Republic of Macedonia.

Marine habitats

■ North and west European seas

Priority status:	A		B		C		D		Total	
	No.	%	No.	%	No.	%	No.	%	No.	%
Belgium	0	0	2	29	2	18	1	4	5	10
Denmark	1	17	2	29	4	36	7	26	14	27
Faroe Islands	5	83	3	43	5	45	8	30	21	41
Greenland	2	33	3	43	3	27	19	70	27	53
Estonia	1	17	3	43	7	64	5	19	16	31
Finland	1	17	3	43	7	64	10	37	21	41
France	4	67	2	29	7	64	6	22	19	37
Germany	1	17	2	29	5	45	8	30	16	31
Iceland	5	83	4	57	6	55	16	59	31	61
Rep. of Ireland	5	83	4	57	6	55	8	30	23	45
Latvia	0	0	1	14	2	18	4	15	7	14
Lithuania	0	0	1	14	2	18	4	15	7	14
Netherlands	0	0	2	29	2	18	5	19	9	18
Norway	4	67	5	71	7	64	12	44	28	55
Svalbard	2	33	2	29	2	18	13	48	19	37
Poland	0	0	2	29	2	18	3	11	7	14
Portugal	0	0	0	0	4	36	4	15	8	16
Russia[1]	3	50	5	71	8	73	18	67	34	67
Spain	2	33	1	14	6	55	4	15	13	25
Sweden	1	17	4	57	7	64	11	41	23	45
United Kingdom	5	83	4	57	9	82	9	33	27	53

[1] Includes species occurring on northern and Baltic seaboards.

■ European Macaronesian seas

Priority status:	A		B		C		D		Total	
	No.	%	No.	%	No.	%	No.	%	No.	%
Portugal										
Azores	2	66	3	75	2	40	0	0	7	54
Madeira	3	100	4	100	2	40	0	0	9	69
Spain										
Canary Islands	1	33	4	100	3	60	0	0	8	61

■ Mediterranean and Black Seas

Priority status:	A		B		C		D		Total	
	No.	%	No.	%	No.	%	No.	%	No.	%
Albania	0	0	0	0	2	40	0	0	2	18
Bulgaria	0	0	1	50	2	40	0	0	3	27
Croatia	0	0	1	50	1	20	0	0	2	18
Cyprus	1	100	0	0	0	0	0	0	1	9
France	1	100	2	100	3	60	1	33	7	64
Greece	1	100	2	100	3	60	1	33	7	64
Italy	1	100	2	100	3	60	2	67	8	73
Malta	0	0	1	50	1	20	0	0	2	18
Romania	0	0	1	50	2	40	0	0	3	27

cont.

Mediterranean and Black Seas (cont.)

Priority status:	A		B		C		D		Total	
	No.	%	No.	%	No.	%	No.	%	No.	%
Russia[1]	0	0	1	50	3	60	4[1]	80	8	73
Slovenia	0	0	0	0	1	20	1	33	2	18
Spain	1	100	2	100	4	80	1	33	8	73
Turkey	1	100	0	0	4	80	1	33	6	55
United Kingdom										
Gibraltar	0	0	0	0	0	0	0	0	0	0
Ukraine	0	0	1	50	3	60	2	67	6	55

[1] Includes Red-throated Diver and Black-throated Diver which only breed in northern Russia.

Coastal habitats

Priority status:	A		B		C		D		Total	
	No.	%	No.	%	No.	%	No.	%	No.	%
Albania	4	44	4	21	14	61	10	43	32	43
Belgium	2	22	7	37	5	22	11	48	25	34
Bulgaria	5	56	4	21	16	70	9	39	34	46
Croatia	3	33	2	11	9	39	8	35	22	30
Cyprus	1	11	2	11	1	4	0	0	4	5
Denmark	3	33	7	37	9	39	13	57	32	43
Faroe Islands	2	22	3	16	3	13	5	22	13	18
Greenland	1	11	3	16	5	22	6	26	15	20
Estonia	3	33	7	37	9	39	15	65	34	46
Finland	2	22	7	37	11	48	15	65	35	47
France	4	44	8	42	15	65	13	57	40	54
Germany	3	33	8	42	13	57	14	61	38	51
Greece	6	67	6	32	14	61	10	43	36	49
Iceland	2	22	3	16	7	30	10	43	22	30
Rep. of Ireland	2	22	6	32	3	13	11	48	22	30
Italy	5	56	6	32	12	52	9	39	32	43
Latvia	1	11	6	32	9	39	13	57	29	39
Lithuania	2	22	6	32	8	35	11	48	27	36
Malta	0	0	0	0	1	4	0	0	1	1
Netherlands	4	44	7	37	9	39	14	61	34	46
Norway	3	33	6	32	10	43	13	57	32	43
Svalbard	1	11	3	16	1	4	5	22	10	14
Poland	3	33	7	37	13	57	13	57	36	49
Portugal	2	22	4	21	10	43	9	39	25	34
Azores	0	0	2	11	0	0	0	0	2	3
Madeira	0	0	2	11	0	0	0	0	2	3
Romania	5	56	7	37	16	70	10	43	38	51
Russia	6	67	14	74	23	100	19	83	62	84
Slovenia	2	22	3	16	3	13	6	26	14	19
Spain	6	67	10	53	16	70	13	57	45	61
Canary Islands	0	0	2	11	1	4	1	4	4	5
Sweden	3	33	9	47	11	48	15	65	38	51
Turkey	6	67	9	47	16	70	12	52	43	58
Ukraine	5	56	8	42	19	83	13	57	45	61
United Kingdom	2	22	7	37	11	48	14	61	34	46
Gibraltar	0	0	1	5	1	4	0	0	2	3
Guernsey	0	0	0	0	0	0	4	17	4	5
Isle of Man	1	11	3	16	2	9	8	35	14	19
Jersey	0	0	0	0	0	0	5	22	5	7

Inland wetlands

Priority status:	A No.	A %	B No.	B %	C No.	C %	D No.	D %	Total No.	Total %
Albania	4	40	10	53	19	58	30	68	63	59
Andorra	0	0	1	5	1	3	2	5	4	4
Austria	1	10	8	42	17	52	30	68	56	53
Belarus	3	30	13	68	21	64	30	68	67	63
Belgium	0	0	7	37	14	42	28	64	49	46
Bulgaria	4	40	13	68	21	64	31	70	69	65
Croatia	2	20	12	63	14	42	29	66	57	54
Cyprus	0	0	1	5	3	9	8	18	12	11
Czech Republic	2	20	9	47	19	58	28	64	58	55
Denmark	0	0	9	47	18	55	32	73	59	56
Faroe Islands	0	0	1	5	5	15	4	9	10	9
Greenland	1	10	0	0	2	6	5	11	8	8
Estonia	1	10	9	47	20	61	31	70	61	58
Finland	2	20	9	47	20	61	30	68	61	58
France	0	0	13	68	23	70	33	75	69	65
Germany	3	30	11	58	21	64	34	77	69	65
Greece	4	40	10	53	21	64	32	73	67	63
Hungary	4	40	14	74	20	61	31	70	69	65
Iceland	1	10	2	11	10	30	7	16	20	19
Rep. of Ireland	0	0	4	21	11	33	19	43	34	32
Italy	1	10	10	53	22	67	30	68	63	59
Latvia	2	20	10	53	19	58	33	75	64	60
Liechtenstein	0	0	1	5	3	9	12	27	16	15
Lithuania	3	30	11	58	17	52	29	66	60	57
Luxembourg	0	0	5	26	5	15	15	34	25	24
Malta	0	0	0	0	0	0	3	7	3	3
Moldova	2	20	12	63	19	58	29	66	62	58
Netherlands	1	10	9	47	20	61	34	77	64	60
Norway	2	20	7	37	18	55	23	52	50	47
Svalbard	0	0	0	0	1	3	1	2	2	2
Poland	3	30	11	58	22	67	33	75	69	65
Portugal	0	0	10	53	16	48	22	50	48	45
Azores	0	0	0	0	0	0	2	5	2	2
Madeira	0	0	0	0	0	0	2	5	2	2
Romania	4	40	14	74	25	76	34	77	77	73
Russia	7	70	18	95	30	91	41	93	96	91
Slovakia	2	20	12	63	18	55	30	68	62	58
Slovenia	2	20	6	32	13	39	29	66	50	47
Spain	3	30	14	74	26	79	29	66	72	68
Canary Islands	0	0	1	5	0	0	5	11	6	6
Sweden	2	20	9	47	20	61	30	68	61	58
Switzerland	1	10	6	32	14	42	25	57	46	43
Turkey	6	60	14	74	27	82	36	82	83	78
Ukraine	5	50	16	84	26	79	36	82	83	78
United Kingdom	1	10	7	37	16	48	29	66	53	50
Gibraltar	0	0	0	0	1	3	2	5	3	3
Guernsey	0	0	0	0	3	9	5	11	8	8
Isle of Man	0	0	2	11	5	15	15	34	22	21
Jersey	0	0	2	11	4	12	9	20	15	14

Tundra, mires and moorland

Priority status:	A No.	A %	B No.	B %	C No.	C %	D No.	D %	Total No.	Total %
Albania	0	0	1	6	3	25	8	17	12	15
Andorra	0	0	0	0	1	8	7	15	8	10
Austria	0	0	1	6	4	33	15	31	20	25
Belarus	0	0	8	44	8	67	13	27	29	36
Belgium	0	0	2	11	4	33	13	27	19	24
Bulgaria	0	0	1	6	4	33	6	13	11	14
Croatia	0	0	1	6	3	25	9	19	13	16

cont.

Tundra, mires and moorland (cont.)

Priority status:	A		B		C		D		Total	
	No.	%	No.	%	No.	%	No.	%	No.	%
Czech Republic	0	0	2	11	6	50	13	27	21	26
Denmark	0	0	3	17	8	67	13	27	24	30
Faroe Islands	0	0	2	11	8	67	10	21	20	25
Greenland	0	0	8	44	6	50	16	33	30	38
Estonia	0	0	10	56	9	75	16	33	35	44
Finland	2	100	14	78	11	92	35	73	62	78
France	0	0	1	6	6	50	16	33	23	29
Germany	0	0	5	28	8	67	15	31	28	35
Greece	0	0	1	6	3	25	8	17	12	15
Hungary	0	0	2	11	3	25	10	21	15	19
Iceland	0	0	8	44	9	75	16	33	33	41
Rep. of Ireland	0	0	4	22	6	50	12	25	22	28
Italy	0	0	1	6	3	25	9	19	13	16
Latvia	0	0	7	39	9	75	15	31	31	39
Liechtenstein	0	0	0	0	0	0	8	17	8	10
Lithuania	0	0	6	33	6	50	14	29	26	33
Luxembourg	0	0	0	0	1	8	6	13	7	9
Moldova	0	0	0	0	3	25	6	13	9	11
Netherlands	0	0	2	11	6	50	12	25	20	25
Norway	2	100	14	78	12	100	36	75	64	80
Svalbard	0	0	4	22	4	33	10	21	18	23
Poland	0	0	6	33	7	58	16	33	29	36
Portugal	0	0	0	0	3	25	6	13	9	11
Romania	0	0	2	11	5	42	13	27	20	25
Russia	2	100	16	89	11	92	40	83	69	86
Slovakia	0	0	1	6	4	33	13	27	18	23
Slovenia	0	0	1	6	3	25	9	19	13	16
Spain	0	0	1	6	7	58	12	25	20	25
Sweden	2	100	14	78	11	92	35	73	62	78
Switzerland	0	0	0	0	3	25	13	27	16	20
Ukraine	0	0	5	28	7	58	13	27	25	31
United Kingdom	0	0	8	44	10	83	22	46	40	50

Lowland Atlantic heathland

Priority status:	A		B		C		D		Total	
	No.	%	No.	%	No.	%	No.	%	No.	%
Belgium	0	0	0	0	4	67	7	70	11	69
Denmark	0	0	0	0	4	67	9	90	13	81
Estonia	0	0	0	0	4	67	8	80	12	75
France	0	0	0	0	6	100	10	100	16	100
Germany	0	0	0	0	5	83	10	100	15	94
Rep. of Ireland	0	0	0	0	2	33	3	30	5	31
Latvia	0	0	0	0	4	67	8	80	12	75
Lithuania	0	0	0	0	4	67	8	80	12	75
Luxembourg	0	0	0	0	4	67	6	60	10	63
Netherlands	0	0	0	0	4	67	8	80	12	75
Norway	0	0	0	0	3	50	8	80	11	69
Poland	0	0	0	0	4	67	10	100	14	88
Portugal	0	0	0	0	5	83	7	70	12	75
Spain	0	0	0	0	6	100	9	90	15	94
Sweden	0	0	0	0	4	67	8	80	12	75
United Kingdom	0	0	0	0	6	100	7	70	13	81
Guernsey	0	0	0	0	0	0	2	20	2	13
Isle of Man	0	0	0	0	2	33	4	40	6	38
Jersey	0	0	0	0	1	17	2	20	3	19

Boreal and temperate forests

■ Boreal forest

Priority status:	A No.	A %	B No.	B %	C No.	C %	D No.	D %	Total No.	Total %
Estonia	0	0	4	67	14	88	32	71	50	72
Finland	1	50	6	100	15	94	40	89	62	90
Iceland	0	0	0	0	2	13	6	13	8	12
Norway	0	0	6	100	14	88	38	84	58	84
Russia	1	50	6	100	16	100	43	96	66	96
Sweden	0	0	6	100	13	81	38	84	57	83
United Kingdom	1	50	3	50	10	63	25	56	39	57

■ Montane forest

Priority status:	A No.	A %	B No.	B %	C No.	C %	D No.	D %	Total No.	Total %
Andorra	0	0	3	50	7	58	24	75	34	68
Austria	0	0	3	50	10	83	31	97	44	88
Bulgaria	0	0	5	83	10	83	28	88	43	86
Croatia	0	0	4	67	9	75	29	91	42	84
Cyprus	0	0	2	33	2	17	5	16	9	18
Czech Republic	0	0	3	50	7	58	30	94	40	80
France	0	0	3	50	12	100	30	94	45	90
Germany	0	0	3	50	10	83	30	94	43	86
Greece	0	0	5	83	10	83	28	88	43	86
Italy	0	0	3	50	11	92	28	88	42	84
Portugal	0	0	3	50	8	67	23	72	34	68
Romania	0	0	4	67	11	92	31	97	46	92
Russia	0	0	5	83	10	83	28	88	43	86
Slovakia	0	0	4	67	11	92	30	94	45	90
Slovenia	0	0	3	50	11	92	31	97	45	90
Spain	0	0	3	50	9	75	28	88	40	80
Switzerland	0	0	3	50	10	83	29	91	42	84
Turkey	0	0	5	83	9	75	27	84	41	82

■ Riverine forest

Priority status:	A No.	A %	B No.	B %	C No.	C %	D No.	D %	Total No.	Total %
Albania	0	0	5	83	14	78	26	84	45	80
Austria	0	0	2	33	12	67	26	84	40	71
Belarus	1	100	2	33	15	83	29	94	47	84
Belgium	0	0	1	17	9	50	26	84	36	64
Bulgaria	0	0	6	100	16	89	24	77	46	82
Croatia	0	0	4	67	14	78	27	87	45	80
Czech Republic	0	0	2	33	15	83	28	90	45	80
Denmark	0	0	1	17	8	44	23	74	32	57
Estonia	0	0	1	17	14	78	24	77	39	70
France	0	0	3	50	15	83	28	90	46	82
Germany	0	0	1	17	15	83	28	90	44	79
Greece	0	0	6	100	14	78	22	71	42	75
Hungary	0	0	5	83	15	83	29	94	49	88
Rep. of Ireland	0	0	0	0	2	11	15	48	17	30
Italy	0	0	3	50	11	61	26	84	40	71
Latvia	1	100	1	17	14	78	26	84	42	75
Liechtenstein	0	0	1	17	6	33	23	74	30	54
Lithuania	0	0	1	17	14	78	25	81	40	71
Luxembourg	0	0	1	17	8	44	23	74	32	57
Moldova	1	100	4	67	14	78	24	77	43	77
Netherlands	0	0	2	33	8	44	24	77	34	61
Poland	1	100	1	17	16	89	28	90	46	82
Portugal	0	0	3	50	11	61	22	71	36	64
Romania	1	100	5	83	17	94	29	94	52	93
Russia	1	100	5	83	18	100	29	94	53	95
Slovakia	0	0	4	67	14	78	29	94	47	84
Slovenia	0	0	2	33	11	61	27	87	40	71

cont.

Riverine forest (cont.)

Priority status:	A No.	%	B No.	%	C No.	%	D No.	%	Total No.	%
Spain	0	0	3	50	14	78	25	81	42	75
Switzerland	0	0	2	33	8	44	27	87	37	66
Turkey	0	0	6	100	15	83	23	74	44	79
Ukraine	1	100	5	83	18	100	29	94	53	95
United Kingdom	0	0	1	17	6	33	22	71	29	52

■ Lowland temperate forest

Priority status:	A No.	%	B No.	%	C No.	%	D No.	%	Total No.	%
Albania	3	60	6	86	24	83	29	91	62	85
Andorra	3	60	2	29	15	52	24	75	44	60
Austria	3	60	4	57	24	83	29	91	60	82
Belarus	4	80	5	71	25	86	29	91	63	86
Belgium	3	60	4	57	18	62	29	91	54	74
Bulgaria	4	80	7	100	23	79	30	94	64	88
Croatia	4	80	6	86	23	79	30	94	63	86
Cyprus	2	40	1	14	7	24	6	19	16	22
Czech Republic	3	60	5	71	24	83	29	91	61	84
Denmark	3	60	3	43	16	55	26	81	48	66
Faroe Islands	0	0	0	0	1	3	7	22	8	11
Greenland	0	0	0	0	2	7	1	3	3	4
Estonia	3	60	5	71	19	66	28	88	55	75
Finland	3	60	4	57	18	62	27	84	52	71
France	3	60	4	57	26	90	29	91	62	85
Germany	3	60	5	71	24	83	30	94	62	85
Greece	4	80	7	100	20	69	29	91	60	82
Hungary	4	80	6	86	24	83	29	91	63	86
Iceland	0	0	0	0	1	3	4	13	5	7
Rep. of Ireland	1	20	2	29	6	21	16	50	25	34
Italy	3	60	4	57	20	69	29	91	56	77
Latvia	4	80	5	71	22	76	29	91	60	82
Liechtenstein	2	40	3	43	15	52	27	84	47	64
Lithuania	3	60	5	71	20	69	27	84	55	75
Luxembourg	3	60	4	57	14	48	27	84	48	66
Malta	0	0	0	0	3	10	4	13	7	10
Moldova	5	100	4	57	21	72	21	66	51	70
Netherlands	3	60	3	43	16	55	27	84	49	67
Norway	3	60	4	57	16	55	26	81	49	67
Svalbard	0	0	0	0	0	0	0	0	0	0
Poland	4	80	5	71	27	93	29	91	65	89
Portugal	3	60	2	29	21	72	23	72	49	67
Azores	0	0	1	14	0	0	3	9	4	5
Madeira	0	0	1	14	1	3	7	22	9	12
Romania	5	100	6	86	28	97	30	94	69	95
Russia	5	100	6	86	28	97	28	88	67	92
Slovakia	4	80	5	71	27	93	29	91	65	89
Slovenia	3	60	4	57	23	79	31	97	61	84
Spain	3	60	3	43	22	76	27	84	55	75
Canary Islands	0	0	0	0	4	14	8	25	12	16
Sweden	3	60	4	57	18	62	27	84	52	71
Switzerland	3	60	4	57	21	72	29	91	57	78
Turkey	4	80	6	86	22	76	27	84	59	81
Ukraine	5	100	6	86	29	100	29	91	69	95
United Kingdom	3	60	3	43	14	48	26	81	46	63
Guernsey	0	0	0	0	6	21	12	38	18	25
Jersey	0	0	0	0	6	21	14	44	20	27
Isle of Man	0	0	1	14	5	17	17	53	23	32
Gibraltar	0	0	0	0	4	14	7	22	11	15

Mediterranean forest, shrubland and rocky habitats

Priority status:	A		B		C		D		Total	
	No.	%	No.	%	No.	%	No.	%	No.	%
Albania	2	22	15	58	24	86	29	78	70	70
Andorra	1	11	9	35	11	39	27	73	48	48
Bulgaria	2	22	18	69	21	75	29	78	70	70
Croatia	3	33	16	62	19	68	29	78	67	67
Cyprus	3	33	12	46	6	21	15	41	36	36
France	3	33	15	58	21	75	32	86	71	71
Greece	4	44	21	81	26	93	29	78	80	80
Italy	3	33	16	62	18	64	31	84	68	68
Portugal	2	22	14	54	19	68	32	86	67	67
Spain	4	44	17	65	22	79	34	92	77	77
Turkey	4	44	21	81	25	89	29	78	79	79

Agricultural and grassland habitats

■ Arable and improved grassland

Priority status:	A		B		C		D		Total	
	No.	%	No.	%	No.	%	No.	%	No.	%
Albania	1	17	17	85	16	67	19	61	53	65
Andorra	2	33	10	50	6	25	15	48	33	41
Austria	3	50	14	70	13	54	23	74	53	65
Belarus	2	33	14	70	17	71	24	77	57	70
Belgium	2	33	12	60	8	33	20	65	42	52
Bulgaria	3	50	17	85	17	71	20	65	57	70
Croatia	3	50	17	85	15	63	23	74	58	72
Cyprus	1	17	8	40	9	38	4	13	22	27
Czech Republic	3	50	12	60	18	75	22	71	55	68
Denmark	2	33	13	65	10	42	23	74	48	59
Faroe Islands	0	0	5	25	2	8	5	16	12	15
Greenland	0	0	1	5	2	8	5	16	8	10
Estonia	2	33	11	55	13	54	23	74	49	60
Finland	2	33	8	40	10	42	22	71	42	52
France	3	50	17	85	18	75	25	81	63	78
Germany	4	67	12	60	15	63	24	77	55	68
Greece	2	33	17	85	17	71	21	68	57	70
Hungary	4	67	15	75	16	67	23	74	58	72
Iceland	0	0	3	15	3	13	7	23	13	16
Rep. of Ireland	1	17	6	30	6	25	17	55	30	37
Italy	3	50	17	85	15	63	21	68	56	69
Latvia	2	33	13	65	14	58	24	77	53	65
Liechtenstein	1	17	7	35	6	25	15	48	29	36
Lithuania	2	33	13	65	13	54	24	77	52	64
Luxembourg	1	17	9	45	10	42	20	65	40	49
Malta	0	0	3	15	3	13	3	10	9	11
Moldova	4	67	15	75	13	54	21	68	53	65
Netherlands	2	33	12	60	10	42	21	68	45	56
Norway	2	33	7	35	8	33	21	68	38	47
Svalbard	0	0	1	5	0	0	3	10	4	5
Poland	3	50	13	65	16	67	23	74	55	68
Portugal	3	50	14	70	15	63	20	65	52	64
Azores	0	0	1	5	1	4	1	3	3	4
Madeira	1	17	3	15	2	8	3	10	9	11
Romania	4	67	17	85	18	75	25	81	64	79
Russia	4	67	18	90	21	88	27	87	70	86
Slovakia	4	67	14	70	16	67	23	74	57	70
Slovenia	2	33	14	70	10	42	20	65	46	57
Spain	4	67	16	80	17	71	25	81	62	77
Canary Islands	0	0	2	10	4	17	6	19	12	15
Sweden	2	33	9	45	11	46	23	74	45	56
Switzerland	2	33	14	70	11	46	22	71	49	60
Turkey	4	67	18	90	17	71	22	71	61	75
Ukraine	4	67	17	85	22	92	23	74	66	81

cont.

Arable and improved grassland (cont.)

Priority status:	A No.	A %	B No.	B %	C No.	C %	D No.	D %	Total No.	Total %
United Kingdom	2	33	10	50	11	46	21	68	44	54
Gibraltar	0	0	6	30	2	8	3	10	11	14
Guernsey	0	0	4	20	2	8	9	29	15	19
Isle of Man	2	33	6	30	6	25	17	55	31	38
Jersey	0	0	5	25	3	13	12	39	20	25

■ Steppic habitats

Priority status:	A No.	A %	B No.	B %	C No.	C %	D No.	D %	Total No.	Total %
Albania	1	14	7	29	15	68	5	42	28	43
Austria	1	14	5	21	11	50	6	50	23	35
Bulgaria	2	29	9	38	16	73				
Slovakia	0	—	6	75	2	25	18	82	26	68
Slovenia	0	—	5	63	2	25	19	86	26	68
Spain	0	—	6	75	7	88	17	77	30	79
Switzerland	0	—	5	63	3	38	16	73	24	63
Turkey	0	—	8	100	6	75	19	86	33	87
Ukraine	0	—	6	75	3	38	18	82	27	71
United Kingdom	0	—	2	25	2	25	15	68	19	50

■ Montane grassland

Priority status:	A No.	A %	B No.	B %	C No.	C %	D No.	D %	Total No.	Total %
Albania	0	0	2	33	9	90	12	75	23	70
Andorra	0	0	0	0	7	70	11	69	18	55
Austria	0	0	2	33	7	70	12	75	21	64
Bulgaria	0	0	2	33	8	80	10	63	20	61
Croatia	0	0	2	33	7	70	10	63	19	58
Cyprus	0	0	0	0	3	30	3	19	6	18
Czech Republic	0	0	1	17	3	30	9	56	13	39
France	0	0	3	50	9	90	13	81	25	76
Germany	0	0	1	17	5	50	12	75	18	55
Greece	0	0	2	33	9	90	12	75	23	70
Italy	0	0	2	33	9	90	13	81	24	73
Norway	0	0	1	17	4	40	7	44	12	36
Portugal	0	0	0	0	9	90	8	50	17	52
Romania	0	0	2	33	7	70	9	56	18	55
Russia	0	0	2	33	6	60	9	56	17	52
Slovakia	0	0	1	17	6	60	9	56	16	48
Slovenia	0	0	2	33	6	60	13	81	21	64
Spain	0	0	2	33	9	90	13	81	24	73
Sweden	0	0	1	17	4	40	7	44	12	36
Switzerland	0	0	2	33	7	70	13	81	22	67
Turkey	1	100	4	67	9	90	14	88	28	85
United Kingdom	0	0	1	17	4	40	6	38	11	33

■ Wet grassland

Priority status:	A No.	A %	B No.	B %	C No.	C %	D No.	D %	Total No.	Total %
Albania	0	0	2	50	3	33	10	56	15	41
Austria	2	33	3	75	4	44	13	72	22	59
Belarus	5	83	3	75	7	78	14	78	29	78
Belgium	2	33	3	75	6	67	12	67	23	62
Bulgaria	1	17	3	75	3	33	12	67	19	51
Croatia	1	17	3	75	4	44	11	61	19	51
Czech Republic	2	33	3	75	6	67	15	83	26	70
Denmark	2	33	3	75	6	67	15	83	26	70
Faroe Islands	0	0	2	50	2	22	5	28	9	24
Estonia	3	50	3	75	6	67	14	78	26	70
Finland	5	83	2	50	5	56	15	83	27	73
France	2	33	3	75	5	56	16	89	26	70

cont.

Wet grassland (cont.)

Priority status:	A No.	A %	B No.	B %	C No.	C %	D No.	D %	Total No.	Total %
Germany	3	50	3	75	7	78	16	89	29	78
Greece	0	0	3	75	4	44	10	56	17	46
Hungary	3	50	4	100	5	56	14	78	26	70
Iceland	0	0	2	50	3	33	6	33	11	30
Rep. of Ireland	2	33	2	50	5	56	9	50	18	49
Italy	2	33	2	50	2	22	11	61	17	46
Latvia	5	83	3	75	7	78	15	83	30	81
Liechtenstein	1	17	0	0	2	22	7	39	10	27
Lithuania	4	67	3	75	7	78	15	83	29	78
Luxembourg	1	17	0	0	3	33	10	56	14	38
Moldova	2	33	4	100	5	56	13	72	24	65
Netherlands	2	33	3	75	6	67	15	83	26	70
Norway	4	67	2	50	5	56	14	78	25	68
Poland	5	83	3	75	7	78	17	94	32	86
Portugal	0	0	2	50	3	33	9	50	14	38
Romania	3	50	3	75	5	56	15	83	26	70
Russia	6	100	4	100	8	89	17	94	35	95
Slovakia	2	33	4	100	5	56	14	78	25	68
Slovenia	1	17	2	50	3	33	10	56	16	43
Spain	2	33	2	50	5	56	14	78	23	62
Sweden	4	67	2	50	5	56	14	78	25	68
Switzerland	1	17	2	50	3	33	13	72	19	51
Turkey	1	17	3	75	4	44	13	72	21	57
Ukraine	5	83	3	75	6	67	16	89	30	81
United Kingdom	2	33	2	50	6	67	14	78	24	65

■ Rice cultivation

Priority status:	A No.	A %	B No.	B %	C No.	C %	D No.	D %	Total No.	Total %
France	0	—	2	100	2	100	14	82	18	86
Greece	0	—	1	50	2	100	13	76	16	76
Italy	0	—	2	100	2	100	11	65	15	71
Portugal	0	—	1	50	2	100	9	53	12	57
Spain	0	—	2	100	2	100	14	82	18	86
Turkey	0	—	1	50	2	100	14	82	17	81

■ Perennial crops

Priority status:	A No.	A %	B No.	B %	C No.	C %	D No.	D %	Total No.	Total %
Albania	0	—	7	88	6	75	19	86	32	84
Andorra	0	—	5	63	2	25	16	73	23	61
Austria	0	—	5	63	2	25	18	82	25	66
Belgium	0	—	3	38	1	13	16	73	20	53
Bulgaria	0	—	8	100	6	75	19	86	33	87
Croatia	0	—	7	88	6	75	19	86	32	84
Cyprus	0	—	5	63	2	25	8	36	15	39
Czech Republic	0	—	5	63	2	25	18	82	25	66
Denmark	0	—	3	38	1	13	15	68	19	50
France	0	—	6	75	6	75	18	82	30	79
Germany	0	—	4	50	3	38	17	77	24	63
Greece	0	—	8	100	6	75	19	86	33	87
Hungary	0	—	5	63	4	50	17	77	26	68
Rep. of Ireland	0	—	0	0	1	13	9	41	10	26
Italy	0	—	6	75	5	63	18	82	29	76
Liechtenstein	0	—	1	13	1	13	15	68	17	45
Luxembourg	0	—	3	38	1	13	16	73	20	53
Malta	0	—	1	13	0	0	7	32	8	21
Moldova	0	—	5	63	2	25	15	68	22	58
Netherlands	0	—	3	38	2	25	15	68	20	53
Poland	0	—	5	63	2	25	18	82	25	66
Portugal	0	—	6	75	7	88	17	77	30	79
Romania	0	—	6	75	5	63	18	82	29	76

cont.

Perennial crops (cont.)

Priority status:	A No.	A %	B No.	B %	C No.	C %	D No.	D %	Total No.	Total %
Russia	0	—	5	63	3	38	15	68	23	61
Slovakia	0	—	6	75	2	25	18	82	26	68
Slovenia	0	—	5	63	2	25	19	86	26	68
Spain	0	—	6	75	7	88	17	77	30	79
Switzerland	0	—	5	63	3	38	16	73	24	63
Turkey	0	—	8	100	6	75	19	86	33	87
Ukraine	0	—	6	75	3	38	18	82	27	71
United Kingdom	0	—	2	25	2	25	15	68	19	50

■ Pastoral woodland

Priority status:	A No.	A %	B No.	B %	C No.	C %	D No.	D %	Total No.	Total %
Albania	1	50	5	100	8	67	16	89	30	81
Andorra	1	50	3	60	3	25	16	89	23	62
Austria	0	0	5	100	5	42	17	94	27	73
Belgium	0	0	3	60	4	33	17	94	24	65
Bulgaria	1	50	5	100	8	67	16	89	30	81
Croatia	1	50	5	100	7	58	17	94	30	81
Cyprus	1	50	3	60	2	17	10	56	16	43
Czech Republic	1	50	4	80	6	50	17	94	28	76
Denmark	0	0	3	60	4	33	16	89	23	62
France	1	50	5	100	11	92	18	100	35	95
Germany	1	50	3	60	6	50	17	94	27	73
Greece	1	50	5	100	9	75	16	89	31	84
Hungary	0	0	5	100	7	58	17	94	29	78
Rep. of Ireland	0	0	0	0	1	8	13	72	14	38
Italy	1	50	5	100	6	50	18	100	30	81
Liechtenstein	0	0	1	20	2	17	16	89	19	51
Luxembourg	1	50	2	40	4	33	17	94	24	65
Malta	1	50	0	0	0	0	7	39	8	22
Moldova	0	0	5	100	7	58	15	83	27	73
Netherlands	0	0	3	60	4	33	17	94	24	65
Poland	1	50	4	80	8	67	17	94	30	81
Portugal	1	50	5	100	10	83	18	100	34	92
Romania	1	50	5	100	8	67	17	94	31	84
Russia	0	0	5	100	9	75	15	83	29	78
Slovakia	1	50	5	100	7	58	17	94	30	81
Slovenia	0	0	5	100	6	50	16	89	27	73
Spain	2	100	5	100	12	100	18	100	37	100
Switzerland	1	50	4	80	5	42	17	94	27	73
Turkey	1	50	5	100	10	83	16	89	32	86
Ukraine	1	50	5	100	9	75	17	94	32	86
United Kingdom	0	0	2	40	2	17	16	89	20	54

APPENDIX 3: Habitat requirements of the priority bird species in Europe

THE TABLES of habitat requirements were produced by the experts of the relevant Habitat Working Groups (see 'Introduction', p. 21), but note that detailed information on habitat requirements is lacking for many species.

Entries in the tables indicate the preference (or otherwise) of particular species for particular features of the various habitats ('preference' indicates that high densities and/or productivity levels of the species are associated with the feature).

Where a requirement is essential to the species, the entry is made in blue; in these cases the feature is always required, and where it is absent the species does not have a long-term viable population in the habitat.

( ) Parentheses indicate a suboptimal habitat feature, i.e. one which is used but is not preferred

[ ] [ ] Square brackets indicate requirements which are alternatives to one another, i.e. one of the two (or more) alternatives is sufficient

Habitat requirements noted here apply throughout the species' life cycle in Europe unless implied or stated otherwise, as follows.

B Breeding season only
W Non-breeding season only
N Nesting requirement only

Marine habitats

KEY

Preferred habitat-zones
• Marginal importance
•• Species occurs
••• Important

Feeding-area characteristics
Principal feeding areas
I Inshore shallow waters (<20 m depth)
S Shelf waters (<200 m depth)
O Oceanic (>200 m depth)
P Pelagic

Salinity
• Brackish (<18–20‰)
•• Moderate salinity (20–30‰)
••• High salinity (>30‰)

Other water-body features
C Clear water
F Fronts (boundaries between water-masses)
U Upwellings (e.g. over sea-mounts and shelf-edges)
R River outflows
I Presence of ice
L Low disturbance from shipping, etc.
S Sheltered waters

Seabed-type
R Rocky
S Sand
M Mud
X Mixed (mosaic)
K Kelp forests present

Food resources
Other food sources
F Fish spawn
C Carcasses
S Sewage
K Kleptoparasitism
B Frequents fishing boats

Nest-site requirements
Nesting location
NB For most species, requirements are highly dependent on absence of mammalian predators.
R Rocky islands and/or sea-cliffs
L Low islands and coasts
I Inland cliffs and ravines
B Beaches and dunes
S Sedimentary islands in coastal wetlands

Nest-site
C Cliffs
B Boulder-slopes or cavities (e.g. caves and burrows)
L Low-lying flat surfaces

Bare ground (mud or sand with <10% cover by low vegetation)
• Required
× Avoided

Low vegetation (<10 cm)
• Required
× Avoided

Moderate vegetation (10–20 cm)
• Required
× Avoided

High vegetation (>20cm)
• Required
× Avoided

Other
U Nest-sites undisturbed by humans
S Presence of other seabirds (e.g. for nest defence)
I Absence of introduced terrestrial mammals
P Absence of introduced mammalian predators
N Absence of native mammalian predators
B Absence of native avian predators
C Colonial breeder

Marine habitats (cont.)

■ **North and west European seas**

Part 1: Preferred habitat-zones

	Breeding season					Outside breeding season				
	Shallow temperate	Shelf temperate	Shallow Arctic	Deep Atlantic	Deep Arctic	Shallow temperate	Shelf temperate	Shallow Arctic	Deep Atlantic	Deep Arctic
Gavia stellata	••	•	••	—	—	•••	••	••	—	—
Gavia arctica	•	•	•	—	—	•••	••	•	—	—
Gavia immer	••	•	••	—	—	•••	••	•	—	—
Gavia adamsii	—	—	••	—	—	•	•	••	—	—
Fulmarus glacialis	••	•••	••	••	••	••	•••	••	••	••
Calonectris diomedea	—	••	—	•••	—	•	••	—	•••	—
Puffinus gravis	—	—	—	—	—	—	•	—	•••	—
Puffinus griseus	—	—	—	—	—	•	•••	•	••	•
Puffinus puffinus	••	•••	—	—	—	••	•••	—	••	—
Puffinus yelkouan	—	—	—	—	—	••	•••	—	—	—
Hydrobates pelagicus	•••	•••	•	•	—	••	••	—	•	—
Oceanodroma leucorhoa	—	••	—	•••	—	—	••	—	•••	—
Oceanodroma castro	—	••	—	•••	—	—	—	—	—	—
Sula bassana	•	•••	•	•	—	•	•••	—	•	—
Phalacrocorax aristotelis	•••	•	•	—	—	•••	•	—	—	—
Aythya marila	•••	—	—	—	—	•••	—	—	—	—
Somateria mollissima	•••	—	••	—	—	•••	—	•	—	—
Somateria spectabilis	•	—	•••	—	—	•	—	•••	—	—
Polysticta stelleri	—	—	••	—	—	••	—	•••	—	—
Histrionicus histrionicus	•	—	—	—	—	•••	—	—	—	—
Clangula hyemalis	•	—	•••	—	—	•••	••	•	—	—
Melanitta nigra	—	—	•••	—	—	•••	—	—	—	—
Melanitta fusca	•	—	•••	—	—	•••	••	—	—	—
Mergus serrator	••	—	•	—	—	•••	••	—	—	—
Phalaropus fulicarius	•	—	•••	—	—	—	•	—	•••	—
Stercorarius pomarinus	—	—	—	—	—	—	••	•••	•••	•••
Stercorarius parasiticus	••	••	•••	—	—	•	•••	—	•••	•
Stercorarius longicaudus	—	—	•••	—	—	—	•	—	•••	•••
Stercorarius skua	••	•••	•	—	—	•	•••	—	•	—
Larus sabini	—	—	•••	—	—	—	•	•••	•••	•••
Larus canus	•••	•	•	—	—	•••	••	•	—	—
Larus fuscus	••	•••	•	—	—	•••	•••	•	•	•
Larus argentatus	•••	••	••	—	—	•••	••	••	—	—
Larus glaucoides	—	—	—	—	—	•••	•	•	—	—
Larus hyperboreus	••	••	••	—	—	••	••	••	—	—
Larus marinus	••	••	••	—	—	••	••	••	—	—
Rhodostethia rosea	—	—	•••	—	—	—	—	•••	—	••
Rissa tridactyla	•	•••	—	—	—	•	•••	•	•	•
Pagophila eburnea	—	—	•••	—	—	•	•	•••	—	•
Sterna caspia	•••	—	—	—	—	•••	—	—	—	—
Stern sandvicensis	•••	—	•	—	—	•••	—	—	—	—
Sterna dougallii	•••	•	—	—	—	•	•	—	—	—
Sterna paradisaea	••	••	••	—	—	••	••	—	•••	••
Sterna albifrons	•••	—	—	—	—	•••	—	—	—	—
Chlidonias niger	—	—	—	—	—	•••	••	—	•	—
Uria aalge	•••	•••	••	—	—	•	•••	•	—	—
Uria lomvia	•	••	••	—	—	—	••	••	••	••
Alca torda	•	•••	•	—	—	•	•••	—	—	—
Cepphus grylle	••	—	••	—	—	••	—	••	—	—
Alle alle	•	—	•••	—	•	—	•	•••	••	—
Fratercula arctica	••	•••	•	—	—	—	••	•	•••	—

cont.

Marine habitats (cont.)

■ North and west European seas (cont.)

Part 2: Feeding-area characteristics and food resources

Species	Feeding-area characteristics				Food resources									
	Principal feeding areas	Salinity	Other water-body features	Seabed-type	Shellfish, echinoderms	Cephalopods	Crustacea	Small bottom fish	Small pelagic fish	Chicks, eggs, small birds	Offal	Discards	Other food sources	Absence of competitive or predatory sea-birds
Gavia stellata	I	≤•^{N}	C,R^N,L^N	S^N	—	—	—	•	•	—	—	—	—	•
Gavia arctica	I	—	C,R^N,L^N	S^N	—	—	—	•^{N}	•^{N}	—	—	—	—	•
Gavia immer	I	••	C,L	S	—	—	—	•	•	—	—	—	—	—
Gavia adamsii	I	•••	C	—	—	—	—	•	•	—	—	—	—	—
Fulmarus glacialis	S,O	•••	F	—	—	—	•	—	•	—	—	—	—	—
Calonectris diomedea	S,O	•••	C,F,U	—	—	•	—	—	•	—	•	—	—	—
Puffinus gravis	S,O	•••	F,U	—	—	—	—	—	•	—	—	—	—	—
Puffinus griseus	S	•••	F	—	—	•	•	—	•	—	—	—	—	—
Puffinus puffinus	S	•••	F	—	—	—	—	—	•	—	—	—	—	—
Puffinus yelkouan	S	•••	F	—	—	—	—	—	•	—	—	—	—	—
Hydrobates pelagicus	S	•••	F	—	—	—	•	—	•	—	—	—	—	•
Oceanodroma leucorhoa	O	•••	—	—	—	—	•	—	•	—	—	—	—	—
Oceanodroma castro	O	•••	C,U	—	—	—	•	—	•	—	—	—	—	—
Sula bassana	S	•••	C	—	—	—	—	•	•	—	—	•	—	—
Phalacrocorax aristotelis	I	••	C,S$^-$	K	—	—	•	•	•	—	—	—	—	—
Aythya marila	I	—	—	R	•	—	—	—	—	—	—	—	—	—
Somateria mollissima	I	—	L^N,S	S	•	—	•	—	—	—	—	—	F	—
Somateria spectabilis	I	•^{N}	—	R^N	•^{N}	—	•	—	—	—	—	—	—	—
Polysticta stelleri	I	•^{N}	R^N,S	S	•	—	•^{N}	—	—	—	—	—	—	—
Histrionicus histrionicus	I	•••	—	R^N	•^{N}	—	•^{N}	—	—	—	—	—	—	—
Clangula hyemalis	I	≤••	—	S^N,M^N,X^N	•	—	•^{N}	—	•	—	•	—	F	—
Melanitta nigra	I	≤•^{N}	L^N	S^N	•^{N}	—	—	—	—	—	—	—	—	—
Melanitta fusca	I	≤•^{N}	—	S^N,M^N,X^N	•	—	•	•	•	—	—	—	—	—
Mergus serrator	I	≤•^{N}	U	—	—	—	•	•	•^{N}	—	—	—	—	—
Phalaropus fulicarius	S,O	•••	—	—	—	—	—	—	—	—	—	—	—	—
Stercorarius pomarinus	S,O	•••	—	—	—	—	—	—	—	•	—	—	K^N	—
Stercorarius parasiticus	S,O	•••	—	—	—	—	—	—	—	—	—	—	K	—
Stercorarius longicaudus	S,O	•••	—	—	—	—	—	—	—	•	—	—	K	—
Stercorarius skua	S	—	—	—	—	—	—	—	—	—	—	—	K	—
Larus sabini	S,O	•^{W}	—	—	—	—	—	—	—	—	—	—	—	—
Larus canus	I,S	•^{W}	F,R,S^N	—	—	—	—	—	•	—	—	•	—	—
Larus fuscus	I,S	—	R^N	—	—	—	—	—	•	—	—	•	—	—

Marine habitats (cont.)

■ North and west European seas (cont.)

Part 2: Feeding-area characteristics and food resources (cont.)

	Feeding-area characteristics				Food resources									
	Principal feeding areas	Salinity	Other water-body features	Seabed-type	Shellfish, echinoderms	Cephalopods	Crustacea	Small bottom fish	Small pelagic fish	Chicks, eggs, small birds	Offal	Discards	Other food sources	Absence of competitive or predatory sea-birds
Larus argentatus	I,S	—	R	—	—	—	—	—	•	—	—	•	S^N	—
Larus glaucoides	S	—	—	—	—	—	—	—	•	—	—	—	S^N	—
Larus hyperboreus	S	—	R^B	—	—	—	—	—	•	•	•	•	S	—
Larus marinus	S,I	—	—	—	—	—	—	—	•	•	•	•	—	—
Rhodostethia rosea	I,S,O	—	—	—	—	—	•	—	•	—	—	—	—	—
Rissa tridactyla	S,O	—	I^N	—	—	—	—	—	•	—	•^{N}	•^{N}	C	—
Pagophila eburnea	S	—	S	—	—	—	—	—	—	—	—	—	—	—
Sterna caspia	I	—	—	—	—	—	—	—	—	—	—	—	—	—
Sterna sandvicensis	I,S	—	—	—	—	—	—	—	•	—	—	—	—	—
Sterna dougallii	I	—	—	—	—	—	—	—	•	—	—	—	—	—
Sterna paradisaea	I,S	—	R	—	—	—	•	—	•	—	—	—	—	—
Sterna albifrons	I	•	F^N,R^N	—	—	—	—	—	•	—	—	—	—	—
Chlidonias niger	I	—	C	—	—	—	—	—	—	—	—	—	—	—
Uria aalge	S	—	C,F,I	—	—	•^{B}	•^{B}	•^{B}	•^{B}	—	—	—	—	—
Uria lomvia	S	—	C	—	—	—	•	—	•	—	—	—	—	—
Alca torda	S	—	C	X^N	—	—	•	—	•^{N}	—	—	—	—	—
Cepphus grylle	I	—	C,F,I	—	—	—	•	•	•	—	—	—	—	•
Alle alle	S	—	C	—	—	—	•	—	•	—	—	—	—	—
Fratercula arctica	S	•••	C	—	—	—	—	—	•^{B}	—	—	—	—	—

Marine habitats (cont.)

■ **North and west European seas** (cont.)

Part 3: Nest-site requirements

	Nesting location	Cliffs	Boulder-slopes	Steep, grassy slopes	Deep, well-drained soil	Pre-existing dry cavities	No vegetation	Vegetation cover	Other
Gavia stellata	—	—	—	—	—	—	—	—	—
Gavia arctica	—	—	—	—	—	—	—	—	—
Gavia immer	—	—	—	—	—	—	—	—	—
Gavia adamsii	—	—	—	—	—	—	—	—	—
Fulmarus glacialis	R	•	—	•	—	—	—	—	—
Calonectris diomedea	R	—	—	—	—	•	—	—	U,I
Puffinus gravis	—	—	—	—	—	—	—	—	—
Puffinus griseus	—	—	—	—	—	—	—	—	—
Puffinus puffinus	R	—	—	—	•	—	—	—	I,N
Puffinus yelkouan	—	—	—	—	—	—	—	—	—
Hydrobates pelagicus	R	—	•	—	—	•	—	—	I,N
Oceanodroma leucorhoa	R	—	•	•	•	•	—	—	I,N
Oceanodroma castro	R	—	—	—	—	•	—	—	I,N
Sula bassana	R	•	—	—	—	—	—	—	U
Phalacrocorax aristotelis	R	•	•	—	—	•	—	—	U,I,N
Aythya marila	—	—	—	—	—	—	—	—	—
Somateria mollissima	R,L	—	—	—	—	—	—	—	I,N
Somateria spectabilis	—	—	—	—	—	—	—	—	U,I,N
Polysticta stelleri	—	—	—	—	—	—	—	—	I,N
Histrionicus histrionicus	—	—	—	—	—	—	—	—	I,N
Clangula hyemalis	—	—	—	—	—	—	—	—	U,I,N
Melanitta nigra	—	—	—	—	—	—	—	—	U,I,N
Melanitta fusca	—	—	—	—	—	—	—	—	—
Mergus serrator	—	—	—	—	—	—	—	—	U
Phalaropus fulicarius	—	—	—	—	—	—	—	—	S
Stercorarius pomarinus	—	—	—	—	—	—	—	—	—
Stercorarius parasiticus	—	—	—	—	—	—	—	—	—
Stercorarius longicaudus	—	—	—	—	—	—	—	—	N
Stercorarius skua	—	—	—	—	—	—	—	—	—
Larus sabini	—	—	—	—	—	—	—	—	—
Larus canus	—	—	—	—	—	—	—	—	—
Larus fuscus	—	—	—	—	—	—	—	—	N
Larus argentatus	—	—	—	—	—	—	—	—	I
Larus glaucoides	—	—	—	—	—	—	—	—	—
Larus hyperboreus	—	•	—	—	—	—	—	—	—
Larus marinus	—	—	—	—	—	—	—	—	I,N
Rhodostethia rosea	—	—	—	—	—	—	—	—	—
Rissa tridactyla	—	•	—	—	—	—	—	—	I
Pagophila eburnea	—	—	—	—	—	—	—	—	—
Sterna caspia	L	—	—	—	—	—	—	—	U,I
Sterna sandvicensis	L	—	—	—	—	—	•	—	S,I,N
Sterna dougallii	R	—	—	—	—	•	—	•	U,S,I,N
Sterna paradisaea	—	—	—	—	—	—	•	—	—
Sterna albifrons	—	—	—	—	—	—	•	—	S,I,N
Chlidonias niger	—	—	—	—	—	—	—	—	—
Uria aalge	R	•	—	—	—	—	—	—	—
Uria lomvia	R	•	—	—	—	—	—	—	—
Alca torda	R	•	•	—	—	•	—	—	—
Cepphus grylle	R,L	•	•	—	—	•	—	—	I,N
Alle alle	—	—	•	—	—	—	—	—	I
Fratercula arctica	R	•	•	•	•	—	—	—	I,N

Marine habitats (cont.)

■ European Macaronesian seas

Part 1: Feeding-area characteristics and food resources

	Feeding-area characteristics		Food resources				
	Principal feeding areas	Other water-body features	Squid	Crustacea (amphipods and copepods)	Pelagic fish (<30 cm)	Mesopelagic fish (<30 cm)	Offal and discards
Pterodroma feae	S?,O	F,U	?	?	?	?	?
Pterodroma madeira	O	—	?	?	?	?	?
Bulweria bulwerii	O	—	●	●	—	●	?
Calonectris diomedea	S,O	F,U	●	●	●	—	●
Puffinus gravis[W]	O	F,U	●	—	●	—	—
Puffinus puffinus	O	—	?	—	●	—	—
Puffinus assimilis	O	?	?	—	●	—	—
Pelagodroma marina	S?,O	F	●	●	●	—	—
Hydrobates pelagicus	O	—	●	●	●	●	—
Oceanodroma castro	O	F	—	●	●	●	●
Sula bassana[W]	S,O	—	—	—	●	●	—
Sterna dougallii	S,O	F,U,L	—	—	●	—	—
Sterna sandvicensis[W]	—	—	—	—	—	—	—
Fratercula arctica[W]	S,O	—	—	—	●	—	—

Part 2: Nest-site requirements

	Nesting location	Barren lava	Sandy soil	Soil plateau	Boulders and crevices	Burrows	Steep, grassy slopes	Caves	Vegetation cover	Other
Pterodroma feae	R,I	—	—	●	—	●	—	—	—	U,P,B
Pterodroma madeira	I	—	—	—	—	—	●	—	—	U,P,B
Bulweria bulwerii	R,L,I	●	—	—	●	●	—	—	—	U,P,B
Calonectris diomedea	R,L,I	●	●	●	●	●	●	●	—	B
Puffinus gravis[W]	—	—	—	—	—	—	—	—	—	—
Puffinus puffinus	I	—	—	—	—	●	●	—	●	U,P
Puffinus assimilis	R,L	—	—	—	●	●	—	—	—	U,P,B
Pelagodroma marina	R	—	●	●	—	●	—	—	●	U,P,N
Hydrobates pelagicus	R,L	—	—	—	●	●	—	—	—	U,P,B
Oceanodroma castro	R,I?	—	—	—	—	●	—	—	—	U,P,B
Sula bassana[W]	—	—	—	—	—	—	—	—	—	—
Sterna dougallii	R	—	—	—	—	—	—	●	●	U,P
Sterna sandvicensis[W]	—	—	—	—	—	—	—	—	—	—
Fratercula arctica[W]	—	—	—	—	—	—	—	—	—	—

■ Mediterranean and Black Seas

Part 1: Preferred habitat-zones

	Breeding			Winter		
	Shallow waters	Shelf seas	Deep waters	Shallow waters	Shelf seas	Deep waters
Gavia stellata	—	—	—	●●●	●●	●
Gavia arctica	—	—	—	●●●	●●	●
Calonectris diomedea	●●	●●●	●	—	●●●	●●
Puffinus yelkouan	●●●	●●●	●	—	●●●	●●
Hydrobates pelagicus	●●●	●●●	—	—	●●●	●●
Larus ichthyaetus	●●●	—	—	●●●	—	—
Larus melanocephalus	●●●	—	—	●●●	—	—
Larus genei	●●●	—	—	●●●	—	—
Larus audouinii	●●●	●●●	—	●●	●●●	●
Larus cachinnans	●●●	●●●	—	●●●	●●●	●
Sterna caspia	●●●	—	—	—	—	—
Sterna sandvicensis	●●●	—	—	●●	—	—
Sterna albifrons	●●●	—	—	—	—	—

Marine habitats (cont.)

■ **Mediterranean and Black Seas** (cont.)

Part 2: Feeding-area characteristics, food resources and nest-site requirements

	Feeding-area characteristics			Food resources		Nest-site requirements						
	Principal feeding areas	Salinity	Other water-body features	Major food resources	Other food sources	Nesting location	Nest-site	Bare ground	Low vegetation	Moderate vegetation	High vegetation	Other
Gavia stellata	I	<•• N	—	Small bottom and pelagic fish	—	—	—	—	—	—	—	—
Gavia arctica	I	—	—	Small bottom and pelagic fish	—	—	—	—	—	—	—	—
Calonectris diomedea	P	••	U	Cephalopods, Clupeidae (esp. anchovies) and Crustacea (Euphausiacea)	B	R	B	•	•	—	—	I
Puffinus yelkouan	P	••	F,U	Little information, probably similar to *Calonectris diomedea*	B	R	C,B	?	?	?	—	P,N
Hydrobates pelagicus	P	••	F	Crustacea, small fish and offal	—	R	C,B	—	—	—	—	P,N
Larus ichthyaetus	I	—	R	Mostly non-marine, but some fish, offal and discards	K	—	—	—	—	—	—	—
Larus melanocephalus	I	••	—	Clupeidae, esp. anchovy and sardine, Gobidae and discards	B	S	L	×	×	•	×	—
Larus genei	I	••	—	Fish and Crustacea	B	—	—	—	—	•	—	—
Larus audouinii	I,P	••	?	Mostly pelagic fish and cephalopods	—	R,B	—	—	—	•	—	B
Larus cachinnans	I	—	R	Generalist; inc. fish and discards, seabird chicks and garbage	B	R,S	B,L	•	—	—	•	—
Sterna caspia	I	—	—	Almost entirely fish	—	B,S	L	•	•	—	—	—
Sterna sandvicensis	I	—	—	Small, surface fish (primarily anchovy and sardine in the Mediterranean)	B	B,S	L	•	•	×	×	C
Sterna albifrons	I	•	R	Small fish and Crustacea	—	B,S	L	•	×	×	×	Prefers small islands, avoids other seabirds

Coastal habitats

KEY

NB Cliff-nesting seabirds are covered under Marine Habitats (p. 59).

Season
The habitat requirements given in this table refer in particular to:
B   Breeding season
N   Non-breeding season
Habitat
Cl   Sea-cliffs
Ro   Rocky shores
Sh   Sand and shingle beaches
Es   Estuaries and intertidal flats
Lg   Lagoons
Sl   Salinas
Ds   Sand-dunes
Sm   Saltmarsh
Dl   Deltas
Open water
•   Required
Salinity
F   Fresh water
B   Brackish water
S   Salt water
Water depth
•   <20 m
••   20–200 m
Tides
○   Absence of regular tidal inundation preferred
•   Large tidal range preferred
Clear water
•   Required

Substrate features
Sediment-type
Sh   Shingle
Sa   Sand
SM   Silt and mud
Exposed wet sediments
•   Required

Vegetation-type and structure
Plant communities
WV   Terrestrial wetland vegetation
AV   Aquatic vegetation
EV   Emergent vegetation
IP   Intertidal aquatic plants (e.g. *Zostera*, *Enteromorpha*)
US   Upper saltmarsh
MS   Mid-saltmarsh
LS   Lower saltmarsh
D   Sand-dune vegetation
DS   Dune-slack vegetation
Plant cover (aquatics and terrestrial)
○   Bare
•   Sparse (<20% cover)
••   Moderate (20–60% cover)
•••   Dense (>60% cover)
Vegetation height (terrestrial)
The scale of height of vegetation below is relative to the plant community in question (e.g. 'tall' saltmarsh is the same absolute height as 'short' emergent vegetation).
•   Short
••   Moderate
•••   Tall
X   Mixed

Food supply
Fi   Fish
AeI   Aerial insects
AqI   Aquatic insects
TI   Terrestrial invertebrates
BI   Benthic invertebrates
Am   Amphibians
MB   Mammals and birds
AqP   Aquatic plants (submerged or floating)
IP   Intertidal aquatic plants
IG   Intertidal grass
TP   Terrestrial and emergent plants
Landscape features
L   Large habitat area
O   Open landscape
Presence of non-coastal habitats
WG   Wet grasslands
NCW   Non-coastal wetlands
DG   Dry grassland
CL   Cultivated land
TW   Trees/woodland
Survival factors
Survival requires low risk from:
Ds   Disturbance
Hu   Hunting
Pl   Powerlines
P   Pollution
Tp   Trampling

Part 1

	Season	Habitat	Open water	Salinity	Water depth	Tides	Clear water	Sediment-type	Exposed wet sediments
Phalacrocorax pygmeus	B	Dl	•	F	••	○	•	—	—
Pelecanus onocrotalus	B	Lg,Dl	•	F/B	••	○	—	—	—
Pelecanus crispus	B	Lg,Dl	•	F/B	••	○	—	—	—
Botaurus stellaris	B	Dl	—	F	≤••	○	•	—	—
Ixobrychus minutus	B	Dl	—	F	≤••	○	—	—	—
Nycticorax nycticorax	B	Dl	—	—	—	—	—	—	—
Ardeola ralloides	B	Dl	—	F	≤••	○	•	—	—
Ardea purpurea	B	Dl	—	F	≤••	○	•	—	—
Plegadis falcinellus	B	Lg,Dl	—	F	≤••	○	—	—	—
Platalea leucorodia	B	Es,Lg,Sm,Dl	•	—	≤••	—	—	SM	—
Phoenicopterus ruber	B	Lg,Sl,Dl	•	S	•	—	—	SM	—
Cygnus columbianus	N	Lg	•	—	≥••	—	—	—	—
Cygnus cygnus	N	Es,Lg	•	—	≥••	—	—	—	—
Anser brachyrhynchus	N	Es	—	—	—	—	—	Sa,SM	—
Branta leucopsis	N	Es,Sm	•	—	≥••	—	—	—	—
Branta bernicla	N	Es,Sm	•	S	≥••	•	—	SM	•
Tadorna ferruginea	B	Lg	•	B/S	≤••	—	—	SM	•
Tadorna tadorna	—	Es,Lg	—	—	—	—	—	—	—
Anas penelope	N	Es,Lg,Dl	•	S	≥••	•	—	SM	•
Anas strepera	—	Es,Lg,Dl	•	F	≤••	—	—	—	—
Anas acuta	N	Es,Lg,Dl	•	S	≥••	•	—	SM	•
Marmaronetta angustirostris	—	Lg,Dl	•	F	≤••	—	—	—	—

cont.

Coastal habitats (cont.)

Part 1 (cont.)

	Season	Habitat	Open water	Salinity	Water depth	Tides	Clear water	Sediment-type	Exposed wet sediments
Netta rufina	N	Lg,Sl,Dl	•	—	≥••	—	—	—	—
Aythya nyroca	N	Lg,Dl	•	—	••	—	—	—	—
Aythya marila	N	Es,Lg	•	B/S	≥••	—	—	—	—
Haliaeetus albicilla	—	Cl,Ro,Es,Lg,Dl	•	—	—	—	—	—	—
Circus cyaneus	N	Ds	—	—	—	—	—	—	—
Aquila adalberti	—	Ds	—	—	—	—	—	—	—
Pandion haliaetus	—	Es,Lg,Dl	•	—	≥••	—	•	—	—
Falco eleonorae	B	Cl	—	—	—	—	—	—	—
Falco peregrinus	N	Cl,Es,Lg,Sm	—	—	—	—	—	—	—
Porzana pusilla	—	Dl	—	F	—	—	—	—	—
Porphyrio porphyrio	—	Lg	—	F/B	—	—	—	—	—
Haematopus ostralegus	—	Ds	—	—	—	—	—	—	—
Recurvirostra avosetta	—	Es,Lg,Sl	—	B/S	•	—	—	SM	—
Glareola pratincola	—	Lg,Dl	—	—	—	—	—	—	—
Glareola nordmanni	—	Lg,Dl	—	—	—	—	—	—	—
Charadrius hiaticula	—	Ro,Sh,ES	—	—	—	—	—	—	—
Charadrius alexandrinus	—	Ds,Es,Lg,Sl,Ds	—	—	—	—	—	Sa,SM	—
Pluvialis squatarola	N	Sh,Es	—	—	—	—	—	—	—
Hoplopterus spinosus	B	Lg,Ds,Sm,Dl	•	F/B	•	—	—	SM	—
Calidris canutus	N	Ds,Es	—	S	—	•	—	Sa,SM	•
Calidris alba	N	Sh	—	—	—	—	—	—	—
Calidris ferruginea	N	Es,Lg,Sl	—	—	—	—	—	—	—
Calidris maritima	N	Ro	—	S	—	•	—	—	—
Calidris alpina	N	Ds,Es,Sl	—	S	—	•	—	SM	•
Lymnocryptes minimus	N	Es,Lg,Sm,Dl	—	F	—	—	—	SM	•
Limosa limosa	N	Es,Lg	—	S	—	•	—	SM	•
Limosa lapponica	N	Ds,Es,Lg	—	S	—	•	—	Sa,SM	•
Numenius phaeopus	N	Es,Lg	—	S	—	•	—	SM	•
Numenius tenuirostris	N	Lg	—	B/S	—	—	—	SM	•
Numenius arquata	N	Es,Lg	—	S	—	•	—	SM	•
Tringa totanus	—	Es,Sl,Sm	—	S	—	•	—	SM	•
Tringa nebularia	N	Es,Lg,Sl	—	—	—	—	—	—	—
Arenaria interpres	N	Ro,Sh,Es	—	—	—	—	—	—	—
Larus ichthyaetus	—	—	—	—	—	—	—	—	—
Larus melanocephalus	B	Ds,Es,Lg,Sl,Dl	•	F	••	—	—	—	—
Larus audouinii	—	Cl,Ro,Sh,Es	—	—	—	—	—	—	—
Larus canus	B	Ds	•[W]	—	••	—	—	Sh[N],Sa[N]	—
Larus fuscus	B	Es,Lg	•[W]	—	••	—	—	—	—
Gelochelidon nilotica	B	Ds,Es,Lg,Sl,Dl	—	—	—	—	—	—	—
Sterna caspia	B	Ds,Es,Lg,Sl,Ds	•	—	≤••	—	—	Sa[N],Sh[N]	—
Sterna sandvicensis	B	Ds,Lg,Ds,Sm,Dl	•	—	≤••	—	•	Sa[N],Sh[N]	—
Sterna dougallii	B	Ro,Ds,Ds	•	—	≤••	—	•	Sa[N],Sh[N]	—
Sterna paradisaea	B	Sh,Ds	•	—	≤••	—	•	—	—
Sterna albifrons	B	Ds,Es,Lg,Sl,Dl	•	—	≤••	—	•	Sa[N],Sh[N]	—
Chlidonias hybridus	B	Lg,Dl	•	—	≤••	—	•	—	—
Chlidonias niger	B	Lg,Dl	•	F/B	••	—	•	—	—
Asio flammeus	B	Lg,Ds,Sm	—	—	—	—	—	—	—
Alcedo atthis	B	Es,Lg,Dl	•	F[B]	≤••	—	•	—	—
Alauda arvensis	—	Ds,Sm	—	—	—	—	—	—	—
Anthus campestris	B	Ds	—	—	—	—	—	Sa,SM	—
Anthus petrosus	B	Cl,Ds	—	—	—	—	—	—	—
Acrocephalus paludicola	B	Sm	—	—	—	—	—	—	—
Pyrrhocorax pyrrhocorax	—	Cl	—	—	—	—	—	—	—

Coastal habitats (cont.)

Part 2

	Plant communities	Plant cover (aquatics) (%)	Plant cover (terrestrial) (%)	Vegetation height (terrestrial)	Food supply	Landscape features	Presence of non-coastal habitats	Survival factors
Phalacrocorax pygmeus	WV	•	•••	•••	Fi	—	TW	P?
Pelecanus onocrotalus	WV	•	•••	•••	Fi	O	—	P?,Pl
Pelecanus crispus	WV	•	•••	•••	Fi	L,O	—	P?,Pl
Botaurus stellaris	EV	—	•••	•••	Fi,Am	—	—	P?
Ixobrychus minutus	WV,EV	—	•••	•••	Fi,Am,AqI	—	—	—
Nycticorax nycticorax	—	—	—	—	—	—	—	—
Ardeola ralloides	WV,EV	—	•••	•••	Fi,Am,AqI	—	—	—
Ardea purpurea	WV,EV	—	•••	•••	Fi,Am	—	—	—
Plegadis falcinellus	WV,EV	—	•••	•••	Fi,Am,AqI	—	—	—
Platalea leucorodia	WV,EV	≤•	≥••	•••	BI	—	TW	—
Phoenicopterus ruber	—	○	≤•	•	AqI,BI	L,O	—	Ds,Pl
Cygnus columbianus	—	—	—	—	—	L,O	WG,NCW,CL	Ds,Hu?,Pl
Cygnus cygnus	—	—	—	—	—	L,O	WG,NCW,CL	Ds,Hu?,Pl
Anser brachyrhynchus	—	—	—	—	—	—	WG,CL	Ds,Hu
Branta leucopsis	US,MS	—	•••	•	IG	—	WG,CL	Ds,Hu
Branta bernicla	IP,US,MS	•••	•••	•	IP,IG	—	WG,CL	Ds,Hu
Tadorna ferruginea	—	≤•	≤•	•	BI	—	DG	—
Tadorna tadorna	—	—	—	—	—	—	—	—
Anas penelope	IP,US,LS	•••	•••	•	IP,IG	—	WG,CL,NCW	Ds,Hu
Anas strepera	AV,EV	≥••	•/••	••	AqP	—	—	—
Anas acuta	—	≤•	—	—	BI	—	—	Hu
Marmaronetta angustirostris	EV,AV	—	≥••	•••	AqP	—	—	—
Netta rufina	AV,EV	•••	•••	•••	AqP	—	—	—
Aythya nyroca	AV,EV,WV	•••	•••	•••	AqP,AqI	—	—	—
Aythya marila	—	—	—	—	BI,AqP	—	—	—
Haliaeetus albicilla	—	—	—	—	Fi,MB	L,O	TW	—
Circus cyaneus	LS,D,DS	—	≥••	•	MB	L,O	NCW,WG,CL	—
Aquila adalberti	D,DS	—	—	X	MB	L	TW	Pl,P?
Pandion haliaetus	—	•	—	—	Fi	L,O	TW	—
Falco eleonorae	—	—	—	—	—	—	—	—
Falco peregrinus	LS,D,DS	—	≥••	•	MB	L,O	NCW,WG,CL	P
Porzana pusilla	EV,WV	—	•••	•	AqI,TI	—	—	—
Porphyrio porphyrio	EV,WV	—	•••	•••	AqP,TP	—	—	—
Haematopus ostralegus	—	—	—	—	—	—	—	—
Recurvirostra avosetta	—	≤•	≤•	—	BI,AqI	—	—	—
Glareola pratincola	—	—	≤•	—	AeI	O	DG,NCW,CL	—
Glareola nordmanni	—	—	≤•	—	AeI	O	DG,NCW,CL	—
Charadrius hiaticula	—	—	—	—	—	—	—	—
Charadrius alexandrinus	—	—	○	—	TI	—	—	—
Pluvialis squatarola	—	—	—	—	—	—	—	—
Hoplopterus spinosus	LS	—	•	•	TI	—	—	—
Calidris canutus	—	—	○	—	BI	—	—	—
Calidris alba	—	—	—	—	—	—	—	—
Calidris ferruginea	—	—	—	—	—	—	—	—
Calidris maritima	—	—	—	—	TI	—	—	—
Calidris alpina	US,MS	—	○	—	BI	—	CL,WG	—
Lymnocryptes minimus	EV,WV	—	≥••	•••	AqI,TI	—	—	—
Limosa limosa	US,MS	—	○	—	BI	—	CL,WG	—
Limosa lapponica	US,MS	—	○	—	BI	—	CL,WG	—
Numenius phaeopus	US,MS	—	○	—	BI	—	CL,WG	—
Numenius tenuirostris	—	—	○?	—	BI,AqI	—	DG,WG,CL	—
Numenius arquata	US,MS	—	○	—	BI	—	CL,WG	Hu
Tringa totanus	LS	—	○/•••[N]	X	BI,AqI,TI	O	CL,WG	Tp
Tringa nebularia	—	—	—	—	—	—	—	—
Arenaria interpres	—	—	—	—	—	—	—	—
Larus ichthyaetus	—	—	—	—	—	—	—	—
Larus melanocephalus	—	—	•/••	•	TI,AqI	O	CL,WG	—
Larus audouinii	—	—	—	—	—	—	—	—
Larus canus	D	—	••	—	TI,AqI,Fi	O	—	—
Larus fuscus	D,DS	—	••	—	varied	O	—	Ds

cont.

Coastal habitats (cont.)

Part 2 (cont.)

	Plant communities	Plant cover (aquatics) (%)	Plant cover (terrestrial) (%)	Vegetation height (terrestrial) (%)	Food supply	Landscape features	Presence of non-coastal habitats	Survival factors
Gelochelidon nilotica	DS,US	—	•/••	•	TI,AeI,Fi	O	CL,WG,DG	—
Sterna caspia	D	—	•/••	•	Fi	O	—	—
Sterna sandvicensis	D	—	•/••	•	Fi	—	—	Ds
Sterna dougallii	D	—	••	X	Fi	—	—	—
Sterna paradisaea	—	—	—	—	Fi	—	—	—
Sterna albifrons	—	—	≤•	—	Fi	—	—	Ds
Chlidonias hybridus	AV,EV,WV	•	≥••	•••	AqI,AeI,Fi,Am	—	—	—
Chlidonias niger	AV,EV	••	•••	•	TI,AqI,Fi,Am	—	—	—
Asio flammeus	D,DS	—	≥••	•••	MB	—	CL,WG	—
Alcedo atthis	—	—	—	—	Fi	—	—	—
Alauda arvensis	D,DS,US,MS	—	≥••	X	TI,TP(seeds)	—	—	—
Anthus campestris	D	—	•	•	TI,TP(seeds)	—	—	—
Anthus petrosus	—	—	—	—	—	—	—	—
Acrocephalus paludicola	EV	—	••	•	TI	—	—	—
Pyrrhocorax pyrrhocorax	—	—	•	•	TI	O	DG	—

Inland wetlands

cont.

KEY

Preferred habitat
R Rivers
L Lakes
M Marshes

Extent of open water
• Small area of open water
•• Large area of open water

Water depth
• <20 m
•• 20–200 m
••• >200m

Water-flow
St Standing water
Sl Slow-flowing water
F Fast-flowing water

Trophic status
O Oligotrophic
M Mesotrophic
E Eutrophic

Sediment-type
M Mud
Sa Sand
Sh Shingle
St Stony

Plant communities
B Bushes within the habitat
E Emergent vegetation
F Floating vegetation
S Submerged vegetation
Te Terrestrial wetland vegetation
Tr Trees within the habitat
Ph *Phragmites* beds

Vegetation height
• <10 cm
•• 10–100 cm
••• >100 cm
X Mix of vegetation heights

Plant cover
o Nil
• <20%
•• 20–60%
••• >60%

Landscape features
C Connected habitat
L Large habitat area
M Mosaic of habitat-types
O Open landscape
T Nearby trees

Food requirements
Ae Aerial invertebrates
Am Amphibians
Aq Aquatic invertebrates
B Birds
EV Emergent vegetation
F Fish
FV Floating vegetation
M Mammals
R Reptiles
SI Soil invertebrates
SV Submerged vegetation
TV Terrestrial vegetation
V Vegetation invertebrates

Survival factors
Survival requires low risk at nest-site from:
D Disturbance
F Flooding
P Predation
T Trampling

	Preferred habitat	Extent of open water	Water-depth	Water-flow	Trophic status	Sediment-type	Plant communities	Vegetation height	Plant cover	Landscape features	Food requirements	Other nesting requirements	Other feeding requirements	Survival factors
Gavia stellata	L	•	••	St	O	—	—	—	•	C	F	Islands	Clear water	P
Gavia arctica	L	—	••	St	E	—	—	—	—	—	F	Islands	Clear water	D,F
Tachybaptus ruficollis	R,L	••	••	St/Sl	E	—	E	(≥••)	≥••	—	Aq	—	—	P
Podiceps cristatus	L	≥••	≥••	St	E	—	E	••	••	—	Aq,F	—	Clear water	D,P
Podiceps grisegena	L	≥••	•	St	E	—	E	••	••	—	Am,Aq	—	Clear water	P
Podiceps auritus	L	•	•	St	M	—	E	••	••	—	Aq	—	Clear water	P
Podiceps nigricollis	L	••	≥••	St	E	—	E	••	••^{B}	T^B	F	—	Clear water	P
Phalacrocorax pygmaeus	L	••	•	—	—	—	Tr^B	•	•	O	F	Islands	—	—
Pelecanus onocrotalus	L	••	••	—	—	—	Te	•	≤•	O	F	—	—	—
Pelecanus crispus	L	•	•	St	E	—	Te	••	≤•	O	F	Islands	—	—
Botaurus stellaris	M	≤•	≤•	St	E	—	Ph	•••	•••	L	Am,F	—	Clear water	D,F
Ixobrychus minutus	R,L,M	≤•	≤•	St	E	—	Ph	•••	•••	—	Am,Aq,F	—	Clear water	D
Nycticorax nycticorax	L,M	≤•	—	St	E	—	E,Tr^B	—	—	—	Am,Aq,F,R	—	Clear water	—
Ardeola ralloides	L,M	•	—	St	E	—	E,Tr^B	—	—	—	Am,Aq,R	—	Clear water	—

cont.

Inland wetlands (cont.)

Species	Preferred habitat	Extent of open water	Water-depth	Water-flow	Trophic status	Sediment-type	Plant communities	Vegetation height	Vegetation plant cover	Landscape features	Food requirements	Other nesting requirements	Other feeding requirements	Survival factors
Egretta garzetta	L,M	—	≤•	St	—	—	E,Tr^B	—	—	—	Am,Aq,F	—	Clear water	—
Egretta alba	L,M	•	≤•	St	—	—	E,Tr^B	—	—	—	F	—	Clear water	—
Ardea cinerea	R,L,M	•	≤•	≤SI	E	—	Te,Tr^B	••	≤•	T	Am,Aq,F,M,R	—	Clear water	—
Ardea purpurea	M	•	•	St	E	—	Ph	≤•	≤•	—	Am,Aq,F,R	—	Clear water	D
Ciconia nigra	R,M	•	•	≤SI	—	M,Sa	—	••	—	L,M,O	Am,Aq,F	—	—	D
Ciconia ciconia	R,L,M	—	•	≤SI	—	M	Te,Tr^B	••	≤•	O	Am,Aq,M,SI	—	—	—
Plegadis falcinellus	L,M	•	≤•	St	—	—	Tr^B	X	••^B	—	Am,Aq	—	—	—
Platalea leucorodia	L,M^N	••	••	St	E	M^B	E,Tr^B	••	≤•^B	—	Aq	—	Clear water	D,P
Phoenicopterus ruber	L	••	••	≤SI	E	—	—	—	○	O	Aq	Islands	Brackish/saline water	—
Cygnus olor	R,L,M	••^w	••	≤SI	≤M	—	E,S,Te	X	≤•	M	SV,TV,V	—	Clear water	—
Cygnus columbianus^w	L	••^w	••	St	—	—	S	—	—	M	TV^w	—	—	D,P
Anser erythropus	L	—	—	St	—	—	—	—	○^w	O^w	TV^w	—	—	—
Branta ruficollis	L	—	—	St	—	—	—	—	○^w	O^w	TV^w	—	—	—
Tadorna ferruginea	L	••^w	—	≤SI	—	M	—	—	—	O	Aq	Cliffs	—	—
Anas strepera	R,L	•	≤•	St	≤M	—	E,Te	••	≤•	—	Aq,SV,seeds	Islands	—	F,P
Anas platyrhynchos	L,M	•	≤•	≤SI	E	—	—	••	••^B	O	Aq,SV,TV,seeds	—	—	P
Anas acuta	L,M	•	≤•	≤SI	≤M	—	E,Te	X	≤•	O	Aq,EV,SV,TV	—	—	P
Anas querquedula	L,M	•	•	St	≤M	—	E,Te	X	••	O	Aq,FV	—	—	P
Anas clypeata	L,M	—	≤•	St	≤M	—	E,Te	X	••	O	Aq	—	—	P
Marmaronetta angustirostris	L,M	•	≤•	St	≤M	—	B,E,F	≤•	≤•	L,M	Aq,FV	—	—	D,P (especially when risk of drying)
Netta rufina	L,M	—	≥•	St	E	—	E,F,S,Te	X	••	L,M	Aq,FV,SV	Islands	—	F,P
Aythya ferina	L	—	≥•	St	E	—	E,F,S	≥•	••	O	Aq,FV,SV	Islands	—	F,P
Aythya nyroca	L,M	—	≤•	St	E	—	E,F,S,Ph	≤•	≥•	M,O	Aq,FV,SV	—	—	F,P
Aythya fuligula	L	—	≥•	St	E	—	E,F,S	X	••	—	Aq,FV,SV	Islands	—	F,P
Bucephala clangula	R,L	••	•	≤SI	M	—	Tr,forest^B	≥•	≥•	L,M,T	Aq,FV	—	Clear water	—
Mergus albellus	R,L	••	≥•	St	M	—	Forest^B	≥•	≥•	L,M,T	Aq,(F),SV	—	Clear water	—
Mergus merganser	R,L	••	≥•	—	≤M	—	Tr^B	••	••	L,T	Aq,F	—	Clear water	P (*M. martes*)
Oxyura leucocephala	L	••	≥•	St	—	—	—	X	X	—	Aq	—	—	—
Milvus migrans	R,L	—	—	St	—	—	Tr	—	—	T	F,M	—	—	—
Haliaeetus albicilla	R,L	—	—	≤SI	E	—	Tr	—	—	T	B,F	—	—	D
Circus aeruginosus	L,M	≤•	•	St	E	—	E,Te	••	••	L,O	B,M,R	—	—	F
Aquila clanga	R,L,M	—	—	—	—	—	Te,Tr/forest^B	X	—	L,O,T	B,M,R,scavenges	—	—	D
Pandion haliaetus	L	—	—	≤SI	—	—	Tr	—	—	L,T	F	—	—	D
Rallus aquaticus	L,M	•	•	M	≤M	M	E,Te	••	••	—	Am,Aq,SI	—	Exposed wet sediment	—

Inland wetlands (cont.)

cont.

	Preferred habitat	Extent of open water	Water-depth	Water-flow	Trophic status	Sediment-type	Plant communities	Vegetation height	Plant cover	Landscape features	Food requirements	Other nesting requirements	Other feeding requirements	Survival factors
Porzana porzana	L,M	●	●	≤Sl	E	M	E,Te	X	●●●	C	Aq,Sl,V	—	Exposed wet sediment	D
Porzana parva	M	●	●	St	E	M	E,F	X	●●●	C	Aq,Sl,V	—	Exposed wet sediment	D
Porzana pusilla	M	●	—	—	—	—	E,F	X	●●●	C	Aq,Sl,V	—	—	D
Crex crex	M	—	—	—	—	—	Te	—	●●●	—	V	—	—	F,P
Gallinula chloropus	R,L,M	●	≤●●	St	E	M	E	X	≤●	M	Aq,V	—	—	
Porphyrio porphyrio	M	●	●	St	—	—	E	X	●●●	—	EV	—	—	F
Fulica atra	R,L	—	●●	St	E	—	E	≥●●	●●	—	EV,SV,seeds	—	—	F
Fulica cristata	L,M	●●	—	St	E	—	E	≥●●	●●	—	EV,SV	—	—	
Grus grus	M	●	●●	St	E	M,Sa	E,Te	X	○	L	Am,M,Sl,TV,V	—	—	D
Himantopus himantopus	M	—	●	St	E	M,Sa	E,Te	≤●●	≤●●	L	Ae,Am,Aq (tadpoles)	Exposed sediment	—	D,F,P,T
Recurvirostra avosetta	L	●●	≤●●	St	E	M,Sa	Te	≤●●	≤●●	L	Ae,Aq	Exposed sediment	Brackish/saline water	D,F,P
Glareola pratincola	L	●●	●	≤Sl	—	M,Sa,Sh	—	○	○	L,O	Ae,Sl	Exposed sediment	—	D,F,P,T
Glareola nordmanni	L	●●	●	≤Sl	—	M,Sa,Sh	Te	≤●●	≤●●	C,L,O	Ae,Sl	Exposed sediment	—	D,F,P,T
Charadrius dubius	R	—	●	≤Sl	E	M,Sa,Sh	Te	≤●●	≤●●	O	Ae,Aq,Sl	Exposed sediment	—	D,F,P
Charadrius alexandrinus	L	—	—	—	—	—	—	≤●	≤●	—	—	—	—	D,P
Charadrius leschenaultii	L	—	—	St	E	M,Sa	Te,*Salicornia*	—	●	—	—	Exposed sediment	—	
Hoplopterus spinosus	L	●●	●	St	E	M,Sa	Te	≤●●	≤●●	L,O	Aq,Sl	—	—	D,F,P,T
Philomachus pugnax	M	●	●	St	E	—	Te	●	●	—	Aq,Sl,V	—	—	P,T
Lymnocryptes minimus	M	—	●	St	—	—	Te	—	●	—	—	—	—	
Limosa limosa	M	●	●	St	—	M	Te	●	●	O	Aq,Sl,V	—	—	P,T
Numenius tenuirostris	L	—	●	St	M	M	—	—	—	—	Sl,V	—	Exposed wet sediment	
Tringa totanus	L,M	●	●	St	E	M	—	≤●	≤●	O	Sl	—	Exposed wet sediment	F,P,T
Tringa stagnatilis	R,L,M	—	—	—	—	—	—	—	—	—	—	—	—	
Tringa glareola	R,M	●	—	—	—	—	—	—	●	—	Sl	—	—	
Actitis hypoleucos	R,L	●	●	F	—	Sh	—	≤●	≤●	O	Aq,Sl	—	Exposed wet sediment	
Larus minutus	L	●●	●●	St	—	—	E	≤●	≤●	O	Ae,Sl	Islands	—	P
Larus ridibundus	L,M[N]	●●	●●	—	≤M	—	E,Te	●	●●	C,O	Ae,Sl	—	—	—
Larus canus	R,L	●●	●●	≤Sl	E	Sa		●	●●	O	Aq,F	Islands	—	D,F

Inland wetlands (cont.)

Species	Preferred habitat	Extent of open water	Water-depth	Water-flow	Trophic status	Sediment-type	Plant communities	Vegetation height	Plant cover	Landscape features	Food requirements	Other nesting requirements	Other feeding requirements	Survival factors
Gelochelidon nilotica	L,M	—	—	—	—	—	—	—	—	—	Aq,F	—	—	—
Sterna albifrons	R	••	≤••	≤Sl	E	Sa,Sh	—	•	•	0	Aq,F	Islands and exposed sediment	—	D,F,P,T
Chlidonias hybridus	L,M	••	••	≤Sl	E	Sa,Sh	F	•	•	0	Aq,F	—	—	F,P,T
Chlidonias niger	L,M	••	≤••	St	E	St	F	•	≤•	0	Ae,Aq,F	Islands	—	F
Chlidonias leucopterus	L,M	≤••	≤••	≤Sl	E	—	F	•	•	0	Ae,Aq,F	Exposed sediment	—	F
Halcyon smyrnensis	R,L,M	≤••	X	—	≥M	Tr	Tr	X	•	M,O	Am,F,R,Sl,V	Sandy cliffs	Clear water, suitable perches	—
Alcedo atthis	R,L	≤••	X	≤Sl	≥M	—	B,E,Tr	X	•	M,T	F	Sandy cliffs	—	—
Ceryle rudis	R,L	••	≤M	—	≥M	—	B,Tr	—	•	T	F	Sandy cliffs	—	—
Merops apiaster	R	••	—	—	—	—	—	•	•	O	F	Sandy cliffs	—	—
Riparia riparia	R	—	—	—	—	—	—	—	•	O	Ae	Sandy cliffs	—	—
Hirundo rustica	R,L,M	≤••	F	—	—	Sh,St	Tr	—	—	M,T	Ae	—	—	—
Motacilla cinerea	R	≤••	X	F	—	St	Tr	X	—	M,T	Ae,Aq,V	—	Exposed sediments, boulders	F
Cinclus cinclus	R	≤•••	X	F	M	St	—	X	•••	M,T	Aq,F	—	Clear water, boulders	—
Cettia cetti	M	•	≤••	≤Sl	E	—	B,E	X	•••	—	Sl,V	—	—	—
Cisticola juncidis	M	•	≤••	≤Sl	E	—	E	≤••	•••	—	Ae,V	—	—	—
Locustella luscinioides	L,M	—	≤••	≤Sl	E	—	E,Ph	•••	•••	T	Ae,Aq	—	—	F,P
Acrocephalus melanopogon	L,M	•	•	St	≤M	—	E,Ph	•••[B]	•••	—	Ae,V	—	—	F
Acrocephalus paludicola	M	—	•	St	—	—	E	≥••	•••	—	Ae,V	—	—	—
Acrocephalus schoenobaenus	R,L,M	•	≤••	≤Sl	≤M	—	E,Te	≥••	•••	C	Ae,V	—	—	—
Acrocephalus agricola	L,M	—	•	St	—	—	E	•	•••	—	Ae,V	—	—	—
Acrocephalus palustris	R,L,M	•	≤••	≤Sl	E	—	B,E,Te	≥••	••	T	Aq,V	—	—	P,T
Acrocephalus scirpaceus	R,L,M	—	≤••	≤Sl	≤M	—	Te,Ph	≥••	≥••	—	Ae,V	—	—	F
Acrocephalus arundinaceus	R,L,M	—	≤••	≤Sl	E	—	E,Ph	•••	•••	—	Aq,V	—	—	—
Panurus biarmicus	M	•	•	St	E	—	E,Ph	•••	•••	T	V	—	—	D,F
Remiz pendulinus	R,L,M	•	•	≤Sl	E	—	B,Tr	X	••	T	SV,V	—	—	D
Emberiza schoeniclus	R,L,M	—	≤••	≤Sl	≤M	—	E,Te,Ph	X	••	M	Ae,Aq	—	—	F,P

Tundra, mires and moorland

KEY

Habitat-types
T Tundra zone
B Boreal montane habitats
M Mires
H Moorland
* >75% of population within these habitats uses this habitat
( ) <10% of population within these habitats uses this habitat

Habitat specialist
Y Only requires tundra, boreal montane habitats, mires and/or moorland when using these habitats
N Also requires other habitats when using these habitats

Topography
F Flat
S Sloping

Habitat extent/variety
L Large habitat area
S Small habitat area
U Uniform habitat
M Habitat mosaic

Vegetation structure
S Scrub (1–2 m)
TS Tall shrubs (0.5–1 m)
MS Medium shrubs (0.2–0.5 m)
DS Dwarf shrubs (<0.2 m)
GS Grass and sedge

ML Moss and lichen
B Bare ground
O Open (treeless)
C Closed

Water-table
W Wet
D Dry

Presence of water-bodies
L Lakes
P Pools
R Rivers
C Coasts

Food source/type
M Marine
F Freshwater
T Terrestrial
W Mire/wetland
P Plant material
V Vertebrates
I Invertebrates

Breeding success factors
W Favourable weather
R Rodent cycles
P Protection provided by presence of other species

Survival factors
Survival requires low risk from:
D Disturbance/tourism
P Persecution
Pr Predation
Po Pollution
H Habitat loss
C Interspecific competition for food, nest-sites, etc.

cont.

	Habitat-types	Habitat specialist	Topography	Habitat extent/variety	Vegetation structure	Water-table	Presence of water-bodies	Food source/type	Breeding success factors	Survival factors
Gavia stellata	T*,(M)	Y	F	S	GS,O	W	L,P	F,M/V	W	D,H
Gavia arctica	T*,M,B	Y	F	—	GS,O	W	L	M.F/V,I	W	D,H
Gavia immer	—	N	F	—	GS,O	W	L,P	M?,F/I,V	W	—
Gavia adamsii	T*	Y	F	—	GS,O	W	L,C	M,F/V	W	D,Po
Cygnus columbianus	T*	—/Y	F	—	GS,O	W	L	F/P,I	W	H
Anser fabalis	M*	Y	F	L,M	TS–GS,O	W	P	W/P	W	H
Anser brachyrhynchus	B,T	Y	—	L,M	GS,ML,O	D,W	R,C	T/P	—	(P),Pr (Alopex),H
Anser albifrons	B,T,M (winter)	Y	F	L,M	MS–ML,O	D,W	R,C	T/P	R,W,P	P,Pr (Alopex),C
Anser erythropus	T*,B	Y	F,S	L,M	MS	D,W	L	T/P	R,W,P	H
Branta leucopsis	T*,B (Baltic)	Y	—	L,M	GS,O	D,W	R,C	T,M/P	W	H
Branta bernicla	T	Y	F	—	ML,O	D,W	P,C	T,F,M/P	—	—
Anas acuta	M,T	Y	F	M	MS–GS	W	L,R,P,C	F/P,I	—	—
Aythya marila	B,T*	Y	F	—	MS–GS	W	L,C	F/I,P	—	—
Clangula hyemalis	T*,B	N	F	—	MS–GS,O	W	L,P	F,M/I	—	P
Melanitta fusca	B,M,T*	N	F	—	MS–GS,C	W	L,P,C	F,M/I	—	—
Melanitta nigra	B,M,T*	N	F	—	MS–GS,C	W	L,P	F,M/I	—	D,P
Bucephala islandica	B*	Y	F	—	MS–GS,C	W	L,P	F/I	W	—
Haliaeetus albicilla	T,M	N	—	L	O (trees)	—	L,R,C	M,F,T/V	—	D,P,Po,H

cont.

Tundra, mires and moorland (cont.)

Habitat-types	Habitat specialist	Topography	Habitat extent/variety	Vegetation structure	Water-table	Presence of water bodies	Food source/type	Breeding success factors	Survival factors
Circus cyaneus — T,H	N	F	L,M	TS,O	D,W	—	T/V	W,R	P,H
Buteo lagopus — B,T	N	S	L,M	O (trees)	D	—	T/V	R	P
Aquila chrysaetos — M,B,H	N	—	L	O (trees)	—	—	T,W/V	—	D,P,H
Falco columbarius — H,M,B,T	N	S	M	TS,MS,O	D	—	T,W/V	—	H
Falco rusticolus — T,B	N	—	L,M	O	—	—	T/V	W,R	P,H,Po,C
Falco peregrinus — T*,B,M,(H)	N	—	L,M	O	—	—	T/V	W	D,P,H,Po
Lagopus lagopus — H,M,T,B	Y	—	S,M	MS	D	—	T,W/P,I	W,R	Pr,H
Lagopus mutus — B*	Y	—	S,M	DS-ML	D	—	T/P,I	W	Pr
Tetrao tetrix — M,H	N	—	S,M	S-MS,C	D	—	T,W/P,I	W	H
Grus grus — M*	N	F	L	GS	W	P	T,W,F/P,V,I	W,P	D,H,Pr
Charadrius morinellus — T,B*	Y	(F)	L,M	ML,O	D	—	T/I,P	R	H
Pluvialis apricaria — B,T,(M),(H)	Y	F	L,M	DS-ML,O	D	—	T/I,P	R	H
Pluvialis squatarola — T*	Y	F	L,M	ML(-GS),O	D	—	T/I,P	W,R	Pr
Calidris canutus — T*	Y	F	L,M	ML,B,O	D	—	T,M/I	W,R	Pr
Calidris alba — T*	Y	F	L,M	ML,B,O	D	—	T,M/I	W,R	Pr
Calidris minuta — T*	Y	F	S,M	DS-ML,O	D,W	C	M,F,W/I	W,R	Pr
Calidris temminckii — B,T,(M)	Y	F	S,M	(MS),GS,(C)	D	C	W,T/I	—	(H)
Calidris maritima — B,T	Y	F	L,M	ML-B,O	D	C	T/I,P	—	Pr
Calidris alpina — T*,M	Y	F	—	GS,ML,O	W	P	T/I	—	H
Limicola falcinellus — T,M*	Y	F	L,M	GS,O	W (stable)	(P)	W/I	W	(H)
Philomachus pugnax — T*,M,(B)	Y	F	—	GS,O	W	(P)	T/I,P	W	H
Lymnocryptes minimus — T,M	Y	F	L,M	GS,ML,O	W	—	T/I	W	H
Gallinago media — B,T	Y	F	(M)	MS-GS,O	D-W	—	T/I	W,R	H
Limosa limosa — M,(B)	Y	F	—	GS	W	—	W/I	—	—
Limosa lapponica — T*,(M)	Y	F	L,M	MS-GS,O	D,W	—	T,W/I,P	W,P	—
Numenius phaeopus — B,T,M	Y	F	L,M	TS-ML,O	D	—	T,W/I,P	—	Pr (*Falco rusticolus*)
Numenius arquata — H,M	N	F	L,M	MS-GS,O	D,W	—	T,W,(M)/V,I,P	W	P
Tringa erythropus — M,T	N	F	L,M	TS (trees)-GS,O	W	P	T,W/F/I	—	—
Tringa totanus — H*,(M)	N	F	(L),M	(MS-GS)/,O	W	Ditches	T,W/I	W	H
Tringa glareola — M,T	N	F	L,U	S-GS,O	W	—	T,M,F/I	—	—
Arenaria interpres — T,B	Y	F	—	ML,B,O	D	—	T,M/I	—	—
Phalaropus lobatus — T*,M,(B)	Y	F	—	GS,O	W	P	F,T/I	W	—
Phalaropus fulicarius — T*	Y	F	—	GS,O	W	P	F,T,M/I	W	—
Stercorarius pomarinus — T*	Y	F	—	GS,ML,O	W	—	T/V	R	—
Stercorarius parasiticus — (H),T*,(B)	Y/N	F	—	DS-ML,O	—	—	T,M/V,I	R	—

Tundra, mires and moorland (cont.)

	Habitat-types	Habitat specialist	Topography	Habitat extent/ variety	Vegetation structure	Water-table	Presence of water bodies	Food source/type	Breeding success factors	Survival factors
Stercorarius longicaudus	T*,(B)	Y	F	—	GS,ML,O	D	—	T/P,V,I	R	—
Stercorarius skua	H,M,B (Iceland)	N	—	—	GS,ML,O	—	—	M/V	—	Po?
Larus sabini	T*	Y	—	—	GS,ML,O	W	L,P,C	M,F/I	—	—
Larus canus	T,(M),B,H	N	—	—	GS	—	—	T,M,F/P,I,V	—	—
Rhodostethia rosea	T*	Y	F	L	GS,O	W	P	F,T/I	W	Pr
Nyctea scandiaca	T*,(B)	Y	—	L	B,O	D	—	T/V	W,R	D,P
Asio flammeus	(T),M,H	Y	—	—	TS,MS,O	D	—	T/V	W,R	—
Alauda arvensis	(M),H,(T),(B)	Y/N	—	S,M?	GS,O	D	—	T/P,I (?)	—	—
Eremophila alpestris	T*,B	Y	—	S,M	GS,ML,O	D	—	T/I,P	—	—
Anthus pratensis	B,T,M,H	Y	—	S,M	GS,O	(W)	—	T/I,P	—	—
Anthus cervinus	T*,(M),(B)	Y	—	—	MS–GS,O	D	—	T/I	—	—
Anthus petrosus	T	Y	F	—	GS	D	C	T/I	—	—
Luscinia svecica	T,(M),B	Y	F	—	TS,MS	—	—	T,W/I	W	H
Saxicola torquata	H*	Y	S	M	TS,O	D	—	T/I	W	H
Turdus torquatus	B*,(H)	N	S	M	S–GS,O	D	—	T/I,P	—	H
Lanius excubitor	M*	Y	F	U	S	D	—	T,W/I,V	—	Pr
Carduelis flavirostris	B*,(H)	Y/N	S	M	MS–GS,O	D	—	T/I,P	—	H
Calcarius lapponicus	B,T*	Y	—	S,M	MS–ML,O	D	—	T/I,P	W	—
Plectrophenax nivalis	B*,T	Y	S	(L),M	ML,B	D	—	T/I,P	W	—
Emberiza pusilla	T,M	Y	F	S,M	TS–ML	W	—	T/I,P	—	—

Lowland Atlantic heathland

KEY

Type of heathland
D Dry
W Wet
V Valley mire

Presence of other habitats
F Forest
A Arable
G Grassland

Vegetation mosaic
• Preferred

Vegetation structure
○ Absence preferred
• Occasional incidence preferred
•• Frequent incidence preferred

Open water
• Presence preferred
X Presence not preferred

Survival factors
Survival requires low risk from:
D Disturbance
P Predation

	Type of heathland	Presence of other habitats	Min. area for 1 pair (ha)	Vegetation mosaic	Trees (>3 m)	Shrubs (1–3 m)	Dwarf shrubs (<1 m)	Grass	Bare ground	Special vegetation needs	Open water	Food requirements	Survival factors	Remarks
Circus cyaneus	D,W	—	10–100?	—	○	••	••	••	○	Dense scrub/short trees for roosting	X	Small birds, mammals	D,P at roosts	Occurs predominantly in winter; needs shrubs for roosting
Tetrao tetrix	W,V	A,G	>100	•	••	••	••	•	•	Fine-grained mosaic, and wet flushes for chicks	—	Insects for chicks	P,D at leks	—
Alectoris rufa	D	A	—	—	○	•	•	•	••	—	—	Insects for chicks	P?	Predominantly a bird of arable land
Perdix perdix	D	A,G	—	—	○	•	••	••	••	Tall veg., open at ground-level for chicks to feed in	—	Insects for chicks	P	Predominantly a bird of arable land
Grus grus	W,V	—	>100	—	—	—	—	•?	—	—	•	Insects for chicks	D	Usually nests in wet situations with pools or small lakes
Burhinus oedicnemus	D	A,G	0–10	•	○	•	•	[••]	[••]	Some bare ground essential for nesting, short, open vegetation for chicks and feeding	X	Invertebrates (worms, beetles)	P,D	—
Asio flammeus	D	—	10–100?	—	•	[••]	[••]	[••]	—	Tall vegetation for nesting	—	Small birds, mammals	D	Nomadic (year-to-year presence not guaranteed)
Caprimulgus europaeus	D	F	0–10	•	•	••	••	—	•	Bare areas, often close to trees, for nesting	•	Flying insects	P	Ultizes wet heath/ meadows/ forest edge for feeding
Picus viridis	D	F	10–100	•	••	—	—	••	••	Very short vegetation and worm patches	X	Insects, esp. ants	—	—
Lullula arborea	D	F	10–100	•	•	—	—	••	••	Bare disturbed ground/short grass for feeding, longer grass for nesting, trees for songposts	X	Insects, and seeds in winter	—	Burning/rabbit-grazing good for creating suitable vegetation

cont.

Lowland Atlantic heathland (cont.)

	Type of heathland	Presence of other habitats	Min. area for 1 pair (ha)	Vegetation mosaic	Vegetation structure					Special vegetation needs	Open water	Food requirements	Survival factors	Remarks
					Trees (>3 m)	Shrubs (1–3 m)	Dwarf shrubs (<1 m)	Grass	Bare ground					
Alauda arvensis	D	—	—	—	○	○	—	••	—	—	X	Invertebrates	—	—
Anthus campestris	D	—	10–100	•	○	○	—	••	••	Very short vegetation/ sandy ground	X	Seeds, invertebrates	—	Needs wide-open areas
Saxicola torquata	D	—	10–100	—	○	•/••	•	•	—	—	X	Invertebrates	—	—
Sylvia undata	D	—	0–10	—	○/•	[••]	[••]	—	—	Continuous areas of scrub/ dwarf scrub; continuity between ground and scrub cover	X	Invertebrates, insects, spiders	Severe winters	—
Lanius collurio	D	G	10–100	•	•	••	••	••	—	Mosaic of long and short grass and elevated perches	X	Large invertebrates, small mammals, birds	—	—
Lanius excubitor	D	—	10–100/>100	—	•	•	••	—	—	Elevated perches essential	X	Large invertebrates, small mammals, birds	D	—

Boreal and temperate forests

KEY

○ Avoided
● Preferred
• Essential

Landscape requirements

Forest cover
L Low
H High

Habitat patchiness
F Fine-grained (i.e. less than home-range size)
C Coarse-grained (i.e. larger than home-range size)

External mosaic
Mosaic of non-forest habitat(s) (with forest) is required within the home-range.

Internal mosaics
Mosaic of the given forest sub-types and/or growth stages is required within the home-range.

Preferred habitats

Tree-growth stages
O <1 m high
I Early and mid-growth stages
II Pole and mature stages
III Old-growth forest

cont.

Part 1: Landscape requirements and preferred habitats

	Landscape requirements			Boreal forest			Montane forest				Lowland temperate forest				Riverine forest			
	Forest cover	Habitat patchiness	External mosaics	Internal mosaics	Pinus (dry)	Mixed Picea/pinus (moist/wet)	Internal mosaics	Krummholz pinus/Larix	Picea	Fagus/Abies	Internal mosaics pinus/Betula	planted Picea & non-native spp. Mixed	Fagus	Mixed deciduous	Internal mosaics	Seasonally flooded	Carrs	Unstable riparian
Phalacrocorax pygmeus	—	—	—	—	—	—	—	—	—	—	—	—	—	—	—	(II–III)	—	—
Nycticorax nycticorax	—	—	—	—	—	—	—	—	—	—	—	—	—	—	—	(I)–III	—	—
Ardeola ralloides	—	—	—	—	—	—	—	—	—	—	—	—	—	—	—	(I)–III	—	—
Ardea purpurea	—	—	—	—	—	—	—	—	—	—	—	—	—	—	—	(I)–III	—	—
Ciconia nigra	H	C	•	—	—	—	•	—	III	II–III?	•	—	II–III	II–III	•	II–III	II–III	—
Plegadis falcinellus	—	—	—	—	—	—	—	—	—	—	—	—	—	—	—	(I)II–III	—	—
Platalea leucorodia	—	—	—	—	—	—	—	—	—	—	—	—	—	—	—	(I)II–III	—	—
Mergus albellus	—	—	—	?	—	III?	—	—	II–III	—	—	—	—	—	—	—	—	—
Pernis apivorus	H	C	•	—	—	II–III	•	—	II–III	II–III	•	?	II–III	II–III	•	II–III	II–III	—
Milvus migrans	L	—	•	—	—	—	—	—	II–III	II–III	?	II–III	II–III	—	—	II–III	II–III	—
Milvus milvus	L	F	•	—	—	—	—	—	II–III	II–III	—	II–III	II–III	II–III	—	II–III	II–III	—

Boreal and temperate forests (cont.)

cont.

Part 1: Landscape requirements and preferred habitats

Species	Landscape requirements			Boreal forest				Montane forest					Lowland temperate forest						Riverine forest			
	Forest cover	Habitat patchiness	External mosaics	Internal mosaics	Pinus (dry)	Mixed Picea/Pinus	Picea (moist/wet)	Internal mosaics	Krummholz	Pinus/Larix	Picea	Fagus/Abies	Fagus/Abies	Internal mosaics Pinus/Betula	Planted Picea & non-native spp.	Mixed	Fagus	Mixed deciduous	Internal mosaics	Seasonally flooded	Cars	Unstable riparian
Haliaeetus albicilla	H	—	•	—	—	(II)III	(II)III	—	—	—	—	—	(II)-III	—	—	(II)-III	(II)-III	(II)-III	III	III	III	—
Circaetus gallicus	—	F	•	—	—	—	—	•	—	—	—	II-III	(0-I)	II-III	—	(0-I) II-III	(0-I) II-III	(0-I) II-III	II-III	II-III	II-III	—
Circus cyaneus	—	—	—	—	—	0-I	0-I	—	—	—	—	—	—	—	—	—	—	—	—	—	—	—
Accipiter brevipes	—	—	—	—	—	—	—	—	—	—	—	—	—	—	—	—	—	—	I-II	I-II	I-II	—
Aquila pomarina	H	F	•	—	—	II-III	II-III	—	—	—	—	II-III	II-III	—	—	II-III	II-III	II-III	II-III	II-III	II-III	—
Aquila clanga	H	F	•	—	—	II-III	II-III	—	—	—	—	—	II-III	—	—	II-III	II-III	II-III	II-III	II-III	II-III	—
Aquila heliaca	—	—	—	—	—	—	—	?	—	—	—	—	?	—	—	II-III	II-III	II-III	II-III	II-III	II-III	—
Aquila chrysaetos	H	—	•	(II)III	—	(II)III	(II)III	(II)III	—	III	III	III	II-III	II-III	—	II-III	II-III	II-III	II-III	II-III	II-III	—
Hieraaetus pennatus	H	C	•	—	—	—	—	—	—	II-III?	II-III	II-III	II-III	—	—	II-III	II-III	II-III	II-III	II-III	II-III	—
Pandion haliaetus	H	—	•	(I)II-III	(I)II-III	(I)II-III	(I)II-III	—	—	—	II-III	(I)-III	(I)-III	(I)-III	—	(I)-III	?	(I)-III	II-III	II-III	II-III	—
Falco tinnunculus	L	—	•	—	—	0-III	0-III	—	—	—	I-III	I-III	(0-I) II-III	(0-I) II-III	(0-I) II-III	(0-I) II-III	(0-I) II-III	(0-I) II-III	II-III	II-III	II-III	—
Falco vespertinus	L	—	•	—	—	0-III	0-III	—	—	—	—	—	—	(II-III)	(II-III)	(I-III)	—	(I)-III	II-III	0-III	0-III	—
Falco cherrug	L	—	—	—	—	—	—	—	—	—	—	—	I-III	—	—	—	I-III	I-III	III	III	—	—
Falco peregrinus	H	C	—	—	—	—	—	•	I-III	I-III	I-III	I-III	I-III	I-III	—	I-III	I-III	I-III	—	—	—	—
Bonasa bonasia	H	C	•	—	—	(I)II-III	(I)II-III	•(•)?	•	I-III	I-III	—	I-III	I-III	—	I-III	I-III	I-III	—	—	—	—
Tetrao tetrix	L	—	•	0-II	0-II	0-II	—	•	0-II	0-II	—	—	—	0-II	0-III	0-II	—	0-I	—	—	—	—
Tetrao urogallus	H	C	—	II-III	II-III	II-III	II-III	—	(O)-II	(O)-III	III	—	(II-III)	(I)II-III	II-III	I-III	I-III	(I)I-III	—	—	—	—

Boreal and temperate forests (cont.)

Part 1: Landscape requirements and preferred habitats (cont.)

cont.

| | Landscape requirements | | | Preferred habitats | | | | | | | | | | | | | | | | | | |
| | Forest cover | Habitat patchiness | External mosaics | Boreal forest | | | | Montane forest | | | | | Lowland temperate forest | | | | | | Riverine forest | | | |
				Internal mosaics	Pinus (dry)	Mixed Picea/Pinus	Picea (moist/wet)	Internal mosaics	Krummholz	Pinus/Larix	Picea	Fagus/Abies	Internal mosaics	Pinus/Betula	Planted Picea & non-native spp.	Mixed	Fagus	Mixed deciduous	Internal mosaics	Seasonally flooded	Cars	Unstable riparian
Grus grus	—	—	—	—	—	O-III	O-III	—	—	—	—	—	• O-II	—	—	O-III	—	O-III	• O-III	O-III	O-III	—
Scolopax rusticola	—	•	—	—	(O)II-III	(O)II-III	O-III	—	—	—	I-III	I-III	O-III	O-III	O-III	O-III	O-III	O-III	II-III	II-III	II-III	—
Tringa glareola	—	—	—	O-I	O-I	O-I	—	—	—	—	—	—	—	—	—	—	—	—	—	—	—	—
Columba oenas	—	—	•	II-III	—	—	—	—	—	—	II-III	II-III	(O)II-III	(O)II-III	(O)II-III	(O)II-III	(O)II-III	(O)II-III	II-III	II-III	II-III	—
Columba palumbus	—	—	—	II-III	II-III	II-III	—	—	—	(II-III)	II-III	II-III	(O)-III	(O)-III	(O)-III	(O)-III	(O)-III	(O)-III	(O)-III	(O)-III	(O)-III	—
Streptopelia turtur	L	•	—	II-III	II-III	II-III	—	—	—	—	I-I	I-I	I-I	I-I	I-I	I-I	I-I	I-I	I-I	I-I	I-I	—
Otus scops	—	—	—	—	—	—	—	—	—	—	—	—	II-III	—	—	II-III	II-III	II-III	II-III	II-III	II-III	—
Bubo bubo	H	C	—	I-III	I-III	I-III	I-III	• ?	O-III	O-III	O-III	O-III	II-III	II-III	II-III	II-III	II-III	II-III	—	—	—	—
Surnia ulula	—	—	—	II-III	II-III	I-III	I-III	—	—	—	—	—	—	—	—	—	—	—	—	—	—	—
Glaucidium passerinum	H	C	—	II-III	II-III	II-III	II-III	—	—	II-III	II-III	II-III	II-III	II-III	II-III	II-III	II-III	II-III	—	—	—	—
Strix aluco	H	C	—	II-III	II-III	—	—	•	II-III?	II-III	II-III	III	(II-III)	II-III	II-III	II-III	II-III	II-III	• II-III	II-III	II-III	—
Strix uralensis	—	—	—	II-III	II-III	II-III	—	—	—	II-III?	II-III	III	II-III	II-III	III	II-III	II-III	II-III	—	—	—	—
Strix nebulosa	—	—	—	—	—	II-III	II-III	—	—	—	III	III	—	—	—	—	—	—	—	—	—	—
Aegolius funereus	H	C	—	II-III	II-III	II-III	II-III	—	—	II-III	II-III	II-III	II-III	II-III	II-III	II-III	II-III	II-III	—	—	—	—
Caprimulgus europaeus	H	F	•	O-III	O-III	—	—	•	O-I(II)	O-I(II)	O-I(II)	—	O-I,III	O-I,III	O-I,III	O-I,III	—	O-I,III	—	—	—	—
Coracias garrulus	—	—	—	—	—	—	—	—	—	—	—	—	II-III	II-III	II-III	II-III	II-III	II-III	• II-III	II-III	II-III	—
Jynx torquilla	—	•	—	—	—	I-III	I-III	• ?	II-III	II-III	II-III	II-III	—	—	II-III	II-III	I-II	II-III	• II-III	II-III	II-III	—

Boreal and temperate forests (cont.)

Part 1: Landscape requirements and preferred habitats (cont.)

Species	Landscape requirements			Preferred habitats — Boreal forest				Montane forest					Lowland temperate forest					Riverine forest		
	Forest cover	Habitat patchiness	External mosaics	Internal mosaics	Pinus (dry)	Mixed Picea/Pinus	Picea/Pinus (moist/wet)	Internal mosaics	Krummholz Pinus/Larix	Picea	Fagus/Abies	Fagus/Abies	Internal mosaics Pinus/Betula	Planted Picea & non-native spp.	Mixed native spp.	Fagus	Mixed deciduous	Internal mosaics Seasonally flooded	Carrs	Unstable riparian
Picus canus	H	F	•	•	—	II-III	—	•	—	II-III	II-III	II-III	II-III	—	II-III	II-III	II-III	II-III	II-III	—
Picus viridis	—	C	—	•	—	II-III	—	•	II-III	II-III	III	III	II-III	—	II-III	III	III	II-III	II-III	—
Dendrocopos medius	H	—	—	—	—	II-III	—	—	—	—	II-III	II-III	—	—	II-III	II-III	II-III	II-III	II-III	—
Dendrocopos minor	—	—	—	—	—	(0)II-III	II-III	—	—	II-III	(0)II-III	(0)II-III	II-III	—	(0)II-III	(0)II-III	(0)II-III	II-III	(0)II-III	—
Picoides tridactylus	H	C	—	—	—	II-III	II-III	—	II-III	II-III	II-III?	0-I	II-III	—	II-III	II-III	—	—	—	—
Lullula arborea	H	F	•	•	0-I	—	—	•	0-I	0-II	0-I	0-I	0-III	0-I	0-III	0-III	0-III	—	—	—
Anthus hodgsoni	—	—	—	—	—	II-III	II-III	?	—	—	—	—	—	—	—	—	—	—	—	—
Bombycilla garrulus	—	—	—	II-III	—	—	II-III	—	—	—	—	—	—	—	—	—	—	—	—	—
Prunella modularis	—	—	•	—	—	I-III	I-III	•	(0)I-III	(0)I-II	(0)I-I	I-III	I-III	I-III	I-III	I-III	I-III	I-III	I-I	—
Prunella atrogularis	—	—	—	•	—	—	II-III	•	I-III	—	—	—	—	—	—	—	—	—	—	—
Erithacus rubecula	—	—	—	II-III	—	I-III	II-III	•	I-III	I-III	I-III	I-III	I-III	—	I-III	I-III	I-III	I-III	I-III	—
Luscinia luscinia	L	•	—	I-I	—	—	—	—	—	—	—	—	—	—	I-I	—	I-I	I-I	I-I	—
Luscinia megarhynchos	—	—	—	—	—	—	—	—	—	—	—	—	—	1	1	—	1	1	1	—
Luscinia calliope	—	—	—	I-III	I-III	—	—	—	—	—	—	—	—	—	—	—	—	—	—	—
Tarsiger cyanurus	—	—	•	II-III	II-III	—	—	—	—	—	—	—	—	—	—	—	—	—	—	—
Phoenicurus phoenicurus	—	—	—	II-III	II-III	—	II-III	?	II-III	II-III	II-III	II-III	II-III	II-III	1	II-III	II-III	II-III	—	•
Saxicola torquata	L	•	•	—	—	—	—	—	—	0-I	0-I	0-I	0-I	0-I	0-I	0-I	0-I	III	—	—

cont.

Boreal and temperate forests (cont.)

Part 1: Landscape requirements and preferred habitats (cont.)

	Landscape requirements		Boreal forest				Montane forest				Lowland temperate forest					Riverine forest		
			Preferred habitats															
	Forest cover	Habitat patchiness / External mosaics	Internal mosaics	Pinus (dry)	Mixed Picea/Pinus	Picea (moist/wet)	Internal mosaics	Krummholz Pinus/Larix	Picea	Fagus/Abies	Internal mosaics	Planted Pinus/Betula	Planted Picea & non-native spp. Mixed	Fagus	Mixed deciduous	Internal mosaics	Seasonally flooded Cars	Unstable riparian
Zoothera dauma	—	—	—	—	II-III	—	—	—	—	—	—	—	—	—	—	—	—	—
Turdus torquatus	—	—	—	—	—	—	•	I-III	I-III	(I-III)	—	—	—	—	—	—	I-III	—
Turdus merula	L	F	II-III	—	II-III	—	—	I-III	I-III	I-III	I-III	I-III	I-III	I-III	I-III	II-III	II-III	—
Turdus pilaris	—	—	II-III	—	II-III	—	—	(I)II-III	(I)II-III	I-III	I-III	I-III	—	I-III	I-III	I-III	I-III	—
Turdus philomelos	—	—	II-III	—	II-III	—	—	—	—	—	II-III	II-III	•	I-III	I-III	•	I-III	—
Turdus iliacus	L	F	II-III	—	II-III	—	—	I-III	II-III	II-III	II-III	II-III	•	II-III	II-III	I-III	I-III	—
Turdus viscivorus	—	F	—	II-III	—	—	—	—	II-III	II-III	I-III	I-III	•	I-II	—	I-II	I-II	—
Locustella fluviatilis	—	—	—	—	—	—	—	—	—	—	—	—	—	—	—	I-II	L	—
Acrocephalus palustris	—	—	—	—	—	—	—	—	—	—	—	—	—	—	—	L	L	I?
Hippolais icterina	F	—	I,II-III	—	—	—	—	—	—	—	—	—	—	—	I-II	I-II	I-II	—
Hippolais polyglotta	—	—	—	—	—	—	—	—	—	—	—	—	—	—	—	—	—	—
Sylvia borin	—	—	I-III	—	—	—	•	(I-III)	I-III	I-III	I-III	(I)	I-II	I-II	I-II	I-II	I-II	•
Sylvia atricapilla	—	—	I-III	—	I-III	—	—	(I-III)	I-III	I-III	I-III	I-III	I-III	I-III	I-III	I-II	I-III	•
Phylloscopus borealis	—	—	I-III	I-III	I-III	—	—	—	—	—	—	I-III	—	—	—	I-III	—	—
Phylloscopus inornatus	—	—	II-III	—	—	—	—	—	—	—	—	II-III	—	—	—	—	—	—
Phylloscopus bonelli	—	—	—	—	—	—	—	—	II-III	II-III	II-III	II-III	II-III	II-III	II-III	—	—	—
Phylloscopus sibilatrix	•	—	II-III	—	II-III	—	—	—	I-III	I-III	I-III	I-III	I-III	I-III	I-III	II-III	—	—

cont.

Boreal and temperate forests (cont.)

Part 1: Landscape requirements and preferred habitats (cont.)

cont.

| | Landscape requirements | | | Preferred habitats | | | | | | | | | | | | | | | | | |
| | | | | Boreal forest | | | | Montane forest | | | | | Lowland temperate forest | | | | | Riverine forest | | | |
Species	Forest cover	Habitat patchiness	External mosaics	Internal mosaics	Pinus (dry)	Mixed Picea/Pinus	Picea (moist/wet)	Internal mosaics	Krummholz	Pinus/Larix	Picea	Fagus/Abies	Internal mosaics Pinus/betula	Planted Picea & non-native spp.	Mixed	Fagus	Mixed deciduous	Internal mosaics	Seasonally flooded	Carr	Unstable riparian
Regulus regulus	—	—	—	—	—	II-III	II-III	—	II-III	II-III	II-III	II-III	II-III	II-III	II-III	II-III	(II-III)	—	—	—	—
Regulus ignicapillus	—	—	—	—	—	II-III	II-III	—	II-III	II-III	II-III	II-III	II-III	II-III	II-III	II-III	II-III	—	—	—	—
Muscicapa striata	—	—	•	—	II-III	II-III	—	—	—	(II-III)?	II-III	II-III	II-III	II-III	II-III	II-III	II-III	II-III	II-III	II-III	—
Ficedula parva	—	—	—	—	—	II-III	II-III	—	—	—	II-III	—	—	—	II-III	II-III	II-III	—	—	II-III	—
Ficedula semitorquata	—	—	—	—	—	—	—	—	—	—	—	—	—	—	(II-III)	—	—	—	—	—	—
Ficedula albicollis	—	—	—	—	—	II-III	—	—	—	—	II-III	II-III	II-III	II-III	II-III	II-III	II-III	II-III	II-III	II-III	—
Ficedula hypoleuca	—	—	—	—	II-III	II-III	II-III	—	—	?	?	II-III	II-III	II-III	I-III	I-III	I-III	II-III	II-III	II-III	—
Parus lugubris	—	—	—	(I)II-III	(I)II-III	—	—	—	—	—	—	—	—	—	—	—	—	—	—	—	—
Parus cinctus	—	—	—	—	—	—	—	—	—	—	—	—	—	—	—	—	—	—	—	—	—
Parus cristatus	—	—	—	II-III	II-III	II-III	—	—	—	(I)II-III	(I)II-III(I)II-III	II-III	II-III	II-III	II-III	II-III	II-III	II-III	II-III	II-III	—
Parus caeruleus	—	—	—	I,II	—	—	—	—	—	(I)II-III)?	II-III	II-III	(ii-III)	—	II-III	II-III	II-III	II-III	II-III	II-III	—
Parus cyanus	—	—	—	—	—	—	—	—	—	—	—	—	—	—	—	I-I	I-I	I-I	I-I	I-I	—
Sitta europaea	—	—	—	II-III	—	—	II-III	—	—	II-III	II-III	II-III	II-III	II-III	II-III	II-III	II-III	II-III	II-III	II-III	—
Certhia familiaris	H	—	—	II-III	—	—	II-III	—	—	II-III	II-III	II-III	II-III	—	II-III	II-III	II-III	—	I-II	I-II	—
Certhia brachydactyla	—	—	—	—	—	—	—	—	—	—	—	—	—	—	II-III	II-III	II-III	II-III	II-III	II-III	—
Oriolus oriolus	—	—	—	II-III	—	—	II-III	—	—	—	—	—	II-III	—	II-III	II-III	II-III	II-III	II-III	II-III	—
Lanius collurio	—	—	—	0-I	0-I	—	—	•	• ?	0-I	0-I	0-I	0-I,III	•	0-I	0-I	0-I	0-I	0-I	0-I	0-I

Boreal and temperate forests (cont.)

Part 1: Landscape requirements and preferred habitats (cont.)

| | Landscape requirements | | Preferred habitats | | | | | | | | | | | | | | | | | | |
| | Forest cover | Habitat patchiness / External mosaics | Boreal forest | | | | Montane forest | | | | | Lowland temperate forest | | | | | | Riverine forest | | | |
Species			Internal mosaics	Pinus (dry)	Mixed Picea/Pinus	Picea (moist/wet)	Internal mosaics	Krummholz	Pinus/Larix	Picea	Fagus/Abies	Internal mosaics	Pinus/Betula	Planted Picea & non-native spp.	Mixed native spp.	Fagus	Mixed deciduous	Internal mosaics	Seasonally flooded	Cars	Unstable riparian
Lanius excubitor	—	—	O-I	—	—	—	—	—	—	—	—	O-I	—	—	—	—	—	O-I	—	—	O-I
Perisoreus infaustus	—	—	—	II-III	II-III	II-III	—	—	—	—	—	—	—	—	—	—	—	—	—	—	—
Corvus monedula	—	•	—	—	—	—	—	—	—	—	II-III	—	—	—	—	II-III	II-III	III	—	—	—
Fringilla coelebs	—	—	II-III	II-III	II-III	II-III	(●)	I-II	I-III	I-III	I-III	I-III	I-III	I-III	I-III	I-III	I-III	II-III	II-III	II-III	—
Fringilla montifringilla	L	•	—	II-III	II-III	II-III	—	—	—	—	—	—	—	—	—	—	—	—	—	—	—
Serinus serinus	F	—	—	—	—	—	•	—	(I-III)?	I-III	I-III	—	—	—	II-III	—	—	II-III	—	—	—
Serinus citrinella	—	—	—	—	—	—	•	—	I-III	I-III	I-III	—	—	—	—	—	—	—	—	—	—
Carduelis chloris	—	•	—	I-I	—	—	—	—	—	—	I-III	I-III	I-III	I-III	I-III	I-III	I-III	I-III	I-III	I-III	—
Carduelis spinus	—	—	—	—	II-III	II-III	—	—	II-III	II-III	II-III	—	—	II-III	—	—	—	II-III	II-III	II-III	—
Loxia leucoptera	—	•	(I)II-III	—	—	—	—	—	I-III	—	—	—	—	—	—	—	—	—	—	—	—
Loxia scotica	—	—	II-III	II-III	—	—	—	—	—	—	—	—	II-III	—	—	—	—	—	—	—	—
Loxia pytyopsittacus	—	—	II-III	II-III	—	—	—	—	—	—	—	—	—	—	—	—	—	—	—	—	—
Pinicola enucleator	—	—	—	—	II-III	—	—	—	—	—	—	—	—	—	—	—	—	—	—	—	—
Coccothraustes coccothraustes	—	—	—	—	—	—	—	—	—	—	—	II-III	II-III	II-III	II-III	II-III	II-III	II-III	II-III	III	—
Emberiza cia	—	—	—	—	—	—	—	—	—	—	—	—	—	—	—	—	—	—	—	—	—
Emberiza hortulana	—	—	O-I	O-I	—	—	—	—	—	—	—	—	—	—	—	—	—	—	—	—	—
Emberiza rustica	—	•	II-III	II-III	II-III	—	—	—	—	—	—	—	—	—	—	—	—	—	—	—	—

Boreal and temperate forests (cont.)

Part 2: Habitat features

Species	Deciduous: Presence	Deciduous: Trees in stands	Deciduous: Snags and dead branches	Coniferous: Presence	Coniferous: Trees in stands	Coniferous: Snags and dead branches	Big trees	Field layer <0.5 m	Shrub layer 0.5–4 m	Canopy layer >4 m	Open canopy	Wet patch	Ecotone	Holes in trees
Phalacrocorax pygmeus														
Nycticorax nycticorax														
Ardeola ralloides														
Ardea purpurea												•		
Ciconia nigra							•(grey)							
Plegadis falcinellus	•									•	•			
Platalea leucorodia														
Mergus albellus														
Pernis apivorus	•						•			•		•		
Milvus migrans	•						•(grey)				•			
Milvus milvus							•				•			
Haliaeetus albicilla														
Circaetus gallicus	•						•			•	•	•		
Circus cyaneus														
Accipiter brevipes							•							
Aquila pomarina		•					•			•	•	•		
Aquila clanga							•			•	?	•		
Aquila heliaca							•							
Aquila chrysaetos							•			•	•			
Hieraaetus pennatus							•							
Pandion haliaetus														
Falco tinnunculus														
Falco vespertinus														
Falco cherrug							•							
Falco peregrinus														
Bonasa bonasia	•						•					•		
Tetrao tetrix				•				•	•		•	•		
Tetrao urogallus					•		•	•(grey)	•				•	•(grey)
Grus grus				•								•		
Scolopax rusticola							•							
Tringa glareola							•	•			•			
Columba oenas									•					•
Columba palumbus														
Streptopelia turtur														
Otus scops													•	•
Bubo bubo											•			•
Surnia ulula														
Glaucidium passerinum					•	•				•				•

cont.

Boreal and temperate forests (cont.)

Part 2: Habitat features (cont.)

Species	Holes in trees	Ecotone	Wet patch	Open canopy	Canopy layer >4 m	Shrub layer 0.5–4 m	Field layer <0.5 m	Big trees	Snags and dead branches (con.)	Trees in stands (con.)	Presence (con.)	Snags and dead branches (dec.)	Trees in stands (dec.)	Presence (dec.)
Strix aluco	(•)				•			•						
Strix uralensis	•			•	•			•						
Strix nebulosa														
Aegolius funereus	•	•		•	•			•						
Caprimulgus europaeus														
Coracias garrulus	•			•	•			•						
Jynx torquilla	•							•				•		
Picus canus				•	•			•			•	(•?)		•
Picus viridis				•	•			•				•	•	•
Dendrocopos medius					•			•				•	•	•
Dendrocopos minor								•				•		
Picoides tridactylus					○				•					
Lullula arborea						○								
Anthus hodgsoni														
Bombycilla garrulus														
Prunella modularis			•			•	•							
Prunella atrogularis							•							
Erithacus rubecula			(•)			•								
Luscinia luscinia			•	•		•							•	
Luscinia megarhynchos			•	•		•							•	
Luscinia calliope							○							
Tarsiger cyanurus														
Phoenicurus phoenicurus	(•)			•		○								
Saxicola torquata														
Zoothera dauma														
Turdus torquatus														
Turdus merula														
Turdus pilaris		•				•								
Turdus philomelos														
Turdus iliacus		•	•				•							
Turdus viscivorus			•	•	•									
Locustella fluviatilis		•												
Acrocephalus palustris			•		•		•							
Hippolais icterina													•	
Hippolais polyglotta														
Sylvia borin				•		•							•	(•)
Sylvia atricapilla						(•)								(•)
Phylloscopus borealis														

cont.

Boreal and temperate forests (cont.)

Part 2: Habitat features (cont.)

Species	Deciduous features			Coniferous features			General features							
	Presence	Trees in stands	Snags and dead branches	Presence	Trees in stands	Snags and dead branches	Big trees	Field layer <0.5 m	Shrub layer 0.5–4 m	Canopy layer >4 m	Open canopy	Wet patch	Ecotone	Holes in trees
Phylloscopus inornatus								●	○					
Phylloscopus bonelli	●							●		●	●			
Phylloscopus sibilatrix	●									●				
Regulus regulus				●						●	●			
Regulus ignicapillus										●				
Muscicapa striata	●									●	●			●
Ficedula parva	●		●						○	●				●
Ficedula semitorquata	●									●	●			●
Ficedula albicollis	●									●				●
Ficedula hypoleuca	●								○	●	●			●
Parus lugubris														●
Parus cinctus				●						●	●			●
Parus cristatus				●						●				●
Parus caeruleus	●									●	●			●
Parus cyanus									●					●
Sitta europaea	●									●				●
Certhia familiaris				●		●	●							●
Certhia brachydactyla	●									●	●			●
Oriolus oriolus	●		●							●				
Lanius collurio									●		●		●	
Lanius excubitor											●		●	
Perisoreus infaustus				●			●							
Corvus monedula							●				●		●	●
Fringilla coelebs										●	●			
Fringilla montifringilla				●						●				
Serinus serinus											●			
Serinus citrinella				●										
Carduelis chloris										●	●			
Carduelis spinus				●						●				
Loxia leucoptera				●										
Loxia scotica				●						●				
Loxia pytyopsittacus				●	●									
Pinicola enucleator										●	●			
Coccothraustes coccothraustes	●									●				
Emberiza cia											●			
Emberiza hortulana														
Emberiza rustica														

Mediterranean forest, shrubland and rocky habitats

KEY

* See 'Notes' column

Habitat-types
C Coniferous forest
B Broadleaved forest
M Maquis
G Garrigue and rocky habitats

Minimum area
Minimum area of habitat required by one breeding pair
• <10 ha
•• 10–100 ha
••• >100 ha
N Nest-site only

Topography
F Flat
S Sloping
V Vertical (cliffs)
M Mixed

Vegetation structure
N None
O Open
D Dense
X Vegetation required, but no preference for open or dense

Vegetation mosaic, Old/dead trees, Tall trees
• Required in home-range

Nest-sites
B Buildings
C Cliffs
G Ground
H Holes in trees
R Rock crevices, ledges, etc.
S Shrubs
T Trees

Survival factors
Survival requires low risk from:
D Disturbance
Pl Powerlines
P Predation
C Competition

	Habitat-types	Minimum area	Topography	Altitude (m)	Vegetation structure						Other habitats required within home-range	Nest-sites	Specific food requirements	Survival factors	Notes
					Herb	Shrub	Tree	Veg. mosaic	Old/dead trees	Tall trees					
Ciconia nigra	C,B,M,G	•••	S,M	<1,000	—	O	X	—	—	•	Rivers, springs, ponds, other wetlands	C,R,T	Fish, tadpoles, large insects	D	Linked to canyons and river valleys
Milvus migrans	C,B	•••	F,S	<1,000	—	—	X	•	—	—	Water, villages, open habitats	T,(C)	Rodents, carrion, rabbits	D,Pl	Linked to rivers, lakes, rubbish-dumps and roads
Milvus milvus	C,B,G	•••	F,S	<1,000	—	—	X	•	—	—	Open habitats, garbage in winter	T	Rodents, rabbits, birds, carrion	D,Pl	Linked to rubbish-dumps and chicken farms in winter
Gypaetus barbatus	G	•••	S,V	300–2,000 (>3,000 for food)	—	N,O	N,O	—	—	—	Mountains, grassland	C	Large-mammal carrion (bones)	D, C (*Gyps fulvus*)	—
Neophron percnopterus	G	•••	S,V	<1,500	—	N,O	N	•	—	—	Open habitats	C,R	Livestock carrion, rabbits	D, C (*Gyps fulvus*)	Linked to disposal sites for dead livestock
Gyps fulvus	G	•••	S,V	200–1,600 (>2,000 for food)	—	N,O	N	•	—	—	Open habitats	C*	Livestock carrion	D	*Colonial; linked to disposal sites for dead livestock
Aegypius monachus	B,M	•••	S	200–1,400	—	D	O	—	—	•	Open habitats, pastoral woodland	T*	Livestock carrion, rabbits	D,C	*Colonial
Circaetus gallicus	C,B,M,G	•••	S	<1,200	—	O	O	•	•	•	—	T	Snakes	D,Pl	—
Accipiter brevipes	C,B*	•••	F,S	<1,200	—	—	O	•	•	•	—	T	Lizards, birds	D	*Moist woods

cont.

Mediterranean forest, shrubland and rocky habitats (cont.)

	Habitat-types	Minimum area	Topography	Altitude (m)	Herb	Shrub	Tree	Veg. mosaic	Old/dead trees	Tall trees	Other habitats required within home-range	Nest-sites	Specific food requirements	Survival factors	Notes
Buteo rufinus	C,M,G	●●●	S,V	<1,000	—	O	N*	●	—	—	Steppe	C,T	Reptiles, susliks	D,Pl,P	*Except isolated trees for nesting
Aquila pomarina	C,B	●●●	S	<1,200	—	—	D	●	—	●	Open habitats and moist areas	T	Amphibians, reptiles	D	—
Aquila heliaca	C,B	●●●	F,S	<1,000	—	—	O	●	—	●	Grasslands and other open habitats	T	Carrion, susliks, other mammals	D,Pl	—
Aquila adalberti	C,B,M	●●●	M	<1,400	—	O	O	●	—	●	Pastoral woodland	T	Rabbits	D,Pl	Nests near birth-place
Aquila chrysaetos	C,M,G	●●●	S,V	200–2,000	—	O	N,O	—	—	—	Open habitats	C,R,T	Rabbits, other mammals, birds, reptiles	D (*Gyps fulvus*)	—
Hieraaetus pennatus	C,B	●●●	F,S	200–2,000	—	—	X	●	—	●	Canyons, river valleys, pastoral woodland	T (pines)	Pigeons, rabbits, lizards, snakes	D,Pl	—
Hieraaetus fasciatus	G	●●●	M	<1,500	O	O	N*	●	—	—	—	C,T	Rabbits, partridges, other birds	D,Pl, P (*Bubo bubo*),C (*Gyps fulvus*)	*Occasionally nests in trees
Falco tinnunculus	C,B,M,G	●●	F,M	<2,000	—	O	O	—	—	—	Open farmland	B,C,T	Insects, rodents, reptiles	D,Pl	—
Falco eleonorae	G	●●	V	<400	—	—	N	—	—	—	—	C*	Passerines	D	*Colonies on sea cliffs
Falco biarmicus	G	●●	S	300–600	—	—	N	—	—	—	Steppe	C	Mostly birds	D	—
Falco peregrinus	M,G	●●	V,M	<1,500	—	—	N	●	—	—	Open habitats	C	Pigeons, other birds	D, P (*B. bubo*)	—
Alectoris chukar	M,G	●●	S	300–800	X	X	N,O	●	—	—	—	G	Insects for chicks; seeds, plant material	P	—
Alectoris graeca	G	●●	S	400–2,000	X	X	N,O	●	—	—	—	G	Insects for chicks; seeds, plant material	P,C*	*Introduced *Alectoris chukar*
Alectoris rufa	M,G	●●	F,M	<2,000	X	X	N	●	—	—	Open farmland	G	Insects for chicks; seeds, plant material	P	—
Alectoris barbara	M,G	●●	S	<900	X	X	N,O	●	—	—	—	G	Insects for chicks; seeds, plant material	P	—

cont.

Mediterranean forest, shrubland and rocky habitats (cont.)

Species	Habitat-types	Minimum area	Topography	Altitude (m)	Herb	Shrub	Tree	Veg. mosaic	Old/dead trees	Tall trees	Other habitats required within home-range	Nest-sites	Specific food requirements	Survival factors	Notes
Francolinus francolinus	M*	••	S	<800	X	X	—	•	—	—	—	G	—	P	*Tamarisk *Tamarix* preferred
Turnix sylvatica	G*	—	F	0	—	O	N,O	•	—	—	—	G	—	—	*In Europe, dwarf palm *Chamaerops* scrub
Scolopax rusticola	C,M	—	S	<2,000	—	O	O	•	—	—	Grassland (winter)	G	Soil invertebrates	—	—
Columba palumbus	C,B	••	F,M	<2,000	—	O	O	•	—	—	—	T	Berries, other plant material; acorns in winter	D	—
Streptopelia turtur	B	••	F,S	<1,000	—	O	O	•	—	—	—	S,T	Seeds, other plant material	D / P (nests)	— / —
Tyto alba	G	••	F,S	<1,400	—	O	N,O	•	—	—	—	B,C	Small mammals	Pl	—
Otus scops	C,B,M	••	F,S	<2,000	—	O	O	•	—	—	—	H	Insects	—	—
Bubo bubo	M,G	•••	S,V	<2,000	X	X	—	•	—	—	—	C,G	Rabbit	D,Pl	Linked to rivers, valleys and canyons
Athene noctua	M,G	•	F,S	<1,500	—	O	O	•	—	—	Farmland	B,C,G,H	Insects, small mammals	D	—
Caprimulgus europaeus	C,G	•	F,S	<2,000	—	—	—	•	—	—	Watercourses and wet areas	G	Moths, other insects	—	—
Caprimulgus ruficollis	C,G	•	F,S	<800	—	—	—	•	—	—	Watercourses and wet areas	G	Insects	—	—
Apus caffer	G	N	V	—	—	—	—	—	—	—	—	—	Insects	—	Occupies nests of *Hirundo daurica* and *Ptyonoprogne rupestris*
Apus affinis	G	N	V	—	—	—	—	—	—	—	—	C	Insects	—	Colonial
Merops apiaster	G	••	F,S	<1,500	—	—	N,O	•	—	—	Sandy banks and open habitats	C,G	Bees	—	Colonial
Coracias garrulus	B,G	••	M	<1,800	—	—	O	•	—	—	—	B,H	Insects	—	—
Jynx torquilla	B	•	F,S	<2,000	—	O	O	—	•	—	—	H	Ants	—	—

cont.

Mediterranean forest, shrubland and rocky habitats (cont.)

Species	Habitat-types	Minimum area	Topography	Altitude (m)	Vegetation structure						Other habitats required within home-range	Nest-sites	Specific food requirements	Survival factors	Notes
					Herb	Shrub	Tree	Veg. mosaic	Old/dead trees	Tall trees					
Picus viridis	C,B	••	S	<1,800	—	O	O	—	•	—	Grassland	H	Insects	—	—
Dendrocopos syriacus	B	•••	F,S	<500	—	—	O	•	•	•	—	H	Insects	D	—
Calandrella brachydactyla	G	•	F	<2,100	X	—	—	•	—	—	Open grassland	G	Insects, seeds and farmland	—	—
Galerida cristata	G	•	F	<1,500	X	—	—	•	—	—	Open habitats	G	Mainly invertebrates	—	—
Galerida theklae	G	••	S	<2,200	O	O	O	—	—	—		G	Mainly invertebrates	—	—
Lullula arborea	C,B,G	•	M	<1,900	X	O	O	•	—	—		G	Mainly invertebrates	—	—
Ptyonoprogne rupestris	G	N	V	<2,000	—	—	—	—	—	—		B,C*	Mainly invertebrates	—	*Colonial
Hirundo daurica	G	N	M	<1,500	—	—	—	—	—	—		B,C,R	Mainly invertebrates	—	Also artificial construction (bridges), houses
Anthus campestris	C*,G	•	F,S	<2,000	O	—	N	—	—	—		G	Mainly invertebrates	—	*Open *Juniperus* forest
Erithacus rubecula	C,B,M	•	F,S	<2,000	—	D	D	—	—	—		S,T	Omnivorous	—	—
Luscinia megarhynchos	C,B,M	•	F,M	<1,500	O	D	O	•	—	—		G,S	Mainly invertebrates	—	—
Phoenicurus phoenicurus	B	•	M	<2,000	O	O	O	•	—	—		H	Mainly invertebrates	—	—
Saxicola torquata	M,G	•	F,M	<1,500	O	O	N	•	—	—		S	Mainly invertebrates	—	Sparse trees, perches
Oenanthe cypriaca	G	•	S	<1,800	O	N,O	N,O	•	—	—		G	Mainly invertebrates	—	—
Oenanthe hispanica	G	•	S	<1,600	O	O	N,O	•	—	—		G	Mainly invertebrates	—	—
Oenanthe leucura	G	•	S,V	<1,800	O	N,O	N,O	—	—	—		R	Insects	—	—
Monticola saxatilis	G	•	S,V	<2,000	O	N,O	N,O	—	—	—		R	Insects	P	—

cont.

387

Mediterranean forest, shrubland and rocky habitats (cont.)

Species	Habitat-types	Minimum area	Topography	Altitude (m)	Herb	Shrub	Tree	Veg. mosaic	Old/dead trees	Tall trees	Other habitats required within home-range	Nest-sites	Specific food requirements	Survival factors	Notes
Monticola solitarius	G	—	V	<1,500	O	N,O	N,O	—	—	—	—	R	Small invertebrates	P	—
Turdus merula	C,B,M,G	•	M	<2,000	—	O,D	O,D	—	—	—	—	S,T	Fruit, invertebrates	—	—
Turdus philomelos	B	•	M	<2,000	—	O	O	—	—	•	—	T	Fruit, invertebrates	—	—
Hippolais pallida	B,M	•	F,S	<1,200	—	O	O	—	—	—	—	S	Insects	—	—
Hippolais olivetorum	B,M	••	S	<500	—	—	O	—	—	—	—	S,T	Insects	—	—
Hippolais polyglotta	B,M,G	•	M	<1,950	—	O,D	O	•	—	—	—	S	Insects	—	—
Sylvia sarda	M,G	•	—	<2,000	—	D	N	—	—	—	—	S	Insects	—	—
Sylvia undata	M	•	F,S	<1,950	—	D	N	—	—	—	—	S	Insects	—	Density affected by shrubland-patch size
Sylvia cantillans	M	••	F,S	300–1,900	—	D	O	•	—	•*	—	S	Insects	—	*Sparse, tall bushes/trees
Sylvia melanocephala	C,B,M	•	F,S	<1,500	—	D	—	•	—	—	—	S	Insects	—	—
Sylvia melanothorax	M	••	F,S	<1,400	—	D	O	•	—	—	Forest edges	S	Insects	—	—
Sylvia rueppelli	M	•	F,S	<1,000	—	D	—	•	—	—	—	S	Insects	—	Coastal zones preferred
Sylvia hortensis	B,G	••	S	<1,300	—	O	O*	—	—	—	—	S	Insects	—	*Sparse trees
Sylvia atricapilla	C,B	•	F,S	<1,600	—	D	D	•	—	—	—	S,T	Omnivorous	—	—
Phylloscopus bonelli	B,M	•	F,S	300–2,000	—	O	D	—	—	—	—	G	Mainly insects	—	—
Regulus ignicapillus	C,B	•	F,S	300–2,000	—	—	D	—	—	—	—	T	Mainly insects	—	—
Muscicapa striata	C,B	•	M	300–1,700	—	—	O	•	•	—	Pastoral woodland, gardens, parks, farms	B,H,R	Mainly insects	—	—

cont.

Mediterranean forest, shrubland and rocky habitats (cont.)

	Habitat-types	Minimum area	Topography	Altitude (m)	Vegetation structure Herb	Shrub	Tree	Veg. mosaic	Old/dead trees	Tall trees	Other habitats required within home-range	Nest-sites	Specific food requirements	Survival factors	Notes
Ficedula semitorquata	B	—	S	<1,000	—	—	O	•	—	•	Spring-fed riparian habitat	H	Mainly insects	—	—
Parus lugubris	C,B	•	S	<1,000	—	O	O	•	—	•	—	H	Mainly insects	—	—
Parus cristatus	C,B	••	S	900–1,500	—	—	D	•	•	•	—	H	Mainly insects, seeds in winter	—	—
Parus caeruleus	C,B	•	S	<2,000	—	—	O,D	—	•	•	—	H	Mainly insects, seeds in winter	—	—
Sitta krueperi	C	—	S	100–1,000	—	—	O	•	•	•	—	—	Seeds, invertebrates	—	—
Sitta whiteheadi	C*	•	M	600–1,800	—	—	D	—	•	—	—	H	Pine seeds, invertebrates	P (*Dendrocopos major*)	*Pinus laricio*
Sitta neumayer	G*	—	V	<1,800	—	—	N	—	—	—	—	B,C,R	Mainly invertebrates	—	*Rocks
Certhia brachydactyla	C,B	•	F,S	<2,000	O	—	O,D	•	—	—	—	H	Insects	—	Abundance affected by tree density, and lichen cover on trees
Lanius collurio	M,G	••	F,S	500–1,800	O	O	N	•	—	—	Farmland	S	Insects, small rodents	—	—
Lanius excubitor	M,G	••	F,S	<1,500	O	O	N,O	•	—	—	—	S	Insects, small rodents	—	—
Lanius senator	B,M	••	F,S	<1,300	O	O	O	•	—	—	—	S	Insects	—	—
Lanius nubicus	B	••	S	<400	—	—	O	—	—	•	Forest edges	S,T	Insects	—	—
Pyrrhocorax pyrrhocorax	G	••	F,V	200–1,800	X	N,O	N	—	—	—	Open grassland	B,C	Grasshoppers	D	—
Sturnus unicolor	C,B,G	•	M	<2,000	—	O,D	O	•	—	—	Open farmland	B,C,H,T	Omnivorous	—	—
Petronia petronia	C,B,G	••	F,V	<1,000	—	O	O	•	—	—	—	B,H,R	Insects	—	Colonial, sometimes with *Merops apiaster*
Fringilla coelebs	C,B,M	•	M	<2,400	—	—	O,D	•	—	—	—	T	Seeds, insects	—	—
Serinus serinus	C,B,G	•	M	<1,800	—	O	O	•	—	—	—	S,T	Seeds, insects	—	—

cont.

Mediterranean forest, shrubland and rocky habitats (cont.)

	Habitat-types	Minimum area	Topography	Altitude (m)	Vegetation structure						Other habitats required within home-range	Nest-sites	Specific food requirements	Survival factors	Notes
					Herb	Shrub	Tree	Veg. mosaic	Old/dead trees	Tall trees					
Serinus citrinella	C,B,M	•	S	<2,000	—	X	O	•	—	—	—	S	Seeds, insects	—	—
Carduelis chloris	C,B,M	•	M	<2,000	—	O	O	•	—	—	—	S,T	Seeds, insects	—	—
Carduelis cannabina	C,M,G	•	S	<2,000	X	O,D	O	•	—	—	—	G,S	Seeds, insects	—	—
Emberiza cirlus	B,M,G	•	S	<1,900	—	X	O	•	—	—	—	S,T	Seeds, insects	—	—
Emberiza cia	M*,G*	•	S	500–2,000	X	X	—	•	—	—	—	G,R	Seeds, insects	—	*Rocky forest edges
Emberiza cineracea	G*	?	S	?	O	N,O	N	?	—	—	—	G	Seeds, insects	—	*Rocky
Emberiza hortulana	B,M,G	••	F,S	<1,500	O	O	—	•	—	•	Mountain pasture and shrubland, vineyards, alpine grassland	G	Seeds, insects	—	—
Emberiza caesia	M*,G	••	S	<1,000	O	N,O	N,O	•	—	—	—	G	Seeds, insects	—	*Low
Emberiza melanocephala	M*,G	•	F,S	<1,200	D	O	O	•	—	—	Farmland	G	Seeds, insects	—	*Low

Agricultural and grassland habitats

KEY

Preferred habitats
∘ Indicates that the species occurs but it is not a priority in the habitat

Arable, improved grassland
A Arable only
M Mixed arable and grass or fodder crops
G Permanent grassland

Steppic habitats
G Grass and shrub steppes
E Extensive cereals with fallow, grass and fodder crops (pseudosteppe)

Montane grassland
H Heath and scrub (subalpine)
S Subalpine grassland
A Alpine grassland
R Rocky habitats

Wet grassland
P Polders
C Coastal grassland
F Flood-plain and lakeside meadows
M Machair
U Upland grassland

Landscape features
Minimum area of agricultural/grassland habitat within home-range
L Low (<10 ha)
M Medium (10–100 ha)
H High (>100 ha)

Habitat mosaic
• Species dependent on presence of other non-agricultural habitats within its home-range

Crop mosaic
• Species dependent on presence of at least two crop-types within its home-range

Openness
Absence of trees, hedges, buildings, pylons, etc.
L Low
M Medium
H High

Topography
F Flat
M Moderate slope
S Steep
V Near-vertical

Vegetation composition and structure
Height
L Low (<10 cm)
M Medium (10–40 cm)
H High (>40 cm)

Plant cover
L Low (<25%)
M Medium (25–75%)
H High (>75%)

Vegetation mosaic
• Species dependent on mosaic of differing vegetation structures at some point in its annual cycle

Other features
R Rank marginal vegetation
H Hedgerows
S Scrub or scattered bushes
T Trees
W Water fringe or emergent vegetation
B Boulders, rocks
P Open pools, ditches

Feeding requirements
Main food
Plants:
PG Grass
PC Crops other than grass (seeds and leaves)
PA Aquatic plants (submerged)
PR Rhizomes
PW Weeds (seeds and leaves)
PF Fruit and seeds, etc., from non-crop vegetation
Invertebrates:
IS Soil invertebrates
IV Vegetation and soil-surface invertebrates
IA Aquatic invertebrates
IF Flying insects

Vertebrates:
VA Amphibians
VR Reptiles
VF Fish
VB Birds
VM Small mammals
VL Large mammals
VC Carrion
Other food sources:
R Refuse

Other habitat requirements
Survival factors
Survival requires low risk from:
D Disturbance (e.g. from farming operations, hunting, recreation)
Pr Predation (e.g. by dogs, foxes, crows, etc.)
S Stock at high densities (due to trampling)
H Harvesting, mowing, etc., during breeding period
Pl Powerlines
T Toxic pollutants

Part 1

	Preferred habitats				Landscape features					
	Arable, improved grassland	Steppic habitats	Montane grassland	Wet grassland	Minimum area	Habitat mosaic	Crop mosaic	Openness	Topography	Specific requirements
Bubulcus ibis	A,M,G	G,E	—	—	—	—	—	—	—	—
Ardea purpurea	—	—	—	F	—	—	—	—	—	Marsh
Ciconia ciconia	M,G	G,E	—	P,F	—	•	—	M–H	—	—
Cygnus columbianus [W]	A,M,G	—	—	P,C,F	—	•	—	H	—	Open water
Cygnus cygnus [W]	—	—	—	P,C,F	—	•	—	H	—	Open water
Anser brachyrhynchus [W]	A,M,G	—	—	P,C,F	—	•	—	H	—	Open water

cont.

Agricultural and grassland habitats (cont.)

Part 1

cont.

Species	Preferred habitats				Landscape features					Specific requirements
	Arable, improved grassland	Steppic habitats	Montane grassland	Wet grassland	Minimum area	Habitat mosaic	Crop mosaic	Openness	Topography	
Anser erythropus [w]	M,G	G,E	—	—	—	●	—	—	—	Open fresh water
Branta leucopsis [w]	A,M,G	—	—	P,C	—	●	—	H	—	Coastal water
Branta bernicla [w]	A,M	—	—	P,C	—	●	—	H	—	Coastal water
Branta ruficollis [w]	—	G[N]	—	—	—	●	—	H	—	Open water
Tadorna ferruginea	—	—	—	F	—	●	—	H	—	Open fresh water
Anas acuta	—	—	—	F	—	—	—	—	—	—
Anas querquedula	—	—	—	—	—	—	—	—	—	—
Elanus caeruleus	A,M,G	—	—	—	M	—	—	M	F-M	Woodland
Milvus migrans	M,G	—	—	o	H	●	—	M-H	F-M	Woodland
Milvus milvus	M,G	G[w],E[w]	—	o	M	●	—	M-H	F-M	—
Gypaetus barbatus	—	—	A,R	—	H	—	—	H	S-V	High-altitude terrain
Neophron percnopterus	—	G,E	A,R	—	H	—	—	H	F-V	—
Gyps fulvus	—	G,E	A,R	—	H	—	—	H	F-V	—
Aegypius monachus	—	G	A,R	—	H	—	—	M-H	F-S	—
Circaetus gallicus	G	G,E	—	C,F,U	H	—	—	M	F-S	—
Circus cyaneus	A,M,G	G	—	—	H	●	—	M-H	F-M	—
Circus macrourus	G	G,E	—	o	H	●	—	H	F	—
Circus pygargus	A,M	G	—	—	H	●	—	M-H	F-M	Fragmented forest
Accipiter brevipes	G	—	—	—	M	●	—	L	F-M	—
Buteo rufinus	M,G	G	—	—	H	—	—	L-M	—	Forest
Aquila pomarina	M,G	—	—	F	H	—	—	M	—	Forest and wetlands
Aquila clanga	—	G	—	F	H	—	—	H	—	—
Aquila rapax	G	G	—	—	H	—	—	H	—	—
Aquila heliaca	G	G	A,H,S	F	H	●	—	M-H	S-V	Forest or scattered trees
Aquila chrysaetos	—	G	A,H,S	—	M	—	—	M-H	—	—
Falco naumanni	M,G	G,E	—	P,C,F,M,U	M	●	●	H	—	—
Falco tinnunculus	A,M,G	G,E	H,S,R,A	—	M	—	—	M-H	F-M	Trees or buildings
Falco vespertinus	A,M,G	G,E	—	—	M	—	—	M-H	—	—
Falco biarmicus	A,M,G	G,E	—	—	M-H	—	—	M-H	—	—
Falco cherrug	A,M,G	G	R	P,C	M-H	●	—	H	—	—
Falco peregrinus	—	—	A,H,S	—	H	—	—	L-M	F-M	—
Tetrao tetrix	—	—	A,H,S	—	M	—	—	L-M	F-M	—
Tetrao mlokosiewiczi	—	—	A,H,S	—	H	—	—	L-M	S	1,700–3,000 m
Tetraogallus caucasicus	—	—	R,A	—	M	—	—	M-H	S-V	1,800–4,000 m
Tetraogallus caspius	—	—	R,A,H	—	H	—	—	M	S-V	1,800–3,000 m

Agricultural and grassland habitats (cont.)

Part 1 (cont.)

Species	Preferred habitats				Landscape features					
	Arable, improved grassland	Steppic habitats	Montane grassland	Wet grassland	Minimum area	Habitat mosaic	Crop mosaic	Openness	Topography	Specific requirements
Alectoris chukar	—	G	—	—	—	—	—	—	—	South-facing slopes
Alectoris graeca	—	—	R,A,H	—	M	—	—	M	M–S	1,200–1,500 m
Alectoris rufa	A,M,G	E	—	—	M	—	—	M–H	—	—
Perdix perdix	A,M,G	G	—	—	M	—	—	M–H	F–M	—
Coturnix coturnix	A,M,G	G	S	F	L	—	—	M–H	F	—
Porzana porzana	—	—	—	F	—	—	—	L–M	F	—
Crex crex	G	—	—	F,M	H	—	—	M–H	—	—
Grus grus	A,M,G	G^w,E^w	—	—	H	•	—	H	F	Water-bodies
Anthropoides virgo	—	G	—	—	H	—	—	H	F	—
Tetrax tetrax	A,M,G	G,E	—	—	M	—	—	H	F–M	—
Chlamydotis undulata	—	G,E	—	—	H	—	—	H	F	—
Otis tarda	A,M,G	G,E	—	—	H	—	—	H	F	—
Burhinus oedicnemus	A,M,G	G,E	—	—	—	—	—	M–H	F–M	—
Cursorius cursor	—	G	—	—	—	—	—	H	F	—
Glareola pratincola	A,M,G	G,E	—	—	—	•	•	H	F	Water-bodies
Glareola nordmanni	M,G	G,E	—	—	—	•	•	H	F	Water-bodies
Charadrius leschenaultii	—	G	—	—	—	•	•	H	F	Water-bodies
Charadrius asiaticus	—	G,E	—	—	—	•	—	H	F–M	Water-bodies
Pluvialis apricaria	G	G	—	○	M	•	—	H	F	—
Chettusia gregaria	—	G	—	—	M	—	—	H	F	Water-bodies
Vanellus vanellus	A,M,G	G,E	—	P,C,F,M,U	M	—	—	H	F–M	—
Calidris alpina	—	—	—	C,M	—	—	—	H	F	—
Philomachus pugnax	—	—	—	P,C,F	M–H	—	—	H	F	—
Gallinago media	—	—	—	F	M	—	—	L	F	—
Scolopax rusticola	G	—	—	—	—	•	—	L	F	Forest
Limosa limosa	—	—	—	P,C,F	M–H	—	—	H	F	—
Numenius tenuirostris	—	—	—	C,M,U	—	—	—	H	—	—
Numenius arquata	—	—	—	—	M–H	—	—	H	F	—
Tringa totanus	A,M,G	—	—	P,C,F,M,U	M–H	—	—	H	F	—
Larus canus	A,M,G	—	—	P,C	—	—	—	H	—	Estuaries
Chlidonias niger	—	E	—	P,F	M	•	—	M–H	—	Marshes
Pterocles orientalis	—	E	—	—	M–H	—	•	H	F–M	—
Pterocles alchata	—	G,E	—	—	M–H	—	•	H	F–M	—
Columba oenas	A,M,G	—	—	—	M	•	—	L–M	—	Woodland
Columba palumbus	A,M,G	—	—	○	M	•	—	L–M	—	Woodland or scrub

cont.

393

Agricultural and grassland habitats (cont.)

Part 1 (cont.)

Species	Preferred habitats				Landscape features					Specific requirements
	Arable, improved grassland	Steppic habitats	Montane grassland	Wet grassland	Minimum area	Habitat mosaic	Crop mosaic	Openness	Topography	
Streptopelia turtur	A,M,G	—	—	—	M	●	—	L–M	—	Woodland or scrub
Tyto alba	G	—	—	P,C,F	M–H	●	—	M	F–M	—
Otus scops	A,M,G	G	—	—	M	●	—	L	—	Open woodland
Bubo bubo	—	—	—	—	H	—	—	—	F–M	—
Athene noctua	A,M,G	G,E	—	P,F	M	●	—	M–H	F–M	—
Strix aluco	A,M,G	—	—	—	M	—	—	L	F	—
Asio flammeus	A,M,G	G,E	—	C,F,M?,U	M	—	—	M–H	F	—
Merops apiaster	A,M,G	G,E	—	—	M	●	—	M–H	F–M	—
Coracias garrulus	M,G	G,E	—	—	M	●	—	L–M	F–M	Woodland, heath
Picus viridis	G	—	—	—	M–H	●	—	L	F–M	Woodland
Chersophilus duponti	—	G,E	—	—	—	—	—	H	F–M	—
Melanocorypha calandra	A,M,G	G,E	—	—	H	—	●	H	F	—
Melanocorypha leucoptera	—	G	—	—	H	—	—	H	F	—
Melanocorypha yeltoniensis	—	G	—	—	—	—	—	H	F–M	—
Calandrella brachydactyla	—	G,E	—	—	—	—	—	H	F–M	—
Calandrella rufescens	A,M,G	G,E	—	—	—	—	—	H	F–M	—
Galerida cristata	A,M,G	G,E	—	—	M–H	—	—	M	F–S	—
Galerida theklae	—	G,E	—	—	M	—	—	M	F–M	—
Lullula arborea	M,G	G,E	S	—	M	—	—	M	F–M	—
Alauda arvensis	A,M,G	G,E	S	P,C,F,M,U	M	—	●	H	F–M	—
Hirundo rustica	A,M,G	—	—	P,C,F,M,U	M	—	●	M	—	—
Anthus campestris	A,M,G	G,E	R,A	—	M	—	—	H	F–M	—
Anthus pratensis	G	G^w,E^w	A,H	P,C,F,M,U	L	—	—	H	M–F	—
Anthus spinoletta	○	—	—	—	L	—	—	M–H	F–M	>1,000 m
Motacilla flava	○	—	—	P,C,F,U	M	—	—	M–H	F–M	—
Prunella ocularis	—	—	R,H	—	L	—	—	L	M	>2,000 m
Prunella collaris	—	—	R,A	—	L	—	—	L	M–V	>1,800 m
Erithacus rubecula	A,M,G	—	—	—	L	—	—	L	F–M	Adjacent woods or hedges
Saxicola rubetra	G	○	H,S	P,C,F,U	L	—	—	M	F–S	—
Saxicola torquata	A,M,G	G,E	H,S	—	M	—	—	M	—	Dry, south-facing slopes
Oenanthe hispanica	M,G	G,E	—	—	M	—	—	M	F–S	—
Oenanthe deserti	—	G	—	—	—	—	—	M	—	—
Monticola saxatilis	—	—	R,A	—	M	—	—	M	M–S	South-facing slopes, <1,000 m
Monticola solitarius	—	—	R	—	M	—	—	M	S–V	South-facing slopes
Turdus merula	M,G	—	—	—	L	—	—	L	F–M	Adjacent woods or hedges

cont.

Agricultural and grassland habitats (cont.)

Part 1 (cont.)

Species	Arable: improved grassland	Steppic habitats	Montane grassland	Wet grassland	Minimum area	Habitat mosaic	Crop mosaic	Openness	Topography	Specific requirements
Turdus pilaris [w]	A,M,G	—	—	∘	L	—	—	L–M	F–M	—
Turdus philomelos	M,G	—	—	—	L	—	—	L	F–M	Adjacent woods or hedges
Turdus iliacus [w]	M,G	—	—	∘	L	—	—	L–M	F–M	—
Turdus viscivorus	M,G	—	—	—	L	—	—	M	F–M	—
Locustella naevia	A,M,G	—	—	F	M	—	—	M–H	F	—
Acrocephalus paludicola	—	—	—	C,F	L	—	—	L–M	F–M	—
Hippolais pallida	A,M,G	G,E	—	—	M	•	—	M	F–M	—
Sylvia nisoria	M,G	—	—	—	L	—	—	M	F–M	—
Sylvia communis	A,M,G	—	—	—	L	—	—	L	F–M	—
Lanius collurio	A,M,G	—	—	—	L	—	—	M	F–M	—
Lanius minor	A,M,G	G,E	—	—	M	—	—	M–H	F–M	—
Lanius excubitor	A,M,G	—	—	—	L	—	—	M	F–M	—
Lanius senator	A,M,G	G,E	—	—	L	—	—	M	F–M	—
Pyrrhocorax graculus	—	—	R,A	—	H	—	—	H	M–V	>1,500 m
Pyrrhocorax pyrrhocorax	A,M,G	G	R,A,S	M	H	—	—	H	F–V	—
Corvus monedula	A,M,G	∘	∘	∘	M	—	—	M–H	F–M	—
Corvus frugilegus	A,M,G	—	—	∘	M	—	—	M–H	F–M	—
Sturnus vulgaris	A,M,G	E	∘	∘	M	—	—	M–H	F–M	Open woodland
Sturnus unicolor	A,M,G	G	—	—	M	•	—	M	M	—
Passer montanus	A,M,G	—	—	—	L–M	—	—	M–H	F–M	—
Montifringilla nivalis	—	—	R,A	—	M	—	—	M	M–S	>1,000 m
Serinus pusillus	—	—	A,H,S	—	L	—	—	M–H	—	>600 m
Serinus citrinella	—	—	R,A,S	—	L	•	—	M–H	M–S	Open pine woods, >700 m
Carduelis chloris	A,M,G	E	—	∘	L	—	—	M	F–M	—
Carduelis cannabina	A,M,G	—	—	—	L	—	—	M–H	F–M	—
Rhodopechys sanguinea	—	—	R,H	—	?	—	—	M–H	M–S	>1,700 m
Bucanetes githagineus	—	G,E	—	—	M	—	—	H	F–M	—
Carpodacus rubicilla	—	—	R,A	—	L	—	—	H	M–S	>2,500 m
Emberiza citrinella	A,M,G	—	—	—	L	—	—	M–H	F–M	—
Emberiza cirlus	A,M,G	—	—	—	L	—	—	M	S	Dry, sunny slopes
Emberiza cia	—	E	R,A,H	—	L	—	—	L–M	S	Dry, sunny slopes
Emberiza hortulana	A,M,G	G	S	—	L	—	•	M	M	—
Emberiza melanocephala	—	—	—	—	—	—	—	M	M	Warm, dry sites
Miliaria calandra	A,M,G	G,E	—	—	M	—	—	M–H	F	Perches

Agricultural and grassland habitats (cont.)

Part 2

cont.

	Vegetation composition and structure					Feeding requirements		Other habitat requirements	
	Specific vegetation requirements	Height	Plant cover	Vegetation mosaic	Other features	Main food	Specific requirements	Nesting requirements	Survival factors
Bubulcus ibis	Grass steppe	—	—	—	—	IV	Presence of livestock	—	—
Ardea purpurea	—	—	—	—	P	VA,VF,IA	—	—	D
Ciconia ciconia	—	L-M	—	—	P	IS,IV,VA	Presence of livestock	Buildings, pylons, trees	PI
Cygnus columbianus w	—	L-M	—	—	—	PG,PC,PR	—	—	D
Cygnus cygnus w	—	L-M	—	—	—	PG,PC,PR	—	—	D
Anser brachyrhynchus w	—	L	—	—	—	PC,PG	—	—	D
Anser erythropus w	Grass	L	—	—	—	PG,PC	Mix of wet and dry grass	—	D,T
Branta leucopsis w	Grass	L	—	—	—	PG	—	—	D
Branta bernicla w	Cereals, temporary grass	L	—	—	—	PG,PC	—	—	D
Branta ruficollis w	Cereals	—	—	—	—	PC	—	—	D
Tadorna ferruginea	—	—	—	—	—	PG,PC,IV	—	—	—
Anas acuta	—	L	—	—	P	PA,IA	—	Inaccessibility to predators	D,S,H
Anas querquedula	—	L-M	—	—	P	PA,IA	—	—	T
Elanus caeruleus	—	L-M	—	—	T	IV,VB,VM	Elevated perches	—	T
Milvus migrans	—	—	—	—	—	VA,VR,VF,VB, VM,VC,R	—	—	—
Milvus milvus	—	L-M	—	—	T	VM,VC,R	—	—	T
Gypaetus barbatus	—	—	—	—	B	VC	Domestic/wild ungulates	Cliffs	D,PI
Neophron percnopterus	—	—	—	—	—	VC,R	—	Cliffs	PI
Gyps fulvus	—	—	—	—	—	VC	Domestic/wild ungulates	Cliffs	PI
Aegypius monachus	—	—	—	—	—	VC	Domestic/wild ungulates	Trees	PI
Circaetus gallicus	—	—	—	—	T	VR	—	Trees	—
Circus cyaneus	—	M-H	H	•	R,W,S	VB,VM	Rough grass or similar	High vegetation	D
Circus macrourus	Natural grass steppe	L-M	—	•	P	VR,VB,VM	Natural grassland	Wetter areas with shrubs	D
Circus pygargus	—	M	M	•	T	VB,VR,VM,IS	—	High vegetation	D,H
Accipiter brevipes	—	—	—	—	—	VR,VB,IF,IV	—	Tall trees	—
Buteo rufinus	—	L	L	•	—	VB,VM,IS,VR	—	Trees or cliffs	D,T
Aquila pomarina	—	L-M	—	—	—	VA,VB,VM	—	—	D
Aquila clanga	—	L	—	—	—	VA,VB,VM	—	—	D
Aquila rapax	Natural grass steppe	L	L	•	—	VM,VB,VL VC	—	—	D,PI
Aquila heliaca	—	L	L	•	—	VM,VL VB,VC	Susliks *Citellus*	Trees	D,PI
Aquila chrysaetos	—	—	—	—	—	VL VC,VR,VB	—	Cliffs	D,PI
Falco naumanni	—	L	M	•	R	IV,VR	—	Normally buildings	T
Falco tinnunculus	—	L-M	L-M	•	R	IV,VM,VR	—	—	T
Falco vespertinus	—	L	L	•	R,T	IV,IF	—	Old nests	D

Agricultural and grassland habitats (cont.)

Part 2 (cont.)

	Vegetation composition and structure					Feeding requirements		Other habitat requirements	
	Specific vegetation requirements	Height	Plant cover	Vegetation mosaic	Other features	Main food	Specific requirements	Nesting requirements	Survival factors
Falco biarmicus	—	L	L	—	—	VB,VR,VM,IF	—	Cliffs	—
Falco cherrug	—	L	L	—	—	VM	Susliks *Citellus*	Old nests	D
Falco peregrinus	—	—	—	—	—	VB	—	Cliffs	T
Tetrao tetrix	Shrubby grassland	M–H	H	•	—	PW,PF,IV	—	—	—
Tetraogallus mlokosiewiczi	—	M–H	H	•	—	PF,IV	Snow patches	—	D,Pr
Tetraogallus caucasicus	—	L	L	—	B	PW,PG,PF	Snow patches	—	—
Tetraogallus caspius	—	L–M	L	—	B	PW,PG,PF	—	—	—
Alectoris chukar	—	L–M	L–M	—	—	PW,IV,PG	—	—	D,Pr
Alectoris graeca	—	L–M	L–M	•	B	IV,PC,PW	—	—	Pr
Alectoris rufa	—	L–M	L–M	—	R	IV,PC,PW	—	—	Pr
Perdix perdix	—	—	—	—	R,H	IV,PC,PW	—	Grassy banks	—
Coturnix coturnix	—	M–H	H	•	—	IV,PW	—	—	H,Pr?
Porzana porzana	—	M–H	M–H	—	W,P	IV	—	—	—
Crex crex	Low-intensity hay meadows	M–H	M	—	R	IS,IV	—	Tall herbaceous vegetation	H
Grus grus [w]	—	—	—	—	—	PC	—	—	—
Anthropoides virgo	—	L	L	•	—	IV,PC,PW	—	—	D
Tetrax tetrax	—	M–H	L	•	R	IV,PC,PW	—	—	D,T,S
Chlamydotis undulata	—	M–H	L	•	—	IV,PC,PW	—	—	D
Otis tarda	—	M–H	L	•	R	IV,PC,PW	—	—	D,Pl, H,S,(T,Pr?)
Burhinus oedicnemus	—	L	L	—	—	IV	Presence of livestock	Bare soil	T,S,H
Cursorius cursor	Shrubby semi-desert	L	L	—	—	IV	—	Bare soil	—
Glareola pratincola	Complex of bare soil and pools	L	L	—	—	IF	—	Bare soil	D,T,Pr S,H
Glareola nordmanni	Complex of bare soil and pools	L	L	—	—	IF	—	Bare soil	D,T,Pr
Charadrius leschenaultii	Halophytic	L	L	—	—	IV	—	Bare soil	—
Charadrius asiaticus	Halophytic	L	L	—	—	IV	—	Bare soil	—
Pluvialis apricaria [w]	—	L	L	—	—	IS,IV	—	—	—
Chettusia gregaria	Natural grass steppe	L	L	•	—	IV	—	Bare saline soil	D,T,Pr
Vanellus vanellus	—	L	M	—	—	IS,IV	—	Rough, patchy grass	H,S
Calidris alpina	—	—	—	—	—		—	—	—
Philomachus pugnax	—	L–M	L–M	•	P	IV,IS	High water-table	—	H,S
Gallinago media	—	M	L–M	•	S	IS	High water-table, soft soil	—	H

cont.

397

Agricultural and grassland habitats (cont.)

Part 2 (cont.)

Species	Vegetation composition and structure — Specific vegetation requirements	Height	Plant cover	Vegetation mosaic	Other features	Feeding requirements — Main food	Specific requirements	Other habitat requirements — Nesting requirements	Survival factors
Scolopax rusticola	—	L	—	—	—	IS	Earthworms	—	—
Limosa limosa	—	L–M	L–M	—	P	IS,IV	High water-table, soft soil, nearby open water	—	H,S
Numenius tenuirostris	—	M	—	—	—	—	Soft soil	—	D,S,H
Numenius arquata	—	L–M	—	•	P	IS,IV	High water-table, soft soil, surface water	Rough, patchy grass	H,S
Tringa totanus	—	L–M	—	—	—	IS,IV,IA	High water-table	—	—
Larus canus [w]	—	L	—	—	P	IS,IV	—	—	—
Chlidonias niger	—	L–M	L	—	P,W	IA,IF,VF	Good water quality	Floating vegetation	T,D
Pterocles orientalis	—	L–M	L	•	—	PC,PW	—	—	D,Pr
Pterocles alchata	—	L–M	L	•	—	PC,PW	—	—	D,Pr
Columba oenas	—	L–M	L	—	T	PC,PW	—	Large trees with holes	—
Columba palumbus	—	L–M	L	—	T,S	PC,PW	—	—	—
Streptopelia turtur	—	M	L	—	T,S	PC,PW	—	—	T
Tyto alba	Rough grass	M	M–H?	—	R,T	VM	—	Old buildings if no suitable trees	—
Otus scops	—	L–M	L–M	—	T	IV	—	Large trees with holes	T
Bubo bubo	—	—	—	—	B	VM,VL,VB	—	Cliffs	D,Pl
Athene noctua	—	L–M	—	—	H,T,B	IV,IS,VM	Elevated perches	Old/pollarded trees, old buildings, rock crevices	T
Strix aluco	Rough grass	L–M	M	—	T,R,H	IV,VM,VB	—	Large trees with holes	T
Asio flammeus	Rough grass	M–H	M	•	R,S	VM	—	—	H
Merops apiaster	—	—	—	—	—	IF	Elevated perches	Sandy cliffs	—
Coracias garrulus	Rough grass	L	L	—	T,H	IV	Elevated perches	Holes in old trees or sandy cliffs	—
Picus viridis	Short infertile grass	L	—	—	T	IV	Ants	Large trees	—
Chersophilus duponti	Shrub steppe	L–M	—	—	—	IS,IV	—	—	—
Melanocorypha calandra	—	L–M	—	•	—	IV,PW,PC	—	—	—
Melanocorypha leucoptera	Primary grass steppe	L	L	•	—	IV,PW	—	—	S
Melanocorypha yeltoniensis	Primary grass steppe	L	L	•	—	IV,PW	—	—	S
Calandrella brachydactyla	—	L–M	L–M	—	—	IV,PW	—	Bare soil with sparse grassy vegetation	S
Calandrella rufescens	Semi-arid stony steppe	L–M	L	—	—	IV,PW	—	—	—
Galerida cristata	—	L	L	•	—	IV,PW	—	—	S
Galerida theklae	Shrubby	L–M	L	—	B	IV,PW	—	—	—

cont.

cont.

Agricultural and grassland habitats (cont.)

Part 2 (cont.)

Species	Vegetation composition and structure					Feeding requirements		Other habitat requirements	
	Specific vegetation requirements	Height	Plant cover	Vegetation mosaic	Other features	Main food	Specific requirements	Nesting requirements	Survival factors
Lullula arborea	—	L–H	L	•	T,S	IV,PW	—	Tall vegetation	—
Alauda arvensis	—	L	L–M	•	P	PC,PW,IV	—	—	S,H
Hirundo rustica	—	—	L	—	—	IF	Livestock	Buildings, mud	—
Anthus campestris	—	—	L	•	B	IV	—	—	S
Anthus pratensis	Rough grass	L–M	—	•	R	IV	—	—	—
Anthus spinoletta	—	L–M	H	•	B	IV	Sandy bogs, pools or snow patches	—	—
Motacilla flava	—	M–H	M–H	•	P	IV,IF	High water-table, livestock	—	H,S
Prunella ocularis	Grass and sparse xerophytic scrub	L–H	M	•	B	IV,PW	—	—	—
Prunella collaris	—	L–M	L–M	•	B	IV,PW	—	—	—
Erithacus rubecula	—	M–H	M	•	H,S,T	IV,IS	—	—	—
Saxicola rubetra	Rank grassland	M–H	M–H	•	R	IV,IF	—	—	H
Saxicola torquata	Sparse scrub	L–M	M	•	S	IV	—	Dense scrub	—
Oenanthe hispanica	Shrubby	L–M	L–M	•	B	IV	—	—	—
Oenanthe deserti	—	—	—	—	—	—	—	—	—
Monticola saxatilis	—	L–M	L–M	•	B,S,T	IV	—	Crags, boulders	—
Monticola solitarius	—	L	L	•	B	IV	—	Crags, boulders	—
Turdus merula	—	L–M	—	•	H,S	IS,PF	—	—	—
Turdus pilaris [w]	—	L	—	—	H,S,T	IS,IV,PF	—	—	—
Turdus philomelos	—	L–M	—	•	H,S,T,R	IS,IV,PF	—	—	—
Turdus iliacus [w]	Short grass	L	—	—	H,S,T	IS,IV,PF	—	—	—
Turdus viscivorus	—	L	—	—	H,S,T	IS,PF,IV	—	—	—
Locustella naevia	—	H	H	—	H,R	IV	—	—	—
Acrocephalus paludicola	Sedge meadows	M	M–H	—	H,R,S	IV	—	—	—
Hippolais pallida	Scrub and shrubs	H	M–H	•	H,R,S	IV	—	Bushes	—
Sylvia nisoria	Hedges or bushes with tall grass	H	H	•	T,H,R,S	IV,PF	—	Bushes	—
Sylvia communis	—	M–H	M–H	•	H,S,R	IV	—	Bushes	—
Lanius collurio	—	L–M	M–H	•	H,R,S	IV,VR,VM,VB	Many perches	Bushes	—
Lanius minor	Bare ground and/or very low herb layer	L–M	M	•	H,R,T,S	IV	Dry, sunny sites, relatively few perches	Tall trees	—
Lanius excubitor	—	L–M	L–M	•	T,S,H	IV,VM,VR,VB	Perches	Trees, bushes	—
Lanius senator	—	L–M	L	•	T,S	IV	Many perches	Trees, bushes	—
Pyrrhocorax graculus	Semi-natural pasture	L	L–M	—	B	IV,IS	—	Cliffs	—

Agricultural and grassland habitats (cont.)

Part 2 (cont.)

	Vegetation composition and structure					Feeding requirements		Other habitat requirements	
	Specific vegetation requirements	Height	Plant cover	Vegetation mosaic	Other features	Main food	Specific requirements	Nesting requirements	Survival factors
Pyrrhocorax pyrrhocorax	Semi-natural pasture	L	L–M		—	IV,IS	—	Cliffs	—
Corvus monedula		L	—		T	IV,IS	—	—	—
Corvus frugilegus		L	—		T	IS,IV,PC	—	Tall trees	—
Sturnus vulgaris		L	—		T,H	IS,IV,PF,PC	—	Buildings, old trees with holes	—
Sturnus unicolor		L	—		T	IS,IV	Insects with livestock	Trees, buildings	—
Sturnus roseus		L	L–M		T,S	IV,PF	—	Rock piles, crevices	—
Passer montanus		—	—		T	IV,PW,PC	—	—	—
Montifringilla nivalis		L	—		B	IV,PW	—	Rock crevices	—
Serinus pusillus	Meadows with scattered dwarf trees, shrubs	—	L	•	T,S,B	PW,IV	—	—	—
Serinus citrinella		L	L–M		T	PW	—	—	—
Carduelis chloris		L–H	M	•	T,H,R	PW,PC,IV	Tall, densely leaved trees	—	—
Carduelis cannabina		L–H	M	•	H,R,S	PW	—	—	—
Rhodopechys sanguinea	Sparse scrub	M–H	L	•	—	PW	—	—	—
Bucanetes githagineus	Shrubby, semi-desert	L–M	L		—	PW	—	—	—
Carpodacus rubicilla	Alpine meadows	L–M	M	•	B	PW,IV,PF	—	—	—
Emberiza citrinella		L	—	•	T,H,S	IV,PW,PC	—	—	—
Emberiza cirlus		L–H	L–H	•	T,H,S	IV,PW	—	—	—
Emberiza cia	Rocky slopes with scattered shrubs, trees	L–H	L		S,T,B	IV,PW	—	—	—
Emberiza hortulana		L–H	L–M	•	T,B	IV,PW	—	—	—
Emberiza melanocephala		—	—		—	—	—	—	—
Miliaria calandra		M–H	—		—	PW,IV	—	—	—

EACH table shows the predicted impact of various threats on the priority birds of a particular habitat. The magnitude of each impact has been scored according to the following scale.

3 **Critical** impact: the species is likely to go extinct in the habitat in Europe within 20 years if current trends continue

2 **High** impact: the species' population is likely to decline by more than 20% in the habitat in Europe within 20 years if current trends continue

1 **Low** impact: likely to have only local effects, and the species' population is not likely to decline by more than 20% in the habitat in Europe within 20 years if current trends continue

The list of threats for each habitat was compiled by the relevant Habitat Working Group. The predicted impacts are mainly the combined expert opinions of members of the relevant Habitat Working Group, and are only rarely based on comprehensive, detailed, quantitative research. Therefore, the predicted impacts on individual species should be treated with caution (especially for poorly monitored or poorly studied species). However, it is considered that the combined assessment across all species is suffi-ciently reliable to give a broad indication of the likely total impact of the threats on priority birds and, by implication, on the habitats themselves. Two points are particularly noteworthy.

- These assessments are at a European scale, and the magnitude of impacts may vary considerably when assessed at smaller scales (regional, local or site populations).

- Where a threat is not listed as having an impact on an individual bird species, this cannot be interpreted as proof that there is no effect or impact, but merely that no harmful effects or impacts have been found by research so far.

In order to assess the relative importance of the various threats within each habitat, a 'weighted threat score' has been derived for each threat in each habitat. This represents the sum of all impact scores for a particular threat within the habitat, each score being weighted (multiplied) according to the priority status of the particular species affected, as follows.

Priority A spp.: multiply relevant impact scores by 4

Priority B: impact × 3
Priority C: × 2
Priority D: × 1

Marine habitats

■ European Macaronesian seas

	Priority score	1. Introduced predators	2. Disturbance	3. Coastal development	4. Increased predators	5. Rubbish	6. Oil pollution	7. Overfishing	8. Vegetation change	9. Reduced discards	10. Fishery bycatch	11. Toxic pollution	12. Increased predatory fish	13. Dredging	14. Bottom-fishing	15. Nutrient pollution	16. Increasing traffic	17. Coastal erosion	18. Overfishing of shellfish
Pterodroma feae	4	–	2	–	2	1	–	–	2	–	–	–	–	–	–	–	–	–	–
Pterodroma madeira	4	3	2	2	–	1	–	–	2	–	–	–	–	–	–	–	–	–	–
Bulweria bulwerii	3	1	1	1	2	1	–	–	–	–	–	–	–	–	–	–	–	–	–
C. diomedea	4	1	2	1	–	1	–	–	1	–	1	–	–	–	–	–	1	–	–
Puffinus gravis	1	–	–	–	–	1	–	–	–	–	–	–	–	–	–	–	–	–	–
Puffinus puffinus	2	2	1	2	–	1	–	–	1	–	–	–	–	–	–	–	–	–	–
Puffinus assimilis	3	1	–	1	1	1	–	–	1	–	–	–	–	–	–	–	–	–	–
Pelagodroma marina	3	–	1	–	1	1	–	–	1	–	–	–	–	–	–	–	–	–	–
H. pelagicus	2	1	–	1	1	–	–	–	–	–	–	–	–	–	–	–	–	–	–
O. castro	3	1	–	1	1	1	–	–	1	–	–	–	–	–	–	–	–	–	–
Sula bassana	1	–	–	–	–	–	–	–	–	–	–	–	–	–	–	–	–	–	–
Sterna dougallii	2	1	2	–	–	–	–	1	1	–	–	–	–	–	–	–	–	–	–
Sterna sandvicensis	2	–	–	–	–	–	–	–	–	–	–	–	–	–	–	–	–	–	–
Fratercula arctica	2	–	–	–	–	–	–	–	–	–	–	–	–	–	–	–	–	–	–
No. of species																			
Critical impact	1	0	0	0	0	0	0	0	0	0	0	0	0	0	0	0	0	0	0
High impact	1	4	2	2	0	0	0	2	0	0	0	0	0	0	0	0	0	0	0
Low impact	6	3	5	4	9	0	1	6	0	1	0	0	0	0	0	1	0	0	
Total	8	7	7	6	9	0	1	8	0	1	0	0	0	0	0	1	0	0	
% of priority spp. threatened	57%	50%	50%	43%	64%	0%	7%	57%	0%	7%	0%	0%	0%	0%	0%	7%	0%	0%	
Weighted threat score	33	36	27	25	27	0	2	33	0	4	0	0	0	0	0	4	0	0	

Marine habitats (cont.)

■ North and west European seas

	Priority score	1. Introduced predators	2. Disturbance	3. Coastal development	4. Increased predators	5. Rubbish	6. Oil pollution	7. Overfishing	8. Vegetation change	9. Reduced discards	10. Fishery bycatch	11. Toxic pollution	12. Increased predatory fish	13. Dredging	14. Bottom-fishing	15. Nutrient pollution	16. Increasing traffic	17. Coastal erosion	18. Overfishing of shellfish
Gavia stellata	3	–	–	1	–	–	1	1	–	–	1	1	1	1	1	1	1	–	–
Gavia arctica	2	–	–	1	–	–	1	1	–	–	1	1	1	1	1	1	1	–	–
Gavia immer	1	–	–	1	–	–	1	–	–	–	1	1	–	1	1	1	1	–	–
Gavia adamsii	1	–	–	–	–	–	–	–	–	–	1	–	–	–	–	–	–	–	–
Fulmarus glacialis	1	1	–	–	–	1	1	–	–	1	–	–	–	–	–	–	–	–	–
C. diomedea	2	1	–	–	–	–	–	–	–	–	–	–	–	–	–	–	–	–	–
Puffinus gravis	1	–	–	–	–	–	–	–	–	–	–	–	–	–	–	–	–	–	–
Puffinus griseus	1	–	–	–	–	–	–	–	–	–	–	–	–	–	–	–	–	–	–
Puffinus puffinus	4	2	–	–	–	1	1	–	–	–	–	–	–	1	–	–	–	–	–
Puffinus yelkouan	1	–	–	–	–	–	–	–	–	–	–	–	–	–	–	–	–	–	–
H. pelagicus	4	2	–	–	1	1	–	–	–	1	–	–	–	–	–	–	–	–	–
O. leucorhoa	3	1	1	–	1	–	–	–	–	–	–	–	–	–	–	–	–	–	–
O. castro	1	–	–	–	–	–	–	–	–	–	–	–	–	–	–	–	–	–	–
Sula bassana	4	–	–	–	1	1	–	–	1	–	–	1	–	1	–	–	–	–	–
P. aristotelis	2	1	1	–	1	1	1	–	1	1	–	1	1	–	1	–	–	–	–
Aythya marila	2	–	–	–	–	–	1	–	–	–	–	–	–	–	–	–	–	–	–
S. mollissima	1	1	–	1	–	–	1	–	1	–	–	1	–	1	1	1	1	–	1
S. spectabilis	1	–	–	–	–	–	1	–	–	–	–	–	–	1	1	–	1	–	–
Polysticta stelleri	4	–	–	1	–	1	2	–	–	–	–	2	–	1	1	–	1	–	–
H. histrionicus	3	–	–	–	–	–	–	–	–	–	–	–	–	–	–	–	–	–	–
Clangula hyemalis	1	–	–	1	–	–	2	–	–	–	–	1	–	2	1	–	1	–	1
Melanitta nigra	1	–	–	1	–	–	2	–	–	–	–	1	–	1	1	–	1	–	1
Melanitta fusca	3	–	–	1	–	–	2	–	–	–	–	1	–	2	1	–	1	–	1
Mergus serrator	1	1	–	1	–	–	1	1	–	–	–	1	–	1	1	–	–	–	–
Phalaropus fulicarius	1	–	–	–	–	–	–	–	–	–	–	–	–	–	–	–	–	–	–
S. pomarinus	1	–	–	–	–	–	–	–	–	–	–	–	–	–	–	–	–	–	–
S. parasiticus	1	–	–	–	1	–	–	1	–	1	–	–	1	–	–	–	–	–	–
S. longicaudus	1	–	–	–	–	–	–	–	–	–	–	–	–	–	–	–	–	–	–
Stercorarius skua	2	–	–	–	–	–	1	1	–	1	–	1	1	–	–	–	–	–	–
Larus sabini	1	–	–	–	–	–	–	–	–	–	–	–	–	–	–	–	–	–	–
Larus canus	3	1	–	–	1	–	–	1	–	1	–	–	–	–	–	–	–	–	–
Larus fuscus	2	–	1	–	–	–	1	1	–	2	1	1	1	–	–	–	–	–	–
Larus argentatus	1	–	1	–	–	–	1	1	–	2	1	1	1	–	–	–	–	–	–
Larus glaucoides	1	–	–	–	–	–	–	–	–	–	–	–	1	–	–	–	–	–	–
Larus hyperboreus	1	–	–	–	–	–	–	1	–	1	–	–	1	–	–	–	–	–	–
Larus marinus	2	–	–	–	–	1	1	1	–	1	1	1	1	–	–	–	–	–	–
Rhodostethia rosea	1	–	–	–	–	–	–	–	–	–	–	–	–	–	–	–	–	–	–
Rissa tridactyla	1	–	–	1	–	1	1	2	–	1	1	–	1	–	–	–	–	–	–
Pagophila eburnea	3	–	–	–	–	–	–	–	–	–	–	–	–	–	–	–	–	–	–
Sterna caspia	2	1	1	2	–	1	–	–	1	–	–	–	–	1	–	–	–	–	–
Sterna sandvicensis	3	1	1	1	–	–	1	1	1	–	–	–	–	1	–	–	1	–	–
Sterna dougallii	2	1	1	1	–	1	1	–	1	1	–	–	–	1	–	–	1	–	–
Sterna paradisaea	1	2	1	1	1	–	1	–	2	1	–	–	–	1	–	–	1	–	–
Sterna albifrons	2	1	2	1	1	1	–	–	–	1	–	–	–	1	–	–	1	–	–
Chlidonias niger	1	–	–	–	–	–	–	–	–	–	–	–	–	–	–	–	–	–	–
Uria aalge	1	1	1	–	–	1	1	2	–	–	1	–	1	–	1	–	–	–	–
Uria lomvia	1	–	–	–	–	–	–	2	–	–	1	–	1	–	1	–	–	–	–
Alca torda	2	1	1	–	–	1	1	2	–	–	1	–	1	–	1	–	–	–	–
Cepphus grylle	4	2	1	1	1	1	1	1	–	–	1	–	1	–	1	–	–	–	–
Alle alle	1	–	–	–	–	–	1	–	–	–	–	–	–	–	–	–	–	–	–
Fratercula arctica	4	2	1	–	1	1	1	2	–	–	1	–	1	–	1	–	–	–	–
No. of species																			
Critical threat	0	0	0	0	0	0	0	0	0	0	0	0	0	0	0	0	0	0	0
High threat	5	5	1	1	0	0	4	6	0	2	1	0	2	0	0	0	0	0	0
Low threat	13	12	15	8	17	20	14	6	10	13	11	20	8	15	8	9	0	4	
Total	18	13	16	8	17	24	20	6	12	14	11	20	10	15	8	9	0	4	
% of priority spp. threatened	35%	25%	31%	16%	33%	47%	39%	12%	24%	27%	22%	39%	20%	29%	16%	18%	0%	8%	
Weighted threat score	59	31	33	21	44	58	49	11	27	34	18	42	23	30	15	17	0	6	

■ Mediterranean and Black Seas

	Priority score	1. Introduced predators	2. Disturbance	3. Coastal development	4. Increased predators	5. Rubbish	6. Oil pollution	7. Overfishing	8. Vegetation change	9. Reduced discards	10. Fishery bycatch	11. Toxic pollution	12. Increased predatory fish	13. Dredging	14. Bottom-fishing	15. Nutrient pollution	16. Increasing traffic	17. Coastal erosion	18. Overfishing of shellfish
Gavia stellata	1	–	–	–	–	–	1	–	–	–	1	1	1	–	1	1	1	–	–
Gavia arctica	1	–	–	–	–	–	1	–	–	–	1	1	1	–	1	1	1	–	–
C. diomedea	3	2	1	1	2	–	1	1	1	1	1	1	1	–	–	–	–	–	–
Puffinus yelkouan	2	2	–	1	2	–	1	1	1	1	1	1	1	–	–	–	–	–	–

cont.

Marine habitats (cont.)

■ Mediterranean and Black Seas (cont.)

	Priority score	1. Introduced predators	2. Disturbance	3. Coastal development	4. Increased predators	5. Rubbish	6. Oil pollution	7. Overfishing	8. Vegetation change	9. Reduced discards	10. Fishery bycatch	11. Toxic pollution	12. Increased predatory fish	13. Dredging	14. Bottom-fishing	15. Nutrient pollution	16. Increasing traffic	17. Coastal erosion	18. Overfishing of shellfish
H. pelagicus	2	2	1	1	2	–	1	–	–	1	–	1	–	–	–	–	–	–	–
Larus ichthyaetus	1	2	–	2	–	–	–	1	–	–	1	–	–	–	–	–	–	–	–
L. melanocephalus	2	1	1	–	1	–	–	1	1	2	–	1	–	1	–	–	–	1	–
Larus genei	1	1	1	–	2	–	–	1	–	1	–	–	–	1	–	–	–	1	–
Larus audouinii	4	2	1	2	2	1	–	1	–	2	–	1	–	1	–	–	–	–	–
Larus cachinnans	1	1	1	–	–	–	–	–	1	–	1	–	–	–	–	–	–	–	–
Sterna caspia	2	1	2	1	–	–	–	–	–	–	1	–	1	–	–	–	–	–	–
S. sandvicensis	3	2	2	–	1	–	–	1	2	–	–	1	–	1	–	1	–	2	–
Sterna albifrons	2	2	2	1	2	–	–	–	–	–	–	1	–	–	–	1	–	2	–
No. of species																			
Critical impact		0	0	0	0	0	0	0	0	0	0	0	0	0	0	0	0	0	0
High impact		7	3	2	6	0	0	0	1	2	0	0	0	0	0	0	0	2	0
Low impact		4	6	5	2	1	5	6	4	5	4	12	0	6	2	4	0	2	0
Total		11	9	7	8	1	5	6	5	7	4	12	0	6	2	4	0	4	0
% of priority spp. threatened		85%	69%	54%	62%	8%	38%	46%	38%	54%	31%	92%	0%	46%	15%	31%	0%	31%	0%
Weighted threat score		40	27	21	33	4	9	15	14	21	7	24	0	10	2	7	0	13	0

Coastal habitats

	Priority score	1. Tourism/recreation	2. Land-claim	3. Nutrient pollution	4. Hunting disturbance	5. Toxic pollution	6. Plant succession	7. Coastal defence	8. Increased predators	9. Sea-level rise	10. Fisheries	11. Inappropriate grazing	12. Military disturbance	13. Barrage construction	14. Afforestation	15. Salina intensification	16. Dredging and extraction	17. Aquaculture
Phalacrocorax pygmeus	3	1	1	1	1	1	1	–	–	–	1	–	–	–	–	–	–	–
Pelecanus onocrotalus	3	1	1	1	1	1	–	–	–	–	1	–	–	–	–	–	–	–
Pelecanus crispus	4	1	–	1	1	1	–	–	–	–	1	–	–	–	–	–	–	1
Botaurus stellaris	1	1	2	2	1	1	1	–	–	–	–	–	–	–	–	–	–	–
Ixobrychus minutus	2	1	2	1	–	1	–	–	–	–	–	–	–	–	–	–	–	–
Nycticorax nycticorax	1	1	1	1	–	1	–	–	–	–	–	–	–	–	–	–	–	–
Ardeola ralloides	2	1	2	1	–	1	–	–	–	–	–	–	–	–	–	–	–	–
Ardea purpurea	2	1	2	1	1	1	1	–	–	–	–	–	–	–	–	–	–	–
Plegadis falcinellus	2	1	1	1	1	1	1	–	–	–	–	–	1	–	–	–	–	–
Platalea leucorodia	4	1	1	2	1	1	1	1	1	1	1	–	–	–	–	–	–	–
Phoenicopterus ruber	3	1	1	1	1	–	–	–	–	1	–	–	–	–	–	1	–	1
Cygnus columbianus	2	–	–	2	1	–	–	–	–	–	–	–	–	–	–	–	–	–
Cygnus cygnus	1	–	–	–	1	–	–	–	–	–	–	–	–	–	–	–	–	–
Anser brachyrhynchus	1	–	–	–	1	–	–	–	–	–	–	–	–	–	–	–	–	–
Branta leucopsis	4	–	–	–	1	–	–	–	–	–	–	–	–	–	–	–	–	–
Branta bernicla	3	1	1	–	1	–	1	–	–	1	–	–	1	1	–	–	1	–
Tadorna ferruginea	2	1	–	1	–	–	–	–	–	–	–	–	–	–	–	–	–	–
Tadorna tadorna	1	2	–	1	–	1	–	1	–	1	1	–	1	–	–	1	–	–
Anas penelope	1	1	–	1	–	1	–	–	–	–	–	–	–	–	–	–	1	–
Anas strepera	2	1	–	1	1	–	1	–	–	–	–	–	–	–	–	–	–	–
Anas acuta	2	1	–	1	1	–	1	–	–	–	–	–	–	–	–	–	–	–
M. angustirostris	4	1	2	1	–	–	–	–	–	–	–	–	–	–	–	–	–	–
Netta rufina	2	–	1	2	1	1	1	–	–	–	–	–	–	–	–	–	–	–
Aythya nyroca	4	–	1	1	1	1	1	–	–	–	–	–	–	–	–	–	–	–
Aythya marila	2	–	1	–	1	–	–	–	–	1	–	–	–	–	–	–	1	–
Haliaeetus albicilla	2	1	–	–	–	1	–	–	–	–	–	–	–	–	–	–	–	–
Circus cyaneus	2	2	–	–	–	–	–	–	–	–	–	–	–	–	1	–	–	–
Aquila adalberti	3	1	–	–	–	1	–	–	–	–	–	–	–	–	1	–	–	–
Pandion haliaetus	2	1	–	–	1	–	1	–	–	–	–	–	–	–	–	–	–	–
Falco eleonorae	4	1	–	–	–	1	–	–	1	–	–	–	–	–	–	–	–	–
Falco peregrinus	2	–	–	–	–	1	–	–	–	–	–	–	–	–	–	–	–	–
Porzana pusilla	1	–	1	–	–	–	1	–	–	–	–	–	1	–	–	–	–	–
Porphyrio porphyrio	3	1	1	–	–	–	–	–	–	–	–	–	–	–	–	–	–	–
Haematopus ostralegus	1	1	1	–	1	–	–	1	1	1	1	1	–	–	–	–	–	–
Recurvirostra avosetta	3	2	1	1	1	1	1	2	1	1	1	–	–	–	–	1	–	1
Glareola pratincola	2	1	1	–	–	–	–	–	–	–	–	–	1	–	–	1	–	–
Glareola nordmanni	2	–	–	–	–	–	–	–	–	–	–	–	–	–	–	1	–	–
Charadrius hiaticula	1	1	1	–	–	–	–	1	1	1	1	–	–	–	1	–	–	–
Charadrius alexandrinus	3	2	1	–	–	–	1	1	1	1	–	1	1	–	1	1	–	–
Pluvialis squatarola	1	–	–	–	1	–	–	1	–	1	–	–	–	1	–	–	–	–
Hoplopterus spinosus	3	1	–	–	–	–	–	–	–	–	–	2	–	–	–	–	–	–
Calidris canutus	3	–	1	–	1	–	–	–	1	1	–	1	1	–	–	–	–	–
Calidris alba	1	–	–	–	–	–	–	1	–	–	–	–	–	–	–	–	–	–
Calidris ferruginea	1	–	–	–	–	–	–	1	–	1	–	–	–	–	–	–	–	–

cont.

Coastal habitats (cont.)

	Priority score	1. Tourism/recreation	2. Land-claim	3. Nutrient pollution	4. Hunting disturbance	5. Toxic pollution	6. Plant succession	7. Coastal defence	8. Increased predators	9. Sea-level rise	10. Fisheries	11. Inappropriate grazing	12. Military disturbance	13. Barrage construction	14. Afforestation	15. Salina intensification	16. Dredging and extraction	17. Aquaculture
Calidris maritima	2	–	–	–	–	–	–	1	–	1	–	–	–	–	–	–	–	–
Calidris ferruginea	1	–	–	–	–	–	–	1	–	–	1	–	–	–	–	–	–	–
Calidris maritima	2	–	–	–	–	–	–	1	–	1	–	–	–	–	–	–	–	–
Calidris alpina	3	1	1	1	1	1	1	1	–	1	–	1	1	1	–	–	–	–
Lymnocryptes minimus	2	–	1	–	–	–	–	1	–	–	–	–	–	–	–	–	–	–
Limosa limosa	3	–	1	–	1	–	2	1	1	1	–	1	–	1	–	–	–	–
Limosa lapponica	3	–	1	1	1	–	–	1	–	1	1	–	1	–	–	–	–	–
Numenius phaeopus	2	–	1	–	–	–	–	1	–	–	–	–	–	–	–	–	–	–
Numenius tenuirostris	4	1	1	–	1	–	–	–	–	–	–	–	–	–	–	–	–	–
Numenius arquata	3	–	1	1	1	1	–	1	–	1	1	–	–	–	–	–	–	–
Tringa totanus	4	1	1	1	1	1	2	1	1	1	1	1	1	1	1	–	–	–
Tringa nebularia	1	–	1	–	–	–	–	–	–	–	–	–	–	–	–	–	–	–
Arenaria interpres	1	–	1	–	–	–	–	1	–	1	–	–	–	–	–	–	–	–
Larus ichthyaetus	1	–	–	1	–	1	–	–	–	–	–	–	–	–	–	–	–	–
Larus melanocephalus	2	1	–	1	–	1	–	–	1	–	–	–	–	–	–	1	–	–
Larus audouinii	4	1	–	–	–	–	–	–	2	–	–	–	–	1	–	–	–	–
Larus canus	3	1	1	–	–	1	–	–	1	–	–	–	–	–	1	–	–	–
Larus fuscus	1	–	–	–	–	1	–	–	–	–	–	–	–	–	–	–	–	–
Gelochelidon nilotica	2	1	2	1	–	1	–	–	–	–	–	1	–	–	–	1	–	–
Sterna caspia	3	1	–	1	–	1	–	–	–	–	–	–	–	–	–	–	–	–
Sterna sandvicensis	4	1	–	1	–	–	2	1	1	–	–	–	–	–	1	1	1	–
Sterna dougallii	3	1	–	–	–	–	1	–	–	–	–	–	–	–	–	–	–	–
Sterna paradisaea	1	1	–	–	–	–	–	–	–	1	1	–	–	–	–	–	–	–
Sterna albifrons	3	2	1	1	–	1	1	1	1	1	–	1	–	–	–	1	1	–
Chlidonias hybridus	2	1	1	1	–	1	1	–	–	–	–	–	–	–	–	–	–	–
Chlidonias niger	2	1	1	1	–	–	1	–	–	–	–	–	–	–	–	–	–	–
Asio flammeus	1	1	–	–	–	–	1	–	–	–	–	1	–	–	1	–	–	–
Alcedo atthis	1	–	–	1	–	–	1	–	–	–	–	–	–	–	–	–	–	–
Alauda arvensis	1	–	–	–	–	–	1	–	–	–	–	1	–	–	1	–	–	–
Anthus campestris	1	1	–	–	–	–	1	–	–	–	–	–	–	–	1	–	–	–
Anthus petrosus	1	–	–	–	–	–	1	–	–	–	–	–	–	–	–	–	–	–
Acrocephalus paludicola	3	–	1	1	–	–	1	–	–	–	–	–	–	–	–	–	–	–
Pyrrhocorax pyrrhocorax	1	1	–	–	–	–	1	–	–	–	–	–	–	–	1	–	–	–
No. of species																		
Critical impact	0	0	0	0	0	0	0	0	0	0	0	0	0	0	0	0	0	0
High impact	5	7	4	0	0	4	0	1	0	0	1	0	0	0	0	0	0	0
Low impact	42	32	32	31	32	24	21	14	19	12	11	8	10	11	7	6	3	
Total	47	39	36	31	32	28	21	15	19	12	12	8	10	11	7	6	3	
% of priority spp. threatened		63%	52%	48%	41%	43%	37%	28%	20%	25%	16%	16%	11%	13%	15%	9%	8%	4%
Weighted threat score		126	113	97	82	80	79	49	47	46	32	30	27	25	23	20	14	10

Inland wetlands

	Priority score	1. Drainage/land-claim	2. Riparian habitat-loss	3. Tourism/recreation	4. Vegetation management	5. Nutrient pollution	6. Water abstraction	7. Toxic pollution	8. Water-level regulation	9. Hunting disturbance	10. Wetland impoundment	11. Canalization	12. Increased predators	13. Angling/Fisheries	14. Acidification	15. Excessive sediment	16. Aquaculture	17. Introduced species
Gavia stellata	2	2	–	1	–	1	–	1	1	–	–	–	2	–	–	–	–	–
Gavia arctica	3	2	–	1	–	1	–	1	1	–	–	–	2	–	–	–	–	–
Tachybaptus ruficollis	1	1	1	–	–	1	1	–	1	–	–	1	–	–	–	–	–	–
Podiceps cristatus	1	1	–	1	–	1	–	1	1	–	–	–	1	–	1	1	–	–
Podiceps grisegena	1	1	1	1	–	1	–	1	1	–	–	–	–	1	1	–	–	–
Podiceps auritus	1	1	1	1	–	1	–	1	1	–	–	–	–	1	1	–	–	–
Podiceps nigricollis	1	1	–	1	–	1	–	1	1	–	–	–	–	1	1	–	–	–
Phalacrocorax pygmeus	3	1	1	1	1	1	1	1	–	1	1	–	1	–	–	–	–	–
Pelecanus onocrotalus	2	1	1	1	–	1	1	1	1	–	–	1	1	1	–	–	–	–
Pelecanus crispus	4	2	1	1	–	1	1	1	1	–	–	1	1	1	–	–	–	–
Botaurus stellaris	3	1	1	1	2	2	1	1	1	–	–	–	–	1	–	–	–	–
Ixobrychus minutus	3	1	1	1	–	1	–	1	1	–	–	–	–	–	–	–	–	–
Nycticorax nycticorax	2	1	–	1	–	1	–	1	–	–	1	–	–	–	–	–	–	–
Ardeola ralloides	3	1	1	1	–	1	–	1	1	–	–	–	–	–	–	–	1	–
Egretta garzetta	1	1	–	–	–	1	–	1	–	–	1	–	–	–	–	–	–	–
Egretta alba	1	1	1	–	–	1	–	–	–	–	–	–	–	–	–	–	–	–
Ardea cinerea	1	–	–	–	–	1	–	1	–	–	1	–	1	1	–	–	–	–
Ardea purpurea	3	2	2	1	2	1	–	1	–	–	–	–	–	–	–	–	–	–
Ciconia nigra	2	1	–	1	–	1	–	1	1	–	–	–	–	–	1	1	–	–
Ciconia ciconia	2	1	–	1	–	1	1	1	–	–	–	–	–	–	1	–	–	–
Plegadis falcinellus	2	1	–	1	1	1	1	1	1	–	1	–	1	–	–	1	–	–

cont.

Inland wetlands (cont.)

	Priority score	1. Drainage/land-claim	2. Riparian habitat-loss	3. Tourism/recreation	4. Vegetation management	5. Nutrient pollution	6. Water abstraction	7. Toxic pollution	8. Water-level regulation	9. Hunting disturbance	10. Wetland impoundment	11. Canalization	12. Increased predators	13. Angling/Fisheries	14. Acidification	15. Excessive sediment	16. Aquaculture	17. Introduced species
Platalea leucorodia	2	2	–	1	1	1	1	1	1	–	1	–	1	–	1	1	1	–
Phoenicopterus ruber	2	1	–	1	–	1	1	1	–	1	–	–	–	–	–	–	–	–
Cygnus olor	1	–	1	–	1	1	–	1	–	–	–	1	–	1	–	–	–	–
Cygnus columbianus	1	–	–	–	2	–	–	–	1	–	–	–	–	–	–	–	–	–
Anser erythropus	4	1	–	–	–	–	–	–	1	–	–	–	–	–	–	–	–	–
Branta ruficollis	4	1	–	–	–	–	1	–	2	–	–	–	–	–	–	–	–	–
Tadorna ferruginea	3	1	–	–	1	–	1	–	1	1	–	–	–	–	1	–	–	–
Anas strepera	2	2	1	1	1	1	1	1	–	1	1	–	–	1	–	–	–	–
Anas platyrhynchos	1	1	–	–	–	–	–	–	–	–	–	–	1	1	–	–	–	–
Anas acuta	2	2	1	1	–	1	1	–	1	1	–	–	–	–	–	–	–	–
Anas querquedula	2	2	–	1	1	–	1	–	1	–	–	–	–	–	–	–	–	–
Anas clypeata	1	1	–	–	–	–	–	–	1	–	–	–	1	–	–	–	–	–
M angustirostris	4	2	–	1	2	1	2	–	1	–	–	–	–	–	1	–	–	1
Netta rufina	2	2	–	1	–	–	1	–	1	1	–	–	–	–	–	–	–	–
Aythya ferina	2	1	–	1	–	–	–	–	1	–	–	–	1	–	–	–	–	–
Aythya nyroca	4	2	–	1	–	1	1	–	1	1	1	–	–	–	–	–	–	–
Aythya fuligula	1	1	–	–	–	1	–	–	1	–	1	–	–	–	–	–	–	–
Bucephala clangula	1	–	1	1	–	1	–	–	–	–	–	–	1	–	–	–	–	1
Mergus albellus	3	2	1	–	–	1	–	1	–	–	–	1	–	1	–	–	–	2
Mergus merganser	1	–	1	1	–	1	–	–	–	–	–	–	–	1	–	–	–	–
Oxyura leucocephala	4	2	1	–	2	1	1	1	–	1	2	–	–	–	1	–	–	3
Milvus migrans	2	–	1	1	–	–	–	1	1	–	–	–	1	–	–	–	–	–
Haliaeetus albicilla	2	–	1	1	–	–	–	1	–	1	–	–	–	1	–	1	–	–
Circus aeruginosus	1	1	1	1	1	–	–	1	–	–	–	–	–	–	–	–	–	–
Aquila clanga	3	2	1	2	–	–	–	1	–	–	–	–	–	–	–	–	–	–
Pandion haliaetus	3	1	1	1	–	1	–	–	–	–	–	–	–	1	1	–	–	–
Rallus aquaticus	1	1	1	–	1	–	–	1	–	–	1	1	–	–	–	–	–	–
Porzana porzana	2	1	1	–	–	–	–	1	–	–	1	1	–	–	–	–	–	–
Porzana parva	2	1	1	–	–	–	–	1	–	–	1	1	–	–	–	–	–	–
Porzana pusilla	3	2	1	–	1	–	1	1	–	–	1	1	–	–	–	1	–	–
Crex crex	3	1	–	–	–	–	–	–	–	1	–	1	–	–	–	–	–	–
Gallinula chloropus	1	–	1	–	–	–	–	1	–	–	1	1	1	–	–	–	–	–
Porphyrio porphyrio	2	–	–	–	–	–	–	–	–	–	–	–	–	–	–	–	–	–
Fulica atra	1	1	–	–	1	–	1	1	1	–	1	1	1	–	–	–	–	–
Fulica cristata	2	–	–	–	–	–	–	–	–	–	–	–	–	–	–	–	–	–
Grus grus	2	–	1	–	–	1	–	1	1	–	–	–	–	–	–	–	–	–
Himantopus himantopus	1	1	–	1	1	–	–	1	–	–	–	–	–	–	1	–	–	–
Recurvirostra avosetta	2	1	–	1	–	1	–	1	–	–	–	–	–	–	1	–	–	–
Glareola pratincola	2	1	–	1	–	–	–	1	–	–	–	1	–	–	–	–	–	–
Glareola nordmanni	2	1	–	–	2	–	–	–	–	–	–	–	–	–	–	–	–	–
Charadrius dubius	1	–	–	1	–	–	–	1	–	–	1	1	1	–	–	–	–	–
Charadrius alexandrinus	1	–	–	1	–	–	–	–	–	–	–	–	–	–	–	–	–	–
Charadrius leschenaultii	3	1	–	–	1	–	–	–	–	–	–	–	–	–	–	–	–	–
Hoplopterus spinosus	2	2	–	1	1	–	1	–	1	–	–	–	1	–	–	–	–	–
Philomachus pugnax	1	2	–	–	1	–	–	1	–	–	–	–	–	–	–	–	–	–
Lymnocryptes minimus	2	1	–	–	–	–	–	–	–	–	–	–	–	–	–	–	–	–
Limosa limosa	3	2	–	–	1	–	–	1	–	–	1	1	–	–	–	–	–	–
Numenius tenuirostris	4	1	–	–	–	–	–	–	1	–	–	–	–	–	–	–	–	–
Tringa totanus	2	2	–	–	1	–	–	1	–	–	2	1	–	–	–	–	–	–
Tringa stagnatilis	1	–	–	–	–	–	–	–	–	–	–	–	–	–	–	–	–	–
Tringa glareola	2	–	–	–	–	–	–	–	–	–	–	–	–	–	–	–	–	–
Actitis hypoleucos	1	–	1	1	–	1	–	1	–	1	1	–	1	1	–	–	–	–
Larus minutus	2	–	–	–	–	–	1	–	–	–	–	–	–	–	–	–	–	–
Larus ridibundus	1	–	–	–	–	–	1	–	–	–	–	–	–	–	–	–	–	–
Larus canus	2	–	–	–	–	–	–	–	–	–	–	–	–	–	–	–	–	–
Gelochelidon nilotica	2	1	–	1	–	–	1	1	–	–	–	–	–	–	–	–	–	–
Sterna albifrons	1	–	1	1	–	1	–	1	1	–	1	1	1	1	–	–	–	–
Chlidonias hybridus	3	2	–	1	–	1	1	1	1	–	–	1	–	–	1	–	1	–
Chlidonias niger	3	1	–	1	–	1	1	1	1	–	–	1	–	–	1	–	1	–
Chlidonias leucopterus	1	2	–	1	–	1	1	1	1	–	–	1	–	–	1	–	1	–
Halcyon smyrnensis	1	1	1	–	–	–	–	–	–	–	–	–	–	–	–	–	–	–
Alcedo atthis	3	1	1	–	–	2	1	2	–	–	–	2	–	1	1	1	–	–
Ceryle rudis	1	1	1	–	–	–	–	1	–	–	–	–	–	–	1	–	–	–
Merops apiaster	1	–	1	1	–	–	–	–	–	–	–	1	–	–	–	–	–	–
Riparia riparia	2	–	1	–	–	–	–	–	–	–	–	1	–	–	–	–	–	–
Hirundo rustica	2	–	–	–	–	–	–	–	–	–	–	–	–	–	–	–	–	–
Motacilla cinerea	1	–	1	–	–	–	–	–	–	–	–	1	–	–	–	–	–	–
Cinclus cinclus	1	–	1	–	–	1	1	–	–	–	–	1	–	1	–	–	–	–
Cettia cetti	1	–	–	–	–	–	–	–	–	1	–	–	–	–	–	–	–	–
Cisticola juncidis	1	–	–	–	–	–	–	–	–	1	–	–	–	–	–	–	–	–
Locustella luscinioides	2	–	1	1	–	1	–	–	–	–	–	–	–	–	–	–	–	–
Acrocephalus melanopogon	1	1	1	–	1	–	–	–	–	–	–	–	–	–	–	–	–	–
Acrocephalus paludicola	4	2	1	–	2	–	–	–	–	–	–	–	–	–	–	–	–	–
Acrocephalus schoenobaenus	1	1	1	–	–	–	–	–	–	–	–	1	–	–	–	–	–	–
Acrocephalus agricola	1	–	1	–	–	–	–	–	–	–	–	–	–	–	–	–	–	–
Acrocephalus palustris	1	–	1	–	1	–	–	–	–	–	–	1	–	–	–	–	–	–
Acrocephalus scirpaceus	2	1	1	–	1	–	–	–	–	–	–	–	–	–	–	–	–	–

cont.

Inland wetlands (cont.)

	Priority score	1. Drainage/land-claim	2. Riparian habitat-loss	3. Tourism/recreation	4. Vegetation management	5. Nutrient pollution	6. Water abstraction	7. Toxic pollution	8. Water-level regulation	9. Hunting disturbance	10. Wetland impoundment	11. Canalization	12. Increased predators	13. Angling/Fisheries	14. Acidification	15. Excessive sediment	16. Aquaculture	17. Introduced species
Acrocephalus arundinaceus	1	1	1	–	1	–	–	–	–	–	–	–	–	–	–	–	–	–
Panurus biarmicus	1	1	1	–	–	–	–	–	–	–	–	–	–	–	–	–	–	–
Remiz pendulinus	1	–	1	–	–	–	–	–	–	–	–	–	–	–	–	–	–	–
Emberiza schoeniclus	1	1	1	–	1	–	–	–	–	–	–	–	–	–	–	–	–	–
No. of species																		
Critical impact	0	0	0	0	0	0	0	0	0	0	0	0	0	0	0	0	0	1
High impact	22	0	1	1	6	3	1	1	0	1	1	2	0	0	2	0	0	1
Low impact	51	50	43	25	34	28	32	37	20	19	24	24	17	17	12	10		2
Total	73	51	44	31	37	29	33	37	21	20	26	24	17	19	12	10		4
% of priority spp. threatened	72%	50%	43%	30%	36%	28%	32%	36%	21%	20%	25%	24%	17%	19%	12%	10%		4%
Weighted threat score	213	94	90	86	79	77	74	69	59	47	49	43	31	40	26	22		23

Tundra, mires and moorland

	Priority score	1. Oil/gas exploitation	2. Afforestation	3. Overgrazing	4. Peat extraction	5. Industrial development	6. Settlement	7. Agricultural drainage	8. Tourism	9. Transport infrastructure	10. Agricultural improvement	11. Energy transport	12. Pollution	13. Mining	14. Hydropower
Gavia stellata	3	1	–	–	1	–	–	1	–	–	–	–	–	–	–
Gavia arctica	3	1	–	–	1	–	–	1	–	–	–	–	1	–	1
Gavia immer	1	–	–	–	–	–	–	–	–	–	–	–	–	–	–
Gavia adamsii	1	–	–	–	–	–	–	–	–	–	–	–	–	–	–
Cygnus columbianus	1	1	–	–	–	1	–	–	–	–	–	–	–	–	–
Anser fabalis	1	1	1	–	–	1	–	–	–	1	–	–	–	–	1
Anser brachyrhynchus	1	–	–	–	–	–	–	–	–	–	–	–	–	–	–
Anser albifrons	1	1	–	–	–	1	–	–	–	–	–	–	–	–	–
Anser erythropus	4	1	–	–	–	1	–	–	1	–	–	–	–	–	–
Branta leucopsis	2	1	–	–	–	–	–	–	–	–	–	–	–	–	–
Branta bernicla	3	1	–	–	–	–	–	–	–	–	–	–	–	–	–
Anas acuta	2	–	1	–	–	–	–	–	–	–	–	–	–	–	–
Aythya marila	1	1	–	–	–	–	–	–	–	–	–	–	–	–	–
Clangula hyemalis	1	1	–	–	–	1	–	–	–	–	–	–	–	–	–
Melanitta fusca	1	1	–	–	–	–	–	–	–	–	–	–	–	–	–
Melanitta nigra	1	1	–	–	–	–	–	–	–	–	–	–	–	–	–
Bucephala islandica	3	–	–	–	–	–	–	–	–	–	–	–	–	–	–
Haliaeetus albicilla	1	–	1	–	–	1	1	–	–	–	–	–	–	–	–
Circus cyaneus	2	–	–	–	–	–	1	–	–	–	–	–	–	–	–
Buteo lagopus	1	–	–	–	–	–	–	–	–	–	–	–	–	–	–
Aquila chrysaetos	1	–	1	–	1	1	1	1	–	–	–	–	–	–	–
Falco columbarius	1	–	1	–	1	–	–	–	–	–	–	–	–	–	–
Falco rusticolus	3	–	–	–	–	1	1	–	–	–	–	–	–	–	–
Falco peregrinus	2	–	1	–	1	1	1	–	–	1	–	–	–	–	–
Lagopus lagopus	1	–	1	1	1	1	1	–	–	1	–	1	–	1	–
Lagopus mutus	1	–	–	1	–	–	–	–	–	–	–	–	–	–	–
Tetrao tetrix	1	–	1	–	1	–	–	1	–	–	–	–	–	–	–
Grus grus	2	–	1	–	1	–	1	1	–	–	–	–	–	–	1
Charadrius morinellus	1	–	–	1	–	–	–	–	–	–	–	–	–	–	–
Pluvialis apricaria	2	–	1	1	1	–	–	1	–	–	–	–	–	–	–
Pluvialis squatarola	1	–	–	–	–	–	–	–	–	–	–	–	–	–	–
Calidris canutus	3	–	–	–	–	–	–	–	–	–	–	–	–	–	–
Calidris alba	1	–	–	–	–	–	–	–	–	–	–	–	–	–	–
Calidris minuta	1	1	–	1	–	–	–	–	–	–	–	–	–	–	–
Calidris temminckii	1	–	–	–	–	–	–	–	–	–	–	–	–	–	–
Calidris maritima	2	1	–	–	–	–	–	–	–	–	–	–	–	–	–
Calidris alpina	3	1	–	1	–	–	–	–	–	–	–	–	–	–	–
Limicola falcinellus	3	–	–	–	–	–	–	–	–	–	–	–	–	–	–
Philomachus pugnax	1	1	–	1	–	–	–	–	–	–	–	1	–	–	–
Lymnocryptes minimus	3	1	–	–	–	–	–	–	–	–	–	–	–	–	–
Gallinago media	4	–	–	1	–	–	–	–	–	–	–	–	–	–	–
Limosa limosa	3	–	1	–	1	–	–	1	–	–	–	–	–	–	–
Limosa lapponica	1	–	–	–	–	–	–	–	–	–	–	–	–	–	–
Numenius phaeopus	2	–	1	–	1	–	–	1	–	–	–	–	–	–	–
Numenius arquata	2	–	1	1	1	–	–	1	–	–	–	–	–	–	–
Tringa erythropus	1	1	–	–	–	–	–	–	–	–	–	–	–	–	–
Tringa totanus	2	–	1	–	–	–	–	1	–	–	1	–	–	–	–
Tringa glareola	3	–	1	1	1	–	–	1	–	–	–	–	–	–	–
Arenaria interpres	1	–	–	–	–	–	–	–	–	–	–	–	–	–	–
Phalaropus lobatus	1	–	–	–	–	–	–	–	–	–	–	–	–	–	–
Phalaropus fulicarius	1	–	–	–	–	–	–	–	–	–	–	–	–	–	–

cont.

Tundra, mires and moorland (cont.)

	Priority score	1. Oil/gas exploitation	2. Afforestation	3. Overgrazing	4. Peat extraction	5. Industrial development	6. Settlement	7. Agricultural drainage	8. Tourism	9. Transport infrastructure	10. Agricultural improvement	11. Energy transport	12. Pollution	13. Mining	14. Hydropower
Stercorarius pomarinus	1	–	–	–	–	–	–	–	–	–	–	–	–	–	–
Stercorarius parasiticus	1	–	–	–	–	–	–	–	–	1	–	–	–	–	–
Stercorarius longicaudus	1	–	–	1	–	–	–	–	–	–	–	–	–	–	–
Stercorarius skua	2	–	–	–	–	–	–	–	–	1	–	–	–	–	–
Larus sabini	1	–	–	–	–	–	–	–	–	–	–	–	–	–	–
Larus canus	2	–	–	–	–	–	–	–	–	–	–	–	–	–	–
Rhodostethia rosea	1	–	–	–	–	–	–	–	–	–	–	–	–	–	–
Nyctea scandiaca	3	–	–	–	–	1	–	–	–	–	–	–	–	–	–
Asio flammeus	1	–	–	–	–	–	–	–	–	–	–	–	–	–	–
Alauda arvensis	1	–	–	–	–	–	–	–	–	–	–	–	–	–	–
Eremophila alpestris	1	–	–	–	–	–	–	–	–	–	–	–	–	–	–
Anthus pratensis	1	–	–	–	–	–	–	–	–	–	–	–	–	–	–
Anthus cervinus	1	–	–	1	–	–	–	–	–	–	–	–	–	–	–
Anthus petrosus	1	–	–	–	–	–	–	–	–	–	–	–	–	–	–
Luscinia svecica	1	–	–	–	–	–	–	–	–	–	–	–	–	–	–
Saxicola torquata	1	–	–	–	–	–	–	–	–	–	–	–	–	–	–
Turdus torquatus	1	–	1	1	–	–	–	–	–	–	–	–	–	–	–
Lanius excubitor	1	–	1	–	1	–	–	1	–	–	–	–	–	–	–
Carduelis flavirostris	1	–	–	–	–	–	–	–	–	–	–	–	–	–	–
Calcarius lapponicus	1	–	–	–	–	–	–	–	–	–	–	–	–	–	–
Plectrophenax nivalis	1	–	–	–	–	–	–	–	–	–	–	–	–	–	–
Emberiza pusilla	1	–	–	–	–	–	–	–	–	–	–	–	–	–	–
No. of species															
Critical impact		0	0	0	0	0	0	0	0	0	0	0	0	0	0
High impact		0	0	0	0	0	0	0	0	0	0	0	0	0	0
Low impact		18	17	4	4	9	9	10	3	3	3	2	1	1	3
Total		18	17	14	14	9	9	10	3	3	3	2	1	1	3
% of priority spp. threatened		25%	23%	19%	19%	12%	12%	14%	4%	4%	4%	3%	1%	1%	4%
Weighted threat score		33	28	23	27	15	16	19	10	4	5	2	3	1	6

Lowland Atlantic heathland

	Priority score	1. Habitat fragmentation	2. Abandonment	3. Afforestation	4. Fires	5. Urbanization	6. Agriculture	7. Nutrient pollution	8. Transport infrastructure	9. Recreation	10. Military activity	11. Drainage	12. Mineral extraction
Circus cyaneus	1	1	–	1	–	1	1	–	1	–	1	–	–
Tetrao tetrix	1	3	–	1	1	1	1	1	1	1	1	1	1
Alectoris rufa	2	1	1	1	–	1	–	–	–	1	1	1	1
Perdix perdix	1	1	1	1	–	1	–	–	–	–	–	–	–
Grus grus	1	1	–	–	–	–	–	–	–	–	–	1	–
Burhinus oedicnemus	1	1	2	1	2	1	1	2	1	1	1	–	1
Asio flammeus	1	1	1	1	1	1	1	–	1	–	–	–	–
Caprimulgus europaeus	2	1	2	1	1	1	1	1	1	1	–	1	1
Picus viridis	2	–	1	1	1	–	1	–	–	–	–	–	–
Alauda arvensis	1	1	2	1	1	1	–	2	–	–	–	–	–
Lullula arborea	2	1	2	1	1	1	–	2	1	1	–	–	–
Anthus campestris	1	1	1	1	1	–	1	1	–	1	–	–	–
Saxicola torquata	1	1	–	1	2	1	1	–	–	–	–	–	–
Sylvia undata	2	1	1	1	2	1	1	–	1	–	–	–	–
Lanius collurio	1	1	1	1	–	1	1	–	–	1	–	–	–
Lanius excubitor	1	2	1	1	–	1	1	–	1	1	1	–	–
No. of species													
Critical impact		1	0	0	0	0	0	0	0	0	0	0	0
High impact		1	4	0	3	0	0	3	0	0	0	0	0
Low impact		13	7	15	7	13	12	3	8	7	5	3	3
Total		15	11	15	10	13	12	6	8	7	5	3	3
% of priority spp. threatened		94%	69%	94%	63%	81%	75%	38%	50%	44%	31%	19%	19%
Weighted threat score		22	22	20	18	17	16	12	11	9	6	4	4

Boreal and temperate forests

■ Boreal forest

	Priority score	1. Intensified forestry	2. Logging	3. Recreation/ disturbance	4. Transport infrastructure	5. Habitat fragmentation	6. Agricultural clearance	7. Water management	8. Overgrazing	9. Air pollution	10. Fires	11. Pesticide use
Mergus albellus	3	–	1	1	–	–	1	1	–	1	–	–
Pernis apivorus	1	–	–	1	–	–	–	–	–	–	–	–
Haliaeetus albicilla	2	1	–	1	1	–	–	1	–	–	–	–
Circus cyaneus	1	–	–	–	–	–	–	1	1	–	–	–
Aquila clanga	4	1	1	2	1	1	–	2	–	–	–	–
Aquila chrysaetos	2	1	1	–	1	–	–	–	–	–	–	–
Pandion haliaetus	2	–	1	1	–	–	–	–	–	–	–	–
Falco tinnunculus	1	–	–	–	–	–	–	–	1	–	–	–
Falco vespertinus	1	–	–	–	–	–	–	1	–	–	–	–
Falco peregrinus	2	–	–	–	–	–	–	–	–	–	–	–
Tetrao tetrix	2	1	1	–	–	1	1	–	1	–	–	–
Tetrao urogallus	1	1	1	–	–	1	1	–	–	–	1	–
Grus grus	1	–	–	1	1	–	–	1	–	–	–	–
Scolopax rusticola	1	1	1	–	–	1	–	–	–	–	–	–
Tringa glareola	1	–	1	–	–	–	1	2	–	–	–	–
Columba oenas	1	–	–	–	–	–	–	–	–	–	–	–
Columba palumbus	1	–	–	–	–	–	–	–	–	–	–	–
Streptopelia turtur	2	–	–	–	–	–	–	–	–	–	–	–
Bubo bubo	2	–	–	2	1	–	–	–	–	–	–	–
Surnia ulula	1	1	1	–	–	–	–	–	–	–	1	–
Glaucidium passerinum	1	1	1	–	–	–	–	1	–	–	–	–
Strix uralensis	1	1	1	–	–	–	–	–	–	–	1	–
Strix nebulosa	1	1	1	–	–	–	–	–	–	–	1	–
Aegolius funereus	1	1	1	–	–	–	–	–	1	–	1	–
Caprimulgus europaeus	3	–	1	–	–	–	–	–	–	–	–	1
Jynx torquilla	2	1	–	–	–	–	–	–	–	–	–	–
Picus canus	2	1	1	–	–	–	–	–	–	–	–	–
Picoides tridactylus	3	2	2	–	–	–	–	–	–	–	–	–
Lullula arborea	3	–	–	–	–	–	–	–	–	–	–	–
Anthus hodgsoni	1	–	–	–	–	–	–	–	–	–	–	–
Bombycilla garrulus	1	1	1	–	–	–	–	–	–	–	1	–
Prunella modularis	1	–	–	–	–	–	–	–	–	–	–	–
Erithacus rubecula	1	–	–	–	–	–	–	–	–	–	–	–
Luscinia luscinia	1	–	–	–	–	1	1	–	–	–	–	–
Luscinia calliope	1	–	–	–	–	1	1	–	–	–	–	–
Tarsiger cyanurus	1	1	1	–	–	1	–	–	–	–	1	–
Phoenicurus phoenicurus	3	1	1	–	–	–	–	–	–	–	–	–
Saxicola torquata	2	–	–	–	–	–	–	–	–	–	–	–
Zoothera dauma	1	–	–	–	–	–	–	–	–	–	–	–
Turdus merula	1	–	–	–	–	–	–	–	–	1	–	–
Turdus pilaris	1	1	1	–	–	–	–	–	–	1	–	–
Turdus philomelos	1	–	–	–	–	–	–	–	–	1	–	–
Turdus iliacus	2	–	–	–	–	–	–	–	–	1	1	–
Turdus viscivorus	1	–	–	–	–	–	–	–	–	1	–	–
Hippolais icterina	1	–	–	–	–	–	–	–	–	–	–	–
Sylvia borin	1	–	–	–	–	–	–	–	1	–	–	–
Sylvia atricapilla	1	–	–	–	–	–	–	–	–	–	–	–
Phylloscopus borealis	1	–	1	–	–	–	–	1	–	–	1	–
Phylloscopus inornatus	1	–	1	–	–	–	–	1	–	–	1	–
Phylloscopus sibilatrix	1	1	1	–	–	1	–	–	1	1	–	–
Regulus regulus	1	–	–	–	–	–	–	–	–	1	–	–
Muscicapa striata	2	1	–	–	–	–	–	–	–	–	–	–
Ficedula hypoleuca	1	1	–	–	–	–	–	–	–	–	–	–
Parus cinctus	1	1	1	–	–	–	–	1	–	1	1	–
Parus cristatus	1	1	–	–	–	–	–	–	–	1	–	–
Lanius collurio	1	1	–	–	–	–	–	–	–	–	–	1
Lanius excubitor	2	–	–	1	–	–	–	–	–	–	–	1
Perisoreus infaustus	3	2	2	–	–	1	–	–	–	1	–	–
Fringilla coelebs	1	–	–	–	–	–	–	–	–	1	–	–
Fringilla montifringilla	1	–	1	–	–	–	–	1	–	1	1	–
Carduelis chloris	1	–	–	–	–	–	–	–	–	1	–	–
Carduelis spinus	1	–	–	–	–	–	–	–	–	–	–	–
Loxia leucoptera	1	–	1	–	–	–	–	1	–	–	1	–
Loxia scotica	4	–	–	–	–	–	–	–	1	–	–	–
Loxia pytyopsittacus	2	1	1	–	–	–	–	–	–	–	–	–
Pinicola enucleator	1	–	1	–	–	–	–	1	–	–	1	–
Emberiza hortulana	2	–	–	–	–	–	–	–	1	–	–	–
Emberiza rustica	1	–	1	–	–	–	–	1	–	–	1	–
No. of species												
Critical impact		0	0	0	0	0	0	0	0	0	0	0
High impact		2	2	2	0	0	0	2	0	0	0	0
Low impact		24	28	6	5	7	6	15	9	13	15	3
Total		26	30	8	5	7	6	17	9	13	15	3
% of priority spp. threatened		38%	44%	12%	7%	10%	9%	25%	13%	19%	22%	4%
Weighted threat score		48	54	23	11	13	9	28	14	18	16	6

Boreal and temperate forests (cont.)

■ Lowland temperate forest

	Priority score	1. Intensified forestry	2. Logging	3. Recreation/ disturbance	4. Transport infrastructure	5. Habitat fragmentation	6. Agricultural clearance	7. Water management	8. Overgrazing	9. Air pollution	10. Fires	11. Pesticide use
Ciconia nigra	2	2	1	1	1	–	1	–	–	–	–	–
Pernis apivorus	1	–	–	1	–	–	–	–	–	–	–	–
Milvus migrans	2	–	–	1	–	–	–	–	–	–	–	–
Milvus milvus	2	1	–	1	–	–	–	–	–	–	–	–
Haliaeetus albicilla	2	1	–	1	1	–	–	–	–	–	–	–
Circaetus gallicus	2	1	–	1	1	–	–	–	–	–	1	–
Accipiter brevipes	3	–	1	1	–	–	1	–	–	–	–	–
Aquila pomarina	3	2	1	1	1	1	1	–	–	–	–	–
Aquila clanga	4	2	1	2	1	1	–	–	–	–	–	–
Aquila heliaca	4	2	2	1	1	–	–	–	–	–	–	–
Aquila chrysaetos	2	1	1	1	1	–	–	–	–	–	–	–
Hieraaetus pennatus	2	1	1	1	1	–	–	–	–	–	1	–
Pandion haliaetus	2	–	1	1	1	–	–	–	–	–	–	–
Falco tinnunculus	1	–	–	–	–	–	–	–	1	–	–	–
Falco vespertinus	1	–	–	–	–	–	–	–	–	–	–	–
Falco cherrug	2	1	1	1	–	–	–	–	–	–	–	–
Falco peregrinus	2	–	–	–	–	–	–	–	–	–	–	–
Bonasa bonasia	1	2	1	–	1	–	–	–	1	–	1	–
Tetrao tetrix	2	1	1	1	1	1	–	–	1	–	–	–
Scolopax rusticola	3	1	1	–	–	–	–	–	1	–	–	–
Columba oenas	1	1	1	–	–	–	–	–	–	–	–	–
Columba palumbus	1	–	–	–	–	–	–	–	–	–	–	–
Streptopelia turtur	2	–	–	–	–	–	–	–	–	–	–	–
Otus scops	2	–	–	–	–	–	–	–	–	–	–	–
Bubo bubo	2	–	–	2	1	–	–	–	–	–	–	–
Strix aluco	2	1	1	–	–	1	–	–	–	–	–	1
Caprimulgus europaeus	3	–	–	1	–	–	–	–	1	–	–	1
Coracias garrulus	2	1	–	–	–	1	–	–	–	–	–	–
Jynx torquilla	3	2	–	–	–	–	–	–	–	–	–	–
Picus canus	3	2	2	–	–	–	–	–	–	–	–	–
Picus viridis	4	2	1	–	–	–	–	–	–	–	–	–
Dendrocopos medius	2	2	1	–	–	–	–	–	–	–	–	–
Dendrocopos minor	1	2	1	–	–	1	–	–	–	–	–	–
Picoides tridactylus	2	2	1	–	–	1	–	–	–	–	–	–
Lullula arborea	4	–	–	1	–	–	–	–	–	–	–	1
Prunella modularis	1	–	–	–	–	–	–	–	1	–	–	–
Erithacus rubecula	1	–	–	–	–	–	–	–	–	–	–	–
Luscinia luscinia	2	–	–	–	–	1	–	1	–	–	–	–
Luscinia megarhynchos	2	1	–	–	–	1	–	1	–	–	–	–
Phoenicurus phoenicurus	4	1	1	–	–	–	–	1	–	–	–	–
Saxicola torquata	2	–	–	–	–	–	–	–	–	–	–	–
Turdus merula	1	–	–	–	–	–	–	–	–	1	–	–
Turdus pilaris	1	1	1	–	–	–	–	–	–	1	–	–
Turdus philomelos	1	–	–	–	–	–	–	–	–	1	–	–
Turdus viscivorus	1	–	–	–	–	–	–	–	–	1	–	–
Hippolais icterina	2	–	–	–	–	–	–	–	–	–	–	–
Hippolais polyglotta	1	–	–	–	–	–	1	–	–	–	–	–
Sylvia borin	2	–	–	–	–	–	–	–	1	–	–	–
Sylvia atricapilla	1	–	–	–	–	–	–	–	1	–	–	–
Phylloscopus bonelli	1	–	–	–	–	–	–	–	1	–	–	–
Phylloscopus sibilatrix	1	1	1	–	–	1	–	–	–	1	–	–
Regulus regulus	1	–	–	–	–	–	–	–	–	1	–	–
Regulus ignicapillus	1	–	–	–	–	–	–	–	–	1	–	–
Muscicapa striata	2	1	–	–	–	–	–	–	–	–	–	–
Ficedula parva	1	1	1	–	–	–	–	–	–	–	–	–
Ficedula semitorquata	3	1	2	–	–	–	1	1	–	–	1	–
Ficedula albicollis	2	1	1	–	–	–	–	–	–	–	–	–
Ficedula hypoleuca	2	1	1	–	–	–	–	–	–	–	–	–
Parus lugubris	1	1	–	–	–	–	–	–	–	–	–	–
Parus cristatus	1	1	1	–	–	–	–	–	–	1	–	1
Parus caeruleus	2	1	–	–	–	–	–	–	–	1	–	1
Sitta europaea	1	1	1	–	–	1	–	–	–	–	–	–
Certhia familiaris	1	1	1	–	–	1	–	–	–	–	–	–
Certhia brachydactyla	2	1	1	–	–	1	–	–	–	–	–	–
Oriolus oriolus	1	–	1	–	–	–	1	–	–	–	–	–
Lanius collurio	1	–	–	–	–	–	–	–	–	–	–	1
Lanius excubitor	1	–	–	1	–	–	–	–	–	–	–	1
Corvus monedula	1	1	–	–	–	–	–	–	–	–	–	1
Fringilla coelebs	1	–	–	–	–	–	–	–	–	1	–	1
Serinus serinus	1	–	–	–	–	–	–	–	–	–	–	–
Carduelis chloris	1	–	–	–	–	–	–	–	–	–	–	–
Carduelis spinus	1	–	–	–	–	–	–	–	–	–	–	–
No. of species												
Critical impact		0	0	0	0	0	0	0	0	0	0	0
High impact		11	3	2	0	0	0	0	0	0	0	0
Low impact		28	28	17	11	12	10	1	11	10	4	8
Total		39	31	19	11	12	10	1	11	10	4	8
% of priority spp. threatened		53%	42%	26%	15%	16%	14%	1%	15%	14%	5%	11%
Weighted threat score		108	75	51	26	23	20	3	19	11	8	15

Boreal and temperate forests (cont.)

■ Montane forest

	Priority score	1. Intensified forestry	2. Logging	3. Recreation/ disturbance	4. Transport infrastructure	5. Habitat fragmentation	6. Agricultural clearance	7. Water management	8. Overgrazing	9. Air pollution	10. Fires	11. Pesticide use
Ciconia nigra	1	1	2	1	1	–	–	–	–	–	–	–
Pernis apivorus	1	–	–	1	–	–	–	–	–	–	–	–
Milvus migrans	2	–	–	1	–	–	–	–	–	–	–	–
Milvus milvus	1	1	–	–	–	–	–	–	–	–	–	–
Circaetus gallicus	2	1	–	1	–	–	–	–	–	–	1	–
Aquila heliaca	3	2	2	1	1	–	–	–	–	–	–	–
Aquila chrysaetos	2	1	1	1	1	–	–	–	–	–	–	–
Hieraaetus pennatus	2	1	1	1	1	–	–	–	–	–	1	–
Tetrao tetrix	2	1	–	1	–	1	–	–	1	–	–	–
Columba palumbus	1	–	–	–	–	–	–	–	–	–	–	–
Bubo bubo	2	–	–	2	1	–	–	–	–	–	–	–
Strix aluco	1	1	–	–	–	1	–	–	–	–	–	–
Caprimulgus europaeus	2	–	–	–	–	–	–	–	–	–	–	–
Jynx torquilla	1	1	–	–	–	–	–	–	–	–	–	–
Picus canus	2	1	2	–	–	–	–	–	–	–	–	–
Picus viridis	3	1	1	–	–	–	–	–	–	–	–	–
Dendrocopos medius	1	1	1	–	–	–	–	–	–	–	–	–
Picoides tridactylus	2	2	1	–	–	–	–	–	–	–	–	–
Lullula arborea	3	–	–	–	–	–	–	–	–	–	–	–
Prunella modularis	1	–	–	–	–	–	–	–	1	–	–	–
Prunella atrogularis	3	–	–	–	–	–	–	–	–	–	–	–
Erithacus rubecula	1	–	–	–	–	–	–	–	–	–	–	–
Phoenicurus phoenicurus	3	1	1	–	–	–	–	–	–	–	–	–
Turdus torquatus	1	–	–	–	–	–	–	–	–	–	–	–
Turdus merula	1	–	–	–	–	–	–	–	–	1	–	–
Turdus pilaris	1	1	1	–	–	–	–	–	–	1	–	–
Turdus philomelos	1	–	–	–	–	–	–	–	–	1	–	–
Turdus viscivorus	1	–	–	–	–	–	–	–	–	1	–	–
Sylvia borin	1	–	–	–	–	–	–	–	1	–	–	–
Sylvia atricapilla	1	–	–	–	–	–	–	–	1	–	–	–
Phylloscopus bonelli	1	–	–	–	–	–	–	–	1	1	–	–
Phylloscopus sibilatrix	1	1	1	–	–	1	–	–	–	1	–	–
Regulus regulus	1	–	–	–	–	–	–	–	–	1	–	–
Regulus ignicapillus	1	–	–	–	–	–	–	–	–	1	–	–
Muscicapa striata	2	1	–	–	–	–	–	–	–	–	–	–
Ficedula semitorquata	3	–	2	–	–	–	–	–	–	–	1	–
Ficedula albicollis	1	1	1	–	–	–	–	–	–	–	–	–
Ficedula hypoleuca	1	1	1	–	–	–	–	–	–	–	–	–
Parus cristatus	1	1	1	–	–	–	–	–	–	1	–	–
Parus caeruleus	1	–	–	–	–	–	–	–	–	1	–	–
Lanius collurio	1	1	–	–	–	–	–	–	–	1	–	–
Fringilla coelebs	1	–	–	–	–	–	–	–	–	1	–	–
Serinus serinus	1	–	–	–	–	–	–	–	–	–	–	–
Serinus citrinella	2	–	–	–	–	–	–	–	1	–	–	–
Carduelis spinus	1	–	–	–	–	–	–	–	–	–	–	–
Emberiza cia	2	–	–	1	–	–	–	–	–	–	–	–
No. of species												
Critical impact	0	0	0	0	0	0	0	0	0	0	0	0
High impact	2	4	1	0	0	0	0	0	0	0	0	0
Low impact	19	11	9	5	3	0	0	6	11	3	0	
Total	21	15	10	5	3	0	0	6	11	3	0	
% of priority spp. threatened	46%	33%	22%	11%	7%	0%	0%	13%	24%	7%	0%	
Weighted threat score	39	36	21	10	4	0	0	8	11	7	0	

■ Riverine forest

	Priority score	1. Intensified forestry	2. Logging	3. Recreation/ disturbance	4. Transport infrastructure	5. Habitat fragmentation	6. Agricultural clearance	7. Water management	8. Overgrazing	9. Air pollution	10. Fires	11. Pesticide use
Phalacrocorax pygmeus	3	–	1	–	–	–	1	1	–	–	–	–
Nycticorax nycticorax	2	–	1	1	–	–	1	1	–	–	–	–
Ardeola ralloides	2	–	1	–	–	–	1	1	–	–	–	–
Ardea purpurea	2	–	1	1	–	–	–	1	–	–	–	–
Ciconia nigra	2	1	2	1	1	–	2	–	–	–	–	–
Plegadis falcinellus	2	–	1	1	–	–	1	1	–	–	–	–
Platalea leucorodia	3	–	1	–	–	–	–	1	–	–	–	–
Milvus migrans	2	–	–	1	–	–	–	–	–	–	–	–
Milvus milvus	1	1	–	–	–	–	–	–	–	–	–	–
Haliaeetus albicilla	2	1	–	1	–	–	–	–	–	–	–	–
Circaetus gallicus	1	1	–	1	1	–	–	–	–	–	1	–
Accipiter brevipes	3	–	1	1	–	–	–	–	–	–	–	–
Aquila pomarina	2	1	1	1	–	–	1	–	–	–	–	–
Aquila clanga	4	1	1	2	–	1	–	–	–	–	–	–

cont.

Boreal and temperate forests (cont.)

■ Riverine forest (cont.)

	Priority score	1. Intensified forestry	2. Logging	3. Recreation/ disturbance	4. Transport infrastructure	5. Habitat fragmentation	6. Agricultural clearance	7. Water management	8. Overgrazing	9. Air pollution	10. Fires	11. Pesticide use
Pandion haliaetus	2	–	1	1	–	–	–	–	–	–	–	–
Falco tinnunculus	1	–	–	–	–	–	–	–	–	–	–	–
Falco vespertinus	1	–	–	–	–	–	–	–	–	–	–	–
Grus grus	2	–	–	1	1	–	–	–	–	–	–	–
Scolopax rusticola	2	–	–	–	–	–	–	–	–	–	–	–
Columba oenas	1	–	–	–	–	1	–	–	–	–	–	–
Columba palumbus	1	–	–	–	–	–	–	–	–	–	–	–
Streptopelia turtur	2	–	–	–	–	–	–	–	–	–	–	–
Otus scops	3	–	1	–	–	–	–	–	–	–	–	–
Strix aluco	1	1	1	–	–	–	–	–	–	–	–	–
Coracias garrulus	2	–	–	–	–	1	1	–	–	–	–	–
Jynx torquilla	2	1	–	–	–	–	–	–	–	–	–	–
Picus canus	2	1	2	–	–	–	–	–	–	–	–	–
Picus viridis	3	1	1	–	–	–	–	–	–	–	–	–
Dendrocopos medius	1	1	1	–	–	–	–	–	–	–	–	–
Prunella modularis	1	–	–	–	–	–	–	–	–	–	–	–
Erithacus rubecula	1	–	–	–	–	–	–	–	–	–	–	–
Luscinia luscinia	1	–	–	–	–	–	1	1	–	–	–	–
Luscinia megarhynchos	1	1	–	–	–	–	1	1	–	–	–	–
Turdus merula	1	–	–	–	–	–	–	–	–	–	–	–
Turdus pilaris	1	1	1	–	–	–	–	–	–	–	–	–
Turdus philomelos	1	–	–	–	–	–	–	–	–	–	–	–
Locustella fluviatilis	2	–	1	–	–	–	1	1	–	–	–	–
Acrocephalus palustris	1	–	–	–	–	–	1	1	–	–	–	–
Hippolais icterina	1	–	–	–	–	–	–	–	–	–	–	–
Hippolais polyglotta	1	–	–	–	–	–	1	1	–	–	–	–
Sylvia borin	1	–	–	–	–	–	–	–	–	–	–	–
Sylvia atricapilla	1	–	–	–	–	–	–	–	–	–	–	–
Muscicapa striata	2	1	–	–	–	–	–	–	–	–	–	–
Ficedula semitorquata	3	–	2	–	–	–	1	1	–	–	1	–
Ficedula albicollis	1	1	1	–	–	–	–	–	–	–	–	–
Ficedula hypoleuca	1	1	1	–	–	–	–	–	–	–	–	–
Parus caeruleus	1	–	–	–	–	–	–	–	–	–	–	–
Parus cyanus	1	1	1	–	–	–	–	–	–	–	1	–
Certhia brachydactyla	1	1	1	–	–	1	–	–	–	–	–	–
Lanius excubitor	1	–	–	–	–	–	–	–	–	–	–	–
Corvus monedula	1	1	–	–	–	–	–	–	–	–	–	–
Fringilla coelebs	1	–	–	–	–	–	–	–	–	–	–	–
Serinus serinus	1	–	–	–	–	–	–	–	–	–	–	–
Carduelis chloris	1	–	–	–	–	–	–	–	–	–	–	–
Carduelis spinus	1	–	–	–	–	–	–	–	–	–	–	–
No. of species												
Critical impact		0	0	0	0	0	0	0	0	0	0	0
High impact		0	3	1	0	0	1	0	0	0	0	0
Low impact		19	20	11	3	4	12	12	0	0	3	0
Total		19	23	12	3	4	13	12	0	0	3	0
% of priority spp. threatened		35%	42%	22%	5%	7%	24%	22%	0%	0%	5%	0%
Weighted threat score		30	54	30	5	8	26	23	0	0	5	0

Mediterranean forest, shrubland and rocky habitats

	Priority score	1. Land abandonment	2. Fires	3. Afforestation	4. Human disturbance	5. Lack of fire	6. Wood-cutting	7. Reservoir construction	8. Urbanization	9. Roads, railways, etc	10. Perennial crops	11. Overgrazing	12. Tree disease	13. Powerlines	14. Habitat fragmentation	15. Arable agriculture	16. Shrub removal	17. Pesticides	18. Mining/ quarrying
Ciconia nigra	2	1	1	–	1	–	1	1	–	1	1	1	1	1	2	–	–	–	1
Milvus migrans	2	1	1	–	1	1	–	–	–	1	1	1	1	1	–	–	–	–	–
Milvus milvus	1	1	1	–	1	1	–	1	–	–	2	–	1	2	–	–	–	–	–
Gypaetus barbatus	2	2	–	1	1	–	–	1	–	–	1	–	1	2	–	–	–	–	1
N. percnopterus	2	1	–	1	2	1	–	1	–	–	–	–	1	–	–	–	–	–	1
Gyps fulvus	2	2	–	1	2	–	1	–	1	–	–	–	1	–	–	–	–	–	1
Aegypius monachus	3	2	1	1	1	–	2	–	1	–	–	1	1	2	–	–	–	–	1
Circaetus gallicus	3	2	1	1	1	1	1	–	1	–	1	1	1	1	1	–	–	–	–
Accipiter brevipes	3	–	–	–	–	–	–	–	–	–	–	–	–	–	–	–	–	–	–
Buteo rufinus	2	1	–	–	1	–	–	–	–	–	–	–	1	1	1	–	–	–	–
Aquila pomarina	1	1	–	–	1	–	–	–	–	–	–	–	1	1	–	–	–	–	–
Aquila heliaca	3	1	1	–	2	–	2	–	–	–	–	–	1	–	–	–	–	–	–
Aquila adalberti	4	1	–	–	1	1	1	1	–	2	–	–	2	2	2	1	–	1	–
Aquila chrysaetos	2	2	–	–	1	1	–	1	–	–	–	–	2	1	–	–	–	–	1

cont.

Mediterranean forest, shrubland and rocky habitats (cont.)

Species	Priority score	1. Land abandonment	2. Fires	3. Afforestation	4. Human disturbance	5. Lack of fire	6. Wood-cutting	7. Reservoir construction	8. Urbanization	9. Roads, railways, etc	10. Perennial crops	11. Overgrazing	12. Tree disease	13. Powerlines	14. Habitat fragmentation	15. Arable agriculture	16. Shrub removal	17. Pesticides	18. Mining/quarrying
Hieraaetus pennatus	3	1	1	–	1	1	1	1	–	–	–	–	1	1	1	–	–	1	–
Hieraaetus fasciatus	3	2	1	–	1	1	1	1	1	2	–	–	1	2	1	1	–	1	2
Falco tinnunculus	1	1	–	–	1	1	–	–	–	–	–	–	1	–	–	–	–	–	–
Falco eleonorae	4	–	–	–	1	–	–	–	–	–	–	–	–	–	–	–	–	–	–
Falco biarmicus	3	–	–	–	1	–	–	1	–	–	–	–	–	–	–	1	–	–	–
Falco peregrinus	2	1	–	–	1	–	–	–	–	–	–	–	–	–	–	–	–	–	1
Alectoris chukar	3	1	–	1	–	–	–	–	–	–	–	–	–	–	–	–	–	–	–
Alectoris graeca	3	1	–	1	1	1	–	1	–	–	–	1	–	–	–	–	–	–	–
Alectoris rufa	3	1	–	1	–	1	–	–	–	–	–	1	–	–	–	–	–	–	–
Alectoris barbara	3	1	1	2	–	1	–	–	–	–	–	1	–	–	–	1	–	–	–
F. francolinus	3	–	1	–	1	–	–	–	1	–	–	–	–	–	–	–	–	–	–
Turnix sylvatica	3	–	–	2	–	–	–	1	2	2	2	1	–	–	2	–	–	–	–
Scolopax rusticola	2	–	1	–	1	–	1	1	–	–	–	–	1	–	–	–	1	–	–
Columba palumbus	1	–	1	–	–	–	1	–	–	–	–	–	1	–	–	–	–	–	–
Streptopelia turtur	2	–	1	–	1	–	1	–	–	–	–	–	1	–	–	–	–	–	–
Tyto alba	1	–	–	–	1	–	–	–	1	1	–	–	–	–	–	–	–	–	–
Otus scops	3	–	1	1	–	–	1	–	–	1	–	–	1	–	–	1	–	1	–
Bubo bubo	2	–	–	–	2	–	–	1	1	2	–	–	–	2	1	–	–	–	1
Athene noctua	1	–	–	–	–	–	–	–	1	–	–	–	–	–	–	–	1	–	–
C. europaeus	3	–	1	1	–	–	1	1	1	1	–	1	–	–	1	–	1	–	–
Caprimulgus ruficollis	1	–	1	1	–	–	–	–	1	1	–	1	–	–	1	–	–	–	–
Apus caffer	1	–	–	–	–	–	–	–	–	–	–	–	–	–	–	–	–	–	–
Apus affinis	1	–	–	–	–	–	–	–	–	–	–	–	–	–	–	–	–	–	–
Merops apiaster	2	1	1	–	1	1	–	–	1	–	–	–	–	–	–	–	–	1	–
Coracias garrulus	3	–	1	–	1	–	–	–	–	1	–	1	–	–	–	–	–	1	–
Jynx torquilla	1	–	1	–	–	–	1	–	–	–	–	–	–	–	1	–	–	–	–
Picus viridis	3	–	1	–	–	–	–	–	–	–	–	–	1	–	–	–	–	–	–
Dendrocopos syriacus	1	–	–	–	–	–	1	–	–	1	–	–	–	–	–	–	–	–	–
Cal. brachydactyla	1	2	–	2	–	1	–	1	1	–	–	–	–	–	–	–	–	–	–
Galerida cristata	1	2	–	2	–	1	–	–	1	–	1	–	–	–	–	–	–	–	–
Galerida theklae	3	1	1	2	–	1	–	1	1	–	1	1	1	–	–	–	–	–	–
Lullula arborea	3	1	1	1	–	–	–	–	–	–	–	–	–	–	–	–	–	–	–
Ptyonoprogne rupestris	1	–	–	–	1	–	–	–	–	–	–	–	–	–	–	–	–	–	1
Hirundo daurica	1	1	1	–	1	–	–	–	–	–	–	–	–	–	–	–	–	–	–
Anthus campestris	2	2	–	1	–	1	–	–	1	–	1	–	–	–	–	1	–	–	–
Erithacus rubecula	1	–	1	–	–	–	1	–	–	–	–	–	–	–	–	–	1	–	–
L. megarhynchos	1	–	1	–	–	–	1	–	–	–	–	1	–	–	–	–	1	–	–
P. phoenicurus	2	–	1	–	–	–	1	–	–	–	–	–	1	–	–	–	1	–	–
Saxicola torquata	2	1	–	–	–	1	–	–	–	–	–	1	–	–	–	1	–	–	–
Oenanthe cypriaca	4	–	–	–	–	–	–	–	–	–	–	–	–	–	–	–	–	–	–
Oenanthe hispanica	4	2	–	2	–	1	–	1	–	–	1	1	–	–	1	–	–	–	–
Oenanthe leucura	3	1	–	1	–	1	–	1	–	–	–	–	–	–	–	–	–	–	1
Monticola saxatilis	2	1	–	1	–	–	–	1	–	–	–	1	–	–	–	–	–	–	1
Monticola solitarius	3	–	–	–	–	–	1	1	–	–	–	–	–	–	–	–	–	–	1
Turdus merula	1	–	1	–	–	–	1	–	–	–	–	–	–	–	–	–	1	–	–
Turdus philomelos	1	–	1	–	–	–	1	–	–	–	–	–	–	–	–	–	–	–	–
Hippolais pallida	2	–	1	–	–	–	–	1	–	–	–	–	–	–	–	–	–	1	–
Hippolais olivetorum	4	–	–	–	–	–	–	–	1	–	1	–	–	–	–	1	1	2	–
Hippolais polyglotta	1	–	–	–	–	–	–	–	–	–	1	–	–	–	–	–	1	–	–
Sylvia sarda	2	–	1	–	–	1	–	–	1	1	–	–	–	–	–	1	–	–	–
Sylvia undata	4	1	1	2	–	1	–	1	1	1	1	1	–	–	–	1	–	–	–
Sylvia cantillans	2	1	1	1	–	1	1	1	1	1	1	1	–	–	–	1	–	–	–
Sylvia melanocephala	2	1	1	1	–	1	1	1	1	1	1	1	–	–	–	1	–	–	–
Sylvia melanothorax	4	–	2	1	2	–	–	–	1	1	1	1	–	–	–	–	2	1	–
Sylvia rueppelli	2	–	1	–	–	–	–	1	–	1	1	–	–	–	1	–	–	–	–
Sylvia hortensis	3	1	1	1	–	1	1	1	1	1	1	1	–	–	–	1	–	–	–
Sylvia atricapilla	1	–	1	–	–	–	–	–	–	–	–	–	–	–	–	–	1	–	–
Phylloscopus bonelli	1	–	1	–	–	–	1	–	–	–	–	–	–	–	–	–	1	–	–
Regulus ignicapillus	1	–	1	–	–	–	1	–	–	–	–	–	–	–	–	–	1	–	–
Muscicapa striata	1	–	1	–	–	–	1	–	–	–	–	–	–	–	–	1	–	–	–
Ficedula semitorquata	3	–	1	–	1	–	–	–	–	–	–	–	–	–	–	1	–	–	–
Parus lugubris	2	–	–	–	–	–	1	–	–	–	–	–	–	–	–	–	1	–	–
Parus cristatus	1	–	1	–	–	–	1	–	–	–	–	–	–	–	–	–	–	–	–
Parus caeruleus	1	–	1	–	–	–	1	–	–	–	–	–	–	–	–	–	–	–	–
Sitta krueperi	2	–	–	–	–	–	1	–	–	–	–	–	–	–	–	–	1	–	–
Sitta whiteheadi	4	–	1	–	–	–	1	–	–	–	–	–	–	–	–	–	1	–	–
Sitta neumayer	2	–	–	–	–	–	–	–	–	–	–	–	–	–	–	–	–	–	–
Certhia brachydactyla	1	–	1	–	–	–	1	–	–	–	–	–	–	1	–	–	–	–	–
Lanius collurio	1	1	1	1	–	1	–	–	1	–	1	–	–	–	–	–	–	–	–
Lanius excubitor	1	1	–	1	1	1	–	–	–	–	–	–	–	–	–	–	–	–	–
Lanius senator	3	1	1	1	–	1	–	–	1	–	1	–	–	1	–	–	–	–	–
Lanius nubicus	4	1	1	–	–	–	–	–	1	–	–	1	–	–	–	1	1	1	–
P. pyrrhocorax	2	2	–	1	1	1	–	1	1	–	1	–	–	–	–	–	–	–	1
Sturnus unicolor	1	1	1	–	–	–	–	–	–	–	–	–	–	1	–	–	–	–	1
Petronia petronia	1	–	–	–	–	–	–	–	–	–	–	1	–	–	–	–	–	–	–
Fringilla coelebs	1	–	1	–	–	–	–	–	–	–	–	–	–	–	–	–	–	–	–
Serinus serinus	1	–	1	–	–	–	–	–	–	–	–	–	–	–	–	–	–	–	–

cont.

Mediterranean forest, shrubland and rocky habitats (cont.)

	Priority score	1. Land abandonment	2. Fires	3. Afforestation	4. Human disturbance	5. Lack of fire	6. Wood-cutting	7. Reservoir construction	8. Urbanization	9. Roads, railways, etc	10. Perennial crops	11. Overgrazing	12. Tree disease	13. Powerlines	14. Habitat fragmentation	15. Arable agriculture	16. Shrub removal	17. Pesticides	18. Mining/quarrying
Serinus citrinella	1	–	1	–	–	–	–	–	–	–	–	–	–	–	–	–	–	–	–
Carduelis chloris	1	–	1	–	–	–	–	–	–	–	–	–	–	–	–	–	–	–	–
Carduelis cannabina	1	1	–	–	–	–	–	–	–	–	1	–	–	–	–	–	–	–	–
Emberiza cirlus	1	–	–	–	–	1	–	–	–	–	–	–	–	–	–	–	–	–	–
Emberiza cia	2	2	–	1	–	1	–	–	–	–	–	–	–	–	–	–	–	–	–
Emberiza cineracea	4	–	–	–	–	–	–	–	1	–	–	–	–	–	–	–	–	–	–
Emberiza hortulana	3	1	–	1	–	–	–	–	–	–	1	1	–	–	–	–	1	–	–
Emberiza caesia	2	–	–	–	–	–	–	–	–	–	1	–	–	–	–	–	–	–	–
E. melanocephala	3	–	–	–	–	–	–	–	–	–	–	1	–	–	–	–	–	–	–
No. of species																			
Critical impact	0	0	0	0	0	0	0	0	0	0	0	0	0	0	0	0	0	0	
High impact	12	1	7	5	0	2	0	1	4	1	1	1	5	5	0	1	1	1	
Low impact	32	51	27	28	33	30	26	25	16	21	24	21	11	9	16	19	11	13	
Total	44	52	34	33	33	32	26	26	20	22	25	22	16	14	16	20	12	14	
% of priority spp. threatened	44%	52%	34%	33%	33%	32%	26%	26%	20%	22%	25%	22%	16%	14%	16%	20%	12%	14%	
Weighted threat score	129	114	104	88	76	70	69	68	61	60	56	54	49	47	46	42	40	32	

Agricultural and grassland habitats

■ Arable and improved grassland

cont.

Species	Priority score	1. Crop improvements	2. Pesticide use	3. Land abandonment	4. High stocking levels	5. Marginal habitat loss	6. Afforestation	7. Cultivation of grassland	8. Farming operations	9. Crop specialization	10. Recreation	11. Overhead structures	12. Increased predators	13. Urbanization/roads	14. Damage/flood control	15. Loss of oak trees	16. Loss of hay meadows	17. Low stocking levels	18. Nutrient pollution	19. Replaced crop/trees	20. Loss of old buildings	21. Reduced carrion	22. Autumn sowing	23. Vegetation clearance	24. Pasture conversion	25. Dry cultiv. rice
Bubulcus ibis	1																									
Ciconia ciconia	3		2												1		1	1								
Cygnus columbianus	1		2																							
Anser brachyrhynchus	3																									
Branta leucopsis	1																									
Branta bernicla	1																									
Branta ruficollis	4			2																						
Elanus caeruleus	2					1																				
Milvus migrans	2		2			1															2	1				
Milvus milvus	1		1			1														1	1	1				
Circaetus gallicus	2	2	2			1										1										
Circus cyaneus	2	2	2			1											1	1								
Circus pygargus	3	2	2			1			2																	
Accipiter brevipes	2																									
Buteo rufinus	2		2			1										1										
Aquila pomarina	4	2	2			2											1	1								
Aquila heliaca	3		2			2						1		1												
Falco naumanni	3		2			2			2	2		1									1					
Falco tinnunculus	2		2			2			2	2		1									1					
Falco vespertinus	2		2													1										
Falco cherrug	2		2			2						1		1												
Alectoris rufa	4	2	2			2			2	2																
Perdix perdix	3	2	2			2			2	2																
Coturnix coturnix	3		2						2	2																
Crex crex	4		2						2								2	2								
Grus grus	3														1											
Tetrax tetrax	3	2	2			2			2	2																
Otis tarda	4	2	2			2			2	2																
Burhinus oedicnemus	2	2	2			2			2	2																
Glareola pratincola	2		2			2			2	2																
Glareola nordmanni	2		2			2			2	2																
Pluvialis apricaria	1																									
Vanellus vanellus	1		1			2											1	1								
Scolopax rusticola	2	2																								
Larus canus	3																									
Columba oenas	2		2																		1					
Columba palumbus	2	2	2																							
Streptopelia turtur	3	2	2			2														1						
Tyto alba	3		2			2						1									1					
Otus scops	3		2			2																				
Athene noctua	3		2			2															1					
Strix aluco	1																									
Asio flammeus	2		2			1											1	2								
Merops apiaster	2	2	2			2																				
Coracias garrulus	2	2	2			2														1	2					

Appendix 4: Predicted Impact of Threats on Priority Bird Species

Agricultural and grassland habitats (cont.)
Arable and improved grassland (cont.)

Species	Priority score	1. Crop improvements	2. Pesticide use	3. Land abandonment	4. High stocking levels	5. Margin habitat loss	6. Afforestation	7. Cultivation of grassland	8. Farming operations	9. Crop specialization	10. Recreation	11. Overhead structures	12. Increased predators	13. Urbanization/roads	14. Drainage/flood control	15. Loss of oak trees	16. Loss of hay meadows	17. Low stocking levels	18. Nutrient pollution	19. Replaced crop	20. Loss of old trees/buildings	21. Reduced carrion	22. Autumn sowing	23. Vegetation clearance	24. Pasture conversion	25. Dry cultiv. rice
Picus viridis	3	–	–	–	–	2	–	–	–	–	–	–	–	–	–	1	–	–	1	–	–	–	–	–	–	–
Melanocorypha calandra	2	2	2	–	–	2	–	–	–	2	–	–	–	–	–	–	–	–	–	–	–	–	–	–	–	–
Galerida cristata	3	2	2	–	–	2	–	–	–	2	2	–	–	–	–	–	–	–	–	–	–	–	–	–	–	–
Lullula arborea	3	2	2	–	–	2	–	–	–	–	–	–	–	–	–	–	–	–	–	–	–	–	–	–	–	–
Alauda arvensis	3	2	2	–	–	2	–	–	–	2	–	–	–	–	–	–	–	–	–	–	–	–	1	–	–	–
Hirundo rustica	3	–	–	–	–	–	–	–	1	–	–	–	–	–	–	–	1	–	–	–	1	1	–	–	–	–
Anthus campestris	1	–	–	–	–	–	–	–	–	–	–	–	–	–	–	–	–	–	–	–	–	–	–	–	–	–
Anthus pratensis	1	–	–	–	–	–	–	–	–	–	–	–	–	–	–	–	–	1	–	–	–	–	–	–	–	–
Saxicola rubetra	1	–	–	2	–	2	–	–	–	–	–	–	–	–	–	–	–	1	–	–	–	–	–	–	–	–
Oenanthe hispanica	3	–	–	2	–	2	–	–	–	–	–	–	–	–	–	–	–	–	–	–	–	–	–	–	–	–
Turdus merula	1	–	–	–	–	–	–	–	–	–	–	–	–	–	–	–	–	–	–	–	–	–	–	–	–	–
Turdus pilaris	2	–	–	2	–	2	–	–	–	–	–	–	–	–	–	–	1	–	–	–	–	–	–	–	–	–
Turdus philomelos	1	–	–	2	–	2	–	–	–	–	–	–	–	–	–	–	–	–	–	–	–	–	–	–	–	–
Turdus iliacus	2	–	–	2	–	2	–	–	–	–	–	–	–	–	–	–	–	–	–	–	–	–	–	–	–	–
Turdus viscivorus	1	–	–	–	–	–	–	–	–	–	–	–	–	–	–	–	–	–	–	–	–	–	–	–	–	–
Locustella naevia	1	–	–	2	–	–	–	–	–	–	–	–	–	–	–	–	–	–	–	–	–	–	–	–	–	–
Hippolais pallida	1	–	–	–	–	–	–	–	–	1	–	–	1	–	–	–	–	–	–	–	–	–	–	–	–	–
Sylvia nisoria	1	–	–	–	–	–	–	–	–	–	–	–	–	–	–	–	–	–	–	–	–	–	–	–	–	–
Sylvia communis	1	–	–	–	–	2	–	–	–	–	–	–	–	–	–	–	–	–	–	–	–	–	–	–	–	–
Lanius collurio	3	–	2	2	–	2	–	–	–	–	–	–	–	–	–	–	–	–	–	–	–	–	–	–	–	–
Lanius minor	3	–	2	2	–	2	–	–	–	1	–	–	–	–	–	–	–	–	–	–	–	–	–	–	–	–
Lanius excubitor	2	–	2	2	–	2	–	–	–	–	–	–	–	–	–	–	–	–	–	–	–	–	–	–	–	–
Lanius senator	2	–	2	2	–	2	–	–	–	2	–	–	–	–	–	–	–	–	–	–	–	–	–	–	–	–
Pyrrhocorax pyrrhocorax	2	–	–	–	–	–	–	–	–	–	–	–	–	–	–	–	–	–	–	–	–	–	–	–	–	–
Corvus monedula	1	–	–	–	–	–	–	–	–	–	–	–	–	–	–	–	–	–	–	–	1	–	–	–	–	–
Corvus frugilegus	1	–	–	–	–	–	–	–	–	–	–	–	–	–	–	–	–	–	–	–	–	–	–	–	–	–
Sturnus vulgaris	1	–	–	–	–	–	–	–	–	–	–	–	–	–	–	–	–	–	–	–	–	–	–	–	–	–
Sturnus unicolor	1	–	–	–	–	–	–	–	–	–	–	–	–	–	–	–	–	–	–	–	–	–	–	–	–	–
Passer montanus	1	–	–	–	–	–	–	–	–	–	–	–	–	–	–	–	–	–	–	–	1	–	–	–	–	–
Carduelis chloris	1	–	–	–	–	–	–	–	–	–	–	–	–	–	–	–	–	–	–	–	–	–	–	–	–	–
Carduelis cannabina	1	–	–	–	–	–	–	–	–	–	–	–	–	–	–	–	–	–	–	–	–	–	–	–	–	–
Emberiza citrinella	1	–	–	–	–	2	–	–	–	–	–	–	–	–	–	–	–	–	–	–	–	–	–	–	–	–
Emberiza cirlus	1	–	–	–	–	2	–	–	–	1	–	–	–	–	–	–	–	–	–	–	–	–	–	–	–	–
Emberiza hortulana	4	–	–	–	–	–	–	–	–	–	–	–	–	–	–	–	–	–	–	–	–	–	–	1	–	–
Emberiza melanocephala	3	–	–	–	–	–	–	–	–	–	–	–	–	–	–	–	–	–	–	–	–	–	–	1	–	–
Miliaria calandra	2	–	–	–	–	–	–	–	–	1	–	–	–	–	–	–	–	1	–	–	1	–	1	1	1	–
No. of species																										
Critical impact		0	0	0	0	0	0	0	0	0	0	0	0	0	0	0	0	0	0	0	0	0	0	0	0	0
High impact		29	21	1	0	17	0	0	3	7	2	9	7	3	6	0	7	0	2	0	4	0	13	0	6	0
Low impact		18	31	26	22	30	18	0	8	11	2	9	7	3	6	2	8	24	2	4	4	13	13	6	6	0
Total		47	52	27	22	47	18	0	11	18	8	22	9	14	4	27	46	6	12	5	22	10				
% of priority spp. threatened		58%	64%	33%	27%	58%	22%	0%	14%	22%	2%	11%	9%	4%	7%	2%	10%	30%	2%	5%	2%	16%	0%	7%	0%	0%
Weighted threat score		183	170	63	48	140	42	0	38	65	8	22	22	9	14	4	27	46	6	12	5	22	10			

cont.

Agricultural and grassland habitats (cont.)

■ Steppe habitats

Species	Priority score	1. Crop improvements	2. Pesticide use	3. Land abandonment	4. High stocking levels	5. Marginal habitat loss	6. Afforestation	7. Cultivation of grassland	8. Farming operations	9. Crop specialization	10. Recreation	11. Overhead structures	12. Increased predators	13. Urbanization/roads	14. Drainage/flood control	15. Loss of oak trees	16. Loss of hay meadows	17. Low stocking levels	18. Nutrient pollution	19. Replaced crop	20. Loss of old trees, buildings	21. Reduced carrion	22. Autumn sowing	23. Vegetation clearance	24. Pasture conversion	25. Dry cultiv. rice
Ciconia ciconia	3	1	2		1	1		2	1	1		1														
Anser erythropus	4		2					2					1													
Tadorna ferruginea	2				2	1	1	2			1															
Milvus milvus	2	1	2	2				2	1	1											2	2				
Neophron percnopterus	1	1	2	2					1				1								1	2				
Gyps fulvus	2			1								1														
Circaetus gallicus	2		2	2			2	2																		
Circus cyaneus	2	1	2	2			2	2	2																	
Circus macrourus	3		2	2				2	2				1													
Circus pygargus	1	1	2	2				2	2																	
Buteo rufinus	3			1								1														
Aquila nipalensis	3			2				2	2	1		1														
Aquila heliaca	3	2		2			2	2				1								2	1					
Falco naumanni	4	2	2	2		2		2	2	2											2	1				
Falco tinnunculus	2	1	2	1		2		1		2																
Falco vespertinus	2		2	1		2		2		2																
Falco biarmicus	3	1	1					1		2		1														
Falco cherrug	3		2	2				2		2		1														
Alectoris chukar	3		2	2	2						2															
Alectoris rufa	3	2	2	2				2		1	2															
Perdix perdix	2	2	2	2				2	2	2			1													
Coturnix coturnix	2	2	2	2				2	2	2																
Grus grus	2		1		1	1	1	2	1	1	1	1	1	1	1											
Anthropoides virgo	1			1				2	1	1			1													
Tetrax tetrax	4	2	2	2	2	1	2	2	2	2	2		1	2												
Chlamydotis undulata	3	2	1	1	2	2		2		2	2		1	2												
Otis tarda	4	2	2	2			2	2	2	2	2	2	1	2												
Burhinus oedicnemus	3	1	2	2			2	2	2	1	2		1	1												
Cursorius cursor	3	2		1		2		2	2				1													
Glareola pratincola	2	2	2			2		2	2		2		1		1											
Glareola nordmanni	2	2	2			1		2	2																	
Charadrius leschenaultii	3		1	1									1													
Charadrius asiaticus	3			2		2		2					1													
Pluvialis apricaria	2				1		2	2	2		1		1													
Chettusia gregaria	4		1	2	2			2	1				1													
Numenius tenuirostris	4		2	2	2	2	2	2	2		2		1													
Pterocles orientalis	3	2	2	2		2		2		2			1													
Pterocles alchata	3	1	1	1		1		2		1																
Bubo bubo	1			1								1									1					
Athene noctua	2	2	2	1		2		2	1																	
Asio flammeus	2		2			2		2	1		1															
Merops apiaster	2	2	2	1		2		1										1								
Coracias garrulus	3	2	2	2		2		1			1										1					
Chersophilus duponti	3	2		1		2	1	2		2				1												
Melanocorypha calandra	2	2	2	1		2	2	2		1																

Agricultural and grassland habitats (cont.)

Steppe habitats (cont.)

Threat columns:
1. Crop improvements
2. Pesticide use
3. Land abandonment
4. High stocking levels
5. Marginal habitat: loss
6. Afforestation
7. Cultivation of grassland
8. Farming operations
9. Crop specialization
10. Recreation
11. Overhead structures
12. Increased predators
13. Urbanization/roads
14. Drainage/flood control
15. Loss of oak trees
16. Loss of hay meadows
17. Low stocking levels
18. Nutrient pollution
19. Replaced crop.
20. Loss of old buildings
21. Reduced carrion
22. Autumn sowing
23. Vegetation clearance
24. Pasture conversion
25. Dry cultiv. rice

Species	Priority score
Melanocorypha leucoptera	2
Melanocorypha yeltoniensis	3
Calandrella brachydactyla	3
Calandrella rufescens	3
Galerida cristata	2
Galerida theklae	2
Alauda arvensis	2
Anthus campestris	2
Anthus pratensis	1
Oenanthe hispanica	3
Oenanthe deserti	2
Hippolais pallida	1
Lanius minor	3
Lanius excubitor	2
Corvus monedula	1
Sturnus unicolor	1
Sturnus roseus	1
Bucanetes githagineus	3
Emberiza melanocephala	2
Miliaria calandra	1

No. of species — summary by threat column (1–25):

	1	2	3	4	5	6	7	8	9	10	11	12	13	14	15	16	17	18	19	20	21	22	23	24	25
Critical impact	0	0	0	0	0	0	0	0	0	0	0	0	0	0	0	0	0	0	0	0	0	0	0	0	0
High impact	27	24	13	7	13	12	13	6	6	1	1	0	0	0	0	0	0	0	1	2	0	0	0	0	0
Low impact	14	23	33	41	12	25	41	10	15	10	11	13	16	0	0	0	0	1	4	1	0	0	0	0	0
Total	41	47	46	48	25	37	54	16	21	11	12	13	16	0	0	0	0	4	5	8	0	0	0	0	0
% of priority spp. threatened	63%	72%	71%	74%	38%	57%	83%	25%	32%	17%	18%	20%	25%	0%	0%	0%	0%	2%	8%	5%	0%	0%	0%	0%	0%
Weighted threat score	159	171	155	143	82	119	168	63	67	34	32	35	44	0	0	0	0	4	14	8	0	0	0	0	0

417

Agricultural and grassland habitats (cont.)

■ Montane grassland

Threat categories: 1. Crop improvements · 2. Pesticide use · 3. Land abandonment · 4. High stocking levels · 5. Marginal habitat loss · 6. Afforestation · 7. Cultivation of grassland · 8. Farming operations · 9. Crop specialization · 10. Recreation · 11. Overhead structures · 12. Increased predators · 13. Urbanization/roads · 14. Drainage/flood control · 15. Loss of oak trees · 16. Loss of hay meadows · 17. Low stocking levels · 18. Nutrient pollution · 19. Replaced crop trees · 20. Loss of old buildings · 21. Reduced carrion · 22. Autumn sowing · 23. Vegetation clearance · 24. Pasture conversion · 25. Dry cultiv. rice

Species	Priority score	1	2	3	4	5	6	7	8	9	10	11	12	13	14	15	16	17	18	19	20	21	22	23	24	25
Gypaetus barbatus	3	–	–	2	–	–	–	–	–	–	–	–	–	–	–	–	–	–	–	–	–	2	–	–	–	–
Neophron percnopterus	2	–	–	2	–	–	–	–	–	–	2	–	–	–	–	–	–	–	–	–	–	1	–	–	1	–
Gyps fulvus	2	–	–	2	–	–	–	–	–	–	–	–	–	–	–	–	–	–	–	–	–	2	–	–	–	–
Aegypius monachus	1	–	–	2	–	–	–	–	–	–	2	–	–	–	–	–	–	–	–	–	–	1	–	–	–	–
Aquila chrysaetos	1	–	–	2	–	–	2	–	–	–	2	–	–	–	–	–	–	–	–	–	–	–	–	–	–	–
Falco tinnunculus	2	1	–	–	–	–	–	–	–	–	–	–	–	–	–	–	–	–	2	–	–	–	–	–	–	–
Falco peregrinus	2	–	–	–	–	–	–	–	–	–	2	–	–	–	–	–	–	–	–	–	–	–	–	–	–	–
Tetrao tetrix	4	–	–	–	–	–	2	–	–	–	2	–	–	–	–	–	1	–	–	–	–	–	–	–	–	–
Tetrao mlokosiewiczi	4	–	–	1	2	–	2	–	–	–	2	–	–	–	–	–	–	–	–	–	–	–	–	–	–	–
Tetraogallus caucasicus	3	–	–	–	2	–	–	–	–	–	–	–	–	–	–	–	–	–	–	–	–	–	–	–	–	–
Tetraogallus caspius	3	–	–	–	2	–	2	–	–	–	–	–	–	–	–	–	–	–	–	–	–	–	–	–	–	–
Alectoris graeca	3	–	–	2	2	–	2	–	–	–	–	–	–	–	–	–	–	–	–	–	–	–	–	–	–	–
Coturnix coturnix	3	2	–	–	2	–	2	–	–	–	2	–	–	–	–	–	–	–	–	–	–	–	–	–	–	–
Crex crex	3	–	–	2	–	–	2	–	–	–	–	–	–	–	–	–	1	–	–	–	–	–	–	–	–	–
Bubo bubo	1	–	–	–	–	–	–	–	–	–	–	2	–	2	–	–	–	–	–	–	–	–	–	–	–	–
Lullula arborea	2	–	–	–	–	–	2	–	–	–	–	–	–	–	–	–	–	–	–	–	–	–	–	–	–	–
Alauda arvensis	2	–	–	–	–	–	2	–	–	–	–	–	–	–	–	–	–	–	1	–	–	–	–	–	–	–
Anthus campestris	1	–	–	2	–	–	2	–	–	–	–	–	–	–	–	–	–	–	–	–	–	–	–	–	–	–
Anthus spinoletta	1	–	–	–	2	–	2	–	–	–	–	–	–	–	–	–	–	–	1	–	–	–	–	–	–	–
Prunella ocularis	3	–	–	–	2	–	2	–	–	–	–	–	–	–	–	–	–	–	–	–	–	–	–	–	–	–
Prunella collaris	1	–	–	–	2	–	–	–	–	–	–	–	–	–	–	–	–	–	–	–	–	–	–	–	–	–
Saxicola torquata	1	–	–	2	2	–	2	–	–	–	–	–	–	–	–	–	–	–	–	–	–	–	–	–	–	–
Monticola saxatilis	2	–	–	2	2	–	2	–	–	–	–	–	–	–	–	–	–	–	–	–	–	–	–	–	–	–
Monticola solitarius	1	–	–	1	–	–	–	–	–	–	–	–	–	–	–	–	–	–	–	–	–	–	–	–	–	–
Pyrrhocorax graculus	2	–	–	–	1	–	–	–	–	–	2	–	–	–	–	–	–	–	2	–	–	–	–	–	–	–
Pyrrhocorax pyrrhocorax	2	–	–	2	2	–	2	–	–	–	2	–	–	–	–	–	–	–	1	–	–	–	–	–	–	–
Montifringilla nivalis	1	–	–	–	–	–	–	–	–	–	1	–	–	–	–	–	–	–	1	–	–	–	–	–	–	–
Serinus pusillus	1	–	–	–	–	–	1	–	–	–	–	–	–	–	–	–	–	–	–	–	–	–	–	–	–	–
Serinus citrinella	1	–	–	–	–	–	1	–	–	–	–	–	–	–	–	–	–	–	1	–	–	–	–	–	–	–
Rhodopechys sanguinea	1	–	–	–	–	–	–	–	–	–	–	–	–	–	–	–	–	–	–	–	–	–	–	–	–	–
Carpodacus rubicilla	3	–	–	1	1	–	–	–	–	–	–	–	–	–	–	–	–	–	–	–	–	–	–	–	–	–
Emberiza cia	2	–	–	1	1	2	1	–	1	–	–	–	–	–	–	–	–	–	1	–	–	–	–	–	–	–
Emberiza hortulana	2	1	–	1	1	2	2	–	1	–	1	–	–	1	–	–	–	–	1	–	–	–	–	–	2	–
No. of species																										
Critical impact		0	0	0	0	0	0	0	0	0	0	0	0	0	0	0	0	0	0	0	0	0	0	0	0	0
High impact		2	0	9	10	1	10	0	1	0	5	0	0	0	0	0	0	0	2	0	0	2	0	0	0	0
Low impact		8	0	13	8	3	13	0	2	0	9	6	8	8	0	0	5	0	15	0	0	2	0	0	2	0
Total		10	0	22	18	4	23	0	3	0	14	6	8	8	0	0	5	0	17	0	0	4	0	0	2	0
% of priority spp. threatened		30%	0%	67%	55%	12%	70%	0%	9%	0%	42%	18%	24%	24%	0%	0%	15%	0%	52%	0%	0%	12%	0%	0%	6%	0%
Weighted threat score		21	0	52	56	8	52	0	8	0	34	11	20	15	0	0	7	0	29	0	0	13	0	0	4	0

Agricultural and grassland habitats (cont.)

■ Wet grassland

Threat columns:
1. Crop improvements
2. Pesticide use
3. Land abandonment
4. High stocking levels
5. Marginal habitat loss
6. Afforestation
7. Cultivation of grassland
8. Farming operations
9. Crop specialization
10. Recreation
11. Overhead structures
12. Increased predators
13. Urbanization/roads
14. Drainage/flood control
15. Loss of oak, trees
16. Loss of hay meadows
17. Low stocking levels
18. Nutrient pollution
19. Replaced crop
20. Loss of old buildings, trees
21. Reduced carrion
22. Autumn sowing
23. Vegetation clearance
24. pasture conversion
25. Dry cultiv. rice

Species	Priority score	1	2	3	4	5	6	7	8	9	10	11	12	13	14	15	16	17	18	19	20	21	22	23	24	25
Ardea purpurea	3	—	—	—	—	—	—	—	—	—	—	—	—	—	2	—	—	—	—	—	—	—	—	—	—	—
Ciconia ciconia	2	—	—	—	—	—	—	—	—	—	—	—	—	—	2	—	—	—	—	—	—	—	—	—	—	—
Cygnus columbianus	2	—	—	—	—	—	—	—	—	—	—	—	—	—	2	—	—	—	—	—	—	—	—	—	—	—
Cygnus cygnus	1	—	—	—	—	—	—	—	—	—	—	—	—	—	2	—	—	—	—	—	—	—	—	—	—	—
Anser brachyrhynchus	1	—	—	—	—	—	—	—	—	—	—	—	—	—	—	—	—	—	—	—	—	—	—	—	—	—
Anser erythropus	4	—	—	—	—	—	—	—	—	—	—	—	—	—	2	—	—	—	—	—	—	—	—	—	—	—
Branta bernicla	2	—	—	—	—	—	—	—	—	—	—	—	—	—	—	—	—	—	—	—	—	—	—	—	—	—
Anas acuta	1	—	—	—	—	—	—	—	—	—	—	—	—	—	2	—	—	—	—	—	—	—	—	—	—	—
Anas querquedula	2	—	—	2	—	—	—	—	—	—	—	—	—	—	2	—	—	—	—	—	—	—	—	—	—	—
Circus cyaneus	2	—	—	1	—	—	—	—	—	—	—	—	—	—	2	—	—	—	—	—	—	—	—	—	—	—
Aquila pomarina	2	—	—	1	—	—	—	—	—	—	—	—	—	—	—	—	—	—	—	—	—	—	—	—	—	—
Aquila clanga	4	—	—	—	—	—	—	—	—	—	—	—	—	—	2	—	—	—	—	—	—	—	—	—	—	—
Aquila heliaca	3	—	—	—	—	—	—	—	—	—	—	—	—	—	—	—	—	—	—	—	—	—	—	—	—	—
Falco tinnunculus	1	—	1	—	—	—	—	—	—	—	—	—	—	—	—	—	—	—	—	—	—	—	—	—	—	—
Coturnix coturnix	1	—	2	—	—	2	—	—	2	—	—	—	—	—	—	—	—	—	—	—	—	—	—	—	—	—
Porzana porzana	2	—	—	—	—	2	—	—	—	—	—	—	—	—	2	—	—	—	—	—	—	—	—	—	—	—
Crex crex	4	—	—	2	2	—	—	—	2	—	—	—	—	—	—	—	2	—	—	—	—	—	—	—	—	—
Grus grus	2	—	—	—	—	—	—	—	—	—	—	—	—	—	2	—	—	—	—	—	—	—	—	—	—	—
Calidris alpina	1	—	—	—	2	—	—	—	—	—	—	—	—	—	—	—	2	—	—	—	—	—	—	—	—	—
Philomachus pugnax	2	—	—	2	2	—	—	—	—	—	—	—	—	—	2	—	2	—	—	—	—	—	—	—	—	—
Gallinago media	4	—	—	2	2	—	—	—	—	—	—	—	—	—	2	—	2	—	—	—	—	—	—	—	—	—
Limosa limosa	4	—	2	2	2	—	—	—	—	—	—	—	—	—	2	—	2	—	—	—	—	—	—	—	—	—
Numenius arquata	2	—	—	—	—	—	—	—	—	—	—	—	—	—	2	—	2	—	—	—	—	—	—	—	—	—
Tringa totanus	3	—	—	—	2	—	—	—	—	—	—	—	—	—	2	—	2	—	—	—	—	—	—	—	—	—
Larus canus	3	—	—	—	—	—	—	—	—	—	—	—	—	—	—	—	—	—	—	—	—	—	—	—	—	—
Chlidonias niger	2	—	—	—	—	—	—	—	—	—	—	—	—	—	1	—	—	—	—	—	—	—	—	—	—	—
Tyto alba	1	—	—	—	—	—	—	—	—	—	—	—	—	1	—	—	—	—	—	—	—	—	—	—	—	—
Athene noctua	2	—	—	—	—	—	—	—	—	—	—	—	—	—	—	—	—	—	—	—	—	—	—	—	—	—
Asio flammeus	1	—	—	—	—	—	—	—	—	—	—	—	—	—	—	—	—	—	—	—	—	—	—	—	—	—
Alauda arvensis	2	—	—	—	—	—	—	—	—	—	—	—	—	—	—	—	—	—	—	—	—	—	—	—	—	—
Hirundo rustica	1	—	—	—	—	—	—	—	—	—	—	—	—	—	—	—	—	—	—	—	—	—	—	—	—	—
Anthus pratensis	1	—	—	—	—	—	—	—	—	—	—	—	—	—	—	—	—	—	—	—	—	—	—	—	—	—
Motacilla flava	1	—	—	—	—	2	—	—	—	—	—	—	—	—	—	—	—	—	—	—	—	—	—	—	—	—
Saxicola rubetra	1	—	—	—	—	2	—	—	—	—	—	—	—	—	—	—	—	—	—	—	—	—	—	—	—	—
Saxicola torquata	1	—	—	—	—	2	—	—	—	—	—	—	—	—	—	—	—	—	—	—	—	—	—	—	—	—
Locustella naevia	1	—	—	—	—	2	—	—	—	—	—	—	—	—	—	—	—	—	—	—	—	—	—	—	—	—
Acrocephalus paludicola	4	—	2	—	—	—	—	—	—	—	—	—	—	—	2	—	—	—	—	—	—	—	—	—	—	—
No. of species																										
Critical impact	0	0	0	0	0	0	0	0	0	0	0	0	0	0	0	0	0	0	0	0	0	0	0	0	0	0
High impact	18	0	2	5	7	0	0	0	2	0	0	0	0	0	7	0	2	0	0	0	0	0	0	0	0	0
Low impact	4	4	8	9	5	17	0	0	8	10	9	7	8	8	16	0	6	0	0	0	2	0	0	1	0	0
Total	22	4	10	14	12	17	0	0	10	10	9	7	8	8	23	0	8	0	0	0	2	0	0	1	0	0
% of priority spp. threatened	59%	11%	27%	38%	32%	46%	0%	0%	27%	27%	24%	19%	22%	22%	62%	0%	22%	0%	0%	0%	5%	0%	0%	3%	0%	0%
Weighted threat score	80	4	31	48	30	36	0	0	25	19	21	19	16	16	56	0	23	0	0	0	3	0	0	2	0	0

Agricultural and grassland habitats (cont.)

▪ Rice cultivation

Species	Priority score	1. Crop improvements	2. Pesticide use	3. Land abandonment	4. High stocking levels	5. Marginal habitat loss	6. Afforestation	7. Cultivation of grassland	8. Farming operations	9. Crop specialization	10. Recreation	11. Overhead structures	12. Increased predators	13. Urbanization/ roads	14. Drainage/flood control	15. Loss of oak trees	16. Loss of hay meadows	17. Low stocking levels	18. Nutrient pollution	19. Replaced crop trees	20. Loss of old buildings	21. Reduced carrion	22. Autumn sowing	23. Vegetation clearance	24. Pasture conversion	25. Dry cultiv. rice
Ixobrychus minutus	3	—	—	—	—	—	—	—	—	—	—	—	—	—	—	—	—	—	—	—	—	—	—	—	—	—
Nycticorax nycticorax	3	—	—	—	—	—	—	—	—	—	—	—	—	—	—	—	—	—	—	—	—	—	—	—	—	—
Ardeola ralloides	—	—	—	—	—	—	—	—	—	—	—	—	—	—	—	—	—	—	—	—	—	—	—	—	—	—
Ardea purpurea	—	—	—	—	—	—	—	—	—	—	—	—	—	—	—	—	—	—	—	—	—	—	—	—	—	—
Plegadis falcinellus	—	—	—	—	—	—	—	—	—	—	—	—	—	—	—	—	—	—	—	—	—	—	—	—	—	—
Platalea leucorodia	2	—	—	—	—	—	—	—	—	—	—	—	—	—	—	—	—	—	—	—	—	—	—	—	—	—
Anas acuta	—	—	—	—	—	—	—	—	—	—	—	—	—	—	—	—	—	—	—	—	—	—	—	—	—	—
Anas querquedula	—	—	—	—	—	—	—	—	—	—	—	—	—	—	—	—	—	—	—	—	—	—	—	—	—	—
Netta rufina	—	—	—	—	—	—	—	—	—	—	—	—	—	—	—	—	—	—	—	—	—	—	—	—	—	—
Milvus migrans	—	—	—	—	—	—	—	—	—	—	—	—	—	—	—	—	—	—	—	—	—	—	—	—	—	—
Falco vespertinus	—	—	—	—	—	—	—	—	—	—	—	—	—	—	—	—	—	—	—	—	—	—	—	—	—	—
Porzana pusilla	—	—	—	—	—	—	—	—	—	—	—	—	—	—	—	—	—	—	—	—	—	—	—	—	—	—
Glareola pratincola	—	—	—	—	—	—	—	—	—	—	—	—	—	—	—	—	—	—	—	—	—	—	—	—	—	—
Glareola nordmanni	—	—	—	—	—	—	—	—	—	—	—	—	—	—	—	—	—	—	—	—	—	—	—	—	—	—
Calidris alpina	3	—	—	—	—	—	—	—	—	—	—	—	—	—	—	—	—	—	—	—	—	—	—	—	—	—
Limosa limosa	—	—	—	—	—	—	—	—	—	—	—	—	—	—	—	—	—	—	—	—	—	—	—	—	—	—
Tringa totanus	2	—	—	—	—	—	—	—	—	—	—	—	—	—	—	—	—	—	—	—	—	—	—	—	—	—
Gelochelidon nilotica	—	—	—	—	—	—	—	—	—	—	—	—	—	—	—	—	—	—	—	—	—	—	—	—	—	—
Chlidonias niger	—	—	—	—	—	—	—	—	—	—	—	—	—	—	—	—	—	—	—	—	—	—	—	—	—	—
Asio flammeus	—	—	—	—	—	—	—	—	—	—	—	—	—	—	—	—	—	—	—	—	—	—	—	—	—	—
Hirundo rustica	—	—	—	—	—	—	—	—	—	—	—	—	—	—	—	—	—	—	—	—	—	—	—	—	—	—
No. of species																										
Critical impact	0	0	0	0	0	0	0	0	0	0	0	0	0	0	0	0	0	0	0	0	0	0	0	0	0	0
High impact	0	0	0	0	0	0	0	0	0	0	0	0	0	0	0	0	0	0	0	0	0	0	0	0	0	0
Low impact	0	10	2	2	0	4	0	0	2	0	0	0	0	0	0	0	0	0	0	0	0	0	0	0	0	10
Total	0	10	2	2	0	4	0	0	2	0	0	0	0	0	0	0	0	0	0	0	0	0	0	0	0	10
% of priority spp. threatened	0%	48%	10%	10%	0%	19%	0%	0%	10%	0%	0%	0%	0%	0%	0%	0%	0%	0%	0%	0%	0%	0%	0%	0%	0%	48%
Weighted threat score	0	16	6	6	0	6	0	0	2	0	0	0	0	0	0	0	0	0	0	0	0	0	0	0	0	15

Agricultural and grassland habitats (cont.)

■ Perennial crops

Threat columns:
1. Crop improvements
2. Pesticide use
3. Land abandonment
4. High stocking levels
5. Marginal habitat loss
6. Afforestation
7. Cultivation of grassland
8. Farming operations
9. Crop specialization
10. Recreation
11. Overhead structures
12. Increased predators
13. Urbanization/roads
14. Damage/flood control
15. Loss of oak contour
16. Loss of hay meadows
17. Low stocking levels
18. Nutrient pollution
19. Replaced crop/trees
20. Loss of old buildings
21. Reduced carrion
22. Autumn sowing
23. Vegetation clearance
24. Pasture conversion
25. Dry cultiv. rice

Species	Priority score
Accipiter brevipes	2
Alectoris rufa	2
Streptopelia turtur	1
Tyto alba	1
Otus scops	3
Athene noctua	1
Merops apiaster	2
Coracias garrulus	3
Jynx torquilla	3
Picus viridis	3
Dendrocopos syriacus	1
Galerida theklae	2
Lullula arborea	3
Erithacus rubecula	1
Phoenicurus phoenicurus	2
Oenanthe hispanica	2
Turdus merula	1
Turdus philomelos	1
Turdus iliacus	1
Turdus viscivorus	1
Hippolais pallida	1
Hippolais olivetorum	3
Sylvia melanocephala	2
Sylvia hortensis	1
Sylvia atricapilla	1
Muscicapa striata	1
Parus lugubris	1
Certhia brachydactyla	1
Lanius collurio	3
Lanius senator	3
Lanius nubicus	3
Sturnus unicolor	1
Fringilla coelebs	1
Serinus serinus	1
Serinus canaria	1
Carduelis chloris	1
Emberiza hortulana	3

Summary by threat

	1. Crop improvements	2. Pesticide use	3. Land abandonment	5. Marginal habitat loss	19. Replaced crop/trees	22. Autumn sowing
Low impact	0	15	0	2	0	0
High impact	6	4	8	4	9	7
Critical impact	0	0	0	0	0	0
Total	6	19	8	6	9	7
% of priority spp. threatened	16%	51%	22%	16%	24%	19%
Weighted threat score	14	43	30	13	32	20

All other threat columns (4, 6–18, 20, 21, 23, 24, 25): Low impact 0, High impact 0, Critical impact 0, Total 0, % of priority spp. threatened 0%, Weighted threat score 0.

Agricultural and grassland habitats (cont.)

■ Pastoral woodland

Species	Priority score	1. Crop improvements	2. Pesticide use	3. Land abandonment	4. High stocking levels	5. Marginal habitat loss	6. Afforestation	7. Cultivation of grassland	8. Farming operations	9. Crop specialization	10. Recreation	11. Overhead structures	12. Increased predators	13. Urbanization/roads	14. Damage/flood control	15. Loss of oak trees	16. Loss of hay meadows	17. Low stocking	18. Nutrient pollution	19. Replaced crop	20. Loss of old trees/buildings	21. Reduced carrion	22. Autumn sowing	23. Vegetation clearance	24. Pasture conversion	25. Dry cultiv. rice
Ciconia nigra	2	—	—	—	—	—	—	—	—	—	—	1	—	—	—	2	—	—	—	—	—	—	—	—	—	—
Ciconia ciconia	3	—	—	3	—	2	—	—	—	—	—	1	—	—	—	—	—	—	—	—	—	—	—	—	—	—
Elanus caeruleus	2	—	—	2	—	2	—	—	—	—	—	—	—	—	—	—	—	—	—	—	—	—	—	—	—	—
Milvus migrans	2	—	—	2	—	—	—	—	—	—	—	—	—	—	—	2	—	—	—	—	—	—	—	—	—	—
Milvus milvus	1	—	—	—	—	—	—	—	—	—	—	—	—	—	—	2	—	—	—	—	—	—	—	—	—	—
Aegypius monachus	2	—	—	—	1	—	—	—	—	—	—	—	—	—	—	2	—	—	—	—	—	—	—	—	—	—
Circaetus gallicus	2	—	—	2	—	—	—	—	—	—	—	—	—	—	—	—	—	—	—	—	—	—	—	—	—	—
Aquila adalberti	4	—	—	—	—	—	—	—	—	—	2	2	—	1	—	2	—	—	—	—	—	—	—	—	—	—
Hieraaetus pennatus	2	—	—	2	—	—	—	—	—	—	—	—	—	—	—	1	—	—	—	—	—	—	—	—	—	—
Falco tinnunculus	2	—	1	2	1	—	—	—	—	—	—	—	—	—	—	1	—	—	—	—	—	—	—	—	—	—
Grus grus	2	—	—	3	—	—	—	—	—	—	—	—	—	—	—	—	—	—	—	—	—	—	—	—	—	—
Columba oenas	1	—	2	2	—	2	—	—	—	—	—	—	—	—	—	2	—	—	—	—	—	—	—	—	—	—
Columba palumbus	1	—	2	2	3	—	—	—	—	—	—	—	—	—	—	1	—	—	—	—	—	—	—	—	—	—
Streptopelia turtur	1	—	2	1	2	—	—	—	—	—	—	—	—	—	—	1	—	—	—	—	—	—	—	—	—	—
Tyto alba	1	—	2	—	—	—	—	—	—	—	—	—	—	—	—	1	—	—	—	—	—	—	—	—	—	—
Otus scops	3	—	2	2	—	—	—	—	—	—	—	—	—	—	—	1	—	—	—	—	—	—	—	—	—	—
Athene noctua	2	—	2	2	2	—	—	—	—	—	—	—	—	—	—	1	—	—	—	—	—	—	—	—	—	—
Merops apiaster	2	—	2	2	—	—	—	—	—	—	—	—	—	—	—	—	—	—	—	—	—	—	—	—	—	—
Coracias garrulus	3	—	—	—	—	—	—	—	—	—	—	—	—	—	—	2	—	—	—	—	—	—	—	—	—	—
Picus viridis	3	—	—	—	—	—	—	—	—	—	—	—	—	—	—	2	—	—	—	—	—	—	—	—	—	—
Galerida theklae	3	—	—	3	—	—	—	—	—	—	—	—	—	—	—	—	—	—	—	—	—	—	—	—	—	—
Lullula arborea	3	—	—	2	—	—	—	—	—	—	—	—	—	—	—	—	—	—	—	—	—	—	—	—	—	—
Erithacus rubecula	2	—	—	—	—	—	—	—	—	—	—	—	—	—	—	1	—	—	—	—	—	—	—	—	—	—
Phoenicurus phoenicurus	2	—	—	—	2	—	—	—	—	—	—	—	—	—	—	1	—	—	—	—	—	—	—	—	—	—
Turdus merula	1	—	—	—	—	—	—	—	—	—	—	—	—	—	—	1	—	—	—	—	—	—	—	—	—	—
Turdus philomelos	1	—	—	—	—	—	—	—	—	—	—	—	—	—	—	1	—	—	—	—	—	—	—	—	—	—
Turdus viscivorus	1	—	—	1	—	—	—	—	—	—	—	—	—	—	—	1	—	—	—	—	—	—	—	—	—	—
Sylvia hortensis	2	—	—	—	—	—	—	—	—	—	—	—	—	—	—	2	—	—	—	—	—	—	—	—	—	—
Sylvia atricapilla	1	—	—	2	—	—	—	—	—	—	—	—	—	—	—	2	—	—	—	—	—	—	—	—	—	—
Muscicapa striata	1	—	—	—	—	—	—	—	—	—	—	—	—	—	—	1	—	—	—	—	—	—	—	—	—	—
Certhia brachydactyla	1	—	—	—	—	—	—	—	—	—	—	—	—	—	—	2	—	—	—	—	—	—	—	—	—	—
Lanius senator	4	—	—	2	—	—	—	—	—	—	—	—	—	—	—	1	—	—	—	—	—	—	—	—	—	—
Corvus monedula	1	—	—	2	—	—	—	—	—	—	—	—	—	—	—	1	—	—	—	—	—	—	—	—	—	—
Sturnus unicolor	1	—	2	2	—	—	—	—	—	—	—	—	—	—	—	1	—	—	—	—	—	—	—	—	—	—
Fringilla coelebs	1	—	—	2	—	—	—	—	—	—	—	—	—	—	—	1	—	—	—	—	—	—	—	—	—	—
Serinus serinus	1	—	2	2	—	—	—	—	—	—	—	—	—	—	—	1	—	—	—	—	—	—	—	—	—	—
Carduelis chloris	1	—	1	2	—	—	—	—	—	—	—	—	—	—	—	1	—	—	—	—	—	—	—	—	—	—
No. of species																										
Critical impact	—	0	0	3	1	0	0	0	0	0	0	0	0	0	0	0	0	0	0	0	0	0	0	0	0	0
High impact	—	0	5	14	3	2	0	0	0	0	0	1	0	0	0	5	0	0	0	0	0	0	0	0	0	0
Low impact	—	0	8	8	1	2	0	0	0	0	8	8	0	1	0	21	0	0	0	0	0	0	0	0	0	0
Total	—	0	13	25	5	4	0	0	0	0	8	9	0	1	0	26	0	0	0	0	0	0	0	0	0	0
% of priority spp. threatened	0%	0%	35%	68%	14%	11%	0%	0%	0%	0%	22%	24%	0%	3%	0%	70%	0%	0%	0%	0%	0%	0%	0%	0%	0%	0%
Weighted threat score	—	0	35	87	15	13	0	0	0	0	17	24	0	4	0	60	0	0	0	0	0	0	0	0	0	0

THE FOLLOWING habitat-types are considered to be of particular interest within the 15 member states of the European Community, and their conservation requires the designation of Special Areas of Conservation under the Habitats Directive (see 'Opportunities for Conserving the Wider Environment', p. 24). This appendix shows which of these 'natural' habitat-types are treated under which of the chapters in this publication (for reasons given in the 'Introduction', p. 17, Macaronesian habitat-types are not covered in this book). The habitat terms are more or less verbatim from the *Official Journal of the European Communities* No. L 206/16 (22.7.92), as amended by No. L 1/136 (1.1.95).

The codes for individual habitats represent the hierarchical classification produced through the Corine programme (Council Decision 85/338/EEC of 27 June 1985, OJ No. L 176, 6.7.1985, p. 14) (Corine biotopes project) as listed in the Technical Handbook, Volume 1, pp. 73–109, Corine/Biotope/89/2.2, 19 May 1988, partially updated 14 February 1989.

× Combination of two or more habitat-types
* Priority habitat-type

Habitat-types are included within this book under chapters as follows:

M   Marine habitats (p. 59)
C   Coastal habitats (p. 93)
W   Inland wetlands (p. 125)
T   Tundra, mires and moorland (p. 159)
H   Lowland Atlantic heathland (p. 187)
B   Boreal and temperate forests (p. 203)
MF   Mediterranean forest, shrubland and rocky habitats (p. 239)
A   Agricultural and grassland habitats (p. 267)

COASTAL AND HALOPHYTIC HABITATS

Open sea and tidal areas

M	11.25	Sandbanks which are slightly covered by seawater all the time
M	11.34	*Posidonia* beds
C	13.2	Estuaries
C	14	Mudflats and sandflats not covered by seawater at low tide
C	21	*Lagoons
M	—	Large shallow inlets and bays
M	—	Reefs
M	—	Marine 'columns' in shallow water made by leaking gases

Sea cliffs and shingle or stony beaches

C	17.2	Annual vegetation of drift lines
C	17.3	Perennial vegetation of stony banks
C	18.21	Vegetated sea cliffs of the Atlantic and Baltic coasts
C	18.22	Vegetated sea cliffs of the Mediterranean coasts (with endemic *Limonium* spp.)
C	18.23	Vegetated sea cliffs of the Macaronesian coasts (flora endemic to these coasts)

Sea-cliffs with nesting colonies of priority seabird species are dealt with under 'Marine Habitats'.

Atlantic and continental salt marshes and salt meadows

C	15.11	*Salicornia* and other annuals colonizing mud and sand
C	15.12	*Spartina* swards (*Spartinion*)
C	15.13	Atlantic salt meadows (*Glauco–Puccinellietalia*)
C	15.14	*Continental salt meadows (*Puccinellietalia distantis*)

Mediterranean and thermo-Atlantic salt marshes and salt meadows

C	15.15	Mediterranean salt meadows (*Juncetalia maritimi*)
C	15.16	Mediterranean and thermo-Atlantic halophilous scrubs (*Arthrocnemetalia fructicosae*)
C	15.17	Iberia halo-nitrophilous scrubs (*Pegano–Salsoletea*)

Salt and gypsum continental steppes

C	15.18	*Salt steppes (*Limonietalia*)
C	15.19	*Gypsum steppes (*Gypsophiletalia*)
A	15.1A	*Pannonic salt steppes and saltmarshes

COASTAL SAND DUNES AND CONTINENTAL DUNES

Sea dunes of the Atlantic, North Sea and Baltic coasts

C	16.211	Embryonic shifting dunes
C	16.212	Shifting dunes along the shoreline with *Ammophila arenaria* (white dunes)
C	16.221–16.227	*Fixed dunes with herbaceous vegetation (grey dunes):
		16.221 *Galio–Koelerion albescentis*
		16.222 *Euphorbio–Helichrysion*
		16.223 *Crucianellion maritimae*
		16.224 *Euphorbia terracina*
		16.225 *Mesobromion*
		16.226 *Trifolio–Gerantietea sanguinei, Galio maritimi–Geranion sanguinei*
		16.227 *Thero–Airion, Botrychio–Polygaletum, Tuberarion guttatae*
C	16.23	*Decalcified fixed dunes with *Empetrum nigrum*
C	16.24	Eu-atlantic decalcified fixed dunes (*Calluno–Ulicetea*)
C	16.25	Dunes with *Hyppophae rhamnoides*
C	16.26	Dunes with *Salix arenaria*
C	16.29	Wooded dunes of the Atlantic coast
C	16.31–16.35	Humid dune slacks
A	1.A	Machairs (* in machairs in Ireland)

Sea dunes of the Mediterranean coast

C	16.223	*Crucianellion maritimae* fixed beach dunes
C	16.224	Dunes with *Euphorbia terracina*
C	16.228	*Malcolmietalia* dune grasslands

C   16.229   *Brachypodietalia* dune grasslands with annuals
C   16.27   *Dune juniper thickets (*Juniperus* spp.)
C   16.28   Dune sclerophyllous scrubs (*Cisto–Lavenduletalia*)
C   16.29 ×   *Wooded dunes with *Pinus pinea* and/or *P. pinaster*
42.8

Continental dunes, old and decalcified
H   64.1 ×   Dry sandy heaths with *Calluna* and *Genista*
31.223
H   64.1 ×   Dry sandy heaths with *Calluna* and *Empetrum nigrum*
31.227
A   64.1 ×   Open grassland with *Corynephorus* and *Agrostis*
35.2   of continental dunes
A   64.71   *Pannonic inland dunes

FRESHWATER HABITATS

Standing water
W   22.11 ×   Oligotrophic waters containing very few minerals
22.31   of Atlantic sandy plains with amphibious vegetation: *Lobelia, Littorella* and *Isoetes*
W   22.11 ×   Oligotrophic waters containing very few minerals
22.34   of west Mediterranean sandy plains with *Isoetes*
W   22.12 ×   Oligotrophic waters in medio European and
(22.31,   perialpine area with amphibious vegetation:
22.32)   *Littorella* or *Isoetes* or annual vegetation on exposed banks (*Nanocyperetalia*)
W   22.12 ×   Hard oligo-mesotrophic waters with benthic
22.44   vegetation of chara formations
W   22.13   Natural eutrophic lakes with *Magnopotamion* or *Hydrocharition*-type vegetation
W   22.14   Dystrophic lakes
W   22.34   *Mediterranean temporary ponds
W   —   *Turloughs (Ireland)

Running water
Sections of watercourses with natural or semi-natural dynamics (minor, average and major beds) where the water quality shows no significant deterioration.
W   24.221,   Alpine rivers and the herbaceous vegetation along
24.222   the banks
W   24.223   Alpine rivers and their ligneous vegetation with *Myricaria germanica*
W   24.224   Alpine rivers and their ligneous vegetation with *Salix elaegnos*
W   24.225   Constantly flowing Mediterranean rivers with *Glaucium flavum*
W   24.4   Floating vegetation of *Ranunculus* of plane, submountainous rivers
W   24.52   *Chenopodietum rubri* of submountainous rivers
W   24.53   Constantly flowing Mediterranean rivers: *Paspalo–Agrostidion* and hanging curtains of *Salix* and *Populus alba*
W   —   Intermittently flowing Mediterranean rivers

TEMPERATE HEATH AND SCRUB

H   31.11   Northern Atlantic wet heaths with *Erica tetralix*
H   31.12   *Southern Atlantic wet heaths with *Erica ciliaris* and *E. tetralix*
H   31.2   *Dry heaths (all subtypes)
H   31.234   Dry coastal heaths with *Erica vagans* and *Ulex maritimus*
—   31.3   *Endemic Macaronesian dry heaths
A   31.4   Alpine and subalpine heaths
A   31.5   *Scrub with *Pinus mugo* and *Rhododendron hirsutum* (*Mugo–Rhododendretum hirsuti*)
T   31.622   Sub-Arctic willow scrub
MF 31.7   Endemic oro-Mediterranean heaths with gorse

SCLEROPHYLLOUS SCRUB (MATORRAL)

Sub-Mediterranean and temperate
A   31.82   Stable *Buxus sempervirens* formations on calcareous rock slopes (*Berberidion p.*)
A   31.842   Mountain *Genista purgans* formations
A   31.88   *Juniperus communis* formations on calcareous heaths or grasslands
H   31.89   *Cistus palhinhae* formations on maritime wet heats (*Junipero–Cistetum palhinhae*)

Mediterranean arborescent matorral
MF 32.131–   Juniper formations
32.135
MF 32.17   *Matorral with *Zyziphus*
MF 32.18   *Matorral with *Laurus nobilis*

Thermo-Mediterranean and pre-steppe brush
MF 32.216   Laurel thickets
MF 32.217   Low formations of euphorbia close to cliffs
MF 32.22–   All types
32.26

Phrygana
MF 33.1   *Astragalo–Plantaginetum subulatae* phrygana
MF 33.3   *Sarcopoterium spinosum* phrygana
MF 33.4   Cretan formations (*Euphorbieto–Verbascion*)

NATURAL AND SEMI-NATURAL GRASSLAND FORMATIONS

Natural grasslands
A   34.11   *Karstic calcareous grasslands (*Alysso-Sedion albi*)
A   34.12   *Xeric sand calcareous grasslands (*Koelerion glaucae*)
A   34.2   Calaminarian grasslands
A   36.314   Siliceous Pyrenean grasslands with *Festuca eskia*
A   36.32   Siliceous alpine and boreal grass
A   36.36   Siliceous *Festuca indigesta* Iberian grasslands
A   36.41–   Alpine calcareous grasslands
36.45
—   36.5   Macaronesian mountain grasslands

Semi-natural dry grasslands and scrubland facies
A   34.31   *Sub-continental steppic grassland
A   34.32–   On calcareous substrates (*Festuco–Brometalia*)
34.34
A   34.5   *Pseudo-steppe with grasses and annuals (*Thero–Brachypodietea*)
A   34.91   *Pannonic steppes
A   34.A1   *Pannonic sand steppes
A   35.1   *Species-rich *Nardus* grasslands, on siliceous substrates in mountain areas (and submountain areas, in continental Europe)

Sclerophyllous grazed forests (dehesas)
A   32.11   With *Quercus suber* and/or *Quercus ilex*

Semi-natural tall-herb humid meadows
A   37.31   Molinia meadows on chalk and clay (*Eu–Molimion*)
A   37.4   Mediterranean tall-herb and rush meadows (*Molinio–Holoschoenion*)
A   37.7,   Eutrophic tall herbs
37.8
A   —   *Cnidion venosae* meadows liable to flooding

Mesophile grasslands
A   38.2   Lowland hay meadows (*Alopecurus pratensis, Sanguisorba officinalis*)
A   38.3   Mountain hay meadows (British types with *Geranium sylvaticum*)

RAISED BOGS AND MIRES AND FENS

Sphagnum acid bogs

T	51.1	*Active raised bogs
T	51.2	Degraded raised bogs (still capable of natural regeneration)
T	52.1, 52.2	Blanket bog (* active only)
T	54.5	Transition mires and quaking bogs
T	54.6	Depressions on peat substrates (*Rhynchosporion*)

Calcareous fens

W	53.3	*Calcareous fens with *Cladium mariscus* and *Carex davalliana*
W	54.12	*Petrifying springs with tufa formation (*Cratoneurion*)
W	54.2	Alkaline fens
—	54.3	Alpine pioneer formations of *Caricion bicoloris-atrofuscae*

Aapa mires

T	54.8	*Aapa mires
T	54.9	*Palsa mires

ROCKY HABITATS AND CAVES

Scree

—	61.1	Siliceous
—	61.2	Eutric
MF	61.3	Western Mediterranean and alpine thermophilous
MF	61.4	Balkan
—	61.5	Medio-European siliceous
—	61.6	*Medio-European calcareous

Chasmophytic vegetation on rocky slopes

—	62.1, 62.1A	Calcareous sub-types
—	62.2	Silicicolous sub-types
—	62.3	Pioneer vegetation of rock surfaces
—	62.4	*Limestone pavements

Other rocky habitats

—	65	Caves not open to the public
—	—	Fields of lava and natural excavations
C	—	Submerged or partly submerged sea caves
T,A	—	Permanent glaciers

FORESTS

(Sub)natural woodland vegetation comprising native species forming forests of tall trees, with typical undergrowth, and meeting the following criteria: rare or residual, and/or hosting species of Community interest.

Boreal forests

F	42.C	*Western taiga

Forests of temperate Europe

F	41.11	*Luzulo–Fagetum* beech forests
F	41.12	Beech forests with *Ilex* and *Taxus*, rich in epiphytes (*Ilici–Fagion*)
F	41.13	*Asperulo–Fagetum* beech forests
F	41.15	Subalpine beech woods with *Acer* and *Rumex arifolius*
F	41.16	Calcareous beech forest (*Cephalanthero–Fagion*)
F	41.24	*Stellatio–Carpinetum* oak–hornbeam forests
F	41.26	*Galio–Carpinetum* oak–hornbeam forests
F	41.2B	*Pannonic oak–hornbeam forest
F	41.4	*Tilio–Acerion* ravine forests
F	41.51	Old acidophilous oak woods with *Quercus robur* on sandy plains
F	41.53	Old oak woods with *Ilex* and *Blechnum* in the British Isles
F	41.7374	*Pannonic white-oak woods
F	41.7A	*Euro-Siberian steppe oak wood
F	41.86	*Fraxinus angustifolia* woods
F	42.51	*Caledonian forest
F	44.A1– 44.A4	*Bog woodland
F	44.3	*Residual alluvial forests (*Alnion glutinoso-incanae*)
F	44.4	Mixed oak–elm–ash forests of great rivers

Mediterranean deciduous forests

MF	41.181	*Apennine beech forests with *Taxus* and *Ilex*
MF	41.184	*Apennine beech forests with *Abies alba* and beech forests with *A. nebrodensis*
MF	41.6	Galicio–Portuguese oak woods with *Quercus robur* and *Q. pyrenaica*
MF	41.77	*Quercus faginea* woods (Iberian peninsula)
MF	41.85	*Quercus trojana* woods (Italy and Greece)
MF	41.9	Chestnut woods
MF	41.1A × 42.17	Hellenic beech forests with *Abies borisii-regis*
MF	41.1B	*Quercus frainetto* woods
MF	42.A1	Cypress forests (*Acero–Cupression*)
MF	44.17	*Salix alba* and *Populus alba* galleries
MF	44.52	Riparian formations on intermittent Mediterranean watercourses with *Rhododendron ponticum, Salix* and others
MF	44.7	Oriental plane woods (*Platanion orientalis*)
MF	44.8	Thermo-Mediterranean riparian galleries (*Nerio–Tamariceteae*) and south-west Iberian peninsula riparian galleries (*Securinegion tinctoriae*)

Mediterranean sclerophyllous forests

MF	41.7C	Cretan *Quercus brachyphylla* forests
MF	45.1	*Olea* and *Ceratonia* forests
MF	45.2	*Quercus suber* forests
MF	45.3	*Quercus ilex* forests
MF	45.5	*Quercus macrolepis* forests
—	45.61– 45.63	*Macaronesian laurel forests (*Laurus, Ocotea*)
MF	45.7	*Palm groves of *Phoenix*
MF	45.8	Forests of *Ilex aquifolium*

Alpine and subalpine coniferous forests

F	42.21, 42.23	Acidophilous forests (*Vaccinio–Piceetea*)
F	42.31, 42.32	Alpine forests with larch and *Pinus cembra*
F	42.4	*Pinus uncinata* forests (* on gypsum or limestone)

Mediterranean mountainous coniferous forests

MF	42.14	*Apennine *Abies alba* and *Picea excelsa* forests
MF	42.19	*Abies pinsapo* forests
MF	42.61– 42.66	*Mediterranean pine forests with endemic black pines
MF	42.8	Mediterranean pine forests with endemic Mesogean pines, including *Pinus mugo* and *P. leucodermis*
—	42.9	Macaronesian pine forests (endemic)
MF	42.A2– 42.A5, 42.A8	*Endemic Mediterranean forests with *Juniperus* spp.
MF	42.A6	*Tetraclinis articulata* forests (Andalusia)
F, MF	42.A71– 42.A73	*Taxus baccata* woods

THE EXPERTS of the Habitat Working Groups played the major role in drafting the habitat conservation strategies that form the eight main chapters of this book; their contribution is explained in more detail in the 'Introduction' (p. 17). Members of the groups are acknowledged at the start of the relevant chapters, and listed here are the titles and institutional affiliations of each.

■ **Marine habitats: north and west European seas**

Viddar Bakken
　Norwegian Polar Institute, Norway

Mardik F. Leopold
　Institute for Forestry and Nature Research, Netherlands

Guillemette Rolland
　SEPNB, France

Rui Rufino
　CEMPA/ICN, Portugal

Kristinn H. Skarphedinsson
　Icelandic Institute of Natural History, Iceland

Henrik Skov
　Ornis Consult, Denmark

Ekaterina Stotskaia
　Laboratory on Rare Bird Conservation, All-Union Research Instute of Nature Conservation and Reserves, Russia

Saulius Švazas
　Institute of Ecology, Lithuania

Mark Tasker
　Senior Marine Advisor, Joint Nature Conservation Committee, United Kingdom

■ **Marine habitats: European Macaronesian seas**

Domingo Concepción García
　Lanzarote, Spain

Felipe Rodríguez Godoy
　Seccíon de Flora y Fauna, Viceconsejería de Medio Ambiente, Spain

Cristina González
　Departamento de Biología Animal, Universidad de La Laguna, Spain

Aurelio Martín Hidalgo
　Departamento de Biología Animal, Universidad de La Laguna, Spain

Juan Luis Rodríguez Luengo
　Centro de Recuperación de Flora y Fauna, Viceconsejería de Medio Ambiente, Spain

Dr Luis Monteiro
　Department of Oceanography and Fisheries, University of Azores, Portugal

Dr Manuel Nogales
　Departamento de Biología Animal, Universidad de La Laguna, Spain

Paulo Oliveira
　Madeira, Portugal

Dr Irene Pereira
　Direcção Regional de Ambiente, Madeira, Portugal

■ **Marine habitats: Mediterranean and Black Seas**

Dr Luca Canova
　Dipartimento di Biologia Animale, Universita' di Pavia, Italy

Dr Mauro Fasola
　Dipartimento di Biologia Animale, Universita' di Pavia, Italy

■ **Coastal habitats**

Dr R. S. K. Barnes
　Department of Zoology, University of Cambridge, United Kingdom

Dr Tricia Bradley
　Royal Society for the Protection of Birds, United Kingdom

Monica Calado
　Parque Natural da Ria Formosa, Portugal

Filiz Demirayak
　Coastal Management Section, Dogal Hayati Koruma Dernegi, Turkey

Dr Pat Doody
　Joint Nature Conservation Committee, United Kingdom

Dr Helena Granja
　Departamento de Ciências da Terra, Universidade do Minho, Portugal

Dr Nathalie Hecker
　Station Biologique de la Tour du Valat, France

Dr R. E. Randall
　Girton College, University of Cambridge, United Kingdom

Cor J. Smit
　Institute for Forestry and Nature Research, Netherlands

John G. Walmsley
　MEDMARAVIS, France

■ **Inland wetlands**

Åke Andersson
　Swedish Hunters' Association, Sweden

Dr Nathalie Hecker
　Station Biologique de la Tour du Valat, France

Dr Verena Keller
　Schweizerische Vogelwarte, Switzerland

Petr Musil
　Institute of Applied Ecology, Agricultural University of Prague, Czech Republic

Dr Steve Ormerod
　School of Pure and Applied Biology, University of Wales, United Kingdom

Tobias Salathé
　Station Biologique de la Tour du Valat, France

Gürdar Sarigül
　Dogal Hayati Koruma Dernegi, Turkey

Dr Valentin Serebryakov
　Biological Department, Kiev State University, Ukraine

Vitas Staneviscius
　Ekologijos Instituas, Lithuania

Dr Janine van Vessem
　Head of Biodiversity Programme, Wetlands International, Netherlands

Dr Maria Wieloch
　Institute of Ecology, Polish Academy of Sciences, Poland

Murat Yarar
　Dogal Hayati Koruma Dernegi, Turkey

■ **Tundra, mires and moorland**

Dr Ingvar Byrkjedal
　Museum of Zoology, Norway

Dr Len Campbell
　Royal Society for the Protection of Birds, United Kingdom

Prof. Vladimir Galushin
　Department of Zoology and Ecology, Moscow State Pedagogical University, Russia

Dr John Atle Kålås
　Division of Terrestial Ecology, Norwegian Institute for Nature Research, Norway

Dr Alexander Mischenko
　All-Russian Research Institute for Nature Conservation, Russia

Vladimir Morozov
　All-Russian Research Institute for Nature Conservation, Russia

Dr Lennart Saari
　Roola, Finland

Dr Karl-Birger Strann
Division of Terrestrial
Ecology, Norwegian
Institute for Nature
Research, Norway

Dr Ivetta Tatarinkova
Kandalaksha State Nature
Reserve, Russia

Dr Des Thompson
Advisory Services,
Scottish Natural Heritage,
United Kingdom

■ Lowland Atlantic
heathland

Dr Herbert Diemont
Instituut voor Bos- en
Natuuronderzoek,
Netherlands

Dr Flemming Pagh Jensen
Ornis Consult, Denmark

Dr Lars Påhlsson
County Administration of
Scania, Department of
Environment, Sweden

Dr Nigel Webb
Institute of Terrestrial
Ecology, United Kingdom

Robin Wynde
Royal Society for the
Protection of Birds, United
Kingdom

■ Boreal and temperate
forests

Dr Per Angelstam
Department of
Conservation Biology
Swedish University of
Agricultural Sciences,
Sweden

Dr Vladimir Anufriev
Institute of Biology,
Russian Academy of
Sciences, Russia

Dr Ian Bainbridge
Head of Research,
Scotland, Royal Society for
the Protection of Birds,
United Kingdom

Dr Olivier Biber
Schweizerische
Vogelwarte, Switzerland

Dr Rob Fuller
British Trust for
Ornithology, United
Kingdom

Dr Ludwik Tomialojc
Museum of Natural
History, Wroclaw
University, Poland

Dr Tomasz Wesolowski
Wroclaw University,
Poland

■ Mediterranean forest,
shrubland and rocky
habitats

Júlia Almeida
Instituto da Conservação
da Natureza, Portugal

Mario Díaz Esteban
Departamento de Ecología,
Universidad Complutense,
Spain

Ben Hallman
Rapsani, Greece

Francesco Petretti
Roma, Italy

Dr Roger Prodon
Observatoire
Oceaniologique, CNRS,
France

Dr Gérard Rocamora
Ligue pour la Protection
des Oiseaux, France

■ Agricultural and
grassland habitats:
arable and improved
grassland

Jim Dixon
Royal Society for the
Protection of Birds, United
Kingdom

Mario Díaz Esteban
Departamento de Ecología,
Universidad Complutense,
Spain

Dr H. P. Kollar
Haringsee, Austria

Dr Elena Lebedeva
Russian Bird Conservation
Union, Russia

Ferenc Markus
WWF-Hungary, Hungary

Michel Métais
Ligue pour la Protection
des Oiseaux, France

Szabolcs Nagy
Magyar Madártani és
Természetvédelmi
Egyesület, Hungary

Dr Deborah J. Pain
International Department,
Royal Society for the
Protection of Birds, United
Kingdom

Dr Juha Tiainen
Finnish Game and Fisheries
Research Institute, Finland

Dr Graham M. Tucker
BirdLife International,
United Kingdom

■ Agricultural and
grassland habitats:
steppic habitats

Javier Alonso
Catedra de Zoología
(Vertebrados), Universidad
Complutense, Spain

Dr Victor Belik
Rostov-on-Don, Russia

Jean Boutin
Ecomusée de la Crau,
France

Dr Gilles Cheylan
Museum d'Histoire
Naturelle, Aix-en-
Provence, France

Jim Dixon
Royal Society for the
Protection of Birds, United
Kingdom

Dr S. Faragó
Department of Wildlife
Management, University of
Sopron, Hungary

Prof. Vladimir Galushin
Department of Zoology
and Ecology, Moscow
State Pedagogical
University, Russia

Dr H. P. Kollar
Haringsee, Austria

Dr Carmen Martínez
Departamento de
Biodiversidad, Museo
Nacional de Ciencias
Naturales, Spain

Miguel Angel Naveso
Sociedad Española de
Ornitología, Spain

Francisco Suárez
Departamento
Interuniversitario de
Ecología, Universidad
Autónoma, Spain

Sylvestre Voisin
Office du Génie
Ecologique, Lagny-sur-
Marne, France

■ Agricultural and
grassland habitats:
montane grassland

Pierandrea Brichetti
Verolavecchia, Italy

Prof. Paolo F. De
Franceschi
Verona, Italy

Johannes Frühauf
BirdLife Österreich,
Austria

László Vasile Kalabér
Reghin, Romania

Barbara Lombatti
Lega Italiana Protezione
Uccelli, Italy

Dr Idris Ogurlu
S. E. S. Biyoloji, Celal
Bayar University, Turkey

Paolo Pedrini
Museo Tridentino di
Scienze Naturali, Italy

Prof. Roald L. Potapov
Zoological Institute,
Russian Academy of
Sciences, Russia

Ueli Rehsteiner
Mörschwil, Switzerland

■ Agricultural and
grassland habitats: wet
grassland

Dr Albert J. Beintema
Instituut voor Bos- en
Natuuronderzoek,
Netherlands

Jean-Jacques Blanchon
Ligue pour la Protection
des Oiseaux, France

Dr Roger S. K. Buisson
Head of Conservation
Management Advice,
Royal Society for the
Protection of Birds, United
Kingdom

Przemek Chylarecki
Institute of Ecology, Polish
Academy of Sciences,
Poland

Hermann Hötker
Forschungs- und
Technologiezentrum
Westküste, Germany

Rein Kuresoo
Estonian Fund for Nature,
Estonia

Dr Jan Plesník
Head, IUCN Project
Coordination Unit, Czech
Republic

Dr Ken W. Smith
Research Department,
Royal Society for the
Protection of Birds, United
Kingdom

Ole Thorup
Ribe, Denmark

GLOSSARY

For **abbreviations**, see p. 433. Terms which are defined separately within this Glossary are given in **bold**.

Agri-environment Regulation EC Council Regulation 2078/92 on 'agricultural production methods compatible with the requirements of the protection of the environment and the maintenance of the countryside'. This regulation updates and broadens the previous EU schemes for supporting environmentally sensitive farming (e.g. ESAs, see below) and extensification. The regulation includes an aid scheme to promote environmental measures including reduction of pollution from agriculture, extensification of crop, sheep and cattle farming (including conversion of arable land into non-intensive grassland), long-term **set-aside**, and, through management agreements, the maintenance of farming practices beneficial to the environment. These measures are implemented at national or regional levels through the development of Zonal Programmes. See p. 267 for more details.

Alluvium Soils formed of fine particles of rock and detritus, washed down by rain or rivers, and deposited in a valley or estuary. Some of the most fertile soils are alluvial.

Alpine The higher regions of a mountain system, usually above the climatic treeline. The actual altitude of the alpine zone differs between mountain-ranges in Europe, depending mainly on the latitude, exposure and maximum height of each range.

Arctic Commonly referred to as the region north of the Arctic Circle (the parallel of latitude 66°33′N). Geographically it comprises the area north of the 10°C July isotherm, where July is the warmest summer month, provided that the mean temperature of the coldest month is not higher than 0°C.

Aspect The position of a plant **community** in relation to its exposure to the sun or to the local climate, e.g. north-facing or south-facing aspect.

Association The basic unit of the science of plant **communities**, defined according to its floristic composition and containing characteristic species; associations are named with the suffix -*etum*, e.g. *Quercetum ilicis*.

Benthos In freshwater and marine **ecosystems**, the assemblage of organisms attached to (or resting on) the bottom sediments (adjective: benthic).

Bern Convention Convention on the Conservation of European Wildlife and Natural Habitats. See 'Opportunities for Conserving the Wider Environment' (p. 24) and Appendix 1 (p. 327) for more information.

Biosphere Reserve Reserve created under the Man and the Biosphere Programme of UNESCO.

Birds Directive Directive and Resolution of the Council of the European Community on the Conservation of Wild Birds (79/409/EEC). See 'Opportunities for Conserving the Wider Environment' (p. 24) and Appendix 1 (p. 327) for more information.

Blanket mire A type of mire (bog) covering the whole surface of the landscape like a blanket, occurring in cool or cold climates with high rainfall and high atmospheric humidity.

Bloom Sudden growth of algae or plankton in an aquatic **ecosystem**.

Bog An acidic, peat-rich, waterlogged soil which develops in regions with high rainfall and high atmospheric humidity, and which is deficient in calcium and other basic minerals. The peat accumulates as a result of the incomplete decay of the vegetation. A bog is an acidic type of **mire**.

Bonn Convention Convention on the Conservation of Migratory Species of Wild Animals. See 'Opportunities for Conserving the Wider Environment' (p. 24) and Appendix 1 (p. 327) for more information.

Boreal A major bioclimatic zone, lying (in Europe) north of the **temperate** zone and south of the **tundra** zone; this region has short, warm summers and long, cold winters with snow, and the characteristic vegetation is the **taiga** (conifer-dominated forest).

Brackish Refers to water which is a mixture of both fresh and salt water.

Bycatch The unwanted catch by fishing gear of aquatic animals other than the target species (e.g. non-target or undersized fish, seabirds, turtles, dolphins, invertebrates).

Calcareous Containing calcium in the form of chalk or lime.

Canopy The uppermost layer of forest vegetation, formed by the branched crowns and massed foliage of trees or shrubs. 'Open-canopy forest' refers to this layer in cases where the tree-crowns are widely spaced and do not form a continuous layer.

Carr A woodland developed over a **fen** or **bog**; typical trees include alder *Alnus* and willow *Salix*.

Catchment (area) The area in which all water

drains into a single river system, separated from other catchment areas by a watershed.

Climax community   A plant **community** which is in equilibrium with the existing natural environmental conditions, as evinced by a more-or-less unchanging species composition.

Community   A general ecological term for any naturally occurring group of organisms inhabiting a common environment. Each community is relatively independent of other communities. Also used (always capitalized) in a more specialist sense in this book to denote the European Community (see **EEC/EC** on p. 433).

Concentrated in Europe   Applied to bird species for which more than 50% of their global breeding or wintering population or range occurs in Europe, according to range maps in Cramp *et al.* (1977–1994) or Harrison (1982) or to global population estimates where available.

Coniferous forest   Forests dominated mainly by cone-bearing trees, e.g. pines *Pinus*, firs *Abies*, larches *Larix*.

Conservation Dependent   A bird species which does not qualify as **Globally Threatened** but is the focus of a continuing conservation programme, the cessation of which would result in the species qualifying as Globally Threatened (Collar *et al.* 1994).

Coppice   Refers to forest vegetation in which the trees or shrubs are regularly cut, e.g. every 10–15 years. Coppiced trees and shrubs include hazel *Corylus*, ash *Fraxinus* and sweet chestnut *Castanea*, and they often occur with widely spaced trees such as oak *Quercus* which are only felled when mature (coppice with standards).

Data Deficient   A bird species for which there is inadequate information to make a direct or indirect assessment of its risk of global extinction (Collar *et al.* 1994).

Deciduous (forest)   Trees which lose their leaves and become dormant during the winter season.

Declining in Europe   Applied to a bird species the European population of which is in **moderate decline** and consists of more than 10,000 breeding pairs or 40,000 wintering individuals.

Dehesa   Spanish name for wooded pastoral habitats in Iberia dominated by oaks *Quercus*. Growth of scrub (**maquis**) is controlled and **pasture** is maintained by extensive rotational cultivation and/or grazing. In Portugal this habitat is called 'montado'.

Demersal   Applied to fish that live close to the sea-floor.

Detritus   Organic debris from decomposing animals and plants.

Dominant   A species in a plant community which has a dominating influence on other members of the community, being larger in size, occupying more space and light, contributing more organic matter, requiring more nutrients, etc.

Ecosystem   All plants, animals and micro-organisms within a community which interact with each other and with the environment.

Ecotone   A transitional or boundary stage between adjacent plant communities.

Effect   When referring in this book to a threat to a bird species or biotic community, used to denote an observed response. See **impact**.

Emergent   Used here to describe aquatic vegetation that emerges substantially above the water surface or **water-table**, e.g. reeds *Phragmites*.

Endangered in Europe   Applied to a bird species the European population of which has any of the following statuses.

- Population in **large decline** and of fewer than 10,000 breeding pairs and not marginal to a larger non-European population; or European wintering and entire flyway population fewer than 40,000 birds.
- Population in **moderate decline** and of fewer than 2,500 breeding pairs and not marginal to a larger non-European population; or European wintering and entire flyway population fewer than 10,000 birds.
- Population neither in **moderate** nor in **large decline** but fewer than 250 breeding pairs and not marginal to a larger non-European population; or European wintering and entire flyway population fewer than 1,000 birds and therefore at risk due to the susceptibility of small populations to factors described under **Rare in Europe**

Endemic   A species only found naturally in one country, region or island.

Environmentally Sensitive Area (ESA)   Area within which farmers received European Community support (under EC Reg. 797/85, as amended by Reg. 2328/91) for management agreements that maintained environmentally beneficial farming practices. This scheme has now been superceded by the **Agri-environment Regulation**.

Ericaceous   Plants, usually shrubs or dwarf shrubs, belonging to the heath family (Ericaceae).

European Threat Status   The demographic status of the European population of a bird species, defined by certain numerical criteria (all population size thresholds refer to minimum population estimates). For further information, see **Secure, Localized, Declining, Rare, Vulnerable** or **Endangered in Europe**.

European Union (EU)   The union whose 15 member states are Austria, Belgium, Denmark, Finland, France, Germany, Greece, Ireland, Italy, Luxembourg, Netherlands, Portugal, Spain, Sweden and the United Kingdom. See also **EEC/EC** on p. 433.

Eutrophic   Of waters or soils: rich in mineral nutrients, either from soil-water or from the decomposition of animal or plant remains. See **trophic status**, also Box 2 on p. 143.

Evergreen   Plants which retain their green leaves throughout the year; see **deciduous**. Semi-evergreen plants shed a large proportion of their leaves in winter or early spring.

Extensification   Used in this book to indicate the process whereby the intensity of land-use practices in an area is reduced (especially with respect to agricultural practices).

Favourable Conservation Status   Applied to bird species whose **European Threat Status** is classed as **Secure**.

Fen   Peatlands (mires) which are alkaline to somewhat acid, and which are relatively well supplied with water and mineral salts via the groundwater.

Formation   A basic unit of vegetation description, referring to a **community** of plants which extends over a very large natural area, and determined primarily by climatic conditions, e.g. tundra, heathland, deciduous forest.

Garrigue   A **community** of low, scattered, often spiny and aromatic **shrubs** found in the Mediterranean region.

Global conservation concern   Applied to bird species whose global status is classified as Globally Threatened, Conservation Dependent or Data Deficient according to IUCN criteria (Collar *et al.* 1994, Baillie and Groombridge 1996). These species are **Species of European Conservation Concern (SPEC)** category 1.

Globally Threatened   A species at risk of global extinction and classed as Critical, Endangered or Vulnerable under IUCN criteria (Collar *et al.* 1994, Baillie and Groombridge 1996).

Habitats Directive   Directive and Resolution of the Council of the European Community on the Conservation of Natural Habitats of Wild Fauna and Flora (92/43/EEC). See 'Opportunities for Conserving the Wider Environment' (p. 24) and Appendix 5 (p. 423) for more information.

Halophytes   Plants able to grow in saline soils.

Heath, heathland   A **community** composed largely of evergreen dwarf shrubs commonly belonging to the heath family (Ericaceae), and usually found on acidic, porous, sandy or gravelly soils. Trees, particularly pines *Pinus* and birches *Betula*, may be present.

Hemi-boreal zone   The southernmost latitudes of the **boreal** zone (around 55°N), at the interface with the **temperate** zone.

Herbs   Non-woody plants, with sappy stems which die back at the end of the growing period. They can be either annuals or biennials, or perennials with overwintering buds.

High forest   Mature woodland of fully-grown trees with a closed **canopy**.

Hydrophytes   Plants adapted to moist or wet conditions.

Impact   When referring in this book to a threat to a bird species or biotic community, used to denote a reduction in survival at the individual or population level. See **effect**.

Important Bird Area (IBA)   A site of particular importance for bird conservation, as listed in *Important Bird Areas in Europe* (Grimmett and Jones 1989).

Insufficiently Known in Europe   Suspected to be **Localized, Declining, Rare, Vulnerable** or **Endangered in Europe**, but insufficient information is available to attribute a **European Threat Status**, even provisionally.

Introduced   Refers to plants or animals which are not native species, but which have been brought into a country or area by human agency, either accidentally or intentionally.

Krummholz   A marginal type of forest, found between the upper altitudinal limit of normal forest and the **treeline**, and usually composed of gnarled, stunted, usually bush forms of trees, typically conifers.

Lacustrine   Of, or pertaining to, a lake or lakes.

Large decline   Applied to the European breeding or wintering population of a bird species which has declined in size or range by at least 20% in at least 66% of the population, or by at least 50% in at least 25% of the population, between 1970 and 1990, and where the total size of (national) populations that declined is greater than the total size of populations that increased. Only wintering populations of waterbirds of the families Anatidae, Haematopodidae, Charadriidae and Scolopacidae are considered in this book because these are typically the only bird species which have well-monitored winter populations in Europe.

Littoral (zone) (1) Of the sea-shore. (2) The shallow water of lakes (usually less than 2 m deep) where light reaches the bottom and rooted plants may grow.

Localized in Europe Applied to a bird species the European population of which consists of more than 10,000 breeding pairs or 40,000 wintering birds, and is neither in **moderate** nor in **large decline**, but has more than 90% of the population occurring at 10 or fewer sites (**Important Bird Areas**), as listed in Grimmett and Jones (1989).

Longline Fishing gear comprising long lines (often many kilometres in length) of baited hooks that are set behind a moving boat.

Macaronesia The biogeographical region comprising the island archipelagos of Madeira, the Canary Islands, the Azores and (outside Europe) the Cape Verde Islands.

Macchia An Italian term for **maquis**.

Machair A grazed, open habitat of coastal northwest Europe, comprising a complex mosaic of dunes, wet hollows, grassland underlain by shell-sand, and peatland.

Macrophyte Larger species of aquatic plant.

Maquis Dense, mostly evergreen shrub community 1–3 m high, characteristic of the Mediterranean region.

Marginal population European population of a bird species that is considered to have adequate potential for repopulation from large non-European populations (the combined total of which is 10,000 pairs or more) and which is therefore not at risk from small population size.

Marsh A general term for flat or gently sloping land with a high water-table (usually adjacent to a water-course or water-body), which remains wet throughout the year and which is periodically flooded. The soil has a mineral source, as opposed to being peat-based. See **fen**, **mire**.

Matorral A Spanish term for **maquis**.

Meadow An enclosed area of usually permanent grassland, traditionally maintained by the grazing of livestock or the cutting of hay. Steppe-meadow is a meadow-like plant community, largely composed of grasses, which occurs on the less arid margins of true **steppe**.

Mesotrophic Of waters or soils: neither poor nor rich in mineral nutrients. See **trophic status**.

Mire Soils and plant communities where there is an accumulation of peat. They may be either alkaline (**fens**), neutral (intermediate or transitional mires), or acid (**bogs**).

Moderate decline Applied to the European breeding or wintering population of a bird species which has declined in size or range by at least 20% in 33–65% of the population, or by at least 50% in 12–24% of the population, between 1970 and 1990, and where the total size of (national) populations that declined is greater than the total size of populations that increased. See also **Large decline** for restriction on species covered.

Montado See **dehesa**.

Moor, moorland A general but not clearly definable term, usually relating to a landscape of acidic, peat-based soils in northern Europe, dominated by heather *Calluna* and other heath-like plant species, or by grasses or sedges. In this book, distinguished from lowland Atlantic **heathland** by having an average total precipitation of more than 1,000 mm/year.

Mosaic Referring (in vegetation) to the patchwork-like distribution of different plant **communities** in an area, resulting from local differences in the environment.

Native species A species occurring naturally in an area or region which has not been introduced by human agency.

Natura 2000 The name given by the European Commission to the 'coherent ecological network' of **Special Protection Areas** (SPAs) and **Special Areas for Conservation** (SACs) that is due to be established within the **European Union** member states by June 2004, under the **Birds and Habitats Directives**.

Natural Referring to plant **communities**, the structure and species composition of which are unaffected by human activities (or those of associated domestic animals), either in the past or present. See **semi-natural**.

Naturalized species Plants or animals of non-native origin which have become fully established in a new area or region through human agency, and are able to reproduce and compete with native species.

Neolithic Of or relating to the later Stone Age.

Oligotrophic Of waters or soils: poor in mineral nutrients. See **trophic status**.

Ombrogenous Referring to mires: those in which the main source of water is from rainfall.

Oxbow (lake) A lake or pond formed by a loop in a river, which eventually becomes cut off by silt-banks from the river itself.

Palearctic One of the major zoogeographical

realms of the world, comprising Europe, Siberia, North Africa, most of the Middle East and northern Eurasia.

Palustrine   Pertaining to, or inhabiting, **marshes**.

Pannonic, Pannonian   Of, or pertaining to, a flat lowland region centred on Hungary.

Pasture   Grassland grazed by livestock.

Peat   Partly decomposed plant remains which accumulate in waterlogged soils, primarily because of lack of oxygen.

Pelagic   In general, applied to organisms that inhabit open water near the surface in aquatic ecosystems (see **demersal**). In ornithology, applied to seabirds that only come ashore to breed and that spend the major part of their lives far out at sea.

Permafrost   Permanently frozen sub-soil; only the surface layer thaws in summer.

Phrygana   A Greek term denoting low scrub developed over dry stony soils in the Mediterranean (Balkan) region. It is, in general, equivalent to the term **garrigue** which is used in the western Mediterranean.

Phytoplankton   The plant plankton and primary producers of aquatic ecosystems, comprising mainly diatoms in cool waters with dinoflagellates being more important in warmer waters.

Primeval forest   Self-regenerating forest whose structure and species composition have been affected only very slightly and temporarily as a result of human use, and that has never experienced any regular or significant removal of biomass by man, and has never been clear-cut.

Priority species   Used in this book to indicate the most important bird species for habitat-conservation measures in Europe, and divided into four classes (A–D, A being of highest priority). Further details are given in the 'Introduction' (p. 19).

Puszta   The flat, almost treeless steppe-grassland of the Hungarian plains.

Raised bog   A type of **bog** (**mire**) which builds up above the general level of the existing peat (becoming convex in cross-section) in climates of high rainfall where drainage is impeded.

Ramsar Convention   Convention on Wetlands of International Importance Especially as Waterfowl Habitat. See 'Opportunities for Conserving the Wider Environment' (p. 24) for more information.

Ramsar Site   Site designated for the Ramsar List of Wetlands of International Importance under the **Ramsar Convention**. The importance of a wetland for waterbirds is a prime criterion for designation, though wetlands may also be designated on the basis of their representativeness or uniqueness, or their importance for plants and animals other than waterbirds (Ramsar Convention Bureau 1990).

Rare in Europe   Applied to a bird species the European population of which is neither in **moderate** nor in **large decline** but consists of fewer than 10,000 breeding pairs and is not marginal to a larger non-European population; or its European wintering population and entire flyway population consists of fewer than 40,000 birds and is therefore at risk due to the susceptibility of small populations to:
- Break-up of social structure.
- Loss of genetic diversity.
- Large-scale population fluctuations and chance events.
- Existing or potential exploitation, persecution, disturbance and interference by man.

Saltmarsh   A marsh which is periodically flooded by sea water, or a marsh whose water-supply is rich in salt.

Sclerophyllous   Referring to plants with thick, leathery, usually relatively small leaves, whose form reduces the loss of water by transpiration in dry climates.

Scrub   A general term referring to low **shrubs** and bushes, usually less than 5 m high.

Secure in Europe   Applied to a bird species the European population of which consists of more than 10,000 breeding pairs or 40,000 wintering birds, and is neither in **moderate** or **large decline** nor is it **Localized in Europe**. Secure species have a **Favourable Conservation Status**.

Semi-natural   Refers to plant **communities** which have been subject to human interference or management, but which retain many of the natural species and some of the physical structure of natural communities.

Set-aside   Land taken out of production through supply-control measures within EC Reg. 1765/92, 'a support system for producers of certain arable crops'. A long-term (20-year) set-aside scheme is available for the promotion of environmental benefits through the **Agri-environment Regulation**.

Shelterbelt   A line or belt of trees planted along the edge of an agricultural field in order to shelter crops from wind.

Shrub   A woody, perennial plant species with well-developed side-branches arising from near its base and usually less than 10 m high. Dwarf shrubs or shrublets refer to woody perennials which are usually less than 0.5 m high.

Species of European Conservation Concern (SPEC) A bird species of conservation concern in Europe, as defined by Tucker and Heath (1994). See 'Introduction' (p. 17) for more information.

Special Protection Area (SPA) A site classified by member states of the European Union for the conservation of wild birds as required under Article 4 of the **Birds Directive**.

Steppe(s) Extensive, usually treeless plains or gently sloping terrain, whose vegetation is commonly dominated by grasses or dwarf shrubs; thus steppe-grasslands or shrub-steppes. Salt-steppes occur where salt is present in the upper layers of the soil. See **meadow**.

Subalpine A zone immediately below the **alpine** zone in mountain ranges, and usually above the treeline. Dwarf shrubs and stunted trees are characteristic of this zone.

Succession The sequence of plant **communities** which replace each other over time in a particular area as the vegetation becomes more integrated with the environment, ultimately reaching a more or less stable state—the **climax community**.

Taiga The coniferous forest of the **boreal** zone of northern Europe and Asia, lying to the south of the **tundra** zone.

Treeline The geographical limit of growth of trees, either in the mountains or in the northern regions. In the mountains, for example, the treeline depends not only on altitude but on aspect, soil and local climate.

Trophic status The status of an ecosystem (usually aquatic) with respect to nutrient levels. See **eutrophic**, **mesotrophic**, **oligotrophic**.

Tundra A treeless zone, lying principally north of the **Arctic** Circle, where winters are long and severe, and summers are short and relatively cool (mean July temperature not above 10°C). The soil is permanently frozen below the surface layers.

Understorey The layer of small or young trees, **shrubs** and/or dwarf shrubs beneath the trees of a mature forest.

Unfavourable Conservation Status Applied to a bird species whose **European Threat Status** is classed as **Endangered**, **Vulnerable**, **Rare**, **Declining**, **Localized** or **Insufficiently Known** (Tucker and Heath 1994).

Upwelling A water current that moves vertically upwards in a water-body, transferring nutrients from bottom sediments to surface waters.

Virgin forest Forest whose structure and species composition are intact and which have never experienced significant human use or disturbance. See **primeval forest**, **natural vegetation**, **semi-natural forest**.

Vulnerable in Europe Applied to a bird species the European population of which has any of the following statuses.

- Population in **large decline** and of more than 10,000 breeding pairs or 40,000 wintering individuals.
- Population in **moderate decline** and of fewer than 10,000 breeding pairs and not marginal to a larger non-European population; or European wintering and entire flyway population fewer than 40,000 birds.
- Population neither in **moderate** nor in **large decline** but fewer than 2,500 breeding pairs and not marginal to a larger non-European population; or European wintering and entire flyway population fewer than 10,000 birds and therefore at risk due to the susceptibility of small populations to the factors described under **Rare in Europe**.

Water-table The uppermost level to which water saturates the soil or substrate. The height of the water-table usually varies with the seasons and with the water-holding capacity of the soil or substrate.

ABBREVIATIONS

AEPS Arctic Environmental Protection Strategy
AEWA Agreement on the Conservation of African–Eurasian Migratory Waterbirds
BSEP Black Sea Environmental Programme
BSPA Baltic Sea Protection Area
CAFF Program for the Conservation of Arctic Flora and Fauna
CAP Common Agricultural Policy
CEC Commission of the European Communities (European Commission)

CBD Convention on Biological Diversity
CFP Common Fisheries Policy
CMS Convention on Migratory Species
CSF Community Support Framework
DDT Dichlorodiphenyltrichloroethane
EAGGF European Agricultural Guidance and Guarantee Fund
ECU European Currency Unit
EEC/EC European (Economic) Community
The EEC first included six member states. This increased

to nine in 1973, 10 in 1981, 12 in 1986 and 15 in 1995. In January 1986, the EEC became the EC. In January 1992, the EU came into being, including the same member states as the EC but embracing wider responsibilities. The terms EU and EC are often used interchangably, although there are differences in technical meaning.

EECONET   European Ecological Network

EEZ   Exclusive Economic Zone

EIA   Environmental Impact Assessment

ERDF   European Rural Development Fund

ESA   Environmentally Sensitive Area

ESF   European Social Fund

EU   European Union (see also EEC/EC)

FAO   United Nations Food and Agriculture Organization

FIFG   Financial Instrument for Fisheries Guidance

FNNPE   Federation of Nature and National Parks of Europe

FSC   Forest Stewardship Council

GATT   General Agreement on Tariffs and Trade

GDP   Gross Domestic Product

GEF   Global Environment Facility

IBA   Important Bird Area

IBRD   International Bank for Reconstruction and Development

ICZM   Integrated Coastal Zone Management

IDA   International Development Association

IMO   International Maritime Organization

IPCC   Intergovernmental Panel on Climate Change

ITQ   Individual Transferable Quota

IUCN   The World Conservation Union

LIEN   Link Inter-European NGOs

LIFE   The European Union's all-encompassing fund for nature conservation

MAB   Man and the Biosphere

MARPOL   International Convention for the Prevention of Pollution from Ships, 1973, as modified by the Protocol of 1978 relating thereto

MedSPA   Mediterranean Specially Protected Area

MSC   Marine Stewardship Council

NGO   Non-governmental organization

OP   Operational Policy/Plan/Programme

PEBLanDS   Pan-European Biological and Landscape Diversity Strategy

PPP   Policies, Plans and Programmes

RAC/SPA   Regional Activity Centre for Specially Protected Areas

RSPB   Royal Society for the Protection of Birds

SAC   Special Area for Conservation

SEA   Strategic Environmental Assessment

SPA   Special Protection Area

SPEC   Species of European Conservation Concern

TAC   Total Allowable Catch

UN   United Nations

UNCLOS   United Nations Convention on the Law of the Sea

UNCSD   United Nations Commission on Sustainable Development

UNDP   United Nations Development Programme

UNECE   United Nations Economic Commission for Europe

UNEP   United Nations Environmental Programme

UNESCO   United Nations Educational, Scientific and Cultural Organization

WCMC   World Conservation Monitoring Centre

WWF   World Wide Fund for Nature

REFERENCES

ABDELLI, C. (1991) *Agreste. La statistique agricole. L'utilisation du territoire en 1990 et son évolution de 1982 à 1990.* Paris: Ministère de l'Agriculture.

ACOPS (1996) *Oil pollution survey around the coasts of the United Kingdom 1995.* London: Advisory Committee on Protection of the Sea.

AEBISCHER, N. J. (1991) Pp.305–331 in L. G. Firbank, N. Carter, J. F. Darbyshire and G. R. Potts, eds. *The ecology of temperate cereal fields.* Oxford, UK: Blackwell Scientific Publications.

AERTS, R. AND HEIL, G. W., EDS. (1993) *Heathlands: patterns and processes in a changing environment.* Dordrecht, Netherlands: Kluwer Academic Publishers (Geobotany 20).

AGUILAR, J. S., MONBAILLIU, X. AND PATERSON, A. M., EDS. (1993) *Status and conservation of seabirds.* Madrid: Sociedad Española de Ornitología/BirdLife/MEDMARAVIS.

ALBANIS, T. (1993) Pesticide residues and their accumulation in wildlife of wetlands in Thermaikos and Amvrakikos Gulfs. WWF Project 4680, Greece (unpublished report).

ALEXANDER, I. AND CRESSWELL, B. (1990) Foraging by nightjars *Caprimulgus europaeus* away from their nesting areas. *Ibis* 132: 568–574.

ALMEIDA, J. (1992) Alguns aspectos dos efeitos do maneio dos montados de sobro *Quercus suber* na avifauna nidificante. *Airo* 3: 69–74.

ALONSO, J. A. AND ALONSO, J. C., EDS. (1990) *Distribución y demografia de la Grulla comun (Grus grus) en España [Distribution and demography of the Common Crane in Spain.]* Madrid: Instituto Nacional para la Conservación de la Naturaleza, CSIC. (In Spanish.)

ALONSO, J. C., ALONSO, J. A. AND MUÑOZ-PULIDO, R. (1994) Mitigation of bird collisions with transmission lines through groundwire marking. *Biol. Conserv.* 67: 129–134.

ALVAREZ, G. AND SANTOS, T. (1992) Efectos de la gestión del monte sobre la avifauna de una localidad mediterránea (Quintos de Mora, Montes de Toledo). *Ecología* 6: 187–198.

ALVERSON, W. S., WALLER, D. M. AND SOLHEIM, S. I. (1988) Forests too deer: edge effects in Northern Wisconsin. *Conserv. Biol.* 2: 348–358.

AMMER, U. (1994) Konsequenzen aus den Ergebnissen der Totholzforschung für die forstliche Praxis. *Forstw. Cbl.* 110: 149–157.

ANDERSEN, K. P. AND URSIN, E. (1977) A multispecies extension to the Beverton and Holt theory of fishing, with accounts of phosphorus circulation and primary production. *Meddelelser fra Fiskeri- og Havunersogelser* 7: 319–435.

ANDRÉN, H. (1992) Corvid density and nest predation in relation to forest fragmentation: a landscape perspective. *Ecology* 73: 794–804.

ANDRÉN, H. AND ANGELSTAM, P. (1988) Elevated predation rates as an edge effect in habitat islands: experimental evidence. *Ecology* 69: 544–547.

ANDREWS, J. (1995) Waterbodies. Pp.121–148 in W. J. Sutherland and D. A. Hill, eds. *Managing habitats for conservation.* Cambridge, UK: Cambridge University Press.

ANDREWS, J. AND REBANE, M. (1994) *Farming and wildlife: a practical management handbook.* Sandy, UK: Royal Society for the Protection of Birds.

ANGELSTAM, P. (1991) Changes in forest landscapes and bird conservation in northern Europe. Pp.2292–2297 in *Proceedings of the XX International Ornithological Congress, Christchurch 1990.*

ANGELSTAM, P. (1992) Conservation of communities—the importance of edges, surroundings and landscape mosaic structure. Pp.9–69 in L. Hansson, ed. *Ecological principles of nature conservation: a boreal perspective.* London: Elsevier.

ANGELSTAM, P. (1996) The ghost of forest past—natural disturbance regimes as a basis for reconstruction of biologically diverse forests in Europe. Pp.287–336 in R. M. de Graaf and R. I. Müller, eds. *Conservation of faunal diversity in forested landscapes.* London: Chapman & Hall.

ANGELSTAM, P. AND MIKUSINSKI, G. (1994) Woodpecker assemblages in natural and managed boreal and hemiboreal forest—a review. *Ann. Zool. Fennici* 31: 157–172.

ANGELSTAM, P., ROSENBERG, P. AND RÜLCKER, C. (1993) [Never, seldom, sometimes, often. Natural fire disturbance as a model for forest management.] *Skog och Forskning* 1: 34–41. (In Swedish.)

ANKER-NILSSEN, T. (1987) The breeding performance of Puffins *Fratercula arctica* on Røst, northern Norway in 1979–1985. *Fauna Norv. Ser. C, Cinclus* 10: 21–38.

ANKER-NILSSEN, T. (1991) [Census of Puffins in the area at risk from oil pollution from the central Norwegian continental shelf.] Pp.13–18 in J. A. Bórresen and K. A. Moe, eds. [*AKUP Annual Report 1990.*] Oslo: Ministry of Oil and Energy.

ANKER-NILSSEN, T. (1992) [Food supply as a determinant of reproduction and population development in Norwegian Puffins *Fratercula arctica.*] Trondheim: University of Trondheim (Ph.D. thesis).

ANKER-NILSSEN, T. AND RØSTAD, O. W. (1993) Census and monitoring of Puffins *Fratercula arctica* on Røst, North Norway 1979–88. *Ornis Scand.* 24: 1–9.

ANON. (1975) [Hydrobiological research on Lithuanian lakes.]

ANON. (1976) Die Seen der Schweiz. *Wasser, Energie, Luft* 68(11/12): 263–266.

ANON. (1978) *Environnement et cadres de vie—dossier statistique—Tome 2.* Paris: Ministère de la Culture et de l'Environnement (La Documentation Française).

ANON. (1983) *Ecological effects of acid deposition.* Stockholm: National Swedish Environment Protection Board.

ANON. (1985) [*The Committee of Threatened Animals and Plants, Committe Report. I. General part.*] Helsinki: Ministry of Environment. (In Finnish.)

ANON. (1988) *De heide heeft toekomst. [Heathlands have a future.]* The Hague, Netherlands: Ministry of Agriculture and Fisheries.

ANON. (1989a) *1899–1989: negentig jaren statistiek in tijdsreeksen.* 's-Gravenhage, Netherlands: SDU Uitgevery/cbs-publikaties.

ANON. (1989b) *Atlas van Nederland: deel 10, Landbouw.* Stichting Wetenschappelijke Atlas van Nederland. 's-Gravenhage, Netherlands: SDU Uitgevery/cbs-publikaties.

ANON. (1990) Bergen Ministerial Declaration on Sustainable Development. *Environmental Policy and Law* 20: 100.

ANON. (1991) [*Report on the monitoring of threatened animals and plants in Finland. Committee report 1991: 30.*] Helsinki: Ministry of the Environment. (In Finnish.)

ANON., ED. (1992a) *European coastal conservation conference 1991: proceedings.* Hague: Dutch Ministry of Agriculture, Nature Management and Fisheries.

ANON. (1992b) *The environment in Europe and North America: annotated statistics 1992.* New York: United Nations Statistical Commission/Economic Commission for Europe Conference of European Statisticians (Statistical Standards and Studies No. 42).

ANON. (1993a) *Forest condition in Europe: results of the 1992 survey.* Brussels/Geneva: Commission of the European Communities/United Nations Economic Commission for Europe.

ANON. (1993b) *Plan national d'actions prioritaires de conservation des zones humides les plus importantes de Bulgarie.* Sofia: Ministry of the Environment.

ANON. (1994a) *Agenda 21: Rio Declaration on Environment and Development, Statement of Forest Principles.* Washington, D.C.: United Nations.

ANON. (1994b) *Biodiversity: the U.K. action plan.* London: HMSO.

ANON. (1994c) *Convention on biological diversity: text and annexes.* Nairobi: United Nations Environment Programme.

ANON. (1994d) *Safer ships, cleaner seas.* Report of Lord Donaldson's inquiry into the prevention of pollution from merchant shipping edition. London: HMSO (CM2560).

ANON. (1994e) *World resources 1994–95: a guide to the global environment.* Oxford: Oxford University Press.

ANON. (1995a) *Action plan to 2010 for central and eastern Europe. Integrating agriculture and the environment.* Report of a conference held at Gödöllö Agricultural University, Hungary, 14–15th September 1995. Sandy, UK: Royal Society for the Protection of Birds.

ANON. (1995b) *Biodiversity: the UK steering group report, 2: action plans.* London: Her Majesty's Stationery Office.

ANON. (1995c) Report of the Working Group on the assessment of Norway pout and sandeel. International Council for the Exploration of the Sea: C.M. 1995/Assess: 5.

ANON. (1996) Natura barometer. *Natura 2000* 1: 6.

ANON. (1997) Vanishing peat in Belarus. *International Mire Conservation Group Newsletter* 2: 10.

ANTOR, R. (1995) The importance of arthropod fallout on snow patches for the foraging of high-alpine birds. *J. Av. Biol.* 26: 81–85.

ARDIZZONE, G. D., CATAUDELLA, S. AND ROSSI, R. (1988) Management of coastal lagoon fisheries and aquaculture in Italy. *FAO Fisheries Technical Paper* 293.

ARGELICH, J., CLAMENS, A. AND DUBOURG, M.-J. (1996) Impact de l'évolution des activités économiques sur l'avifaune en région de montagnes: le cas de l'Andorre (Pyrénées). *Alauda* 64: 163–168.

ARLETTAZ, R. (1990) La population relictuelle du Hibou petit-duc *Otus scops* en Valais central: dynamique, organisation spatialle, habitat et protection. *Nos Oiseaux* 40: 321–343.

ATHANASIOU, H. (1987) Past and present importance of the Greek wetlands for wintering waterfowl. International Waterfowl Research Bureau, Slimbridge. EU/GR/17.

ATLANTIC CONSULTANTS (1996) Attitude survey of the value of heathlands. Peterborough, UK: English Nature (unpublished report).

AULITZKY, H. (1994) Musterbeispiele vermeidbarer Erosions-, Hochwasser- und Lawinen schäden. *Österreichische Akademie der Wissenschaften, Veröff. Komm. Humanökologie* 5: 105–148.

AVERY, M. I. AND LESLIE, R. (1990) *Birds and forestry.* London: Poyser.

AZZARELLO, M. Y. AND VAN VLEET, E. S. (1987) Marine birds and plastic pollution. *Mar. Ecol. (Prog. Ser.)* 37: 295–303.

BAILEY, R. S., FURNESS, R. W., GAULD, J. A. AND KUNZLIK, P. A. (1991) Recent changes in the population of the sandeel (*Ammodytes marinus*) Raitt at Shetland in relation to estimates of seabird predation. *ICES Mar. Sci. Symp.* 193: 209–216.

BAILLIE, J. AND GROOMBRIDGE, B., EDS. (1996) *1996 IUCN Red List of threatened animals.* Cambridge, UK: International Union for Conservation of Nature and Natural Resources.

BAINES, D. (1988) The effects of improvement of upland grassland on the distribution and density of breeding wading birds (Charadriiformes) in northern England. *Biol. Conserv.* 45: 221–236.

BAINES, D. (1989) The effects of improvement of upland, marginal grasslands on the breeding success of Lapwings *Vanellus vanellus* and other waders. *Ibis* 131: 497–506.

BAINES, D. (1990) The roles of predation, food and agricultural practice in determining the breeding success of the Lapwing (*Vanellus vanellus*) on upland grasslands. *J. Anim. Ecol.* 59: 915–929.

BALDOCK, D. (1984) *Wetland drainage in Europe.* London: Institute for European Environmental Policy, International Institute for Environment and Development.

BALDOCK, D. (1990a) *The CAP Structures Policy.* Gland, Switzerland: WWF International (CAP Discussion Paper 2).

BALDOCK, D. (1990b) *Agriculture and habitat loss in Europe.* Gland, Switzerland: WWF International (CAP Discussion Paper 3).

BALDOCK, D., BEAUFOY, G., BENNETT, G. AND CLARK, J. (1993) *Nature conservation and new directions in the EC Common Agricultural Policy.* London: Institute for European Environmental Policy.

BÁRCENA, F., TEIXEIRA, A. M. AND BERMEJO, A. (1984) Breeding seabirds populations in the Atlantic sector of the Iberian Peninsula. Pp.335–345 in J. P. Croxall, P. G. H. Evans and R. W. Schreiber, eds. *Status and conservation of the world's seabirds.* Cambridge, UK: International Council for Bird Preservation (Techn. Publ. 2).

BARIS, Y. S. (1989) Turkey's bird habitats and ornithological importance. *Sandgrouse* 11: 42–51.

BARIS, Y. S. (1991) Conservation problems of steppic avifauna in Turkey. Pp.93–96 in P. D. Goriup, L. A. Batten and J. A. Norton, eds. *The conservation of lowland dry grassland birds in Europe.* Peterborough, UK:

Joint Nature Conservation Committee.

BARNES, R. S. K. (1980) *Coastal lagoons*. Cambridge, UK: Cambridge University Press.

BARNES, R. S. K. (1989) The coastal lagoons of Britain: an overview and conservation appraisal. *Biol. Conserv.* 49: 295–313.

BARNES, R. S. K. (1994) The coastal lagoons of Europe. *Coastline* 3: 3–8.

BARNES, R. S. K. AND HUGHES, R. N., EDS. (1982) *An introduction to marine ecology*. Oxford, UK: Blackwell.

BARR, C., BUNCE, R., CLARK, R., FULLER, R., FURSE, M., GILLESPIE, M., GROOM, G., HALLAM, C., HORNING, M., HOWARD, D. AND NESS, M. (1993) *Countryside Survey 1990: main report*. London: Department of the Environment.

BARRETT, R. T. (1979) Small oil spill kills 10–20,000 seabirds in north Norway. *Mar. Pollut. Bull.* 10: 253–255.

BARRETT, R. T. AND VADER, W. (1984) The status and conservation of breeding seabirds in Norway. Pp.323–333 in J. P. Croxall, P. G. H. Evans and R. W. Schreiber, eds. *Status and conservation of the world's seabirds*. Cambridge, UK: International Council for Bird Preservation (Techn. Publ. 2).

BARTLE, J. A. (1995) Tuna long-lining in the Southern Ocean and its effects on New Zealand seabird populations. Pp.9 in M. L. Tasker, ed. *Threats to seabirds: Proceedings of the 5th International Seabird Group conference*. Sandy, UK: Seabird Group.

BARTRAM, H. (1995) *A North Sea view*. Cambridge, UK: BirdLife International.

BATTEN, L. A., BIBBY, C. J., CLEMENT, P., ELLIOTT, G. D. AND PORTER, R. F., EDS. (1990) *Red data birds in Britain: action for rare, threatened and important species*. London: T. and A. D. Poyser.

BATTISTI, A. *ET AL.* (1996) Organochlorine compounds, heavy metals and radionuclides in raptors from central Italy. In *Proceedings 2nd International Conference on Raptors*. Urbino, Italy.

BÄTZING, W. (1994) Nachhaltige Naturnutzung im Alpenraum. *Österreichische Akademie der Wissenschaften, Veröff. Komm. Humanökologie* 5: 15–52.

BAUER, H.-G. AND BERTHOLD, P. (1996) *Die Brutvögel Mitteleuropas: bestand und Gefährdung*. Wiesbaden, Germany: AULA.

BAYLE, P. (1994) Preventing bird of prey problems at transmission lines in western Europe.

BEAUBRUN, P.-C. (1983) Le Goéland d'Audouin (*Larus audouinii* Payr.) sur les côtes du Maroc. [Audouin's Gull *Larus audouinii* Payr. on the Moroccan coasts.] *L'Oiseau et R.F.O.* 53: 209–226.

BEAUFOY, G., BALDOCK, D. AND CLARK, J. (1994) *The nature of farming*. Peterborough, UK: Joint Nature Conservation Committee.

BEGON, M., HARPER, J. L. AND TOWNSEND, C. R. (1990) *Ecology*. Oxford, UK: Blackwell.

BEHRE, K. E. (1988) The role of man in European vegetation history. Pp.633–672 in B. Huntley and T. Webb, eds. *Vegetation history*. Amsterdam: Kluwer Academic Publishers.

BEINTEMA, A. J. (1988) Conservation of grassland bird communities in the Netherlands. Pp.105–111 in P. D. Goriup, ed. *Ecology and conservation of grassland birds.*

Cambridge: International Council for Bird Preservation (Techn. Publ 7).

BEINTEMA, A. J. (1991) Status and conservation of meadow birds in the Netherlands. *Wader Study Group Bull.* 61 (Suppl.): 12–13.

BEINTEMA, A. J. AND MÜSKENS, G. J. D. M. (1987) Nesting success of birds breeding in Dutch agricultural grasslands. *J. Appl. Ecol.* 24: 743–758.

BEINTEMA, A. J., DUNN, E. AND STROUD, D. A. (1997) Birds and wet grasslands. Pp.269–296 in D. J. Pain and M. W. Pienkowski, eds. *Farming and birds in Europe: the Common Agricultural Policy and its implications for bird conservation*. London: Academic Press.

BEINTEMA, A. J., MOEDT, O. AND ELLINGER, D. (1995) *Ecologische Atlas van de Nederlandse Wiedevogels*. Haarlem, Netherlands: Schuyt and Co.

BEKSHTREM, E. A. (1927) O faune zverej i ptits Rjazanskoi Mesheri. [About mammals and birds fauna in Rjazan Meschera.] In *Materiali k izucheniju flori i fauni Centralno-Promishlennoj oblasti. [Materials to the study of flora and fauna of Central-Industrial Region, Moscow.]* Moscow.

BELL, D. V. AND OWEN, M. (1990) Shooting disturbance: a review. Pp.159–171 in G. V. T. Matthews, ed. *Managing waterfowl populations*, 12. Slimbridge, UK: International Waterfowl and Wetlands Research Bureau.

BELLROSE, F. C. (1958) Lead poisoning as a mortality in waterfowl populations. *Ill. Nat. Hist. Surv. Bull.* 27: 235–288.

BENNETT, G. (1991) *Towards a European ecological network*. Arnhem, Netherlands: Institute for European Environmental Policy.

BERG-SCHLOSSER, G. (1984) Zoogeographische und faunenhistoriche Bemerkungen zur Vogelwelt der Alpen—ein Überblick. *Monticola* 6 (54): 42–60.

BERNES, C. (1993) *The Nordic environment: present state, trends and threats*. Copenhagen: Nordic Council of Ministers.

BERRY, R. (1979) Nightjar habitats and breeding in East Anglia. *Brit. Birds* 72: 207–218.

BEVANGER, K. (1994) Bird interactions with utility structures: collision and electrocution, causes and mitigating measures. *Ibis* 136: 412–425.

BIBBY, C. J. (1979) Breeding biology of the Dartford Warbler *Sylvia undata* in England. *Ibis* 121: 41–52.

BIBER, J.-P. (1996) International action plan for the Lesser Kestrel *Falco naumanni*. Pp.191–203 in B. Heredia, L. Rose and M. Painter, eds. *Globally threatened birds in Europe: action plans*. Strasbourg: Council of Europe, and BirdLife International.

BIBER-KLEMM, S. (1991) International legal instruments for the protection of migratory birds: an overview for the west Palearctic–African flyways. Pp.315–344 in T. Salathé, ed. *Conserving migratory birds*. Cambridge, UK: International Council for Bird Preservation (Techn. Publ. no. 12).

BIGNAL, E., BIGNAL, S. AND CURTIS, D. J. (1989) Functional unit systems and support ground for Choughs: the nature conservation requirements. Pp.102–109 in E. Bignal and D. J. Curtis, eds. *Choughs and land-use in Europe*. Clachan, UK: Scottish Chough Study Group.

BINA, O. AND BRIGGS, B. (1996) Transport and regional development: an uncertain economic impact. Sandy, UK:

Royal Society for the Protection of Birds.

BINA, O., BRIGGS, B. AND BUNTING, G. (1995) The impact of Trans-European Networks on nature conservation: a pilot project. Sandy, UK: RSPB/WCMC-BT/BirdLife International (unpublished report).

BINDING, T. AND FRANSDEN, B. L. (1995) Long-term successional patterns in Nørholm Heath: investigations on permanent plots 1921–1995. Pp.19–20 in *Proceedings 5th European Heathland Workshop, Santiago de Compostela, Spain, September 1995*. Spain: University of Santiago de Compostela.

BINK, R. J., BAL, D., VAN DEN BERK, V. M. AND DRAAIJER, L. J. (1994) *Toestand van de natuur 2*. Wageningen, Netherlands.

BIRD, E. C. F. (1984) *Coasts. An introduction to coastal geomorphology*. Third edition. Oxford, UK: Basil Blackwell.

BIRDLIFE INTERNATIONAL (1995a) IBA criteria: categories and thresholds. Cambridge, UK: BirdLife International (unpublished report).

BIRDLIFE INTERNATIONAL (1995b) *The Structural Funds and biodiversity conservation*. Sandy, UK: Royal Society for the Protection of Birds.

BIRKS, H. H., BIRKS, H. J., KALAND, P. E. AND MOE, D., EDS. (1988) *The cultural landscape: past, present and future*. Cambridge, UK: Cambridge University Press.

BIRÓ, P. (1984) Lake Balaton: a shallow Pannonian water in the Carpathian Basin. Pp.231–245 in F. B. Taub, ed. *Ecosystems of the world—lakes and reservoirs*. Amsterdam, Oxford, Tokyo: Elsevier.

BJÖRK, S. (1994) The evolution of lakes and wetlands. Pp.6–15 in M. Eiseltová, ed. *Restoration of lake ecosystems: a holistic approach*. Slimbridge, UK: International Waterfowl and Wetlands Research Bureau (Publ. 32).

BLACK, J. M. AND MADSEN, J. (1993) Red-breasted Goose: conservation and research needs. *IWRB Goose Res. Group Bull.* 4: 8–15.

BLONDEL, J. (1986) *Biogéographie évolutive*. Paris: Massen.

BLONDEL, J. (1988) Biogéographie évolutive à différentes échelles: L'histoire des avifaunes méditerranéennes. Pp.155–188 in H. Ouellet, ed. *Acta XIX Congressus Internationalis Ornithologici*. Ottawa: University of Ottawa Press.

BLONDEL, J. AND FARRÉ, H. (1988) The convergent trajectories of bird communities in European forests. *Oecologia (Berlin)* 75: 83–93.

BLONDEL, S. AND ARONSON, J. (1995) Biodiversity and ecosystem function in the Mediterranean basin: human and non-human determinants. Pp.43–119 in G. W. Davis and D. M. Richardson, eds. *Mediterranean-type ecosystems: the function of biodiversity*. Berlin and London: Springer-Verlag (Ecological Studies, Vol. 109).

BLUM, W. E. H. (1990) The challenge of soil protection in Europe. *Environ. Conserv.* 17: 72–74.

BOADA, M. (1994) *Breu historia de Catalunya*. Barcelona: Generalitat de Catalunya.

BOBBINK, R. ET AL. (1992) *Nitrogen eutrophication and critical load for nitrogen based upon changes in flora and fauna in (semi-) natural terrestrial ecosystems*. New York: United Nations.

BODDEKE, R. AND HAGEL, P. (1995) Eutrophication, fisheries and productivity of the North Sea Continental Zone. Pp.290–315 in N. B. Armantrout, ed. *Condition of the world's aquatic habitats*. Oxford and New Delhi: IBH Publishing Co. PVT. Ltd.

BONAN, G. B. AND SHUGART, H. H. (1989) Environmental factors and ecological processes in boreal forests. *Annual Review of Ecology and Systematics* 20: 1–28.

BORODIN, Y. W. AND KUYLENSTIERNA, J. C. J. (1992) Acidification and critical load in Nordic countries: a background. *Ambio* 21: 332–338.

BOSSERT, A. (1980) [Winter nutrition of Alpine Rock Ptarmigan *Lagopus mutus* and the influence of snow cover and the melting period.] *Orn. Beob.* 77: 121–166. (In German with English summary.)

BOURNE, W. R. P. (1976) Seabirds and pollution. Pp.403–502 in R. Johnston, ed. *Marine pollution*. London: Academic Press.

BOWDEN, C. G. R. (1990) Selection of foraging habitats by Woodlarks (*Lullula arborea*) nesting in pine plantations. *J. Appl. Ecol.* 27: 410–419.

BOYD, H. AND PIROT, J.-Y., EDS. (1989) *Flyways and reserve networks for waterbirds*. Slimbridge, UK: International Waterfowl and Wetlands Research Bureau (Spec. Publ. 9).

BRAAE, L., NOHR, H. AND PETERSEN, B. S. (1988) [*The bird fauna of conventional and organic farmland*.] Copenhagen: Miljoministeriet, Miljostyrelsen (Miljoprojekt 12). (In Danish.)

BRADSHAW, R. AND HANNON, G. (1992) The disturbance dynamics of Swedish boreal forest. Pp.528–535 in A. Teller, P. Mathy and J. N. R. Jeffers, eds. *Responses of forest ecosystems to environmental changes*. London: Elsevier.

BRASIER, C. M. (1992) Oak tree mortality in Iberia. *Nature* 360: 539.

BRAUN-BLANQUET, J., ROUSSINE, N. AND NÈGRE, R. (1951) *Les groupements végétaux de la France Méditerranéene*. Montpellier, France.

VAN BREEMEN, N. AND VAN DIJK, H. F. G. (1988) Ecosystem effects of atmospheric deposition of nitrogen in the Netherlands. *Environ. Pollut.* 54: 249–274.

BRIGGS, B. AND HOSSELL, J. (1995) The implications of global climate change for biodiversity. *RSPB Conservation Review* 9: 41–47.

BRINKMANN, R., KÖHLER, B., HEINS, J. U. AND RÖSLER, S. (1991) *Menderes Delta: Zustand und Gefährdung eines ostmediterranen Flußdeltas*. Germany: Gesamthochschule Kassel.

BROCKMAN, U., BILLEN, G. AND GIESKES, W. W. C. (1988) North Sea nutrients and eutrophication. Pp.348–389 in W. Salomons, B. L. Bayne, E. K. Duursma and U. Förstner, eds. *Pollution of the North Sea*. Heidelberg, Germany: Springer-Verlag.

BROOKES, A. (1987) The distribution and management of channelized streams in Denmark. *Regulated Rivers* 1: 3–16.

BROOKES, A. (1988) *Channelized rivers: perspectives for environmental management*. New York: Wiley.

BROTHERS, N. P. (1991) Albatross mortality and associated bait loss in the Japanese longline fishery in the Southern Ocean. *Biol. Conserv.* 55: 255–268.

BUCKLEY, G. P. (1992) *Ecology and management of coppice woodlands*. London: Chapman and Hall.

BUISSON, R. AND WILLIAMS, G. (1991) RSPB action for lowland wet grassland. *RSPB Conservation Review* 5:

60–64.

BULL, K. R., EVERY, W. J., FREESTONE, P., HALL, J. R. AND OSBORN, D. (1983) Alkyl lead pollution and bird mortalities on the Mersey estuary, UK, 1979–1981. *Environ. Pollut.* (Series A) 31: 239–259.

BURD, F. (1989) *The saltmarsh survey of Great Britain.* Peterborough, UK: Nature Conservancy Council.

BURGER, J. AND GOCHFELD, M. (1994) Predation and effects of humans on island-nesting seabirds. Pp.39–67 in D. N. Nettleship, J. Burger and M. Gochfeld, eds. *Seabirds on islands: threats, case studies, and action plans.* Cambridge, UK: BirdLife International (BirdLife Conservation Series no. 1).

BURGESS, N., WARD, D., HOBBS, R. AND BELLAMY, D. (1995) Reedbeds, fens and acid bogs. Pp.149–196 in W. J. Sutherland and D. A. Hill, eds. *Managing habitats for conservation.* Cambridge, UK: Cambridge University Press.

BURGIS, M. J. AND MORRIS, P. (1987) *The natural history of lakes.* Cambridge, UK: Cambridge University Press.

BURNESS, G. P. AND MORRISS, R. D. (1993) Direct and indirect consequences of mink presence in a Common Tern colony. *Condor* 95: 708–711.

BYRKJEDAL, I. AND THOMPSON, D. B. A. (in press) *Golden and grey plovers: birds of the tundra.* London: Poyser.

CABRAL, M. T., LOPES, F. AND SARDINHA, R. A. (1993) Determinação das causas de morte do sobreiro nos concelhos de Santiago do Cacém, Grandola and Sines. Relatório síntese. *Silva Lusitana* 1: 7–24. (In Portuguese.)

CADBURY, C. J. (1981) Nightjar census methods. *Bird Study* 28: 1–4.

CADBURY, C. J. (1989) What future for lowland heaths in southern Britain? *RSPB Conservation Review* 3: 61–67.

CADENAS, R. (1992) Informe de resultados de las campañas de reproducción del aquila imperial en el Parque Nacional de Doñana. [Report on the Spanish Imperial Eagle breeding population in Doñana National Park.] Huelva, Spain: ICONA (unpublished report).

CAFF (1994) *The state of protected areas in the circumpolar Arctic 1994.* Trondheim: Directorate for Nature Management (Conservation of Arctic Flora and Fauna Habitat Conservation Report 1).

CAFF (1996) *Proposed protected areas in the Circumpolar Arctic 1996.* Trondheim: Conservation of Arctic Flora and Fauna (CAFF Habitat Conservation Report No. 2).

CAGNOLARO, L., DI NATALE, A. AND NOTARBARTOLO DI SCIARA, G. (1983) *Cetacei.* Rome: Guida CNR (No. 9).

CALDERÓN, J., CASTROVIEJO, J., GARCÍA, L. AND FERRER, M. (1988) El Aquila Imperial *Aquila adalberti*: dispersión de los jovenes, estructura de edades y mortalidad. [The Imperial Eagle *Aquila adalberti*: juvenile dispersion, age structure and mortality.] *Doñana Acta Vert.* 15: 79–98. (In Spanish.)

CALOW, P. AND PETTS, G. E. (1992) *The rivers handbook*, 1. Oxford, UK: Blackwell Science.

CAMPBELL, L. H. (1978) Patterns of distribution and behaviour of flocks of seaducks wintering at Leith and Musselburgh, Scotland. *Biol. Conserv.* 14: 111–123.

CAMPBELL, L. H. (1984) The impact of changes in sewage treatment on seaducks wintering in the Firth of Forth, Scotland. *Biol. Conserv.* 28: 173–180.

CAMPBELL, L. H., AVERY, M. I., DONALD, P., EVANS, A. D., GREEN, R. E. AND WILSON, J. D. (in press) *A review of the indirect effects of pesticides on birds.* Peterborough, UK: Joint Nature Conservation Committee (JNCC Research Report).

CAMPHUYSEN, C. J. (1989) Beached bird surveys in the Netherlands, 1915–1988: seabird mortality in the southern North Sea since the early days of oil pollution. Amsterdam: Vogelbescherming Werkgroep Noordzee (Technisch Rapport 1).

CAMPHUYSEN, C. J. (1991) Beached bird surveys and the assessment of total mortality in the case of oil incidents. *Sula* 5: 41–42.

CAMPHUYSEN, C. J. AND VAN FRANEKER, J. A. (1992) *The value of beached bird surveys in monitoring marine oil pollution.* Zeist, Netherlands: Vogelbescherming Nederland (Vogelbescherming report 10).

CAMPHUYSEN, C. J., CALVO, B., DURINCK, J., ENSOR, K., FOLLESTAD, A., FURNESS, R. W., GARTHE, S., LEAPER, G., SKOV, H., TASKER, M. L. AND WINTER, C. J. N. (1995) *Consumption of discards by seabirds in the North Sea.* Final report EC DG XIV research contract BIOECO/93/10. NIOZ Rapport 1995–5 edition. Texel, Netherlands: Netherlands Institute for Sea Research.

CAMPOS, P. (1984) *Economía y energía en la dehesa extremeña.* Madrid: Pesqueros y Alimentarios, Instituto de Estudios.

CAMPOS, P. (1992) Spain. Pp.165–200 in S. Wibe and T. Jones, eds. *Forests: market and intervention failures. Five case studies.* London: Earthscan.

CAMPOS, P. (1993) Valores comerciales y ambientales de las dehesas españolas. *Agricultura y Sociedad* 66: 9–41.

CAMPOS, P. (1994) The total economic value of agroforestry systems. Pp.33–47 in N. E. Koch, ed. *The scientific basis for sustainable multiple-use forestry in the European Community.* Brussels: Commission of the European Communities.

CAMPOS, P. (1995a) Dehesa forest economy and conservation in the Iberian peninsula. Pp.112–117 in D. I. McCracken and E. M. Bignal, eds. *Farming on the edge: the nature of traditional farmland in Europe.* Peterborough, UK: Joint Nature Conservation Committee.

CAMPOS, P. (1995b) Conserving commercial and environmental benefits in the western Mediterranean forest. Pp.301–310 in L. M. Albisu and C. Romero, eds. *Environmental and land-use issues: an economic perspective.* Kiel, Germany: Wissenschaftsverlag Vauk Kiel KG.

CANOVA, L. AND FASOLA, M. (1993) Evoluzione della popolazione nidificante di spatola *Platalea leucorodia* in Italia. Atti VII Convegno Associazione A. Ghigi per la biologia e la conservazione di vertebrati. *Suppl. Ric. Biol. Selvaggina* 21: 525–528.

CARTER, I. C., WILLIAMS, J. M., WEBB, A. AND TASKER, M. L. (1993) *Seabird concentrations in the North Sea: an atlas of vulnerability to surface pollutants.* Aberdeen, UK: Joint Nature Conservation Committee.

CDPE (1995) *Pan-European Biological and Landscape Diversity Strategy. 11th Meeting, Strasbourg, 7–9 June 1995.* CDPE(95)9 edition. Strasbourg, France: Steering Committee for the Protection and Management of the Environment and Natural Habitats, Council of Europe.

CEC (1991) *Rapport d'activité de la Commission dans le secteur régi par le Règlement (CEE) no 3521/86 du Conseil du 17 novembre 1986, et par le Règlement (CEE) no 1641/89 du Conseil du 29 Mai 1989, relatifs à la protection des forêts contre les incendies.* Brussels: Commission of the European Communities (DG VI).

CEC (1992) *Council Regulation instituting a Community aid scheme for forestry measures in agriculture.* Luxembourg: Commission of the European Communities (EEC2080/92).

CEC (1993) *Towards sustainability. The European Community programme of policy and action in relation to the environment and sustainable development.* Brussels: Commission of the European Communities (OJ C138, 17 May 1993).

CEC (1994a) *Promoting biodiversity in the European Community: the ACE-Biotopes programme 1984–1991.* Luxembourg: Commission of the European Communities.

CEC (1994b) *Proposal for a European Parliament and Council Decision on Community guidelines for the development of the trans-European transport network.* Brussels: Commission of the European Communities (COM(94)106 fin., 7.4.94).

CEC (1994c) *Special Protection Areas.* Brussels: Commission of the European Communities.

CEC (1995a) Agricultural situation and prospects in the Central and Eastern European countries: summary report (European Commission, Directorate for Agriculture). Brussels: Commission of the European Communities.

CEC (1995b) *Communication from the Commission to the Council and the European Parliament. Wise use and conservation of wetlands. COM(95)189 final.* Brussels: Commission of the European Communities.

CEC (1995c) *Communication from the Commission to the Council of the European Parliament on the integrated management of coastal zones.* COM(95)511 final edition. Brussels: Commission of the European Communities.

CEC, ED. (1995d) *Europe's environment: statistical compendium for the Dobrís assessment.* Luxembourg: Commission of the European Communities.

CEC (1995e) Study on alternative strategies for the development of relations in the field of agriculture between the EU and the associated countries with a view to future accession of these countries. Draft for discussion by EU Heads of Government. Brussels: Commission of the European Communities.

CENTRAL STATISTICAL OFFICE (1995) *Agriculture Statistical Almanac of Hungary.* Budapest: Orient Press.

CERNUSCA, A. (1977) Ökologische Veränderungen im Beriech von Skipisten. Pp.81–150 in R. Sprung and B. König, eds. *Das österreichische Skirecht.* Innsbruck, Austria: Wagner.

CHADWICK, L. (1982) *In search of heathland.* Durham, UK: Dobson.

CHAMBERLAIN, D., EVANS, J., FULLER, R. AND LANGSTON, R. (1995) Where there's muck, there's birds. *BTO News* 200: 15–17.

CHANTREL, C. (1984) Eléments d'étude pour un bilan économique de la transformation des zones humides par l'agriculture. Olonne-sur-Mer, France: Office National de la Chasse.

CHAPMAN, V. J. (1976) *Coastal vegetation.* Oxford, UK: Pergamon Press.

CHAUVET, C. (1988) Manuel sur l'aménagement des pêches dans les Lagunes Côtières: la Bordique Méditerranéenne. *FAO Fisheries Technical Paper* 290.

CHERNOV, Y. I. (1975) [*Natural zonation and animal life on the continent.*] Moscow. (In Russian.)

CHERNOV, Y. I. (1985) *The living tundra.* Cambridge, UK: Cambridge University Press (Studies in Polar Research).

CHEYLAN, G. (1991) Patterns of Pleistocene turnover, current distribution and speciation among Mediterranean mammals. Pp.227–262 in R. H. Groves and F. di Castri, eds. *Biogeography of Mediterranean invasions.* Cambridge, UK: Cambridge University Press.

CHEYLAN, M. (1984) The true status and future of Hermann's tortoise *Testudo hermanni robertmertensi* in western Europe. *Amphibia. Reptilia* 5: 17–26.

CHRISTENSEN, M. AND EMBORG, J. (1996) Biodiversity in natural versus managed forest in Denmark. *Forest Ecology and Management* 85: 47–51.

CLARK, R. B. (1982) The impact of oil pollution on marine populations, communities and ecosystems: a summing up. *Phil. Trans. R. Soc. Lond. B* 297: 433–443.

CLARK, R. B. (1984) Impact of oil pollution on seabirds. *Environ. Pollut. A* 33: 1–22.

CLUTTON-BROCK, T. H. AND ALBON, S. D. (1989) *Red deer in the Highlands.* Oxford, UK: BSP Professional Books.

COLLAR, N. J. AND ANDREW, P. (1988) *Birds to watch: the ICBP world check-list of threatened birds.* Cambridge, UK: International Council for Bird Preservation (Techn. Publ. 8).

COLLAR, N. J., CROSBY, M. J. AND STATTERSFIELD, A. J. (1994) *Birds to watch 2: the world list of threatened birds.* Cambridge, UK: BirdLife International (BirdLife Conservation Series no. 4).

COMIN, F. A. AND PARAREDA, X. F. (1979) Les llacunes litorals. *Quaderns d'Ecologia Aplicada* 4: 51–68.

COMÍNS, J. S., SANZ, J. M., NAVALPOTRO, P. AND SANTAOLALLA, A. (1994) El meio ambiente en la PAC: impactos recientes en la agricultura española. *El Campo* 131: 9–30.

CONSTANT, P., EYBERT, M.-C. AND MAHEO, R. (1973) Breeding bird populations in conifer plantations in Paimpont Forest, Brittany. *Alauda* 41: 371–384. (English title, original in French.)

COOK, R. T. (1992) Trace metals and organochlorines in the eggs and diets of Dalmatian Pelicans (*Pelecanus crispus*) and Spoonbills (*Platalea leucorodia*) in northern Greece. Manchester, UK: University of Manchester (M.Sc thesis).

COSGROVE, P. AND TURNER, R. (1995) Implementation in the United Kingdom of the Helsinki Guidelines for Sustainable Forest Management. Edinburgh, UK: Royal Society for the Protection of Birds (unpublished report).

COSTANZA, R. (1991) *Ecological economics: the science and management of sustainability.* New York, USA: Columbia University Press.

COULSON, J. C. (1988) The structure and importance of invertebrate communities on peatlands and moorlands, and effects of environmental and management changes. Pp.365–380 in M. B. Usher and D. B. A. Thompson, eds. *Ecological changes in the uplands.* Oxford, UK: Blackwell Scientific Publications (Special Publication of the British Ecological Society 7).

COULSON, J. C., GOODYER, A. AND FIELDING, S. (1992) *The management of moorland areas to enhance their nature conservation interest.* Peterborough, UK: Joint Nature Conservation Committee (JNCC Report No. 134).

COVENEY, J., MERNE, O., WILSON, J., ALLEN, D. AND THOMAS, G. (1993) A conservation strategy for birds in Ireland. Monkstown, Ireland: Irish Wildbird Conservancy (unpublished report).

CPRE (1995) *What are they doing to our land?* London: Council for the Protection of Rural England.

CRAIK, J. C. A. (1995) Effects of North American mink on British seabirds. Pp.17–18 in M. L. Tasker, ed. *Threats to seabirds: Proceedings of the 5th International Seabird Group conference.* Sandy, UK: Seabird Group.

CRAMP, S., ED. (1988) *The birds of the western Palearctic*, 5. Oxford, UK: Oxford University Press.

CRAMP, S. AND SIMMONS, K. E. L., EDS. (1980) *The birds of the western Palearctic*, 2. Oxford, UK: Oxford University Press.

CRAMP, S. *ET AL.* (1977–1994) *The birds of the western Palearctic*, 1–9. Oxford, UK: Oxford University Press.

CRICK, H. Q. P., DUDLEY, C., EVANS, A. D. AND SMITH, K. W. (1994) Causes of nest failure among buntings in the UK. *Bird Study* 41: 88–94.

CRIVELLI, A. J. (1996) Action plan for the Dalmatian Pelican (*Pelecanus crispus*. Pp.53–66 in B. Heredia, L. Rose and M. Painter, eds. *Globally threatened birds in Europe: action plans.* Strasbourg: Council of Europe, and BirdLife International.

CRIVELLI, A. J., FOCARDI, S., FOSSI, C., LEONZIO, C., MASSI, A. AND RENZONI, A. (1989) Trace elements and chlorinated hydrocarbons in eggs of *Pelecanus crispus* a world endangered bird species nesting at Lake Mikri Prespa, north-western Greece. *Environ. Pollut.* 61: 235–247.

CRIVELLI, A. J., NAZIRIDES, T. AND JERRENTRUP, H. (1996) Action plan for the Pygmy Cormorant (*Phalacrocorax pygmeus*. Pp.41–52 in B. Heredia, L. Rose and M. Painter, eds. *Globally threatened birds in Europe: action plans.* Strasbourg: Council of Europe, and BirdLife International.

CROCKFORD, N., GREEN, R., ROCAMORA, G., SCHÄFFER, N., STOWE, T. AND WILLIAMS, G. (1996) Action plan for the Corncrake *Crex crex* in Europe. Pp.205–243 in B. Heredia, L. Rose and M. Painter, eds. *Globally threatened birds in Europe: action plans.* Strasbourg: Council of Europe, and BirdLife International.

CROXALL, J. P. (1975) The effect of oil on nature conservation, especially birds. Pp.93–101 in H. A. Cole, ed. *Petroleum and the continental shelf of north-west Europe*, 2. London: Petroleum and Halstead Press.

CROXALL, J. P. (1977) The effects of oil on seabirds. *Rapports et Procès-verbaux des Réunions, Conseil International pour l'Exploration de la Mer* 171: 191–195.

CROXALL, J. P., ED. (1987) *Seabirds: feeding ecology and role in marine ecosystems.* Cambridge, UK: Cambridge University Press.

CROXALL, J. P., EVANS, P. G. H. AND SCHREIBER, R. W., EDS. (1984) *Status and conservation of the world's seabirds.* Cambridge, UK: International Council for Bird Preservation (Techn. Publ. 2).

CUFF, J. AND RAYMENT, M. (1997) Working with nature: economies, employment and conservation in Europe. A RSPB Policy Research Department Report on behalf of BirdLife International.

CURRIE, F. (1995) Lowland heathlands and forestry in Great Britain. Pp.62 in *Proceedings 5th European Heathland Workshop, Santiago de Compostela, Spain, September 1995.* Spain: University of Santiago de Compostela.

CUSTODIO, E. (1991) Characteristics of aquifer overexploitation: comments on hydrogeological and hydrochemical aspects: the situation in Spain. IAH XXIII Int. Cong., Puerto de la Cruz, Tenerife, Spain.

DAGET, P. (1980) Un élément actuel de la caractérisation du monde méditerranéen: le climat. *Naturalia Monspelensia* No. Hors Série: 101–126.

DAHL, A. H. (1991) *Regional seas: island directory. UNEP regional seas directories and bibliographies.* Nairobi: United Nations Environment Programme (Vol 35).

DAHL, E., ELVEN, R., MOEN, A. AND SKOGEN, A. (1986) *Vegetasjonsregionkart over Norge 1:1,500,000. Nasjonal atlas for Norge.* Oslo: Statens Kartverk. (In Norwegian.)

DAMMAN, A. W. H. (1957) The south Swedish *Calluna*-heath and its relation to the Calluneto-Genistetum. *Botaniska Notiser* 110: 363–398.

DANIELS, R. E. (1978) Floristic analyses of British mires and mire communities. *J. Ecol.* 66: 773–802.

DANISH MINISTRY OF ENVIRONMENT AND ENERGY (1995) Declaration signed at 4th International Conference on the Protection of the North Sea, Esbjerg, Denmark, 8–9 June 1995.

DAVIDSON, N. C. AND EVANS, P. R. (1986) The role and potential of man-made and man-modified wetlands in the enhancement of the survival of overwintering shorebirds. *Col. Waterb.* 9: 176–188.

DAVIDSON, N. C. AND EVANS, P. R. (1988) Pre-breeding accumulation of fat and muscle protein by arctic-breeding shorebirds. *Acta XIX Congr. Int. Orn.*: 342–352.

DAVIDSON, N. C. AND ROTHWELL, P. I., EDS. (1993) Disturbance to waterfowl on estuaries. *Wader Study Group Bull.* 68 (Special Issue): 97–106.

DAVIDSON, N. C., D'ALAFFOLEY, D., DOODY, J. P., WAY, L. S., GORDON, J., KEY, R., DRAKE, C. M., PIENKOWSKI, M. W., MITCHELL, R. AND DUFF, K. L. (1991) *Nature conservation and estuaries in Great Britain.* Peterborough, UK: Nature Conservancy Council.

DAVIS, T. J., ED. (1993) *Towards the wise use of wetlands.* Gland, Switzerland: Ramsar Convention Bureau.

DEBUSSCHE, M. AND ISENMANN, P. (1990) Introduced and cultivated fleshy-fruited plants: consequences on a mutualistic Mediterranean plant-bird system. Pp.399–416 in F. di Castri, A. Hansen and M. Debussche, eds. *Biological invasions in Europe and the Mediterranean Basin.* Dordrecht, Netherlands: Kluwer Publishers.

DEGANGE, A. R. AND NEWBY, T. C. (1980) Mortality of seabirds and fish in a lost salmon driftnet. *Mar. Pollut. Bull.* 11: 322–323.

DE LATTIN, G. (1967) *Grundriß der Zoogeographie.* Jena, Germany: Gustav Fischer.

DEVILLIERS, P. AND DEVILLIERS-TERSCHUREN, J. (1996) *A classification of Palaearctic habitats.* Strasbourg, France: Council of Europe (Nature and Environment No. 78). HA/AA/12.

DGF (1990) *Inventário florestal do sabreiro 1990.* Lisbon: DGF (Estudos e Informação 300).

DGF (1993) *Distribuição da floresta em Portugal Continental. Areas florestais por distritos. Informação disponível em 1992.* Lisbon: DGF (Estudos e Informação 303).

DÍAZ, M., ASENSIO, B. AND TELLERÍA, J. L. (1996) *Aves*

Ibéricas I. No paseriformes. Madrid: J. M. Reyero.

DIAS, M., GAUDÊNCIO, M. J. AND GUERRA, M. T. (1994) Economically important bivalves in the coastal waters adjacent to the Tagus and Sado estuaries. Publ. avulsas do IPIMAR No. 1, pp.47–48.

DÍAZ, M., CAMPOS, P. AND PULIDO, F. J. (1997) The Spanish dehesas: a diversity in land-use and wildlife. Pp.178–209 in D. J. Pain and M. W. Pienkowski, eds. *Farming and birds in Europe: The Common Agricultural Policy and its implications for bird conservation*. London: Academic Press.

DÍAZ, M., GONZÁLEZ, E., MUÑOZ-PULIDO, R. AND NAVESO, M. A. (1993) Abundance, seed predation rates, and body condition of rodents wintering in Spanish holm-oak *Quercus ilex* L. dehesas and cereal croplands: effects of food abundance and habitat structure. *Z. Säugetierkunde* 58: 302–311.

DI CASTRI, F. AND MOONEY, H. A. (1973) *Mediterranean-type ecosystems: origin and structure*. London: Chapman and Hall.

DICKSON, K. L., MAKI, A. W. AND BRUNGS, W. A. (1987) *Fate and effects of sediment bound chemicals in aquatic systems*. UK: Pergamon Press.

DIEMONT, H., WEBB, N. AND DEGN, H. J. (1996) Pan European view on heathland conservation. Pp.21–32 in *Proceedings of the National Heathland Conference, 18th-20th September 1996*. UK.

DIEMONT, W. H. (1996) *Survival of Dutch heathlands*. Wageningen, Netherlands: DLO Institute for Forestry and Nature Research (IBN–DLO) (IBN Scientific Contributions 1).

DIERSSEN, K. (1982) *Die wichtigsten Pflanzengesellschaften der Moore NW-Europas*. [*The major plant communities of north-west European mires.*] Geneva: Conservatoire et Jardin Botaniques.

VAN DIJK, G. (1991) The status of semi-natural grasslands in Europe. Pp.15–36 in P. D. Goriup, L. A. Batten and J. A. Norton, eds. *The conservation of lowland dry grassland birds in Europe*. Peterborough, UK: Joint Nature Conservation Committee.

DIJKEMA, K. S., ED. (1984) *Saltmarshes in Europe*. Strasbourg, UK: Council of Europe (Nature and Environment Sciences No. 30).

DILL, W. A. (1990) *Inland fisheries of Europe*. Rome: Food and Agricultural Organization of the United Nations (EIFAC Techn. Pap. No. 52).

DIMBLEBY, G. W. (1962) The development of British heathlands and their soils. *Forestry Memoirs* 23: 120.

DIXON, J. (1997) European agriculture: threats and opportunities. Pp.389–421 in D. J. Pain and M. W. Pienkowski, eds. *Farming and birds in Europe: The Common Agricultural Policy and its implications for bird conservation*. London: Academic Press.

DONALD, P. F. AND EVANS, A. D. (1994) Habitat selection by Corn Buntings *Miliaria calandra* in winter. *Bird Study* 41: 199–210.

DONAUBAUER, E., KREHAN, H. AND SCHADAUER, K. (1995) Wie gesund oder wie krank ist Österreichs Wald? Pp.67–77 in *Ökobilanz Wald Österreich 1995*. Vienna: Österreichisches Statistisches Zentralamt.

DONÁZAR, J. A. AND FERNÁNDEZ, C. (1990) Population trends of the Griffon Vulture (*Gyps fulvus*) in northern Spain between 1969 and 1989 in relation to conservation measures. *Biol. Conserv.* 53: 83–91.

DONÁZAR, J. A., NAVESO, M. A., TELLA, J. E. AND CAMPIÓN, D. (1997) Extensive grazing and raptors in Spain. Pp.117–149 in D. J. Pain and M. W. Pienkowski, eds. *Farming and birds in Europe: the Common Agricultural Policy and its implications for bird conservation*. London: Academic Press.

DOODY, J. P. (1989) Dungeness—a national nature conservation perspective. *Botanical J. Linn. Soc.* 101: 163–171.

DOODY, J. P., ED. (1991) *Sand dune inventory of Europe*. Peterborough, UK: Joint Nature Conservation Committee/European Union for Coastal Conservation.

DOODY, P., ED. (1984) *Spartina anglica in Great Britain*. Peterborough, UK: Nature Conservancy Council.

VAN DORP, D. AND OPDAM, P. F. M. (1987) Effects of patch size, isolation and regional abundance on forest bird communities. *Landscape Ecol.* 1: 59–73.

DRABLØS, D. AND TOLLAN, A., EDS. (1980) *Ecological impact of acid precipitation*. Proceedings of the International Conference on the Ecological Impact of Acid Precipitation edition. Oslo: SNSF (project report).

DRENT, P. J. AND WOLDENDORP, J. W. (1989) Acid rain and eggshells. *Nature* 339: 431.

DUDLEY, N. (1992) *Forests in trouble: a review of the status of temperate forests worldwide*. Gland, Switzerland: World Wide Fund for Nature.

DUFFEY, E. (1976) Breckland. In *Nature in Norfolk: a heritage in trust*. Norwich, UK: Jarrold.

DUGAN, P., ED. (1993) *Wetlands in danger: a Mitchell Beazley world conservation atlas*. London: Mitchell Beazley.

DUGAN, P. J., ED. (1990) *Wetland conservation: a review of current issues and required action*. Gland, Switzerland: International Union for the Conservation of Nature and Natural Resources.

DUINKER, J. C. AND KOEMAN, J. H. (1978) Summary report on the distribution and effects of toxic pollutants (metals and chlorinated hydrocarbons) in the Wadden Sea. Pp.45–54 in K. Essink and W. J. Wolff, eds. *Pollution of the Wadden Sea area*, Report 8. Balkema, Rotterdam: Wadden Sea Working Group.

DUNN, E. (1994a) *Case studies of farming and birds in Europe: lowland wet grasslands in the Netherlands and Germany*. Sandy, UK: Royal Society for the Protection of Birds (Studies in European agriculture and environmental policy no. 10).

DUNN, E. (1994b) *Interactions between fisheries and marine birds: research recommendations*. Sandy, UK: Royal Society for the Protection of Birds.

DUNN, E. (1995) Global impacts of fisheries on seabirds: a paper prepared by BirdLife International for the London Workshop on Environmental Science, Comprehensiveness and Consistency in Global Declines on Ocean Issues. Sandy, UK: Royal Society for the Protection of Birds.

DUNN, E. (1996) Fisheries and the development of sustainability indicators. Pp.57–62 in B. Earll, ed. *Marine Environmental Management: review of events in 1995 and future trends*. Kempley, UK: Candle Cottage.

DUNN, E. AND HARRISON, N. (1995) *RSPB's vision for sustainable fisheries*. Sandy, UK: Royal Society for the Protection of Birds.

DUNNET, G. M. (1982) Oil pollution and seabird populations.

Phil. Trans. R. Soc. Lond. B 297: 413–427.

DUNNET, G. M. (1987) Seabirds and North Sea oil. *Phil. Trans. R. Soc. Lond. B* 316: 513–524.

DUNNET, G. M., CRICK, H. Q. P. AND BAHA EL DIN, S. (1986) Bardawil lagoon baseline environmental study and vulnerability to oil-pollution. Pp.335–358 in MEDMARAVIS and X. Monbailliu, eds. *Mediterranean Marine Avifauna.* Berlin: Springer-Verlag.

DUNNET, G. M., FURNESS, R. W., TASKER, M. L. AND BECKER, P. H. (1990) Seabird ecology in the North Sea. *Netherlands Journal of Sea Research* 26: 387–425.

DURINCK, J., SKOV, H., JENSEN, F. P. AND PIHL, J. (1994) *Important marine areas for wintering seabirds in the Baltic Sea.* Copenhagen: Ornis Consult.

ECGB (1992) *Climate change: a threat to global development: Enquete Commission of the German Bundestag.* Berlin: Economia Verlag.

EDWARDS, C. A. AND LOFTY, J. R. (1975) The influence of soil cultivation on soil animal populations. Pp.399–407 in J. Vanek, ed. *Progress in soil zoology.* The Hague, Netherlands: Junk.

EDWARDS, C. A. AND LOFTY, J. R. (1982a) The effect of direct drilling and minimal cultivation on earthworm populations. *J. Appl. Ecol.* 19: 723–734.

EDWARDS, C. A. AND LOFTY, J. R. (1982b) Nitrogenous fertilizers and earthworm populations in agricultural soils. *Soil Biology & Biochemistry* 14: 515–521.

EGDELL, J. M. (1993) *Impact of agricultural policy on Spain and its steppe regions.* Sandy, UK: Royal Society for the Protection of Birds (Studies in European agriculture and environmental policy no. 2).

EISELTOVÁ, M., ED. (1994) *Restoration of lake ecosystems: a holistic approach.* Slimbridge, UK: International Waterfowl and Wetlands Research Bureau (IWRB Spec. Publ 32).

ELENA, M., LÓPEZ, J. A., CASAS, M. AND SÁNCHEZ DEL CORRAL, A. (1987) *El carbon de encina y la dehesa.* Madrid: Ministerio de Agricultura, Pesca y Alimentación.

ELLENBERG, H. (1986a) *Vegetation Mitteleuropas mit den Alpen.* Stuttgart, Germany: Eugen Ulmer.

ELLENBERG, H. (1986b) Warum gehen die Neuntöter (*Lanius collurio*) in Mitteleuropa in Bestand zurück? *Corax* 12: 34–46.

ELLENBERG, H. (1988) *Vegetation ecology on Central Europe.* Cambridge, UK: Cambridge University Press.

ELLENBERG, H., RÜGER, A. AND VAUK, G. (1989) [*Eutrophication: the most serious threat in nature conservation.*] Germany: Norddeutsche Naturschutzakademie (Berichte 2). (In German.)

ELLIOTT, C. (1996) *WWF guide to forest certification.* Godalming, UK: World Wide Fund for Nature–UK.

ELTRINGHAM, S. K., ED. (1971) *Life in mud and sand.* London: The English Universities Press Ltd.

ENGLISH NATURE (1996) *Impact of water abstraction on wetland SSSIs.* Peterborough, UK: English Nature.

ENOKSSON, B., ANGELSTAM, P. AND LARSSON, K. (1995) Deciduous forest and resident birds: the problem of fragmentation within a coniferous forest landscape. *Landscape Ecol.* 10: 267–275.

EPPLE, W. (1988) Das Braunkelchen—Jahresvogel 1987—im Brennpunkt der Extensivierungdebatte in der Landwirtschaft. Einführung in das Artenschutzsymposium Braunkehlchen. *Beih. Veröff. Naturschutz Landschafts-*

pflege Bad.-Württ. 51: 15–31.

ERIKSSON, M. O. G. (1987) Some effects of freshwater acidification on birds in Sweden. Pp.183–190 in A. W. Diamond and F. L. Filion, eds. *The value of birds.* Cambridge: International Council for Bird Preservation (Techn. Publ. 6).

ERIKSSON, M. O. G. (1994) Susceptibility of freshwater acidification by two species of loon: Red-throated Loon *Gavia stellata* and Arctic Loon *Gavia arctica* in southwest Sweden. *Hydrobiologia* 279/280: 439–444.

ERIKSSON, M. O. G., ARVIDSSON, B. L. AND JOHANSSON, I. (1988) [Habitat characteristics of Red-throated Diver, *Gavia stellata*, breeding lakes in south-west Sweden.] *Vår Fågelv.* 47: 122–132. (In Swedish, English summary.)

ERIKSSON, M. O. G., JOHANSSON, I. AND AHLGREN, C.-G. (1992) Levels of mercury in eggs of Red-throated Diver *Gavia stellata* and Black-throated Diver *Gavia arctica* in south-west Sweden. *Ornis Svecica* 2: 29–36.

ERKAMO, V. (1952) On plant-biological phenomena accompanying the present climate change. *Fennia* 75: 25–37.

ERTAN, A., KILIÇ, A. AND KASPAREK, M. (1989) *Turkiye'nin onemli kus alanlari.* [*Important Bird Areas in Turkey.*] Istanbul: Dogal Hayati Koruma Dernegi and the International Council for Bird Preservation.

ESCODA, R. (1996) *Memoriès d'un poble: la Vilella d'Amunt.* Tarragona, Spain: Institut d'Estudis Tarraconense.

ESPEUT, M. (1984) Avifaune nicheuse du massif des Madres et du Mont Coronat. Montpellier (unpublished thesis) (Thèse USTL).

ESSEEN, P.-A., EHNSTRÖM, B., ERICSON, L. AND SJÖBERG, K. (1992) Boreal forest—the focal habitats of Fennoscandia. Pp.252–325 in L. Hansson, ed. *Ecological principles of nature conservation—a boreal perspective.* London: Elsevier.

ESSINK, K. (1984) The discharge of organic waste into the Wadden Sea—local effects. *Netherlands Inst. Sea Res. Publ. Ser.* 10: 165–177.

EUROSTAT (1992) *Europa in Zahlen.* Luxembourg: Commission of the European Communities.

EVANS, A. (1997) The importance of mixed farming for seed-eating birds in the UK. Pp.331–357 in D. Pain and M. Pienkowski, eds. *Farming and birds in Europe: the Common Agricultural Policy and its implications for bird conservation.* London: Academic Press.

EVANS, A. D. AND SMITH, K. W. (1994) Habitat selection of Cirl Bunting *Emberiza cirlus* wintering in Britain. *Bird Study* 41: 81–87.

EVANS, A., PAINTER, M., WYNDE, R. AND MICHAEL, N. (1994) An inventory of lowland heathland: a foundation for an effective conservation strategy. *RSPB Conservation Review* 8: 24–30.

EVANS, P. R. (1981) Reclamation of intertidal land: some effects on Shelduck and wader populations in the Tees estuary. *Verh. Orn. Ges. Bayern* 23: 147–168.

EVANS, P. R. AND PIENKOWSKI, M. W. (1983) Implications of coastal engineering projects from studies, at the Tees estuary, on the effects of reclamation of intertidal land on shorebird populations. *Water Science and Technology* 16: 347–354.

EVANS, P. R., UTTLEY, J. D., DAVIDSON, N. C. AND WARD, P. (1987) Shorebirds (S.Os Charadrii and Scolopaci) as agents of transfer of heavy metals within and between estuarine ecosystems. Pp.337–352 in P. J. Coughtrey, M. H. Martin and M. H. Unsworth, eds. *Pollutant transport and fate in ecosystems*. Oxford, UK: Blackwell (British Ecological Society Special Publ. No. 6).

FAO (1993) *Recent trends in the fisheries and environment of the GFCM area*. General fisheries for the Mediterranean twentieth session, Valletta, Malta, 5–9 July 1993. Rome: Food and Agriculture Organization of the United Nations.

FARAGÓ, S. (1992) [Studies on animal food resource of game-birds in agriculture areas in Hungary III.] *Erd. Faip. Tud. Közl.* 1990(1): 5–161. (In Hungarian.)

FARALLI, U. (1995) Effetti della riforestazione sulle comunità ornitiche di una brughiera dell'Appennino settentrionale, Toscana. *Suppl. Ric. Biol. Selvaggina* 22: 299–306.

FARALLI, U. AND LAMBERTINI, M. (1989) Ecologia del Fuoco. Lega Italiana Protezione Uccelli (unpublished report).

FARRELL, L. (1993) *Lowland heathland: the extent of habitat change*. Peterborough, UK: English Nature (English Nature Science Series No. 12).

FASOLA, M. (1986) Resource use of foraging herons in agricultural and non-agricultural habitats in Italy. *Col. Waterb.* 9: 139–148.

FASOLA, M. AND BARBIERI, F. (1978) Factors affecting the distribution of heronries in northern Italy. *Ibis* 120: 337–340.

FASOLA, M. AND CANOVA, L. (1991) Colony site selection by eight species of gulls and terns breeding in the Valli di Comacchio (Italy). *Boll. Zool.* 58: 261–266.

FASOLA, M. AND CANOVA, L. (1992) Nest habitat selection by eight syntropic species of Mediterrnaean gulls and terns. *Col. Waterb.* 15: 169–178.

FASOLA, M. AND CANOVA, L. (1996) Conservation of gull and tern colony sites in north-eastern Italy, an internationally important bird area. *Col. Waterb.* 19: 59–67.

FASOLA, M. AND GHIDINI, M. (1983) Use of feeding habitat by breeding Night Heron and Little Egret. *Avocetta* 7: 29–36.

FASOLA, M. AND RUÍZ, X. (1996) The value of rice fields as substitutes for natural wetlands for waterbirds in the Mediterranean region. *Col. Waterb.* 19 (Suppl. 1): 122–128.

FASOLA, M. AND RUÍZ, X. (1997) Rice farming and waterbirds: integrated management in an artificial landscape. Pp.210–235 in D. J. Pain and M. W. Pienkowski, eds. *Farming and birds in Europe: the Common Agricultural Policy and its implications for bird conservation*. London: Academic Press.

FASOLA, M., BOGLIANI, G., SAINO, N. AND CANOVA, L. (1989) Foraging, feeding and time activity niches of eight species of breeding birds in the coastal wetlands of the Adriatic Sea. *Boll. Zool.* 56: 61–62.

FAWCETT, D. AND VAN VESSEM, J., EDS. (1995) *Lead poisoning in waterfowl: international update 1995*. Peterborough, UK: Joint Nature Conservation Committee (JNCC Report No. 252).

FERNÁNDEZ, C. (1988) Inventariación y valoración de los muladares para las aves carroñeras. Pamplona, Spain: Gobierno de Navarra (unpublished report).

FERNÁNDEZ, J. A. AND MONTERO, G. (1993) Prospección de secas en *Quercus* de Extremadura y La Mancha. *Montes* 32: 32–36.

FERNÁNDEZ-ALÉS, R., MARTÍN, A., ORTEGA, F. AND ALÉS, E. E. (1992) Recent changes in landscape structure and function in a Mediterranean region of south-west Spain (1950–1984). *Landscape Ecol.* 7: 3–18.

FERNS, P. N. (1992) *Bird life of coasts and estuaries*. Cambridge, UK: Cambridge University Press.

FERRER, M. AND CALDERÓN, J. (1990) The Spanish Imperial Eagle *Aquila adalberti* Brehm 1861 in Doñana National Park (south west Spain): a study of population dynamics. *Biol. Conserv.* 51: 151–161.

FERRER, M., GUYONNE, J. AND CHACÓN, M. L. (1993) Mortalidad de aves en tendidos eléctricos: situación actual en España. *Quercus* 94: 20–29.

FERRER, X., MUNTANER, J. AND MARTINEZ, A. (1984) *Atlas d'els ocells nidificants de Catalunya i d'Andorra*. Barcelona: Ketres.

FERRY, C. AND FROCHOT, B. (1970) L'avifaune nidificatrice d'une forêt des chênes pédonculés en Bourgogne: étude de deux successions écologiques. *La Terre et la Vie.* 1970: 153–250.

FERRY, C. AND FROCHOT, B. (1974) L'influence du traîtment forestier sur les oiseaux. Pp.309–326 in P. Pesson, ed. *Ecologie forestière*. Paris: Gauthier-Villars.

FINLAYSON, C. M. AND LARSSON, T., EDS. (1991) *Wetland management and restoration: proceedings of a workshop, Sweden, 12–15 September 1990*. Solna, Sweden: Swedish Environmental Protection Agency. HA/WE/21.

FINLAYSON, M. AND MOSER, M., EDS. (1991) *Wetlands*. New York: Facts on File.

FINLAYSON, M., ED. (1992) *Integrated management and conservation of wetlands in agricultural and forested landscapes*. Slimbridge, UK: International Waterfowl and Wetlands Research Bureau.

FINLAYSON, M., HOLLIS, T. AND DAVIS, T., EDS. (1992) *Managing Mediterranean wetlands and their birds: proceedings of an International Waterfowl and Wetlands Research Bureau international symposium, Grado, Italy, February 1991*. Slimbridge, UK: International Waterfowl and Wetlands Research Bureau (IWRB Spec. Publ. 20).

FISHER, J. (1952) *The Fulmar*. London: Collins.

FLADE, M. (1991) Die Habitate des Wachtelkönigs während der Brutsaison in drei europäischen Stromtälern (Aller, Save, Biebrza). [The habitat of Corncrakes in the breeding season in three European valleys (Aller, Save, Biebrza).] *Vogelwelt* 112: 16–40. (In German.)

FLADE, M. (1994) *Die Brutvogelgemeinschaften Mittel- und Norddeutschlands*. Eching, Germany: IHW.

FLOUSEK, J. (1989) Impact of industrial emissions on bird populations breeding in mountain spruce forests in central Europe. *Ann. Zool. Fennici* 26: 255–263.

FOLKESTAD, A. O. (1982) The effect of mink predation on some seabird species. *Viltrapport* 21: 42–49.

FORESTRY COMMISSION (1989) *Native pinewoods grants and guidelines*. Edinburgh, UK: Forestry Commission.

FOSTER, D. R. (1983) The history and pattern of fire in the boreal forest of southeastern Labrador. *Can. J. Botany* 61: 2459–2471.

VAN FRANEKER, J. A. (1985) Plastic ingestion in the North Atlantic fulmar. *Mar. Pollut. Bull.* 16: 367–369.

FRANK, G. (1995) Naturwaldreservate. Pp.37–41 in *Ökobilanz Wald Österreich 1995*. Vienna: Österreichisches

Statistisches Zentralamt.

FRANZ, H. (1979) *Ökologie der Hochgebirge*. Stuttgart, Germany: Eugen Ulmer.

FRAZIER, S. (1996) *An overview of the world's Ramsar sites*. Slimbridge, UK: Wetlands International.

FREEDMAN, B. (1989) *Environmental ecology: the impacts of pollution and other stresses on ecosystem structure and function*. London: Academic Press.

FREEMARK, K. E. AND MERRIAM, H. G. (1986) Importance of area and habitat heterogeneity to bird assemblages in temperate forest fragments. *Biol. Conserv.* 36: 115–141.

FREMSTAD, E. AND KVENILD, L. (1993) [Poor heath vegetation in Norway: a distribution map.] *NINA Oppdragsmelding* 188: 1–17. (In Norwegian.)

FRID, C. L. J. AND EVANS, P. R. (1995) Coastal habitats. Pp.59–83 in W. J. Sutherland and D. A. Hill, eds. *Managing habitats for conservation*. Cambridge, UK: Cambridge University Press.

FULLER, R. J. (1982) *Bird habitats in Britain*. Calton, UK: T. and A. D. Poyser.

FULLER, R. J. (1990) Responses of birds to lowland woodland management in Britain: opportunities for integrating conservation with forestry. *Sitta* 4: 39–40.

FULLER, R. J. (1995) *Bird life of woodland and forest*. Cambridge, UK: Cambridge University Press.

FULLER, R. J. AND WARREN, M. S. (1990) *Coppiced woodlands: their management for wildlife*. Peterborough, UK: Nature Conservancy Council.

FULLER, R. J., GREGORY, R. D., GIBBONS, D. W., MARCHANT, J. H., WILSON, J. D., BAILLIE, S. R. AND CARTER, N. (1995) Population declines and range contractions among farmland birds in Britain. *Conserv. Biol.* 9: 1425–1442.

FULLER, R., HILL, D. AND TUCKER, G. (1991) Feeding the birds down on the farm: perspectives from Britain. *Ambio* 20: 232–237.

FULLER, R. J., REED, T. M., BUXTON, N. E., WEBB, A., WILLIAMS, T. D. AND PIENKOWSKI, M. W. (1986) Populations of breeding waders Charadrii and their habitats on the crofting lands of the Outer Hebrides, Scotland. *Biol. Conserv.* 37: 333–361.

FURNESS, R. W. (1985) Plastic particle pollution: accumulation by procellariform seabirds at Scottish colonies. *Mar. Pollut. Bull.* 16: 103–106.

FURNESS, R. W. (1987a) Seabirds as monitors of the marine environment. Pp.217–230 in A. W. Diamond and F. L. Filion, eds. *The value of birds*. Cambridge: International Council for Bird Preservation (Techn. Publ. 6).

FURNESS, R. W. (1987b) *The skuas*. Calton, UK: T. and A. D. Poyser.

FURNESS, R. W. (1992) *Implications of changes in net mesh size, fishing effort and minimum landing size regulations in the North Sea for seabird populations*. Peterborough, UK: Joint Nature Conservation Committee (JNCC Rep. No. 133).

FURNESS, R. W. (1993) *An assessment of human hazards to seabirds in the North Sea*. Godalming, UK: World Wide Fund for Nature.

FURNESS, R. W. (1995) Are industrial fisheries a threat to seabirds? Pp.22 in M. L. Tasker, ed. *Threats to seabirds: Proceedings of the 5th International Seabird Group conference*. Sandy, UK: Seabird Group.

FURNESS, R. W. AND MONAGHAN, P. (1987) *Seabird ecology*. Glasgow, UK: Blackie.

FURNESS, R. W., GREENWOOD, J. J. D. AND JARVIS, P. J. (1993) Can birds be used to monitor the environment? Pp.1–41 in R. W. Furness and J. J. D. Greenwood, eds. *Birds as monitors of environmental change*. London: Chapman and Hall.

FURNESS, R. W., MUIRHEAD, S. J. AND WOODBURN, M. (1986) Using bird feathers to measure mercury in the environment: relationships between mercury content and moult. *Mar. Pollut. Bull.* 17: 27–30.

GAGGINO, G. F., CAPPELLETTI, E., MARCHETTI, R. AND CALCAGNINI, T. (1987) The water quality of Italian lakes. European Water Polluttion Control Association. International Congress. *Lake pollution and recovery*, Rome 15–18 April, 1985.

GAMAUF, A. (1991) [*Birds of prey in Austria: populations, threats, laws.*] Vienna: Umweltbundesamt (Monogr. 29). (In German.)

GAMLIN, L. (1988) Sweden's factory forests. *New Scientist* 28 January.

GARCÍA, A. (1992) Conserving the species-rich meadows of Europe. *Agric. Ecosyst. Environ.* 40: 219–232.

GASCÓ, J. M. (1987) Condicionamientos del medio natural de las dehesas extremeñas desde las perspectivas de su desarrollo compatible con el mantenimiento de su capacidad productiva. Pp.19–35 in P. Campos and M. Martín, eds. *Conservación y desarrollo de las dehesas portuguesa y española*. Madrid: Ministerio de Agricultura.

GATAULINA (1992) In C. J. Pearson, ed. *Field crop ecosystems*. Amsterdam: Elsevier (Ecosystems of the World 18).

GÉHU, J. M. (1960a) Une site célèbre de la côté Nord Bretonne: le sillon de Talbert (C-du-N). Observations phytosociologique et ecologiques. *Bull. Lab. Marit. Dinard.* 46: 93–113.

GÉHU, J. M. (1960b) Une site célèbre de la côté Nord Bretonne: le sillon de Talbert (C-du-N). Observations phytosociologiques. *Bull. Lab. Marit. Dinard.* 46: 78–92.

GÉHU, J. M. (1960c) La végétation des levées des galets du littoral français de la Manche. *Bull. Soc. Bot. Nord France* 13: 141–152.

GÉHU, J. M. (1991) Végétations et paysages littoraux de type Cantabro-atlantique. *Berichte der Reinhold-Tuxen Gesellschaft* 3: 59–128.

GEORGE, K. (1996) Deutsche Landwirtschaft im Spiegel der Vogelwelt. *Vogelwelt* 117: 187–198.

GEORGII, B., ZEITLER, A. AND HOFER, D. (1994) Hängegleiten, Gleitsegeln und Wildtiere—Eine Umfrage unter Piloten, Berufsjägern und Bergsteigern. *Verh. Ges. Ökol.* 23: 262–268.

GILBERT, G. AND GIBBONS, D. W. (1996) A review of habitat, land cover and land-use survey and monitoring in the United Kingdom. Royal Society for the Protection of Birds (unpublished report).

GILES, N. (1992) *Wildlife after gravel: 20 years practical research by the Game Conservancy at the ARC Wildfowl Centre*. Fordingbridge, UK: The Game Conservancy.

GIMINGHAM, C. H. (1972) *Ecology of heathlands*. London: Chapman & Hall.

GIMINGHAM, C. H. (1978) *Calluna* and its associated species: some aspects of coexistence in communities. *Vegetatio* 36: 179–186.

GIMINGHAM, C. H. (1992) *The lowland heathland manage-*

ment book. Peterborough, UK: English Nature.

GIMINGHAM, C. H., CHAPMAN, S. B. AND WEBB, N. R. (1979) European heathlands. In R. L. Specht, ed. *Ecosystems of the world heathlands and related dwarf shrublands.* Amsterdam: Elsevier.

GIMINGHAM, C. H., HOBBS, R. J. AND MALLIK, A. U. (1981) Community dynamics in relation to management of heathland vegetation in Scotland. *Vegetatio* 46–7: 149–155.

GLEIK, P. H. (1989) Global climatic changes and regional hydrology: impacts and responses. *Internatn. Assoc. Scient. Hyd. Publ.* 168: 389–402.

GLUE, D. (1973) The breeding birds of a New Forest valley. *Brit. Birds* 66: 461–473.

GLUTZ VON BLOTZHEIM, U. N., BAUER, K. M. AND BEZZEL, E. (1973) *Handbuch der vögel Mitteleuropas: Galliformes–Gruiformes [Handbook of the birds of Central Europe]*, 5. Wiesbaden, Germany: Akademische Verlagsgesellschaft. (In German.)

GLUTZ VON BLOTZHEIM, U. N. AND BAUER, K. M. (1985) *Handbuch der vogel Mitteleuropas: Motacillidae–Prunellidae*, 10/II. Wiesbaden: Aula-Verlag.

GLUTZ VON BLOTZHEIM, U. N. *ET AL.* (1966–1997) *Handbüch der Vögel Mitteleuropas [Handbook of the birds of Central Europe]*, 1–13. Wiesbaden, Germany: Aula-Verlag.

GOMEZ-CAMPO, C. (1985) The conservation of Mediterranean plants: principles and problems. Pp.3–8 in C. Gomez-Campo, ed. *Plant conservation inf the Mediterranean area.* Dordrecht, Netherlands: Junk.

GONZÁLEZ, L. M. (1996) Action plan for the Spanish Imperial Eagle *Aquila adalberti.* Pp.175–189 in B. Heredia, L. Rose and M. Painter, eds. *Globally threatened birds in Europe: action plans.* Strasbourg: Council of Europe, and BirdLife International.

GONZÁLEZ, L. M., BUSTAMANTE, J. AND HIRALDO, F. (1990) Factors influencing the present distribution of the Spanish Imperial Eagle *Aquila adalberti. Biol. Conserv.* 51: 311–320.

GOODWILLIE, R. (1980) *European peatlands.* Strasbourg: Council of Europe. EU/AA/130.

GORDIENKO, N. S. (1991) [Biology and number of the Sociable Lapwing *Chettusia gregaria* in Kustanai Steppe, Northern Kazakhstan.] *Ornitologiya* 25: 54–61. (In Russian.)

GORDON, D. (1992) Joint logging ventures in the Russian Far East. *Taiga News* 2.

GORDON, J., GODDARD, A., LEAPER, R., LEAPER, L., STEINER, L. AND WHITMORE, C. (1989) *Cetacean research program in the Azores.* Cambridge, UK: International Fund for Animal Welfare.

GORDON, N. D., MCMAHON, T. A. AND FINLAYSON, B. L. (1992) *Stream hydrology: an introduction for ecologists.* Chichester, UK: John Wiley & Sons.

GORIUP, P. D., ED. (1988) *Ecology and conservation of grassland birds.* Cambridge, UK: International Council for Bird Preservation (Techn. Publ. 7).

GORIUP, P. D., BATTEN, L. A. AND NORTON, J. A., EDS. (1991) *The conservation of lowland dry grassland birds in Europe: proceedings of an international seminar held at the University of Reading 20–22 March 1991.* Peterborough, UK: Joint Nature Conservation Committee.

GOSS-CUSTARD, J. D. (1977a) The ecology of the Wash III. Density-related behaviour and the possible effects of a loss of feeding grounds on wading birds (Charadrii). *J. Appl. Ecol.* 14: 721–739.

GOSS-CUSTARD, J. D. (1977b) Predator responses and prey mortality in Redshank, *Tringa totanus* (L.), and a preferred prey, *Corophium volutator* (Pallas). *J. Appl. Ecol.* 46: 21–35.

GOSS-CUSTARD, J. D. AND DURELL, S. E. A. LE V. DIT (1990) Bird behaviour and environmental planning: approaches in the study of wader populations. *Ibis* 132: 273–289.

GOSS-CUSTARD, J. D. AND MOSER, M. E. (1988) Rates of change in the numbers of Dunlin *Calidris alpina*, wintering in British estuaries in relation to *Spartina anglica. J. Appl. Ecol.* 25: 95–109.

GOUDRIAAN, J., VAN KEULEN, H. AND VAN LAAR, H. H., EDS. (1990) *The greenhouse effect and primary productivity in European agro-ecosystems.* Wageningen, Netherlands: Pudoc.

GRABHERR, G. (1982) The impact of trampling by tourists on a high-altitudinal grassland in the Tyrolean Alps, Austria. *Vegetatio* 48: 209–217.

GRABHERR, G. (1987) Ökologische Probleme des alpinen Raumes. *Schriftenreihe des Deutschen Rates für Landschaftspflege* 52: 124–131.

GRABHERR, G., MAIR, A. AND STIMPFL, H. (1987) Wachstums- und Reproduktionsstratagien von Hochgebirgspflanzen und ihre Bedeutung für die Begrünung von Skipisten und anderen alpinen Erosionsflächen. *Verh. Ges. Ökol.* 15: 183–188.

GRAJETZKY, B. (1993) Breeding success of the Robin *Erithacus rubecula* in hedgerows. *Vogelwelt* 114: 232–240.

GRANSTRÖM, A. (1991) [Fire and its associated plants in southern Sweden.] *Skog och Forskning* 91: 22–28.

GREEN, A. (1996) International action plan for the Marbled Teal (*Marmaronetta angustirostris*. Pp.99–117 in B. Heredia, L. Rose and M. Painter, eds. *Globally threatened birds in Europe: action plans.* Strasbourg: Council of Europe, and BirdLife International.

GREEN, A. J. AND ANSTEY, S. (1992) The status of the White-headed Duck *Oxyura leucocephala. Bird Conserv. Internatn.* 2: 185–200.

GREEN, A. AND HUGHES, B. (1996) Action plan for the White-headed Duck (*Oxyura leucocephala.* Pp.119–145 in B. Heredia, L. Rose and M. Painter, eds. *Globally threatened birds in Europe: action plans.* Strasbourg: Council of Europe, and BirdLife International.

GREEN, B. (1990) Agricultural intensification and the loss of habitat, species and amenity in British grasslands. *Grass and Forage Sciences* 45: 365–372.

GREEN, P. C. (1989) The use of Trichoptera as indicators of conservation value. *J. Environ. Mgmt* 29: 95–104.

GREEN, R. E. (1986) The management of lowland wet grassland for breeding waders. Sandy, UK Royal Society for the Protection of Birds (internal report).

GREEN, R. E. (1988) Stone Curlew conservation. *RSPB Conserv. Rev.* 2: 30–33.

GREENPEACE (1995) *Driftnets: walls of death.* London: Greenpeace.

GREENWOOD, J. J. D., BAILLIE, S. R., GREGORY, R. D., PEACH, W. J. AND FULLER, R. J. (1995) Some new approaches to conservation monitoring of British breeding birds. *Ibis* 137: 516–528.

GREIF, F. AND SCHWACKHÖFER, W. (1979) *Die Sozialbrache im Hochgebirge am Beispiel des Außerferns. Agrarwissenschaftliches Institut des Bundesministeriums für Land- und Forstwirtschaft*. Wien: Österreicheischer Agrarverlag.

GRETTON, A. (1996) International action plan for the Slender-billed Curlew *Numenius tenuirostris*. Pp.271–288 in B. Heredia, L. Rose and M. Painter, eds. *Globally threatened birds in Europe: action plans*. Strasbourg: Council of Europe, and BirdLife International.

GRIGORYEV, A. (1992) Status report of the forest situation in Russia. *Taiga News* 2.

GRIMMETT, R. F. A. AND JONES, T. A. (1989) *Important Bird Areas in Europe*. Cambridge, UK: International Council for Bird Preservation (Techn. Publ. 9).

GROOMBRIDGE, B. (1982) *The IUCN Amphibia–Reptilia Red Data Book: Testudines, Crocodylia, Rhynchocephalia*. Gland, Switzerland: International Union for Conservation of Nature and Natural Resources.

GROOMBRIDGE, B., ED. (1993) *1994 IUCN Red List of threatened animals*. Gland, Switzerland and Cambridge, UK: International Union for Conservation of Nature and Natural Resources.

GUÉLORGET, O. AND PERTHUISOT, J.-P. (1992) Paralic ecosystems: biological organization and functioning. *Vie et Milieu* 42: 215–251.

GUNNARSSON, B. (1990) Vegetation structure and the abundance and size distribution of spruce-living spiders. *J. Anim. Ecol.* 59: 743–752.

GUYOT, I. (1993) Breeding distribution and number of Shag (*Phalacrocorax aristotelis* in the Mediterranean. Pp.37–45 in J. S. Aguilar, X. Monbailliu and A. M. Paterson, eds. *Status and conservation of seabirds: ecogeography and Mediterranean action plan*. Madrid: Sociedad Española de Ornitología.

HAGEMEIJER, E. J. M. AND BLAIR, M. J. (1997) *The EBCC atlas of European breeding birds: their distribution and abundance*. London: T. and A. D. Poyser.

HAILA, Y. AND JÄRVINEN, O. (1990) Northern conifer forests and their bird species assemblages. Pp.61–85 in A. Keast, ed. *Biogeography and ecology of forest bird communities*. The Hague, Netherlands: SPB Academic Publishing.

HAILA, Y., HANSKI, I. K. AND RAIVIO, S. (1987) Breeding bird distribution in fragmented coniferous taiga in Southern Finland. *Ornis Fenn.* 64: 90–106.

HÅKANSON, L. (1980) The quantitative impact of pH, bioproduction and Hg-contamination on the Hg-content of fish (pike). *Environ. Pollut. Serv. Bull.* 1: 285–304.

HAKE, M. (1991) The effects of needle loss in coniferous forests in south-west Sweden on the winter foraging behaviour of willow tits *Parus montanus. Biol. Conserv.* 58: 357–366.

HALLER, H. (1978) Zur Populationsökologie des Uhus *Bubo bubo* im Hochgebirge: Bestand, Bestandsentwicklung und Lebensraum in den Rätuschen Alpen. *Orn. Beob.* 75: 237–265.

HALLER, H. (1979) Raumorganisation und Dynamik einer Population des Steinadlers *Aquila chysaetos* in den Zentralalpen. *Orn. Beob.* 79: 163–211.

HALLMANN, B. (1985) Status and conservation problems of birds of prey in Greece. Pp.55–59 in I. Newton and R. D. Chancellor, eds. *Conservation studies on raptors*. Cam-

bridge, UK: International Council for Bird Preservation (Techn. Publ. 5).

HALPERN, C. B. AND SPIES, T. A. (1995) Plant species diversity in natural and managed forests of the Pacific Northwest. *Ecol. Applicat.* 5: 913–934.

HAMPICKE, U. (1978) Agriculture and conservation—ecological and social aspects. *Agriculture and the Environment* 4: 25–42.

HANCOCK, J. AND ELLIOT, H. (1978) *The herons of the world*. London: Harper and Row.

HANSKI, I. (1982) Dynamics of regional distribution: the core and satellite species hypothesis. *Oikos* 38: 210–221.

HANSKI, I. AND TIAINEN, J. (1988) Populations and communities in changing agro-ecosystems in Finland. *Ecological Bulletin* 39: 159–168.

HANSSON, L. (1983) Bird numbers across edges between mature conifer forest and clearcuts in central Sweden. *Ornis Scand.* 14: 97–103.

HARDING, P. T. AND ROSE, F. (1986) *Pasture woodlands in lowland Britain*. Cambridge, UK: Institute of Terrestrial Ecology.

HARDY, A. R., STANLEY, P. I. AND GREIG-SMITH, P. W. (1987) Birds as indicators of the intensity of use of agricultural pesticides in the U.K. Pp.119–132 in A. W. Diamond and F. L. Filion, eds. *The value of birds*. Cambridge, UK: International Council for Bird Preservation (Techn. Publ. 6).

HARIO, M. (1993) First incident of mass seabird deaths from presumed paralytic shellfish posioning in Finland. *Suomen Riista* 39: 7–20.

HARPER, D. (1992) *Eutrophication of freshwaters: principles, problems and restoration*. London: Chapman & Hall.

HARRIS, L. D. (1984a) *The fragmented forest: island biogeography theory and the preservation of biotic diversity*. Chicago, USA: University of Chicago Press.

HARRIS, M. P. (1970) Rates and causes of increases of some British gull populations. *Bird Study* 17: 325–335.

HARRIS, M. P. (1984b) *The Puffin*. Calton, UK: T. and A. D. Poyser.

HARRISON, C. (1981) Recovery of lowland grassland and heathland in southern England from disturbance by seasonal trampling. *Biol. Conserv.* 19: 119–130.

HARRISON, C. (1982) *An atlas of the birds of the western Palearctic*. London: Collins.

HASKINS, L. E. (1978) The vegetational history of southeast Dorset. Southampton, UK: University of Southampton (unpublished Ph.D. thesis).

HAVELKA, P. AND RÜGE, K. (1993) [Population trends of woodpeckers (Picidae) in the Federal Republic of Germany.] *Veröff. Natursch. Landschaftspfl. Bad.-Wurtt.* 67: 33–38. (In German.)

HAWKE, C. J. AND JOSÉ, P. V. (1996) *Reedbed management for commercial and wildlife interest*. Sandy, UK: Royal Society for the Protection of Birds.

HAWKES, H. A. (1979) Invertebrates as indicators of river water quality. Pp.1–45 in A. James and L. Evison, eds. *Biological indicators of water quality*. Chichester, UK: John Wiley & Sons.

HAWKINS, S. J. AND JONES, H. D. (1992) *Rocky shores*. London: Immel Publishing.

HEATH, M. F. AND PAYNE, A. J. (in prep.) *Important Bird Areas in Europe*. Cambridge, U.K.: BirdLife International (BirdLife Conservation Series).

HEATHWAITE, A. L. (1993) *Mires. Process, exploitation and conservation*. Chichester, UK: John Wiley.

HEINIGER, P. (1991) [Adaptions of the Snow Finch *Montifringilla nivalis* to his cold-dominated alpine environment.] *Orn. Beob.* 88: 193–207. (In German with English summary.)

HELLE, P. AND JÄRVINEN, O. (1986) Population trends of north Finnish landbirds in relation to their habitat selection and changes in forest structure. *Oikos* 46: 107–115.

HELLE, P. AND MÖNKKÖNEN, M. (1990) Forest successions and bird communities: theoretical aspects and practical implications. Pp.299–318 in A. Keast, ed. *Biogeography and ecology of forest bird communities*. The Hague, Netherlands: SPB Academic Publishing.

HENRIKSEN, A., LIEN, L., TRAAEN, T. J., SEVALDRUD, I. S. AND BRAKKE, D. F. (1988) Lake acidification in Norway: present and predicted chemical status. *Ambio* 17: 259–266.

HEREDIA, B. (1996a) Action plan for the Cinereous Vulture *Aegypius monachus* in Europe. Pp.147–158 in B. Heredia, L. Rose and M. Painter, eds. *Globally threatened birds in Europe: action plans*. Strasbourg: Council of Europe, and BirdLife International.

HEREDIA, B. (1996b) International action plan for the Imperial Eagle *Aquila heliaca*. Pp.159–174 in B. Heredia, L. Rose and M. Painter, eds. *Globally threatened birds in Europe: action plans*. Strasbourg: Council of Europe, and BirdLife International.

HEREDIA, B. (1996c) Action plan for the Aquatic Warbler *Acrocephalus paludicola* in Europe. Pp.327–338 in B. Heredia, L. Rose and M. Painter, eds. *Globally threatened birds in Europe: action plans*. Strasbourg: Council of Europe, and BirdLife International.

HEREDIA, B., ROSE, L. AND PAINTER, M., EDS. (1996) *Globally threatened birds in Europe: action plans*. Strasbourg, France: Council of Europe and BirdLife International.

HERMELINE, M. AND REY, G. (1994) *L'Europe et la forêt*. Tome I et II edition. EUROFOR/ONF/Parlement Européen.

HERRERA, C. M. (1982a) Defense of ripe fruits from pests: its significance in relation to plant-disperser interactions. *Amer. Nat.* 120: 218–241.

HERRERA, C. M. (1982b) Seasonal variation in the quality of fruits and diffuse coevolution between plants and avian dispersers. *Ecology* 63: 773–785.

HERTZMAN, T. AND LARSSON, T. (1991) Lake Hornborga: a case study. Pp.154–160 in C. M. Finlayson and T. Larsson, eds. *Wetland management and restoration: proceedings of a workshop, Sweden, 12–15 September 1990*. Solna, Sweden: Swedish Environmental Protection Agency.

HEUTZ DE LEMPS, A. (1970) *La végétation de la terre*. Paris: Masson.

HEY, PFEIFFER AND TOPAN (1996) The economic impact of motorways in the peripheral regions of the EU—a literature survey. Sandy, UK: Royal Society for the Protection of Birds.

HICKS, J. R. (1946) *Value and capital*. Second edition. Oxford, UK: Oxford University Press.

HIDALGO DE TRUCIOS, S. J. AND CARRANZA ALMANSA, J. C. (1990) *Ecología y comportamiento de la Avutarda (*Otis tarda *L.). [The ecology and behaviour of Great Bustard*

Otis tarda.] Cáceres, Spain: Universidad de Extremadura.

HILL, D. A. (1985) The feeding ecology and survival of pheasant chicks on arable farmland. *J. Appl. Ecol.* 22: 645–654.

HILL, D. A. (1994) An assessment of the population consequences to birds of the proposed compensation measures associated with the Cardiff Bay barrage scheme. Cambridge, UK: Ecoscope Applied Ecologists (unpublished report).

HILL, D. A. AND STREET, M. (1987) Survival of mallard ducklings and competition with fish for invertebrates on a flooded gravel quarry in England. *Ibis* 129: 159–167.

HILL, D. A., HOCKIN, D., PRICE, D., TUCKER, G. M. AND MORRIS, R. (1997) Bird disturbance: improving the quality and utility of disturbance research. *J. Appl. Ecol.* 34: 275–288.

HILL, D., RUSHTON, J. P., CLARK, N., GREEN, P. AND PRYS-JONES, R. (1993) Shorebird communities on British estuaries: factors affecting community composition. *J. Appl. Ecol.* 30: 220–234.

HOBBS, R. J., RICHARDSON, D. M. AND DAVIS, G. W. (1995) Mediterranean-type ecosystems: opportunities and constraints for studying the function of biodiversity. Pp.1–42 in G. W. Davis and D. M. Richardson, eds. *Mediterranean-type ecosystems: the function of biodiversity*. Berlin and London: Springer-Verlag (Ecological Studies Vol. 109).

HOCKIN, D., OUNSTED, M., GORMAN, M., HILL, D., KELLER, V. AND BAKER, M. (1992) Examination of the effects of disturbance on birds with reference to the role of environmental impact assessments. *J. Environ. Mgmt* 36: 253–286.

HODGE, C. A. H., BURTON, R. G. O., CORBETT, W. M., EVANS, R. AND SEALE, R. S. (1984) *Soils and their use in eastern England*. UK: Soil Survey of England and Wales.

HODGSON, J. G. (1987) Growing rare in Britain. *New Scientist* 1547: 38–39.

HÖGBOM, A. G. (1934) *Om skogseldar förr och nu: och deras null i skogarnas utvecklingshistoria*. Stockholm: Almquist och Wiksells förlag.

HOLLIS, T. (1992) The causes of wetland loss and degradation in the Mediterranean. Pp.83–90 in M. Finlayson, T. Hollis and T. Davis, eds. *Managing Mediterranean wetlands and their birds: proceedings of an IWRB international symposium, Grado, Italy, February 1991*. Slimbridge, UK: International Waterfowl and Wetlands Research Bureau (Spec. Publ. 20).

HOLMES, N. T. H. AND HONBURY, R. G. (1995) Rivers, canals and dykes. Pp.84–120 in W. J. Sutherland and D. A. Hill, eds. *Managing habitats for conservation*. Cambridge, UK: Cambridge University Press.

HOLMES, W. N. AND CRONSHAW, J. (1977) Biological effects of petroleum on marine birds. Pp.359–398 in D. C. Malins, ed. *Effects of petroleum on arctic and subarctic marine environments and organisms*. New York: Academic Press.

HÖLZINGER, J., ED. (1987) *Avifauna Baden-Württemberg. 1: Gefährdung und Schutz. [Birds of Baden-Württemberg. 1: Threats and conservation.]* Karlsruhe: E. Ulmer Verlag. (In German.)

HOMSKIS, V. (1969) [Dynamic and thermic regime of small Lithuanian lakes.]

HOOPER, T. D., VERMEER, K. AND SZABO, I. (1987) Oil

pollution of birds: an annotated bibliography. *Can. Wildl. Serv. Techn. Rep. Ser.* 34: 1–180.

HOPKINS, J. H. (1991) Management of semi-natural lowland dry grasslands. Pp.119–124 in P. D. Goriup, L. A. Batten and J. A. Norton, eds. *The conservation of lowland dry grassland birds in Europe. Proceedings of a seminar, Reading 20–22 March 1991.* Peterborough, UK: Joint Nature Conservation Committee.

HOPKINS, J. J. (1983) Studies of the historical ecology, vegetation and flora of the Lizard district, Cornwall, with particular reference to heathland. University of Bristol, UK (Ph.D. Dissertation).

HOS (1992) Greek threatened areas project. Athens: Hellenic Ornithological Society (unpublished annual report).

HOSSELL, J. (1994) The implications of global climate change for biodiversity. Oxford, UK: Environmental Change Unit, University of Oxford (unpublished report to the Royal Society for the Protection of Birds).

HÖTKER, H., ED. (1991) *Waders breeding on wet grasslands.* Peterborough, UK: Wader Study Group (Wader Study Group Bulletin No. 61, Suppl.).

HOUSDEN, S. D., THOMAS, G., BIBBY, C. J. AND PORTER, R. F. (1991) Towards a habitat conservation strategy for bird habitats in Britain. *RSPB Conservation Review* 5: 9–16.

DEL HOYO, J., ELLIOT, A. AND SARGATAL, J., EDS. (1992) *Handbook of the birds of the world,* 1: Ostrich to ducks. Barcelona, Spain: Lynx Edicions.

HRBÁCEK, J. (1994) Food web relations. Pp.44–58 in M. Eiseltová, ed. *Restoration of lake ecosystems: a holistic approach.* Slimbridge, UK: International Waterfowl and Wetlands Research Bureau (IWRB Spec. Publ. 32).

HUBOLD, G. (1994) Massnahmenkatalog für eine ausgewogenere und rationellere Bewirtschaftung der von der deutschen Fischerei genutzten Fischereiressourcen im EG Meer. *Inform. Fischwirt.* 41: 3–18.

HUDSON, P. J. (1992) Grouse in space and time: the population biology of a mmanaged gamebird: the report of the Game Conservancy's Scottish Grouse Research Project and North of England Grouse Research Project. Fordingbridge, UK: Game Conservancy.

HUDSON, R., TUCKER, G. M. AND FULLER, R. J. (1994) Lapwing *Vanellus vanellus* populations in relation to agricultural changes: a review. Pp.1–33 in G. M. Tucker, S. M. Davies and R. J. Fuller, eds. *The ecology and conservation of lapwings* Vanellus vanellus. Peterborough, UK: Joint Nature Conservation Committee (UK Nature Conservation Series 9).

HUNTER, J. M. AND BLACK, J. M. (1996) International action plan for the Red-breasted Goose (*Branta ruficollis.* Pp.79–98 in B. Heredia, L. Rose and M. Painter, eds. *Globally threatened birds in Europe: action plans.* Strasbourg: Council of Europe, and BirdLife International.

HUNTER, M. L. (1990) *Wildlife, forests and forestry.* Englewood Cliffs, USA: Prentice Hall.

HUNTINGS SURVEY (1986) *Moorland landscape change,* 1: *Main report.* Borehamwood, UK: Huntings Survey.

HUNTLEY, B. (1988) Europe. Pp.341–383 in B. Huntley and T. Webb, eds. *Vegetation history.* Amsterdam: Kluwer Academic Publishers.

HUNTLEY, B. (1990) European post-glacial forests: compositional changes in response to climatic change. *Journal of Vegetation Science* 1: 507–518.

HUNTLEY, B. (1994) Plant species' response to climate change: implications for the conservation of European birds. *Ibis* 137 (Suppl.): 127–138.

HUNTLEY, B. AND BIRKS, H. J. B. (1983) *An atlas of past and present pollen maps for Europe: 0–13,000 years ago.* Cambridge, UK: Cambridge University Press.

HUSBY, E. (1995) *Wilderness quality mapping in the Euro-Arctic Barents region.* Trondheim, Norway: Direktoratet for Naturforvaltning (DN-rapport 1995–4).

IBERO, C. (1994) *The status of old-growth and semi-natural forests in western Europe.* Gland, Switzerland: World Wide Fund for Nature.

IBERO, C. (1996) *Ríos de vida. El estado de conservación de las riberas fluviales en España.* Madrid: SEO/BirdLife.

ICBP (1992) *Putting biodiversity on the map: priority areas for global conservation.* Cambridge, UK: International Council for Bird Preservation.

ICES (1994) Report of the working group on ecosystem effects of fishing activities. International Council for the Exploration of the Sea Doc. C.M. 1994/ASSESS/ENV.1.

ICONA (1991) [*Forest oak decline in Spain (1990–1991).*] Madrid: Informe Interno del ICONA (MAPA). (In Spanish.)

IMBODEN, E., ED. (1987) *Riverine forests in Europe—status and conservation, 15th Conference of the European Continental Section, Rapperswil, Switzerland: 20th–25th February 1985.* Cambridge, UK: International Council for Bird Preservation.

VAN IMPE, J. (1985) Estuarine pollution as a probable cause of increase of estuarine birds. *Mar. Pollut. Bull.* 16: 271–276.

INE (1993) *Portugal Agrícola.* Lisbon: Instituto Nacional de Estatística.

INSTITUTO FLORESTAL (1993a) Relatório periódico de incêndios florestais relativo ao período de 93/10/10. Unpublished.

INSTITUTO FLORESTAL (1993b) *Inventário pinheiro-bravo 1992.* Lisbon: Instituto Florestal (Estudos e Informação 305).

IPCC (1990) *Climate change: the IPCC scientific assessment.* Houghton, J.T., Callender, B.A. and Varney, S.K. (eds.). Cambridge, UK: Cambridge University Press.

IPCC (1992) *The supplementary report to the IPCC scientific assessment of climate change 1992.* Houghton, J.T., Callender, B.A. and Varney, S.K. (eds.). Cambridge, UK: Cambridge University Press.

IREMONGER, S., KAPOS, V., RHIND, J. AND LUXMOORE, R. (in press) A global overview of forest conservation. Cambridge, UK: World Conservation Monitoring Centre (unpublished paper written for presentation at the World Forestry Congress, October 1997).

IRVING, R. (1993) *Too much of a good thing: nutrient enrichment in the U.K.'s inland and coastal waters.* Godalming, UK: World Wide Fund for Nature.

IUCN (1991) *The lowland grasslands of Central and Eastern Europe.* Cambridge, UK: International Union for Conservation of Nature and Natural Resources.

IUCN (1993) *The wetlands of central and eastern Europe.* Gland, Switzerland: International Union for Conservation of Nature and Natural Resources.

IUCN (1995) *The mountains of Central and Eastern Europe.* Cambridge, UK: International Union for Nature Conservation and Natural Resources.

IUCN–CNPPA (1994) *Parks for life: action plan for protected areas in Europe.* Gland, Switzerland: International Union for Nature Conservation and Natural Resources (Commission on National Parks and Protected Areas).

IVERSEN, T. M., KRONVANG, B., MADSEN, B. L., MARKMAN, P. AND NIELSEN, M. B. (1993) Re-establishment of Danish streams, restoration and maintenance measures. *Aquatic Conserv.* 3: 1–20.

JAMES, F. C. AND WAMER, N. O. (1982) Relationships between temperate forest bird communities and vegetation structure. *Ecology* 63: 159–171.

JANSEN, P. B. AND DE NIE, H. W. (1986) Thirty years of passerine breeding bird monitoring in a mixed wood. *Limosa* 59: 127–134.

JEDRZEJEWSKI, W., JEDRZEJEWSKA, B., OKARMA, H. AND RUPRECHT, A. L. (1992) Wolf predation and snow cover as mortality factors in the ungulate community of the Bialowieza National Park, Poland. *Oecologia* 90: 27–36.

JEDRZEJEWSKI, W., SCHMIDT, K., MILKOWSKI, L., JEDRZEJEWSKA, B. AND OKARMA, H. (1993) Foraging by lynx and its role in ungulate mortality: the local (Bialowieza Forest) and Palearctic viewpoints. *Acta Theriologica* 38: 385–403.

JELGERSMA, S. (1994) Examples of the geological past: evolution of coastal sedimentation sequences during Holocene sea level rise. Pp.45–57 in J. Pernetta, R. Leemans, D. Elder and S. Humphrey, eds. *Impacts of climate change on ecosystems and species: marine and coastal ecosystems.* Gland, Switzerland: IUCN.

JENKINS, D. (1984) *Agriculture and the environment.* Cambridge, UK: Natural Environment Research Council (Proceedings of ITE symposium No. 13).

JENNI-EIERMANN, S., BÜHLER, U. AND ZBINDEN, N. (1996) Vergiftungen von Greifvögeln durch Carbofurananwendung um Ackerbau. *Orn. Beob.* 93: 69–77.

JENNY, M. (1990) Nahrungsökologie der Feldlerche *Alauda arvensis* in einer intensiv genutzten Agrarlandschaft des schweizerischen Mittellandes. [Diet of the Skylark *Alauda arvensis* in an area of intensive agriculture in central Switzerland.] *Orn. Beob.* 87: 31–53.

JIMÉNEZ, J. (1992) [*Census of Spanish Imperial Eagle and Black Vulture populations in Ciudad Real.*] Ciudad Real, Spain: Consejeria de Agricultura. (In Spanish.)

DE JUANA, E. (1984) The status and conservation of seabirds in the Spanish Mediterranean. Pp.347–361 in J. P. Croxall, P. G. H. Evans and R. W. Schreiber, eds. *Status and conservation of the world's seabirds.* Cambridge, UK: International Council for Bird Preservation (Techn. Publ. 2).

DE JUANA, E., SANTOS, T., SUÁREZ, F. AND TELLERÍA, J. L. (1988) Status and conservation of steppe birds and their habitats in Spain. Pp.113–123 in P. D. Goriup, ed. *Ecology and conservation of grassland birds.* Cambridge, UK: International Council for Bird Preservation (Techn. Publ. 7).

DE JUANA, E., VARELA, J. AND WITT, H.-H. (1984) The conservation of seabirds at the Chafarinas islands. Pp.363–370 in J. P. Croxall, P. G. H. Evans and R. W. Schreiber, eds. *Status and conservation of the world's seabirds.* Cambridge, UK: International Council for Bird Preservation (Techn. Publ. 2).

DE JUANA, F. (1989) Situación actual de las rapaces diurnas. *Ecología* 3: 237–292.

JOHANSSON, M. (1994) Quantification of mycorrhizal infection in roots of *Calluna vulgaris* (L.) Hull from Danish heathland. *Soil Biology & Biochemistry* 26: 763–766.

JOHNSGARD, P. A. (1978) *Ducks, geese and swans of the World.* Lincoln and London: University of Nebraska Press.

JONES, T. A. AND HUGHES, J. M. R. (1993) Wetland inventories and wetland loss studies: a European perspective. Pp.164–169 in M. Moser, R. C. Prentice and J. van Vessem, eds. *Waterfowl and wetland conservation in the 1990s: a global perspective.* Slimbridge, UK: International Waterfowl and Wetlands Research Bureau (Proc. Symp., St Petersburg Beach, Florida, U.S.A).

DE JONG, F., BAKKER, J. F., DAHL, K., DANKERS, N., FARKE, H., JÄPPELT, W., KOßMAGK-STEPHAN, K. AND MADSEN, P. B. (1993) *Quality Status Report of the North Sea. Subregion 10: the Wadden Sea.* Wilhelmshaven, Germany: Common Wadden Sea Secretariat.

JÖNSSON, P. E. (1990) [Mink—a serious threat for Black Guillemots on Hallands Väderö.] *Anser* 29: 278–281. (In Swedish with English summary.)

JORDANO, P. (1985) [The annual cycle of frugivorous passerines in southern Spanish Mediterranean shrubland: the wintering season and between-year variations.] *Ardeola* 32: 69–94. (In Spanish.)

JORDHØY, P., STRAND, O., SKOGLAND, T. AND GAARE, E. (1996) *Monitoring of cervids. Wild reindeer 1991–95.* Trondheim, Norway: Norsk Institutt for Naturforskning (NINA-Fagrapport).

JURKOVSKAYA, T. K. (1980) Sphagnovie bolota. [Sphagnum bogs.] Pp.303–328 in S. A. Gribova, T. I. Isachenko and E. M. Lavrenko, eds. *Rastitelnost evropejskoj chasti SSSR. [Vegetation of European part of the USSR.]* Leningrad: Nauka.

KAESER, G. AND SCHMID, H. (1989) Bestand und Bruterfolg des Turmfalken *Falco tinnunculus* und der Schleiereule *Tyto alba* in der Region Rheinfelden 1951–1988. *Orn. Beob.* 86: 199–208.

KALAND, P. E. (1995) Traditional land-use methods in the coastal heathlands of Norway. How can we encourage the farmers to maintain the tradition within a modern society? Pp.29b–30b in *Proceedings 5th European Heathland Workshop, Santiago de Compostela, Spain, September 1995.* Spain: University of Santiago de Compostela.

KÅLÅS, J. E., FISKE, P. AND HÖGLUND, J. (in press) Food supply and breeding occurrences: the west European population of the lekking Great Snipe (Aves). *J. Biogeogr.*

KARAPETKOVA, M., ZHIVKOV, M. AND ALEXANDROVA-KOLEMANOVA, K. (1993) The freshwater fishes in Bulgaria. Pp.515–546 in M. Sakalian, ed. *National strategy for conservation of the biological diversity*, 1. Sofia: Bulvest 2000.

KASPAREK, M. (1985) *Die Sultanssumpfe: naturgeschichte eines vogelparadieses in Anatolien. [The Sultan marshes: natural history of a bird paradise in Anatolia.]* Heidelberg, Germany: Kasparek. (In German.)

KATS, N. Y. (1948) [*Types of bogs of USSR and west Europe, and their geographic distribution.*] Moscow: OGIZ. (In Rusian.)

KELLER, V. E. (1991) Effects of human disturbance on Eider ducklings *Somateria mollissima* in an estuarine habitat in Scotland. *Biol. Conserv.* 58: 213–228.

KEMPE, G., TOET, H., MAGNUSSON, P.-H. AND BERGSTEDT, J. (1992) *Riksskogtaxeringen 1983–87.* [*The national forest survey 1983–87.*] Stockholm: Institutionen för skogstaxering (Rapport 51).

KENCE, A. (1987) *Biological diversity in Turkey.* Ankara: Environmental Problems Foundation of Turkey.

KILIÇ, A. AND KASPAREK, M. (1990) The Eregli marshes. Unpublished report.

KING, C. A. M. (1972) *Beaches and coasts.* London: Arnold.

KIRBY, K. (1995) *Rebuilding the English countryside: habitat fragmentation and wildlife corridors as issues in practical conservation.* Peterborough, UK: English Nature (Science Series no. 10).

KISIOV, H., KOLAROV, P., DIKOV, T., ZLATANOVA, S., BOJADZHIEV, A. AND PETROV, P. (1994) Status and assessment of the fish resources in Bulgaria. Pp.153–174 in M. Sakaljan and C. Meine, eds. *National strategy for conservation of the biological diversity in Bulgaria,* 2. Sofia, Bulgaria:

KLEIN, J. (1920) *The mesta: a study of Spanish economic history, 1273–1836.* Cambridge, USA: Harvard University Press.

KOHH, E. (1975) Studier över skogsbränder och skenhälla i älvdalsskogarna. [Studies of forest fires and hardpan in the Älvdalen forests.] *Svenska skogsvårdsförbundets tidskrift* 71: 299–360.

KOLLAR, H. P. (1996) Action plan for the Great Bustard *Otis tarda.* Pp.245–260 in B. Heredia, L. Rose and M. Painter, eds. *Globally threatened birds in Europe: action plans.* Strasbourg: Council of Europe, and BirdLife International.

KOLLAR, H. P. AND WURM, H. (1996) Zur Bestandssituation der Großtrappe (*Otis t. tarda* L., 1758). *Österreich. Naturschutz und Landschaftspflege in Branderburg* 5: 7–9.

KOPANEVA, L. M. AND STEBAJEV, I. V. (1985) [*Life of locusts.*] Moscow. (In Russian.)

KORNAS, J. (1983) Man's impact on flora and vegetation in Central Europe. *Geobotany* 5: 277–286.

KOVALENKO, O. A., ED. (1994) *Main indices of agro-industrial complex development in 1993.* Moscow: Informagrotekh.

KRAL, F. (1994) Der Wald im Spiegel der Waldgeschichte. Pp.11–40 in *Österreichischer Forstverein [Hrsg.] Österreichs Wald. Vom Urwald zur Waldwirtschaft,* 2. Vienna: Aufl. Österreichischer Forstverein.

KRAUSE, G. H. M. (1989) Forest decline in central Europe: the unravelling of multiple causes. Pp.377–399 in P. J. Grubb and J. B. Whittaker, eds. *Toward a more exact ecology.* Oxford, UK: Blackwell Scientific Publications.

KRUG, H. M. S. AND SILVA, H. M. (1990) Avaliação do stock açoreano de goraz. *Relatório da IX Semana das Pescas dos Açores* 9 (1989): 231–238.

KUUSELA, K. (1994) *Forest resources in Europe.* Cambridge, UK: Cambridge University Press (European Forest Institute Research Report 1).

LACK, D. (1935) The breeding bird populations of British heaths and moorland. *J. Anim. Ecol.* 4: 43–51.

LAIST, D. W. (1987) Overview of the biological effects of lost and discarded plastic debris in the marine environment. *Mar. Pollut. Bull.* 18: 319–326.

LAMBECK, R. H. D. (1990) Changes in abundance, distribution and mortality of wintering Oystercatchers after habitat loss in the Delta area, SW Netherlands. *Acta XX*

Congressus Internationalis Ornithologici 4: 2208–2218.

LAMBECK, R. H. D., SANDEE, A. J. J. AND DE WOLF, L. (1989) Long-term patterns in the wader usage of an intertidal flat in the Oosterchelde (SW Netherlands) and the impact of the closure of an adjacent estuary. *J. Appl. Ecol.* 26: 419–431.

LAMBERTINI, M. (1996) International action plan for Audouin's Gull (*Larus audouinii*). Pp.289–301 in B. Heredia, L. Rose and M. Painter, eds. *Globally threatened birds in Europe: action plans.* Strasbourg, France: Council of Europe and BirdLife International.

LAMBERTINI, M. AND FARALLI, U. (1991) Ecologia del Fuoco in un ambiente a macchia Mediterranea (Monte Argentario, Toscana): struttura ed evoluzione delle comunità orniche. *Suppl. Ric. Biol. Selvaggina* XVII: 1–552.

LAMBERTINI, M. AND LEONZIO, C. (1986) Pollutant levels and their effects on Mediterranean seabirds. Pp.359–378 in MEDMARAVIS and X. Monbailliu, eds. *Mediterranean marine avifauna.* Berlin: Springer-Verlag.

LANDMANN, A. AND WINDING, N. (1993) Niche segregation in high-altitude Himalayan chats (Aves, Turdidae): does morphology match ecology? *Oecologia* 95: 505–519.

LANG, J. T. AND COCKER, M. (1991) A nest of Caucasian Black Grouse *Tetrao mlokosiewiczi* in Turkey. *Sandgrouse* 13: 102–103.

LANGEVELD, M. (1991) *Important Bird Areas in the European Community. A shadow list of special protection areas.* Cambridge, UK: International Council for Bird Preservation.

LANGEVELD, M. J. AND GRIMMETT, R. F. A. (1990) *Important Bird Areas in Europe: wetlands for the shadow list of Ramsar sites.* Cambridge, UK: International Council for Bird Preservation.

LARSEN, T. AND SANDVIK, H. (1994) [Status of Lapwing in Norway 1994.] *Vår Fuglefauna* 17: 196–204. (In Norwegian.)

LARSSON, T. (1976) Composition and density of the bird fauna in Swedish shore meadows. *Ornis Scand.* 7: 1–12.

LAURSEN, K., GRAM, I. AND FRIKKE, J. (1984) Traekkende vandfugle ved det fremskudte dige ved Højer, 1982. [Migratory waterfowl and waders at a new dike in the Danish Wadden Sea, 1982.] *Danske Vildtundersøgelser* 37: 1–36. (In Danish.)

LE HOUEROU, H. N. (1980) *Forêt méditerranéenne.* Tome 2, no. 1 et 2 edition. Aix-en-Provence, France.

LECOMTE, P. AND VOISIN, S. (1991) Dry grassland birds in France: status, distribution and conservation measures. Pp.59–68 in P. D. Goriup, L. A. Batten and J. A. Norton, eds. *The conservation of lowland dry grassland birds in Europe. Proceedings of a seminar, Reading 20–22 March 1991.* Peterborough, UK: Joint Nature Conservation Committee.

LEE, J. (1987) European land use and resources: an analysis of future EEC demands. *Land Use Policy* 4: 179–199.

LEFRANC, N. (1997) Shrikes and the farmed landscape in France. Pp.236–268 in D. Pain and M. Pienkowski, eds. *Farming and birds in Europe: the Common Agricultural Policy and its implications for bird conservation.* London: Academic Press.

LEOPOLD, M. F. (1993) *Spisula*, scoters and cockle fisheries: a new environmental problem. *Sula* 7: 24–28.

LEPART, J. AND DEBUSSCHE, M. (1992) Human impact on

landscape patterning: Mediterranean examples. Pp.76–106 in A. J. Hansen and F. di Castri, eds. *Landscape boundaries: consequences for biotic diversity and ecological flows*. New York: Springer-Verlag.

LICHTENBERGER, E. (1994) Die Alpen in Europa. *Österreichische Akademie der Wissenschaften, Veröff. Komm. Humanökologie* 5: 53–86.

LILJELUND, L.-E. (1990) Effects of climate change on species diversity and zonation in Sweden. In J. I. Holten, ed. *Effects of climate change on terrestrial ecosystems*, Notat 4. Trondheim: NINA.

LILJELUND, L.-E., PETTERSSON, B. AND ZACKRISSON, O. (1992) [Forestry and biodiversity.] *Svensk Bot. Tidskr.* 86: 227–232. (In Swedish.)

LINDSAY, R. (1995) *Bogs: the ecology, classification and conservation of ombrotrophic mires*. Battleby, UK: Scottish Natural Heritage.

LITZBARSKI, B. AND LITZBARSKI, H. (1996) Zur Situation der Großtrappe *Otis tarda* in Deutschland. *Vogelwelt* 117: 213–224.

LITZBARSKI, H. (1993) Das Schutzprojekt 'Großtrappe'in Brandenburg. *Ber. Vogelschutz* 31: 61–66.

LLOYD, C. S., TASKER, M. L. AND PARTRIDGE, K. E. (1991) *The status of seabirds in Britain and Ireland*. London: T. and A. D. Poyser.

LONG, S. P. AND MASON, C. F., EDS. (1983) *Saltmarsh ecology*. Glasgow, UK: Blackie and Son Ltd.

LÓPEZ-BELLIDO (1992) In C. J. Pearson, ed. *Field crop ecosystems*. Amsterdam: Elsevier (Ecosystems of the World 18).

LOVARI, S. (1976) Population trends and seasonal flock size variation of Alpine Choughs, Choughs and Ravens in the Abruzzo National Park, Italy. *Gerfaut* 66: 207–219.

LOWDAY, J. E., MARRS, R. H. AND NEVISON, G. B. (1983) Some of the effects of cutting bracken (*Pteridium aquilinum* (L.) Kuhn) at different times during the summer. *J. Environ. Mgmt* 17: 373–380.

LUDER, R. (1993) Vogelbestände und -lebensräume in der Gemeinde Lenk (Berner Oberland): Veränderungen im Laufe von 12 Jahren. *Orn. Beob.* 90: 1–34.

LUDIKHUIZE, D. (1992) *Environmental management and protection of the Black Sea*. Background document prepared for Technical Experts Meeting, 20–21 May 1992. Constanta, Romania: UNDP/UNEP/World Bank.

VAN LYNDEN, G. W. J. (1995) *European soil resources: current status of soil degradation, causes, impacts and need for action*. Wageningen, Netherlands: Council of Europe Press.

LYSTER, S. (1985) *International wildlife laws*. Cambridge, UK: Grotius.

MACDONALD, A. (1990) *Heather damage: a guide to types of damage and their causes*. Edinburgh, UK: Nature Conservancy Council (Research and Survey in Nature Conservation No. 28).

MACGARVIN, M. (1990) *The North Sea*. London: Collins & Brown.

MACKENZIE, D. (1994) Where has all the carbon gone?*New Scientist* 19 (8 January): 30–33.

MADSEN, J. (1996) International action plan for the Lesser White-fronted Goose *Anser erythropus*. Pp.67–78 in B. Heredia, L. Rose and M. Painter, eds. *Globally threatened birds in Europe: action plans*. Strasbourg: Council of Europe, and BirdLife International.

MADSEN, J. AND FOX, A. D. (1995) Impacts of hunting disturbance on waterbirds: a review. *Wildl. Biol.* 1: 193–207.

MADSEN, J. AND PIHL, S. (1993) Jagt- og forstyrrelsesfrie kerneområder for vandfugle I Danmark. Danmarks Miljøundersøgelser, Faglig Rapport fra DMU No. 132.

MADSEN, J., FOX, A. D., MOSER, M. AND NOER, H. (1995) The impact of hunting disturbance on the dynamics of waterbird populations: a review. A report from the National Environment Research Institute and the International Waterfowl and Wetlands Research Bureau to the Commission of the European Communities.

MAFF (1994) *Water level management plans*. London: Ministry of Agriculture, Food and Fisheries.

MAGNIN, G. (1991) Hunting and persecution of migratory birds in the Mediterranean region. Pp.63–75 in T. Salathé, ed. *Conserving migratory birds*. Cambridge, UK: International Council for Bird Preservation (Techn. Publ. 12). LIBREP

MAJORAL, R. (1987) La utilización del suelo agrícola en España. Aspectos evolutivos y locacionales. *El Campo* 104: 13–26.

MALO, J. E., LEVASSOR, C., JIMÉNEZ, B., SUÁREZ, F. AND PECO, B. (1994) La sucesión en cultivos abandonados en zonas agropastorales: semejanzas y diferencias entre tres localidades peninsulares. Pp.131–136 in SEEP, ed. *Recursos pastables. Hacia una gestión de calidad*. Santander, Spain: Sociedad Española para el Estudio de los Pastos.

MALTBY, E., DUGAN, P. AND LEFEUVRE, J. C., EDS. (1992) *Conservation and development: the sustainable use of wetland resources*. Cambridge, UK: IUCN Wetlands Programme.

MAÑOSA, S., REAL, J. AND SOLÉ, J. (1996) L'impacte de les línies elèctriques sobre l'avifauna a Catalunya: electrocució i collisió. Internal report to Universitat Central de Barcelona/Fundació Miguel Torres, Spain (unpublished).

MARCHAND, M. AND UDO DE HAES, H. A. (1989) *Abstracts: International Conference on Wetlands; The people's role in wetland management, Leiden 5–8 June 1989*. Leiden, Netherlands: Leiden University.

MARCHANT, J. H., HUDSON, R., CARTER, S. P. AND WHITTINGTON, P. (1990) *Population trends in British breeding birds*. Tring, UK: British Trust for Ornithology.

MARÉCHAL, P. (1993) [On the external factors influencing the habitat quality of the Red-backed Shrike.] *Vogeljaar* 41: 34–48. (In German.)

MARTI, C. AND BOSSERT, A. (1985) [Notes on summer activity and breeding biology of Rock Ptarmigan *Lagopus mutus* in the Aletsch region (Swiss Alps).] *Orn. Beob.* 82: 153–168. (In German with English summary.)

MARTÍN, A. R. (1986) Feeding associations between dolphins and shearwaters around the Azores Islands. *Canadian Journal of Zoology* 64: 1372–1374.

MARTÍNEZ, C. (1991) Patterns of distribution and habitat selection of a Great Bustard *Otis tarda* population in north-western Spain. *Ardeola* 38: 137–146.

MARTÍNEZ, C. AND DE JUANA, E. (1996) Breeding bird communities of cereal crops in Spain: habitat requirements. Pp.99–106 in J. Fernández Gutierrez and J. Sanz-Zuasti, eds. *Conservación de las aves esteparias y su hábitat*. Valladolid, Spain: Junta de Castilla y León.

References

MAYER, H. (1994a) Der Bergwald als Garant zur Erhaltung der Gebirgslandschaft und seine Gefährdung. *Österreichische Akadamie der Wissenschaften, Veröff. Komm. Humanökologie* 5: 87–105.

MAYER, H. (1994b) Geschichte des Waldbaus. Pp.243–260 in *Österreichischer Forstverein [Hrsg.] Österreichs Wald. Vom Urwald zur Waldwirtschaft*, 2. Vienna: Aufl. Österreichischer Forstverein.

MAYOL SERRA, J. (1986) Human impact on seabirds in the Balearic Islands. Pp.379–396 in MEDMARAVIS and X. Monbailliu, eds. *Mediterranean marine avifauna*. Berlin and Heidelberg: Springer-Verlag (NATO ASI Ser. G. 12).

MAYR, C. (1993) Vierzehn Jahre EG-Vogelschutzrichtlinie. Bilanz ihrer Umsetzung in der Bundersrepublik Deutschland. [Fourteen years of the EC Wild Birds Directive. Balance of its application in the Federal Republic of Germany.] *Ber. Vogelschutz* 31: 13–22.

MCCANN, B. AND APPLETON, B. (1993) *European water: meeting the supply challenges*. London: Financial Times (Management Report).

MCGINN, A. P. (1996) A private-sector sustainable fishing initiative. *Worldwatch* 9: 8.

MEADOWS, B. S. (1972) Kingfisher numbers and stream pollution. *Ibis* 110: 443.

MEE, L. D. (1992) The Black Sea in crisis: a need for concerted international action. *Ambio* 21: 278–286.

MEIRE, P. M. (1990) Effects of a substantial reduction in intertidal area on numbers and densities of waders. *Acta XX Congressus Internationalis Ornithologici* IV: 2219–2227.

MEIRE, P. M. AND KUIJKEN, E. (1984) Relations between the distribution of waders and the intertidal benthic fauna of the Oosterchelde, Netherlands. Pp.57–68 in P. R. Evans, J. D. Goss-Custard and W. G. Hale, eds. *Coastal waders and wildfowl in winter*. Cambridge, UK: Cambridge University Press.

MELTOFTE, H. (1982) Jaglige forstyrrelser af svømme- og vadefugle. [Shooting disturbance of waterfowl.] *Dansk Orn. Foren. Tidsskr.* 76: 21–35.

MERILEHTO, K., KENTTÄMIES, K. AND KÄMÄRI, J. (1988) *Surface water acidification in the EEC region*. Copenhagen: Nordic Council of Ministers (Miljörapport 1988: 14).

MERMET, L. (1984) France. Pp.24–47 in D. Baldock, ed. *Wetland drainage in Europe*. London: Institute for European Environmental Policy and International Institute for Environment and Development.

MERRIAM, G. (1988) Landscape dynamics in farmland. *Tree* 3: 16–20.

MESÓN, M. AND MONTOYA, J. M. (1993) Factores desencadenantes de la seca de los *Quercus* en España. *Quercus* 92: 30–31.

MESTRE, P. (1984) Variaciones sobre el status de algunas especies de aves durante los veinticinco ultimos años en la comarca del Penedès. *Alytes* 2: 191–223.

MIKOLA, J., MIETTINEN, M., LEHIKOINEN, E. AND LEHTILA, K. (1994) The effects of disturbance caused by boating on survival and behaviour of Velvet Scoter *Melanitta fusca* ducklings. *Biol. Conserv.* 67: 119–124.

MINISTERIAL CONFERENCE ON THE PROTECTION OF FORESTS IN EUROPE (1993) *Report on the follow-up of the Strasbourg Resolutions. Helsinki, 16–17 June 1993*. Helsinki: Ministry of Agriculture and Forestry.

MINISTERIO DE AGRICULTURA (1979) *Las coníferas en el Primer Inventario Forestal Nacional*. Madrid: Ministerio de Agricultura.

MINISTERIO DE AGRICULTURA (1980) *Las frondosas en el Primer Inventario Forestal Nacional*. Madrid: Ministerio de Agricultura.

MINISTERO DELL'AMBIENTE (1992) *Relazione sullostato dell'ambiente*. Rome: Ministero dell'Ambiente (Report of the State of the Environment). (In Italian.)

MIQUET, A. (1990) Mortality in Black Grouse *Tetrao tetrix* due to elevated cables. *Biol. Conserv.* 54: 349–355.

MITCHELL, B., STAINES, B. W. AND WELCH, D. (1977) *Ecology of red deer*. Cambridge, UK: Institute of Terrestrial Ecology.

MITCHELL, F. J. G. AND KIRBY, K. J. (1990) The impact of large herbivores on the conservation of semi-natural woods in the British uplands. *Forestry* 63: 333–353.

MÖCKEL, R. (1992) Effect of 'Waldsterben' (forest damage due to airborne pollution) on the population dynamics of Coal Tit *Parus ater* and Crested Tit *Parus cristatus* in the western Erzgebirge. *Ökol. Vögel* 14: 1–100.

MOLDAN, B. AND SCHROOR, J. L. (1992) Czechoslovakia: examining a critically ill environment. *Environmental Science and Technology* 26: 14–21.

MØLLER, H. S. (1975) Danish salt-marsh communitites of breeding birds in relation to different types of management. *Ornis Scand.* 6: 125–133.

MOLODAN, G. N. (1980) [The breeding of the Pratincole in north-east Priazovye.] *Vestnik Zool.* 4: 96–97. (In Russian.)

MONAGHAN, P. (1992) Seabirds and sandeels: the conflict between exploitation and conservation in the northern North Sea. *Biodiv. & Conserv.* 1: 98–111.

MONAGHAN, P., UTTLEY, J. D. AND BURNS, M. D. (1992) Effect of changes in food availability on reproductive effort in Arctic Terns *Sterna paradisaea*. *Ardea* 80 (Special Publ.): 71–81.

MONDAIN-MONVAL, J. Y. AND SALATHÉ, T. (1993) Actions of the European Community for wetland conservation: an assessment of the wetland conservation projects committed by the European Community under the ACE regulations (EEC) No. 1872/84 and No. 2242/87. Doc. XI/060/94. Brussels: Commission of the European Communities (DG XI.B.2).

MONTEIRO, L. R. (1996) Seabirds as monitors of mercury contamination in the Portuguese Atlantic. University of Glasgow (Ph.D. thesis).

MONTEIRO, L. R., RAMOS, J. A. AND FURNESS, R. W. (1996) Past and present status and conservation of the seabirds breeding in the Azores archipelago. *Biol. Conserv.* 78: 319–328.

MONTERO, G. (1988) *Modelos para cuantificar la producción de corcho en alcornocales* Quercus suber L. *en función de la calidad de estación y de los tratamientos selvícolas*. Madrid: INIA.

MONTES, C. (1990) Estudio de las zonas húmedas de la España peninsular. Inventario y tipificación. INITEC. Madrid: Dirección General de Obras Hidráulicas, Ministerio de Obras Públicas y Urbanismo (unpublished).

MONTEVECCHI, W. A. (1993) Birds as indicators of changes in marine prey stocks. Pp.217–266 in R. W. Furness and J. D. D. Greenwood, eds. *Birds as monitors of environ-*

mental change. London: Chapman and Hall.

MONTSERRAT, P. AND FILLAT, F. (1987) *Ecosystems of the world: the systems of grassland management in Spain*. Amsterdam: Elsevier.

MOORE, J. J. (1968) A classification of the bogs and wet heaths of northern Europe. Pp.306–320 in R. Tüxen, ed. *Pflanzensoziologische Systematik*. The Hague, Netherlands: Junk.

MOORE, N. W. (1962) The heaths of Dorset and their conservation. *J. Ecol.* 50: 369–391.

MOORE, N. W. (1983) Ecological effects of pesticides. Pp.159–179 in A. Warren and F. B. Goldsmith, eds. *Conservation in perspective*. Chichester, UK: John Wiley & Sons.

MOORE, N. W. AND HOOPER, M. D. (1975) On the number of bird species in British woods. *Biol. Conserv.* 8: 239–250.

MOORE, P. D. (1984) The classification of mires: an introduction. Pp.1–10 in P. D. Moore, ed. *European mires*. London: Academic Press.

MOORS, P. J. AND ATKINSON, I. A. E. (1984) Predation on seabirds by introduced animals, and factors affecting its severity. Pp.667–690 in J. P. Croxall, P. G. H. Evans and R. W. Schreiber, eds. *Status and conservation of the world's seabirds*. Cambridge, UK: International Council for Bird Preservation (Techn. Publ. 2).

MORENO, J. M. AND OECHEL, W. C., EDS. (1994) *The role of fire in Mediterranean-type ecosystems*. New York: Springer-Verlag.

MORGAN, R. A. (1978) Changes in the breeding bird communities at Gibraltar Point, Lincolnshire, between 1965 and 1974. *Bird Study* 25: 51–58.

MORMAT, J. Y. AND GUERMEUR, Y. (1979) *L'Amoco Cadiz et les oiseaux*. Paris: Ministère de l'Environment et du Cadre de Vie.

MOSKÁT, C. AND WALICZKY, Z. (1992) Bird–vegetation relationships along ecological gradients: ordination and plexus analysis. *Ornis Hungarica* 2: 45–60.

MÜLLER, M. J. (1982) *Selected climate data for a global set of standard stations for vegetation science*. The Hague, Netherlands: Junk.

MUNDY, J. (1992) *Land redistribution and nature conservation in central and eastern Europe*. Cambridge, UK: International Union for Conservation of Nature and Natural Resources.

MUNIZ, J. P. (1991) Freshwater acidification: its effects on species and communities of freshwater microbes, plants and animals. *Proc. R. Soc. Edinburgh* 97B: 227–254.

MUNTANER, J., FERRER, X. AND MARTÍNEZ-VILLALTA, A. (1983) [*Atlas of the breeding birds of Cataluña and Andorra*.] Barcelona: Ketres. (In Catalan.)

MUNTEANU, D. (1990) Les peuplements d'oiseaux des écrans forestiers de Ceanu Mare (Plaine de Transylvanie). [Bird populations in the forest shelterbelts of Ceanu Mare (Transylvanian Plain).] *Studia Univ. Babes-Bolyai, Biologia* 35: 18–21.

MUÑOZ-COBO, J. (1992) Breeding bird communities in the olive tree plantations of southern Spain: the role of the age of trees. *Alauda* 60: 118–122.

MUÑOZ-COBO, J. AND PURROY, F. J. (1980) Wintering bird communities in the olive tree plantations of Spain. Pp.185–189 in *Proc. VI Int. Conf. Bird Census Work*. Gottingen, Germany: University Gottingen.

MUSTERS, C. J. M., NOORDERVLIET, M. A. W. AND TERKEURS, W. J. (1996) Bird casualties caused by a wind energy project in an estuary. *Bird Study* 43: 124–126.

MUUS, B. J. (1967) The fauna of Danish estuaries and lagoons. *Meddelelser fra Danmarks Fiskeriog Havundersøgelser (Ny Serie)* 5: 1–316.

MYRBERGET, S. (1963) [The Corncrake in Norway.] *Sterna* 5: 289–305. (In Norwegian.)

NACHURISVILI, G. (1983) Untersuchungen der ökologischen Auswirkungen intensiver Schafbeweidung im Zentral-Kaukasus. *Verh. Ges. Ökol.* 10: 183–191.

NAIRN, R. G. W. AND SHEPPARD, J. R. (1985) Breeding waders of sand dune machair in north-west Ireland. *Irish Birds* 3: 53–70.

NAIRN, R. G. W. AND WHATMOUGH, J. A. (1978) Breeding bird communities of a sand-dune system in north-east Ireland. *Irish Birds* 1: 160–170.

NALBORCZYK AND ZEBOR (1992) In C. J. Pearson, ed. *Field crop ecosystems*. Amsterdam: Elsevier (Ecosystems of the World 18).

NATIONAL COMMISSION FOR STATISTICS (1996) *Romanian Statistical Yearbook 1995*. Bucharest.

NATIONAL RIVERS AUTHORITY (1991) *NRA facts, 1990*. Bristol, UK: National Rivers Authority.

NCC (1984) *Nature conservation in Great Britain*. Nature Conservancy Council: Peterborough, UK.

NEGRO, J. J. AND FERRER, M. (1995) Mitigating measures to reduce electrocution of birds on power lines: a comment on Bevanger's review. *Ibis* 137: 136–137.

NEHLS, G. (1996) Der Kiebitz in der Agrarlandschaft—Perspektiven für den Erhalt des Vogels des Jahres 1996. *Ber. Vogelschutz* 34: 123–132.

NERC (1983) *Contaminants in top predators*. Swindon, UK: Natural Environment Research Council.

NETTLESHIP, D. N. AND BIRKHEAD, T. R., EDS. (1985) *The Atlantic Alcidae: the evolution, distribution and biology of the auks inhabiting the Atlantic Ocean and adjacent water areas*. Orlando, USA: Academic Press.

NETTLESHIP, D. N. AND EVANS, P. G. H. (1985) Distribution and status of the Atlantic Alcidae. Pp.53–154 in D. N. Nettleship and T. R. Birkhead, eds. *The Atlantic Alcidae: the evolution, distribution and biology of the auks inhabiting the Atlantic Ocean and adjacent water areas*. Orlando, USA: Academic Press.

NEWTON, I. (1979) *Population ecology of raptors*. Berkhamsted, UK: T. and A. D. Poyser.

NILSSON, S. G., OLSSON, O., SVENSSON, S. AND WIKTANDER, U. (1992) Population trends and fluctuations in Swedish woodpeckers. *Ornis Svecica* 2: 13–21.

NISBET, I. C. T. (1994) Effects of pollution on marine birds. Pp.8–25 in D. N. Nettleship, J. Burger and M. Gochfeld, eds. *Seabirds on islands: threats, case studies, and action plans*. Cambridge, UK: BirdLife International (BirdLife Conservation Series no. 1).

NISBET, I. C. T. (1995) Seabirds off the east coast of the USA: status, trends and threats. Pp.34 in M. L. Tasker, ed. *Threats to seabirds: Proceedings of the 5th International Seabird Group conference*. Sandy, UK: Seabird Group.

NOIRFALISE, A. AND VANESSE, R. (1976) *Heathlands of western Europe*. Strasbourg, France: Council of Europe.

VAN NOORDEN, B. (1986) *Dynamiek en dichtheid van bosvogels in geïsoleerde loofbosfragmenten*. Leersum,

Netherlands: Research Institute for Nature Management (Report 86/19).

NORDIC COUNCIL OF MINISTERS (1993) The Nordic environment: present state, trends and threats. *Nord* 12: 1–211.

NORTH SEA TASK FORCE (1993) *North Sea Quality Status Report 1993*. London: Oslo and Paris Commissions.

O'BRIEN, M. AND SELF, M. (1994) Changes in the numbers of breeding waders on lowland wet grasslands in the UK. *RSPB Conserv. Rev.* 8: 38–44.

O'CONNOR, R. J. AND SHRUBB, M. (1986) *Farming and birds*. Cambridge, UK: Cambridge University Press.

ØIEN, I. J. AND FOLVIK, A. (1995) [The Corncrake—the voice that became silent.] *Vår Fuglefauna* 18: 105–111. (In Norwegian with English summary.)

OLDÉN, B., PETERZ, M. AND KOLLBERG, B. (1985) Seabird mortality in the gill net fishery, southeast Kattegat, south Sweden. *Anser* 24: 159–180.

OPDAM, P. AND HELMRICH, V. R. (1984) Vogelgemeenschappen van heide en hoogveen: een typologische beschrijving. [Bird communities of heath and peat moors: a typological description.] *Limosa* 57: 47–63.

OPDAM, P., FOPPEN, R., REIJEN, R. AND SCHOTMAN, A. (1995) The landscape ecological approach in bird conservation: integrating the metapopulation concept into spatial planning. *Ibis* 137 (Suppl. 1): 139–146.

OPDAM, P., RIJSDIJK, G. AND HUSTINGS, F. (1985) Bird communities in small woods in an agricultural landscape: effects of area and isolation. *Biol. Conserv.* 34: 333–352.

OPDAM, P., VAN APELDOORN, R., SCHOTMAN, A. AND KALKHOVEN, J. (1993) Population responses to landscape fragmentation. Pp.147–171 in C. C. Vos and P. Opdam, eds. *Landscape ecology of a stressed environment*. London: Chapman & Hall.

ORIA, J. AND CABALLERO, J. (1992) [Monitoring and conservation of Spanish Imperial Eagle in central Spain.] Madrid: Instituto Nacional para la Conservación de la Naturaleza (unpublished manuscript).

ORME, S. E. (1990) *Nature conservation in the European Community: the present and the future*. Cirencester, UK: Royal Agricultural College (Centre for Rural Studies Occ. Pap 10).

ORMEROD, S. J. AND TYLER, S. J. (1987) Dippers (*Cinclus cinclus*) and grey wagtails (*Motacilla cinerea*) as indicators of stream acidity in upland Wales. Pp.191–208 in A. W. Diamond and F. Filion, eds. *The value of birds*. Cambridge: International Council for Bird Preservation.

ORMEROD, S. J. AND TYLER, S. J. (1993) Birds as indicators of changes in water quality. Pp.179–216 in R. W. Furness and J. D. D. Greenwood, eds. *Birds as monitors of environmental change*. London: Chapman and Hall.

ORÓ, D. AND MARTÍNEZ, A. (1994) Migration and dispersal of Audouin's Gull *Larus audouinii* from the Ebro Delta colony. *Ostrich* 65: 225–230.

ORTUÑO, F. (1990) El plan para la repoblación forestal de España del año 1939. Análisis y comentarios. *Ecologia frera de serie* 1: 373–392.

OSPARCOM (1992a) *Nutrients in the Convention Area*. London: Oslo and Paris Commissions.

OSPARCOM (1992b) *Dumping and incinerating at sea*. London: Oslo and Paris Commissions.

ÖSTERREICHISCHE RAUMORDNUNGSKONFERENZ (1992) *Österreichisches Raumordnungskonzept 1991, Schriftenreihe 96*. Wien: Österreichische Raumordnungskonferenz.

OSVALD, H. (1925) Zue Vegetation der ozeanischen Hochmoore in Norvegen. [On the vegetation of oceanic raised bogs in Norway.] *Svenska Växtsociologiska Sällskapets Handlingar* VII: 1–106.

OVERREIN, L. N., SEIP, H. M. AND TOLLAN, A. (1980) *Acid precipitation—effects on forests and fish*. Final report of SNSF project, 1972–1980, FR/80 edition. Oslo: Norwegian Council for Scientific and Industrial Research.

OWEN, M., ATKINSON-WILLIES, G. AND SALMON, D. (1986) *Wildfowl in Great Britain*. Cambridge, UK: Cambridge University Press.

OZENDA, P. (1964) *Biogéographie végétale*. Paris: Doin.

OZENDA, P. (1979) *Vegetation map of the Council of Europe Member States*. Council of Europe: Strasbourg, France (Nature and Environment Series 16).

PÅHLSSON, L. (1994) Vegetationstyper i Norden. Pp.665 in *Tema Nord 1994*. Copenhagen: Nordic Council of Ministers.

PAIN, D. J. (1990) Lead shot ingestion by waterbirds in the Camargue, France: an investigation of levels and interspecific differences. *Environ. Pollut.* 66: 273–285.

PAIN, D. J. (1991) Lead shot densities and settlement rates in Camargue marshes, France. *Biol. Conserv.* 57: 273–286.

PAIN, D. J., ED. (1992) *Lead poisoning in waterfowl. Proc. IWRB Workshop, Brussels, Belgium, 1991*. Slimbridge, UK: International Wetlands Research Bureau (IWRB Spec. Publ. 16).

PAIN, D. J. (1994a) Olive farming in Portugal. Pp.38 in D. J. Pain, ed. *Case studies of farming and birds in Europe*. Sandy, UK: RSPB unpublished report.

PAIN, D. J. (1994b) Rice farming in Italy. Pp.40 in D. J. Pain, ed. *Case studies of farming and birds in Europe*. Sandy, UK: RSPB unpublished report.

PAIN, D. AND DUNN, E. (1995) The effects of agricultural intensification upon pastoral birds: lowland wet grasslands (the Netherlands) and transhumance (Spain). Pp.90–98 in D. I. McCracken and E. M. Bignal, eds. *Farming on the edge: the nature of traditional farmland in Europe*. Peterborough, UK: Joint Nature Conservation Committee.

PAIN, D. J. AND BAVOUX, G. (in press) Seasonal blood lead concentrations in Marsh Harriers *Circus aeruginosus* from Charante-Maritime, France: relationship with the hunting season. *Biol. Conserv.*

PAIN, D. J. AND HANDRINOS, G. I. (1990) The incidence of ingested lead shot in ducks of the Evros Delta, Greece. *Wildfowl* 41: 167–170.

PAIN, D. J. AND PIENKOWSKI, M. W., EDS. (1997) *Farming and birds in Europe: the Common Agricultural Policy and its implications for bird conservation*. London: Academic Press.

PAIN, D. J., AMIARD-TRIQUET, C., BAVOUX, C., BURNELEAU, G., EON, L. AND NICOLAU-GUILLAUMET, P. (1994) Lead poisoning in wild populations of Marsh Harriers *Circus aeruginosus* in the Camargue and Charente-Maritime, France. *Ibis* 135: 379–386.

PARR, S. J., NAVESO, M. A. AND YARAR, M. (1997) Habitat and potential prey surrounding Lesser Kestrel *Falco naumanni* colonies in Central Turkey. *Biol. Conserv.* 79: 309–312.

PARUNKSTIS, A. (1975) [Sedimentation in Lithuanian lakes.]

PASTOR, J., NAIMAN, R. J., DEWEY, B. AND MCINNES, P. (1988) Moose, microbes and the boreal forest. *BioScience* 38: 770–777.

PATERSON, A. M., MARTÍNEZ-VILALTA, A. AND DIES, J. I. (1992) Partial breeding failure of Audouin's Gull in two Spanish colonies in 1991. *Brit. Birds* 85: 97–100.

PAYETTE, S., MORNEAU, C., SIROIS, L. AND DESPONTS, M. (1989) Recent fire history of the northern Quebec biomes. *Ecology* 70: 656–673.

PEAKALL, D. B. AND BOYD, H. (1987) Birds as bio-indicators of environmental conditions. Pp.113–118 in A. W. Diamond and F. L. Filion, eds. *The value of birds.* Cambridge, UK: International Council for Bird Preservation (Techn. Publ. 6).

PEARCE, F. (1995) How the Soviet seas were lost. *New Scientist* 11 November: 39–42.

PEARSON, C. J., ED. (1992) *Field crop ecosystems.* Amsterdam: Elsevier (Ecosystems of the World 18).

PEDRINI, P. (1991) Ecologia riproduttiva e problemi di conservazione dell'Aquila Reale (*Aquila chrysaetos*) in Trentino (Alpi Centro Orientali). Pp.365–369 in A. Montemaggiori, ed. *Atti V Convegno Italiano di Ornitologia.* Bologna, Italy: Instituto Nazionale di Biologia della Selvaggina "Alessandro Ghigi".

PENTERIANI, V. (1994) Electrocution as a limiting factor for Eagle Owl in Abruzzo, central Italy. In *Proceedings VI Italian Ornithological Conference.* Torino, Italy.

PÉREZ, A. (1988) *Cambios y problematica de la dehesa (el suroeste de Badajóz).* Badajóz, Spain: Universidad de Extremadura.

PERTHUISOT, J.-P. AND GUÉLORGET, O. (1992) Morphologie, organisation hydrologique, hydrochimie et sédimentologie des bassins paraliques. *Vie et Milieu* 42: 93–109.

PETERKEN, G. F. (1993) *Woodland conservation and management.* Second edition. London: Chapman and Hall.

PETERSEN, B. S. (1994) Interactions between birds and agriculture in Denmark: from simple counts to detailed studies of breeding success and foraging behaviour. Pp.49–56 in E. J. M. Hagemeijer and T. J. Verstrael, eds. *Bird numbers 1992, distribution, monitoring and ecological aspects: proceedings of the 12th International Conference of IBCC and EOAC, Noordwijkerhout, Netherlands, September 14–18 1992.* Beek-Ubbergen: Statistics Netherlands, Voorburg/Heerlen and SOVON.

PETRETTI, F. (1988) An inventory of steppe habitats in southern Italy. Pp.125–143 in P. D. Goriup, ed. *Ecology and conservation of grassland birds.* Cambridge, UK: International Council for Bird Preservation (Techn. Publ. 7).

PETRETTI, F. (1995) Examples of extensive farming systems in Italy. Pp.38–42 in D. I. McCracken and E. M. Bignal, eds. *Farming on the edge: the nature of traditional farmland in Europe.* Peterborough, UK: Joint Nature Conservation Committee.

PETTERSSON, B. (1991) Bevarande av faunans och florans mångfald vid skogsbruk. [Conservation of animal and plant diversity in forestry.] Rapport från skogsstyrelsens arbetsgrupp för översyn av föreskrifter och allmänna råd till 21 # skogsvårdslagen. Stockholm: Skogsstyrelsen (unpublished report).

PETTERSSON, R. B., BALL, J. P., RENHORN, K.-E., ESSEEN, P.-A. AND SJÖBERG, K. (1995) Invertebrate communities in boreal forest canopies as influenced by forestry and lichens with implications for passerine birds. *Biol. Conserv.* 74: 57–63.

PETTY, S. J. AND AVERY, M. I. (1990) *Forest bird communities.* Edinburgh: Forestry Commission (Occasional paper 26).

PFISTER, H. P., NAEF-DAENZER, B. AND BLUM, H. (1986) Qualitative und quantitative Beziehungen zwischen Heckenvorkommen im Kanton Thurgau und ausgewählten Heckenbrütern: Neuntöter, Goldammer, Dorngrasmücke, Mönchsgrasmücke und Gartengrasmücke. *Orn. Beob.* 83: 7–34.

PHILIP, P. E. G. AND MACLEAN, N. I. G. (1985) The invertebrate fauna of Dungeness. In B. W. Ferry and S. Waters, eds. *Ecology and conservation.* Peterborough, UK: Nature Conservancy Council, Royal Holloway and Bedford New Colleges.

PHILLIPS, G. (1992) A case-study in restoration: shallow eutrophic lakes in the Norfolk Broads. Pp.251–278 in D. Harper, ed. *Eutrophication of freshwaters: principles, problems and restoration.* London: Chapman & Hall.

PIATT, J. F. AND NETTLESHIP, D. N. (1995) Diving depths of four alcids. *Auk* 102: 293–297.

PIATT, J. F., CARTER, H. R. AND NETTLESHIP, D. N. (1991) Effects of oil pollution on marine bird populations. Pp.125–141 in J. White, ed. *The effects of oil on wildlife: research, rehabilitation, and general concerns. Proceedings of International Wildlife Rehabilitation Council Symposium, Herndon, Virginia, 16–18 October 1990.* Hanover, USA: Sheridan Press.

PIATT, J. F., LENSINK, C. J., BUTLER, W., KENDZIOREK, M. AND NYSEWANDER, D. R. (1990) Immediate impact of the *Exxon Valdez* oil spill on marine birds. *Auk* 107: 387–397.

PIERSMA, T., DEKKINGA, A. AND KOOLHAAS, A. (1993) Modder, nonnetjes en kanoeten bij Griend. [Mud, "nuns" and Knots at Griend.] *Waddenbulletin* 28: 144–149.

PINA, J. P., RUFINO, R., ARAÚJO, A. AND NEVES, R. (1990) Breeding and wintering passerine densities in Portugal. Pp.273–276 in K. Stasny and V. Bejcek, eds. *Bird census and atlas studies. Proceedings of the XIth International Conference on bird census and atlas work, Prague.* Prague: Institute of Apllied Ecology and Ecotechnology, Agriculture University.

PLATTEEUW, M. (1986) Effecten van geluidhinde door militaire activiteiten op-gedrag en ecologie van wadvogels. RIN-rapport 86/13: 50pp.

POKORNÝ, J. (1994) Development of aquatic macrophytes in shallow lakes and ponds. Pp.36–43 in M. Eiseltová, ed. *Restoration of lake ecosystems: a holistic approach.* Slimbridge, UK: International Waterfowl and Wetlands Research Bureau (Publ. 32).

POLUNIN, O. AND WALTERS, M. (1985) *A guide to the vegetation of Britain and Europe.* New York: Oxford University Press.

PONS, A. AND QUÉZEL, P. (1985) The history of the flora and vegetation and past and present human disturbance in the Mediterranean region. Pp.25–43 in C. Gomez-Campo, ed. *Plant conservation in the Mediterranean area.* The Hague, Netherlands: Junk Publishers.

PONS, P. (1991) Biogeografia i ecologia de l'avifauna nidificant en les suredes de la conca mediterranea occidental. Universitat Central de Barcelona, Spain (upublished thesis).

POTAPOV, R. L. (1985) [*Fauna of the U.S.S.R.: Family Tetraonidae.*] Leningrad: Nauka. (In Russian.)

POTTER, C. (1997) Europe's changing farmed landscape.

Pp.25–42 in D. J. Pain and M. W. Pienkowski, eds. *Farming and birds in Europe: The Common Agricultural Policy and its implications for bird conservation*. London: Academic Press.

POTTS, D. (1997) Cereal farming, pesticides and grey partridges. Pp.150–177 in D. J. Pain and M. W. Pienkowski, eds. *Farming and birds in Europe: The Common Agricultural Policy and its implications for bird conservation*. London: Academic Press.

POTTS, G. R. (1970) Recent changes in the farmland fauna with special reference to the decline of the Grey Partridge (*Perdix perdix*). *Bird Study* 17: 145–166.

POTTS, G. R. (1986) *The Partridge: pesticides, predation and conservation*. London: Collins.

POTTS, G. R. (1990) Causes of the decline in partridge populations and effect of the insecticide dimethoate on chick mortality. Pp.62–71 in J. T. Lumeij and Y. R. Hoogeveen, eds. *The future of wild Galliformes in the Netherlands*. The Hague, Netherlands: Gegevens Koninklijke Bibliotheek.

POULSEN, J. G. (1993) Comparative ecology of Skylarks (*Alauda arvensis* L.) on arable farmland (unpublished thesis).

POWER, S. A., ASHMORE, M. R., COUSINS, D. A. AND AINSWORTH, N. (1995) Long-term effects of enhanced nitrogen deposition on a lowland dry heath in soutern Britain. *Water, Air and Soil Polln.* 85: 1701–1706.

PRATER, A. J. (1981) *Estuary birds of Britain and Ireland*. Calton, UK: T. and A. D. Poyser.

PRINCE, P. A. AND CROXALL, J. P. (1995) Threats to albatrosses at South Georgia. Pp.41 in M. L. Tasker, ed. *Threats to seabirds: Proceedings of the 5th International Seabird Group conference*. Sandy, UK: Seabird Group.

PRINS, F. (1993) Grazing on heathland: the Netherlands experience. Pp.13–18 in Anon., ed. *Heathland Conference 1992 Proceedings*. Kingston-upon-Thames, UK: Surrey County Council.

PRINZ, B. (1987) Causes of forest damage in Europe. *Environment* 29: 11–37.

PRODON, R. (1987) Incendies et protection des oiseaux en France mediterranéene. *Oiseau Rev. Fr. Ornithol.* 57: 1–12.

PRODON, R. (1992) Animal communities and vegetation dynamics: measuring and modelling animal community dynamics along forest successions. Pp.126–141 in A. Teller, P. Mathy and J. N. R. Jeffers, eds. *Responses of forest ecosystems to environmental changes*. London: Elsevier Applied Science.

PRODON, R. (1993) Une alternative aux types biogéographiques de Voous: la mesure des distributions latitudinales. *Alauda* 62: 83–90.

PRODON, R. AND LEBRETON, J. D. (1981) Breeding avifauna of a Mediterranean succession: the holm oak and cork oak series in the eastern Pyrenees, 1. Analysis and modelling of the structure gradient. *Oikos* 37: 21–38.

PRODON, R. AND LEBRETON, J. D. (1983) Prediction of bird census from vegetation structure. Pp.190–194 in F. J. Purroy, ed. *Bird census and Mediterranean landscapes (Proc. 7th Int. Conf. on bird census works, 8–12 Sept. 1981)*. Leon, Spain: IBCC Univ.

PRODON, R., FONS, R. AND ATHIAS-BINCHE, F. (1987) The impact of fire on animal communities in Mediterranean area. Pp.121–157 in L. Trabaud, ed. *The role of fire in ecological systems*. The Hague, Netherlands: SPB Academic Publishing.

PRODON, R., FONS, R. AND PETER, A. M. (1984) L'impact du feu sur la végétation, les oiseaux et les micro-mammifières dans diverses formations méditerranéennes des Pyrénées Orientales: premiers résultats. *Rev. Ecol. (Terre Vie)* 39: 129–158.

PTUSHENKO, E. S. AND INOZEMTSEV, A. A. (1968) *Biologiya i khozyaistvennoye znacheniye ptits Moskovskoy oblasti i sopredelnykh territoriy.* [*Biology and practical value of the birds in the Moscow oblast and adjacent territories.*] Moscow: Izdat. Moskovskogo Universiteta.

PULIDO, F. J. AND DÍAZ, M. (1992) Relaciones entre la estructura de la vegetación y las comunidades de aves en las dehesas: influencia del manejo humano. *Ardeola* 39: 63–72.

PUTMAN, R. J., EDWARDS, P. J., MANN, J. C. E., HOW, R. C. AND HILL, S. D. (1989) Vegetational and faunal changes in an area of heavily grazed woodland following relief of grazing. *Biol. Conserv.* 47: 13–32.

QUÉZEL, P. (1976) Les forêts du pourtour méditerranéen. Pp.9–34 in P. Quézel, R. Morandini and R. Tomazelli, eds. *Forêts et maquis méditerranéens: écologie, conservation et aménagement*. Paris: UNESCO.

QUÉZEL, P. (1978) Analysis of the flora of Mediterranean and Saharan Africa. *Ann. Missouri Bot. Gard.* 65: 479–534.

QUÉZEL, P. (1985) Definition of the Mediterranean region and origin of its flora. Pp.9–24 in C. Gomez-Campo, ed. *Plant conservation in the Mediterranean area*. Dordrecht, Netherlands: Junk.

RABAÇA, J. E. (1983) [Contribution to the study of the avifauna of Cork Oak Forest *Quercus suber*.] Lisbon: Universidade de Lisboa (B.Sc. thesis).

RABAÇA, J. E. (1994) Bird communities of olive tree (*Olea europaea*) plantations in Portugal: a preliminary approach. Pp.97–100 in E. J. M. Hagemeijer and T. J. Verstrael, eds. *Bird numbers 1992. Distribution, monitoring and ecological aspects. Poster appendix of the proceedings of the 12th International Conference of IBCC and EOAC, Noordwijkerhout, Netherlands*. Beek-Ubbergen, Netherlands: Statistics Netherlands, Voorburg/Heerlen and SOVON.

RACKHAM, O. (1986) *The history of the countryside*. London: J. M. Dent & Sons.

RACKHAM, O. (1990) *Trees and woodland in the British landscape*. Second edition. London: Dent.

RAIVIO, S. AND HAILA, Y. (1990) Bird assemblages in silvicultural habitat mosaics in southern Finland during the breeding season. *Ornis Fenn.* 67: 73–83.

RAKONCZAY, Z. (1990) *Természetvédelem* [*Nature conservation*.] Sopron, Hungary: University Press.

RAMSAR CONVENTION BUREAU (1990) *Proceedings of the Fourth Meeting of the Conference of the Contracting Parties, Montreux, Switzerland, 1990*. Gland, Switzerland: Ramsar Convention Bureau.

RANDALL, R. E. (1989) Shingle habitats in the British Isles. *Botanical J. Linn. Soc.* 101: 3–18.

RANNER, A. (1990) Ein Brutvorkommen des Ziegenmelkers (*Caprimulgus europaeus*) am Ruster Hügelzug (Burgenland). *Vogelkundliche Nachrichten aus Österreich* 1/2: 2–3.

RATCLIFFE, D. (1993) *The Peregrine Falcon*. 2nd edition.

London: T. and A. D. Poyser.

RATCLIFFE, P. R. AND PETTY, S. J. (1986) The management of commercial forests for wildlife. Pp.177–187 in D. Jenkins, ed. *Trees and wildlife in the Scottish uplands*. Huntingdon, UK: Institute of Terrestrial Ecolocy (ITE Symposium 17).

RAYMONT, J. E. G. (1980) *Plankton and productivity in the oceans*, 1: *Phytoplankton*. 2nd edition. Oxford, UK: Pergamon Press.

REAL, J. AND MAÑOSA, S. (1992) *La conservació de l'Aguila perdiguera a Catalunya. [The conservation of Bonelli's Eagle in Catalonia.]* Barcelona: Memoria Universitat de Barcelona/Miguel TORRES SA. (In Spanish.)

REIJEN, M. J. S. M. AND THISSEN, J. B. M. (1987) Effects from road traffic on breeding-bird populations in woodland. Pp.121–132 in *Annual Report 1986*. Leersum, Netherlands: Research Institute for Nature Management.

REMMERT, H. (1973) Über die Bedeutung warmblütiger Pflanzenfresser für den Energiefluß in terrestrischen Ökosystem. *J. Orn.* 114: 227–249.

REMMERT, H. (1980) *Ökologie. Ein Lesebuch*. Berlin: Springer.

REVIER, H. (1992) The Wadden Sea: the threat from fisheries. *North Sea Monitor* 12: 11–12.

REY, P. J. (1993) The role of olive orchards in the wintering of frugivorous birds in Spain. *Ardea* 81: 151–160.

REY, P. J. (1995) Spatio-temporal variation in fruit and frugivorous bird abundance in olive orchards. *Ecology* 76: 1625–1635.

REYNOLDS, J. AND TAPPER, S. C. (1995) Ecology of Red Fox (*Vulpes vulpes*) in relation to small game in rural southern England. *Wildl. Biol.* 1: 105–119.

RICH, T. (1996) Wildlife corridors. *In Practice* 11: 1, 4–5.

RIIS-NIELSEN, T. (1995) The agricultural use of Danish heathland in the past. In *Proceedings 5th European Heathland Workshop, Santiago de Compostela, Spain, September 1995*. Spain: University of Santiago de Compostela.

RIKLI, M. (1943) *Das Pflanzenklied der Mittelmeerländer*. Bern, Switzerland: Verlag Huber.

RIPPEY, B. H. R. T. (1973) The conservation of British heaths: a factual study. London: University College of London (unpublished M.Sc. thesis).

RITCHIE, W. AND O'SULLIVAN, M., EDS. (1994) *The environmental impact of the wreck of the Braer*. Edinburgh, UK: The Scottish Office.

ROBERTS, T. M., SKEFFINGTON, R. A. AND BLANK, L. W. (1989) Causes of Type 1 spruce decline in Europe. *Forestry* 62: 179–222.

ROBINS, M. (1991) Synthetic gill nets and seabirds. Sandy, UK: WWF-UK and RSPB (unpublished report).

ROBINS, M. AND BIBBY, C. J. (1985) Dartford Warblers in 1984 Britain. *Brit. Birds* 78: 269–280.

ROBREDO, F. AND SÁNCHEZ, A. (1983) Lucha química contra la lagarta verde de la encina, *Tortrix viridana* L. (Lep.: Tortricidae). Evolución de las técnicas de applicación desde los primeros ensayos y trabajos realizados hasta el momento actual. *Boletín del Servicio de Sanidad Vegetal. Plagas* 9: 253–272.

ROBSON, N. (1997) The evolution of the Common Agricultural Policy and the incorporation of environmental considerations. Pp.43–78 in D. J. Pain and M. W. Pienkowski, eds. *Farming and birds in Europe: The Common Agri-*

cultural Policy and its implications for bird conservation. London: Academic Press.

ROCAMORA, G. (1987) Biogéographie et écologie de l'avifaune nicheuse des massifs périméditerranéens d'Europe occidentale. Thèse ENSAM. Université de Montpellier, France (unpublished thesis).

ROCAMORA, G. (1990) Reflexions sur la mise en place des avifaunes dans les massifs forestiers péri-méditerranéens d'Europe occidentale. Pp.122–130 in Anon., ed. *Segon colloqui de naturalistes vallesans*. Barcelona, Spain: Museu de Granollers.

ROCAMORA, G. (1994) *Les zones importantes pour la conservation des oiseaux en France*. Paris: Ministère de l'Environnement/LPO-BirdLife.

ROCAMORA, G., HOTTE, J. F. AND MAILLET, N. (1995) *La conservation des ZICO en France: recherche de priorités en fonction de l'intérêt ornithologique et des niveaux de menaces*. France: Rapport Ministère de l'Environnement/LPO-BirdLife.

RODRIGUES, J. F. (1995) Evolução dos montados de sobro. *Informação Florestal* 9: 10–19.

RODRÍGUEZ, J. M. AND DE JUANA, E. (1991) Land-use changes and the conservation of dry grassland birds in Spain: a case study of Almería Province. Pp.49–58 in P. D. Goriup, L. A. Batten and J. A. Norton, eds. *The conservation of lowland dry grassland birds in Europe: proceedings of an international seminar held at the University of Reading 20–22 March 1991*. Peterborough, UK: Joint Nature Conservation Committee.

RODWELL, J. S., ED. (1991) *British plant communities, 2: mires and heaths*. Cambridge, UK: Cambridge University Press.

ROSBERG, I., OVSTEDAL, D. O., SELJELID, R., SCHREINER, O. AND GOKSOYR, J. (1981) Estimation of carbon fluid in a *Calluna* heath system. *Oikos* 37: 295–305.

ROSE, P. M. AND SCOTT, D. A. (1994) *Waterfowl population estimates*. Slimbridge, UK: International Waterfowl and Wetlands Research Bureau (IWRB Spec. Publ. 29).

RÖSLER, S. AND WEINS, C. (1996) Aktuelle Entwicklungen in der Landwirtschaftspolitik und ihre Auswirkungen auf die Vogelwelt. *Vogelwelt* 117: 169–186.

ROTENBERRY, J. T. (1985) The role of habitat in community composition: physiognomy or floristics? *Oecologia (Berlin)* 67: 213–217.

RSPB (1993) Lowland heathland habitat action plans. Sandy, UK: Royal Society for the Protection of Birds (unpublished report).

RSPB (1995) *The future of the Common Agricultural Policy*. Sandy, UK: Royal Society for the Protection of Birds.

RSPB/ITE/EN (in press) *UK wet grassland management (floodplain and coastal)*. Sandy, UK: Royal Society for the Protection of Birds.

RSPB/NRA/RSNC (1994) *The new rivers and wildlife handbook*. Sandy, UK: Royal Society for the Protection of Birds.

RUBIO, J. L. AND MARTÍNEZ, C. (1992) *Valle de Alcudia*. Madrid: ICONA (Cuadernos de la Trashumancia 2).

RUGGIERI, L., MANFREDO, I. AND BLONDIN, M. (1996) The importance of electrical lines as a cause of mortality of the Eagle Owl in the North-Western Alps (Val d'Aosta, Italy). In *Proceedings 2nd International Conference on Raptors*. Urbino, Italy.

RUÍZ, M. AND RUÍZ, J. P. (1986) Ecological history of

transhumance in Spain. *Biol. Conserv.* 37: 73–86.

RUTSCHKE, E. (1989) *Die wildenten Europas,* [*Ducks of Europe.*] Berlin: VEB Deutscher Landwirtschaftsverlag.

RYAN, P. G. (1987) The effects of ingested plastic on seabirds: correlations between plastic load and body condition. *Environ. Pollut.* 46: 119–125.

RYAN, P. G. (1990) The effects of ingested plastic and other marine debris on seabirds. Pp.623–634 in R. S. Shomura and M. L. Godfrey, eds. *Proceedings 2nd international conference on marine debris, 2–7 April, Honolulu, Hawaii,* 1. Honolulu: U.S. Department of Commerce NOAA-TM-NMFS-SWFC-154.

SACCHI, C. F. (1979) The coastal lagoons of Italy. Pp.593–601 in R. L. Jefferies and A. J. Davy, eds. *Ecological processes in coastal environments.* Oxford, UK: Blackwell Scientific.

SAGE, B. (1986) *The Arctic and its wildlife.* London: Croom Helm.

SÁNCHEZ, A. AND TELLERÍA, J. L. (1988) Influencia de la presión humana sobre la comunidad de aves de un encinar ibérico (*Quercus rotundifolia*). *Misc. Zool.* 12: 295–302.

SANTOS, R. S., HAWKINS, S. J., MONTEIRO, L. R., ALVES, M. AND ISIDRO, E. J. (1995) Marine research, resources and conservation in the Azores. *Aquatic Conserv.* (Marine and Freshwater Ecosystems) 5: 311–354.

SANTOS, T. AND ALVAREZ, G. (1990) Efectos de las repoblaciones con eucaliptos sobre las comunidades de aves forestales en un maquis mediterráneo (Montes de Toledo). [Effects of eucalypt plantations on forest bird communities in a Mediterranean maquis (Toledo mountains, west-central Spain.] *Ardeola* 37: 319–324. (In Spanish.)

SANTOS, T. AND TELLERÍA, J. L. (1991) Effects of leafing and position on nest predation in a Mediterranean fragmented forest. *Wilson Bull.* 103: 675–682.

SANTOS, T. AND TELLERÍA, J. L. (1992) Edge effects on nest predation in Mediterranean fragmented forests. *Biol. Conserv.* 60: 1–5.

SCHERZINGER, W. (1995) *Naturschutz im Wald. Qualitätsziele einer dynamischen Waldentwicklung.* Stuttgart, Germany: Ulmer.

SCHIMMEL, J. (1993) *Fire behavior, fuel succession and vegetation response to fire in Swedish boreal forest.* Umeå, Sweden: Swedish University of Agricultural Sciences (Dissertation in forest vegetation ecology 5).

SCHLÄPFER, A. (1988) [Population studies of the Skylark *Alauda arvensis* in an area of intensive agriculture.] *Orn. Beob.* 85: 309–371. (In German.)

SCHOENER, T. W. (1976) The species-area relation within archipelagos: models and evidence from island land birds. Pp.629–642 in H. J. Frith and J. H. Calaby, eds. *Proceedings 16th International Ornithological Conference.* Canberra: Australian Academy of Sciences.

SCHOENER, T. W. (1986) Patterns in terrestrial vertebrate versus arthropod communities: do systematic differences in regularity exist? Pp.556–586 in J. Diamond and T. J. Case, eds. *Community Ecology.* New York: Harper & Row.

SCHULZ, H. (1994) Zur Bestandssituation des Weißstorchs—Neue Perspektiven für den "Vogel des Jahres 1994"? *Ber. Vogelschutz* 32: 7–18.

SCOTT, D. A., ED. (1982) *Managing wetlands and their birds.* Slimbridge, UK: International Waterfowl and Wetlands Research Bureau.

SCOTT, D. A. AND ROSE, P. M. (1996) *Atlas of Anatidae populations in Africa and western Eurasia.* Wageningen, Netherlands: Wetlands International (Publ. No. 41).

SCOTTISH NATURAL HERITAGE (1994) *Red deer and the natural heritage.* Perth, UK: Scottish Natural Heritage (SNH Policy Paper).

SCOTTISH NATURAL HERITAGE (1995) *The natural heritage of Scotland: an overview.* Perth, UK: Scottish Natural Heritage.

SEGERSTRÖM, U., HÖRNBERG, G. AND BRADSHAW, R. (1996) The 9000-year history of vegetation development and disturbance patterns of a swamp-forest in Dalarna, northern Sweden. *Holocene* 6: 37.

SELLIN, D. (1989) Vergleichende Untersuchungen zur Habitatstruktur des Seggesrohrsängerss *Acrocephalus paludicola*. [Comparative investigation of habitat structure in the Aquatic Warbler.] *Vogelwelt* 110: 198–208. (In German.)

SEO (1992) Proposals to declare Madrigal-Peñaranda, Villafáfila and Tierra de Campos as Environmentally Sensitive Areas. Madrid: Sociedad Española de Ornitología (unpublished reports).

SEQUEIRA, M. AND FERREIRA, C. (1994) Coastal fisheries and cetacean mortality in Portugal.*Rep. Int. Whal. Comm.* (Special Issue) 15: 165–181.

SÉRIOT, J. AND ROCAMORA, G. (1992) *Les rapaces et le réseau électrique aérien; analyse de la mortalité et solutions.* [*Raptors and aerial electric power-line network in France: analysis of mortality and solutions.*] Rochefort, France: Electricité de France and Ligue pour la Protection des Oiseaux. (In French.)

SGOP (1983) *Estudio de la explotación de aquas subterráneas en las proximidades del Parque Nacional de las Tablas de Damiel y su influencia sobre el soporte hidrico del ecosistema.* Madrid: Ministerio de Obras Públicas y Urbanismo (Report 12/83).

SHEPHERD, J. G. (1993) *Key issues in the conservation of fisheries,* MAFF Laboratory Leaflet No. 71. Lowestoft, UK: Directorate of Fisheries Research.

SHEPHERD, K. B. AND STROUD, D. A. (1991) Breeding waders and their conservation on the wetlands of Tiree and Coll, Inner Hebrides. *Wildfowl* 42: 108–117.

SHERMAN, K., JONES, C., SULLIVAN, L., SMITH, W., BERRIEU, P. AND EJSYMENT, L. (1981) Congruent shifts in sand eel abundance in western and eastern North Atlantic ecosystems. *Nature* 291: 486–489.

SHRUBB, M. AND LACK, P. C. (1991) The numbers and distribution of Lapwings *Vanellus vanellus* nesting in England and Wales in 1987. *Bird Study* 38: 20–37.

SILVA, H. M. AND KRUG, H. M. (1992) Virtual population analysis of the forkbeard,*Phycis phycis* (Linnaeus, 1766), in the Azores Archipelago. *Life and Earth Sciences* 10: 5–12.

SIMBERLOFF, D. (1995) Habitat fragmentation and population extinction of birds. *Ibis* 137 (Suppl. 1): 105–111.

SJÖRS, H. (1983) Pp.69–94 in A. J. P. Gore, ed. *Mires: swamp, bog, fen and moor. Regional studies.* Amsterdam: Elsevier Scientific (Ecosystems of the World 4B).

SKJELKVÅLE, B. AND WRIGHT, R. F. (1990) *Overview of areas sensitive to acidification: Europe.* Oslo: Norwegian Institute for Water Research (Acid Rain Research Rep. 20/1990).

SKOV, H., DURINCK, J., LEOPOLD, M. F. AND TASKER, M. L. (1995) *Important Bird Areas for seabirds in the North Sea*. Cambridge, UK: BirdLife International.

SLOTTA-BACHMAYR, L. AND WERNER, S. (1992) Bestandssituation und Ökologie felsbrütender Vogelarten im Bundesland Salzburg. *Salzburg Vogelkdl. Ber.* 4: 30–43.

SMAL, C. (1991) Feral American mink in Ireland: a guide to the biology, ecology, pest status and control of feral American mink *Mustela vison* in Ireland. *Irish Office of Public Works, Wildlife Service* 1: 32.

SMIT, C. J. AND PIERSMA, T. (1989) Numbers, midwinter distribution, and migration of wader populations using the East Atlantic flyway. Pp.24–63 in H. Boyd and J.-Y. Pirot, eds. *Flyways and reserve networks for waterbirds*. Slimbridge: International Waterfowl and Wetlands Research Bureau (Spec. Publ. 9).

SMIT, C. J., LAMBECK, R. H. D. AND WOLFF, W. J. (1987) Threats to coastal wintering and staging areas for waders. *Wader Study Group Bull.* 49 (Suppl.): 105–113.

DE SMIDT, J. T. AND VAN REE, P. (1991) The decrease of bryophytes and lichens in Dutch heathland since 1975. *Acta Botanica Neerlandica* 40: 379.

SMITH, K. W. (1988) Breeding bird communities of commercially managed broadleaved plantations. *RSPB Conservation Review* 2: 43–46.

SNEDDON, N. P. AND RANDALL, R. E. (1993) *Vegetated shingle structures of Great Britain: main report*. Peterborough, UK: Joint Naturem Conservation Committee.

SOBRAL, M. P., CARVALHO, M., SOBRAL, M., MONTEIRO, C. C. AND DIAS, D. (1989) [Short reference to the exploitation of bivalves in the Portuguese continental coast.] *Açoreana* 7: 175–207. (In Portuguese.)

SOIKKELI, M. AND SALO, J. (1979) The bird fauna of abandoned shore pastures. *Ornis Fenn.* 56: 124–132.

SOLOMON, A. M. (1991) *Predicting boreal and temperate forest ecosystems response to climate change: ICCES symposium*. Amersfoot: Abstracts Global Reviews.

SOLOMON, A. M. AND CRAMER, W. (1993) Biospheric implications of global environmental change. Pp.25–52 in A. M. Solomon and H. H. Shugart, eds. *Vegetation dynamics and global change*. New York: Chapman and Hall.

SOLOMON, A. M., PRENTICE, I. C., LEEMANS, R. AND CRAMER, W. P. (1993) The interaction of climate and land use in future terrestrial carbon storage and release. *Water, Air and Soil Polln.* 70: 595–614.

SOLONEN, T. (1985) Birds and birdlife in Finland: a review. *Ornis Fenn.* 62: 47–55.

SOU (1992) *Skogspolitiken inför 2000-talet. [The forestry policy for the 21st century.]* Stockholm: Swedish Government (Official Report 1992: 76).

SOUTHWOOD, T. R. E. AND CROSS, D. J. (1969) The ecology of the partridge. III Breeding success and the abundance of insects in natural habitats. *J. Anim. Ecol.* 38: 497–509.

SPIERTZ *ET AL.* (1992) In C. J. Pearson, ed. *Field crop ecosystems*. Amsterdam: Elsevier (Ecosystems of the World 18).

SPITZENBERGER, F. (1988) [*Protection of species in Austria, 8.*] Vienna: Ministry for the environment, young people and families (Green Series). (In German.)

SPOOR, G. AND CHAPMAN, J. M. (1992) Comparison of the hydrological conditions of the Nene Washes and West Sedgemoor reserves in relation to their suitability for breeding waders. Silsoe College, Silsoe (unpublished report to the Royal Society for the Protection of Birds).

STANEVISCIUS, V. (1992) [*Numbers, structure and space distribution of bird communities on south Lithuanian lakes.*] Moscow. (In Russian.)

STANNERS, D. AND BOURDEAU, P., EDS. (1995) *Europe's environment: the Dobrís Assessment*. Copenhagen: European Environment Agency.

STASTNY, K. AND BEJCEK, V. (1985) Bird communities of spruce forests affected by industrial emmissions in the Krusne Hory (Ore Mountains). Pp.243–253 in K. Taylor, R. J. Fuller and P. C. Lack, eds. *Bird census and atlas studies*. Tring, UK: British Trust for Ornithology.

STATISTISK SENTRALBYRÅ (1996) *Official Statistics of Norway 1996*. Oslo: Kongsvinger.

STATTERSFIELD, A. J., CROSBY, M. J., LONG, A. J. AND WEGE, D. C. (in press) *Endemic bird areas of the world: priorities for bird conservation*. Cambridge, UK: BirdLife International (BirdLife Conservation Series no. 7).

STEINBERG, C. E. W. AND WRIGHT, R. F., EDS. (1994) *Acidification of freshwater ecosystems: implications for the future*. Chichester, UK: John Wiley & Sons.

STEMPNIEWICZ, L. (1993) Waterfowl mortality in fishing nets in the Gulf of Gdansk, 1972–1990. *Proceedings of Baltic-birds Conference 7. Phalanga*.

STOCK, M., KIEHL, K. AND REINKE, D. (1994) Salzwiesenschutz in Schleswig-Holstein. Landesamt für den Nationalpark Schleswig-Holteinisches Wattenmeer. Tönning.

STORTENBEKER, C. W. (1987) Ecologische voorwaarden voor het behoud van de heide als levensgemeenschap. [Ecological conditions for the conservation of heath as a natural community.] *Levende Nat.* 88: 98–103.

STOWE, T. J., NEWTON, A. V., GREEN, R. E. AND MAYES, E. (1993) The decline of the Corncrake *Crex crex* in Britain and Ireland in relation to habitat. *J. Appl. Ecol.* 30: 53–62.

STRANDGAARD, S. (1982) *Factors affecting the moose population in Sweden during the 20th century with special attention to silviculture*. Stockholm: Swedish University of Agricultural Science (Dept. Wildlife Ecol. Report 8).

STRANN, K. B., VADER, W. AND BARRETT, R. (1990) Auk mortality in fishing nets in north Norway. Pre-publication manuscript.

STRESEMANN, E. (1920) Die Herkunft der Hochgebirgsvögel Europas. *Jber. Cl. Nederland. Vogelkundig.* 10: 149–172.

SUÁREZ, F., NAVESO, M. A. AND DE JUANA, E. (1997a) Farming in the drylands of Spain: birds of the pseudo-steppes. Pp.297–330 in D. J. Pain and M. W. Pienkowski, eds. *Farming and birds in Europe: The Common Agricultural Policy and its implications for bird conservation*. London: Academic Press.

SUÁREZ, F., HERRANZ, J. AND YANES, M. (1997b) *Conservación y gestión de las estepas en la España peninsular. Congreso International de Aves Esteparias. [Spanish peninsular steppe: conservation and management.]* Valladolid, Spain: Sociedad Española de Ornitología. (In Spanish with English summary.)

SUÁREZ, F., SAINZ-OLLERO, H., SANTIS, T. AND GONZÁLEZ BERNÁLDEZ, F. (1992) *Las estepas ibéricas*. Madrid: MOPT, Unidades Temáticas de la Secretaría de Estado para las Políticas del Agua u el Medio Ambiente.

SUGAL, C. (1996) Labeling wood. How timber certification may reduce deforestation. *Worldwatch* 9: 29–34.

SUHONEN, J., NORRDAHL, K. AND KORPIMAKI, E. (1994) Avian predation risk modifies breeding bird community on a farmland area. *Ecology* 75: 1626–1634.

SUTHERLAND, W. J. AND HILL, D. A. (1995) *Managing habitats for conservation*. Cambridge, UK: Cambridge University Press.

SVAZAS, S. (1992) The threat of an ever increasing oil activity in Lithuanian coastal waters for the waterfowl areas of international importance. *IWRB Seaduck Bull.* 1: 43–44.

SVAZAS, S. (1995) Threats to seabirds in the Eastern Baltic. Pp.46–47 in M. L. Tasker, ed. *Threats to seabirds: Proceedings of the 5th International Seabird Group conference*. Sandy, UK: Seabird Group.

SVAZAS, S. AND PAREIGIS, V. (1992) The significance of Lithuanian Baltic coastal waters for the wintering population of the Velvet Scoter. *IWRB Seaduck Bull.* 1 (Abstract): 41–42.

SVAZAS, S. AND VAITKUS, G. (1995) Oil terminals and seabirds in Lithuanian coastal waters. Pp.104–118 in L. Lazauskiene, ed. *The Butinge oil terminal (ecological conditions)*. Vilnius: Lithuania.

SWCL (1993) All at sea? Coastal zone management—the case for Scotland. Perth, UK: Scottish Wildlife and Countryside Link.

SZIJJ, J. (1957) [Ecological and biogeographical studies on the Treecreepers of the Carpathian basin.] *Aquila* 63/64: 119–144. (In Hungarian.)

TASKER, M. L. AND BECKER, P. H. (1992) Influences of human activities on seabird populations in the North Sea. *Netherlands Journal of Aquatic Ecology* 26: 59–73.

TASKER, M. L., WEBB, A., HALL, A. J., PIENKOWSKI, M. W. AND LANGSLOW, D. R. (1987) *Seabirds in the North Sea*. Peterborough, UK: Nature Conservancy Council.

TEIXEIRA, A. M. (1986) Winter mortality of seabirds on the Portuguese coast. Pp.409–419 in MEDMARAVIS and X. Monbailliu, eds. *Mediterranean marine avifauna: population studies and conservation*. Paris: Springer-Verlag (NATO ASI Series G Ecological Sciences, 12).

TELLERÍA, J. L. (1992) Gestión forestal y conservación de las aves en España peninsular. *Ardeola* 39: 99–114.

TELLERÍA, J. L. (1993) Criteria for the optimal design of new forests as faunal biotopes. EC Scientific Workshop: scientific criteria for the establishment of zonal afforestation plans. November 25–26, 1993, Brussels.

TELLERÍA, J. L. AND SANTOS, T. (1992) Spatiotemporal patterns of egg predation in forest islands: an experimental approach. *Biol. Conserv.* 62: 29–33.

TELLERÍA, J. L., DÍAZ, M. AND ASENSIO, B. (1996) *Aves Ibéricas II. Paseriformes*. Madrid: J. M. Reyero.

TELLERÍA, J. L., SANTOS, T., ALVAREZ, G. AND SAEZ-ROYUELA, C. (1988) [Birds of cropland habitats in central Spain.] Pp.173–319 in F. Bernis, ed. [*Birds of urban and agricultural areas in the Spanish plains.*] Madrid: Sociedad Española de Ornitología.

TELLERÍA, J. L., SANTOS, T. AND ALCÁNTARA, M. (1991) Abundance and food-searching intensity of wood mice *Apodemus sylvaticus* in fragmented forests. *J. Mammal.* 72: 183–187.

TELLERÍA, J. L., SANTOS, T. AND DÍAZ, M. (1994) Effects of agricultural practices on bird populations in the Mediter-

ranean region: the case of Spain. Pp.57–75 in E. J. M. Hagemeijer and T. J. Verstrael, eds. *Bird Numbers 1992: distribution, monitoring and ecological aspects*. Beek-Ubbergen, Netherlands: Statistics Netherlands & SOVON.

THALER, R. (1994) Alpentransit. Entwicklungen, Einflußfaktoren und Handlungsstrategien. *Österreichische Akademie der Wissenschaften, Veröff. Komm. Humanökologie* 5: 149–172.

THIBAULT, J. C. (1993) Breeding distribution and numbers of Cory's Shearwater (*Calonectris diomedea*) in the Mediterranean. Pp.25–35 in J. J. Aguilar, X. Monbailliu and A. M. Paterson, eds. *Status and conservation of seabirds. Proceedings of the 2nd Mediterranean Seabird Symposium*. Madrid, Spain: Sociedad Española de Ornitología.

THOMPSON, D. B. A. AND BROWN, A. (1992) Biodiversity in montane Britain: habitat variation, vegetation diversity and some objectives for conservation. *Biodiv. & Conserv.* 1: 179–209.

THOMPSON, D. B. A., HESTER, A. J. AND USHER, M. B., EDS. (1995a) *Heaths and moorland: cultural landscapes*. Edinburgh and London, UK: Her Majesty's Stationery Office.

THOMPSON, D. B. A., MACDONALD, A. J., MARSDEN, J. H. AND GALBRAITH, C. A. (1995b) Upland heather moorland in Great Britain: a review of international importance, vegetation change and some objectives for nature conservation. *Biol. Conserv.* 71: 163–178.

THORUP, O. (1991) Population trends and studies on breeding waders at the Nature-Reserve Tipperne. *Wader Study Group Bull.* 61 (Suppl.): 78–81.

THORUP, O. (1992) Ynglefuglene på Tipperne 1928–1992. Copenhagen: Ministry of Environment, Skov- og Naturstyrelsen (unpublished manuscript).

TIAINEN, J. (1985) [Changing avifauna of Finnish farmland.] *Suomen Luonto* 44: 22–27. (In Finnish with English summary.)

TICKLE, A. K. (1992) Critical loads for nitrogen. Pp.18–23 in C. Ågren and P. Elvingson, eds. *Critical loads for air pollutants: report of the third international NGO strategy seminar on air pollution*. Göteborg, Sweden: Swedish NGO Secretariat on Acid Rain.

TOMASELLI, R. (1976) La dégradation du maquis méditerranéen. Pp.35–76 in *Notes techniques du MAB, 2 Forêts et Maquis Méditerranées: écologie, conservation et aménagement*. Paris: UNESCO.

TOMIALOJC, L. (1995) The birds of the Bialowieza Forest—additional data and summary. *Acta Zool. Cracov.* 38: 363–397.

TOMIALOJC, L. AND WESOLOWSKI, T. (1994) [Bird community stability in a primeval forest: 15-year data from Bialowieza National Park (Poland).] *Orn. Beob.* (In German.)

TRABAUD, L. (1994) Postfire plant community dynamics in the Mediterranean basin. Pp.1–15 in J. M. Moreno and W. C. Oechel, eds. *The role of fire in Mediterranean-type ecosystems*. New York: Springer-Verlag (Ecological Studies Vol. 107).

TRABAUD, L. AND PRODON, R., EDS. (1992) *Fire in Mediterranean ecosystems*. Brussels: Commission of the European Communities (Ecosystems Research Report 5).

TREWEEK, J. (1996) Ecology and environmental impact assessment. *J. Appl. Ecol.* 33: 191–199.

TUBBS, C. R. (1968) *The New Forest: an ecological history.*

Newton Abbot, UK: David and Charles.

TUCKER, G. (1997) Priorities for bird conservation in Europe: the importance of the farmed landscape. Pp.79–116 in D. J. Pain and M. W. Pienkowski, eds. *Farming and birds in Europe: the Common Agricultural Policy and its implications for bird conservation*. London: Academic Press.

TUCKER, G. M. (1992) Effects of agricultural practices on field use by invertebrate-feeding birds in winter. *J. Appl. Ecol.* 29: 779–790.

TUCKER, G. M. AND HEATH, M. F. (1994) *Birds in Europe: their conservation status*. Cambridge, UK: BirdLife International (BirdLife Conservation Series no. 3).

UNECE (UNITED NATIONS ECONOMIC COMMISSION FOR EUROPE) (1991a) *Forest damage and air pollution. Report of the 1990 forest damage survey in Europe*. Geneva: United Nations Economic Commission for Europe/United Nations Environment Programme.

UNECE (1991b) *Interim report on cause-effect relationships in forest decline*. Geneva: United Nations Economic Commission for Europe/United Nations Environment Programme.

UNECE (1995) *Workshop on the ECE database on environmental impact assessment*. Oslo: Norwegian Institute for Urban and Regional Research.

UNECE/FAO (1992a) *The forest resources of temperate zones: the UNECE/FAO 1990 Forest Resource Assessment*, 1. New York: United Nations.

UNECE/FAO (1992b) *The forest resources of temperate zones: the UNECE/FAO 1990 Forest Resource Assessment*, 2. New York: United Nations.

UNEP–ROE (1996) *Pan-European Biological Diversity and Landscape Conservation Strategy. Action Theme 5: Coastal and Marine Ecosystems. Proposal for the programme of work, 1996/1997. Prepared by meeting of Working Party, Geneva, 24–25 September 1996.* Geneva: United Nations Environment Programme Regional Office for Europe.

UNITED NATIONS (1991) *European red list of globally threatened animals and plants*. New York: United Nations.

USFWS (1986) *Use of lead shot for hunting migratory birds in the United States. Final supplement environmental impact statement*. Washington, DC: Department of the Interior, Fish & Wildlife Service, Government Printing Office.

USHER, M. B. AND THOMPSON, D. B. A. (1993) Variation in the upland heathlands of Great Britain: conservation importance. *Biol. Conserv.* 66: 69–81.

UTSCHICK, H. (1991) Beziehungen zwischen Totholzreichtum und Vogelwelt in Wirtschaftswäldern. *Forstwiss. Centralbl.* 110: 135–148.

VAITKUS, G., MEISSNER, W. AND KUROCHKIN, A. (1995) Oil and death of seabirds in the eastern Baltic. *WWF Bulletin* 2–3: 28–34.

VALKANOV, A. (1966) Lakes. Pp.308–320 in I. P. Gerassimov and Z. Galabov, eds. *Geography of Bulgaria*, 1: *Physical geography*. Sofia: Bulgarian Academy of Sciences Publishing House.

VAN WAGNER, C. E. (1978) Age-class distribution and the forest fire cycle. *Canad. J. For. Res.* 8: 220–227.

VASILIANSKIENE, M. (1975) [Littoral characteristics: sedimentation of south Lithuanian small lakes.]

VASSILAKIS, K. AND VASSILAKOPOULOU, R. (1993) *Dam construction at the Nestos River*. Athens: Hellenic Ornithological Society.

VAUGHAN, R. (1992) *In search of Arctic birds*. Calton, UK: Poyser.

VÉLEZ, R. (1990) Los incendios forestales en España. *Ecología* fuera de serie 1: 213–221.

VERNET, J. L. (1990) Man and vegetation in the Mediterranean area during the last 20,000 years. Pp.161–168 in F. di Castri, A. Hansen and M. Debussche, eds. *Biological invasions in Europe and the Mediterranean basin*. Dordrecht: Kluwer Publishers.

VIADA, C., CRIADO, J. AND NAVESO, M. Á. (1995) Áreas Importantes para las Aves y ZEPAs en España. *Áreas* 5: 11.

VICECONSEJERÍA DE MEDIO AMBIENTE, ED. (1994) *Medio Ambiente en Canarias*. Canarias: Viceconsejería de Medio Ambiente.

VIEIRA, L. M. AND EDEN, P. (1995) The Portuguese montados. Pp.99–102 in D. I. McCracken, E. M. Bignal and S. E. Wenlock, eds. *Farming on the edge: the nature of traditional farmland in Europe*. Peterborough, UK: Joint Nature Conservation Committee.

VIRKKALA, R. (1990) Ecology of the Siberian Tit *Parus cinctus* in relation to habitat quality: effects of forest management. *Ornis Scand.* 21: 139–147.

VIRKKALA, R., ALANKO, T., LAINE, T. AND TIAINEN, J. (1993) Population contraction of the White-backed Woodpecker *Dendrocopos leucotos* in Finland as a consequence of habitat alteration. *Biol. Conserv.* 66: 47–53.

VIRKKALA, R., RAJASÄRKKÄ, A., VÄISÄNEN, R. A., VICKHOLM, M. AND VIROLAINEN, E. (1994) Conservation value of nature reserves: do hole-nesting birds prefer protected forests in southern Finland. *Ann. Zool. Fennici* 31: 173–186.

VIVIAN, H. (1989) Hydrological changes of the Rhone River. Pp.57–77 in G. E. Petts, H. Moller and A. L. Roux, eds. *Historical change of large alluvial rivers*. Chichester, UK: Wiley.

VMPI (1980) [*Peat resource of the Latvian SSR. Report of the State Drainage Planning Institute. Situation on 01.01.1980.*] Riga: State Drainage Planning Institute. (In Latvian.)

VOOUS, K. H. (1960) *Atlas of European birds*. London: Nelson.

WAHLSTRÖM, E., REINIKAINEN, T. AND HALLANARO, E.-L. (1992) *Miljöns tillstand Finland*. Helsingfors, Finland: Gaudeamus.

WALICZKY, Z. (1994) An interim list of sites qualifying as Special Protected Areas under the Directive 79/409/EEC. Important Bird Areas in the European Union. Cambridge, UK: BirdLife International (unpublished report).

WALTER, H. (1968) [*The vegetation of the earth.*] Jena, Germany: Gustaf Fischer. (In German.)

WALTER, H. AND BRECKLE, S.-W. (1986a) *Ökologie der Erde*, 3. Stuttgart, Germany: Gustav Fischer.

WALTER, H. AND BRECKLE, S.-W. (1986b) *Spezielle Ökologie der Gemäßigten und Arktischen Zonen Euro-Nordasiens*. Stuttgart, Germany: Gustav Fischer.

WALTER, R. AND BRECKLE, S.-W. (1991) *Ökologie der Erde*, 4. Stuttgart, Germany: Gustav Fischer.

WARD, D. (1990) Recreation on inland lowland waterbirds: does it affect birds? *RSPB Conservation Review* 4: 62–68.

WARD, R. S. AND AEBISCHER, N. J. (1994) *Changes in Corn Bunting distribution on the South Downs in relation to agricultural land-use and cereal invertebrates*. Peterborough, UK: English Nature (English Nature Research Report No. 134).

WATT, A. S. (1955) Bracken versus heather: a study in plant sociology. *J. Ecol.* 43: 490–506.

WCED (WORLD COMMISSION ON ENVIRONMENT AND DEVELOPMENT) (1987) *Our common future*. Oxford, UK: Oxford University Press.

WEBB, A., STRONACH, A., TASKER, M. L. AND STONE, C. J. (1995) *Vulnerable concentrations of seabirds south and west of Britain*. Joint Nature Conservation Committee: Peterborough, UK.

WEBB, N. R. (1986) *Heathlands*. London: Collins.

WEBB, N. R. (1989) Studies on the invertebrate fauna of fragmented heathland in Dorset, UK, and the implications for conservation. *Biol. Conserv.* 47: 153–165.

WEBB, N. R. (1990) Changes on the heathlands of Dorset, England, between 1978 and 1987. *Biol. Conserv.* 51: 273–286.

WEBB, N. R. AND HASKINS, L. E. (1980) An ecological survey of heathlands in the Poole Basin, Dorset, England in 1978. *Biol. Conserv.* 17: 281–296.

WEBB, N. R. AND HOPKINS, P. J. (1984) Invertebrate diversity on fragmented *Calluna*-heathland. *J. Appl. Ecol.* 21: 921–933.

WELSH, D. A. (1987) Birds as indicators of forest stand condition in boreal forests of eastern Canada. Pp.259–267 in A. W. Diamond and F. L. Filion, eds. *The value of birds*. Cambridge, UK: International Council for Bird Preservation (Techn. Publ. 6).

WESOLOWSKI, T. (1995) Value of Bialowieza Forest for the conservation of White-backed Woodpecker *Dendrocopos leucotos* in Poland. *Biol. Conserv.* 71: 69–76.

WESTERHOFF, D. V. (1992) *The New Forest heathlands, grasslands and mires: a management review and strategy*. Peterborough, UK: Nature Conservancy Council.

WHILDE, A. (1979) Auks trapped in salmon drift nets. *Irish Birds* 1: 370–376.

WHITFIELD, D. P. AND TOMKOVICH, P. S. (1996) Mating system and timing of breeding in Holarctic waders. *Biol. J. Linn. Soc.* 57: 277–290.

WIENS, J. A. (1989a) *The ecology of bird communities, 1: foundations and patterns*. Cambridge, UK: Cambridge University Press.

WIENS, J. A. (1989b) *The ecology of bird communities, 2: processes and variations*. Cambridge, UK: Cambridge University Press.

WIENS, J. A. (1995a) Habitat fragmentation: island v. landscape perspectives on bird conservation. *Ibis* 137 (Suppl. 1): 97–104.

WIENS, J. A. (1995b) Is oil pollution a threat to seabirds? Lessons from the Exxon Valdez oil spill. Pp.50–51 in M. L. Tasker, ed. *Threats to seabirds: Proceedings of the 5th International Seabird Group conference*. Sandy, UK: Seabird Group.

WIETFELD, J. (1981) Bird communities of Greek olive plantations in the breeding season. Pp.127–128 in F. J. Purroy, ed. *Bird census and the Mediterranean landscape. Proceedings 7th Int. Con. Bird Census IBCC/5th Meeting EOAC, León, España, 8–12/9/1981*. León, Spain: Universidad de León.

WILCOVE, D. S. (1985) Nest predation in forest tracts and the decline of migratory songbirds. *Ecology* 66: 1211–1214.

WILCOVE, D. S., CHARLES, H., McLELLAN AND DOBSON, A. P. (1986) Habitat fragmentation in the temperate zone. Pp.237–256 in M. E. Soulé, ed. *Conservation biology: the science of scarcity and diversity*. Sunderland, USA: Sinauer Associates, Inc.

WILLIAMS, G. AND BOWERS, J. (1987) Land drainage and birds in England and Wales. *RSPB Conservation Review* 1: 25–30.

WILLIAMS, G., NEWSON, M. AND BROWNE, D. (1988) Land drainage and birds in Northern Ireland. *RSPB Conservation Review* 2: 72–77.

WILLIAMS, J. M., TASKER, M. L., CARTER, I. C. AND WEBB, A. (1995) A method of assessing seabird vulnerability to surface pollutants. *Ibis* 137 (Suppl.): 147–152.

WILSON, J. D. (1993) The BTO birds and organic farming project—one year on. *BTO News* 185: 10–12.

WILSON, J. D. AND BROWNE, J. J. (1993) *Habitat selection and breeding success of Skylarks* Alauda arvensis *on organic and conventional farmland*. Thetford, UK: British Trust for Ornithology (BTO Research Rep. No. 129).

WILSON, J. D., EVANS, A. D., POULSEN, J. G. AND EVANS, J. (1995) Wasteland or oasis? The use of set-aside by wintering and breeding birds. *Brit. Wildlife* 6: 214–223.

WINDING, N., WERNER, S., STADLER, S. AND SLOTTA-BACHMAYR, L. (1993) Die struktur von Vogelgemeinschaften am alpinen Hohengradienten: Quantitative Brutvogel-Bestandsaufnahmen in den Hohen Tauern (Österreichische Zentralalpen). *Wiss. Mitt. Nationalpark Hohe Tauern* 1: 106–124.

WINKELMAN, J. E. (1995) Bird/wind turbine investigations in Europe. Proceedings of national avian–wind power planning meeting, Denver, Colorado, 20–21 July 1994.

WOLDHEK, S. (1980) *Bird killing in the Mediterranean*. Zeist, Netherlands: European Committee for the Prevention of Mass Destruction of Migratory Birds.

WOLFF, W. J. (1989) Getijdewateren. Pp.92–96 in W. J. Wolff, ed. *De internationale betekenis van de Nederlandse natuur. Een verkenning*. 's-Gravenhage.

WOLFF, W. J. AND BINSBERGEN, M. (1985) *Het beheer van de wadden. De visie van de Werkgroep Waddengebied*. Arnhem, Netherlands: Stichting veth tot Steun aan Waddenonderzoek.

WOLKINGER, F. AND PLANK, S. (1981) *Dry grasslands of Europe*. Strasbourg, France: Council of Europe.

WRI (WORLD RESOURCES INSTITUTE) (1994) *World resources 1994–95. A guide to the global environment*. Oxford, UK: Oxford University Press.

WRIGHT, P. J. AND BAILEY, M. C. (1993) *Biology of sandeels in the vicinity of seabird colonies at Shetland*. Aberdeen, UK: SOAFD.

WTO (1993) *Yearbook of tourism statistics, 45th edition*. Madrid: World Tourism Organisation.

WWF (1994) *Status of old growth and semi-natural forests in western Europe*. Gland, Switzerland: World Wide Fund for Nature.

WWF (1996) Forests for life. WWF Annual Forest Report no. 1.

WWF/IUCN (1994) *Centres of plant diversity: a guide and strategy for their conservation, 1: Europe, Africa, South West Asia and the Middle East*. Cambridge, UK: World

Wide Fund for Nature and International Union for Conservation of Nature and Natural Resources.

WYNDE, R. AND BURROWS, J. (1991) *Caledonian pine forests action plan*. Sandy, UK: Royal Society for the Protection of Birds.

WYNNE, G., AVERY, M., CAMPBELL, L., GUBBAY, S., HAWKSWELL, S., JUNIPER, T., KING, M., NEWBERY, P., SMART, J., STEEL, C., STUBBS, A., TAYLOR, J., TYDEMAN, C. AND WYNDE, R. (1995) *Biodiversity challenge*. Second edition. Sandy: Royal Society for the Protection of Birds.

YANES, M. (1994) The importance of land management in the conservation of birds associated with the Spanish steppes. Pp.34–40 in E. D. Bignal, D. I. McCracken and D. J. Curtis, eds. *Nature conservation and pastoralism in Europe*. Peterborough, UK: Joint Nature Conservation Committee.

YANES, M. AND SUÁREZ, F. (1996) Incidental nest predation and lark conservation in an Iberian semiarid shrubsteppe. *Conserv. Biol.* 10: 881–887.

YON, D. AND TENDRON, G. (1981) *Alluvial forests of Europe*. Strasbourg, France: Council of Europe.

ZACKRISSON, O. (1977a) Influence of forest fires on the north Swedish boreal forest. *Oikos* 29: 22–32.

ZACKRISSON, O. (1977b) Forest fire and vegetation pattern in the Vindelälven valley, N. Sweden during the past 600 years. The river valleys as a focus of interdisciplinary research. In *Proc. Int. Conf. Commem. Maupertuis' Exp. to the river Tornio, Northern Finland, 1736–37.*

ZAMORA, R. (1990) [Snow-field use by high mountain passerines in Sierra Nevada (SE of Spain).] *Doñana Acta Vert.* 17: 57–66. (In Spanish with English summary.)

ZANG, H. (1990) Population decrease of Coal Tit *Parus ater* in the Harz Mountains due to forest damage ('Waldsterben'). *Vogelwelt* 111: 18–28. (English title.)

ZEITLER, A. J. (1994) Skilauf und Rauhfußhühner. *Verh. Ges. Ökol.* 23: 289–294.

ZINO, F., HEREDIA, B. AND BISCOITO, M. J. (1996a) Action plan for Fea's Petrel (*Pterodroma feae*. Pp.25–31 in B. Heredia, L. Rose and M. Painter, eds. *Globally threatened birds in Europe: action plans*. Strasbourg, France: Council of Europe and BirdLife International.

ZINO, F., HEREDIA, B. AND BISCOITO, M. J. (1996b) Action plan for Zino's Petrel (*Pterodroma madeira*. Pp.33–39 in B. Heredia, L. Rose and M. Painter, eds. *Globally threatened birds in Europe: action plans*. Strasbourg, France: Council of Europe and BirdLife International.

ZINO, P. A. AND ZINO, F. (1986) Contribution to the study of the petrels of the genus *Pterodroma* in the archipelago of Madeira. *Bol. Mus. Mun. Funchal* 38: 141–165.

ZUNA-KRATKY, T. AND FRÜHAUF, J. (1996) Brutzeitbericht für die March/Thaya-Auen im Jahr 1995. Ramsar-Gebietsbetreuung March/Thaya-Auen, Distelverein, Orth/Donau (unpublished report).

ZWARTS, L. (1974) *Vogels van het brakke getijgebied*. Amsterdam: Bondsuitgeverij.

ZWARTS, L. AND DRENT, R. H. (1981) Prey depletion and the regulation of predator density; Oystercatchers (*Haematopus ostralegus*) feeding on mussels (*Mytilus edulis*). Pp.193–216 in N. V. Jones and W. J. Wolff, eds. *Feeding and survival strategies of estuarine organisms*. New York: Plenum Press.